THE TECHNICAL DESIGN GRAPHICS PROBLEM SOLVER

Staff of Research and Education Association,

Dr. M. Fogiel, Director

Research and Education Association

THE TECHNICAL DESIGN GRAPHICS PROBLEM SOLVER

Printed in the United States of America

Library of Congress Catalog Card Number 81-86648

International Standard Book Number 0-87891-534-6

WHAT THIS BOOK IS FOR

For as long as technical design graphics has been taught in schools, students have found this subject difficult to understand and learn. Despite the publication of hundreds of textbooks in this field, each one intended to provide an improvement over previous textbooks, students continue to remain perplexed, and the subject is often taken in class only to meet school/departmental requirements for a selected course of study.

In a study of the problem, REA found the following basic reasons underlying students' difficulties with this subject as taught in schools:

(a) No systematic rules of analysis have been developed which students may follow in a step-by-step manner to solve the usual problems encountered. This results from the fact that the numerous different conditions and principles which may be involved in a problem, lead to many possible different methods of solution. To prescribe a set of rules to be followed for each of the possible variations, would involve an enormous number of rules and steps to be searched through by students, and this task would perhaps be more burdensome than solving the problem directly with some accompanying trial and error to find the correct solution route.

(b) Textbooks currently available will usually explain a given principle in a few pages written by a professional who has an insight in the subject matter that is not shared by students. The explanations are often written in abstract manners which leave the students confused as to the application of the principle. The explanations given are not sufficiently detailed and extensive to make the student aware of the wide range of applications and different aspects of the principles being studied. The numerous possible variations of principles and their applications are usually not discussed, and it is left for the students to discover these for themselves while doing exercises. Accordingly, the average student is expected to rediscover that which has been long known and practiced, but not published or explained extensively.

(c) The illustrations usually following the explanation of a principle in design graphics are too few in number and too simple to enable the student to obtain a thorough grasp of the principle involved. The illustrations do not provide sufficient basis to enable a student to solve problems that may be subsequently assigned for homework or given on examinations.

The illustrations are presented in abbreviated form which leaves out much material between steps, and requires that students derive the omitted material themselves. As a result, students find the illustrations difficult to understand-contrary to the purpose of the illustrations.

Illustrations are, furthermore, often worded in a confusing manner. They do not state the problem and then present the solution. Instead, they pass through a general discussion, never revealing what is to be solved for.

Illustrations, also, do not always include diagrams, wherever appropriate, and students do not obtain the training to draw diagrams to simplify and organize their thinking.

(d) Students can learn the subject only by doing the exercises themselves and reviewing them in class, to obtain experience in applying the principles with their different ramifications.

In doing the exercises by themselves, students find that they are required to devote considerably more time to their courses in design graphics than to other subjects of comparable credits, because they are uncertain with regard to the selection and application of the principles involved. It is also often necessary for students to discover those "tricks" not revealed in their texts (or review books), that make it possible to solve problems easily. Students must usually resort to methods of trial-and-error to discover these "tricks," and as a result they find that they may sometimes spend several hours to solve a single problem.

(e) When reviewing the exercises in classrooms, instructors usually request students to take turns in writing solutions on the boards and explaining them to the class. Students often find it difficult to explain in a manner that holds the interest of the class, and enables the remaining students to follow the material written on the boards.

The remaining students seated in the class are, furthermore, too occupied with copying the material from the boards, to listen to the oral explanations and concentrate on the methods of solution.

This book is intended to aid students in technical design graphics to overcome the difficulties described, by supplying detailed illustrations of the solution methods which are usually not apparent to students. The solution methods are illustrated by problems selected from those that are most often assigned for class work and given on examinations. The problems are arranged in order of complexity to enable students to learn and understand a particular topic by reviewing the problems in sequence. The problems are illustrated with detailed step-by-step explanations of the principles involved, to save the students the large amount of time that is often needed to fill in the gaps that are usually omitted between steps of illustrations in textbooks or review/outline books.

The staff of REA considers technical design graphics a subject that is best learned by allowing students to view the methods of analysis and solution techniques themselves. This approach to learning the subject matter is similar to that practiced in the medical fields, for example, and various scientific laboratories.

In using this book, students may review and study the illustrated problems at their own pace; they are not limited to the time allowed for explaining problems on the board in class.

When students want to look up a particular type of problem and solution, they can readily locate it in the book by referring to the index which has been extensively prepared. It is also possible to locate a particular type of problem by glancing at just the material within the boxed portions. To facilitate rapid scanning of the problems, each problem has a heavy border around it. Furthermore, each problem is identified with a number immediately above the problem at the right-hand margin.

To obtain maximum benefit from the book, students should familiarize themselves with the section, "HOW TO USE THIS BOOK," located in the front pages.

To meet the objectives of this book, staff members of REA have selected problems usually encountered in assignments and examinations, and have solved each problem meticulously to illustrate the steps which are usually difficult for students to comprehend. Gratitude for their patient work in this area is due to Lewis Stern, Simon Halapir, Carole Livingston, Susan Mileaf, E. Polanco, and Valerie Skleros.

The manuscript that was evolved with its endless inserts, changes, modifications to the changes, and editorial remarks, must have been an arduous typing task for Louise Baggot, Yvette Fuchs, Sophie Gerber, and Sarah Nicoll. These ladies typed the manuscript expertly with almost no complaints about the handwritten material and the numerous symbols that require much patience and special skill.

For their efforts in the graphic-arts required in the layout arrangement, and completion of the physical features of the book, gratitude is expressed to Vivian Lopez, R. Colon Puig, Irene Schaub, Fran Wolfson, and Catherine Fusco. They also helped in the training and supervision of other artists that were needed to prepare the book for printing.

The difficult task of coordinating the efforts of all persons was carried out by Carl Fuchs. His conscientious work deserves much appreciation. He also trained and supervised art and production personnel in the preparation of the book for printing.

Finally, special thanks are due to Helen Kaufmann for her unique talents to render those difficult border-line decisions and constructive suggestions related to the design and organization of the book.

Max Fogiel, Ph. D.
Program Director

HOW TO USE THIS BOOK

This book can be an invaluable aid to students in technical design graphics as a supplement to their textbooks. The book is subdivided into 17 chapters, each dealing with a separate topic. The subject matter is developed beginning with orthographic projection of lines and extending through planes, auxiliary views, sectional views, surfaces and solids and their intersections. Included also, are sections on developments, fasteners, cams and gears, as well as vector analysis and dimensioning.

TO LEARN AND UNDERSTAND A TOPIC THOROUGHLY

1. Refer to your class text and read there the section pertaining to the topic. You should become acquainted with the principles discussed there. These principles, however, may not be clear to you at that time.

2. Then locate the topic you are looking for by referring to the "Table of Contents" in front of this book, "The Technical Design Graphics Problem Solver."

3. Turn to the page where the topic begins and review the problems under each topic, in the order given. For each topic, the problems are arranged in order of complexity, from the simplest to the more difficult. Some problems may appear similar to others, but each problem has been selected to illustrate a different point or solution method.

To learn and understand a topic thoroughly and retain its contents, it will be generally necessary for students to review the problems several times. Repeated review is essential in order to gain experience in recognizing the principles that should be applied, and to select the best solution technique.

TO FIND A PARTICULAR PROBLEM

To locate one or more problems related to a particular subject matter, refer to the index. In using the index, be certain to note that the numbers given there refer to problem numbers, not to page numbers.

This arrangement of the index is intended to facilitate finding a problem more rapidly, since two or more problems may appear on a page.

If a particular type of problem cannot be found readily, it is recommended that the student refer to the "Table of Contents" in the front pages, and then turn to the chapter which is applicable to the problem being sought. By scanning or glancing at the material that is boxed, it will generally be possible to find problems related to the one being sought, without consuming considerable time. After the problems have been located, the solutions can be reviewed and studied in detail. For this purpose of locating problems rapidly, students should acquaint themselves with the organization of the book as found in the "Table of Contents."

In preparing for an exam, it is useful to find the topics to be covered in the exam from the Table of Contents, and then review the problems under those topics several times. This should equip the student with what might be needed for the exam.

CONTENTS

CHAPTER 1

ORTHOGRAPHIC PROJECTION OF LINES

TRUE LENGTH, SLOPE, POINT VIEW, AND TRIANGULATION

• PROBLEM 1-1

Explain the meaning of, and method for determining, the true length of a line.

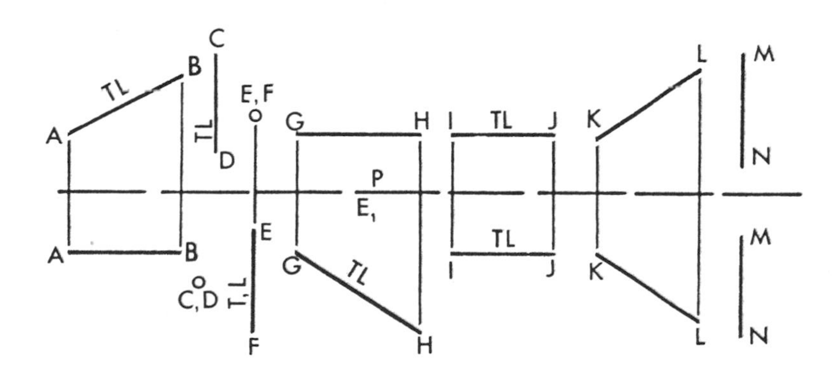

Fig. 1

Solution: (1) The true length of a line will be seen only when the lines of sight are at right angles to the line. In other words, if a line is to be true length in any view, it must be parallel to the image plane for that view. (2) A line will appear shorter than true length in any view in which the lines of sight are not perpendicular to the line. (3) Fig. 1 illustrates the above principles: lines A-B and C-D are true length in the plane (or horizontal) view; lines E-F and G-H are true length in the elevation (or vertical) view; and line I-J is true length in both views. (4) Note that the line appears parallel to the line of intersection between the plan and elevation viewing planes (a rotation or border line) in the view adjacent to the true-length view. A border line (also called a rotation line) is the line of intersection between true perpendicular planes. In addition, the line appears true-length and perpendicular to the rotation line in the view adjacent to the view that shows the line as a point. (5) Since lines K-L and M-N do not appear parallel to the border line in either the plan or elevation views, lines K-L and M-N are not true length in either given view. A new view must be drawn in order to find the true length of K-L or M-N. (6) As shown in Figs. 2 and 3, the true length of the given line may be obtained in either an elevation view or an inclined view.

LINES AND PLANES

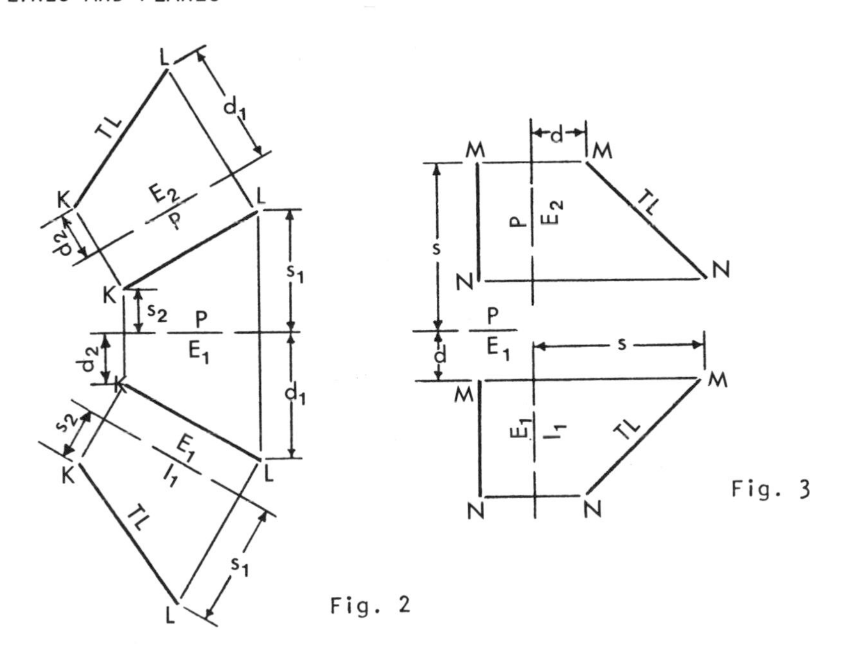

Fig. 2

Fig. 3

An elevation view shows the line in a plane drawn perpendicular to the horizontal plane. The frontal plane and the side plane are specific examples of elevation planes. An elevation plane is one in which true distances below (or above) a horizontal plane can be either found or measured off. An inclined plane is one drawn perpendicular to the frontal plane. Measurements in an inclined plane are true distances on any horizontal plane. Note that the elevation view is obtained when the lines of sight are horizontal and the image plane is vertical. When the lines of sight are inclined (i.e., neither vertical nor horizontal), the image plane is inclined, and the view obtained is an inclined view. An infinite number of elevation and inclined views of an object can be drawn. (7) In the case of either line K-L or M-N, a new rotation line is drawn parallel to the given line, and projection lines are drawn perpendicular to both the rotation line and the given line. (8) Measurements for view E_2, the second elevation view drawn, are obtained in view E_1; for example, the measurements d_1 and d_2 in Fig. 2. Measurements for view I_1, the primary inclined view, are obtained in the plan (or horizontal) view; for example, the measurements S_1 and S_2 in Fig. 2. (9) In general, any point on an object must appear the same distance from the rotation line in all views connected to a given view by projection lines. (10) As an illustration of step (9), refer to Fig. 2: Point L is a distance d_1 from the rotation lines in views E_1 and E_2, each of which is connected to view P by projection lines. The distance from the rotation line to point L is S_1 in the views connected to E_1. (11) For future reference, the letters TL signify that a line is true length.

PRIMARY AUXILIARY VIEW OF A LINE

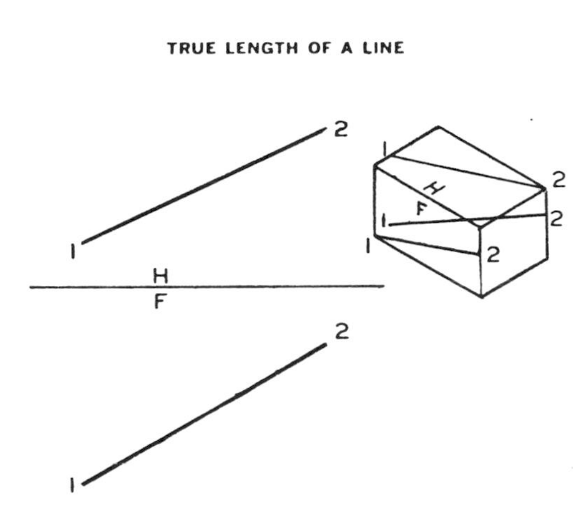

Given: The top and front views of line 1-2.
Required: Find the true length view of line 1-2 by the auxiliary-view method.

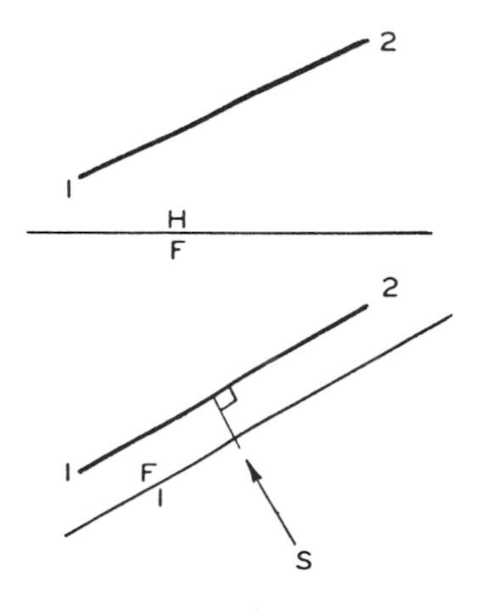

Step 1: A line will appear true length when viewed in a perpendicular direction. Therefore we shall establish a line of sight perpendicular to the front view of 1-2 and draw reference plane F-1 parallel to line 1-2 and perpendicular to the line of sight. Label the reference plane F-1 since it is perpendicular to the frontal plane.

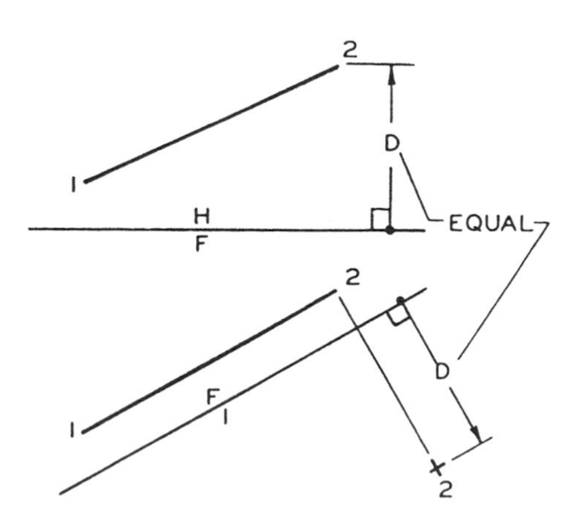

Step 2: We shall use the frontal plane as a reference plane for measurements since the auxiliary view is projected from the front view. Point 2 is distance *D* from the frontal plane in the top view. Measure this distance, which is perpendicular to the F-1 plane, in the auxiliary view along the projector from the front view of point 2.

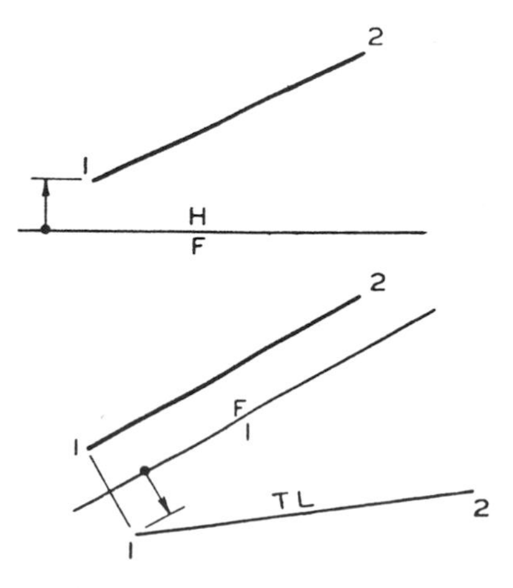

Step 3: Locate point 1 in the auxiliary view in the same manner as we did point 2. Connect points 1 and 2 to establish the true-length view of line 1-2. We could also find the true length of line 1-2 by projecting from the top view and using an H-1 reference plane.

• **PROBLEM 1-2**

Explain the meaning and determination of the slope of a line.

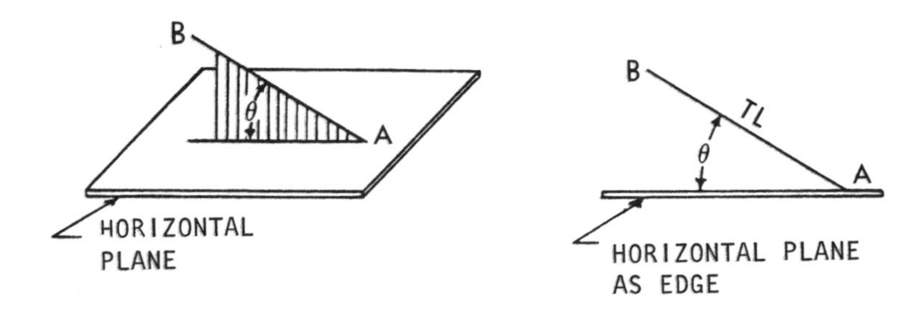

Fig. 1

Solution: (1) The slope of a line may be defined as the angle the line makes with a horizontal plane. (2) As shown in Fig. 1, the angle θ is the slope of line A-B. The true size of angle θ is seen when the line is in its true length and the horizontal plane is an edge in the same view. Since a horizontal plane can be seen as an edge only in an elevation view, the slope is found in an elevation view showing the line in true length.

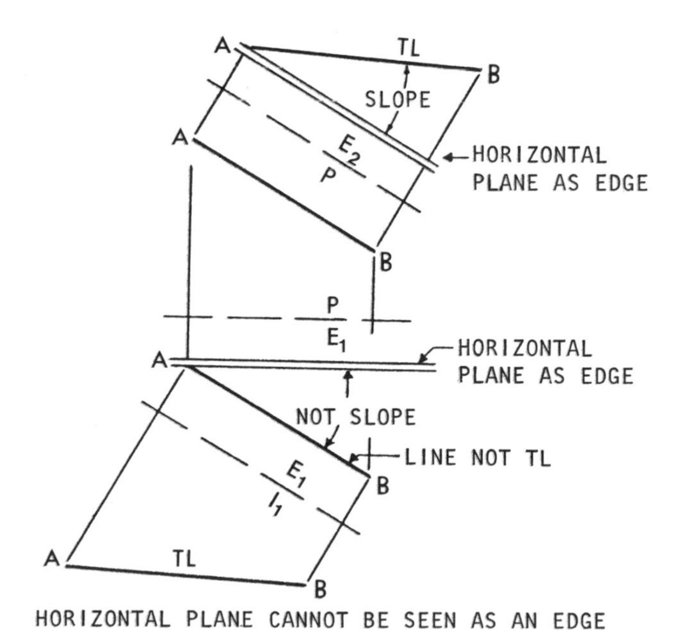

HORIZONTAL PLANE CANNOT BE SEEN AS AN EDGE Fig. 2

(3) In Fig. 2, line A-B is given in the P- (plan or horizontal) and E_1- (primary elevation or frontal) views. (4) In order to show the slope of A-B, the secondary elevation view, E_2, is drawn to show the true length of A-B. In this view, a horizontal plane appears as an edge and

the slope angle is seen in its true size. (5) Note that the primary auxiliary view, I_1, also shows the true length of A-B, but that the slope cannot be seen because a horizontal plane cannot appear as an edge in an inclined view. Consequently, the slope of a line can be seen only in an elevation view that shows the true length of the line. (6) Referring to Fig. 2, since point B is lower than A in view E_1, line A-B has a downward, or negative slope. If the line had been designated as line B-A, the slope would be upward, or positive.

• PROBLEM 1-3

Obtain the point view of a line.

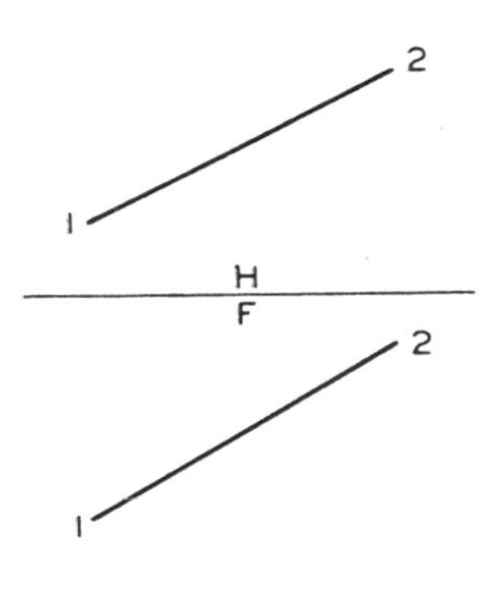

Fig. 1

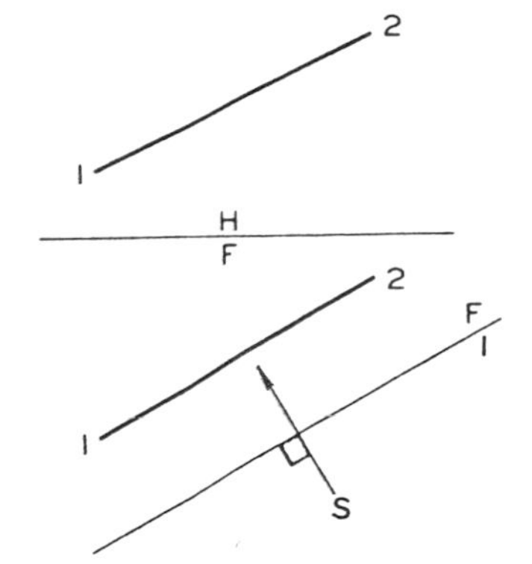

Fig. 2

Solution: (1) Fig. 1 shows the given line, 1-2, in the H- and F- planes of projection. (2) Problem analysis reveals that a point view of a line can be projected only from a true-length view of the line. (3) Project an auxiliary view from one of the principal views, as shown in Fig. 2, to obtain a true length view of the line. In this particular case, the reference plane is drawn parallel to the front view of line 1-2. (4) The projectors will be parallel to the given line of sight, which is perpendicular to the F-1 plane. (5) The primary auxiliary view, view I, could have also been projected from the horizontal view. (6) Line 1-2 is true length in view I, as shown in Fig. 3, since the line 1-2 is parallel to the reference plane in the preceding view. (7) Referring to Fig. 4,draw a secondary plane, 1-2, perpendicular to the true length view of line 1-2. (8) Transfer measurement L from view F to view 2 to find the point view. Note that this measurement is preserved since view F and view 2 are connected to view I by projection lines. In other words, measurement L will appear true length in these positions since it is perpendicular to the primary auxiliary plane, which appears as an edge in the front and secondary auxiliary views.

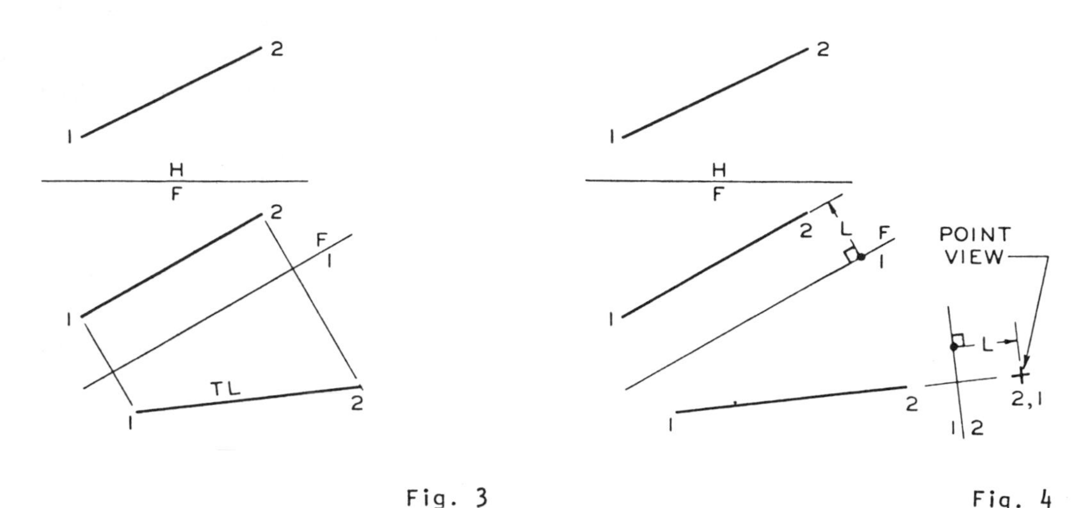

Fig. 3 Fig. 4

• PROBLEM 1-4

> By using the triangulation method, find the true length and slope of a given line KL.

Solution: (1) Fig. 1 shows the given line, K-L, in the plan view, P, and the primary elevation view, E_1. The primary elevation view is also known as the frontal plane.

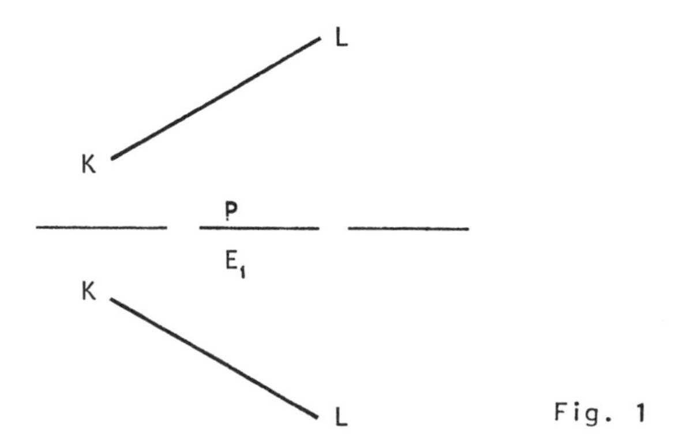

Fig. 1

(2) The true length and slope of a line can be found by considering the line to be the hypotenuse of a right triangle whose base and altitude are equal to the horizontal and vertical projections of the line. The true length of the horizontal projection, or plan length, can be measured in the plan view. The true length of the

vertical projection, or difference in elevation between the ends of the line, can be measured in any elevation view.

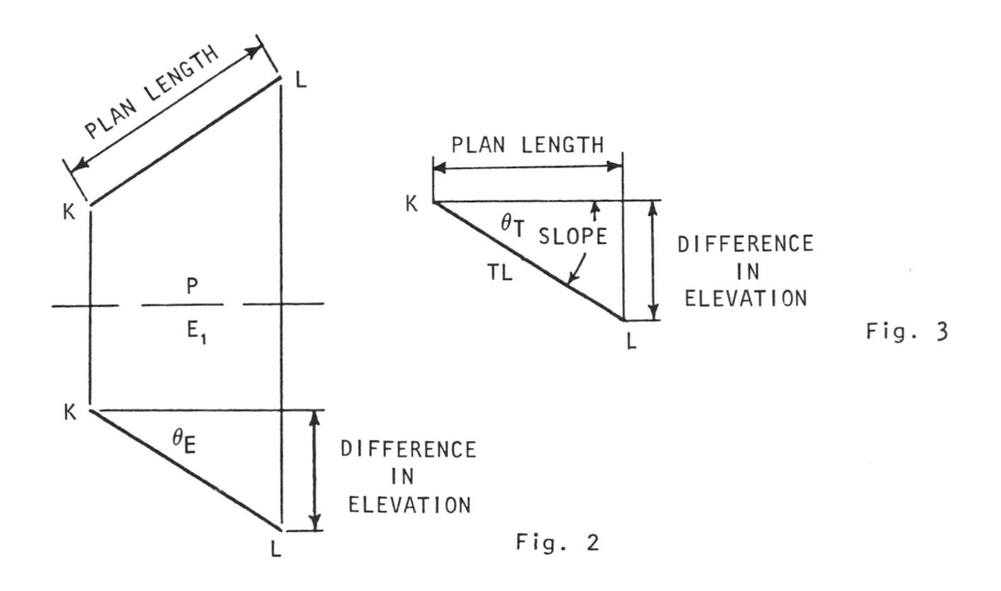

Fig. 3

Fig. 2

(3) In Figure 2, the given line is not true length in either view. The plan length and the difference in elevation are measured in the given views and laid out perpendicular to each other, as shown in Figure 3. The true length and slope of KL can then be measured. Note that the slope of KL is negative since L is lower than K. Note that θ_T, in Fig. 3, the true angle, or slope, between line K-L and the horizontal plane appears larger than true size when line KL appears less than true length, as shown by θ_E in Fig. 2.

• PROBLEM 1-5

Explain the meaning and determination of the bearing of a line. Describe the meaning of an azimuth.

Solution: The bearing of a line measures the angle of the horizontal projection of the line. Bearings are essentially directions on a map relating to directions on the earth's surface. The earth's surface is considered to be a horizontal plane. Thus, it is the horizontal projection of the line which is given by the bearing regardless of the true angle actually made by the line with the frontal plane. The plan or horizontal view is sometimes called the map view.

Bearings are given with reference to the directions of a compass. By convention they are measured down or up from the north or south line. (The north or south line is called a meridian.) The bearing angles are always less than 90°. The angle may start from either the north or south end of the meridian and may be measured either toward the east or toward the west. Consequently, in giving a bearing one must specify both the end of the meridian used and the direction of measurement of the angle except in cases where the line is reported as Due north, Due south, Due east, or Due west.

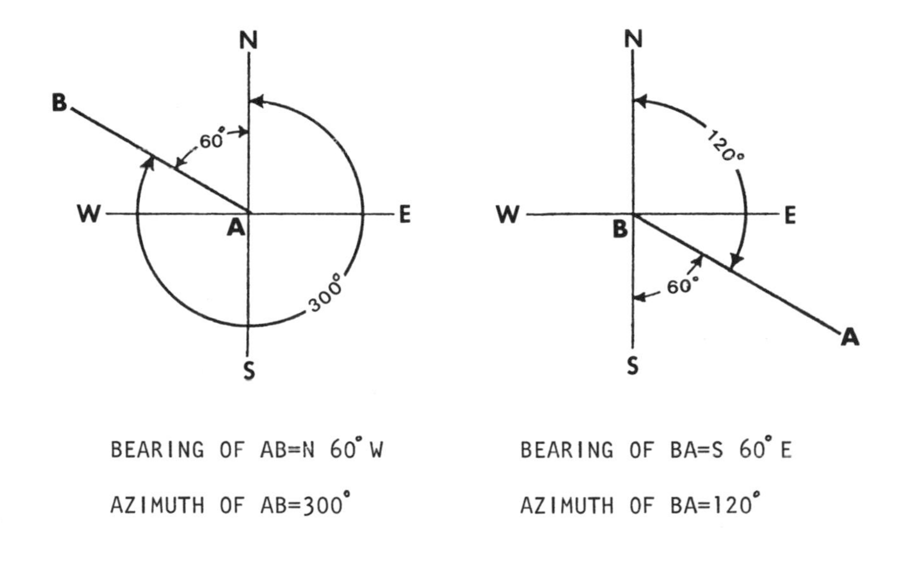

BEARING OF AB=N 60° W

AZIMUTH OF AB=300°

Fig. 1

BEARING OF BA=S 60° E

AZIMUTH OF BA=120°

Fig. 2

In Fig. 1 the bearing of line AB is measured from the north end of a meridian, 60° toward the west, and is written N60°W. The bearing of line BA in Fig. 2 is measured from the south end of a meridian, 60° toward the east, and thus is written S60°E. The bearing of a line may also be given as an azimuth instead of a bearing angle. An azimuth also measures a line's horizontal projection; however, the angle is measured clockwise from north through 360°. Thus, line AB of Fig. 1 has an azimuth of 300°. Line BA of Fig. 2 has an azimuth of 120°. A line bearing Due east has an azimuth of 90°. A line bearing Due north has an azimuth of 0°.

LENGTH, BEARING, AND SLOPE

• **PROBLEM 1-6**

Lay out a line of given length, bearing, and slope.

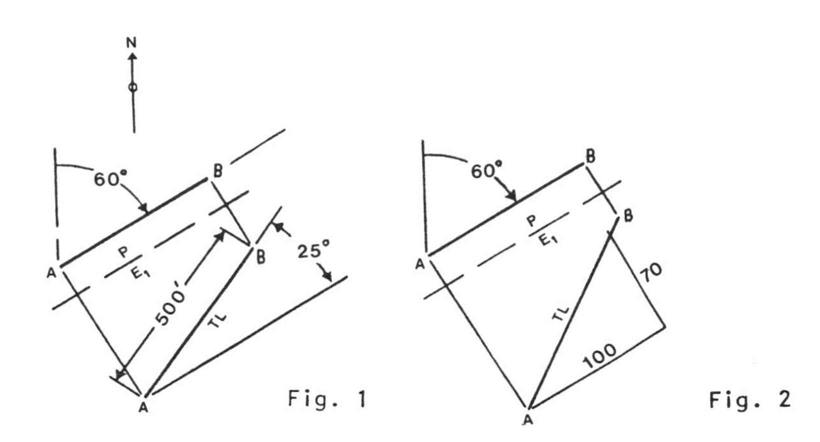

Solution: (1) Problem analysis calls for an understanding of the following principles: The slope of a line is defined as the angle the line makes with a horizontal plane. The bearing of a line is the horizontal angle of 90° or less measured between the horizontal projection of the line and the north or south meridian. Consequently, the bearing of a line must be measured in the plan view.

(2) Let AB be the given line, with a true length of 500 feet, a bearing of N60°E, and a slope of +25°. Fig. 1 illustrates the following method of solution for the stated problem. (3) Locate point A at a convenient point in the plan view. (4) Through A, draw a line of indefinite length, making an angle of 60° toward the east from north. (5) Draw rotation line $\frac{P}{E_1}$ parallel to this line, giving a view in which true-length measurements can be made. (6) Locate A at some convenient point in view E_1. (7) From A, in view E_1, draw a line of indefinite length at an angle of 25° upward from the horizontal. (8) On this line, locate point B at a distance of 500 feet to scale from A. (9) Project B back to the plan view.

(10) Note: If the slope is given as a grade of +70 percent, follow the same procedure as listed above, except step (7). To lay out the +70 percent grade, measure 100 units horizontally using any convenient scale, and 70 units vertically using the same scale. This solution is illustrated in Fig. 2.

Lay out a Line from Given Rectangular Coordinates.

GIVEN
The line AB with coordinates of A and B as follows:

Point	X	Y	Z
A	36,745	12,147	4821
B	38,896	11,261	3116

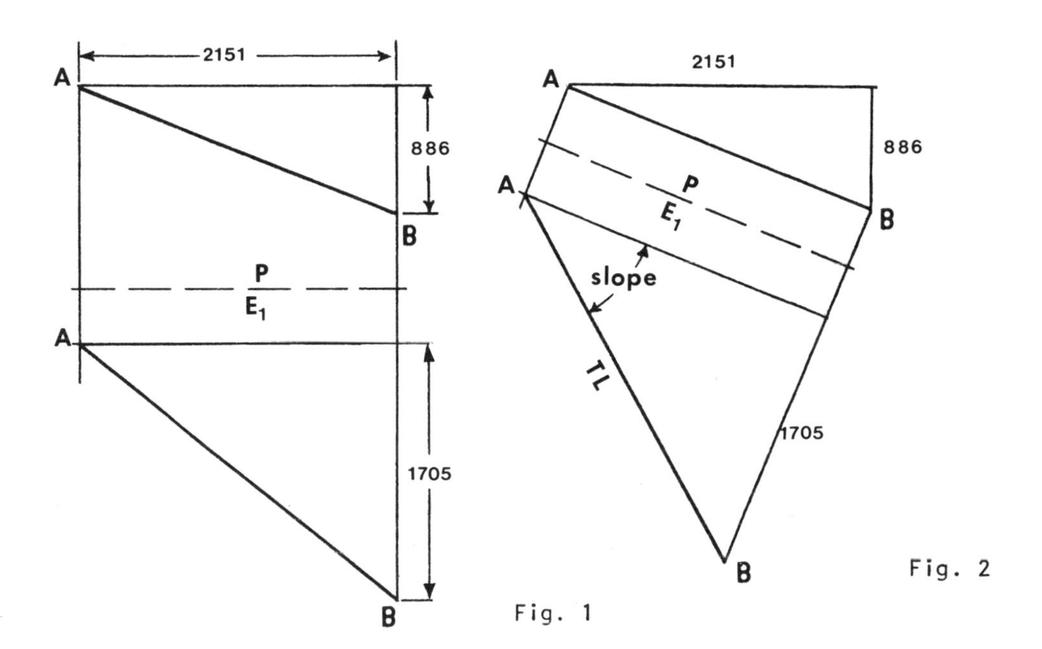

Solution: (1) Subtract the coordinates of A from the coordinates of B, observing the direction indicated by the signs.

X: 38,896 - 36,745 = +2151 (B east of A)

Y: 11,261 - 12,147 = -886 (B south of A)

Z: 3116 - 4821 = -1705 (B lower than A)

(2) Locate A in the plan view at any convenient point, as in fig. 1.

(3) The X and Y coordinates are shown in the plan view. Establish the position of B in the plan view by laying off (to scale) 2151 feet eastward and 886 feet southward.

(4) Draw line AB in the plan view.

(5) Draw rotation line $\frac{P}{E_1}$ between the plan and the desired

elevation view. In Figure 1, the lines of sight for the elevation view are due north.

(6) Draw projection lines into view E_1 and select a position for point A.

(7) The X and Z coordinates are shown in this elevation view. From a horizontal line through A, scale 1705 feet downward along the projection line from B, thereby establishing the position of B in the E_1 view. Draw AB.

If the true length and slope of AB are desired, the above procedure is followed except that the $\frac{P}{E_1}$ rotation line is drawn parallel to AB in the plan view; Figure 2. The lines of sight for the elevation view are therefore perpendicular to line AB, so that its true length and slope are obtained in view E_1.

Note that the origin of the coordinate system need not be available within the limits of the drawing, since the difference in coordinates is used to plot the points.

The information for line AB might have been given using difference in coordinates; for example: Point B is 2151 feet east of point A, 886 feet south of point A, and 1705 feet lower than A.

LINE TRACES AND PARALLEL LINES

• PROBLEM 1-8

Define and determine the traces of a line.

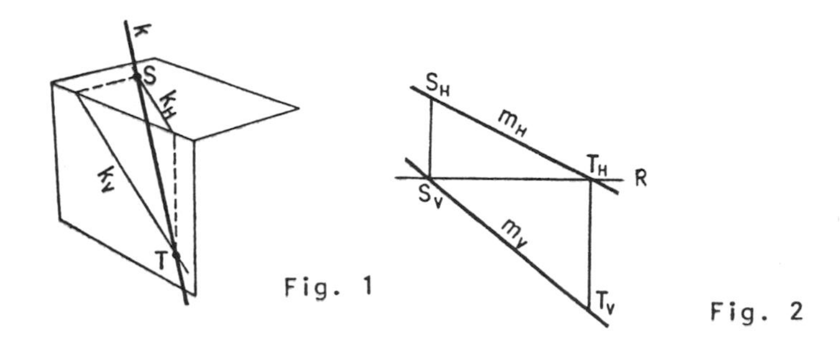

Fig. 1 Fig. 2

Solution: (1) The point in which a line intersects a projection plane is the trace of a line. (2) The point S (Fig. 1) in which line K pierces the H-plane is the horizontal trace of K; the point T in which K pierces the

V-plane is the vertical trace of K. (3) If a line is parallel to a projection plane, it will have no trace on that plane.

(4) Fig. 2 illustrates the method for determining the traces of the given line m, in the horizontal, H, and vertical, V, planes of projection. (5) The point S_V, in which m_V meets the reference line, R, is the vertical projection of the point in which line m pierces the H-plane. The projection S_H is the H-trace of m and lies on the projector through S_V. (6) The point T_H in which m_H meets the reference line is the horizontal projection of the point in which line m pierces the V-plane. The projection T_V is the V-trace of m and lies on the projector through T_H. (7) The projections S_V and T_H are always on the reference line.

• PROBLEM 1-9

Explain parallel lines with reference to orthographic views.

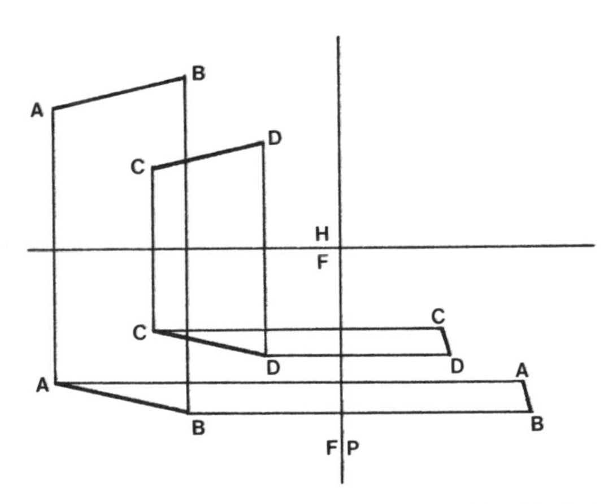

PARALLEL LINES APPEAR PARALLEL IN ALL ORTHOGRAPHIC VIEWS.

Fig. 1

Solution: (1) Lines that are parallel in space will appear parallel in any orthographic view (except in a view in which one line completely hides the other line or lines, and only one line appears; it is also possible for two or more parallel lines to appear as points.) See Fig. 1 (2) Lines are parallel in space if they appear parallel in two adjacent views, provided that they are not at right angles to the rotation line between the two views. (3) If the latter case is true, a third view will show whether or not the lines are parallel. See Fig. 2.

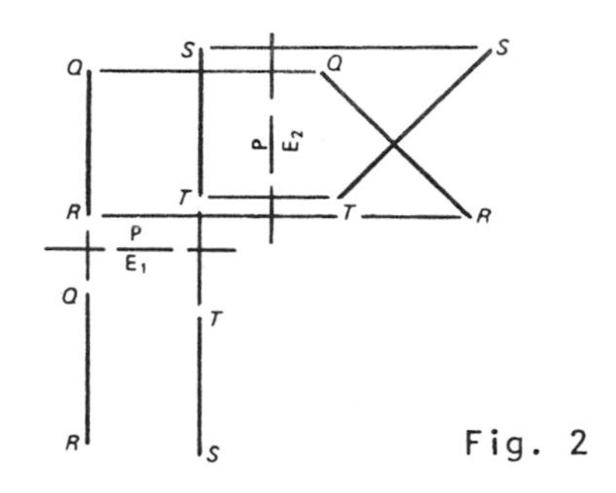

Fig. 2

• PROBLEM 1-10

Construct a line, using projection, equal in length and parallel to a given line AB, with a midpoint at a given point, O.

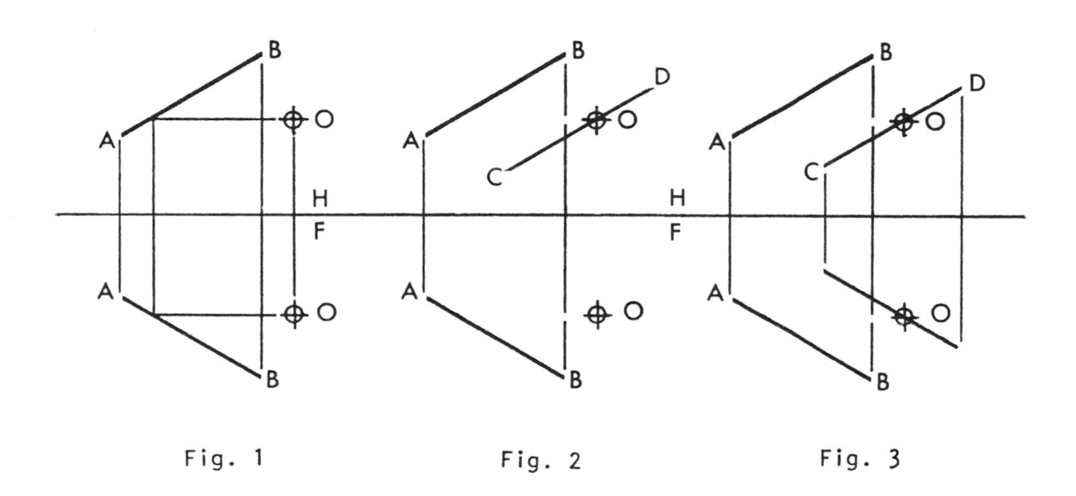

Fig. 1 Fig. 2 Fig. 3

Solution: To determine whether or not two lines are parallel, it is necessary to project more than one orthographic view of the lines. Parallel lines in space will appear parallel in all orthographic views (except in the views where both lines appear as a single line or as two points). Applying this concept to the problem renders a solution quite readily. The midpoint of a line will also be the midpoint of any projection of that line in an orthographic view. Project the given point, O, to the front view, F, from the given horizontal view, H, along with the given line, AB. See Fig. 1. Draw the top view of the required line, CD, making it parallel and equal in length to the top view of line AB, with its midpoint at O. See Fig. 2. Draw the front view of line CD, making it parallel to the front view of line AB, with O as its midpoint. Determine the endpoints of line CD by projection of the line from the top view. See Fig. 3. The front views of the lines AB and CD will then be equal in length, also. Thus the required line is line CD.

• PROBLEM 1-11

Explain intersecting and non-intersecting lines, and the method of determining intersection of lines.

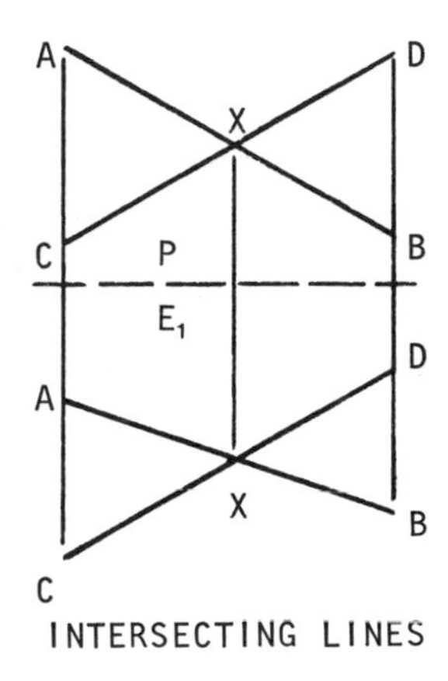

Fig. 1

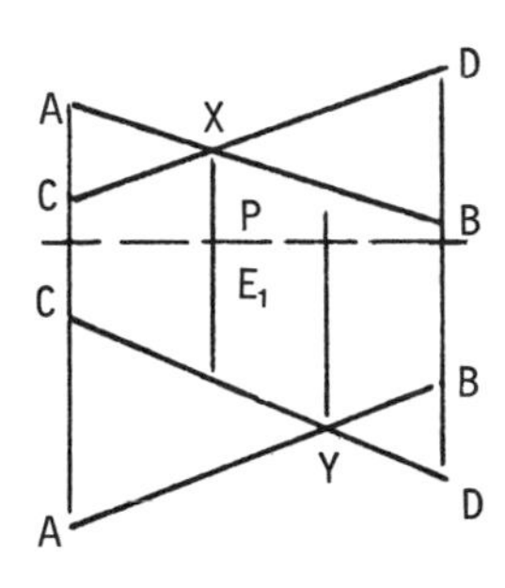

NONINTERSECTING LINES

Fig. 2

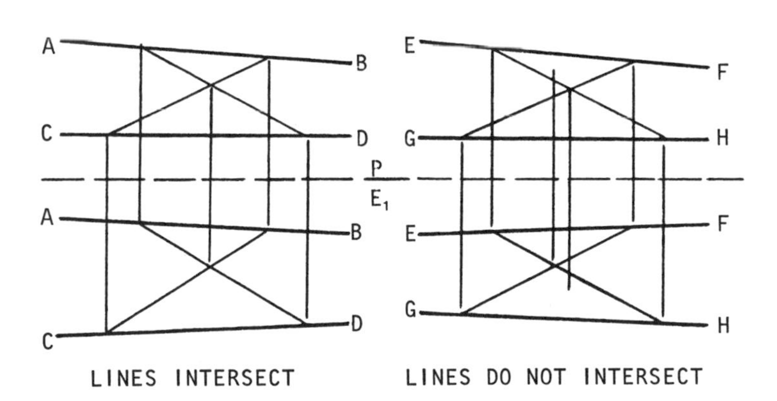

Fig. 3

Fig. 4

Solution: (1) By definition, lines intersect if they have a common point. (2) If two crossing lines apparently intersect in each of two adjacent views, they actually do intersect only if the apparent points of intersection lie on the same projection line. (3) As illustrated in Fig. 1, the lines cross at a common point along the projection line in the P- and E_1- views, and thus are intersecting lines. The lines in Fig. 2 are non-intersecting. (4)

If two lines that are not parallel have no apparent point of intersection within the limits of the drawing, the method illustrated in Figs. 3 and 4 will determine whether or not the lines intersect. (5) These given lines intersect if any two intersecting lines, drawn between the given lines in one view and projected to the adjacent view, cross at a common point along the projection line in the two adjacent views. (6) The lines in Fig. 3 intersect, while those in Fig. 4 do not.

• PROBLEM 1-12

Explain the method for determining the visibility of nonintersecting lines at the points of crossing.

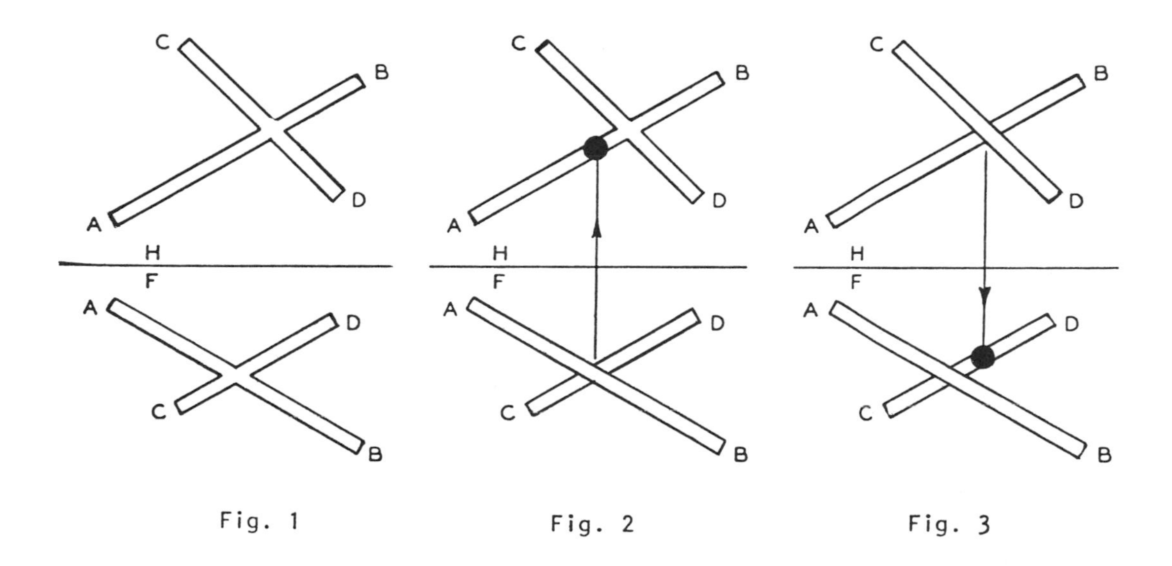

Fig. 1 Fig. 2 Fig. 3

Solution: (1) Two lines, A-B and C-D are shown crossing in the top and front views of Fig. 1. By applying the method of determining the intersection of lines, as described in the foregoing problem, it is obvious that these lines do not intersect.

(2) To determine the visibility in the front view, the crossing point in the front view is projected to the top view in Fig. 2. (3) This projector intersects line A-B before line C-D, indicating that line C-D is further back. (4) This establishes line A-B as being visible in the front view, since the horizontal view depicts true distances from the frontal plane and since the line that is closest to the front view would therefore be the one that is visible.

(5) The visibility of the top view is determined by projecting the crossing point from the top view to the front view, shown in Fig. 3. (6) This projector inter-

sects line C-D first, establishing line C-D as being higher than or above line A-B, and therefore visible in the top view.

(7) Note: Visibility in a given view cannot be established by that view alone. It is necessary to determine visibility by inspecting the preceding view, as outlined in this example.

• PROBLEM 1-13

Find the shortest connection from a given point, C, to a given line, A-B, using the line method.

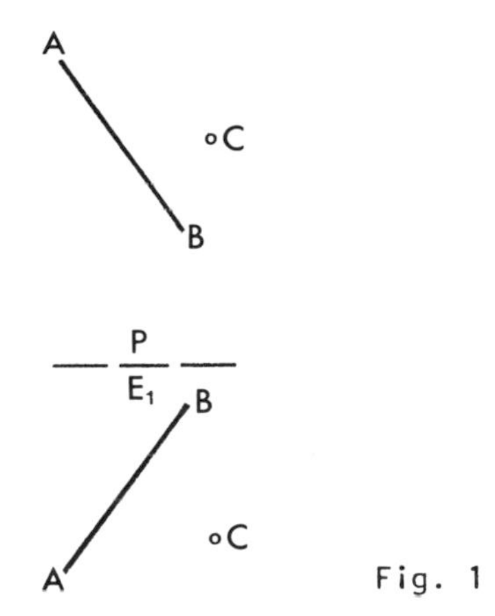

Fig. 1

Solution: (1) Fig. 1 shows the given line and point in the plan view, P, and the primary elevation view, E_1. (2) The shortest connection from a point to a line is perpendicular to the line. This connection must appear perpendicular to the line in any view showing the true length of the line. Since the true length of a line can be seen only if the lines of sight are perpendicular to the line, the true length of the shortest connection must appear in a view showing the line as a point. (3) In Fig. 2, view E_2, the secondary elevation view, is drawn to show the true length of A-B. (4) The shortest connection, C-X, is located in view E_2 by drawing a line from C perpendicular to A-B. (5) View I_1, the primary inclined view, shows A-B as a point and the true length (but not the true slope) of the shortest connection, C-X. (6) The true slope is found in view E_3, the third elevation view, rotation line P-E_3 being parallel to C-X. This is based on the fact that the slope of a line is the angle the line makes with the horizontal, and it can be seen only when the line appears in true length in an elevation view.

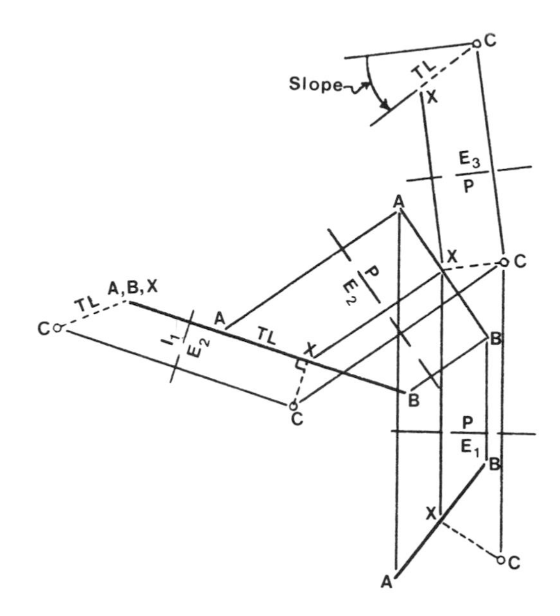

Fig. 2

• PROBLEM 1-14

Find the shortest distance between two nonintersecting, nonparallel (skew) lines, A-B and C-D, using the line method.

Solution: (1) Fig. 1 shows the given lines in the plan view, P, and the primary elevation view E_1. (2) Through spacial analysis, one recognizes that when the lines are viewed so that one appears as a point, the common perpendicular to both lines will show in true length, from the point view perpendicular to the other line. This is known as the line method of solving the problem. (3) In Fig. 2, lines A-B and C-D are to be connected with the shortest possible line. (4) The secondary elevation view, E_2, is projected from the plan view to show A-B in true length. (5) The primary inclined view, I_1, is drawn having lines of sight parallel to the true length projection of A-B, in order to produce a point view of A-B. Furthermore, because the lines of sight are parallel to A-B, view I_1 shows the true length of any line that is perpendicular to A-B. Since the shortest connection is perpendicular to both A-B and C-D, it is drawn at right angles to C-D in view I_1 to locate point R. (6) Point R is then projected into view E_2, where a line is drawn at 90° to A-B to locate point S. Note that R-S is parallel to the $\frac{E_2}{I_1}$ rotation line in view E_2 since it is true length in view I_1.

(7) The shortest connection, R-S, may then be projected into the given views, P and E_1, and its bearing can be measured in the plan. (8) If the slope of this connection is required, another view, E_3, must be drawn. E_3 is the necessary elevation view showing R-S in true length.

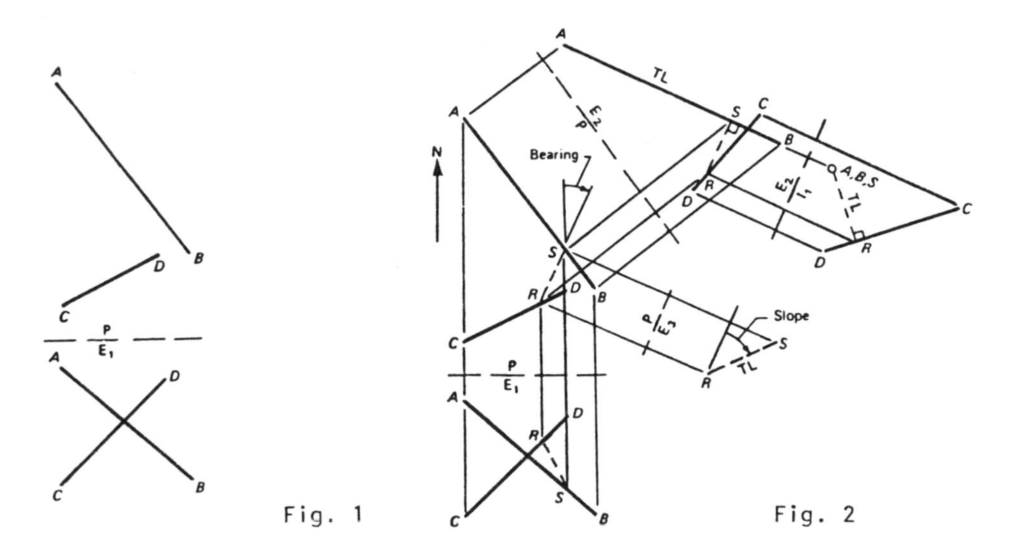

Fig. 1 Fig. 2

APPLICATIONS

• PROBLEM 1-15

> Complete the front view of line AB, Fig. 1, if the line slopes upward from point A and makes an angle of 25° with a frontal plane.

<u>Solution</u>: If the required front view of the line A-B were made to appear parallel to the border line between the horizontal and frontal planes, A^H-C, shown in Fig. 2, then the horizontal view of the line would be a true length view and would appear to be 25° from the same border line.

a) Construct the border line A^H-C as shown in Fig. 2.

b) Measure 25° up from this line from point A and draw a line. If the true length of the line AB were to appear in the horizontal view it would appear along this 25° line.

c) Construct a line parallel to the border line AC at B^H and locate the intersection point, B_r^H, between this line and the 25° line. This line, B^H-B_r^H, represents the horizontal view of the arc length traveled by point B of line A-B. The projection line from B_r^H then locates the point B_r^F. Line $A^F - B_r^F$ is the desired line seen rotated into a horizontal position in the frontal plane.

d) Counter-rotate line $A^F - B_r^F$ back to intersect the projection of line AB in the frontal plane at points $A^F - B^F$. Line $A^F - B^F$ is the required line.

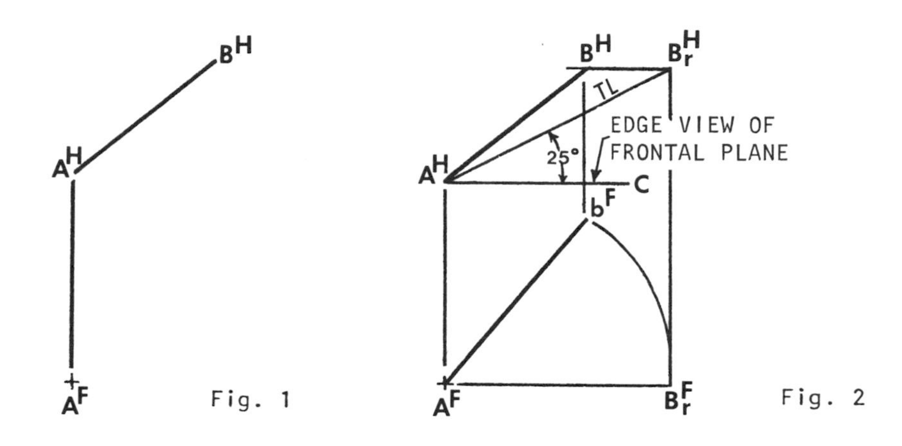

Fig. 1 Fig. 2

• PROBLEM 1-16

1) Complete the front view of line EG, Fig. 1, if the line slopes upward from E at an angle of 30° with a horizontal plane.

2) Find the true angle which line EG makes with the frontal plane.

Solution: 1) To solve this problem one must first recognize what information one has been given. First, a line in space exists sloping outward from the frontal plane at an unknown angle and upward from the given horizontal plane at a known angle of 30°. Second, the location of one point, E^F, is known in the front view and the vertical line representing the possible locations of point G in the frontal plane is also known - it is shown as the line through G^H perpendicular to the border line, B-B, between the top (horizontal) plane and the frontal plane. This line is known as a projection line. Note that the angle between line $E^H - G^H$ and line B-B in the top view is not the unknown angle which the actual line EG makes with the frontal plane since $E^H - G^H$ is not the true length of line EG.

Suppose the actual line EG were to rotate back through the unknown angle with the frontal plane until it was parallel with the frontal plane. If this were to happen the angle between line EG and the horizontal plane would remain 30°, the line E^H- G^H would appear parallel to the border line, B-B, as shown by line E^H- G'^H in Fig. 2, and the frontal view would show the true length of the line EG, as shown by line E^F-G'^F in Fig. 2.

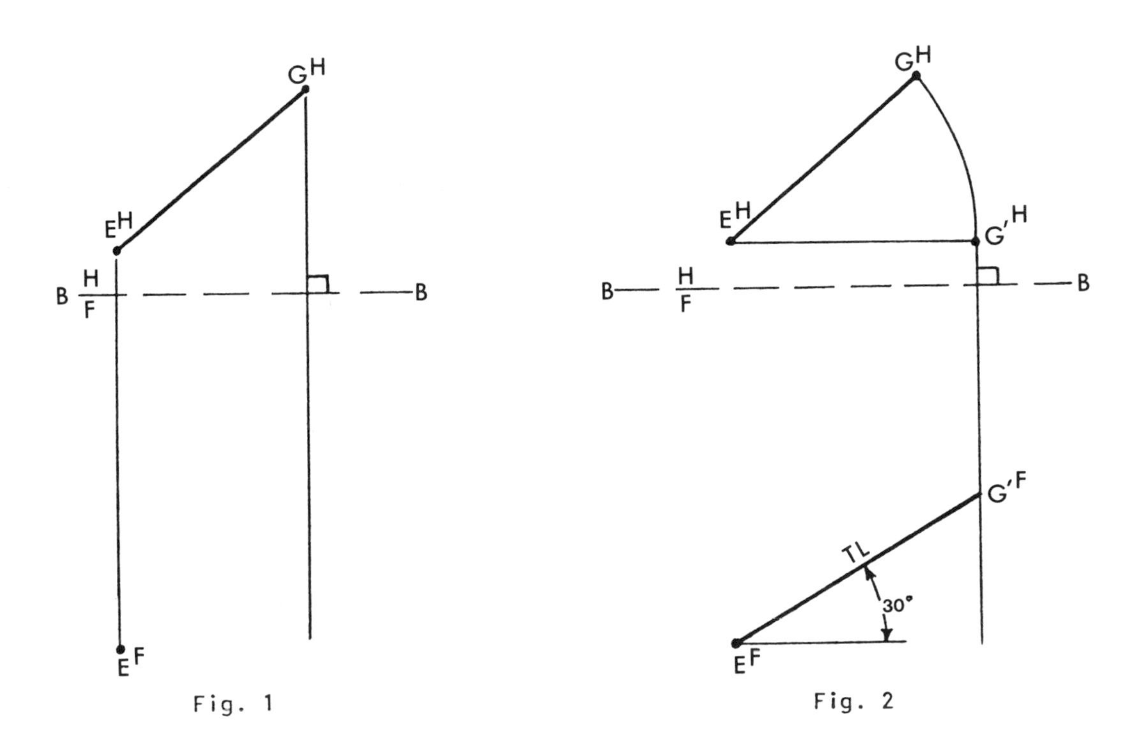

Fig. 1

Fig. 2

The next step is to realize that arc G^H-G'^H, which is the top view of the arc turned by point G through the unknown frontal angle must appear as a straight line parallel to the horizontal plane in the frontal view since any angle made with the frontal plane must be made in a plane parallel with the horizontal plane. Line G^F-G'^F, in Fig. 3, represents this arc in the frontal view. The intersection of the projection line, G^H-G^F, and the arc line, G'^F-G^F, is the location of point G in the frontal view. Line E^F-G^F is the required frontal view of line EG.

2) In order to view the true angle between line FG and the frontal plane in the horizontal view the true length of line E-G must be shown in the horizontal view. In order to show line E-G in true length in the horizontal view line E-G must be made to rotate back through the 30° which it makes with the horizontal plane. The path of this rotation is shown as a dotted line through G^H, parallel to the fron-

tal plane, in the top view of Fig. 4.

The intersection of this dotted line and the line representing the true length of EG measured from point E^H is shown in Fig. 4 as point G_{TL}. The line E^H-G_{TL} shows line EG in its true length and also shows the true angle between line E-G and the frontal plane.

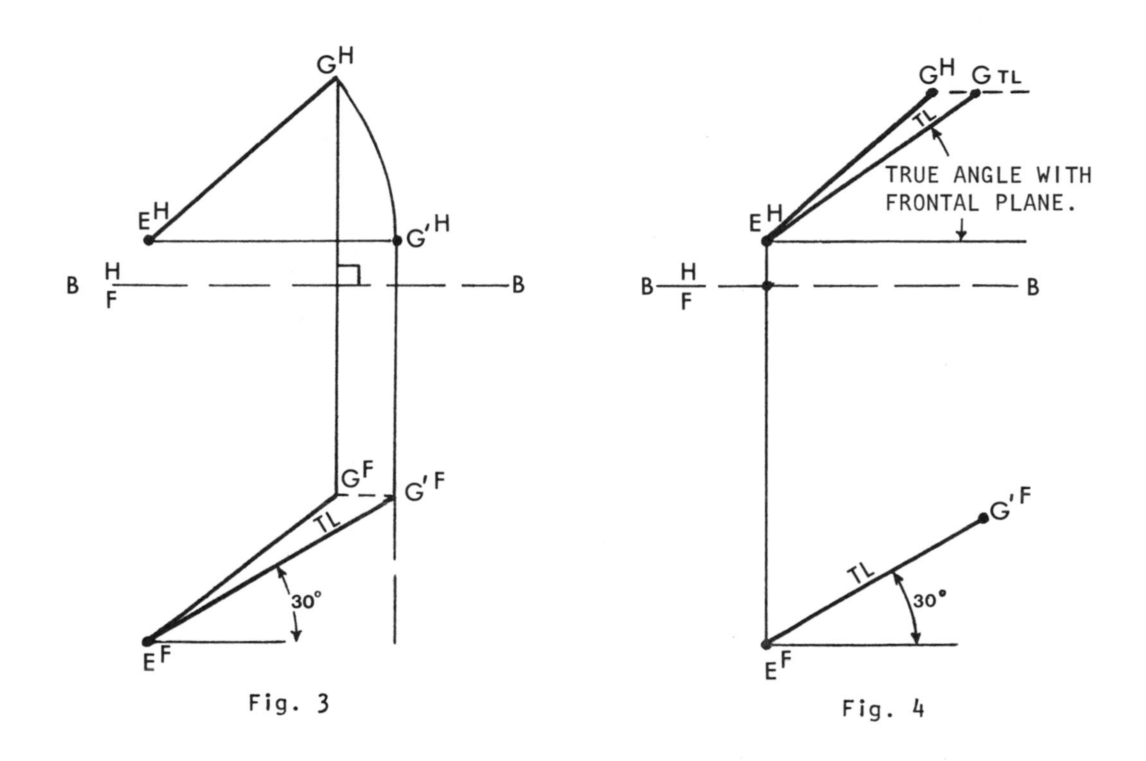

Fig. 3

Fig. 4

• **PROBLEM 1-17**

Find the true clearance between the spherical tank with its center at point O and the cylindrical pipe with center line XY. The scale of the figure is 3/4" = 1'.

Solution: The true distance between the pipe and the tank can be measured in a view which shows the point view of center line, X-Y. a) Construct a line, B_1-B_1' in the figure, which shows the intersection line between the horizontal plane and a plane perpendicular to the horizontal plane and parallel to the pipe's center line XY. By measuring the distances D and D_1 found in the frontal view, the true length of center line X-Y is found in this first auxiliary view. Locate the center of the spherical tank in auxiliary view 1 by using D_3 from the frontal view.

b) Construct auxiliary view 2 by drawing borderline B_2-B_2' perpendicular to the center line X^1-Y^1. Measure distances D_2 and D_4 found in auxiliary view 1. Note that the true circular shape of the pipe is also shown in this point view of center line XY. The point to point distance between the pipe and the tank is shown in the figure as 8 inches.

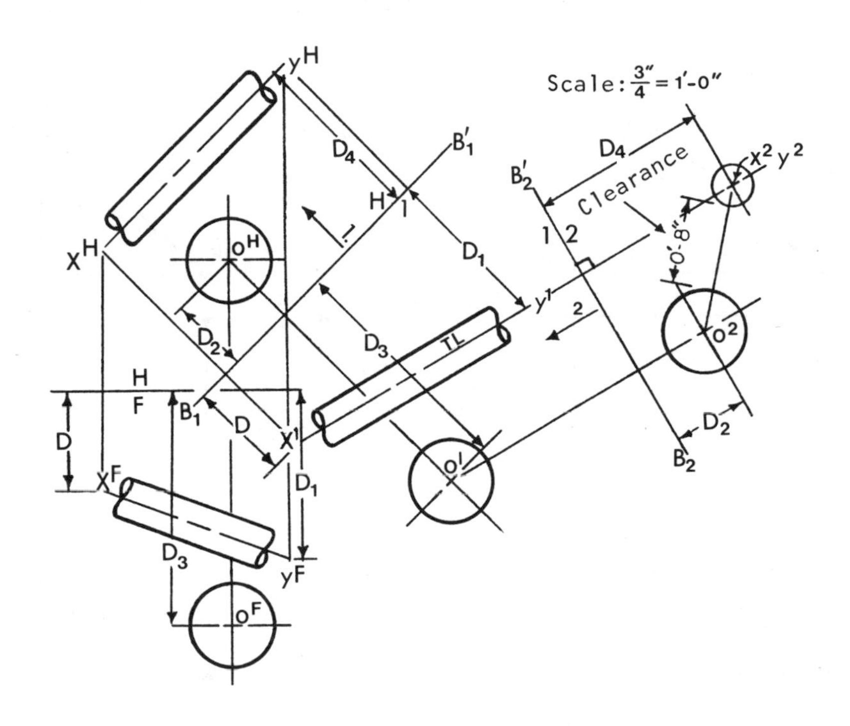

• **PROBLEM 1-18**

Locate in both the frontal view and in the horizontal view a point K which lies 1/4" above a given horizontal line AB and 3/16" in front of a given frontal line AC and in the plane ABC.

Solution: Only one point, K, exists which meets all the stated requirements; namely, that it be in a given plane, ABC, that it be in a second plane representing all points 1/4" above a line, AB, and that it be in a third plane 3/16" in front of a line, AC. Fig. 1 illustrates the problem in perspective.

The problem will be solved in the following way. First, the two given lines, AB and AC, will be shown in both the frontal and horizontal views. Second, the two planes which represent the loci of points at the required distances from the lines will be shown. Third, a line, H-H, will be established in both views which represents the intersection of the plane ABC with the plane of points 1/4" from line AB. Point K must be on line H-H at the intersection of line H-H with the plane which shows the 3/16" distance from the frontal line, AC. Line H-H must be parallel to line AB since it is the intersection by a plane, ABC, with two parallel planes; namely, the horizontal plane containing line AB and the plane of points 1/4" from the horizontal plane. Finally, point K will be located in the top (horizontal) view and then projected to the frontal view.

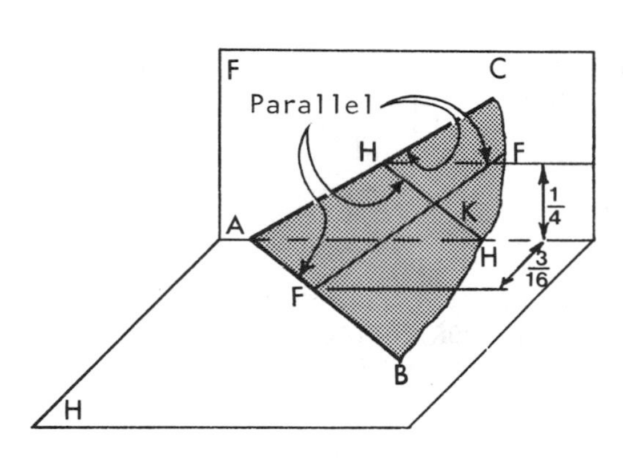

Fig. 1

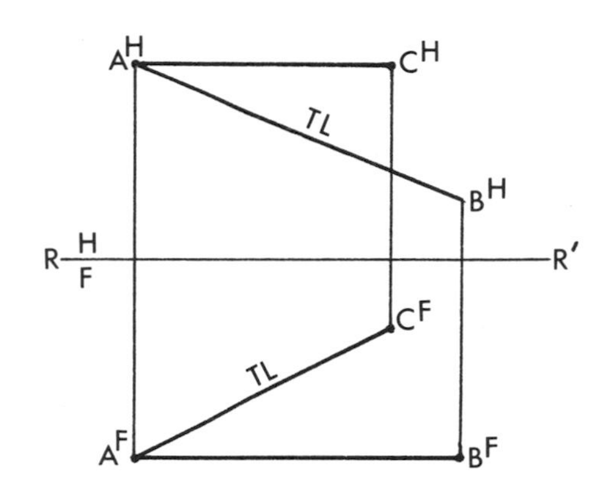

Fig. 2

a) Draw the frontal view of line AC as shown in Fig. 2 sloping generally upward as shown in the perspective sketch, Fig. 1. Line AC by definition lies in the frontal plane; therefore, it is seen true length in this view.

b) Construct a top view of line AC. Note that line AC must appear parallel to the borderline R-R' in the top view because it is shown true length in the frontal view.

c) Draw a line in the top view to represent the horizontal line AB sloping generally downward as indicated by the sketch of Fig. 1. Note that the horizontal view shows the horizontal line, AB, in true length. Note also that lines AC and AB will appear to intersect at A in all views which show both lines.

d) Construct the frontal view of line AB.

e) Construct the two planes which will each represent the locus of points at the specified distances as shown in Fig. 2. Note that the horizontal plane 1/4" from the line AB does not appear in the horizontal view and appears as a line in the frontal view. Similarly, the frontal plane at a distance of 3/16" from line AC appears only as a line in the top view and does not appear in the frontal view.

f) Construct the line, H-H, which is the intersection of plane ABC with the horizontal measurement plane. Note that line H-H must be parallel to line AB and must intersect line AC (since it is not parallel to line AC and is in the same plane as line AC). Line H-H also appears to lie coincidentally with the edge view of the horizontal plane as shown in Fig. 3.

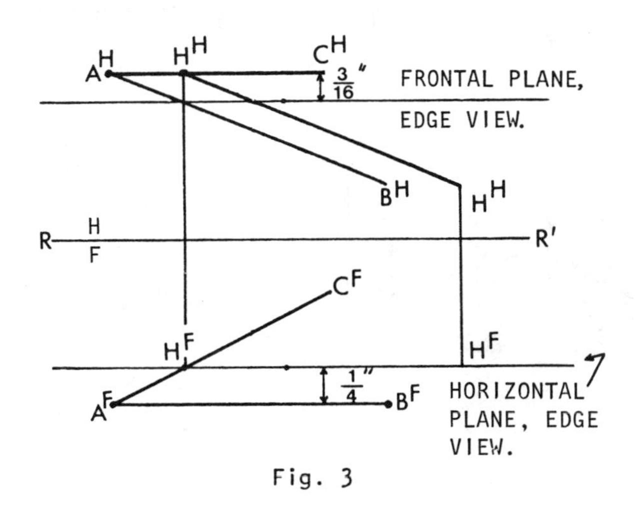

Fig. 3

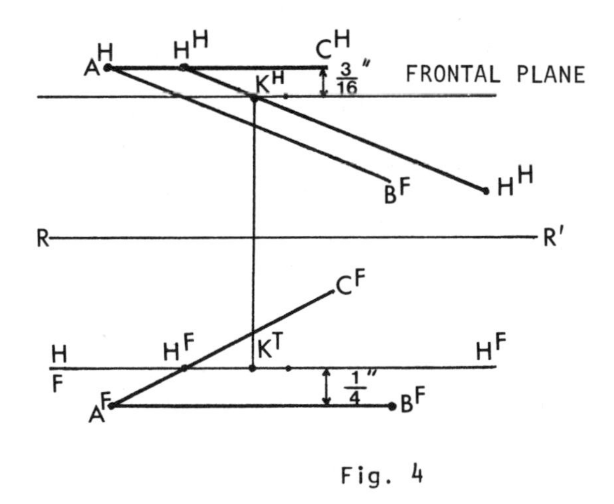

Fig. 4

g) Locate point K in the top view at the intersection of line H-H with the frontal measurement plane as shown in Fig. 4.

h) Project point K into the frontal view and locate point K on line H^F-H^F.

Complete the top view of the hoist frame, Fig. 1, given the information that the braces CE and GE intersect AB at point E.

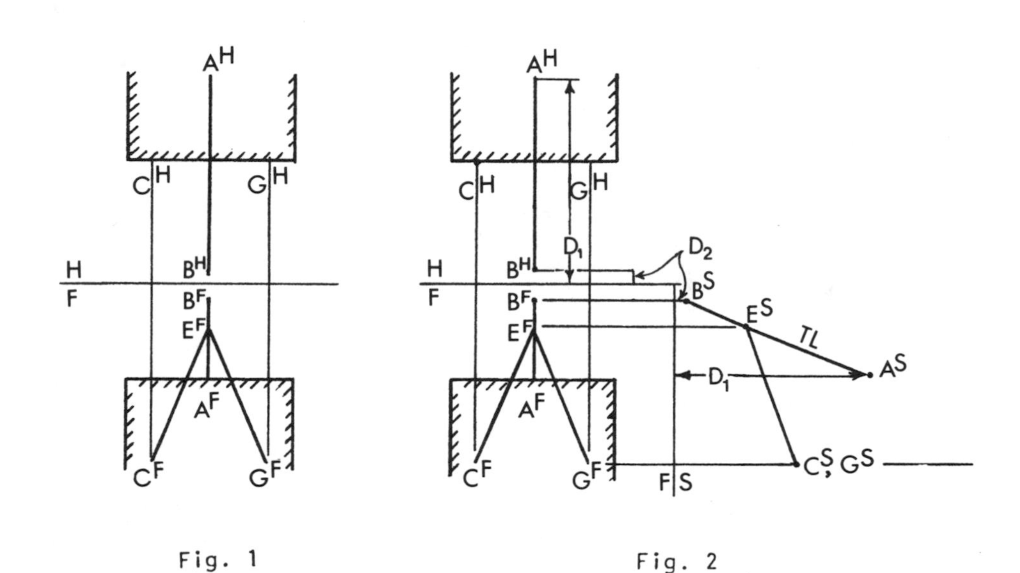

Fig. 1

Fig. 2

Solution: The top (or horizontal) view of the three lines, A-B, C-E, and G-E is incomplete because it is missing point E. Since A^H-B^H and A^F-B^F appear perpendicular to the borderline of the horizontal and frontal planes, point E^F cannot be simply located by projection into the top view. Therefore, a third view is necessary to locate point E.

A side view of line AB will locate point E.

a) Construct a side viewing plane to the right of the frontal plane as shown in Fig. 2.

b) Locate points B and A in the side view by measuring off distances D_2 and D_1 from the horizontal plane into the side plane. Note that line AB appears true length in the side view. Points C^S and G^S (which appear to coincide in the side view) are located in a similar way although the locations of points C and G in the side view are not necessary to solve this problem.

c) Project point E into the side viewing plane. Line E^S- C^S,G^S can also be drawn for completeness.

d) Once E^S is located in the side plane its location in the horizontal plane can be obtained simply by measuring its distance, D, as shown in Fig. 3, from the frontal plane and locating it at this distance along line A^H-B^H in the top view. Fig. 3 shows the correct location of point E in the top view.

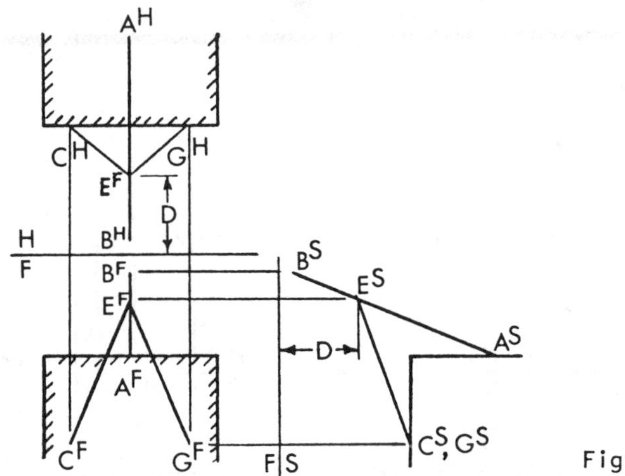

Fig. 3

• PROBLEM 1-20

Given the horizontal and vertical projections of point A, determine the horizontal and vertical projections of line AB having a bearing of N30°W, a slope of 45° down from A, and a length of 1.5 inches. Construct to the given scale.

Solution: By definition, the bearing of a line is the direction the horizontal projection of the line has relative to the points of a compass located in the horizontal plane. Thus, regardless of what angle line AB actually makes with a frontal plane it will appear at an angle of 60° from the frontal plane in the horizontal view.

a) Construct a line, A^H-X^H, through point A in the horizontal plane bearing N30°W as shown in the figure.

b) Construct a vertical viewing plane, V'-V', parallel to line A^H-X^H in order to draw line A-B in true length. Line V'-V' represents the intersection of the horizontal plane with a vertical plane. (By definition, any vertical plane is perpendicular to the horizontal plane, and any distance measured vertically down in such a plane will be a true distance in any other vertical plane, such as the frontal plane.)

c) Measure the vertical distance, D_1, in the frontal plane and transfer this distance along the projection of point A, locating point A^V as shown. Construct line A^V-B^V in true length and at a 45° slope from the horizontal plane. Note that a line shown in true length will appear in true angle.

d) Project point B^V back into the horizontal plane locating B^H along the bearing line, A^H-X^H, as shown in the figure.

e) Project point B^H into the frontal plane locating B^F at the true vertical distance, D_2, as shown.

f) Draw line A^F-B^F to complete the required frontal view.

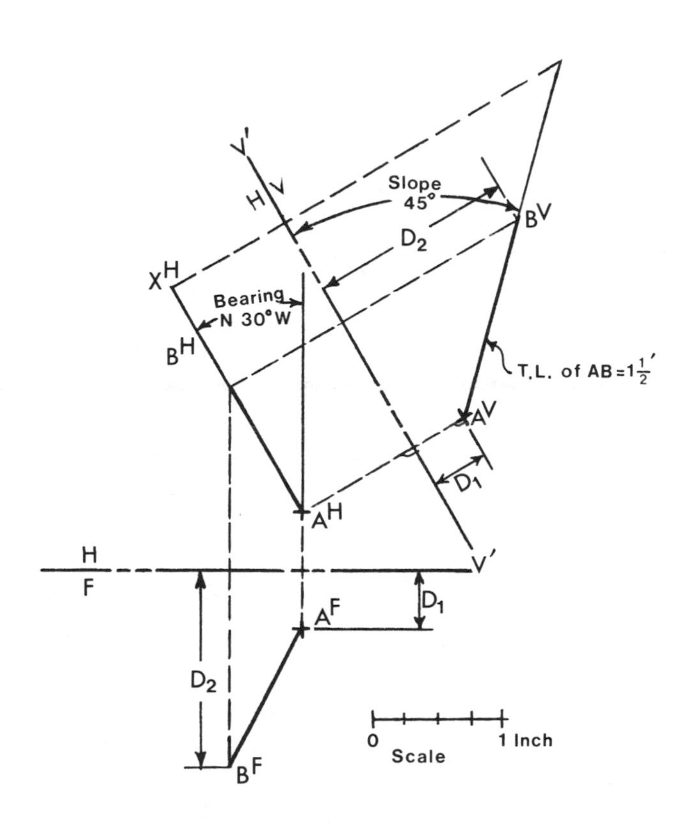

CHAPTER 2

PLANES AND PLANAR VIEWS

EDGE VIEW, SLOPE, TRUE SHAPE

• PROBLEM 2-1

Explain plane surfaces.

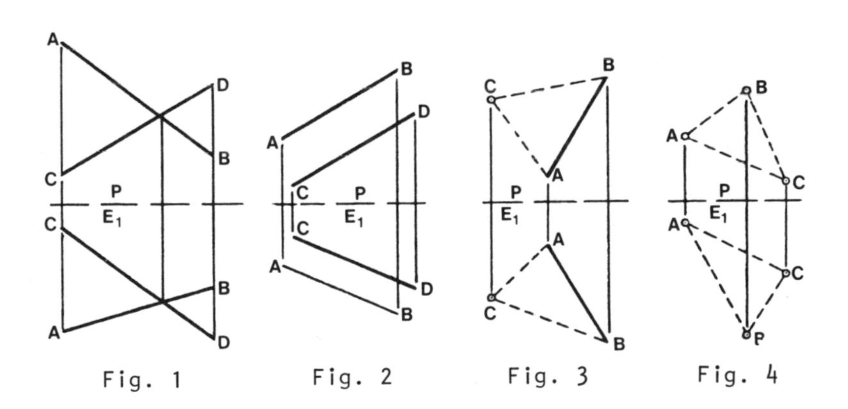

Fig. 1 Fig. 2 Fig. 3 Fig. 4

Solution: (1) A plane is a surface such that any two points on the surface may be connected by a straight line that lies entirely on the surface. The locations of planes in space may be designated by the locations of two intersecting lines (Fig. 1), two parallel lines (Fig. 2), a line and an external point (Fig. 3), or three points not in a straight line (Fig. 4).

(2) In solving problems, planes are considered to be unlimited in extent, and the plane, or any line in the plane, may be extended beyond the limits given for the plane. Furthermore, any number of lines may be drawn in a plane to

aid in the solution of the problem.

(3) Planes are grouped into four main categories; oblique planes and three principal planes. Planes parallel to one of the three principal projection planes--horizontal, frontal, or profile--are principal planes that will appear true size in a principal view.

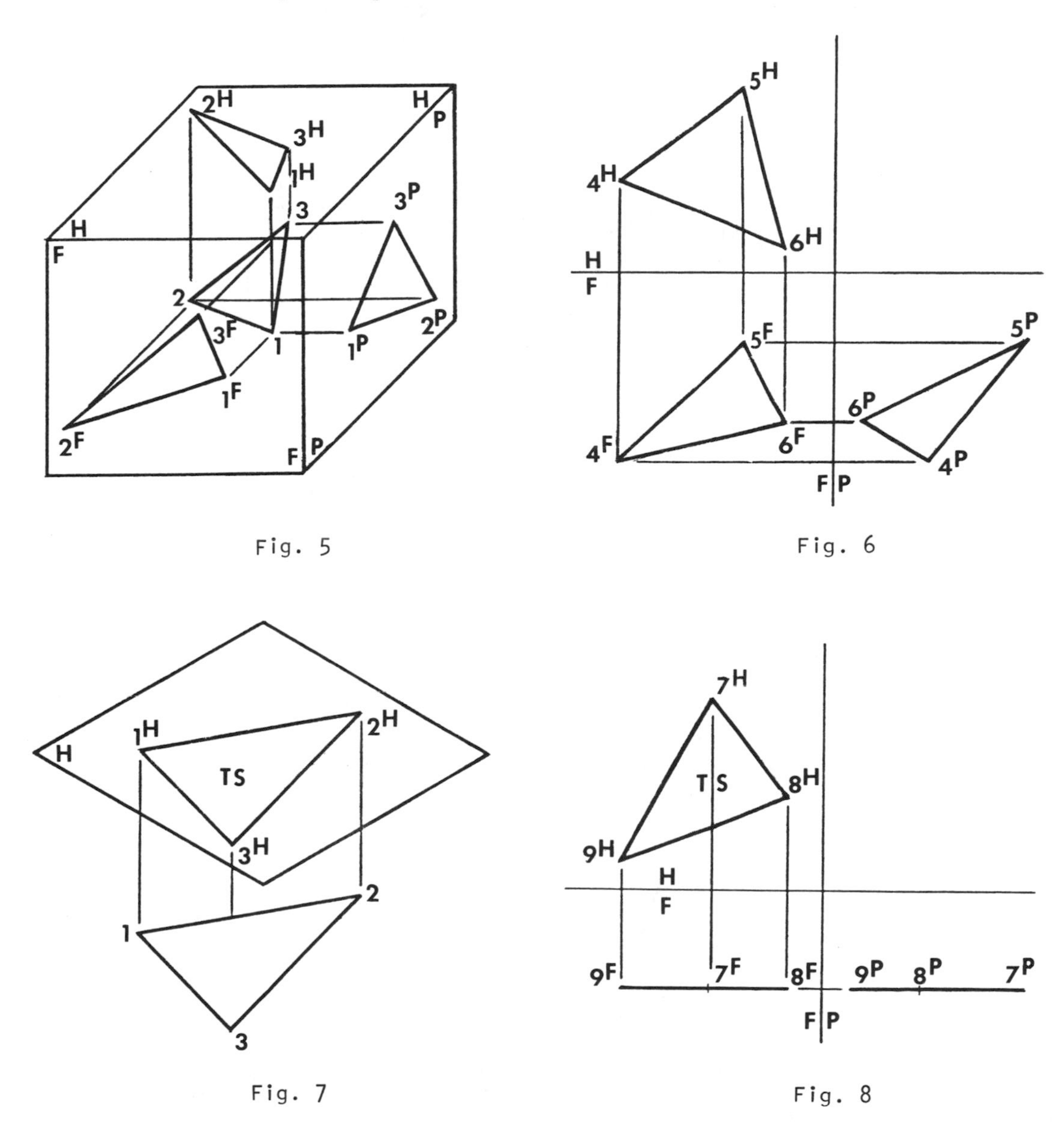

Fig. 5

Fig. 6

Fig. 7

Fig. 8

(4) The oblique plane is the general case of a plane. It is a plane that is not parallel to a principal projection plane in any view, as shown in Fig. 5. Its projections may appear as lines or as foreshortened areas which are smaller than its true size. Fig. 6 represents an oblique plane, 4-5-6, in three views. Each of the vertex points is found in the same manner in each view as though it were an individual point.

(5) A horizontal plane is parallel to the horizontal projection plane, as shown in Fig. 7. This plane is a principal plane and it appears true size in the top view. Three orthographic views of a horizontal plane, 7-8-9, are shown in Fig. 8. If the plane appears as an edge in both the front and side views, or as an edge in either the front or side view and is parallel to the H-F fold line (i.e., the rotation line) in the same view, it is a horizontal plane. Observation of the top view of a plane is not sufficient to determine whether it is a horizontal plane.

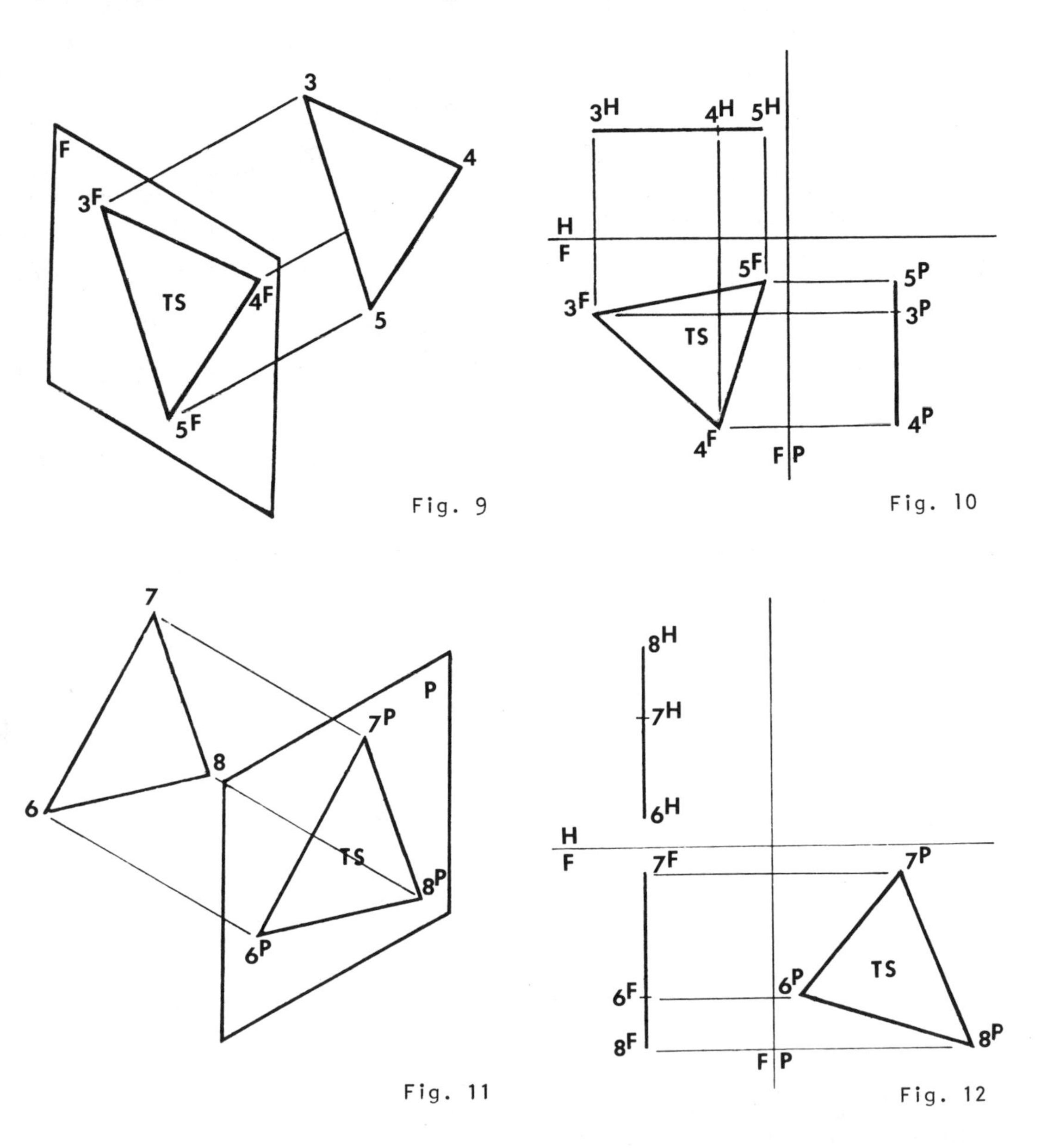

Fig. 9

Fig. 10

Fig. 11

Fig. 12

(6) By definition, a frontal plane is parallel to the frontal projection plane, as shown in Fig. 9. This principal plane appears true size in the front view and as an edge in the top and side views. The edge views of plane 3-4-5 are shown

parallel to the frontal plane in Fig. 10. There are an infinite number of shapes a frontal plane may have in the front view, but the top view and side views must be edges that are parallel to the frontal plane.

(7) The third principal plane is the profile plane, which is parallel to the profile projection plane (Fig. 11).

Plane 6-7-8 is true size in the profile view, or side view, as shown in Fig. 12. Note that the plane appears as an edge in the top and front views and that these edges are parallel to the edge view of the profile plane.

• PROBLEM 2-2

> Define and determine the edge view of a plane and the slope of a plane.

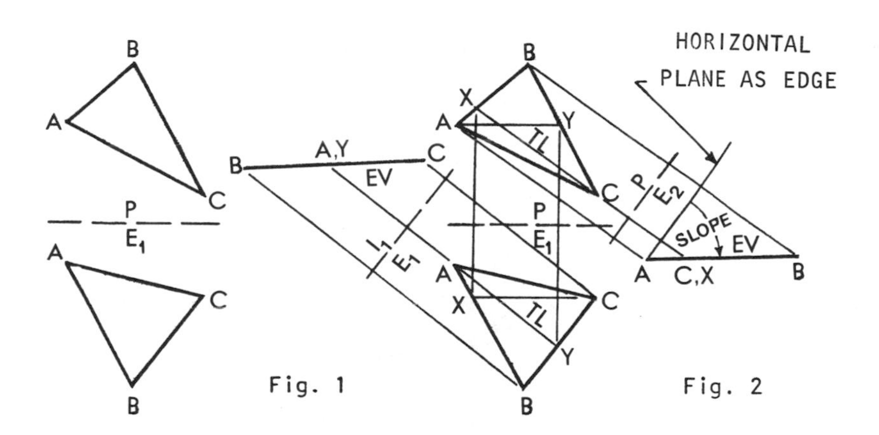

Fig. 1 Fig. 2

Solution: (1) If a plane is to appear as an edge (i.e., a line) in any view, the lines of sight for that view must be parallel to the plane. A line appears as a point when the lines of sight are parallel to the line. Obviously, if the lines of sight are parallel to any line on a plane, the lines of sight are also parallel to the plane. As a consequence, and a general rule, a plane must appear as an edge, in any view in which a line on the plane appears as a point.

(2) Since the point view of a line must be projected from a view showing its true length, a line on the plane must appear in the true length before an edge view of the plane can be drawn. (3) In Fig. 1, the P and E_1 views of plane A-B-C are given. (4) In Fig. 2, the horizontal line C-X is drawn in view E_1 and projected to the plan, where it appears in its true length. (5) View E_2, the secondary elevation, is then drawn having lines of sight parallel to the true length line C-X, showing C-X as a point and the plane A-B-C as an edge. (6) Similarly, the edge view is found in an inclined view by drawing, in the plan view, line A-Y

parallel to the rotation line $\frac{P}{E_1}$. (7) Line A-Y then appears in its true length in view E_1, and the inclined view I_1 shows A-Y as a point and A-B-C as an edge.

(8) Edge views could also be obtained by views with lines of sight at an angle of 180 degrees to those used for the E_2 and I_1 views; that is, looking at the true length lines C-X and A-Y from the opposite end. (9) Note that all true-length lines on a plane appear parallel in any view except the view that shows the true shape and size of the plane. (10) Consequently, only two edge views of a plane can be projected from any view except a true-shape-and-size view. All views projected from a true-shape-and-size view are edge views (marked EV in the diagram).

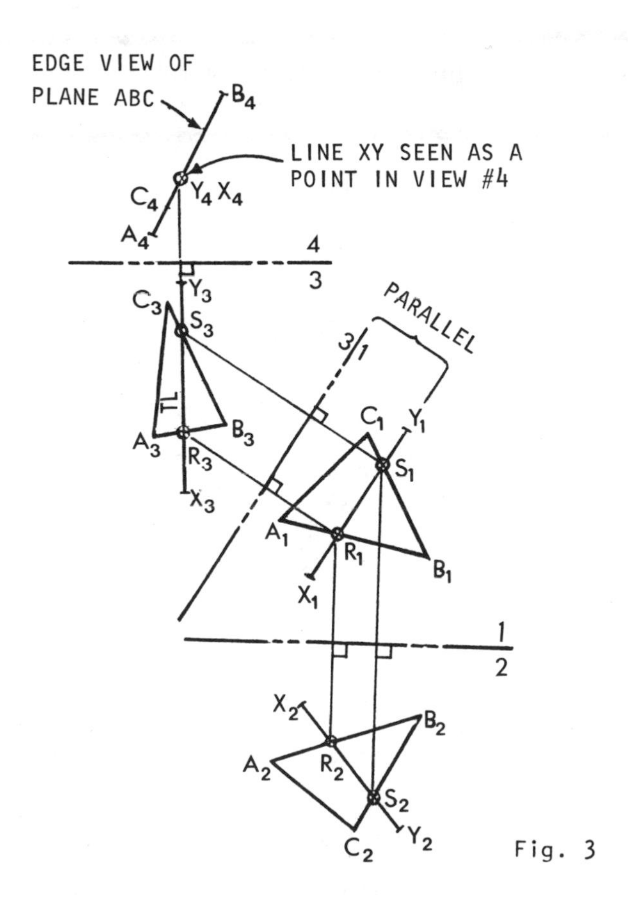

Fig. 3

(11) In general, an edge view of a plane is obtained by constructing an arbitrary line as a point, as shown in Fig. 3. (12) First, construct any line X-Y on plane A-B-C in the frontal view, view #2. (13) Determine the horizontal projection, X_1Y_1 of line X-Y (view #1). (14) Pass a projection plane, plane #3, parallel to X_1Y_1 in order to determine the true length projection of X-Y, which is noted as X_3Y_3 in view #3. (15) Pass a projection plane (view #4) perpendicular to the true length projection, X_3Y_3, in order to see line X-Y as a point X_4Y_4 in view #4. (16) View #4 shows the edge view of plane A-B-C. Note that measurements

for view #4 were obtained from view #1.

(17) The slope of a plane is the angle that the plane makes with a horizontal plane. This angle can be measured only if both the given plane and the horizontal plane are seen as edges. Since the horizontal plane appears as an edge only in an elevation view, the slope of a plane can be seen only in an elevation view that shows the plane as an edge. The slope of a plane can be expressed in the same manner as the slope of a line.

(18) In Fig. 2, the slope of the plane A-B-C is shown in the elevation view, E_2, in which a horizontal plane appears as an edge. The slope cannot be seen in the inclined view.

• PROBLEM 2-3

Define and determine the true-shape-and-size of a plane using the edge-view method.

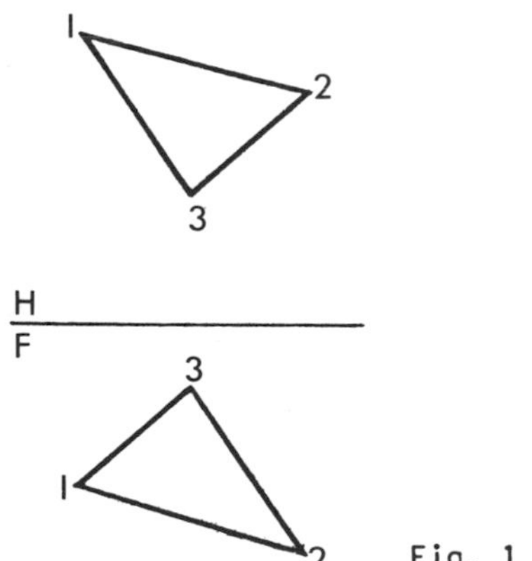

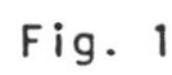
Fig. 1

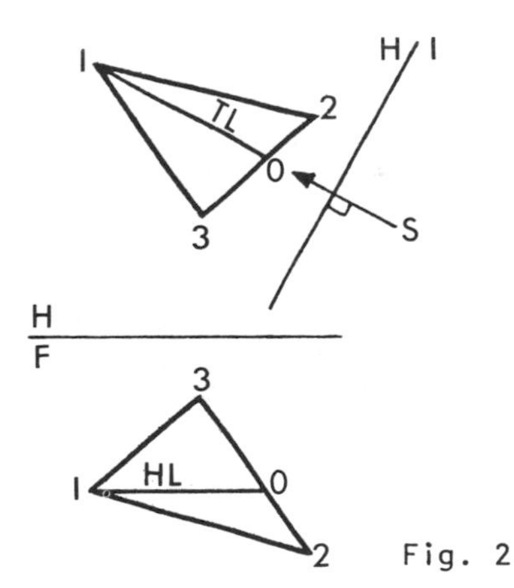

Fig. 2

Solution: (1) In Fig. 1, plane 1-2-3 is given in the horizontal and frontal planes of projection, planes H and F. (2) Problem analysis reveals that a plane will be seen in its true size and shape in any view for which the lines of sight are perpendicular to the plane. Consequently, as a rule, the view showing the true shape and size of a plane must be projected from a view that shows the plane as an edge.

(3) As illustrated in Fig. 2, draw horizontal line 1-0 in the front view of plane 1-2-3 and project it to the top view, where the line appears true length. (4) Project a primary auxiliary view from the top view parallel to the direction of line 1-0. The H-I reference plane is perpendicular to 1-0 and the line of sight.

(5) Referring to Fig. 3, the point view of line 1-0 is found in the primary auxiliary view. (6) Project points 2 and 3 to this view where the plane will appear as an edge.

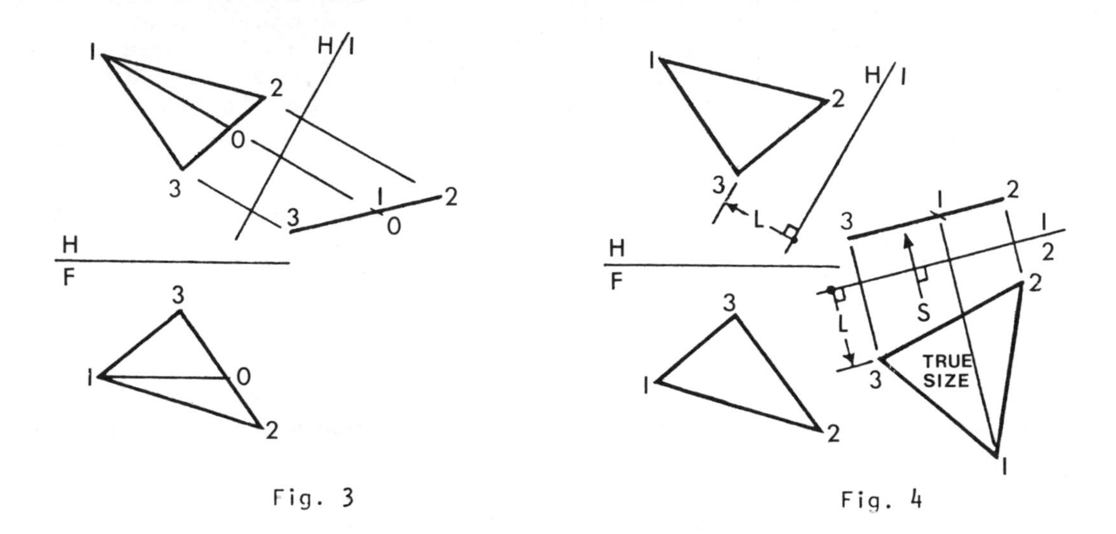

Fig. 3 Fig. 4

(7) A secondary auxiliary view plane 1-2 is drawn parallel to the edge view of plane 1-2-3, as shown in Fig. 4. (8) The line of sight is drawn perpendicular to the 1-2 plane. (9) The true-size view of the plane is found by locating each point with measurements taken perpendicularly from the edge view of the primary auxiliary plane, as indicated.

(10) The new view thus obtained shows the plane in its true shape and size. In this view, any line on the plane is true length, and any angle on the plane is true size.

• PROBLEM 2-4

Find the true shape of a given triangle, ABC, using triangulation.

Solution: Fig. 1 shows the horizontal and frontal plane views of triangle ABC. Since the triangle's true shape can only be represented if the legs of the triangle are true length, then the true lengths of AB, BC, and AC must be found first.

Using the auxiliary-view method, as shown in figs. 2,3, and 4, find the true lengths of the sides of the triangle ABC. Then construct a triangle using the true lengths of AB, BC, and AC. This triangle will show the true shape of the given triangle ABC, as shown in fig. 5.

The true shape of any polygon can be found by dividing the figure into triangles and then laying off these triangles in true size. This process is appropriately known as triangulation, and it has many important applications in the layout and development of surfaces.

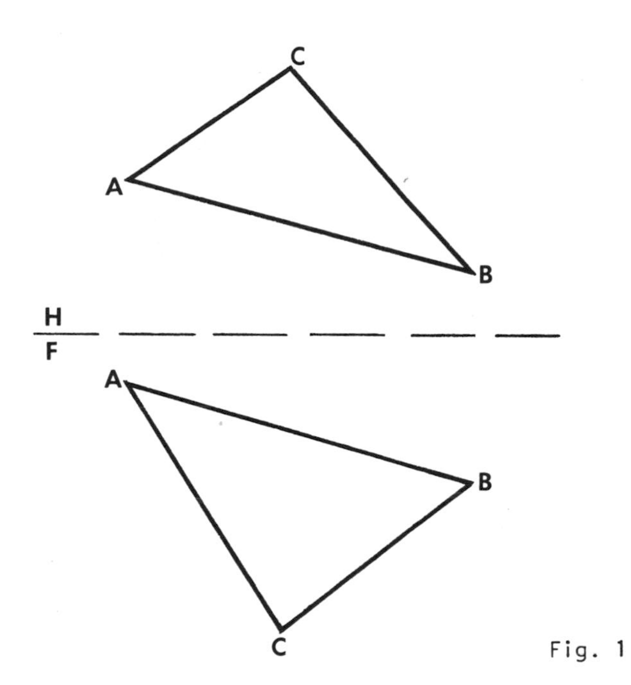

Fig. 1

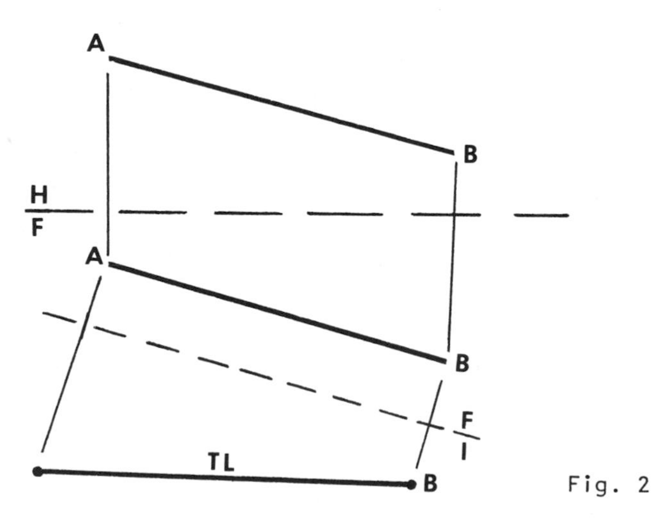

Fig. 2

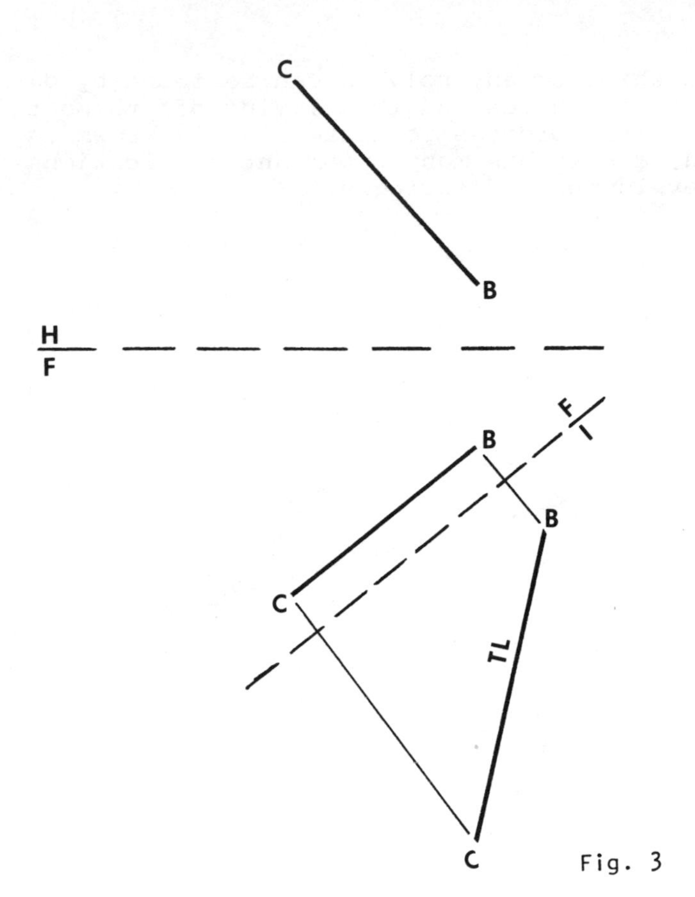

Fig. 3

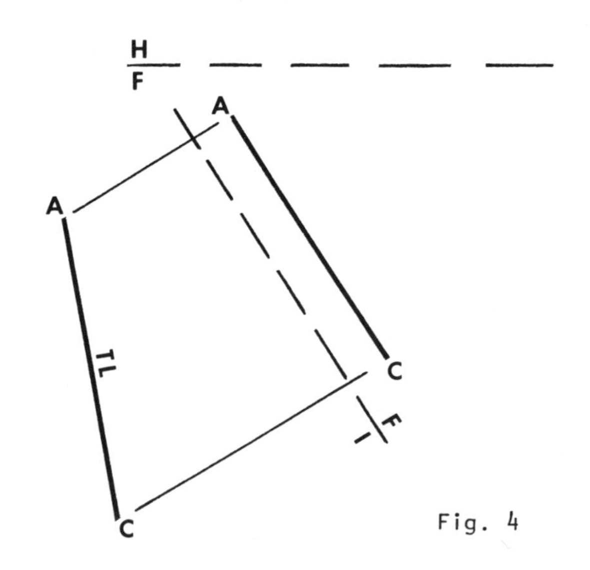

Fig. 4

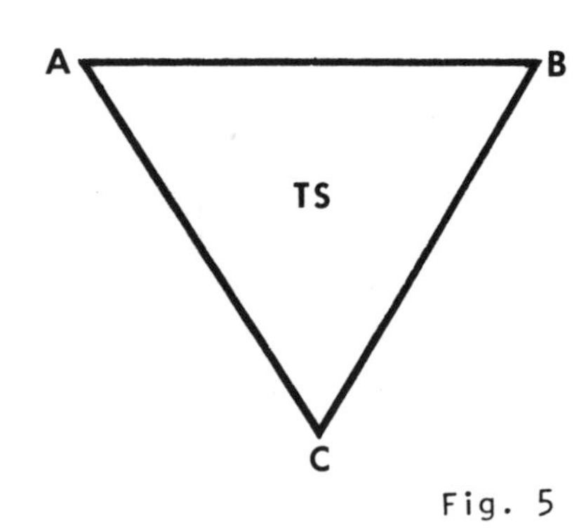

Fig. 5

POINT IN A PLANE

• **PROBLEM 2-5**

Determine the position of a point in a plane surface in a primary plane of projection, given its location in the other primary plane of projection.

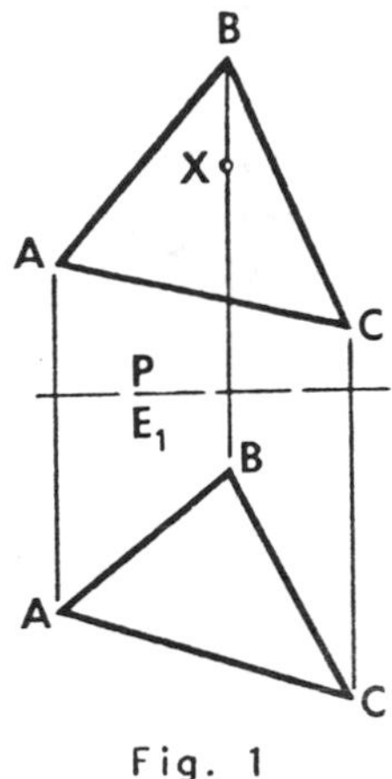

Fig. 1

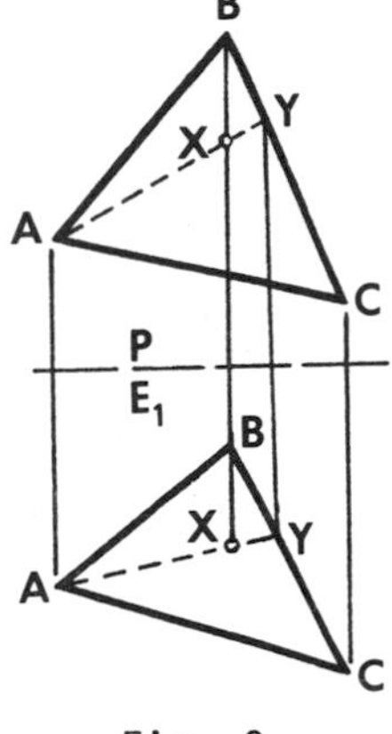

Fig. 2

Solution: (1) Two views of a plane are required to show the location of a point in a plane in space. The position of a point on the plane is fixed, however, if it is shown in one view of the plane. (2) In general, the point can be located in another view if a line is drawn on the plane through the point and is projected into the other view, as illustrated in Figs. 1 and 2.

(3) In Fig. 1 the plan view, P, and the primary elevation view, E_1, of plane A-B-C are given. Point X, the given point on the plane, is shown only in the plan view. (4) In Fig. 2, line A-Y is drawn through X in the plan view. (5) Point Y, being on line B-C, is easily located in the elevation view by projection. Line A-Y is then drawn in this view. (6) Since point X is on this line, it is located by projection from the plan. (7) Note that although line A-Y was chosen for convenience in projection, any line through X could have been used.

(8) In Figs. 3 and 4, a special case of locating a point on a plane is illustrated. (9) In Fig. 3, plane A-B-C is again given with a point X shown only in the plan. (10) In this case, because point X is on line A-C, which is perpendicular to the rotation line, it cannot be located in E_1 by projection. (11) One of the ways of locating point X in the elevation view is illustrated in Figs. 4, 5, and 6. (12) Line AB is extended in both views. (Fig. 4.) (13) Point X is projected horizontally in the plan view until it intersects

with AB extended, at point Z, locating line XZ in the plan view. (Fig. 5.) (14) Line XZ intersects line BC in the plan view at point Y, which can be projected vertically to the elevation view, along with point Z. Line YZ is then drawn in the elevation view, which locates point X on line AC, as in Fig. 6.

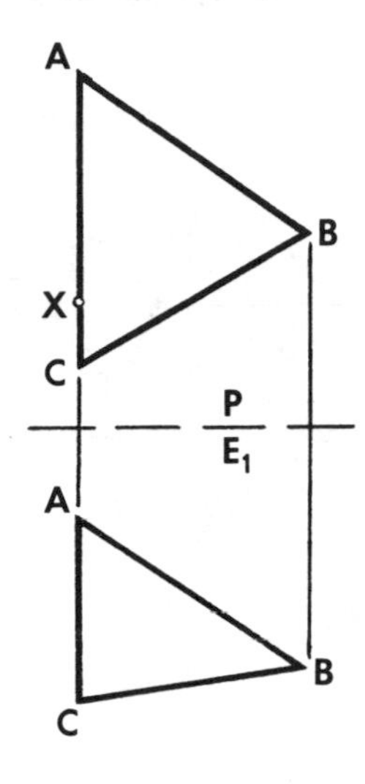

Fig. 3

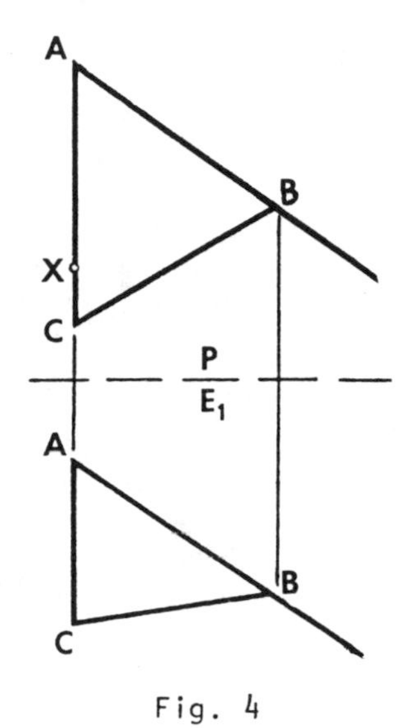

Fig. 4

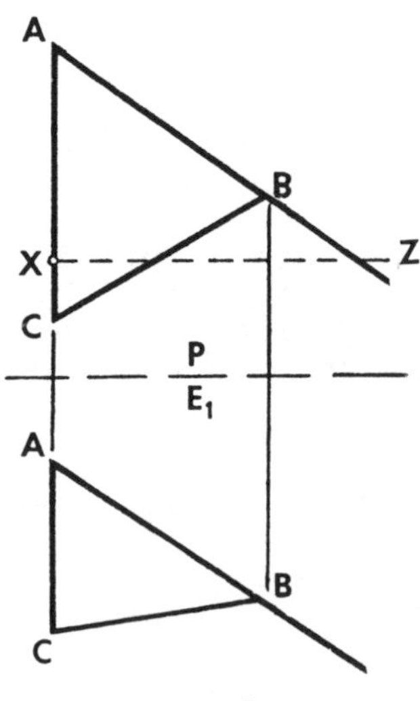

Fig. 5

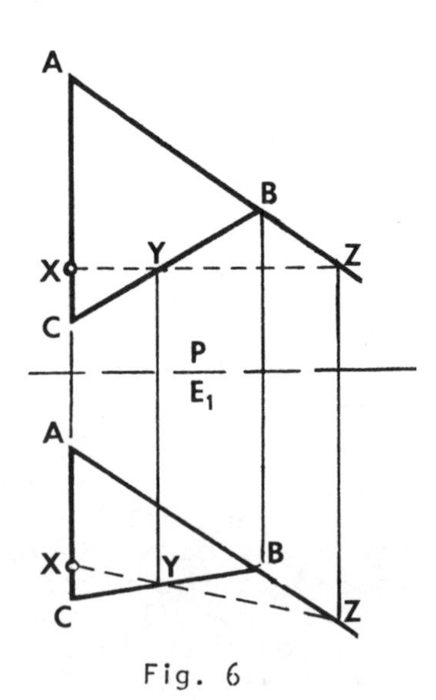

Fig. 6

INTERSECTION OF PLANES

• PROBLEM 2-6

Define and determine the traces of a plane.

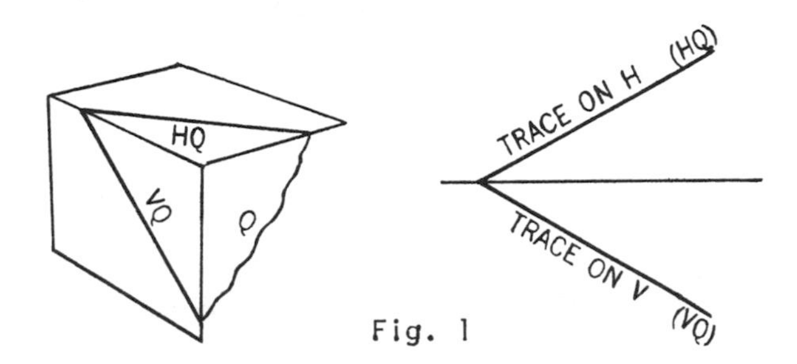

Fig. 1

Solution: (1) The line-of-intersection formed when one plane intersects another is called its trace on that plane. The intersections of a plane with the H- and V- projection planes are its horizontal and vertical traces, respectively, as shown in Fig. 1.

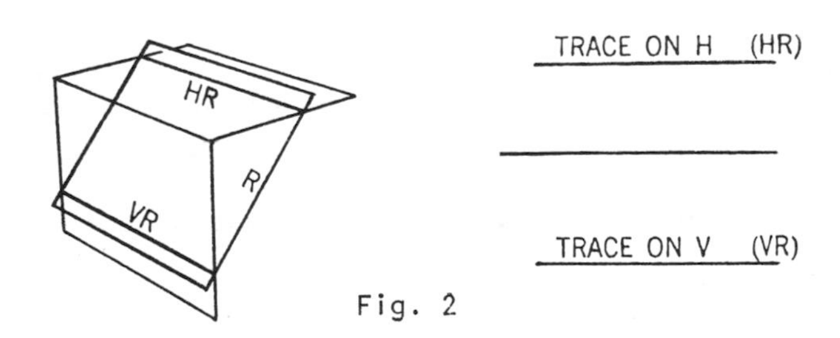

Fig. 2

The traces of a plane are particular principal lines of the plane. The H- and V- traces of a given plane are simply the principal lines which lie in the H- and V- projection planes.

(2) In general, the horizontal and vertical traces of a plane meet at a point on the reference axis. If the plane is parallel to the reference axis, the traces are parallel, as illustrated in Fig. 2.

A plane parallel to the H-plane will have no H-trace; the V-trace will be an edge view of the plane. Similarly, a plane parallel to the V-plane will have no V-trace; the H-trace will be an edge view of the plane.

(3) Traces of planes are commonly denoted by the names of the two determining planes. Thus, a plane Q may have traces H-Q, V-Q, and P-Q on the H-, V-, and P- (profile) planes respectively.

(4) Let the plane Q be passed through points A, B, and C, as shown in Figs. 3 and 4. (5) The intersection of Q and the H-plane is determined by the points 1, 2, 3 in which the lines A-B, B-C, and A-C of plane Q intersect the H-plane. (6) Similarly, the intersection of plane Q and the V-plane is determined by points 4, 5, and 6 where these same lines pierce the V-plane. Thus, the two required points, needed for the location of each trace, are obtained. (7) Note: Since points 1 and 4 are the H- and V- traces of the line A-B, it is evident that when a line lies in a plane, a trace of the line must lie on a trace of the plane.

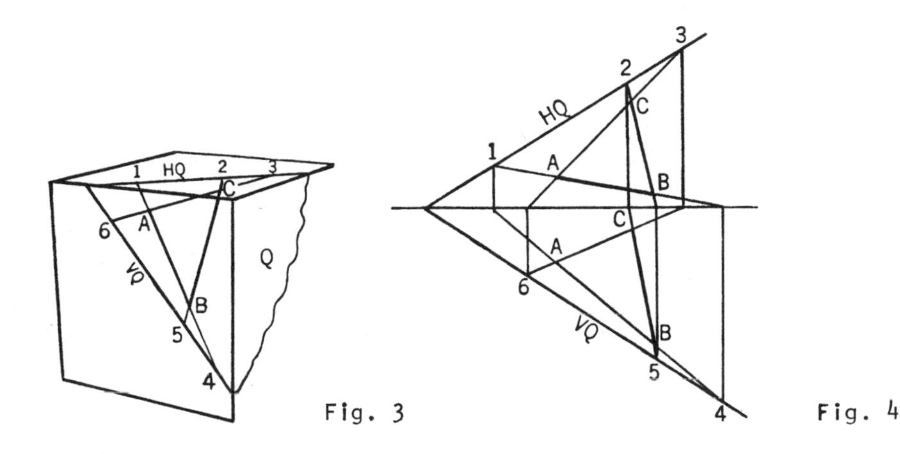

Fig. 3 Fig. 4

• PROBLEM 2-7

> Find the intersection of two planes, which are represented by their traces.

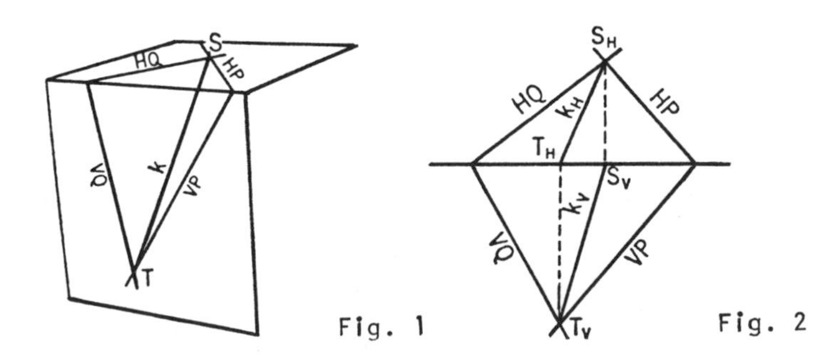

Fig. 1 Fig. 2

Solution: (1) Let P and Q be the given planes. Fig. 1 shows the horizontal traces, H-P and H-Q, and the vertical traces, V-P and V-Q, of the given planes. (2) Problem analysis reveals that if in any plane of projection, such as H (horizontal), V (vertical), or P (profile), the traces of two planes intersect, then the planes must intersect. This follows quite logically from the fact that the intersection of two plane traces (or lines) is a point, and this point is itself the trace of a line. This line is the line-of-intersection of the two planes whose traces intersect. The line of intersection is, in general, determined by the two points in which two pairs of traces intersect.

(3) The horizontal traces H-P and H-Q intersect at point S. (4) The vertical traces V-P and V-Q intersect at point T. (5) The points S and T are the traces of the required intersection K: K_H is determined by S_H-T_H; K_V is determined by S_V-T_V . See Fig. 2.

SURFACES LYING WITHIN PLANES

• PROBLEM 2-8

Construct a circle of given diameter lying in a given parallelogram, A-B-C-D, and centered at point O.

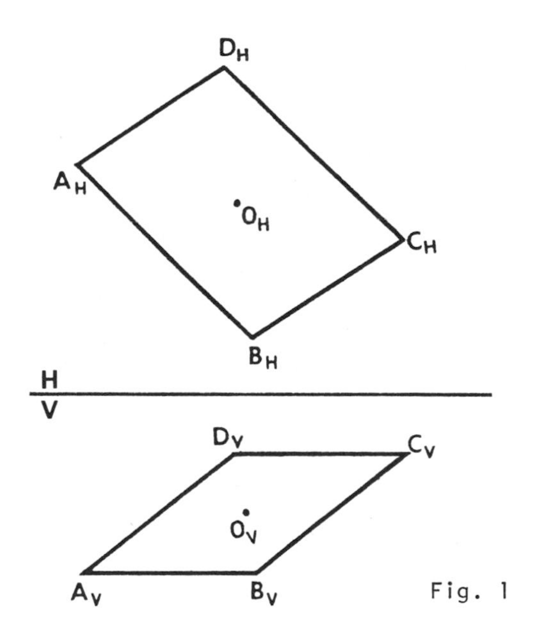

Fig. 1

Solution: (1) Fig. 1 shows the given parallelogram, A-B-C-D, in the H-, horizontal, and V-, vertical, planes of projection, and the location of point O. (2) Problem analysis calls for the realization of the fact that a circle having its plane inclined to a projection plane projects as an ellipse on that plane. The diameter of the circle which is parallel to the projection plane projects in true length and becomes the major axis of the ellipse. A second diameter at right angles to the first projects as the minor axis.

(3) Fig. 2 illustrates the following method of construction. (4) Draw the horizontal line E_V-F_V and lay off the diameter of the circle on the true length projection, $E_H F_H$. This diameter is the major axis of the top view of the ellipse. (5) The minor axis is drawn through O_H and at right angles to E_H-F_H. This is based on the fact that when two lines intersect at right angles, the true size of the right angle will appear in any view showing either of the lines in true length (unless the other line appears as a point). (6) The minor axis in the top view is a projection of the diameter of the circle constructed on the edge view of A-B-C-D, as shown in view 3. (7) The ellipse is constructed on the axes thus determined by the methods of plane geometry.

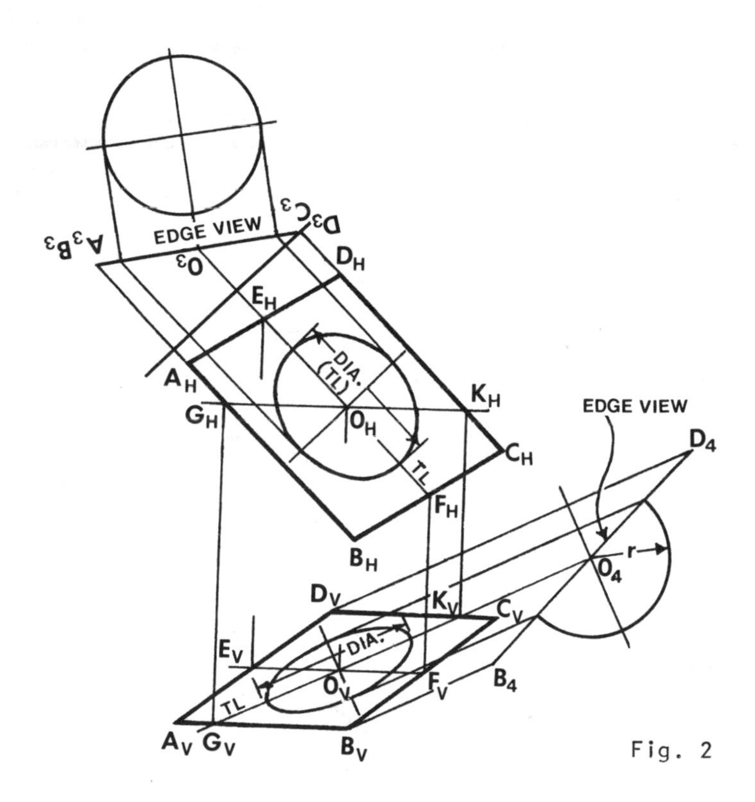

Fig. 2

(8) The major axis of the ellipse which represents the front view of the oblique circle lies along the true length frontal line $G_V K_V$, and is equal in length to the diameter of the given circle. (9) The minor axis is the projection of the diameter of the circle constructed on the edge-view of A-B-C-D, as shown in view 4. The ellipse in the frontal view is constructed on these axes thus determined.

• PROBLEM 2-9

Construct a hexagon of diameter BC lying in the given plane ABC.

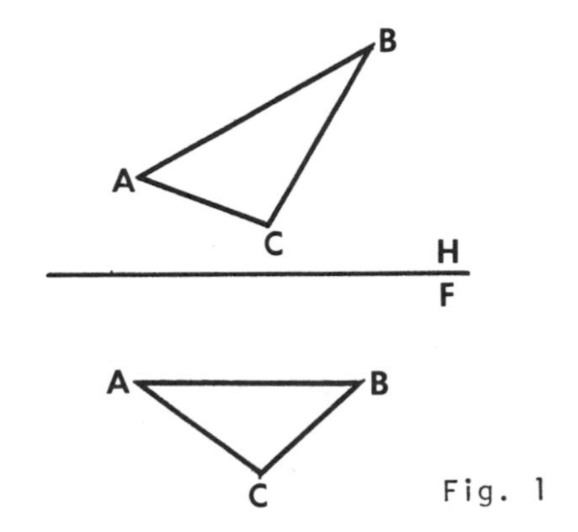

Fig. 1

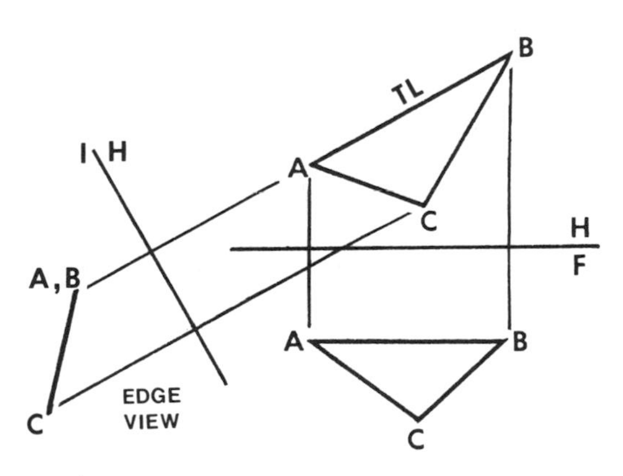

Fig. 2

Solution: Fig. 1 shows the front and top views of the plane ABC. Problem analysis reveals that a polygon to be constructed in a plane must be drawn in a view that shows the true size and shape of the plane. The true size and shape view of a plane must be drawn from an edge view of the plane. The first auxiliary view to be drawn, then, is an edge view of plane ABC. Line AB in the front view is parallel to the reference plane H-F, and therefore it projects to true length in the top view. With the line of sight parallel to line AB in the top view, make the reference plane H-1 perpendicular to line AB and project plane ABC to view 1 as an edge, as shown in fig. 2. Now the true size and shape view of plane ABC can be drawn from this edge view. Make the reference plane 1-2 parallel to the edge view and project plane ABC to view 2, where it appears in true size and shape, as in fig. 3. Locate the midpoint, M_1 of line BC and construct a circle of radius MB centered at M. Divide the circle into six equal arcs using a protractor, and connect the divisions sequentially to form the required hexagon of diameter BC, as in fig. 4. Project the hexagon back to the first auxiliary, top, and front views, as in fig. 5.

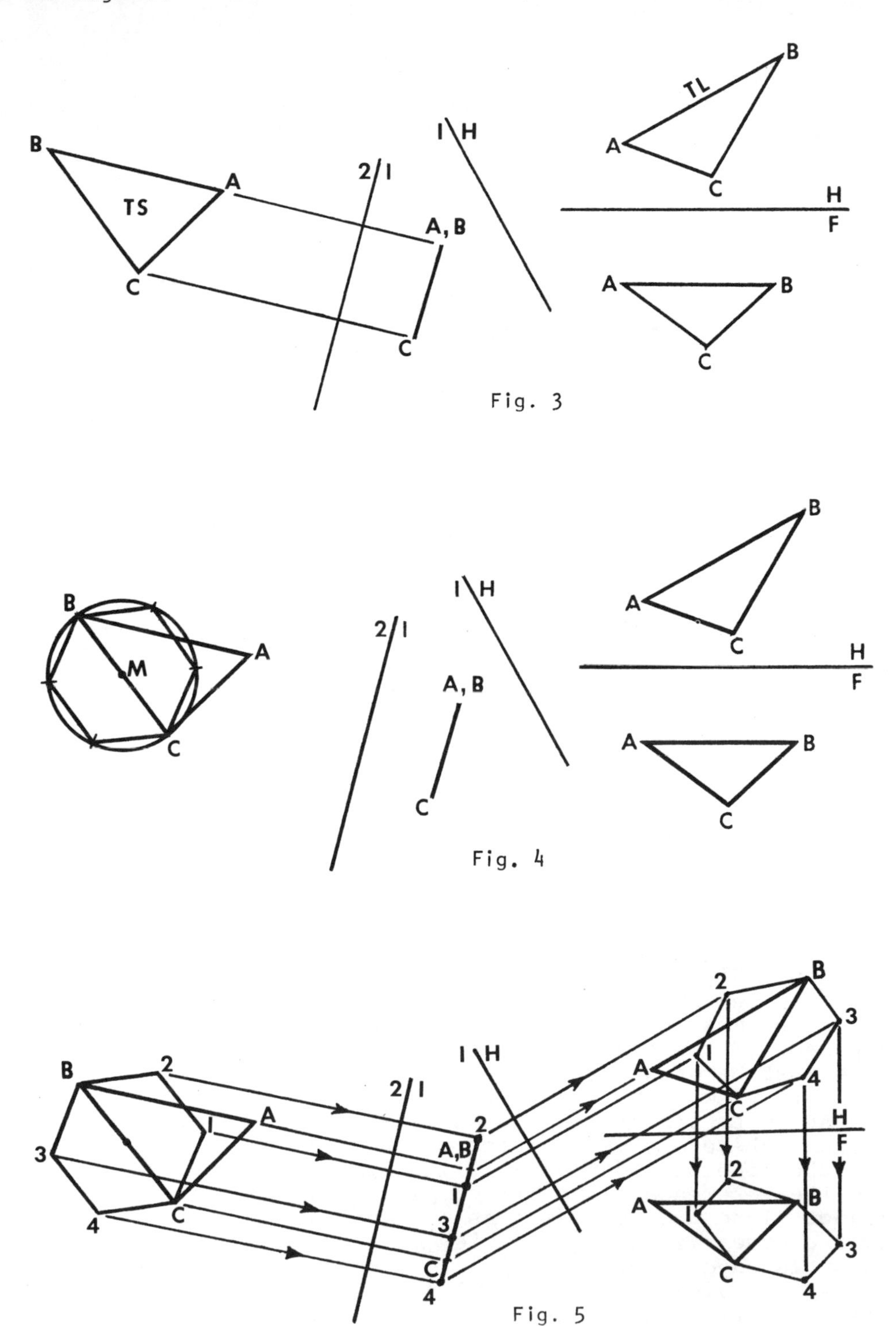

Fig. 3

Fig. 4

Fig. 5

CHAPTER 3

ORTHOGRAPHIC SKETCHING AND PROJECTION

FUNDAMENTAL PRINCIPLES

• PROBLEM 3-1

Describe various types of orthographic views.

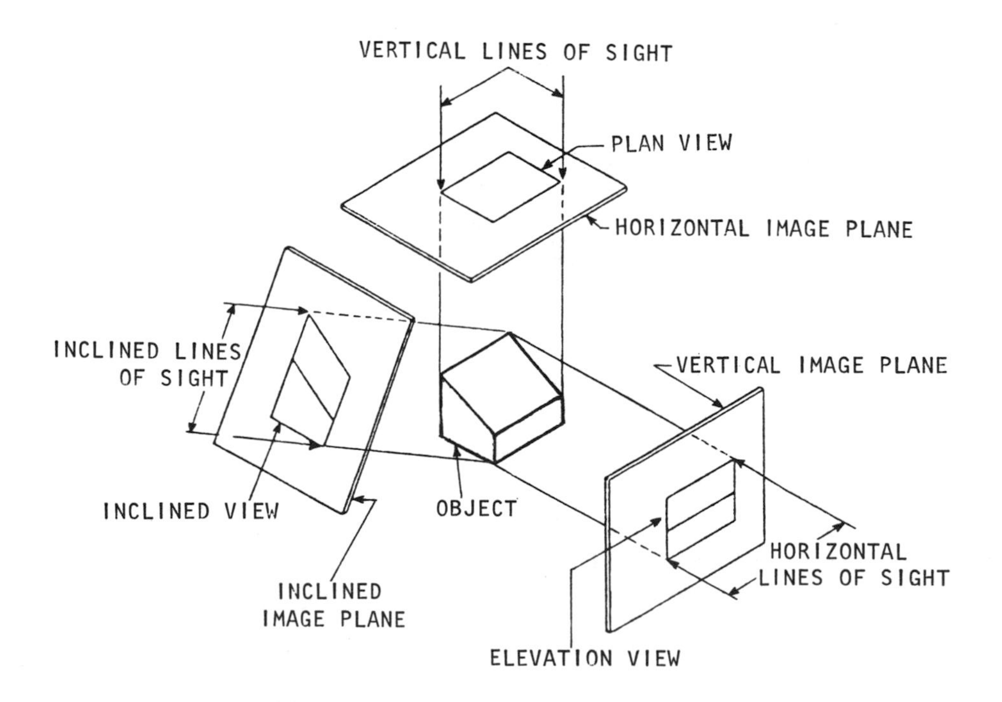

Solution: In orthographic projection, the object is considered to be drawn on a transparent image plane that lies between the observer and the object. All lines of sight from the observer to the object are at right angles to the image plane for the particular view.

The object is assumed to remain in a fixed position whereas the observer moves around the object to obtain different views. The term "change of position" therefore refers to the change in the position of the observer and not to a change in the position of the object being drawn.

When the observer is looking downward vertically, the image plane is horizontal and the view obtained is the plan view. Since the object remains in a fixed position only one plan view is possible.

If the lines of sight are horizontal, the image plane is vertical, and the view obtained is an elevation view. An infinite number of elevation views of an object can be drawn.

If the lines of sight are inclined, that is, neither vertical nor horizontal, the image plane is inclined, and the view obtained is an inclined view. An infinite number of inclined views of an object can be drawn.

• PROBLEM 3-2

Describe the nature of orthographic projection.

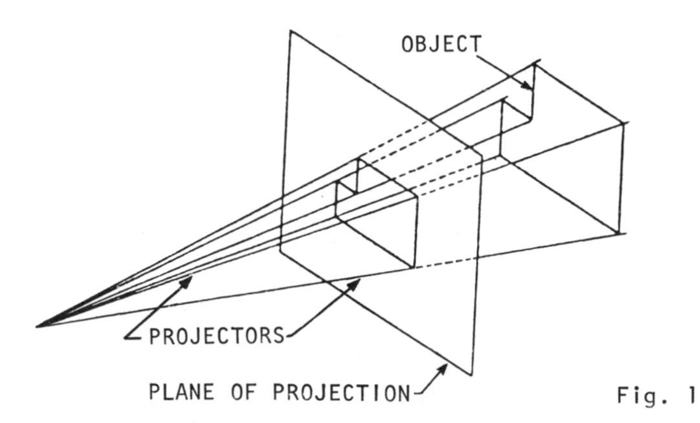

Fig. 1

Solution: The problem continually confronting the engineering draftsman is to represent a three-dimensional object on a sheet of paper which has only two dimensions. Such drawings must give the mechanic or shopman a complete description of the object to enable him to construct it with facility and without error. There are several methods of representing a three-dimensional object on a sheet of paper - namely, by photographs, by picture drawings, and by orthographic projections.

Photographs are not always practicable because the object must be made before it can be photographed. Moreover, the interior of the object cannot be seen and the back view can be shown only by another photograph. Photographs also have a tendency to distort the object somewhat, so that true relationships between the different views of the object cannot be fully expressed.

Picture drawings also are limited in their usefulness because, as in photographs, interior details are not shown. During World War II, however, many factories made use of pictorial drawings of simple objects to enable the untrained person with limited skill to perform minor operations on these objects.

Regardless of the type of representation used to show an object, all methods of representation make use of lines of sight, or projectors.

If the projectors converge to a point or are inclined to a plane placed between the observer and the object (Fig. 1), the resulting projection of the object on this plane will be distorted. This type of representation is known as perspective projection and is used principally by architects to show how a proposed building will appear when completed. Fig. 1 shows how projectors converge. This kind of projection is more nearly what the eye actually sees than is any other type of graphic representation. The plane which is inserted between the observer and the object is known as the plane of projection.

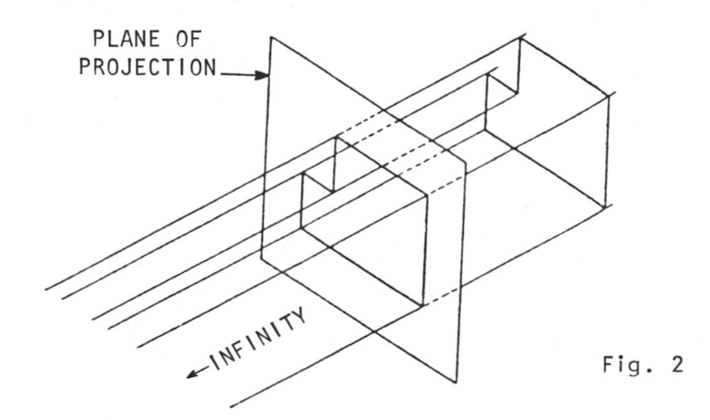

Fig. 2

If a plane, such as a pane of glass, is placed parallel to the side of an object to be drawn, the projectors may be drawn perpendicular to the pane of glass, as shown in Fig. 2. The resulting projection on the glass will be an orthographic projection of that side of the object and will show that side in its true size and shape.

• PROBLEM 3-3

Given the isometric drawing shown in Fig. 1, draw the orthographic projections of the front, top, and right side views of the object.

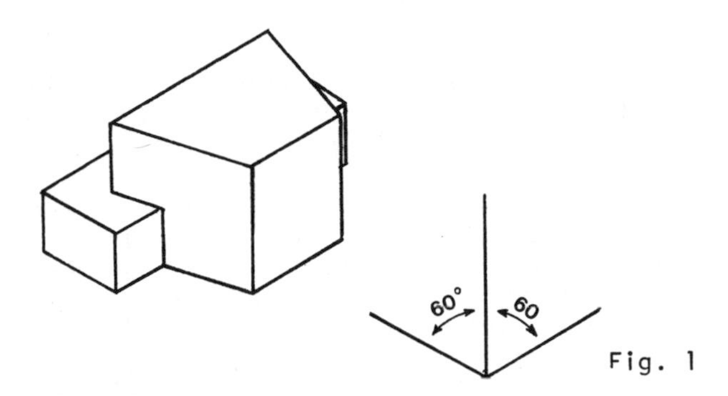

Fig. 1

Solution: Notice that no dimensions are required in this problem because it was stated that Fig. 1 is an isometric drawing. Recall that the measurements needed for drawing the required orthographic projections can be taken directly from the isometric lines of the given figure; thus, the height, width, or depth of each of the object's surfaces are taken directly from lengths of those lines of Fig. 1 which are parallel with any of the isometric axes. Note that measurements cannot be taken from a line which is not parallel to the isometric axes.

In drawing the orthographic projections of the views called for in the problem,always picture yourself as an observer looking at the object at right angles to the specific view. Another way to determine each principal view is to identify all the vertical and horizontal planes and determine inclined planes by using the vertical and horizontal planes.

When all the required views have been determined, proceed to draw them as follows: first, box in the required views as shown in Fig. 2. Include all the details in each view as shown in Fig. 3. When all the features of the object have been fully described, darken all the object lines, erase the construction lines, and include all the hidden lines. (See Fig. 4.) Notice that a hidden line was included in the right side view. This hidden line represents the edge view of the top surface of the rectangular block.

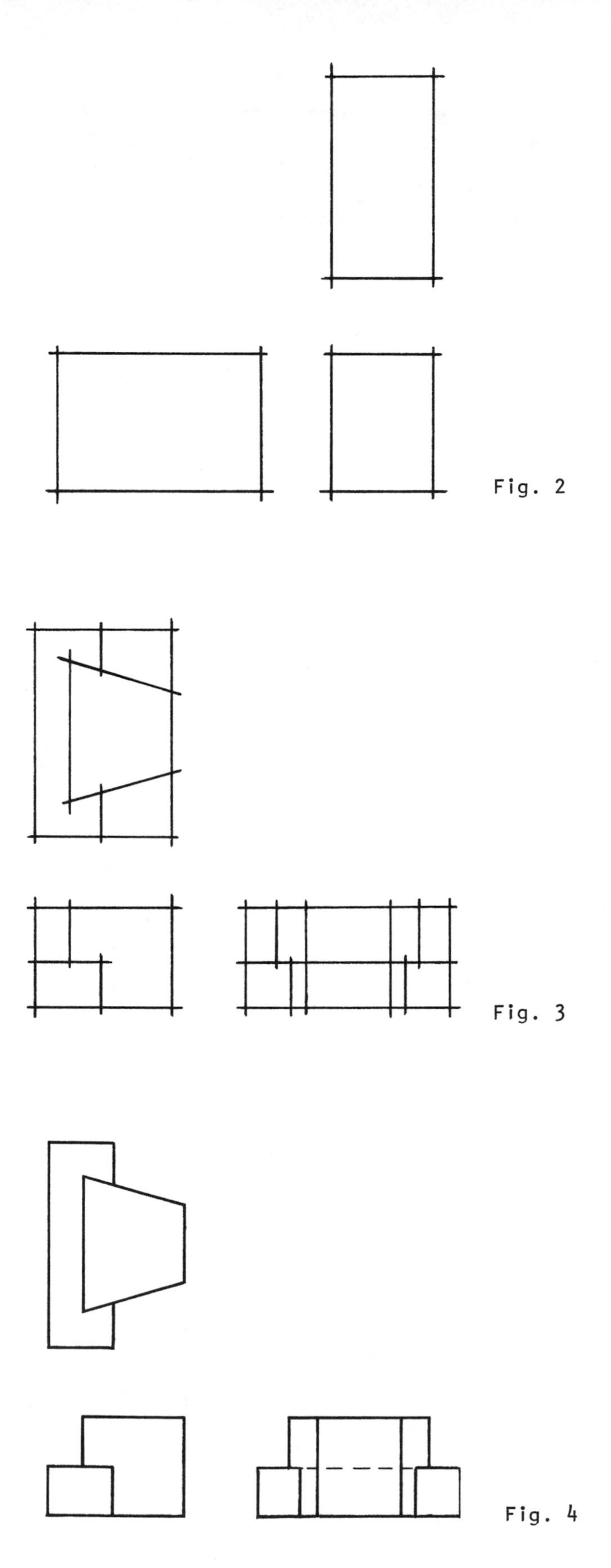
Fig. 2

Fig. 3

Fig. 4

• **PROBLEM 3-4**

From the isometric drawing given in Fig. 1 draw the orthographic projections of the front, top, and right side views of the object.

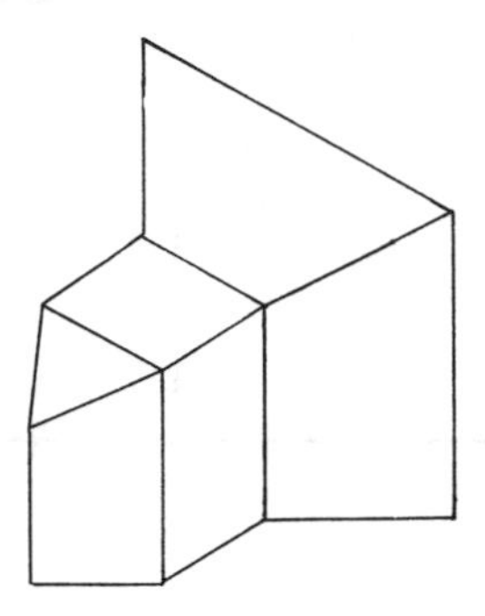

Fig. 1

Solution: The dimensions needed in order to draw the required orthographic projections can be taken directly from the isometric lines of the figure. Recall that an isometric line is any line which is parallel to any of the isometric axes. The two axes for an isometric drawing are: (a) vertical and (b) 30 degrees up from the horizontal. Make sure that all the measurements are taken from these lines. Any inclined line; that is, any line not parallel with any of the three isometric lines, must be determined by using references to isometric lines.

The steps in drawing the orthographic projections of the given object are as follows: first, box in each of the required views as shown in Fig. 2. Proceed to lay out all the details as shown in Fig. 3. Finally, darken all the lines and erase the construction lines (see Fig. 4).

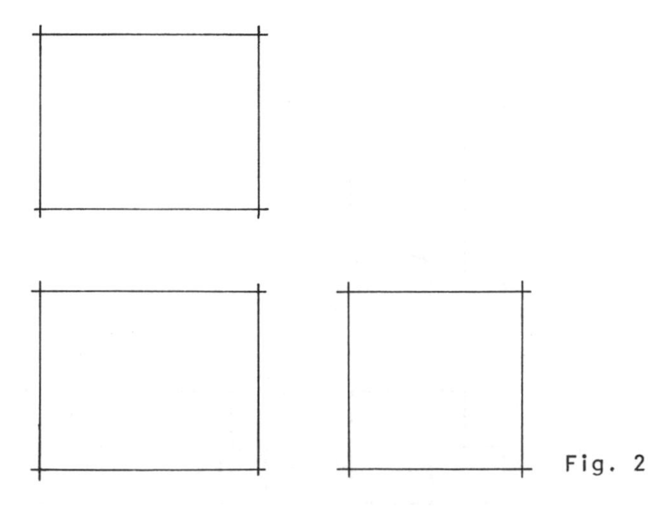

Fig. 2

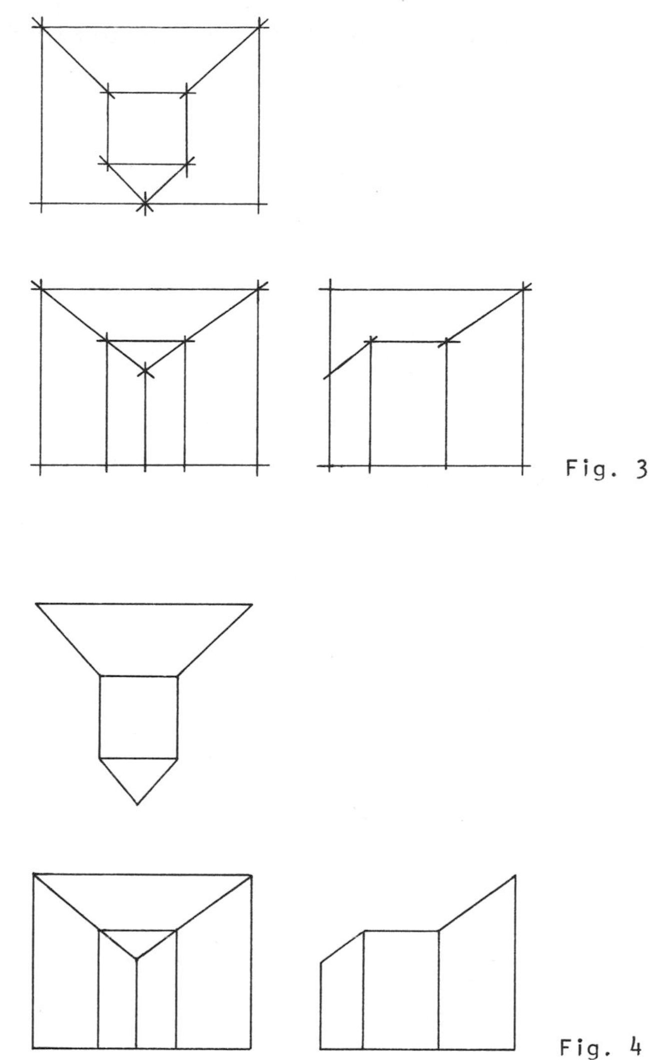

• PROBLEM 3-5

From the isometric drawing of the object shown in Fig. 1 draw the orthographic projections of the front, top, and right side views of the object.

Solution: Notice that the drawing in Fig. 1 is not dimensioned. Dimensions are unnecessary because the problem states that the given figure is an isometric drawing; that is, the dimensions needed in order to draw the required orthographic projections can be taken directly from the isometric lines of the drawing. Recall that an isometric line is any line parallel to the principal axes of an isometric drawing. Keep in mind that measurements can be taken only along such lines and, therefore, no inclined line will appear true length in an isometric drawing.

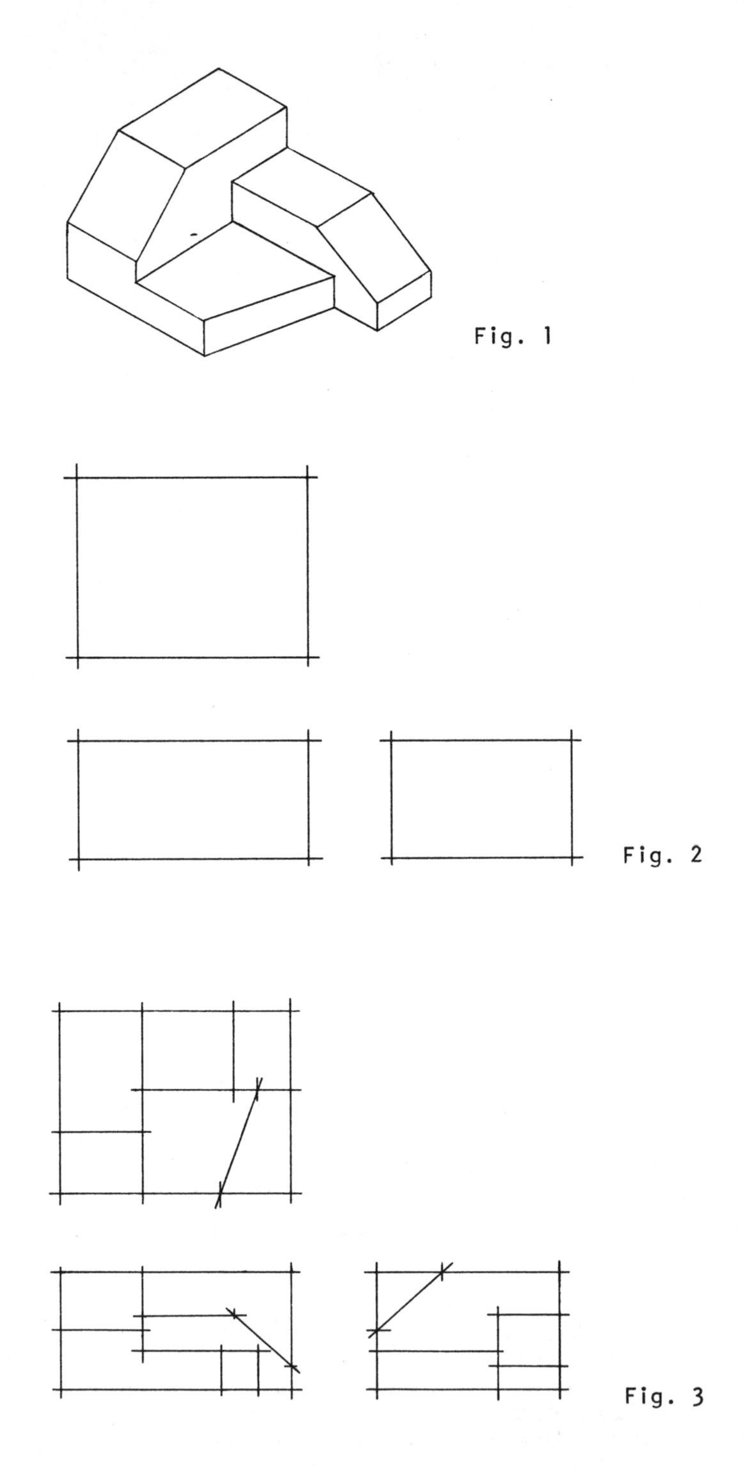

Fig. 1

Fig. 2

Fig. 3

In drawing the required views of the given object, these steps should be followed: first, box in each of the required views as shown in Fig. 2. Then proceed to lay out all the visible lines in each of the views. These lines should be drawn very lightly as in Fig. 3. When all the lines in the object have been drawn in each of the views and they have been verified to be correct, proceed to darken all the lines and erase all the construction or unnecessary lines. The result is shown in Fig. 4. Notice that the inclined planes are determined in each view by the use of planes parallel to the principal planes of projection. Avoid measuring along inclined planes unless absolutely sure that such a measurement is valid.

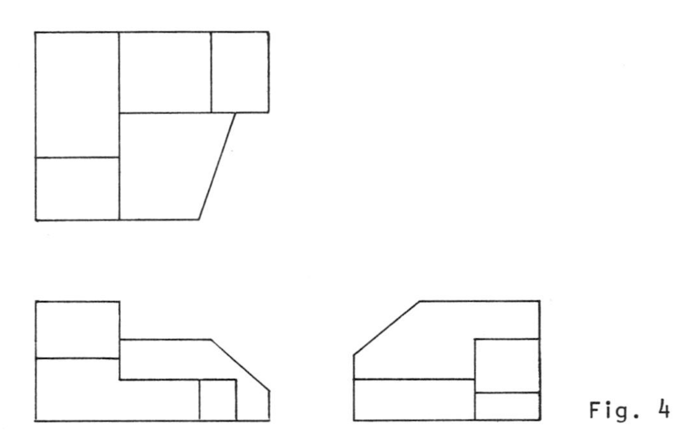

Fig. 4

• PROBLEM 3-6

From the isometric drawing shown in Fig. 1, draw the orthographic projections of the front, top, and right side views of the given object.

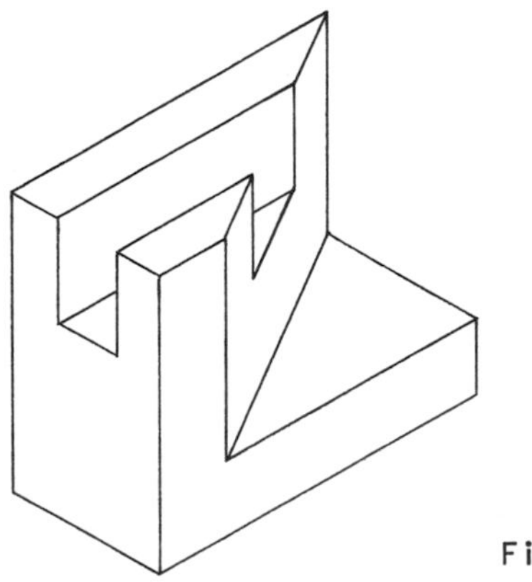

Fig. 1

Solution: Because the problem states that figure 1 is an isometric drawing, we can take the measurements we need for the orthographic projections directly from the isometric lines of the figure. If the problem had read "isometric projection " this would not have been the case.

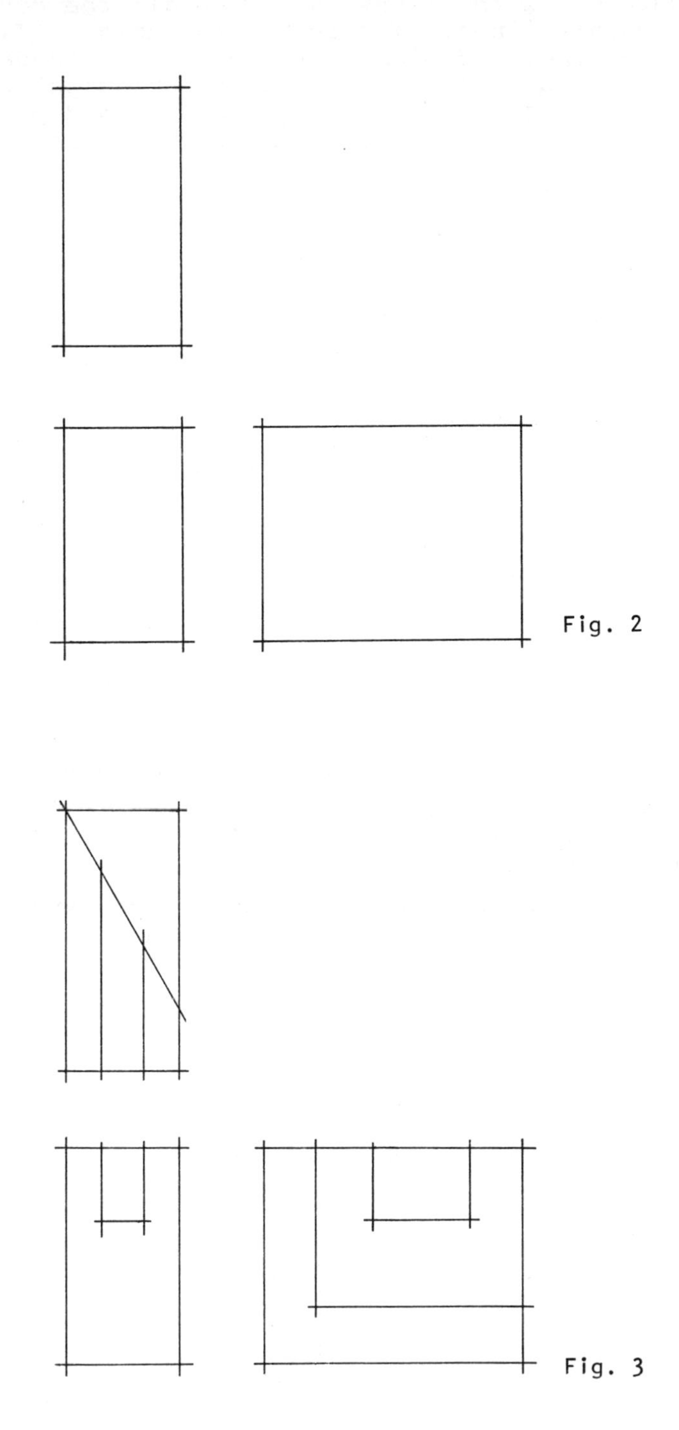

Fig. 2

Fig. 3

In order to draw the required orthographic projections first notice that all the planes of the object are either horizontal or vertical. Every vertical plane will appear as a line in the top view and any horizontal plane will appear as a line in the front and side views. Draw the views in the following order: first, box in all the required views as shown in Fig. 2. Lay out all the visible lines as shown in Fig. 3. Finally, darken all the lines, include all the hidden lines, and erase the construction lines. Fig. 4 shows the three required orthographic projections completed.

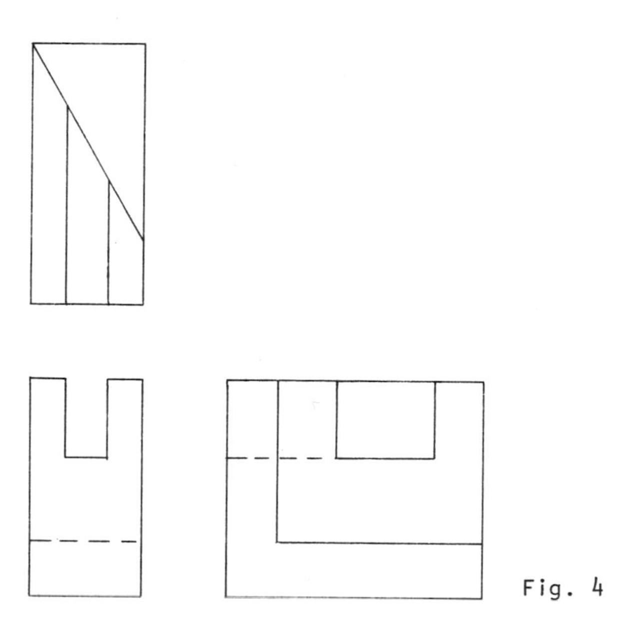

Fig. 4

• PROBLEM 3-7

From the isometric drawing shown in Fig. 1, draw the orthographic projections of the front and top views of the object.

Solution: The measurements needed to draw the required orthographic projections can be taken directly from the isometric lines of Fig. 1. This can be done because the given figure is an isometric drawing. In the case of an isometric projection, this cannot be done. Recall that an isometric line is any line which is parallel to any of the isometric axes. The isometric axes for an isometric drawing are defined as follows: one vertical axis and two other axes, each one 30 degrees up from the horizontal on each the left and right sides.

In drawing the required orthographic projections, first box in the desired views very lightly as shown in Fig. 2. Locate all the center lines and proceed to draw all the circles and rounds present in the object (see Fig. 3). Make sure that in drawing the front and top views of this object the top view is drawn first. The reason for drawing the top view first is that, when drawing circles or rounds, those lines which have to be projected should be projected from the circles and rounds and not vice versa. When all the circles and rounds have been drawn, proceed to include all the other details, as shown in Fig. 3. Finally, darken all the object lines, erase all the construction lines, and make sure the lines that should appear hidden appear as such. (See Fig. 4.) Notice that the 45 degree incline, marked in Fig. 1, refers to the circular surface at which the bigger hole becomes smaller.

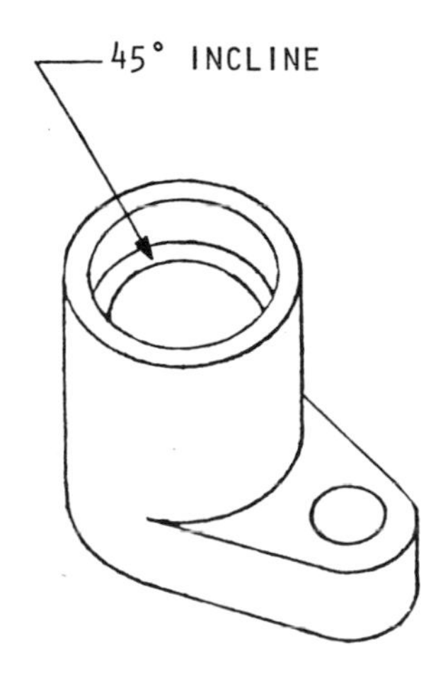

Fig. 1

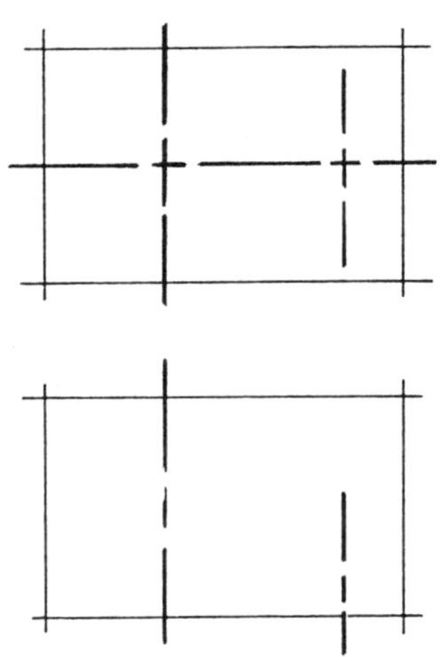
Fig. 2

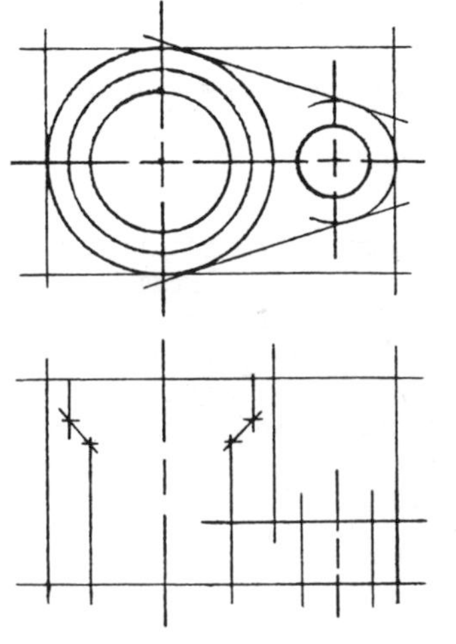
Fig. 3

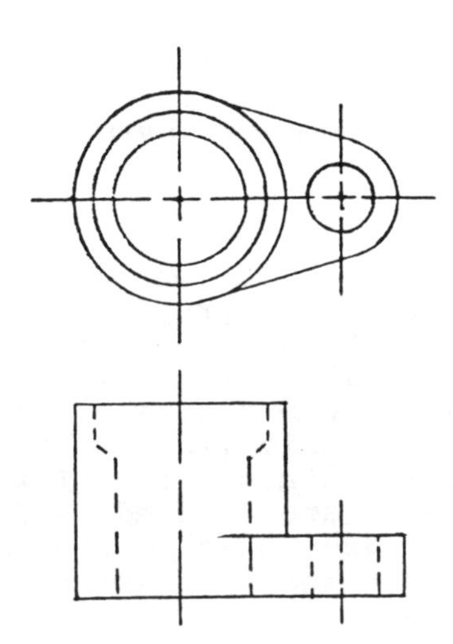
Fig. 4

From the isometric drawing shown in Fig. 1 draw the orthographic projections of the front and top views.

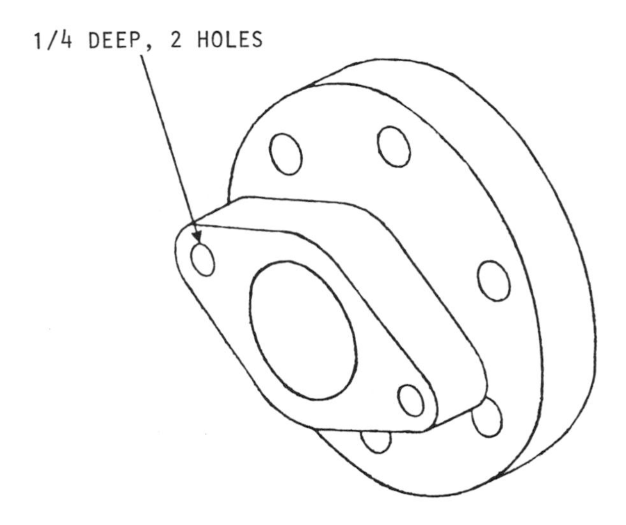

Fig. 1

Solution: The measurements needed to draw the orthographic projections of the front and top views can be taken directly from the isometric lines of the isometric drawing in figure 1. Recall that the isometric lines are all the lines parallel to the axes of the isometric drawing which are drawn (a) vertical and (b) 30 degrees up from both sides of the horizontal.

In order to draw the required orthographic projections, first establish all the center lines. When all these center lines have been located proceed to include all the circles and rounds of the object (see Fig. 2). Notice that since all the holes in the bottom plate of the object are equidistant from the center of the front view, the center line of these holes is a circle going through the centers of the six circles. When drawing the top plate of the object in the front view notice that parts of circles must first be drawn in the center and extreme left and right of the top plate. When the circles have been established join the tangency points with straight lines. Make sure to project the top view from the front rather than the other way around. Notice that the isometric drawing calls for the two holes in the top plate to be a quarter inch deep. The points of these holes which were drawn in the top view of Fig. 2 represent the tips of the drills used to make the holes. The other six holes on the object are through holes since no limitation on their depth was indicated.

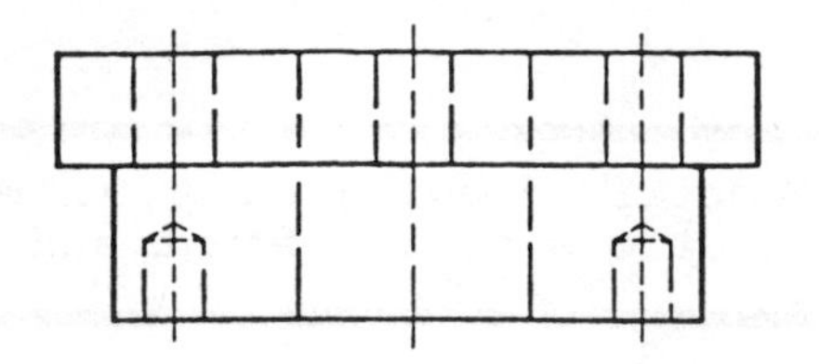

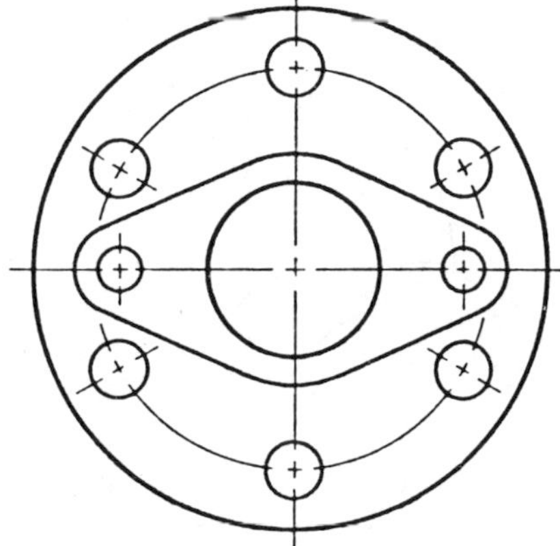

Fig. 2

• PROBLEM 3-9

From the isometric drawing shown in Fig. 1, draw the orthographic projections of the front, top, and right side views.

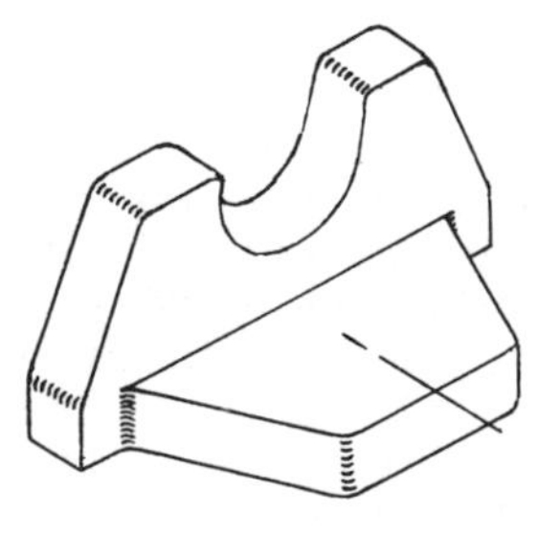

Fig. 1

Solution: The measurements needed to draw the required orthographic projections can be taken directly from the isometric lines of the isometric drawing in figure 1. Recall that all the lines parallel to the isometric axes are true length in an isometric drawing.

The steps in drawing the orthographic projections of the object in Fig. 1 are as follows: first, box in all the required views and find all the center lines as shown in Fig. 2. Draw all the circles and rounds which are centered around the already determined center lines. Proceed to draw all the visible lines and include the fillets and the rounds (see Fig. 3). Finally, darken all the lines, include the hidden lines, and erase the construction lines as shown in Fig. 4.

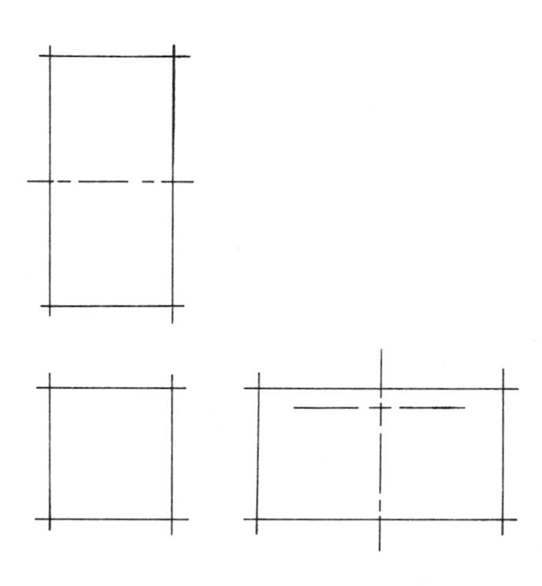

Fig. 2

Fig. 3

Fig. 4

Draw the orthographic projections of the front, top, and right side views of the object represented by the isometric drawing shown in Fig. 1.

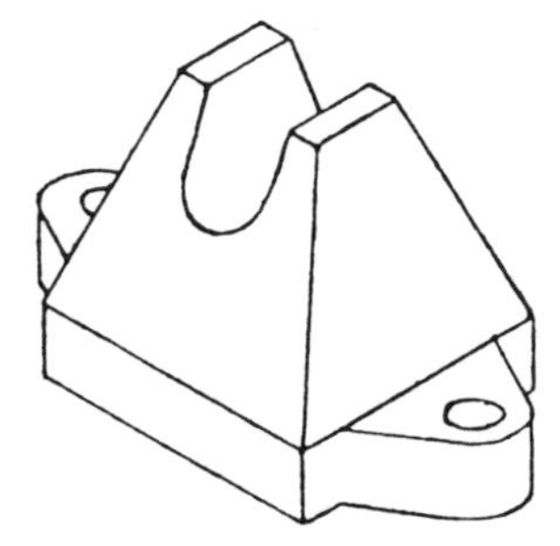

Fig. 1

Solution: The measurements needed to draw the required orthographic projections can be taken directly from the isometric lines of figure 1. This can be done because the problem states that the figure is an isometric drawing. For this and any other object it is necessary to keep in mind that the isometric lines are not the object's lines. Recall that the isometric lines are the two lines 30 degrees above the horizontal axis and the vertical line . All the lines parallel to these three lines are also isometric lines. In taking the measurement from this isometric drawing, it is very important to measure from isometric lines only, because there are inclined planes which contain non-isometric lines.

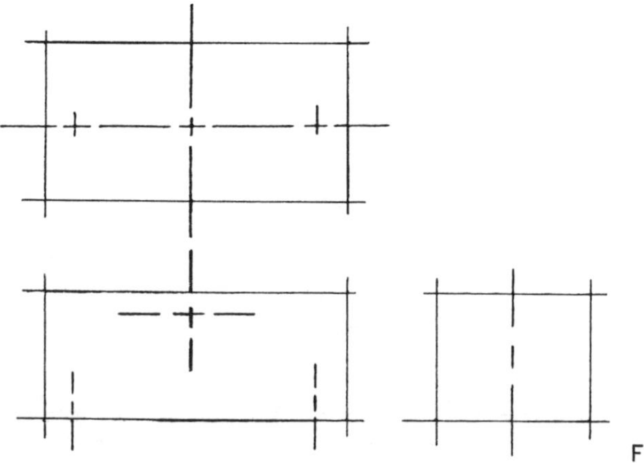

Fig. 2

The procedure in drawing the orthographic projections of this object is as follows: first, box in all the required views as shown in Fig. 2. Locate all the center lines to be used in the object and then proceed to draw all the circles and rounds. After all the circles and rounds have been drawn, lay out the other visible lines in each view (see Fig. 3). Finally, darken all the lines, erase the construction lines, and include all the hidden lines. Fig. 4 shows the required completed orthographic projections. When drawing the two rounds at each end of the slot, as shown in the top view, project points from the inclined plane vertically into the top plane. When enough points have been located for each round, use a french curve to connect the points. (For most practical purposes these rounds can be approximated by using an ellipse template.)

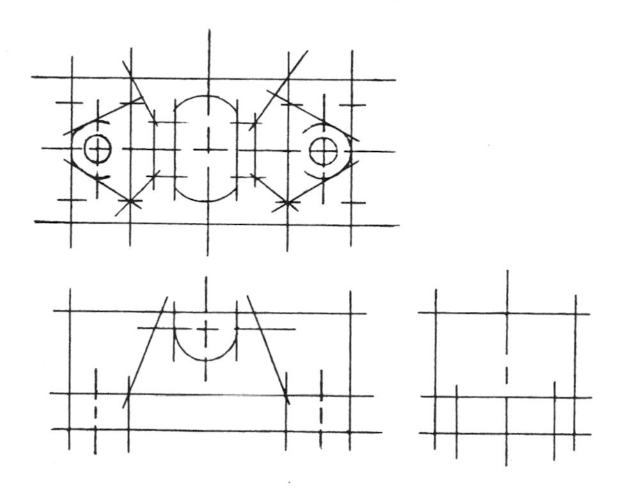

Fig. 3

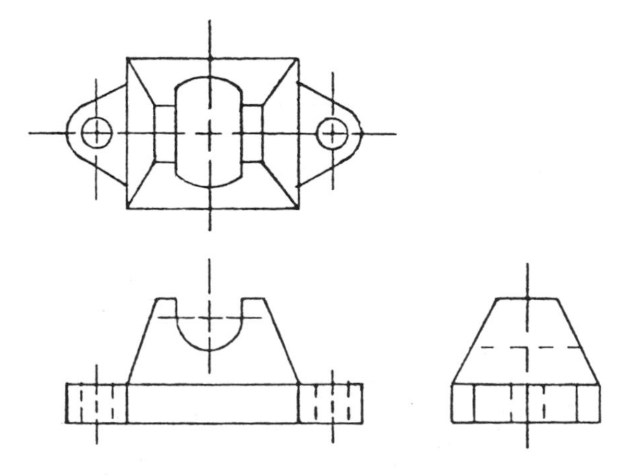

Fig. 4

Given the isometric drawing of the object shown in Fig. 1, draw the orthographic projections of the front, top, and right side view of the object.

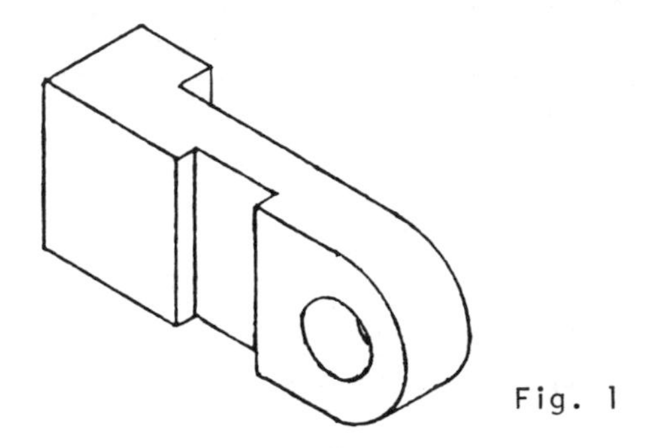

Fig. 1

Solution: The measurements necessary to draw the required orthographic projections can be taken directly from the isometric lines of figure 1 because the problem states that this is an isometric drawing. Had the problem stated that the figure was an isometric projection, this would have not been the case.

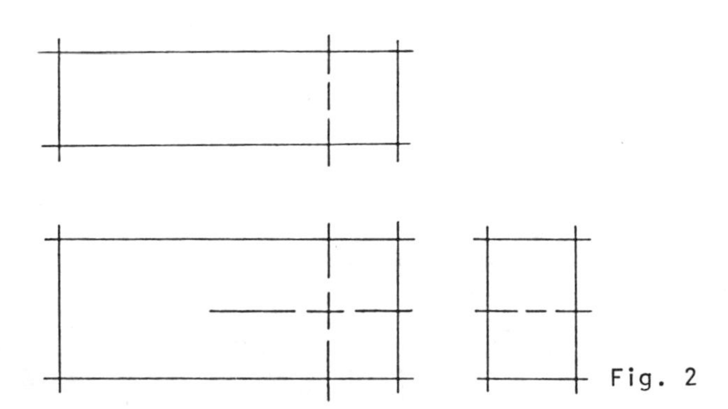

Fig. 2

To draw the orthographic projections of the object, first box in all the necessary views as shown in Fig. 2. When the views have been boxed in, proceed to find all the center lines in the object. Then draw all the circles and rounds. Outline all the other details as shown in Fig. 3. Finally, darken all the lines, include the hidden lines, and erase construction lines. Fig. 4 shows the completed projections. Remember to draw all construction lines very lightly since they are temporary. Work on all views at the same time and project from rounds and circles rather than fitting circles and rounds to projection lines.

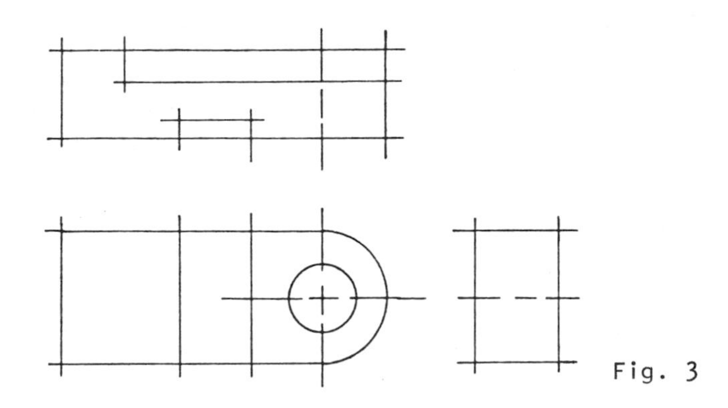

Fig. 3

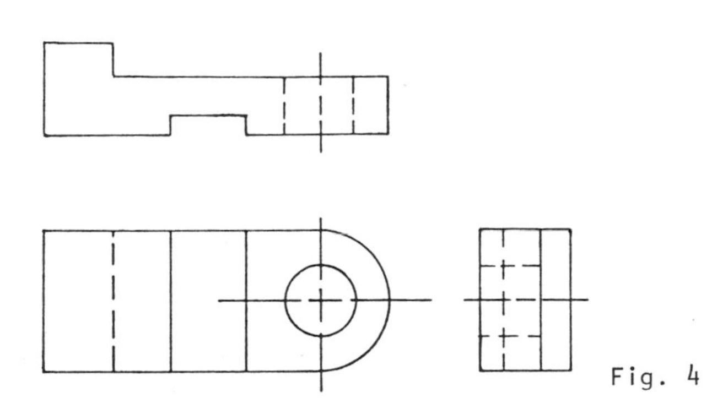

Fig. 4

• PROBLEM 3-12

Given the isometric drawing shown in Fig. 1, draw the orthographic projections of the front, top, and right side views of the object.

Solution: Figure 1 shows an isometric drawing of the given object; therefore, we can obtain all the measurements we need for the orthographic projections directly from this drawing.

To draw the required orthographic projections of the given object, first box in all the desired views as shown in Fig. 2. Proceed to sketch the rest of the visible lines appearing in each of the views. Next, include all the rounds in the object. Finally, include all the hidden lines, darken all the lines, and erase the construction lines. Fig. 3 shows the completed projections. A concept to keep in mind when drawing orthographic projections is that for every vertical plane in the object, there is a line representing the edge view of this plane in the top view, and for every horizontal plane there is a line representing the edge view of this plane in each of the front and side views. Notice that an inclined plane cannot be represented by a single line in any of the principal views.

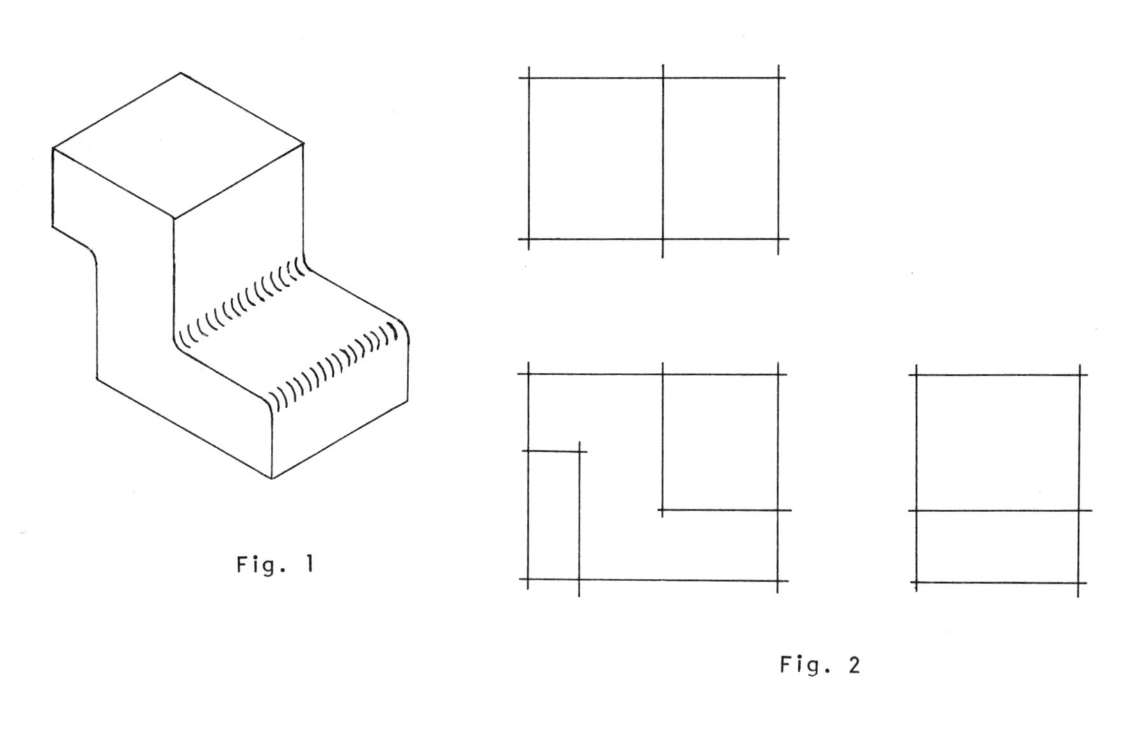

Fig. 1

Fig. 2

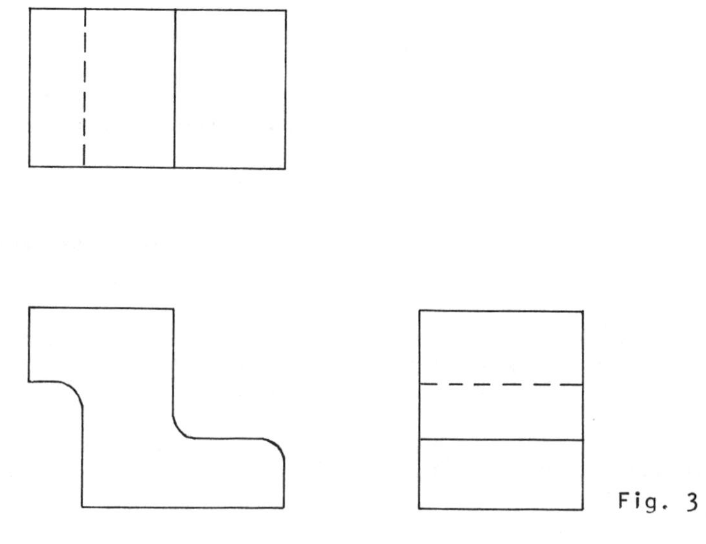

Fig. 3

• **PROBLEM 3-13**

Given the isometric drawing of the object shown in Fig. 1, draw the orthographic projections of the front, top, and right side views.

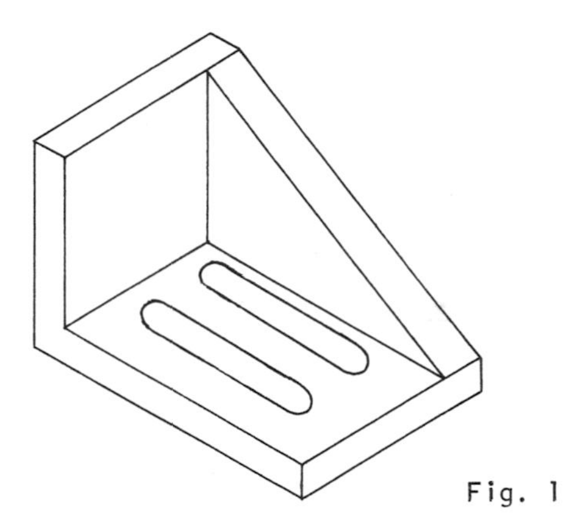

Fig. 1

Solution: Notice that the drawing shown in figure 1 is not dimensioned. The reason is that the problem states that this figure is an isometric drawing. We can, therefore, take the necessary measurements for the orthographic projections directly from the isometric lines of the drawing. Recall that this can only be done in the case of an isometric drawing.

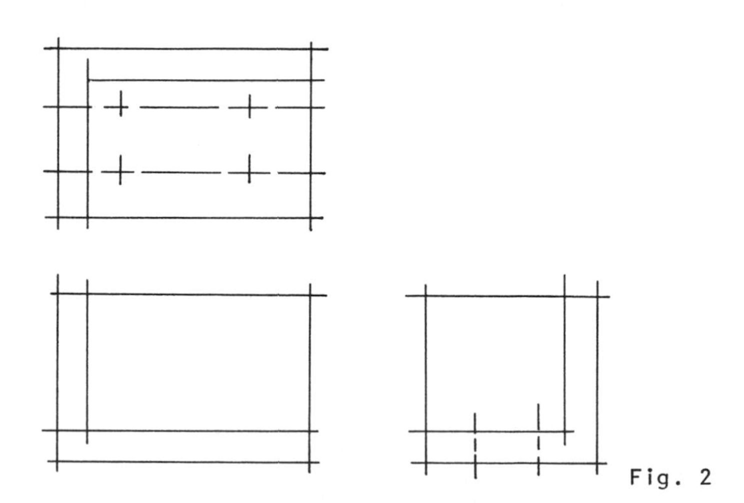

Fig. 2

To draw the required orthographic projections imagine yourself looking at the object at right angles to the front, the top, and the right side. Once you have established what is seen in each view, proceed to draw the views in the following manner: first, box in each of the three views and locate all the center lines (see Fig. 2). When all the center lines have been located draw all the circles and rounds present in the object (see Fig. 3). Next, draw the rest of the visible lines. Finally, darken all the object lines, include all the hidden lines, and remove the construction lines. Fig. 4 shows the three completed orthographic projections.

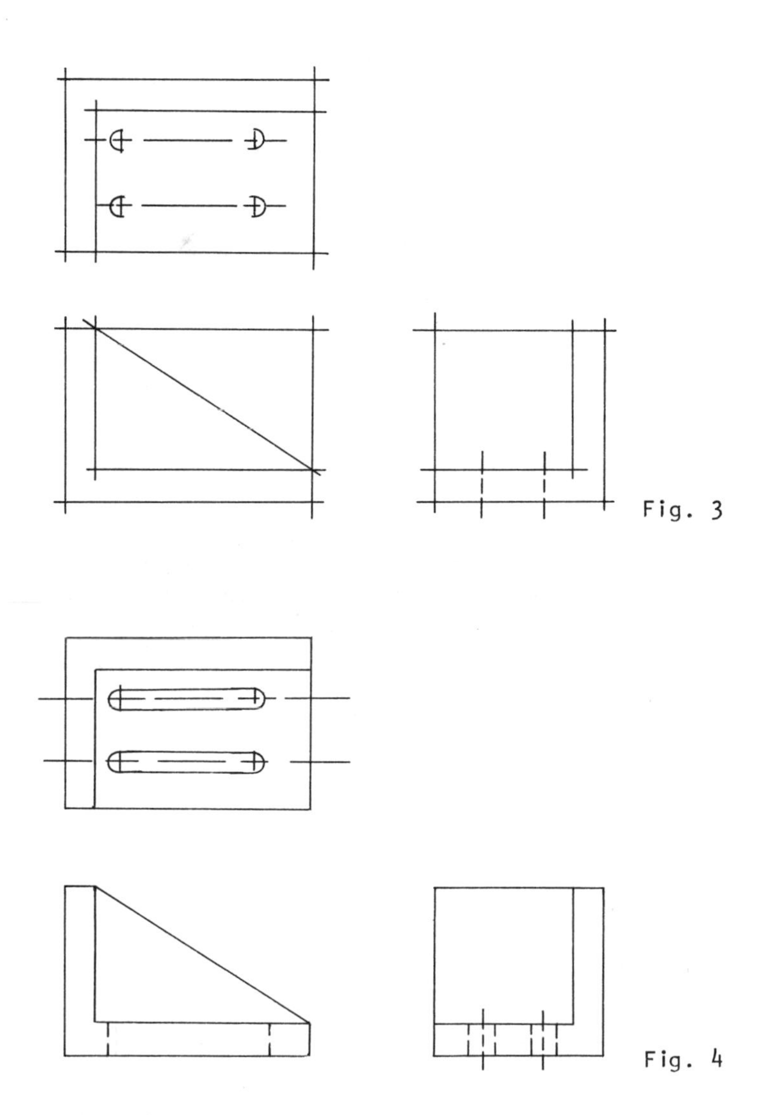

Fig. 3

Fig. 4

Whenever you are drawing the orthographic projections of an object keep in mind that a vertical plane will appear as a line in the top view, and any horizontal plane will appear as a line in the front and side views. Any type of circular hole within the object will appear as two hidden lines in the views in which it doesn't appear as a circle. Any plane that ends behind a visible plane will also appear hidden.

• PROBLEM 3-14

Given the isometric drawing shown in Fig. 1, draw the orthographic projections of the front and top views of the object.

Solution: Notice that in this problem no dimensions are shown in the given figure. This is because the problem states that the figure is an isometric drawing. This means that the measurements needed for drawing the orthographic projections can be taken directly from the isometric lines of the given figure.

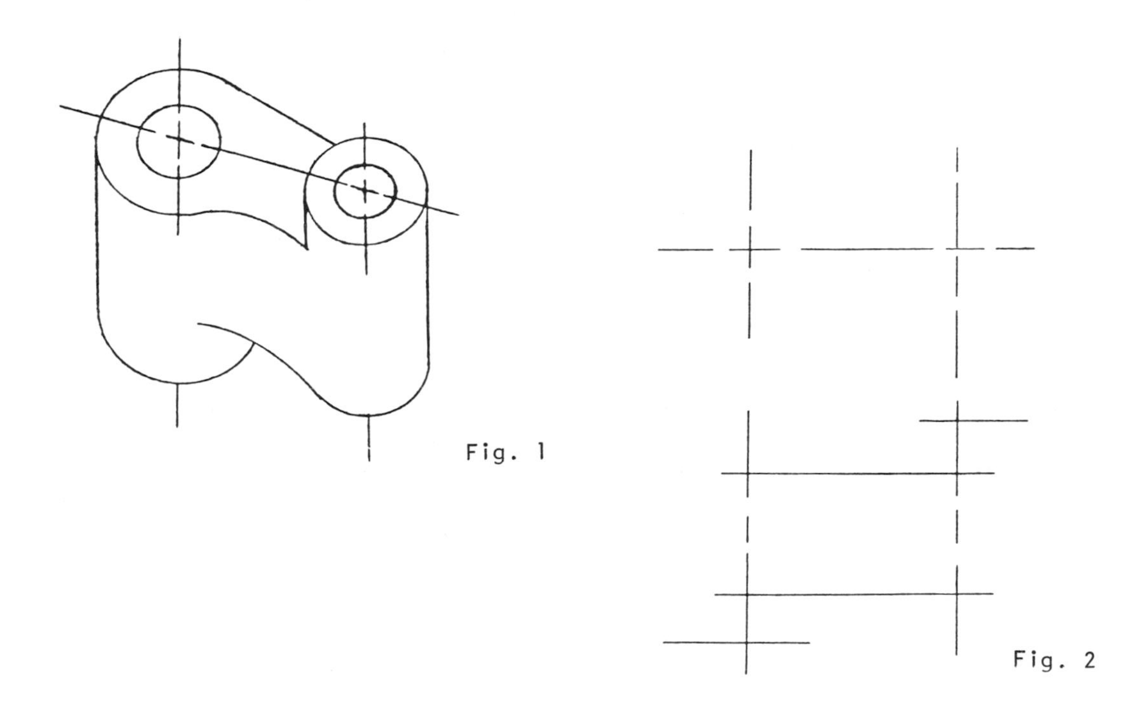

Fig. 1

Fig. 2

In drawing the isometric projections of this object, first locate the center lines in both the front and top views (see Fig.2). At this point the heights in the front view can be outlined. Make sure that when the heights are being located there is an adequate space between the top and front views. Once all the center lines have been drawn include all the circles and rounds, as shown in Fig. 3. Next, project the vertical lines corresponding to the circles and rounds. Finally, darken all the lines, include the hidden lines, and remove the construction lines. The completed drawing is shown in Fig. 4.

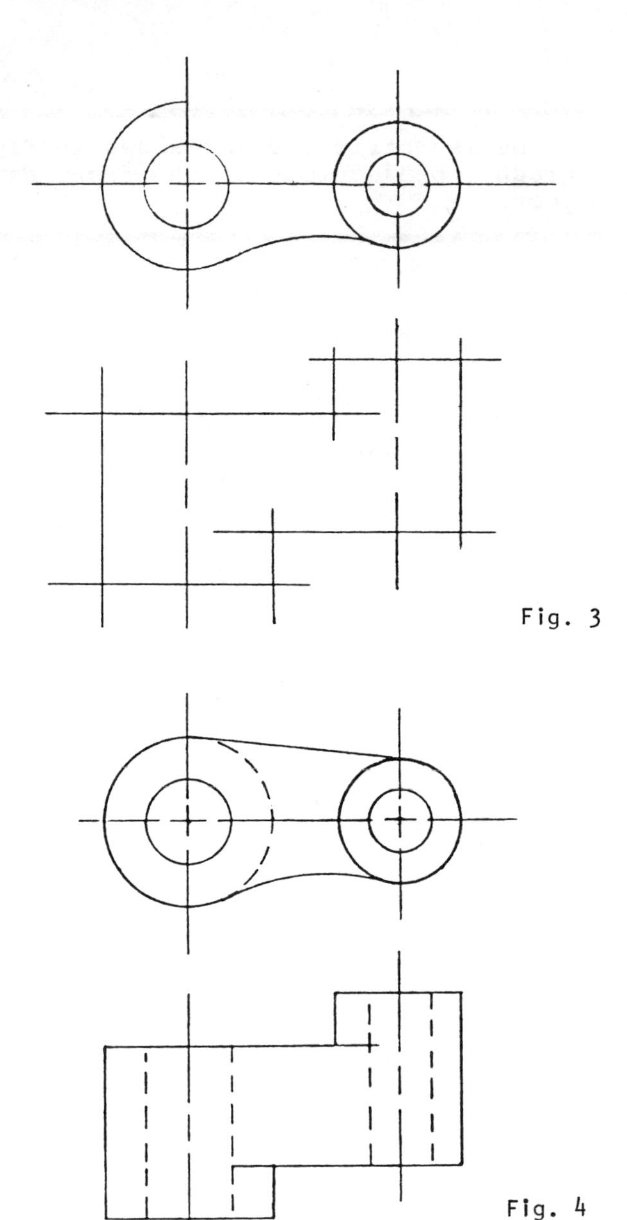

Fig. 3

Fig. 4

• PROBLEM 3-15

> Given the isometric drawing shown in Fig. 1, find the orthographic projections of the object's front, top, and right side views.

Solution: The first thing to notice in this problem is that figure 1 is an isometric drawing. This means that measurements can be taken directly from the isometric lines. Had this been an isometric projection this would have not been the case.

To draw the required views the first step is to box in all the views and draw all the center lines (see Fig. 2). After all the center lines have been drawn include all the circles, rounds, and other visible lines (see Fig. 3). In the last stage of the drawing, darken all the lines, erase the construction lines, and include all the hidden lines. Fig. 4 shows the completed drawing.

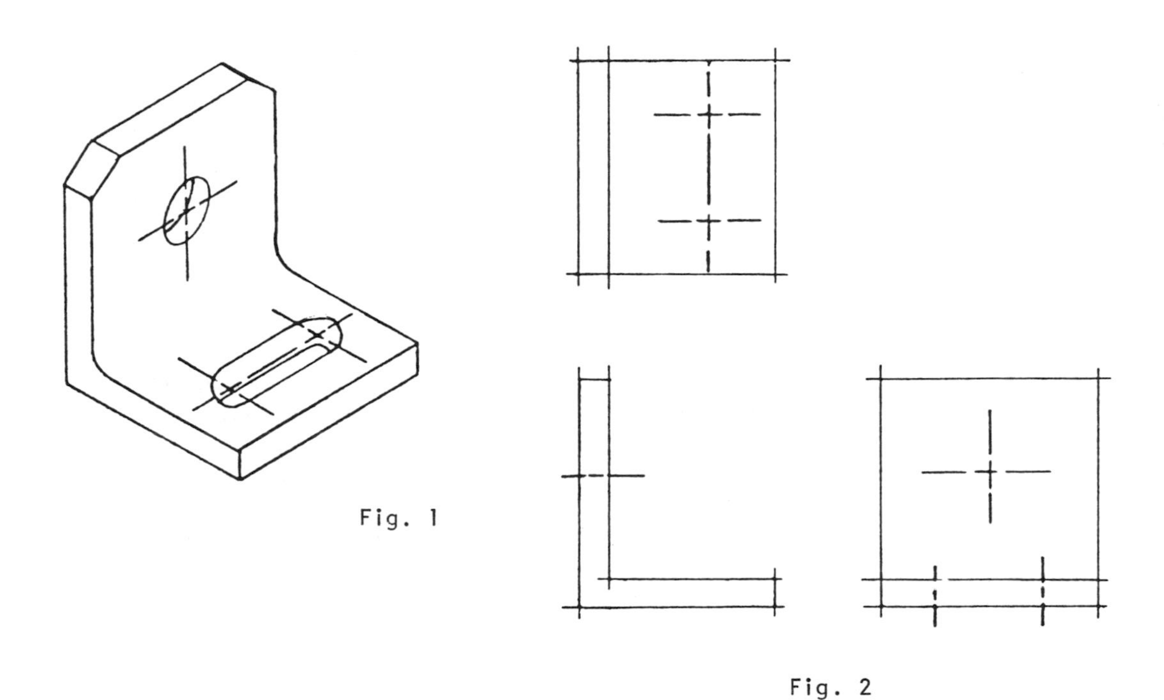

Fig. 1

Fig. 2

Fig. 3

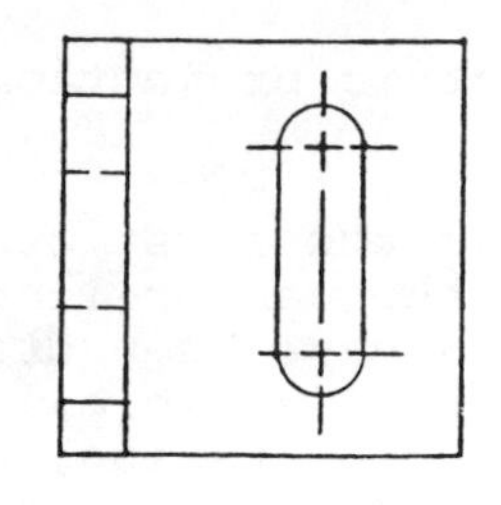

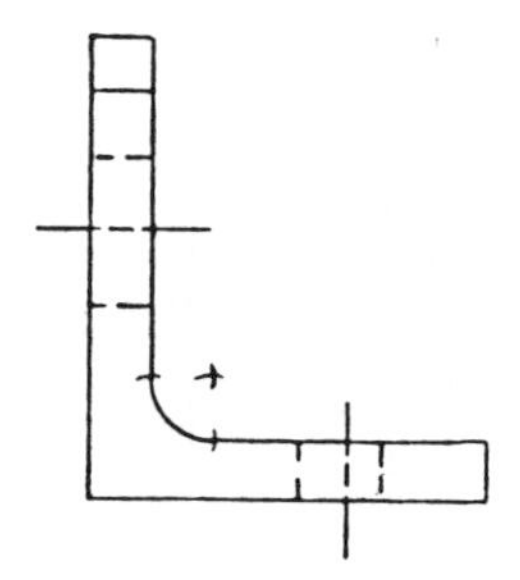

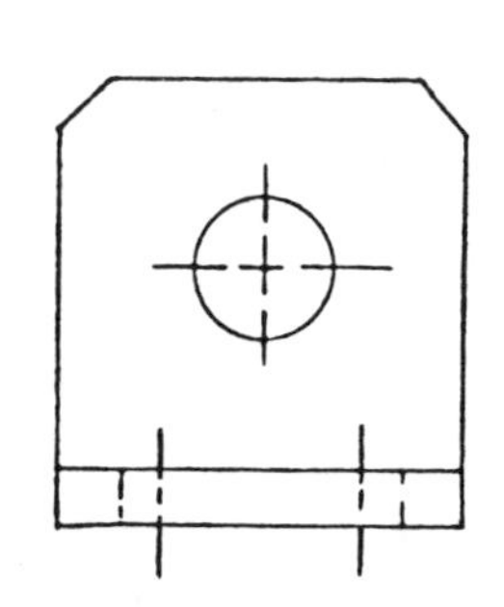

Fig. 4

• PROBLEM 3-16

Given the isometric drawing shown in Fig. 1, find the orthographic projections of the front, top, and right side views.

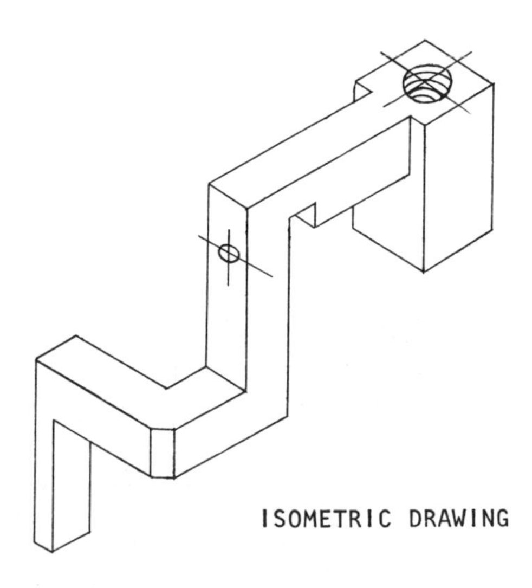

Fig. 1

Solution: Notice that no dimensions are specified in the given figure. The problem states, however, that it is an isometric drawing; therefore, all the isometric lines are in true length. This means that all the measurements can be made directly from the isometric drawing. Unlike an isometric projection, an isometric drawing contains all the isometric lines in true length.

In order to draw the orthographic projections of the required views it is necessary to act as an observer looking at the object from the directions specified. For example, an observer looking at the object from the top would see the threaded hole as a regular circle, the vertical planes as lines, and would not see the other hole in the object.

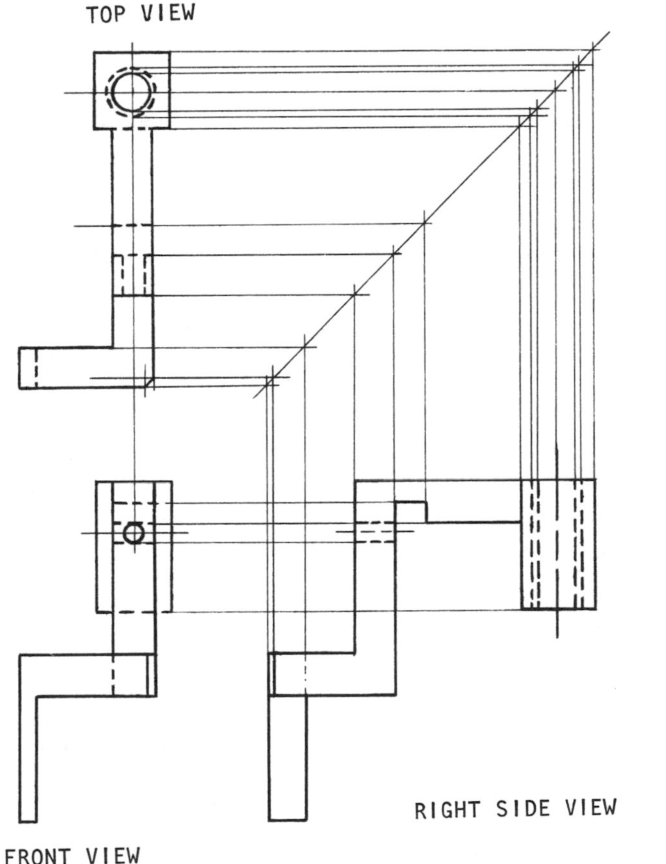

Fig. 2

When actually drawing the orthographic projections, it is advisable to draw the extreme edges of all the required views and then fill in the details. For instance, to draw the top view of this object outline all the rectangles first and then draw the circles for the threaded hole. Draw the hidden lines and check every vertical plane to make sure that for every plane appearing as an edge there is a line. If a vertical surface ends below a horizontal plane then the edge view of this plane is hidden; otherwise it is shown as a solid line. This same procedure is employed for all the other views. Fig. 2 also illustrates another aspect of orthographic projections; namely, given two orthographic projections a third projection can be constructed.

In Fig. 2, the right side view represents the intersections of projection lines from the top and front views. The projection lines are projected to the right from the front view. From the top view the projection lines are projected across to a 45° line and then down into the side view. The 45° line and the projection lines are customarily not shown in the completed orthographic projections.

• PROBLEM 3-17

> Given the isometric drawing of the object shown in Fig. 1, find the orthographic projections of the front, top, and side views.

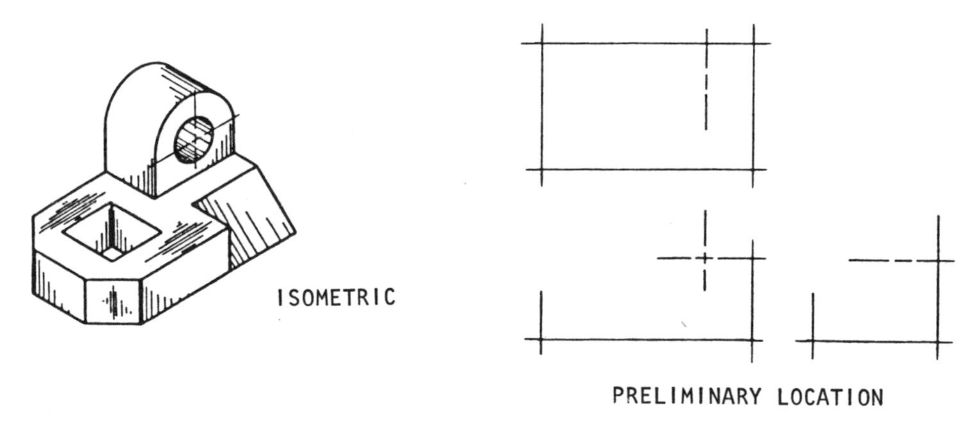

Fig. 1 Fig. 2

Solution: The procedure in this type of problem is to first outline the outside edges of the object in each of the views by (a) drawing the straight edges, (b) locating all the center lines, and (c) completing the round edges (see Fig. 2). It is better to work on all the views at one time rather than to finish one before starting another. Draw the circles and rounds where they appear as such, and project from them rather than first drawing the lines and later fitting circles and rounds to the projecting lines. The latter is an unrecommended practice because it is easier to project lines from circles or rounds than it is to project lines first and then attempt to fit circles or rounds tangent to them.

When all the outside edges have been outlined the other details are then drawn (see Fig. 3). When all the details are included and the construction lines have not yet been removed, the drawing looks like a skeleton of the finished drawing as shown in Fig. 3.

The front and top views can be drawn just by looking at the given isometric drawing and measuring from it. But since the right side view of the object is not seen in the isometric drawing it must be projected from both the top and front views. To project the right side view, first draw a 45 degree line and project all the points on the top view to this line. Then from the line drop vertical lines and project all the points in the front view horizontally to the right. The corresponding projection lines determine points which can be connected to construct the required view. See Fig. 4. Always keep in mind that any point in a view is fully determined by projecting it from any two adjacent views. Also notice that any line which is hidden is drawn as a series of dashes in order to distinguish it from those which are actually seen.

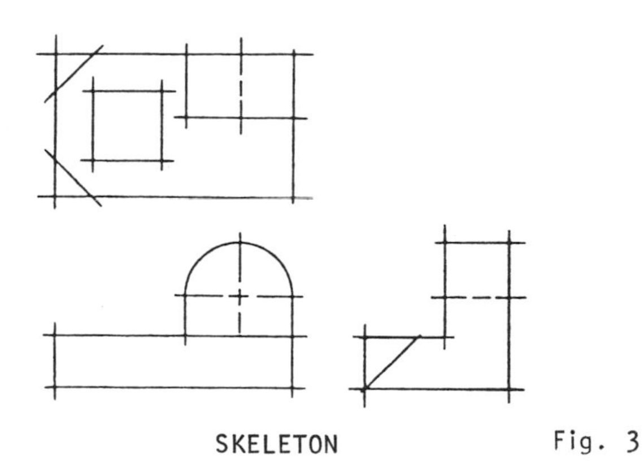

Fig. 3

Fig. 4

• **PROBLEM 3-18**

Given the isometric drawing of the anchor nut shown in Fig. 1, find the orthographic projections of the top and front views.

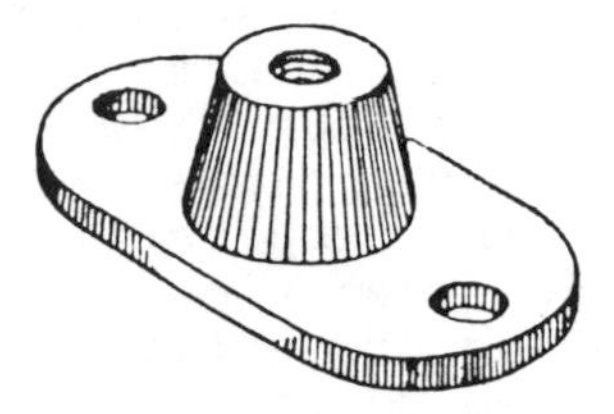

Fig. 1

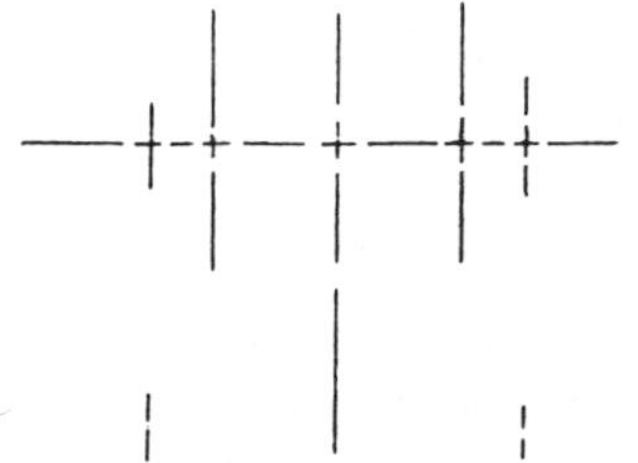

Fig. 2

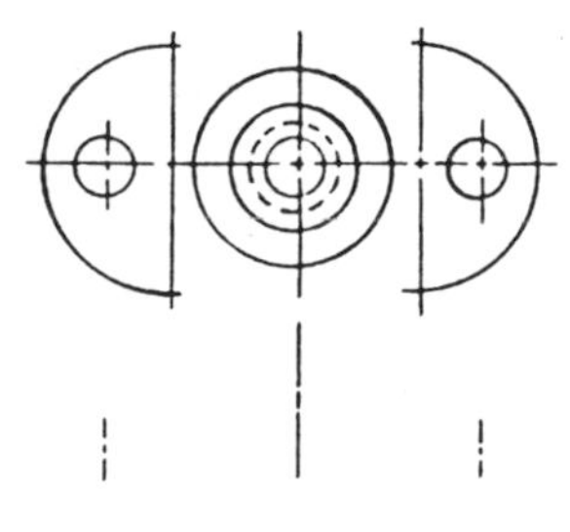

Fig. 3

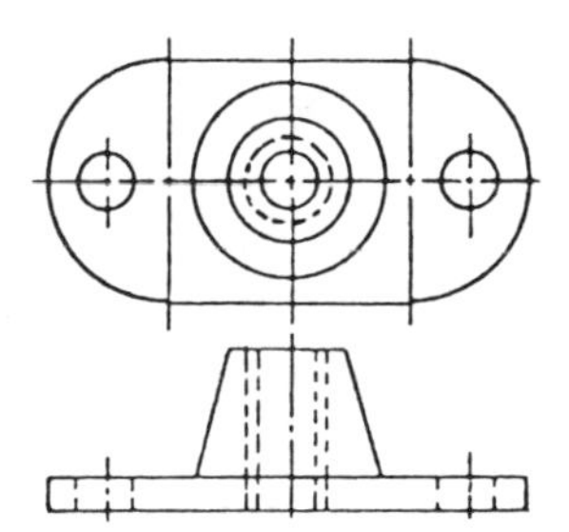

Fig. 4

Solution: For cylindrical objects like this one it is advisable to first draw all the center lines needed for each hole and round off the object (see Fig. 2). After locating all the center lines, draw all the circles and rounds and then add the straight lines. Notice that in the top view there is a dashed circle. This circle is hidden because the middle hole is threaded and the outside diameter of an internal thread is always shown as hidden. The diameter of the hidden circle is referred to as the outside diameter of the thread because it is father away from the center than the diameter of the inner solid circle.In the front view the hidden lines are shown for the inside and outside diameter of the thread. Also keep in mind that rounds and circles are always drawn first before drawing their projections in any other view.

From the isometric drawing given in Fig. 1, draw the orthographic projections of the front, top, and right side views.

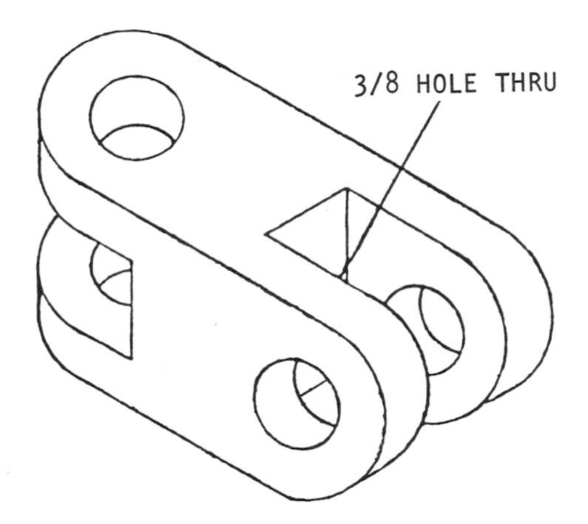

Fig. 1

Solution: The problem states that the pictorial view shown in Fig. 1 is an isometric drawing. Therefore, all the measurements required to construct the orthographic projections can be taken directly from the isometric lines in the isometric drawing. Had the pictorial view shown in Fig. 1 been an isometric projection, the measurements could not be taken from the drawing.

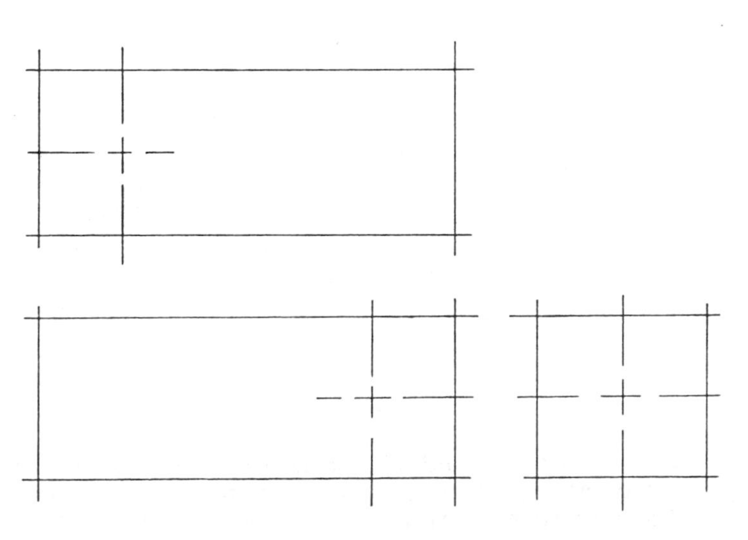

Fig. 2

In order to draw the orthographic projections of the required views imagine an observer viewing the object at right angles to the particular view which is being drawn. Enclose each of the views within rectangles as shown in Fig. 2, and include the center lines. The rectangles should be drawn very lightly. Using the center lines already determined, draw all the circles and rounds in all the views (see Fig. 3). Include all the visible lines and make sure that the lines that appear tangent to the circles or rounds are indeed tangent. Finally, include all the hidden lines and erase the construction lines.

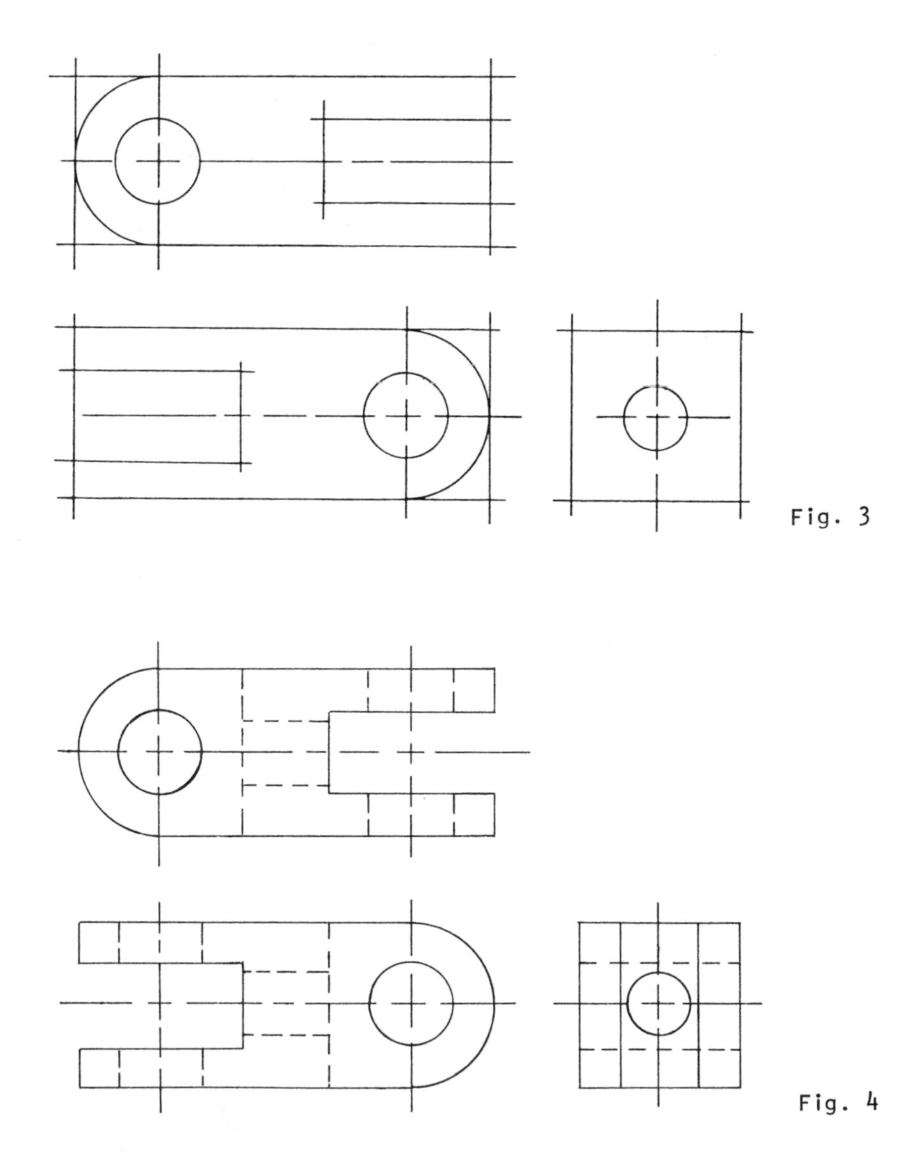

Fig. 3

Fig. 4

Fig. 4 shows the completed orthographic drawing. Notice that the 3/8 diameter hole, which is not actually seen in the given isometric drawing, is centered and that it goes through the link.

• PROBLEM 3-20

Given the numbered isometric drawing shown in Fig. 1, draw and number the orthographic projections of the front, top, and right side views of the object.

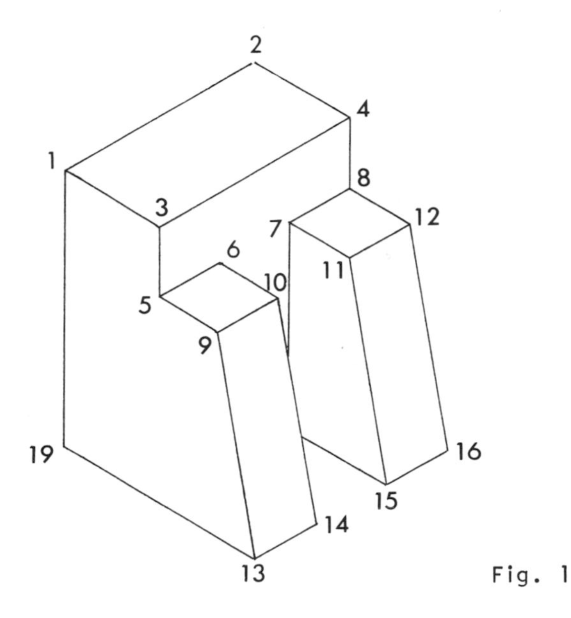

Fig. 1

Solution: Since figure 1 is an isometric drawing the measurements needed for drawing the orthographic projections can be taken from the isometric lines of this figure.

To determine each of the views asked for in the problem, first determine which planes are vertical and which are horizontal. Refer to Fig. 2. Inclined planes can be determined using vertical and horizontal planes even if the planes used are not part of the object. Keep in mind that all vertical planes appear as lines in the top view and all horizontal planes appear as lines in the front and side views. To draw the top view first draw the horizontal plane 1-2-4-3. Then using line 3-4 for reference, draw planes 5-6-10-9 and 7-8-12-11. To construct the inclined planes measure the length of line 19-13 and lay it out in the top view. Follow the same procedure in locating the planes necessary to draw the front and right side views. Notice that the hidden line in the front view represents the section of the middle vertical plane which is not visible to a viewer in the front.

In numbering the orthographic projections it is a good idea to label each plane as soon as it is drawn to save time. Notice that in numbering the planes in Fig. 2 the visible corners are labeled on the outside of the figure while those that are not visible are drawn on the inside.

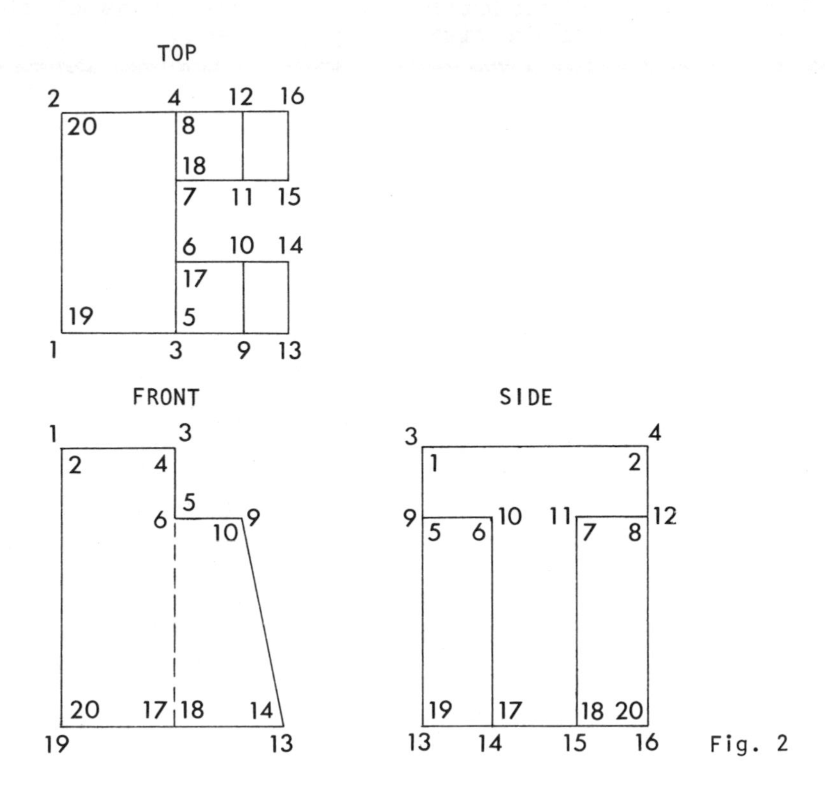

Fig. 2

• PROBLEM 3-21

> Given the numbered isometric drawing shown in Fig. 1, draw and number the orthographic projections of the top, front, and right side views of the object.

<u>Solution</u>: As stated in the problem Fig. 1 is an isometric drawing. Therefore, measurements can be taken from the isometric lines of the given object to draw the required orthographic projections.

In order to draw the different views asked for in the problem, first determine which planes are horizontal and which planes are vertical (refer to Fig. 2). For example, planes 1-2-15-12-13-14 and 7-8-9-10 are horizontal planes while planes 6-7-10-13-12-11 and 4-1-14-9-8-5 are vertical. Keep in mind that a vertical plane will appear as a line in the top view while a horizontal plane will appear as a line in the front and side views. To draw the top view first draw the horizontal plane 1-2-15-12-13-14. The horizontal plane 7-8-9-10 lies directly below it. This means that lines 8-9 and 9-10 of the lower plane appear to be on the same lines as lines 1-14 and 14-13 as seen from the top. Therefore, plane 7-8-9-10 can be drawn by using lines 1-14 and 14-13 as references. The same type of analysis can be made for both the front and right side views. Notice that this problem would be made more complicated had inclined planes been included in the given drawing. Had this been the case, the lines representing the inclined planes would have to be located by using horizontal and vertical planes for reference.

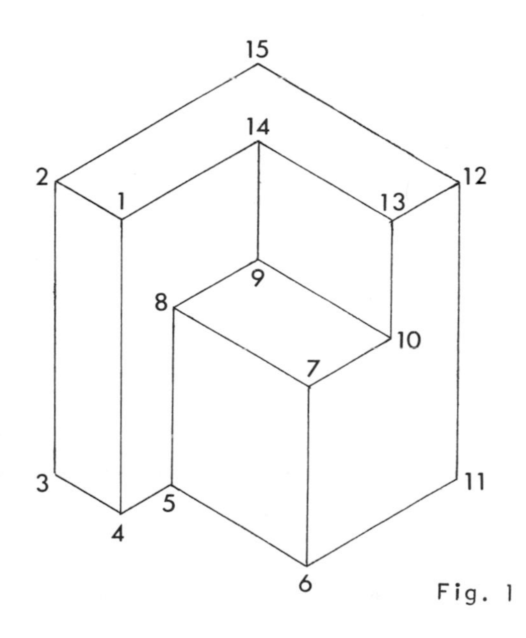

Fig. 1

In numbering each of the views number each plane as soon as it is drawn. This will make locating the other planes easier. The system that was used in numbering the views shown in Fig. 2 is: points which are visible are numbered on the outside corners while those which are not visible are drawn on the inside corners. Another system (seldomly employed) uses commas to separate the numbers or letters. The visible point is written first, followed by a comma and the invisible point. When there are more than one invisible point write the points as seen by the viewer in descending order and separate each by a comma.

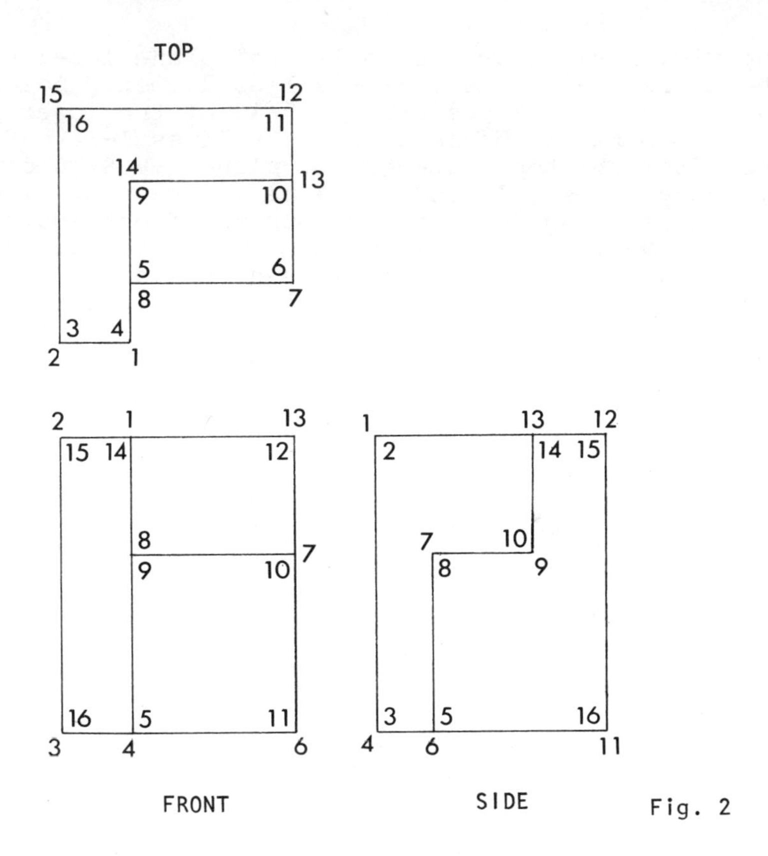

Fig. 2

• PROBLEM 3-22

Draw the orthographic projections of the front, top, and right views of the object represented by the isometric drawing in Fig. 1. Use the "Glass Box" method.

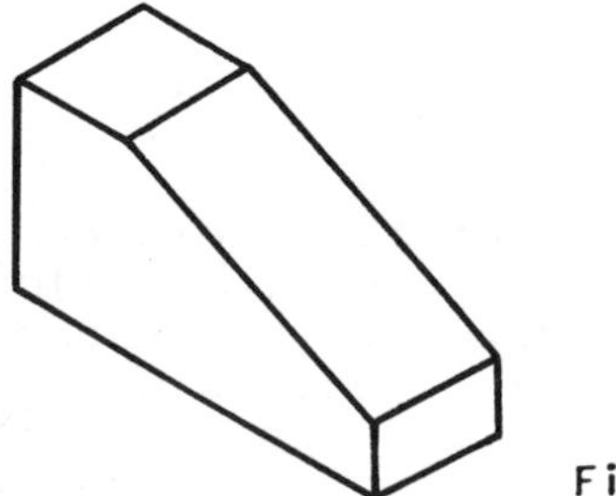

Fig. 1

Solution: The "Glass Box" method of drawing the orthographic projections of an object from an isometric drawing makes use of an imaginary box which surrounds the given object.

The box is placed so that the faces of the "glass box" are parallel to the principal sides of the given object. When the "glass box" has been properly placed, each of the faces of the object is projected perpendicularly to the corresponding face of the "glass box". Figure 2 shows what the "glass box" looks like after all the sides of the object have been projected. Notice that for clarity's sake Fig. 2 is not drawn to scale. After all the sides of the object have been projected, the "glass box" is opened until all the sides coincide in one plane (see Fig. 3). Figure 4 shows how the three required views are pictured as orthographic projections.

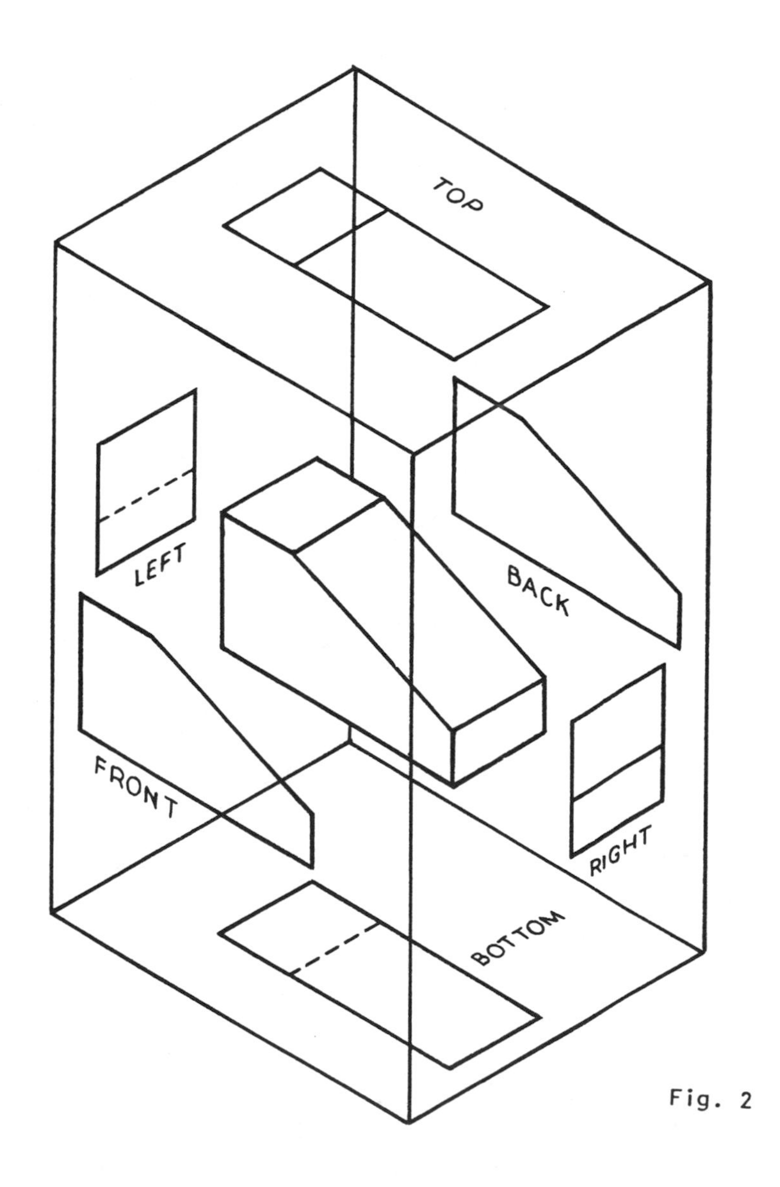

Fig. 2

The "glass box" method can be very helpful to a student in understanding orthographic projections. For example, by examining the "glass box" method we can see that the front view describes all the heights of the object while the top and side views fully describe the depths. Because

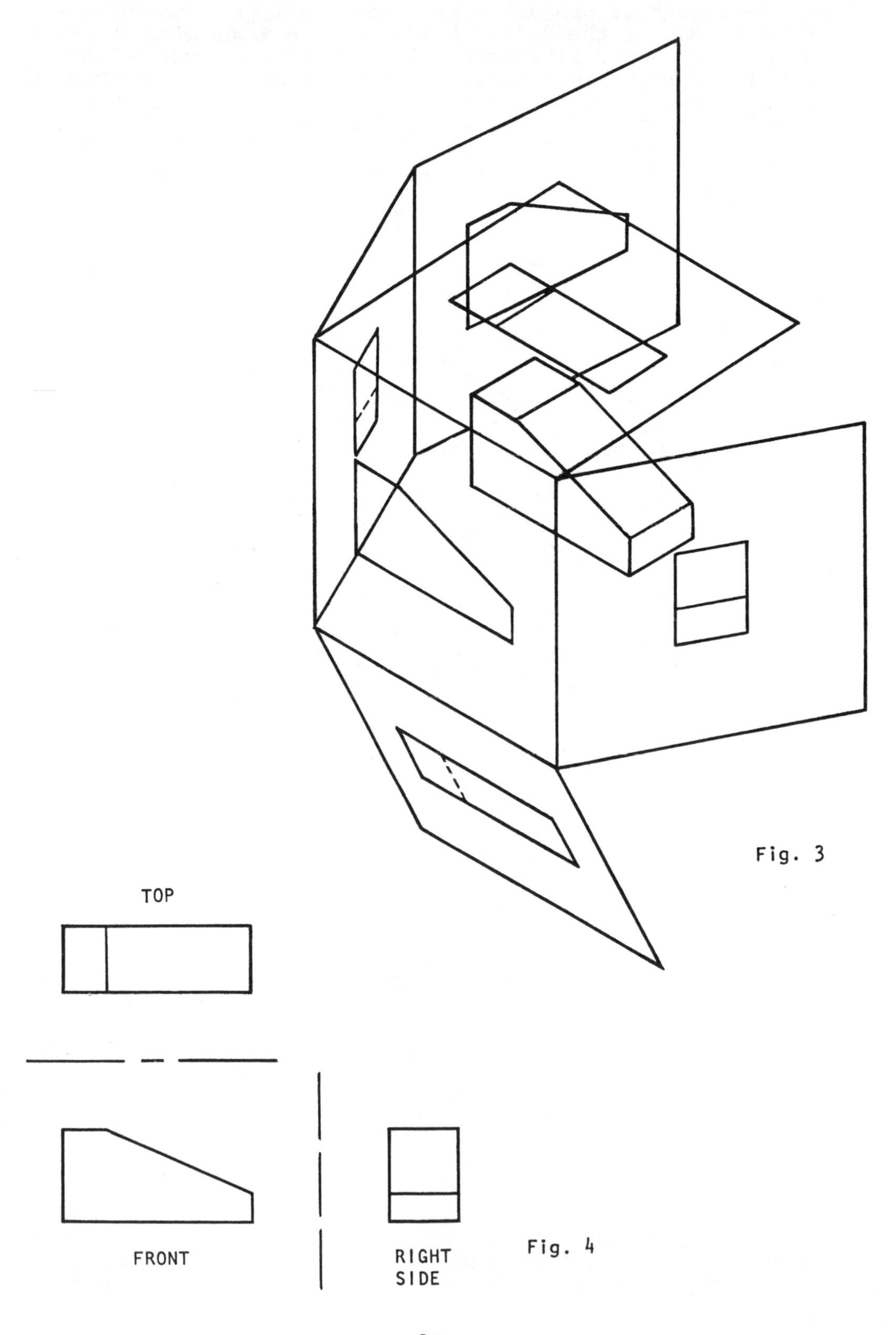

Fig. 3

Fig. 4

the top and side views both represent depths we can use a 45 degree line to project the side view when the front and top views are given, or the top view when the front and side views are given.

THIRD VIEWS DERIVED FROM TWO ORTHOGRAPHIC VIEWS

• PROBLEM 3-23

From the orthographic projections of the front and right side views shown in Fig. 1, draw the orthographic projection of the top view of the object.

Solution: The right side view of the object reveals an inclined plane in the upper right hand corner. From both given views we can see that there is a trapezoidal slot in the middle of the front view whose depth is indicated by the vertical hidden line of the side view.

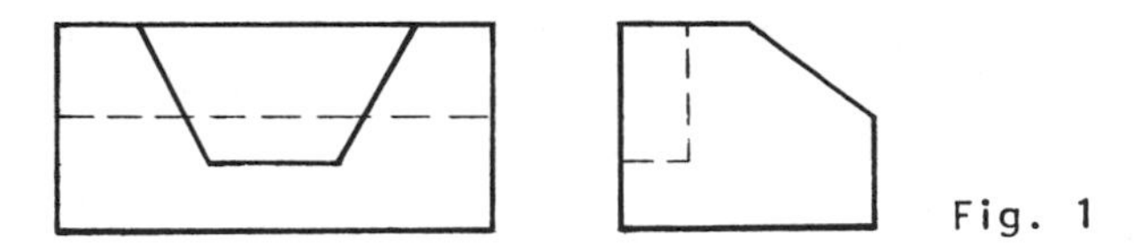

Fig. 1

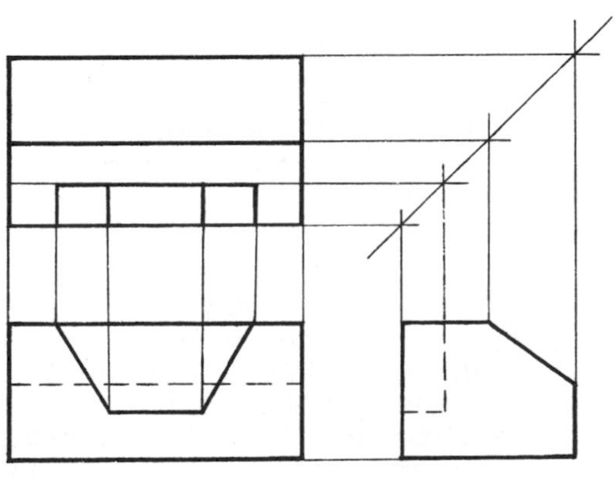

Fig. 2

In drawing the required orthographic projection of the top view first draw a 45 degree line at a convenient distance above the right side view. (See Fig. 2.) Onto this line project all the points in the side view vertically up. Then project these intersections horizontally to the left. Project all the points in the front view vertically up until all the projection lines cross. From the given views we can see that the given object is a block; therefore, darken the outside rectangle formed by the projection lines. For the inclined plane darken the corresponding projection line and for the trapezoidal slot darken the line corresponding to the depth. Then darken the four vertical lines that represent the corners of the slot. Keep in mind that the projection lines that come from the slot are darkened until they reach the depth line of the slot.

• PROBLEM 3-24

From the orthographic projections of the front and right side views shown in Fig. 1, draw the orthographic projection of the top view of the object.

Solution: Before drawing the required top view, notice that the two given views reveal that the object is composed of a 90 degree cut from a hollow cylinder and a rectangular block at the bottom which contains a rectangular slot. Knowing all the parts that make up the object makes it unnecessary to place identifying symbols on the given views.

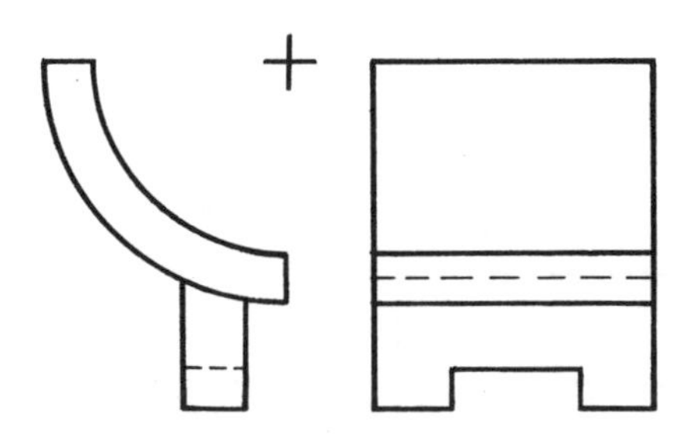

Fig. 1

In order to draw the orthographic projection of the top view, first draw a 45 degree line at a convenient distance from the right side view. Refer to Fig. 2. Onto this line project all the points in the side view vertically up and project these intersections horizontally to the left. Then proceed to project all the points in the front view vertically up until all the projection lines cross. Now determine and draw all the object lines as follows: knowing that the top part of the object is a 90 degree cut from a cylinder which can be represented by a rectangle on the outside edges of the top view, darken the outside rectangle formed by the projection lines. Darken the second vertical line going from the left to the right of the top view. This line determines the thickness of the top part. Now draw the hidden lines for the bottom block and the slot in it. These lines are hidden because the circular part is on top of the block and, therefore, the lines are not seen by an observer looking perpendicularly from the top.

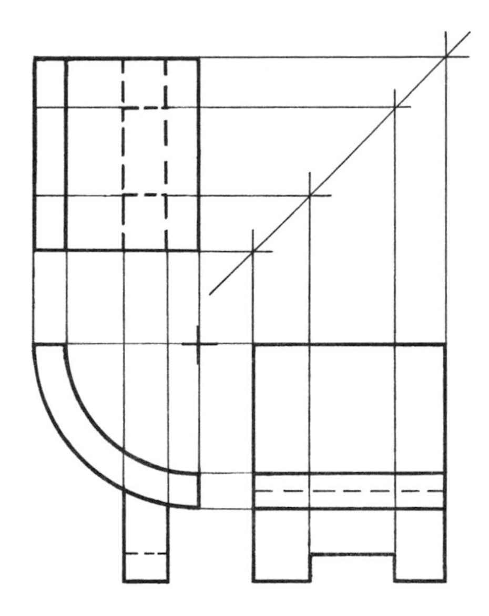

Fig. 2

• PROBLEM 3-25

> From the orthographic projection of the top and right side views shown in Fig. 1, draw the orthographic projection of the front view of the object.

Solution: The two given views show that the object is composed of two rectangular blocks, one vertical and the other horizontal. Notice that the horizontal block has a trapezoidal indentation on each of the left and right sides of the right side view.

In drawing the orthographic projection of the front view first project all the points in the top view vertically down and all the points in the right side view horizontally to the left. See Fig. 2. Notice that the vertical block is inside the horizontal block, and therefore, the horizontal block is completely visible as viewed from the front. The part of the vertical block that goes into the horizontal block is hidden. The trapezoidal indentations appear as double rectangles joined at each corner. This object is called a "Try Square."

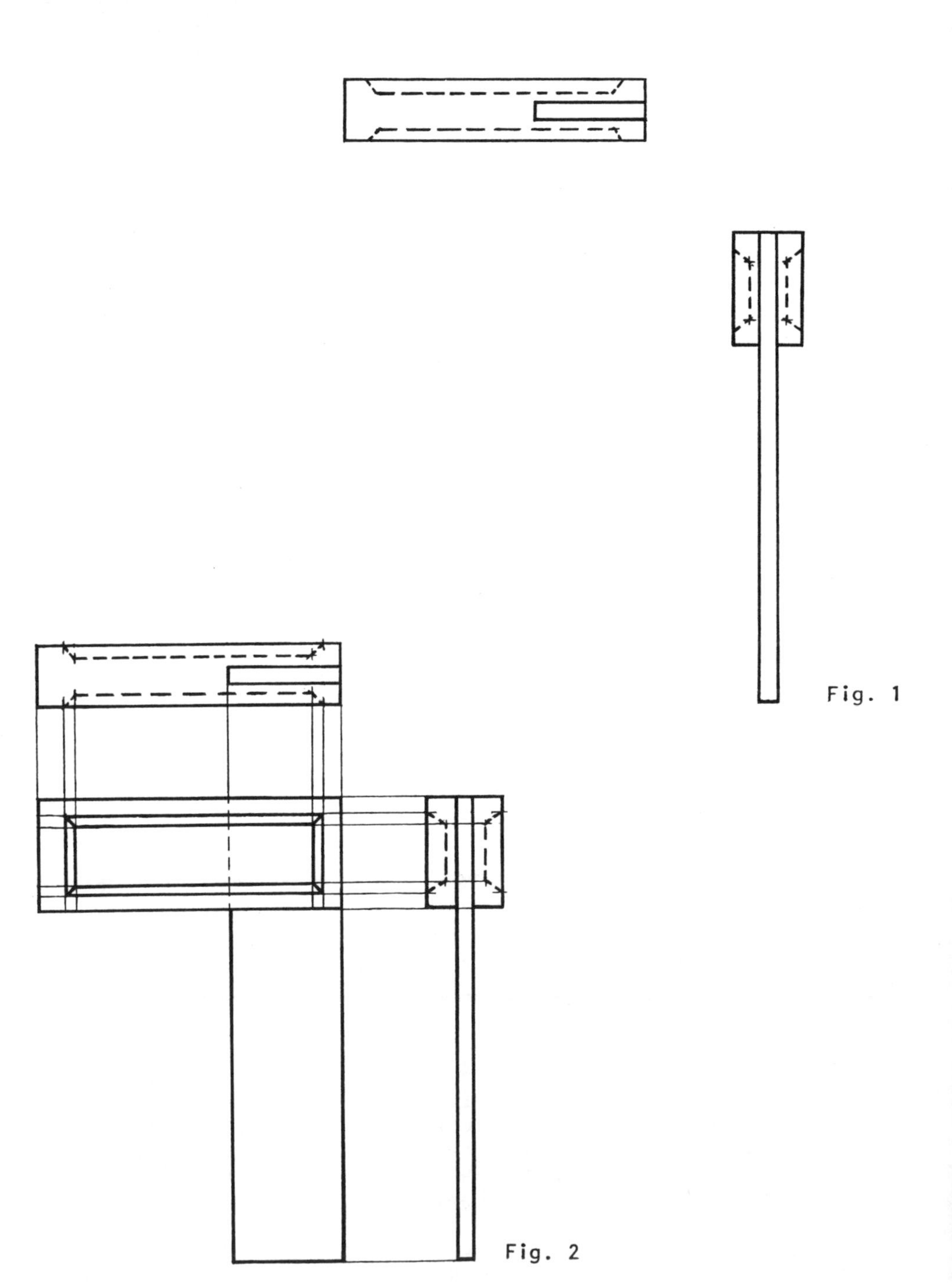
Fig. 1

Fig. 2

• **PROBLEM 3-26**

From the orthographic projections of the front and top views shown in Fig. 1, draw the orthographic projection of the right side view adjacent to the top view.

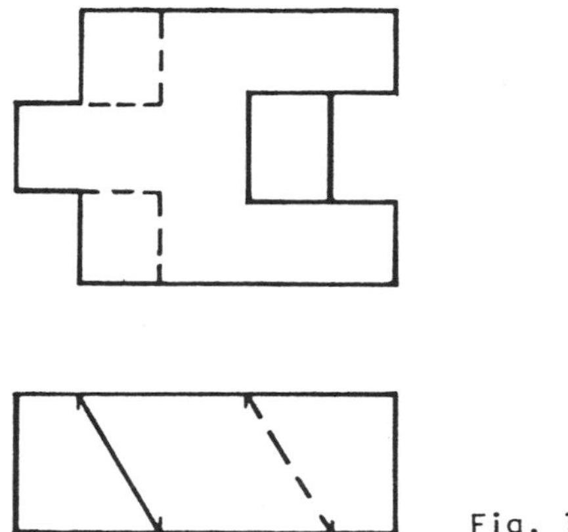

Fig. 1

Solution: The problem calls for the right side view to be drawn adjacent to the top view. The choice of this location was made to save drawing space. Recall that the right side view of this object would be the same no matter where it would be drawn.

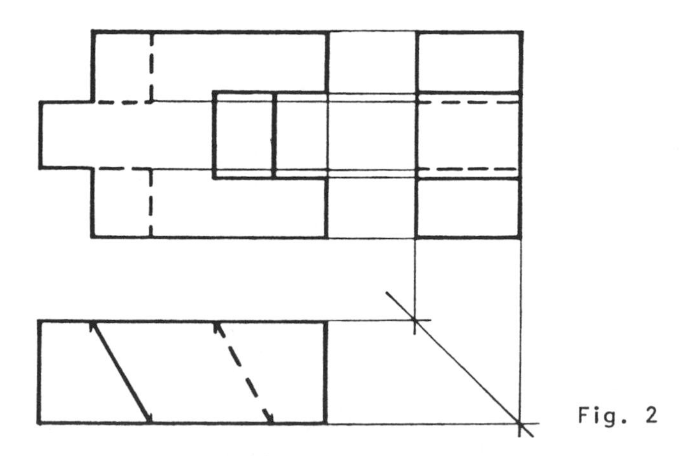

Fig. 2

In order to draw the required side view draw a 45 degree line at a convenient distance to the right of the front view. (Refer to Fig. 2.) Onto this line project horizontally all the points in the front view. Notice that only the two horizontal lines are projected. Proceed to project the intersections of the 45 degree line vertically up and then project all the points in the top view horizontally to the right until all the projection lines cross. Now proceed to determine and draw the object lines. Because of the lines in the extreme right of both the given views the outside edges of the side view form a rectangle. Therefore, darken the outside rectangle formed by the constructed projection lines. Then notice the slot to the right of the top view. The slot goes all the way through and it has an inclined plane. Therefore, darken the two projection lines coming from the slot. Finally, draw hidden lines for the block coming out of the extreme left in the top view. These lines are hidden because an observer looking at this object from the right side would not see this detail.

• PROBLEM 3-27

From the orthographic projections of the front and top views of the object shown in Fig. 1, draw the orthographic projection of the left side view.

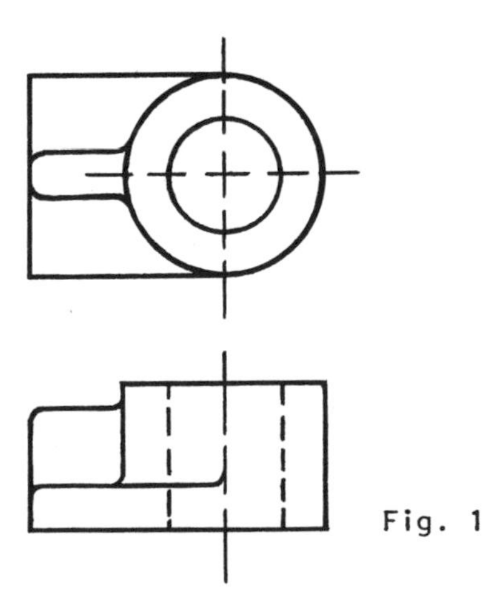

Fig. 1

Solution: The best way to solve this specific problem is by carefully analyzing the two given views. For example, from the front and top views it can be noticed that the outer circle in the top view represents a cylinder. The inner circle represents a hole going through the cylinder and the rectangle to the left of the center line represents a horizontal plate. Again by looking at both of the given views it can be noticed that the inner rectangle in the top view represents a vertical plate joined to the vertical cylinder.

After carefully analyzing the given views, draw a 45 degree line at a convenient distance to the left of the top view. Onto this line project horizontally all the significant points in the top view. Then project these intersections vertically down. Now project all the significant points in the front view horizontally to the left until all the projection lines cross (see Fig. 2). Then proceed to darken the object lines with the aid of the analysis done to the two given views in the above discussion. For example, the cylinder as viewed from the side view is represented by the outer rectangle. The vertical and horizontal plates are represented by the inner vertical rectangle and the inner horizontal rectangle, respectively. Notice that no line is drawn at the intersection of the inner rectangles. No line is drawn because both the vertical and horizontal plates start from the same surface. The object is rounded at all the inner corners. Therefore, include rounds in all the inner corners of the left side view.

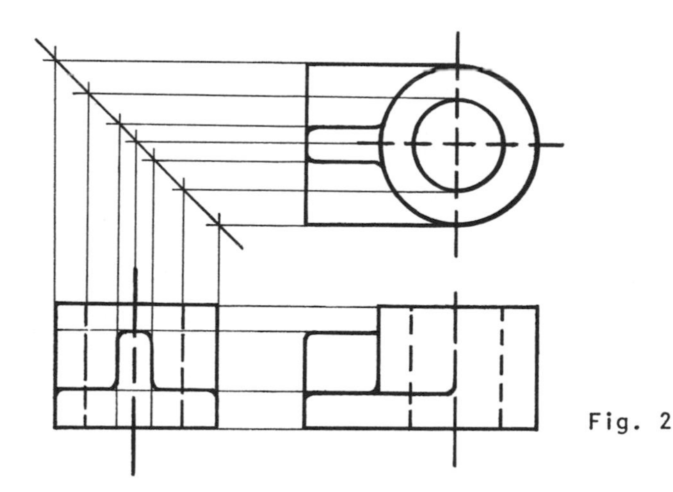

Fig. 2

• PROBLEM 3-28

From the orthographic projections of the front and top views shown in Fig. 1, draw the orthographic projection of the right side view of the object.

Solution: In this type of problem it is sometimes a good idea to label all the significant points in both of the given views. In Fig. 2 the front and top views were labeled as follows: first we notice that the middle vertical line in the front view (line 2-5) lines up with the middle "vertical" line in the top view (line 2-3). Also, in the front view the top plane lies to the left of the middle vertical line (line 2-5). Therefore, label the plane to the left of the middle vertical line in the top view with the numbers 1-2-3-4. From the front view it is seen that the plane to the right is the next lower plane. Therefore, label it 5-6-7-8. Notice that the visible points are labeled on the outside corners while the hidden points are labeled in the inside corners. For example, the top line in the front view represents the edge view of the top plane 1-2-3-4. The visible points in the front view are 1 and 2 while the hidden points are 3 and 4. Notice that 3 and 4 were labeled in the inside corners. Plane 5-6-7-8 was labeled in the front view by using the same method. To finish labeling the significant points in the top view label the lowest visible plane (Plane 9-10-11-12-13-14-15-16) and locate the visible points of this plane in the front view. The bottom line in the front view was not labeled because at this point we can see that all the planes below line 16-11 are vertical.

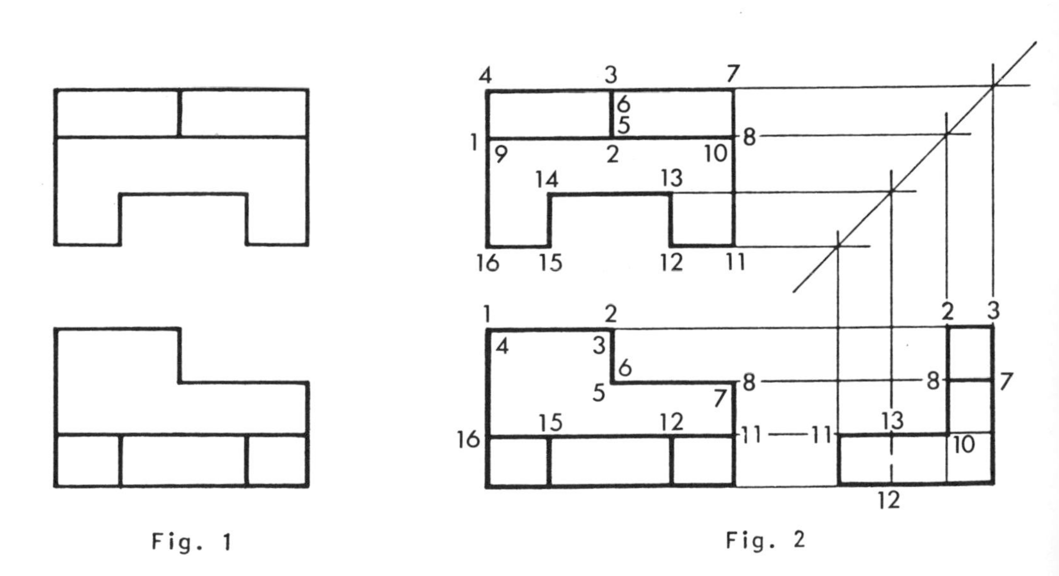

Fig. 1

Fig. 2

In order to construct the right side view, draw a 45 degree line at a convenient distance to the right of the top view. Project all the points in the top view horizontally to the right until the projection lines intersect the 45 degree line. Project these intersections vertically down and proceed to project all the points in the front view horizontally to the right until all the projection lines cross. Determine and label all the visible points in the side view as follows: the visible points in the right side view are located in the extreme right of the front view. For example, points 2, 8 and 11 are located in the extreme right of the front view. Therefore, they are visible in the right side view. Notice that the visible points in the extreme right of the front view are visible in the extreme left of the right side view. Also, the hidden points in the extreme right of the front view, such as points 3 and 7, are visible at the right of the side view. Recall that each point in the side view is determined by two projection lines, one coming from the top view and another from the front view.

• PROBLEM 3-29

> From the orthographic projections of the top and right side views of the object in Fig. 1, draw the orthographic projection of the object's front view.

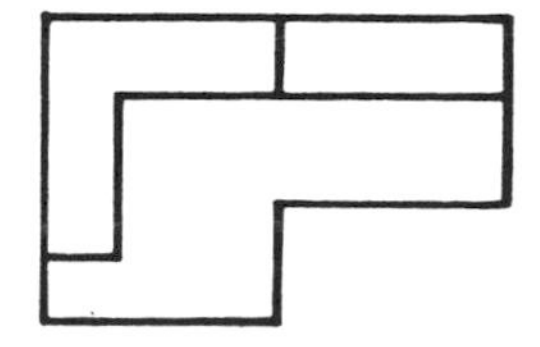

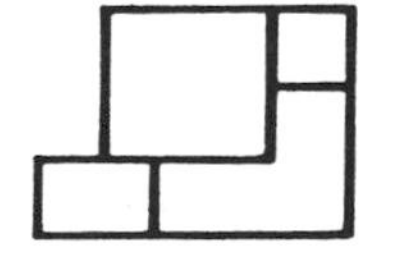

Fig. 1

Solution: When faced with a problem like this which has a comparatively large number of planes, it is best to recognize each plane in the given views and label them. The views in Fig. 2 are labeled as follows: the points that are visible in a view are labeled on the outside corners while those that are not visible are labeled on the inside corners. In labeling the two given views in this problem the following analysis is employed. First, to determine which plane is the top plane, take a close look at the top and side views. Notice that line 9-10 in the side view is solid. Therefore, the plane to the left of line 5-4 in the top view is the top plane and was labeled 1-2-3-4-5-6. These points are then located in the top edge of the side view. Notice that points 1, 3 and 6 are not visible to an observer looking perpendicularly to the right side. Therefore, the points were labeled on the inside corners. Proceed to label the next highest plane which is the plane whose edge view is line 9-10 in the side view. Finally, to complete labeling all the visible planes in the top view label the lowest plane seen from the top. Notice that this plane is represented by line 13-16 in the side view. Whenever a plane is labeled on one view locate the corresponding points in the other views.

After all the visible points in the top view have been labeled and located in the side view, proceed to label the planes which are unlabeled in the side view and locate their corresponding points in the top view.

Once the given views have been labeled, project all the points in the top view vertically down, and all the points in the side view horizontally to the left until all the projection lines cross. Locate all the points that were labeled in the top and side views in the front view and proceed to connect any two points that appear connected in both of the given views. Also connect any two points which on one view appear connected and on the other appear as a point. For example, points 9 and 7 appear connected in the top view and appear as a point in the side view. Therefore, they are connected in the front view. The rest of the front view is drawn in the same manner (see Fig. 2).

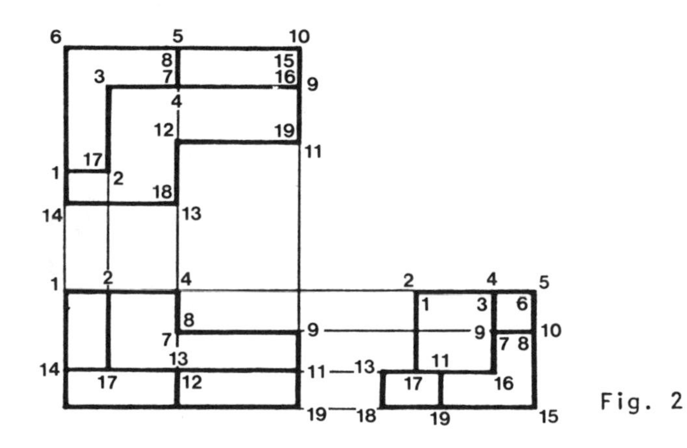

Fig. 2

• PROBLEM 3-30

From the numbered orthographic projections of the front and top view of the object in Fig. 1, draw the orthographic projection of the right side view. The views have been numbered in such a way that the numbers on the outside corners are visible while the ones in the inside are not.

Solution: In drawing the right side view from the given orthographic projections first draw a 45 degree line at a convenient distance to the right of the top view.

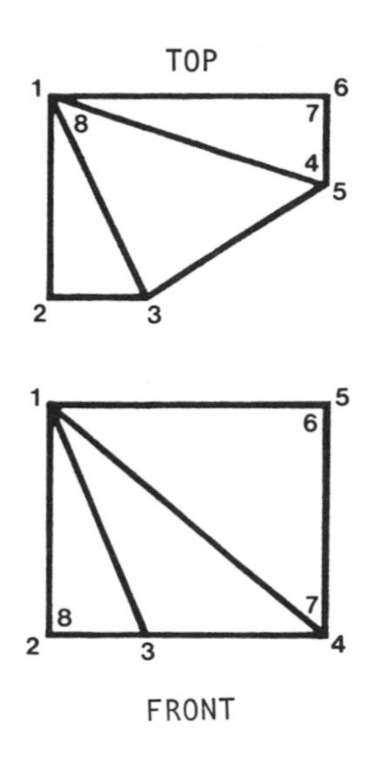

Fig. 1

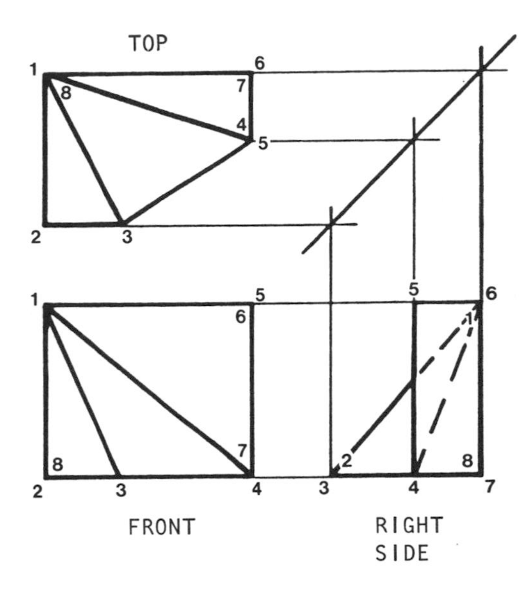

Fig. 2

Project all the points in the top view horizontally onto this line. Then from the 45 degree line project all the intersections vertically down. Proceed to project all the points in the front view horizontally to the right until all the projection lines cross. Label all the points in the side view by determining the intersections of every two projection lines that come from the same specific number. For example, the intersection of the projection line for point "5" coming from the top view and the projection line for point "5" coming from the front view determine point "5" in the side view. When all the points in the side view have been labeled, connect the object lines in the following manner. Notice that line 5-6 in the top view represents the edge view of the vertical plane 4-5-6-7. Line 5-4 in the front view represents the edge view of the same plane. Both of these edge views appear to the right of their respective drawings. Therefore, this plane must be visible in the right side view. Connect every two points which appear connected in both of the given views. For example, points 1 and 2, 1 and 3, 1 and 4, 1 and 6, and 3 and 4 all appear connected in both of the given views; therefore, they are connected in the side view. Keep in mind that plane 4-5-6-7 is visible over any other plane (see Fig. 2); any line that enters the interior of this plane must appear hidden unless it is on the same plane.

Given the orthographic projections of the front and top views of the object shown in Fig. 1, find the orthographic projection of the right side view.

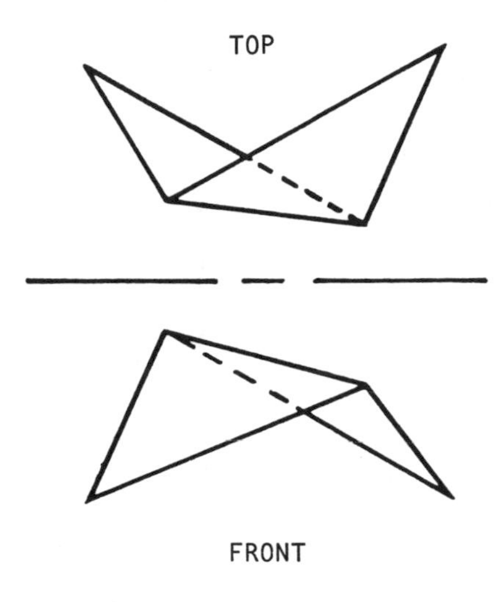

Fig. 1

Solution: This problem illustrates the fact that in many problems of this type first attempting to visualize the desired view is often the least effective approach. Instead, by first establishing the location of as many points as possible by orthographic projection, most, if not all, of the required view will become apparent. First label all the points in the two given views. In this case it is easy to label each point because each point lines up vertically. Note that the place where lines AB and CD appear to cross is not labeled. Since these locations do not line up vertically they are not the same point in the top and front views. This point cannot be projected into the side view. (Actually lines AB and CD are skewed lines; they do not intersect. Where they appear to cross is not actually a point at all.) When all the points have been labeled in both views, construct a 45 degree line at a convenient distance to the right of the top view. Onto this line project horizontally all the points in the top view. Then project these intersections vertically down. Project all the points in the front view horizontally to the right until all the projection lines cross. Label the intersections of the projection lines that come from the same point. For example, the intersection of the two projection lines coming from point "C" is labeled "C". When all the points have been labeled, connect every two points which appear connected in the two given views. For example, points "B" and "C" appear connected in both given views; therefore, they must be connected in the side view. After all the lines present in the object have been drawn, the next problem is to determine which lines are hidden.

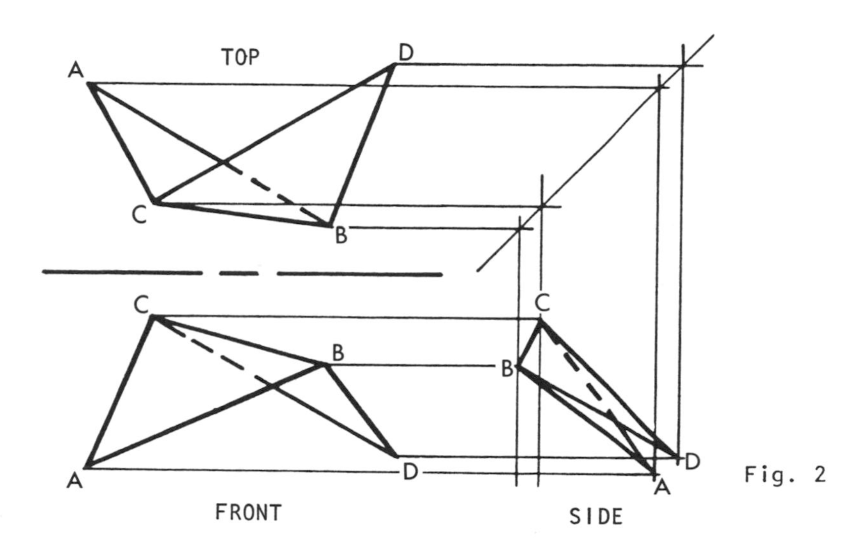

Fig. 2

To determine which lines are hidden first notice that all the outside edges of any view must be visible. Therefore, only the inside segments of lines AC and BD could be hidden. The way to determine which of the lines is visible is to look at the front view. Whichever one of the two lines appears to be closer to the side view is the one that is visible in the side view. In the front view of Fig. 2, line BD appears closer to the side view than line AC; therefore, line BD is visible in the side view.

• PROBLEM 3-32

Given the orthographic projections of the front and top views of the rectangular plate shown in Fig. 1, draw the orthographic projection of the right side view.

Solution: To draw the right side view of the given object, first draw a 45 degree line at a convenient distance to the right of the top view. Project all the points in the top view horizontally onto this line. Then from the 45 degree line project the intersections vertically down. Project all the points in the front view horizontally to the right until all the projection lines cross. Make sure that the projection lines are thin and very light. When all the projection lines have been drawn, draw and darken all the center lines. Keep in mind that the presence of center

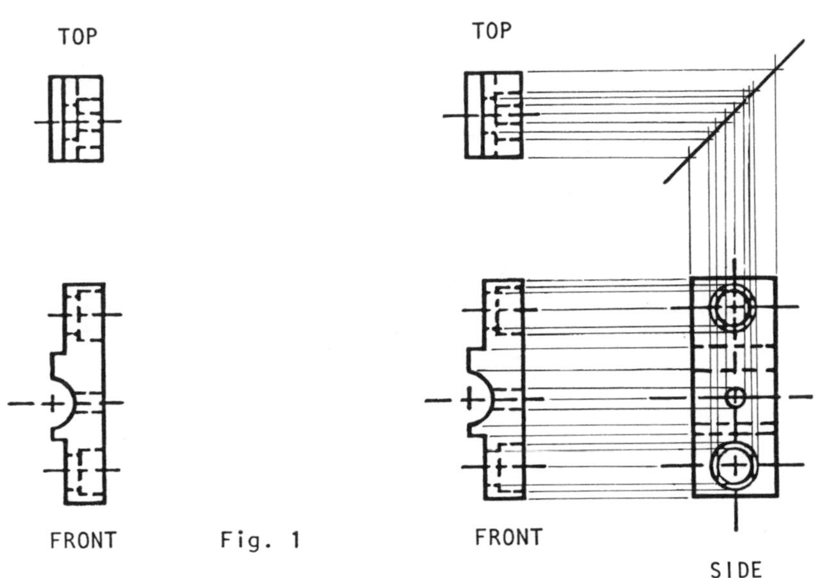

lines on a drawing indicates that there are circles or rounds centered around the intersections of these center lines. After all the center lines have been drawn, include all the circles and rounds present in the object. Now draw the outside rectangle. Notice the problem states that the object is a rectangular plate. This statement is necessary because the right side view may have contained rounded corners which would not necessarily show in the front or top view. (For this reason if the object is to be described by only two views it would be better to describe it by the front and side views.) Finally, draw the four hidden lines which represent the edges on the left side of the front view.

• PROBLEM 3-33

Given the orthographic projections of the top and front views of the object shown in Fig. 1, draw the orthographic projection of the right side view.

Solution: To construct the orthographic projection of the given object, first draw a 45 degree line at a convenient distance to the right of the top view. Project all the points in the top view horizontally onto this line. Then project the intersections of the 45 degree line vertically down. Project all the points in the front view horizontally to the right until all the projection lines meet. Make sure that all the projection lines are thin and very light.

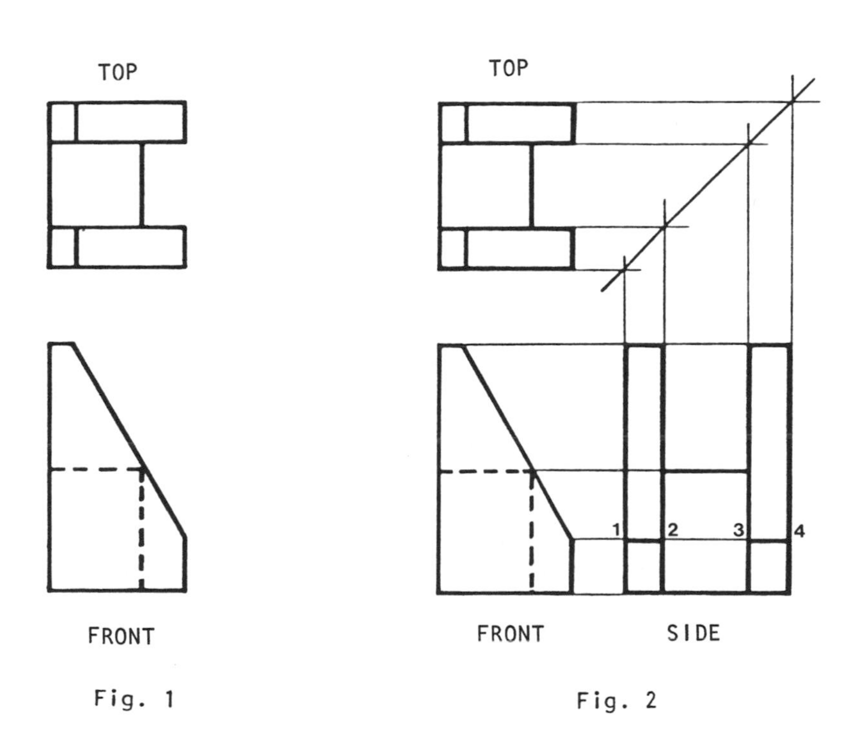

Fig. 1

Fig. 2

The lines that are part of the side view can be determined by analyzing both the top and front views. For example, it can be noticed from the top and front views that the object has two vertical plates of the exact same shape, each of which contains an inclined surface. Therefore, the rectangles can be drawn on the left and right of the side view (see Fig. 2). The lines drawn inside these rectangles, lines 1-2 and 3-4, represent the edges at which the inclined surfaces end. By analyzing the front view it is seen that the hidden lines represent the edge views of two planes, one vertical and one horizontal. The horizontal plane is seen in the middle of the top view. Therefore, draw the vertical plane in the middle of the side view.

Given the orthographic projections of the front and right side views of the object shown in Fig. 1, draw the orthographic projection of the top view.

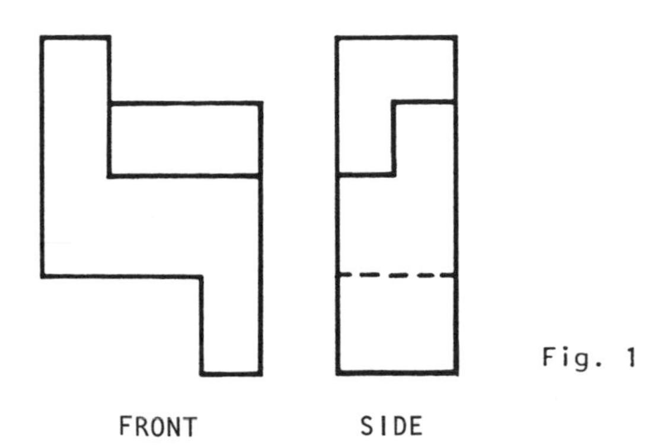

Fig. 1

Solution: A very important concept to keep in mind whenever a third view is being drawn from two given views is the fact that every line in the interior of a drawing represents a break, change, or end of a plane and unless an inclined plane is being represented, every line is the edge view of a plane parallel to the line of sight. The top view of this object can be determined by using the above mentioned concept. However, in order to better illustrate this concept, the front and side views have been numbered (see Fig. 2).

In determining a third view from two given views it is advisable to label every corner point of the object. To label every point, first determine the points that are on the same plane and then label all such points. Do this for every plane on the object. For example, there is a frontal plane on the object which is enclosed by six lines. Every corner point of this plane has been labeled in Fig. 2 (plane 1-2-3-4-5-6-7-8).

By inspecting the right side view it can be seen that points 1,2,3,4,5,6,7,8 must be located on the left vertical edge of the right side view. Thus, point 8 is located on the left edge of the right side view at the same height as it is in the front view. Now notice point 7 which is also seen to be on the front surface. If points 7 and 8 are on the same front surface then point 7 must be hidden by point 8 in the right side view and so line 7-8 appears as a point in the right side view. In the middle of the right side view there is another point. This point, labeled point 10 in Fig. 2, is hidden in the front view.

Now notice line 8-10 in the right side view. This is the edge view of a horizontal plane. Another edge view of this same plane is represented by line 7-8 in the front view. The fourth corner of this horizontal plane, point 9, can then be located. Point 9 is hidden in both the front and right side views. This point by point analysis reveals that it must be located behind point 7 in the front view.

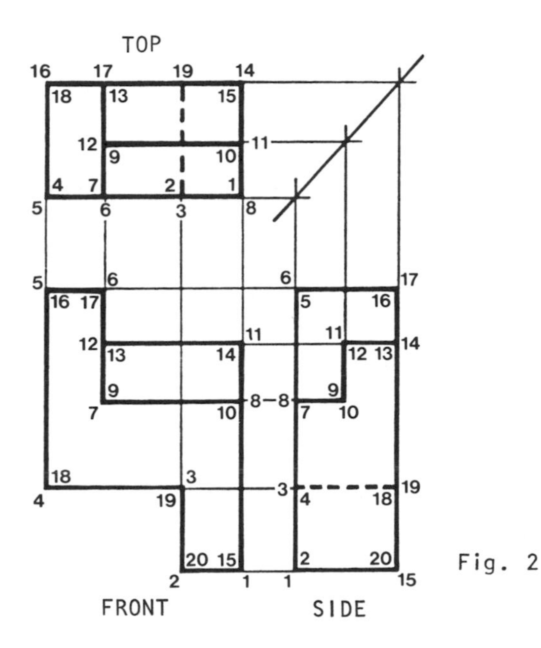

Fig. 2

With points 9 and 10 located, it is now easier to see that there is a second plane visible from the front. This plane is recessed back from the front surface of the object. The edge view of this plane, line 10-11, is visible in the right side view. The right view shows that this plane is in the middle of the block, halfway between the front and back surfaces. This same procedure is used to label all the other points. Notice that in numbering the two given views the visible points were placed on the outside of each corner while the points that are not visible were labeled in the inside of each corner.

After all the points have been labeled, draw a 45 degree line at a convenient distance above the right side view and project every point on the side view vertically up to this 45 degree line. Then project these intersections horizontally to the left. Project every point in the front view vertically up until each projection line meets the corresponding projection line previously drawn for that particular point. For example, point 11 in the side view is projected to the 45 degree line and then horizontally to the left. This same point is also projected up from the front view, thus determining point

11 in the top view. Every other point is determined in the same manner. Once all the points have been determined and labeled in the top view they are connected to form the correct top view. In connecting two points make sure that those points appear to be connected in every other view.

• PROBLEM 3-35

Given the orthographic projections of the top and right side views of the object shown in Fig. 1, determine the orthographic projection of the front view.

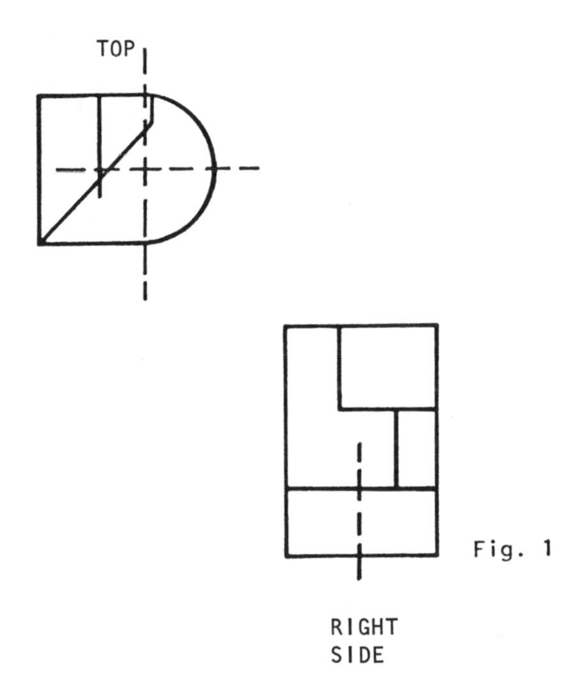

Fig. 1

Solution: In this problem it is recommended that the corner points of the object be labeled in order to make the recognition of points easier. In labeling these points it is suggested that the two given views be labeled plane by plane. For example, in Fig. 2 plane 1-2-3-4 was determined to be the top plane of the object. This fact was arrived at by looking at the right side view which shows that the plane containing points 5, 6 and 7 is lower than plane 1-2-3-4. Had it been the other way around line 1-2-3-4 in the side view would have been hidden. Once the top plane has been established the next lower plane is also labeled in the top view and then the points are located in the side view. Notice that plane 1-2-3-4 and the plane which contains points 6, 5 and 7 are both horizontal. This can be deduced by noticing that there are no trapezoidal planes in the side view.

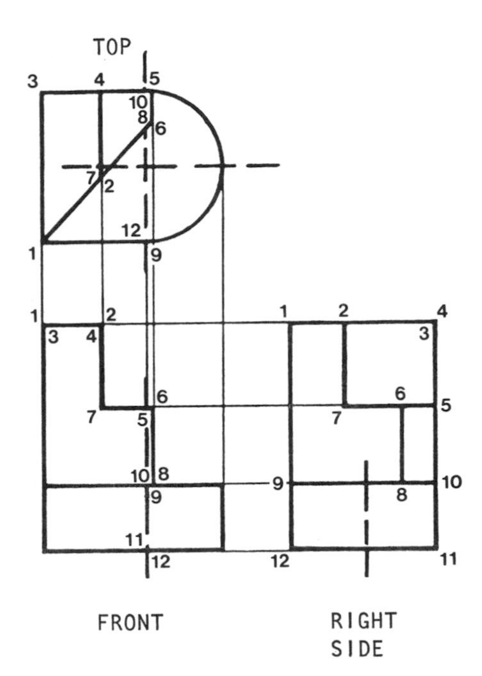

Fig. 2

The fact that planes 1-2-3-4 and 5-6-7 appear as horizontal lines in the side view indicates that they are horizontal planes. The same is true of the section on the lower right side of the object.

Once all the planes and points have been labeled in the manner described above in both the top and right side views, the front view is then projected from the given views as follows: project every point on the side view horizontally to the left. Then, project every point in the top view vertically down until all the projection lines cross. Identify each point in the front view by first spotting each point on the top and side views then following their projection lines until they meet at a point. Label this point with the same symbol as the points from which the projection lines were taken. After all the points have been identified in this manner, connect every two points by a line in order to draw the correct front view. When connecting these points it is a good idea to first label the planes which are recognized. For example, plane 1-2-7-6-8 can be recognized in the front view by looking at the side view. Then the rest of the points are connected by using the same procedure.

Given the front and right side views of the object shown in Fig. 1, construct the top view of the object.

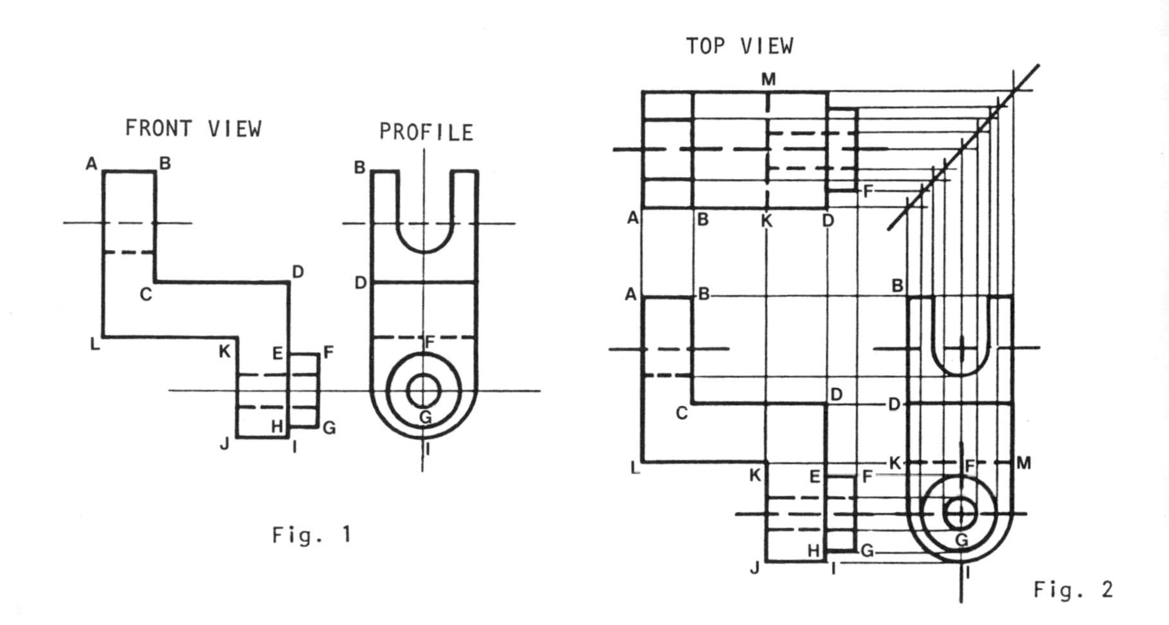

Fig. 1

Fig. 2

Solution: The usual approach for problems of this type is to first draw a 45 degree line at a convenient distance from the side view. Project the side view vertically up to the 45 degree line and from this line project all the points horizontally to the left. Now project the front view vertically up to meet the projection lines coming from the 45 degree line (see Fig. 2). Each corner point of the top view is determined by two projection lines, one taken from the 45 degree line and the other coming from the front view. It is always a good idea to label the corner points in the front and profile views so that the projections of each point can be easily followed. For example, point B is easily identified in the front and side views. Point B on the profile view is projected up to the 45 degree line and then to the top view. Next, point B on the front view is projected up to where it meets the other projection line coming from the side view. This point is labeled B on the top view also. This procedure is followed for all other points on the drawing.

To determine the visibility of a line on the top view simply follow the projection lines down to the front view. If the projection line crosses a solid line that is not part of the line being analyzed, the line is hidden, otherwise, it is solid. For example, when line KM in the top view is projected down to the front view it crosses line CD first and since line CD is not part of KM, line KM is hidden in the top view.

Given the front and top view shown in Fig. 1 find the right side view of the object.

Solution: Whenever a top or side view is to be constructed from the front and a top or a side view, first construct a 45 degree line at a convenient distance from either the

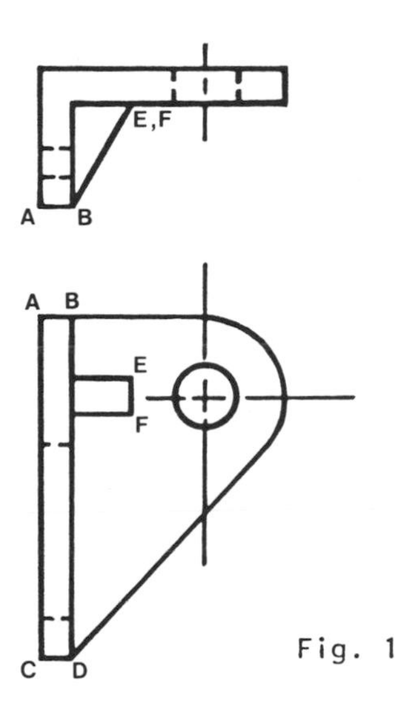

Fig. 1

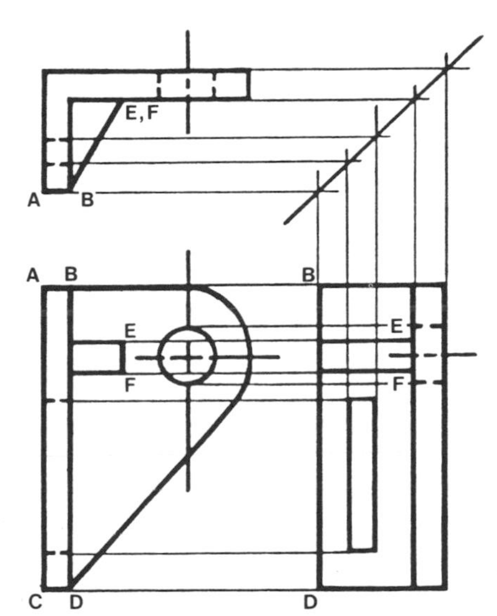

Fig. 2

top or side view (see Fig. 2). For this problem draw the 45 degree line to the right of the top view. Project all the horizontal lines from the top view onto this 45 degree line. From this line project the points of intersection vertically down to the right of the front view. Then go to the front view and project all the horizontal lines to the right until they meet the projection lines coming from the 45 degree line. Each point on the side view is determined by two projection lines. After all the points have been identified they are connected to form the side view. For example, points E and F are determined by their corresponding projection lines and once they have been identified, line EF can be drawn. The same procedure follows for all the lines in the side view. Notice that point F appears directly beneath point E in the top view.

Given the front and top views of the object shown in Fig. 1, find the orthographic projection of the right side view.

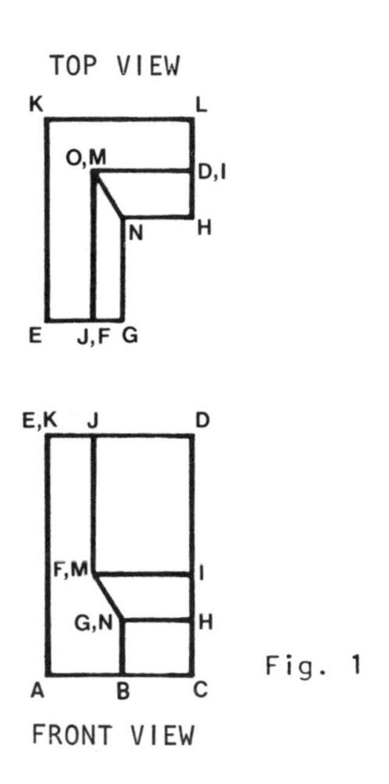

Fig. 1

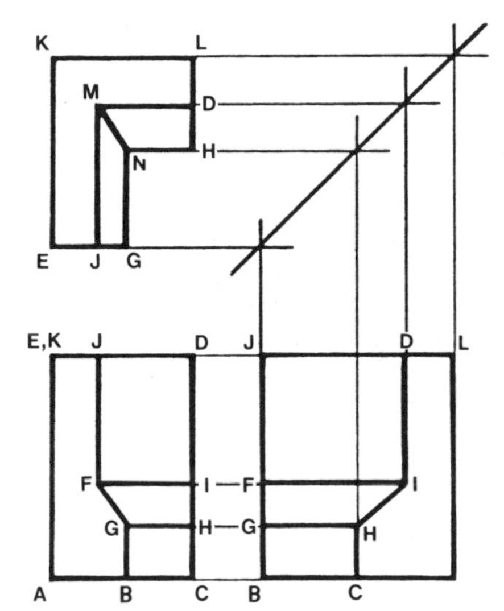

Fig. 2

Solution: The front and top views of the given object were conveniently labeled in order to make the problem a little easier. If the given views had been unlabeled, they could have been labeled as follows: Recall that every straight line in any view represents a change in plane. For example, in the top view of the given object the points K-L-D-O-J-E all lie in the same plane (the top plane). This can be reasoned because lines O-D and J-O both represent a change in plane. It is worth noticing that since line M-N is inclined, both planes F-G-N-M and I-H-N-M (shown in the top view) are also inclined. Once the top plane is determined and labeled, the top line in the front view (which shows the edge view of the top plane) can be labeled. This same procedure can be used to label all the other points. It is an accepted practice to label all the points on every view by separating the points with commas (as shown in the upper left hand corner of the front view).

Once all the points have been labeled on both the front and top views, draw a 45 degree line at a convenient distance to the right of the top view. On this line project horizontally all the corner points shown in the top view. Then project the intersection points of the 45 degree line and the projection lines vertically down to the right of the front view. Now project all the corner points in the front view horizontally to the right to meet the projection lines coming from the 45 degree line. Follow each projection line and label the points that they determine. For example, the projection line of point "D" in the top view meets the projection line of point "D" in the front view at a point labeled "D". Do the same with every point. Not all the points have been labeled. Once the points have been labeled they can be connected. To connect the points simply notice the locations of the planes. For example, points M-N-H-I represent an inclined plane in the front view. Therefore, join the points that define this plane. Points F-G-N-M in the front view represent another inclined plane. These points can be joined also. Follow this procedure until all the planes have been determined.

• PROBLEM 3-39

Given the orthographic projections of the top and right side views of the object shown in Fig. 1, find the orthographic projection of the object's front view.

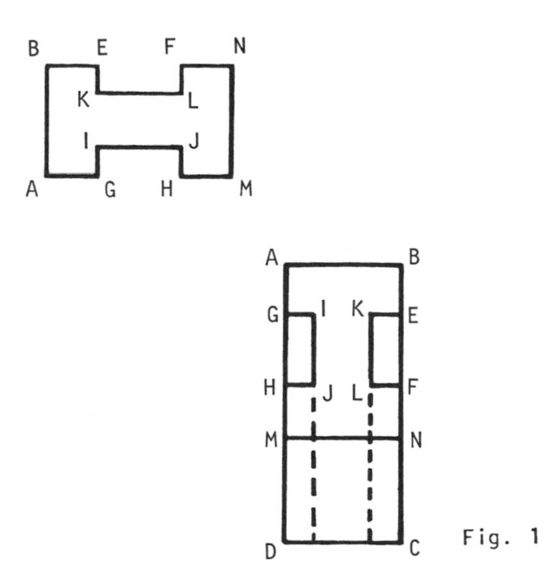

Fig. 1

Solution: Notice that the given front and top views have been completely labeled. These views were labeled by noticing that the plane A-B-E-K-L-F-N-M-H-J-I-G appeared in both views. This means that the plane must be inclined and therefore, the points are easily determined and labeled.

To construct the front view of the given object, project all the corner points shown in the right side view horizontally to the left. Then project all the corner points shown in the top view vertically down. By using the corresponding projection lines determine and label all the points in the front view. For example, the two projection lines of point "H" intersect at a point labeled "H", in the front view. Follow the same method to determine and label all the points in the front view.

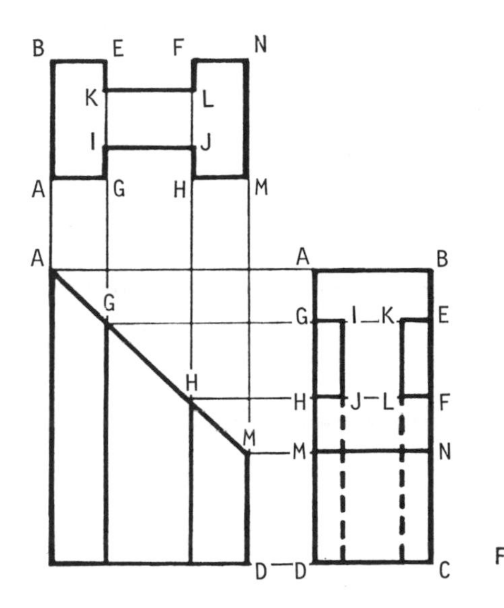

Fig. 2

After all the points in the front view have been determined, connect these points for the correct front view. Recall that if two points are connected on a plane to form a line they must be connected in all views. For example, points "A" and "G" are on the inclined plane so they must appear connected in all views. The same procedure may be followed to connect the rest of the points.

• PROBLEM 3-40

Given the orthographic projections of the front and right side views of the object shown in Fig. 1, find the orthographic projection of the top view.

Solution: When a draftsman cannot readily picture an object from two given views and wishes to construct a third view, the first thing he does is letter or number each significant corner point of the two given views. In Fig. 2 both the front and right side views were labeled as follows: remember that each solid line in the interior of a drawing represents a change in plane. For example, in the front view the line CB represents a change or break of the plane A-B-C-D. This means that the "apparent" plane C-B-G is not on the same plane of A-B-C-D. (The word apparent is used because in the front view point C,B and G appear as if they were on the same plane. But, the real plane is I-K-G, whose edge view is shown in the side view.) When analyzing the side view, notice that lines FE, IK, IG and GH all represent a break in a plane and by this determine and label all the planes shown in the given views.

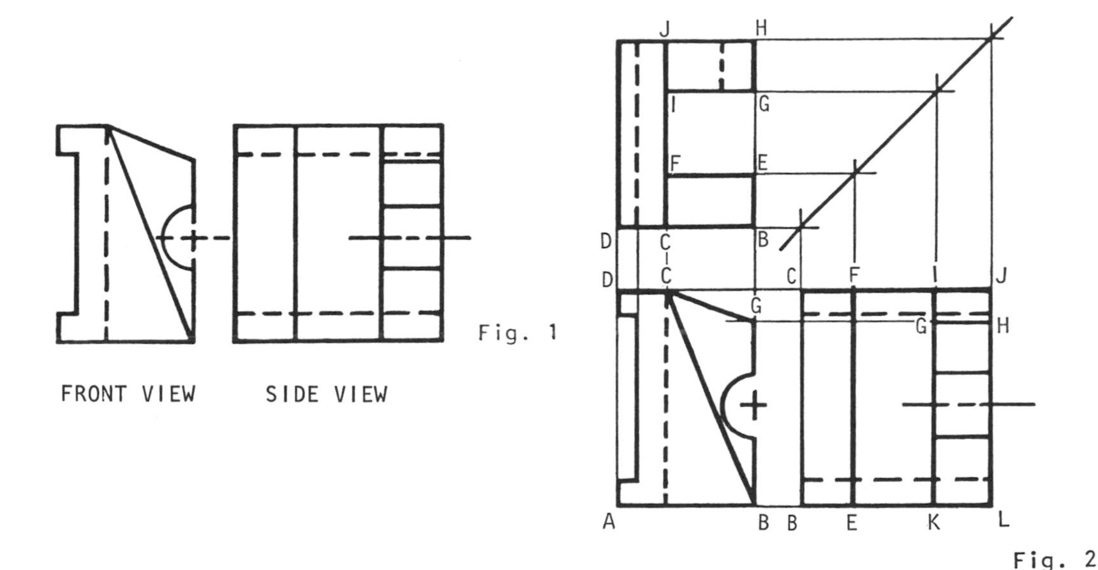

Fig. 1

Fig. 2

To construct the top view after labeling the significant corner points, draw a 45 degree line at a convenient distance from the right side view. On this line project vertically up all the corner points from the side view. Then project horizontally to the left, the intersections of the projection lines and the 45 degree line. Now project all the corner points in the front view vertically up to meet the projection lines previously drawn. Proceed to identify each corner point in the following manner: take each point on one of the views (for example, the side view) and follow its projection line to the top view until it intersects with the corresponding projection line coming from the other view (the front view). For example, the projection line of point "G" is followed from the side view to the 45 degree line and then to the top view until it meets the projection line of point "G" coming from the front view. This point is labeled "G" in the top view. This procedure is followed for all the points in the front and side views. Notice that the horizontal lines in the side view will appear as a point in the front view and that for this reason point "C" in the front view also represents points F, I and J. Connect the points so as to draw all the existing lines in the drawing. For instance, line G-H is known to exist in the side view; therefore, points G and H should be connected in the top view. Be aware that whenever a line being projected is behind a visible plane, it should be drawn as a hidden line.

• PROBLEM 3-41

Given the front and side views of the object shown in Fig. 1, find the object's top view.

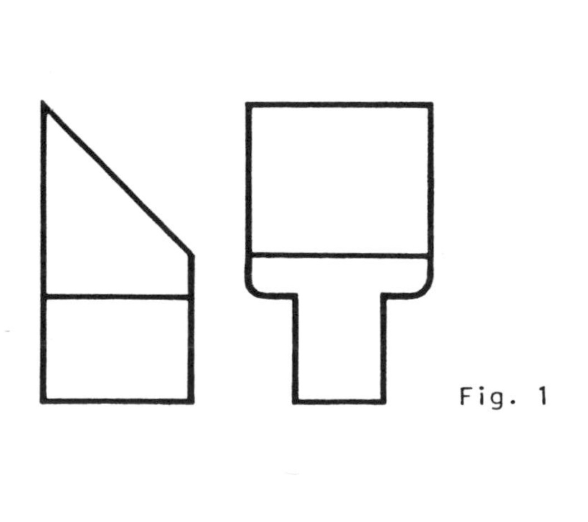

Fig. 1

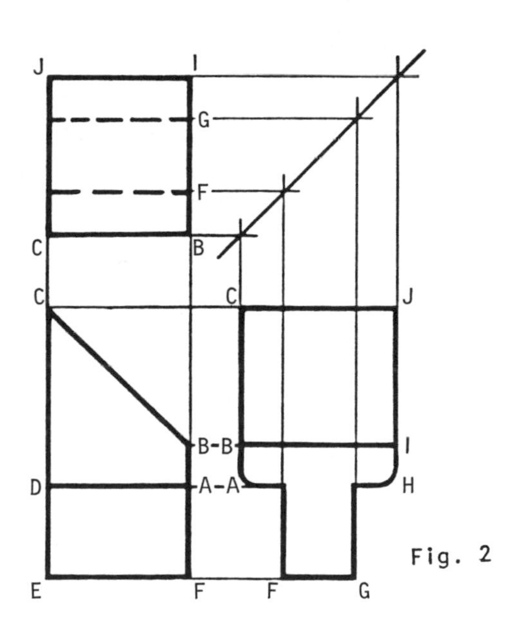

Fig. 2

Solution: This is a rather simple object that can be easily pictured from the two given views. However, it is a good practice to label the corner points of any object when determining the third view. When labeling this object notice that line BC in the front view represents the edge view of the inclined plane C-B-I-J and that line D-A represents the end of plane A-B-C-D (see Fig. 2). Identifying each plane in the given view is important because it aids the draftsman in labeling all the corner points of the object.

In order to construct the top view of this object, first draw a 45 degree line at a convenient distance from the right side view. Onto this line project all the corner points in the side view, vertically up and then project these points of intersection horizontally to the left. Project all the corner points in the front view vertically up until each projection line intersects the corresponding projection line that was taken from the side view. For example, to determine point "C" project point "C" from the side view onto the 45 degree line and from this line project it horizontally to the left. Now take point "C" in the front view and project it vertically up until it intersects the projection line for point "C" previously drawn. This procedure is followed for all the other points. Each pair of points is then connected to form the existing lines (see Fig. 2). Notice that for the round corners in the side view, only a straight line is projected from it onto the front view because this is how the round corner is viewed by an observer looking perpendicularly to the frontal plane. The two dashed lines shown in the top view are hidden because the visible plane C-B-I-J is on top of the lines being represented by these dashes.

• PROBLEM 3-42

Given the orthographic projections of the front and right side views of the object shown in Fig. 1, draw the orthographic projection of the top view.

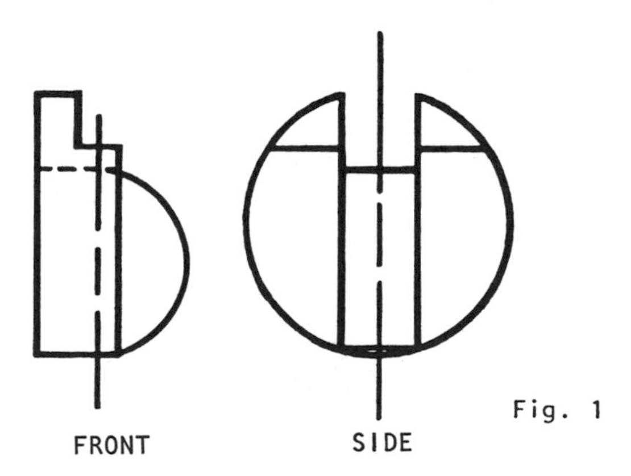

Fig. 1

Solution: It is usually convenient to label the corner points of the given views before attempting to draw the third view. In this case, however, the labeling of points may confuse the diagram because there are two circular discs. So rather than labeling the two given views, simply analyze each given view and recognize the geometrical figures they represent.

To draw the top view, first draw a 45 degree line at a convenient distance from the right side view. Project all the points in the side view vertically up to the 45 degree line and from this line project the intersections horizontally to the left. Then, project all the points in the front view vertically up until the projection lines cross.

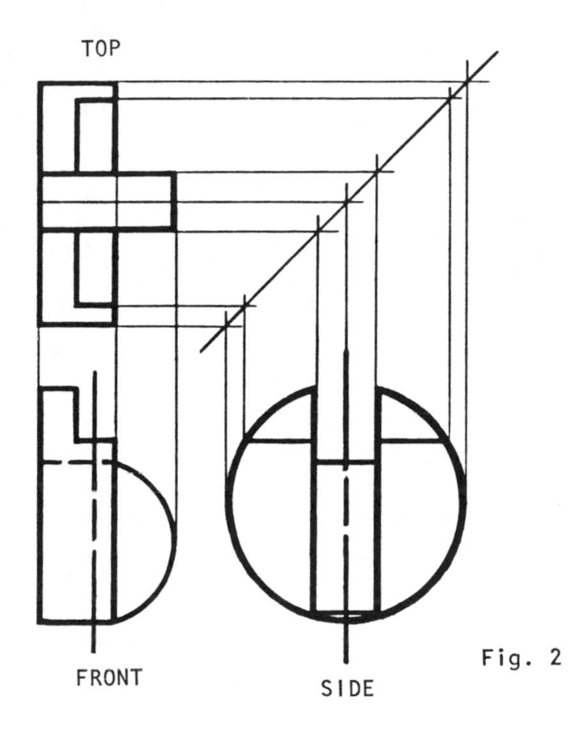

Fig. 2

To determine which lines are part of the object, go through the analysis of the two given views. For example, by looking at the side view it can be noticed that the left plate shown in the front view is a circular disc. Therefore, draw a rectangle in the top view to represent the circular disc. From the front view it is seen that the circular plate has a slot in the upper right hand corner from which a centered rectangle can be drawn. Finally notice that there is another slot in the upper center of the side view from which another rectangle can be drawn. These rectangles determine the whole top view. Notice that all the different components of the object have been fully described.

CHAPTER 4

PRIMARY AND SUCCESSIVE AUXILIARY VIEWS

PRIMARY AND SECONDARY AUXILIARY VIEWS

• **PROBLEM** 4-1

Find the projections of a point contained in the interior of the pentagonal box shown in figure 1 onto all the planes surrounding it.

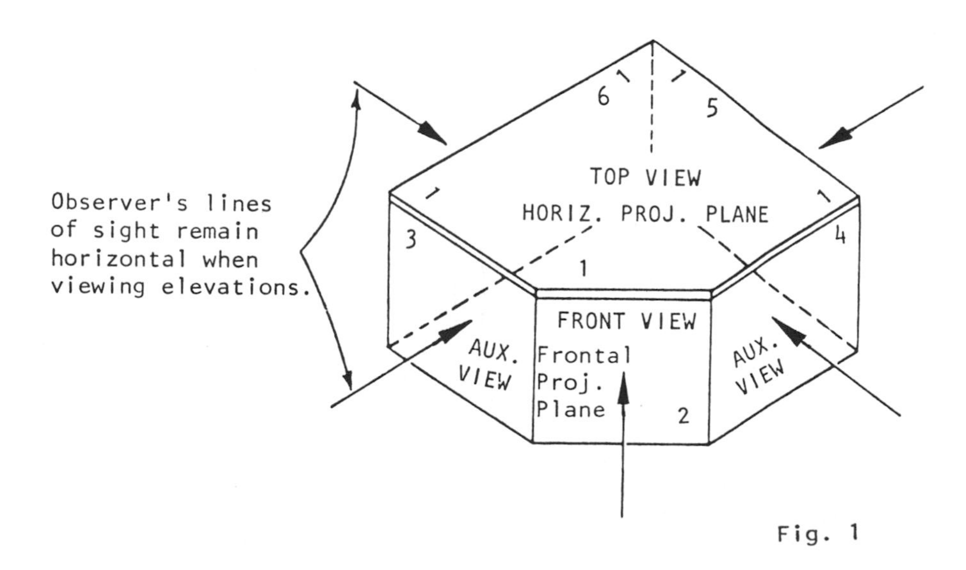

Fig. 1

Solution: In this problem all that is needed is to project the location of the point perpendicularly to every

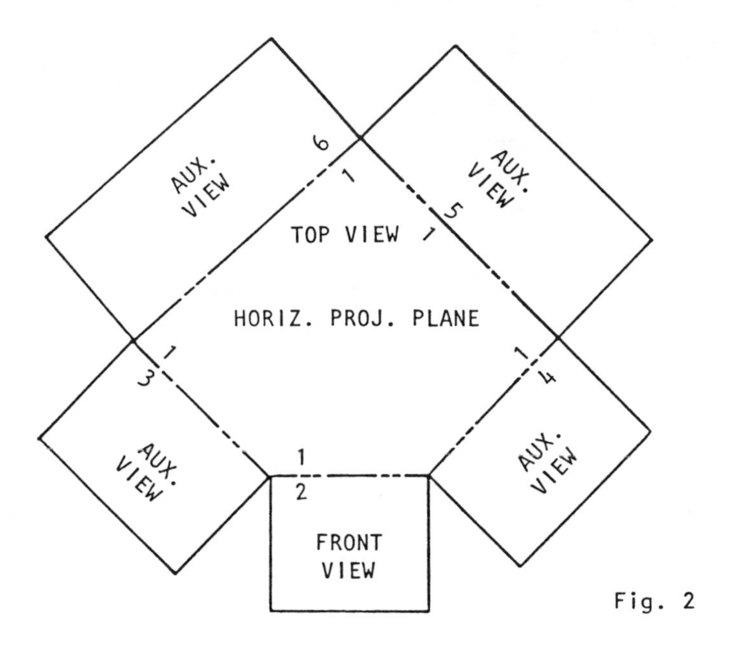

Fig. 2

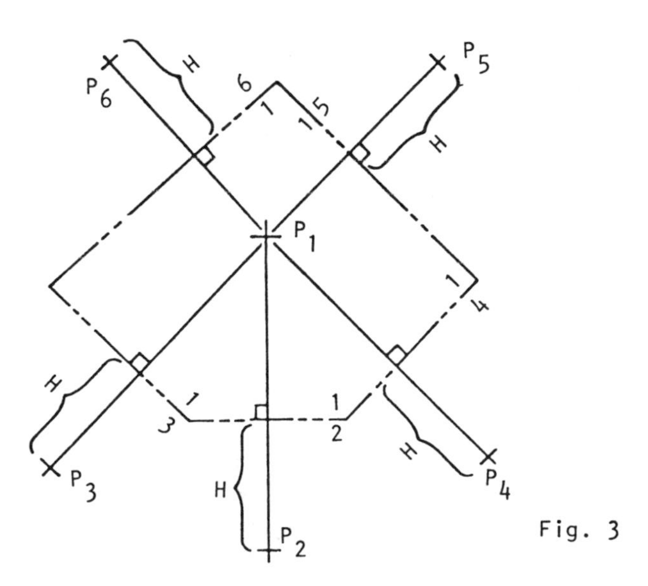

Fig. 3

side of the box. This is equivalent to viewing the point at right angles from the faces of the box as illustrated by the arrows in Fig. 1. To show the different auxiliary views of the point open up the box so all the sides of the box lie on the same plane, as shown in Fig. 2. Now, locate the point on the horizontal plane and from this top view, project the point perpendicularly to every edge of the box. Notice that distance "H" is the same on all the auxiliary views because the depth of the point is the same throughout. This is so because upon viewing the point at right angles to each face, the horizontal plane appears in edge.

• PROBLEM 4-2

Draw an auxiliary view of the tetrahedron as seen in the direction DE.

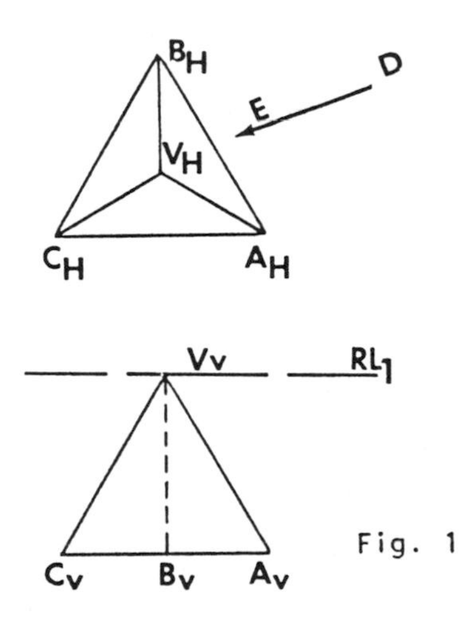

Fig. 1

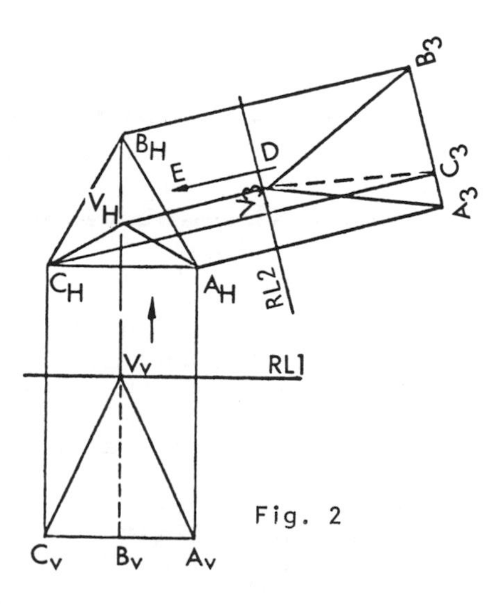

Fig. 2

Solution: The desired view is the projection of the figure on a plane taken perpendicular to the given direction of sight DE. Notice that the direction of DE is perpendicular to the line A_HB_H of the tetrahedron. Also be aware that the direction of DE is not perpendicular to the face of the tetrahedron to which it is pointing but instead it is parallel to the base of the tetrahedron.

Draw RL2 perpendicular to DE allowing a convenient space between RL2 and the top view. Draw RL1 between the top and front views, and at right angles to the projectors, from the top view to the front. Draw projectors from V_H, A_H, B_H, C_H perpendicular to RL2. Locate V_3, A_3, B_3, C_3 on the projectors from RL2 at distances equal to the distances of V_V, A_V, B_V, C_V from RL1 respectively. These distances can be transferred because both the frontal and auxiliary planes are vertical while the top plane is horizontal. For this reason the corresponding heights in the front and auxiliary views are the same, since both reference lines RL1 and RL2 lie on the horizontal plane.

The speed and accuracy with which an auxiliary view can be drawn is determined to some extent by the location of the reference line. It can be seen in Figure 2 that placing RL1 through V_V, instead of above it, eliminates one measurement and allows the auxiliary view to be constructed by using three equal distances. It can further be seen that if RL1 had been taken through C_VA_V, RL2 placed at the distance of A_3B_3, and the measurement to V laid off toward the top view, only one measurement need have been made. In general, it is convenient to place the reference lines through at least one point of a figure.

• PROBLEM 4-3

Determine the secondary auxiliary view of a solid: the view of the solid indicated by the line of sight.

Solution: Fig. 1 shows the given solid, a rectangular prism, in the H, horizontal, and F, frontal, planes of projection. The desired view, the secondary auxiliary, is projected from the primary auxiliary. Recall that a secondary auxiliary plane is perpendicular to the primary auxiliary plane, and the plane of a successive auxiliary view projected from a secondary auxiliary view is perpendicular to the secondary auxiliary plane. In general, all sequential auxiliary planes are perpendicular to the preceding plane from which the projection was made.

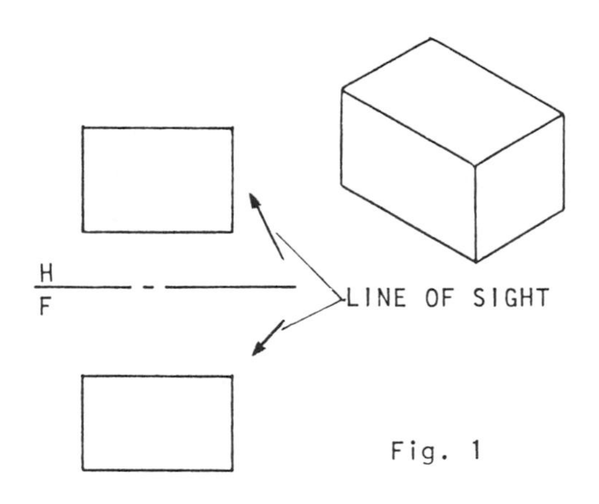

Fig. 1

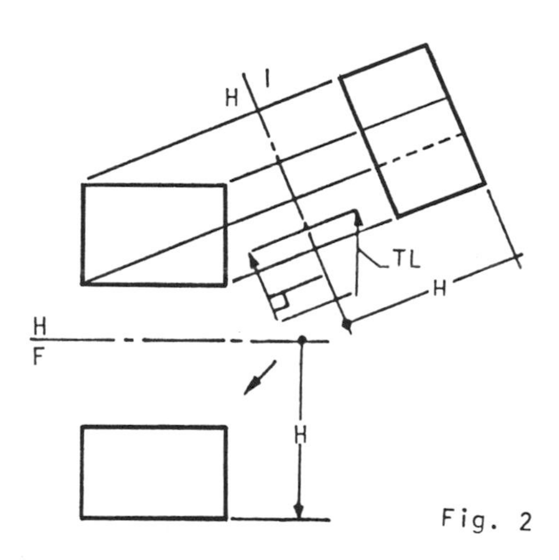

Fig. 2

The given line of sight was arbitrarily chosen in the top and front views. Since it is required to view the prism in the direction of the line of sight, a view must be constructed in which the line of sight appears as a point: the secondary auxiliary.

As shown in Fig. 2, project a primary auxiliary view, from one of the given views so that it is perpendicular to the line of sight. The primary auxiliary will give the true length of the line of sight since it is viewed perpendicularly. Likewise, project the prism to view 1. Transfer the dimension H from the front view, where the horizontal plane appears as an edge, to the primary auxiliary, where the horizontal plane appears as an edge also, in order to establish points.

Since the line of sight is true length in the primary auxiliary, a point view of the line can be found, in a secondary projection plane, which is drawn perpendicular to that true length line. This will give the required view

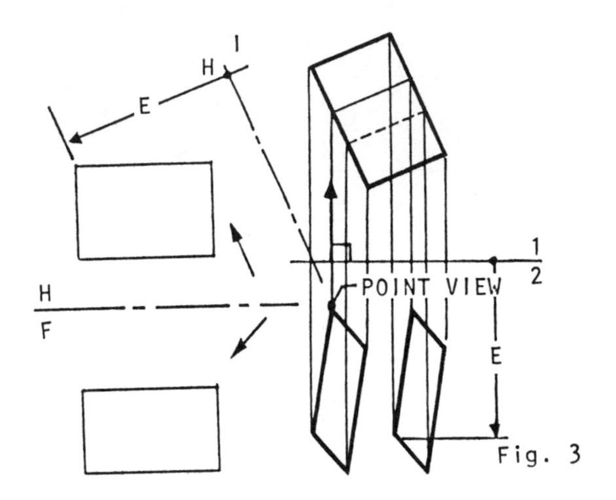

Fig. 3

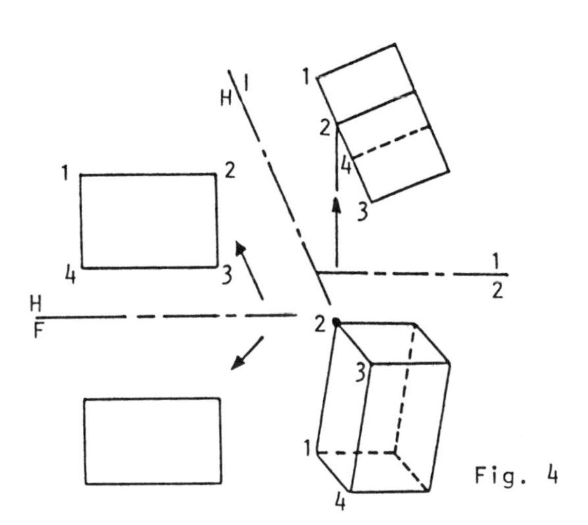

Fig. 4

of the prism. Referring to Fig. 3 project the upper and lower planes to the secondary auxiliary by transferring all dimensions from the H-1 plane in the form of the measurement E: the perpendicular distance from the auxiliary plane of projection.

Complete the view 2 projection of the prism by connecting the respective corners with the missing lines. Plane 1-2-3-4 will appear visible in the secondary auxiliary view since the line of sight gives an unobstructed view of the plane in the primary auxiliary view. Since plane 1-2-3-4 is visible in view 2, all lines crossing the plane must be behind it and consequently are hidden. The remaining outlines of the solid are visible.

Draw the auxiliary view of the hexagonal pyramid whose top and front views are given in Fig. 1, in the direction of P-Q.

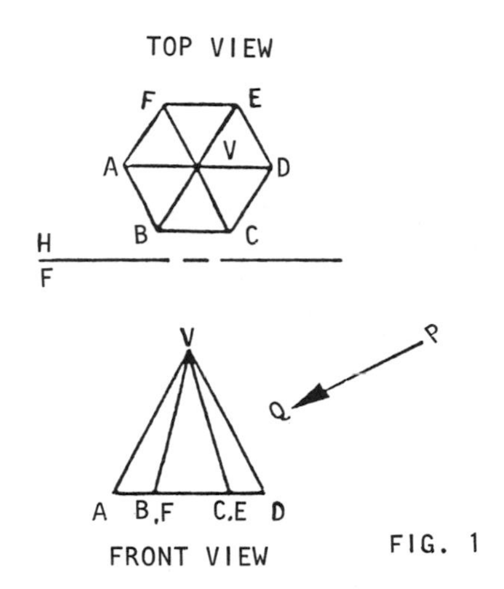

FIG. 1

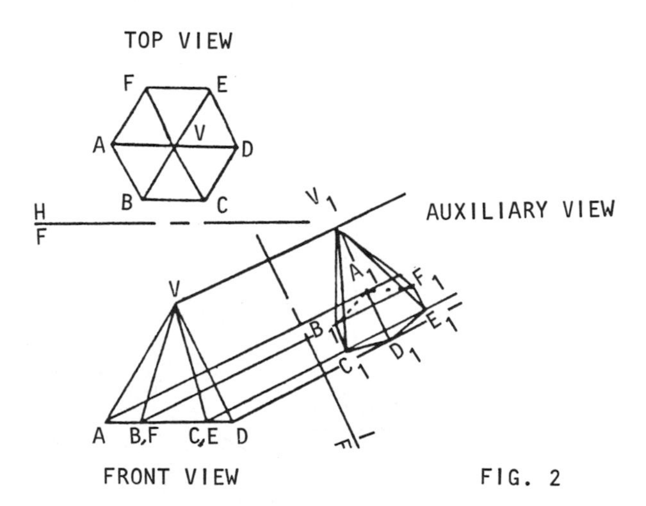

FIG. 2

Solution: The direction of line P-Q is perpendicular to line V-D and the required auxiliary plane has to be perpendicular to line P-Q. Notice that when an auxiliary view is drawn for a specific direction, the auxiliary plane has to be perpendicular to the given direction.

To draw the auxiliary view choose a reference line

F-1 at a convenient distance from the front view and parallel to line V-D. This line represents the edge view of the auxiliary plane. Project all the corner points in the front view perpendicularly to reference line F-1. Measure the distances of all the corner points in the top view from reference line F-H and transfer these distances to the corresponding projections in the auxiliary view as shown in Fig. 2. Notice that the distance of the corner points in the auxiliary view is measured from reference line F-1 after this distance has been measured from reference line F-H in the top view.

A convenient choice of auxiliary plane can save time when drawing an auxiliary view. For example, if reference line F-H had been drawn through point B and C, the measurements of these two points would have been unnecessary because these points would have been on reference line F-1. In the same way reference line F-H could have been chosen to go through points A,V, and D and in this way the measurements of these three points would have been avoided.

• PROBLEM 4-5

> Given the front and top views of a right cylinder, construct the auxiliary view.

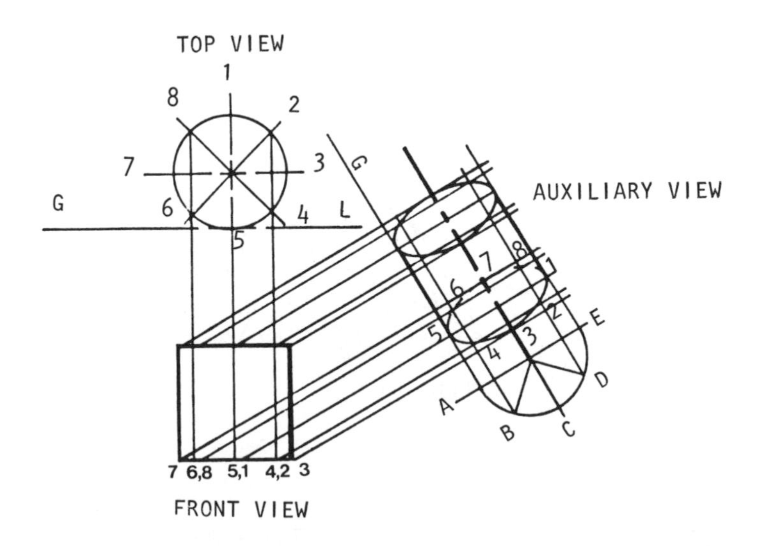

Solution: As shown in the accompanying figure, the auxiliary view is projected from the front view. In order to construct the front view, simply project the extreme elements from the front elevation, down to the front view. The length and width of the plan are equal to the diameter of the cylinder.

In order to produce an auxiliary view of the cylinder, divide the top view into eight equal parts. Project these divisions downward to the front view. Establish a new ground line, G-L, for the auxiliary view, and draw the center line parallel to it at a distance equal to the radius of the cylinder. Draw a half-end view (i.e., a semi-circle which is bisected by the center line) and equally divide it as before: A-B, B-C, C-D, and D-E. Project points A through E parallel to the center line. The intersections of these lines and the points projected from the front view will complete the auxiliary view.

• PROBLEM 4-6

Given the front and top views of figure 1 draw the right front and left front auxiliary views.

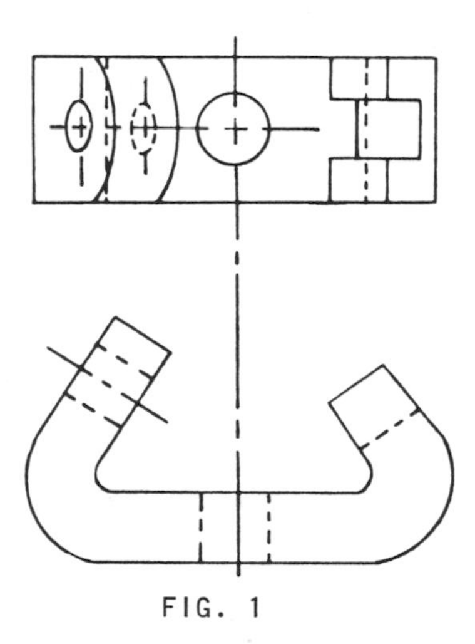

FIG. 1

Solution: In this problem the terms right front and left front auxiliary views refer to the projections at right angles of the left and right sides onto auxiliary planes which are parallel to the sides and thought of as being hinged to the plane that contains the front view. The edge views of these auxiliary planes have been denoted by F-1 and F-2, in figure 2.

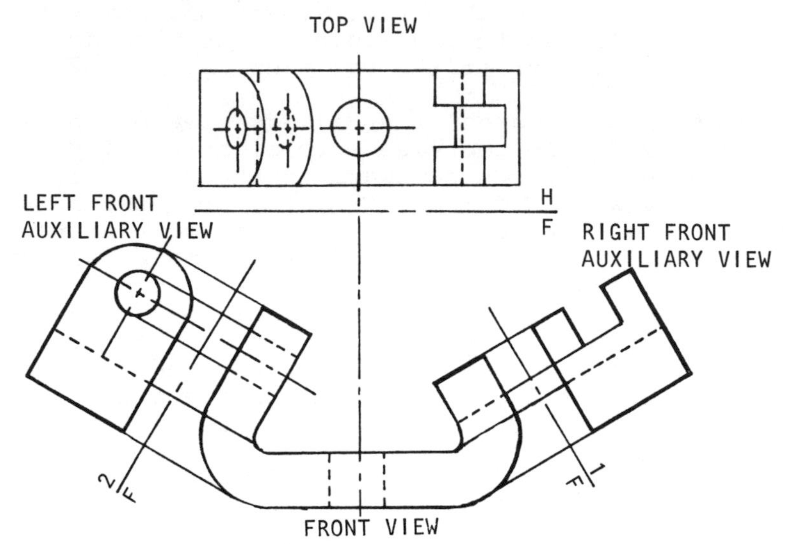

FIG. 2. Right Front and Left Front Auxiliary views

To obtain the right front and left front auxiliary views first project all the corner points in each side perpendicularly to the parallel auxiliary planes. Then measure the distance of these points from the reference line F-H to the top view and lay out this distance from reference line F-1 or F-2, depending on which auxiliary view you are working, to the corresponding projection lines. Finally, join the points to complete the drawing.

• PROBLEM 4-7

Draw the secondary elevation view, E_2, of an object given in the P, plan, and E_1, primary elevation views.

Solution: (1) Fig. 1 shows the given object in the P and E_1 views. (2) Given two orthographic views, any other view of the object, such as the elevation view indicated by the rotation line $\frac{P}{E_2}$ are drawn one surface at a time. Each surface is drawn one line at a time to connect points on the surface that are projected separately into the new view.

(3) Each point on the object may be designated by a symbol (i.e., letter, number, etc.) to aid in identifying and keeping track of the surface being drawn. To avoid

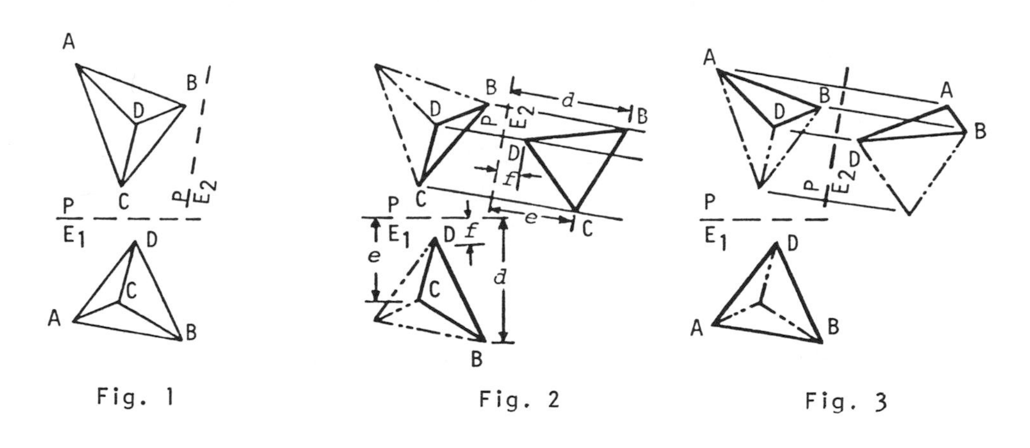

Fig. 1 Fig. 2 Fig. 3

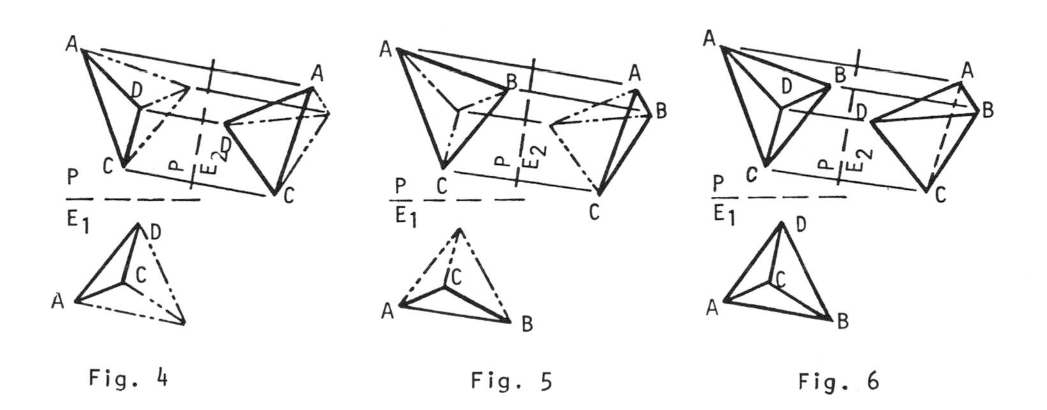

Fig. 4 Fig. 5 Fig. 6

confusion, the accompanying figures will show the particular surface being drawn in E_2 by solid lines and the remainder of the object by phantom lines.

(4) First, as illustrated in Fig. 2, the single surface B-C-D is selected and identified in the given views. (5) Next, from each point of intersection in the plan view, projection lines perpendicular to the rotation line are drawn into view E_2. (6) As shown by measurements d, e, and f, each point is located on its projection in view E_2 at the proper distance from the rotation line by transferring the measurement from view E_1 with dividers. (7) Draw lines to connect B to C, C to D, and D to B in the same order in which they are connected in the given views. (8) In general, if two points are connected by a line in space, they are connected by a line in all views.

(9) The same procedure is followed for surface A-B-D (Fig. 3), A-C-D (Fig. 4), and A-B-C (Fig. 5). (10) Note: No new lines are actually drawn in Fig. 5 to locate surface A-B-C, since all lines bounding this surface are lines previously drawn for the other surfaces.

(11) Fig. 6 shows both the given views and the com-

pleted view E_2. (12) Lines A-B, B-C, C-D, and D-A are visible because they are boundaries of the view. (13) The visibility of lines A-C and B-D can be determined by the visibility methods previously illustrated and employed.

• PROBLEM 4-8

Draw the primary inclined view, I_1, of an object given in the P, plan, E_1, primary elevation, and E_2, secondary elevation, views.

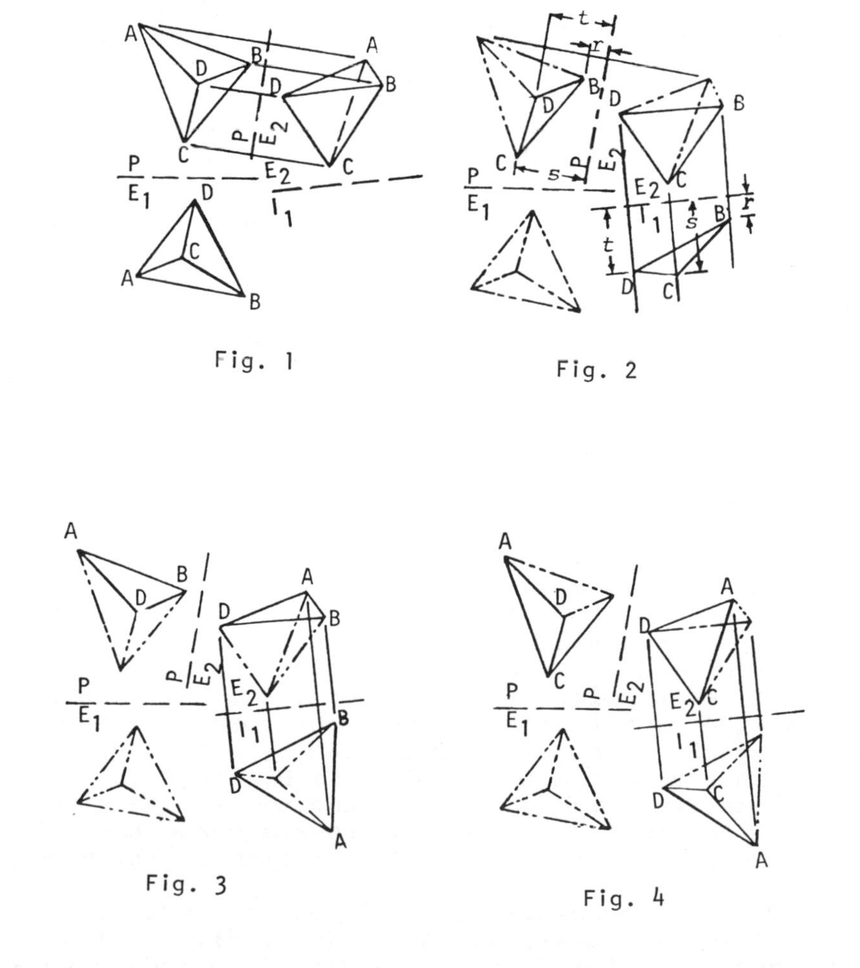

Fig. 1

Fig. 2

Fig. 3

Fig. 4

Solution: (1) Fig. 1 shows the given object in the P, E_1, and E_2 views, which are the same as the final views

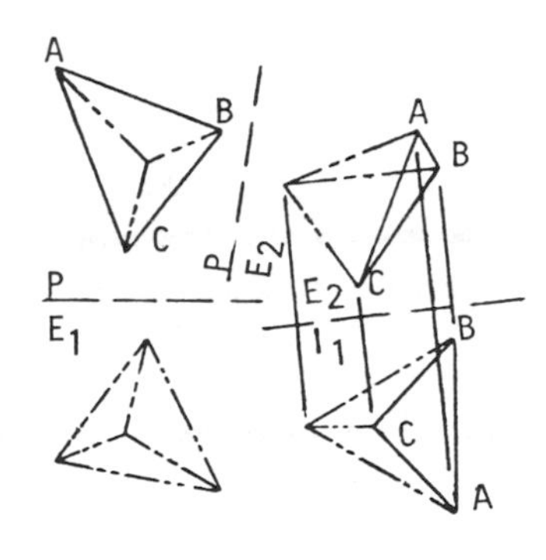

Fig. 5

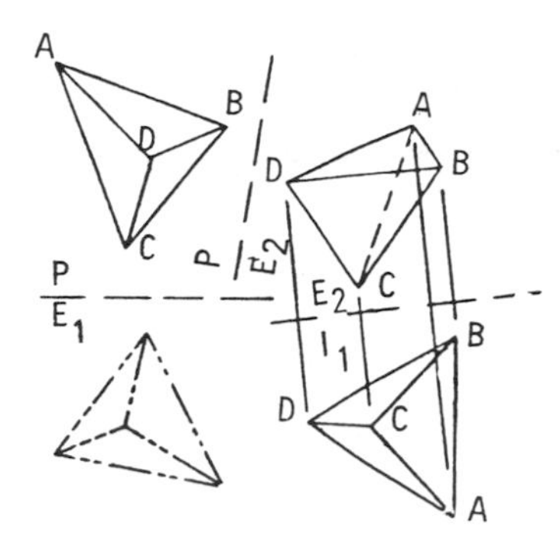

Fig. 6

in the preceding problem. (2) Having given two orthographic views (in this case, three), any other views of the object, such as the inclined view indicated by the rotation line E_2-I_1 are drawn one surface at a time. Each surface is drawn one line at a time to connect points on the surface that are projected separately into the new view.

(3) Each point on the object may be designated by a symbol (i.e., letter, number, etc.) to aid in identifying and keeping track of the surface being drawn. To avoid confusion, the accompanying figures will show the particular surface being drawn in I_1 by solid lines and the remainder of the object by phantom lines.

(4) First, as illustrated in Fig. 2, the single surface B-C-D is selected and identified in the given views. (5) Next, from each point of intersection in view E_2, projection lines perpendicular to the rotation line are drawn into view I_1. (6) As shown by measurements r, s, and t, each point is located on its projection in view I_1 at the proper distance from the rotation line by transforming the measurement from view P with dividers. (7) Draw lines to connect B to C, C to D, and D to B in the same order in which they are connected in the given views. (8) In general, if two points are connected by a line in space, they are connected by a line in all views. (9) Note: The entire E_1 view is shown by phantom lines since it is not used at all in drawing view I_1. (10) The same procedure is followed for surface A-B-D (Fig. 3), A-C-D (Fig. 4), and A-B-C (Fig. 5). (11) The completed view is shown in Fig. 6. The visibility of the interior lines in the I_1 view is determined by the fact that point C is nearest the rotation line E_2-I_1 in view E_2; therefore, it must be visible in view I_1.

TRUE SIZE AND SHAPE OF OBLIQUE PLANES

• **PROBLEM** 4-9

Given the front and top views in the accompanying figure, find the true size and shape of the oblique face 1-2-3.

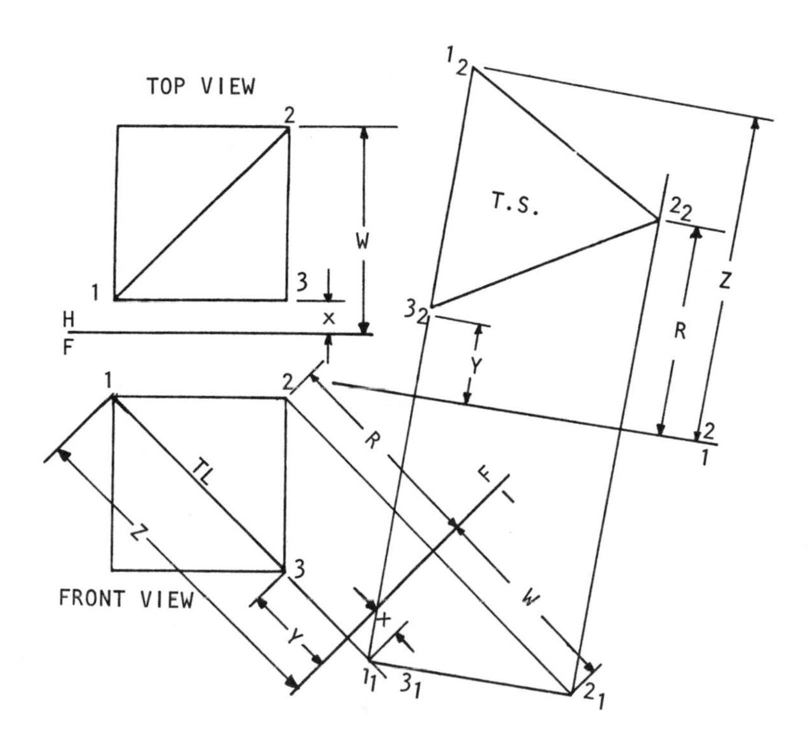

Solution: Since neither the top nor front views show the plane on which face 1-2-3 lies in edge, face 1-2-3 does not appear in true shape on either of the given views. Therefore in order to find the true shape of face 1-2-3, construct an edge view of said plane. To do this notice that in the top view line 1-3 is parallel to the horizontal plane represented by H-F. Therefore line 1-3 appears in true length in the front view. Now draw an auxiliary view whose plane of projection is perpendicular to the true length line 1-3, shown in the front view. The distance of points 1 and 3 from the reference line F-1, is denoted by "X". Notice that this distance is taken directly from the top view. The distance of point 2 is represented by "W" also taken from the top view. Now that an edge view of the plane that contains face 1-2-3 has been obtained, construct an auxiliary view whose plane is parallel to face 1-2-3. Now take the three points that lie on the edge view and project them perpendicularly onto the parallel plane. On the corresponding projections lay off the distances Y, R and Z which are measured from the front view as shown in the diagram. The resulting figure is true shape because it was projected on a parallel plane and at right angles.

• PROBLEM 4-10

Find the true size view of the inclined surface of a block.

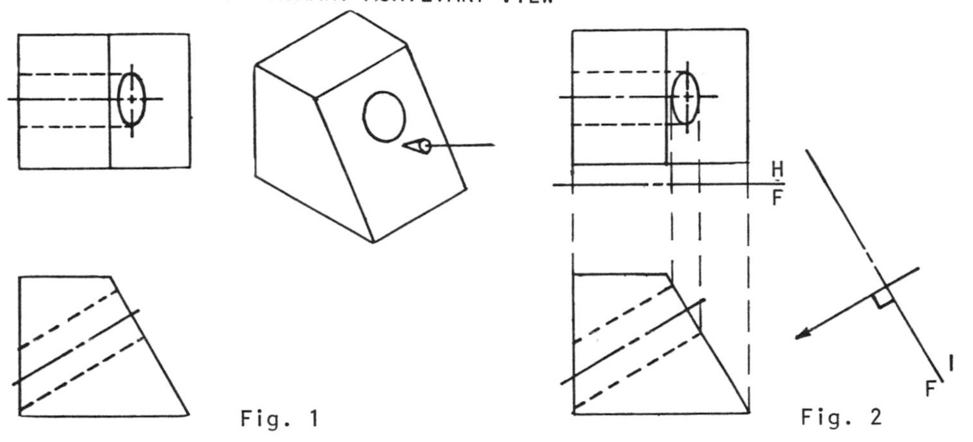

Solution: Fig. 1 shows the given top and front views of a block with an inclined surface which appears as an edge in the front view. Construct the H-F reference plane, as shown in Fig. 2, between the top and front views that is perpendicular to the projectors between the views. Draw a line of sight perpendicular to the edge view of the inclined surface. Draw the edge view of the auxiliary plane perpendicular to the line of sight and parallel to the inclined surface. The auxiliary view, view 1, is the normal view of the inclined surface, and, therefore, shows the inclined surface in true shape and the other faces, foreshortened.

Number points in the top and front views on the inclined surface, as illustrated in Fig. 3. Project these points perpendicularly from the inclined edge to the auxiliary plane. Transfer the dimensions from the H-F plane to the auxiliary view (i.e., dimension D). Note: This dimension can be transferred since the auxiliary

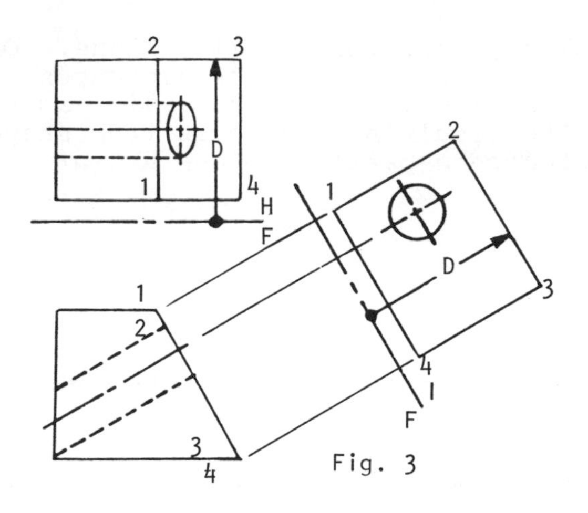

Fig. 3

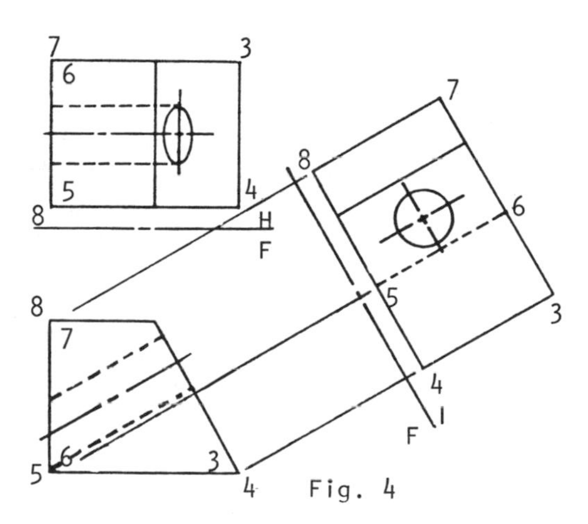

Fig. 4

plane is perpendicular to the frontal plane; consequently, the observer will see the frontal plane as an edge when his line of sight is perpendicular to the auxiliary plane. Distances that are perpendicular to the frontal plane will appear true length when the frontal plane appears as an edge. Therefore, line 2-3 can be located in the auxiliary view by measuring its distance, D, from the frontal plane in the top view and transferring this distance to the auxiliary plane, where the frontal plane appears as an edge also. Connect the projected points, located by the transferred distances, to yield a true-size view of the surface.

Number the remaining points of the object in the top and front views and project them to the auxiliary view to complete the auxiliary view of the object. Line 5-6, a base line of the cube, will be a hidden line. Be sure to measure all dimensions perpendicularly to the reference planes used.

Given the object in Fig. 1, find the auxiliary view in which the surface sloping down appears in true shape.

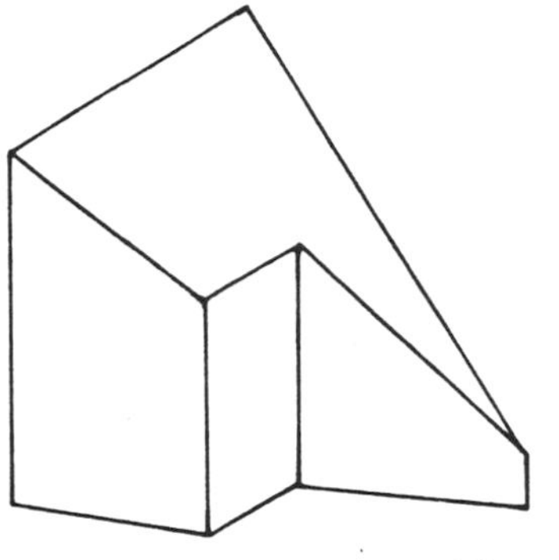

Fig. 1

Solution: The inclined surface referred to in the problem will appear in true shape only if the surface is projected perpendicularly onto a plane parallel to the given surface.

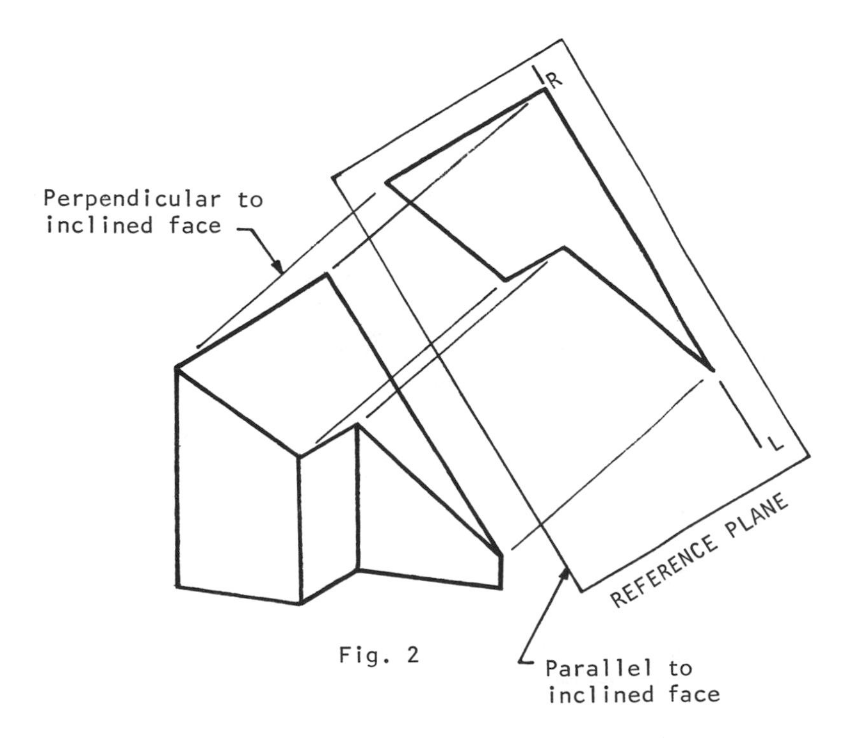

Fig. 2

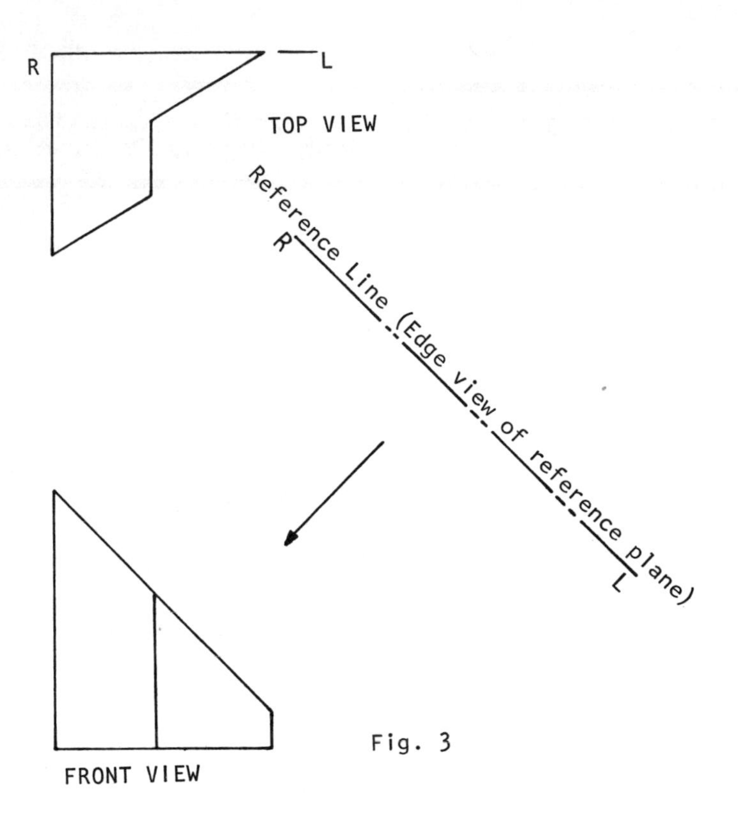
R
L
TOP VIEW
Reference Line (Edge view of reference plane)
R
L
FRONT VIEW
Fig. 3

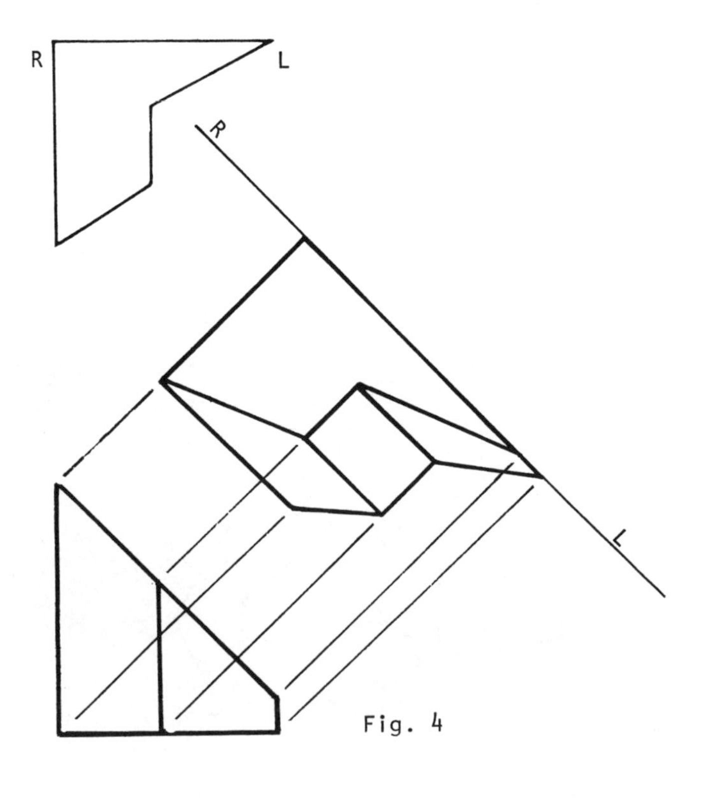
R
L
R
L
Fig. 4

(See Fig. 2.) To show it on the drawing take R-L, a line on the reference plane, to be the edge view of the reference plane on the front view as shown in Fig. 3. Project all the corner points from the front view normal to the reference line R-L. Notice that reference line R-L was taken to be the back edge of the object as shown in the top view in figures 3 and 4. Now measure the distances for the depths from the top view and transfer them onto the corresponding projections made from the front view. Notice that each depth is used twice in the auxiliary view, one for the top and one for the bottom of the object. Also be aware that in figures 1, 2 and 4 there are missing hidden lines which were omitted for simplicity.

• PROBLEM 4-12

Given the front and top views of the object shown in figure 1, find the auxiliary view that shows the inclined surface in true shape.

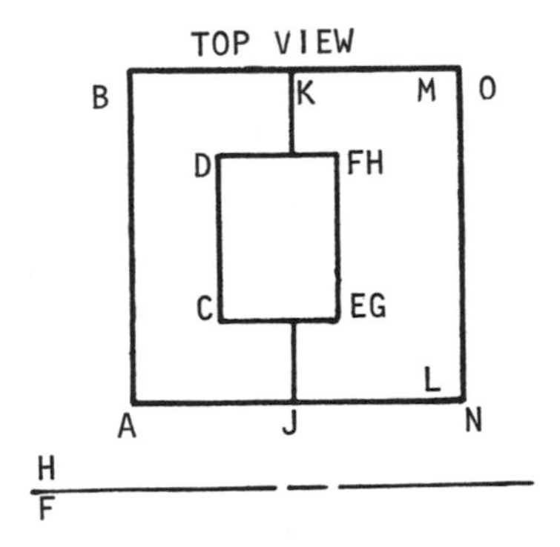

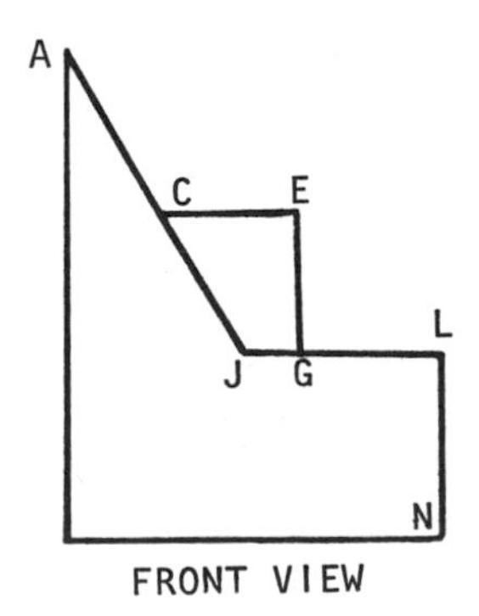

Fig. 1

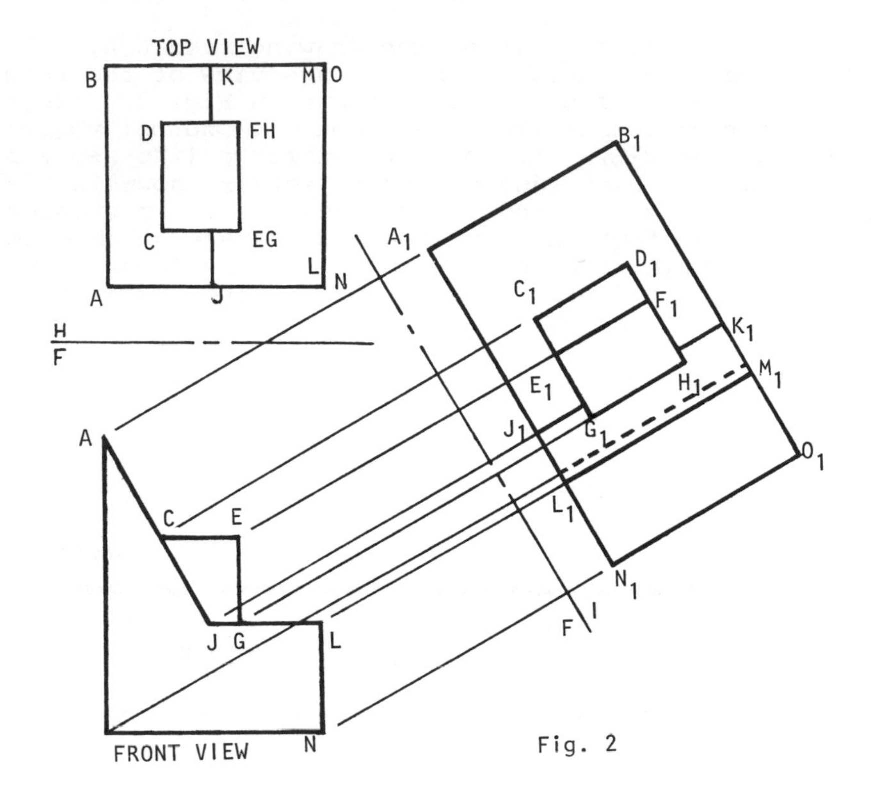

Fig. 2

Solution: To obtain the required auxiliary view we have to project the inclined surface perpendicularly to a parallel plane as shown in Fig. 2. First draw reference line F-1 at a convenient distance from the front view. Measure the distance of all the corner points in the top view from reference line F-H and transfer these distances to the auxiliary view, using F-1 as the new reference line. Notice that rectangle $A_1B_1K_1J_1$ in the auxiliary view is the true shape of the slanted surface and that all the other parts of the drawing are foreshortened. The reference lines F-H and F-1 could have been made to coincide with lines A-N and A_1N_1 respectively, to avoid the two measurements required to determine each of the lines, A-N and A_1N_1.

• PROBLEM 4-13

Given the front and top view of plane ABC, find the angle it makes with the frontal plane. Also find its true shape.

Solution: In order to find the angle plane ABC makes with the frontal plane, a view is necessary in which both the

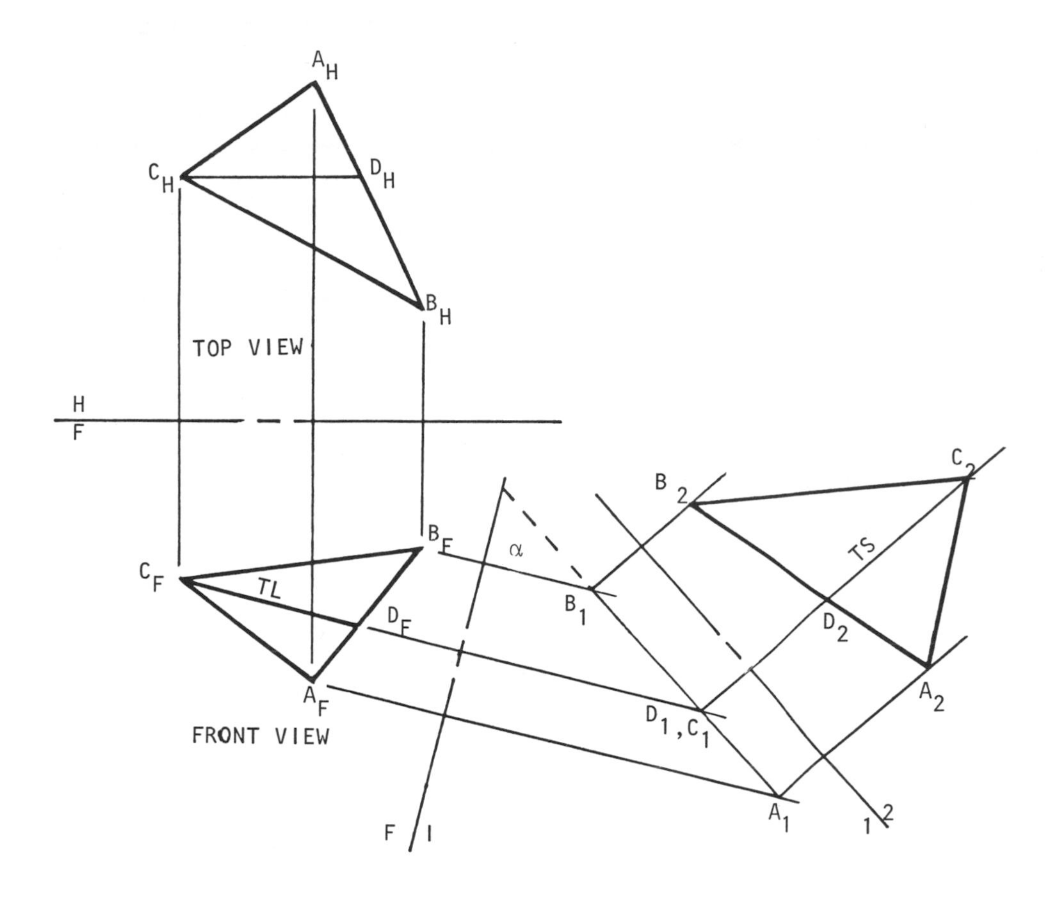

frontal and given plane ABC show in edge. From this same edge view of plane ABC its true shape can be found by projecting the plane onto a parallel plane.

The procedure of finding the edge view is as follows; first find a true length line, which lies on the plane. In the accompanying figure line CD was drawn parallel to the cutting plane H-F (the horizontal plane). This line, when projected down to the front view, shows in true length. Now find an auxiliary view of plane ABC which projects perpendicularly to the true length line shown in the front view. Notice that cutting plane F-1 represents a plane which is perpendicular to the front view where both plane ABC and frontal planes appear in edge. The angle plane ABC makes with the frontal plane is therefore the acute angle "α".

To find the true shape of plane ABC, project the edge view shown in the first auxiliary view onto a parallel plane (cutting plane 1-2). This second auxiliary view is drawn by measuring the distance each point in the front view is from reference line F-1 and transferring these measurements onto the second auxiliary plane, using 1-2 as the new reference line.

AUXILIARY VIEWS IN OBLIQUE DIRECTIONS

• PROBLEM 4-14

Points 1, 2, and 3 in fig. 1 are three points in space. Construct a circle connecting these three points, and project the circle to the front and top views shown.

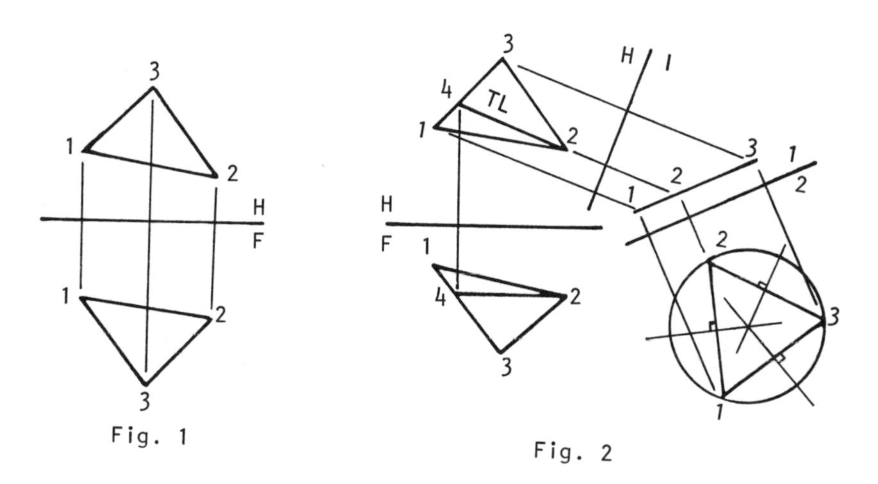

Solution: Points 1, 2, and 3 determine a plane, and in order to construct the circle going through the three points, the true shape of the plane is required. To construct the true shape diagram of plane 1-2-3, draw a horizontal line, 2-4, through point 2 in the front view. Project this line to the top view where it appears true length, as in fig. 2. With the line of sight parallel to the true length line 2-4, make the reference plane H-1 perpendicular to line 2-4 and construct auxiliary view 1. Because line 2-4 appears true length in the top view, it appears as a point in the primary auxiliary view, and therefore view 1 is an edge view of the plane, as shown in fig. 2.

To obtain the true shape of the plane, take another auxiliary view with the line of sight perpendicular to the edge view and the reference plane 1-2 parallel to it. The resulting plane is then true shape. On this true shape diagram of the plane construct the circle going through the vertices of the plane by constructing the perpendicular bisectors on each of the sides of the plane, and taking their intersection as the center, O, of the circle. (See fig. 2.)

To find the projections of this circle on the front and top views, theoretically project back all the points on the circle. In practice, project back a number of points and then connect them with a smooth curve. The technique used to project back is that illustrated by the distance "E" of the top and secondary auxiliary views in fig. 3. A more common method of projecting a circle is by the construction of diameters AB and CD, parallel and perpendicular to the reference plane 1-2, respectively, in view 2, as shown in fig. 3. Project diameters AB and CD and center O from view 2 to the top view, and now use an ellipse template to draw the best ellipse with major diameter CD and minor diameter AB. (See fig. 3.)

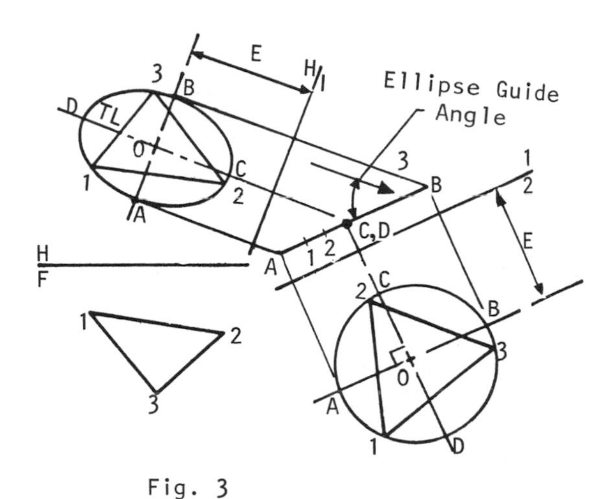

Fig. 3

To construct the ellipse on the front view, an edge view of the plane projected from the front view is needed. To find this edge view, draw a horizontal line, EF, on the plane 1-2-3 in the top view, passing through the center O of the ellipse. Project line EF to the front view where it appears true length, as shown in fig. 4. Pro-

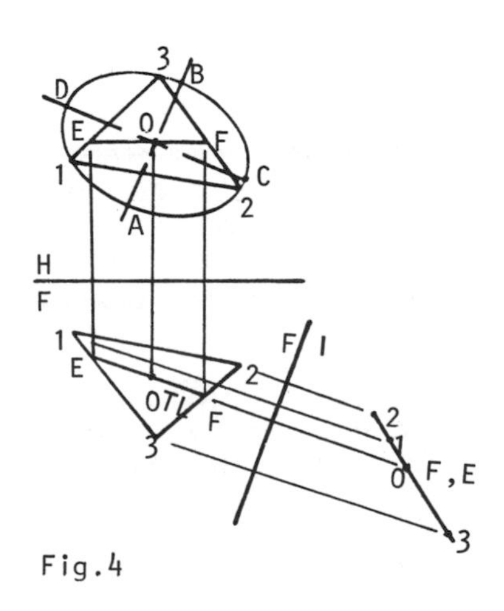

Fig.4

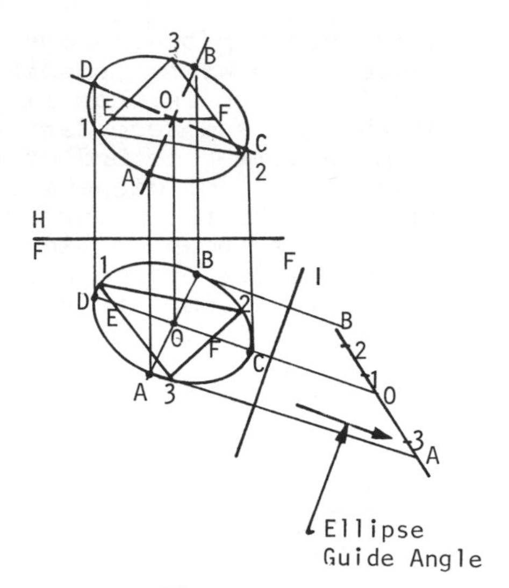

Fig. 5

ject center O to the front view. Construct the auxiliary view of the plane with the line of sight parallel to the true length line EF and the reference plane F-1 perpendicular to it. Line EF will appear as a point, and therefore the auxiliary view is an edge view of the plane. Extend the edge view of plane 1-2-3 to points A and B, with OA equal to OB and AB equal to the diameter of the circle through points 1, 2, and 3, as in fig. 5. This edge view now shows the true length of the minor diameter, AB, of the elliptical projection of the circle. Project points A and B back to the front view to where they intersect their respective projections from the top view. Draw the major diameter true length in the front view, through point O and parallel to line EF. Measure the ellipse angle in the edge view and draw the ellipse in the front view, as shown in fig. 5.

• PROBLEM 4-15

Given the top and front views of the object shown in Fig. 1, find a series of auxiliary views that will show the object from five different angles.

Solution: All four of the required auxiliary views can be drawn from either the front or top views. But for practice and clarity in the drawing, they were drawn from one another.

As shown in Fig. 2, the first auxiliary view was

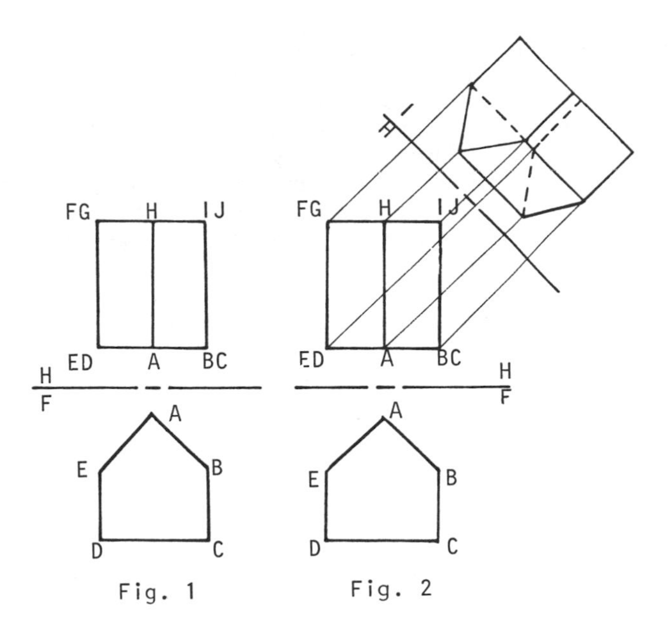

Fig. 1 Fig. 2

drawn from the top view in the direction of the arrow shown. In order to draw an auxiliary view in the indicated direction, all the corner points in the top view must be projected parallel to the line of sight. A reference line H-1, which represents the beginning of the auxiliary plane, is chosen at a convenient distance from the top view. Be aware that all the auxiliary planes are perpendicular to the plane they are drawn from. In this case then, the first auxiliary plane is perpendicular to

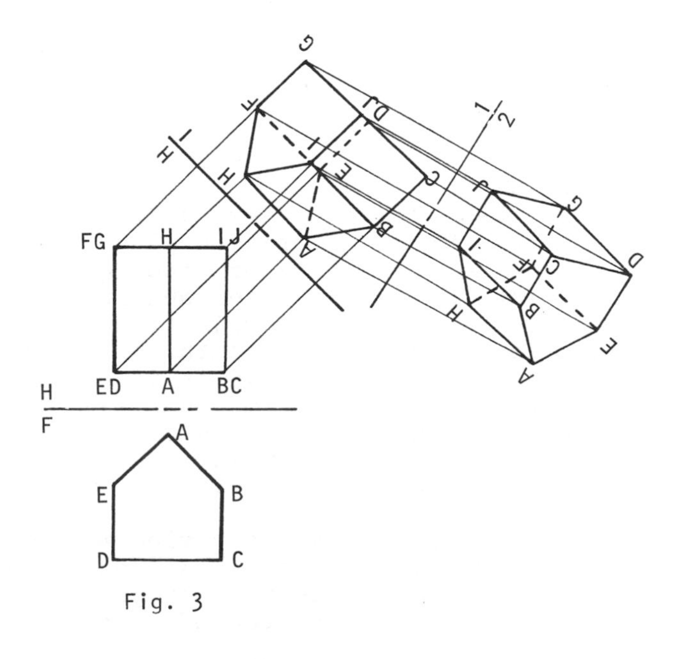

Fig. 3

the horizontal plane (in the top view), the second auxiliary plane is perpendicular to the first and so on. After reference line H-1 is chosen, the distance that every point in the front view is from reference line H-F is measured and transferred to the auxiliary plane by measuring from reference line H-1 on the corresponding projecting line. When all the corner points have been located on the auxiliary plane draw lines between the points just as they appear in the principal views. For example, points "E" and "A" are connected by a line on both the top and front views, therefore they must be connected in any other view. After connecting the points the problem then lies in the visibility of the lines.

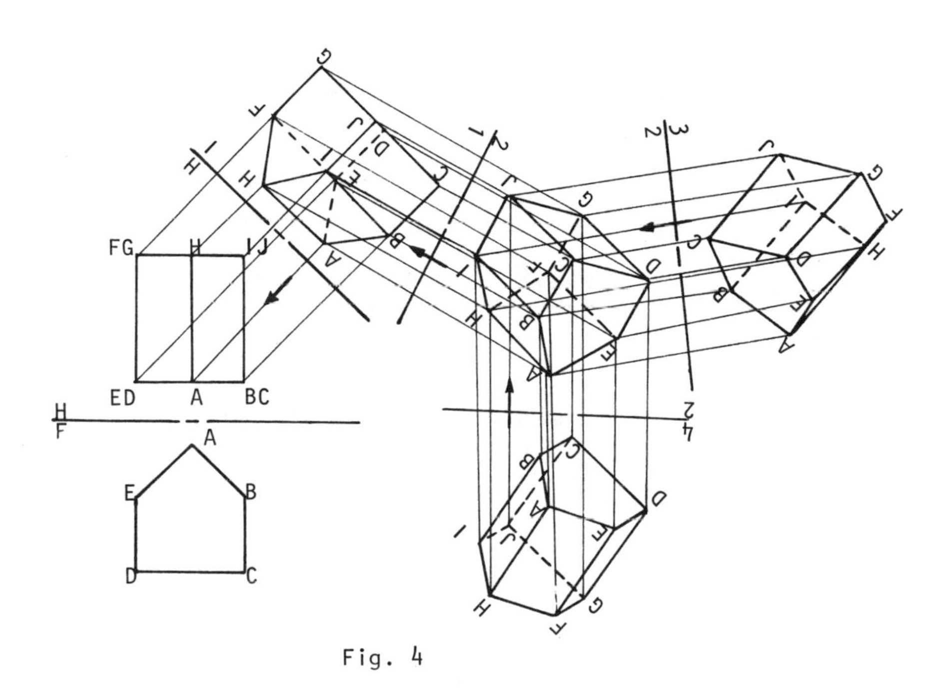

Fig. 4

To determine the visibility of a line, it is necessary to analyze two adjacent views and treat them in the following manner: First, isolate a plane and a line as in Fig. 5 where the visibility of line FE is sought. Then pick a point "p" on the plane and project it back to the previous view. If the projection line meets the plane in question before it meets the line, then the plane is visible over the line as it is the case in Fig. 5. Notice that the visibility of the plane need not be known in order to find the relative visibility between the line and the plane. This relative visibility is only settled in the view from where point "p" was projected.

In Fig. 3, a second auxiliary view was drawn off the first auxiliary view and as explained in the previous problem all the corner points were projected parallel to the line of sight. The distances of each corner point

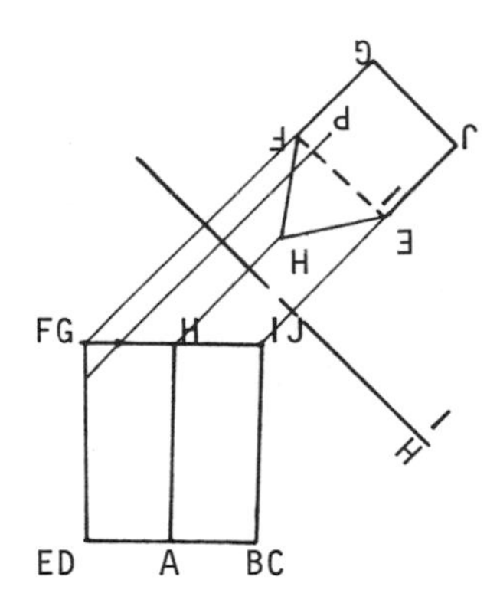

Fig. 5

in the first auxiliary view are measured from reference line H-1 and transferred to the second auxiliary view. The visibility of each line is determined by the method described in the previous paragraph. In Fig. 4 two auxiliary views were constructed from the second auxiliary view, employing the same methods used for the other auxiliary views.

GLASS-BOX METHOD

• PROBLEM 4-16

> Find the auxiliary view of the object shown in Fig. 1, whose auxiliary plane is parallel to the top inclined surface. Do it by the "glass box" method.

Solution: As shown in Fig. 2, the given object is enclosed by a "glass box" having a vertical plane parallel to the front, a profile plane parallel to the right side and a plane parallel to the inclined surface. The object is then projected perpendicularly to each of the three above mentioned phases of the box. Notice that the plane to which the inclined surface was projected is an auxiliary plane and not a principal plane. This is because even though this plane is perpendicular to the vertical plane,

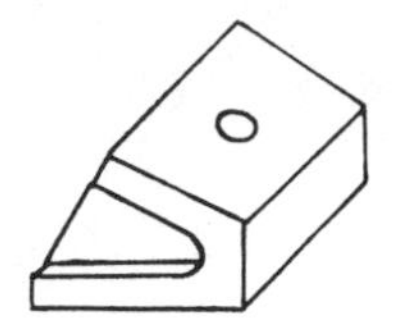

Fig. 1

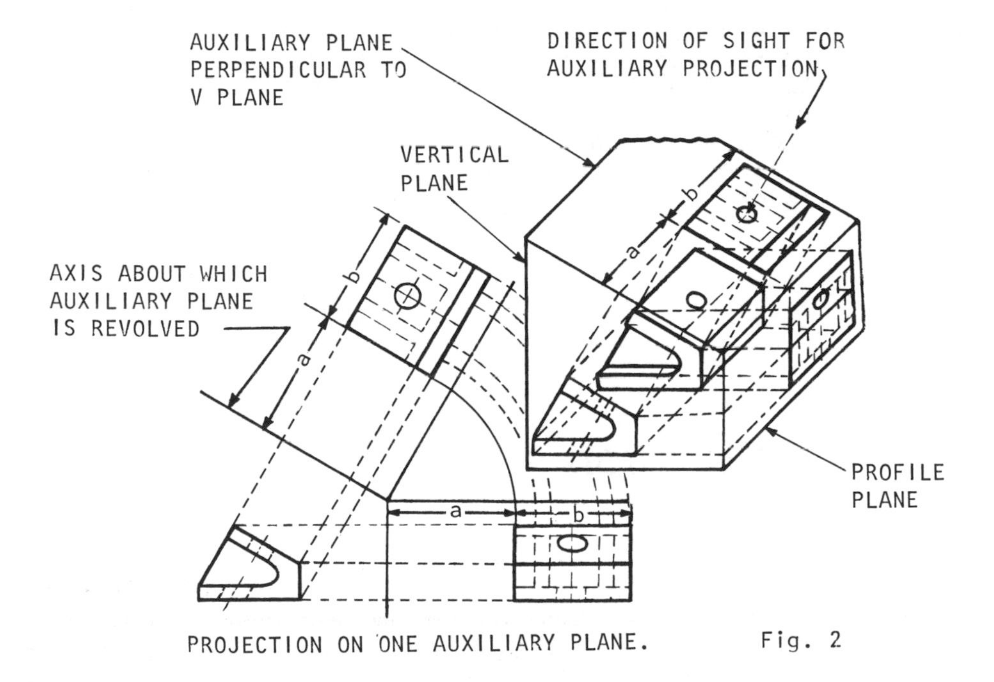

Fig. 2

it is not perpendicular to the profile plane.

After all the projections have been made, the three sides of the box to which the object was projected are then unfolded until they are co-planar. (See Fig. 2.) It is important to notice that in this case the two front edges of the "glass box" can be thought of as axes of rotation and that these axes serve as reference lines from which measurements such as "a" are taken.

• PROBLEM 4-17

Given the front, top, and side views of a line (see figure 1) none of which is parallel to any of the principal planes, find its normal view.

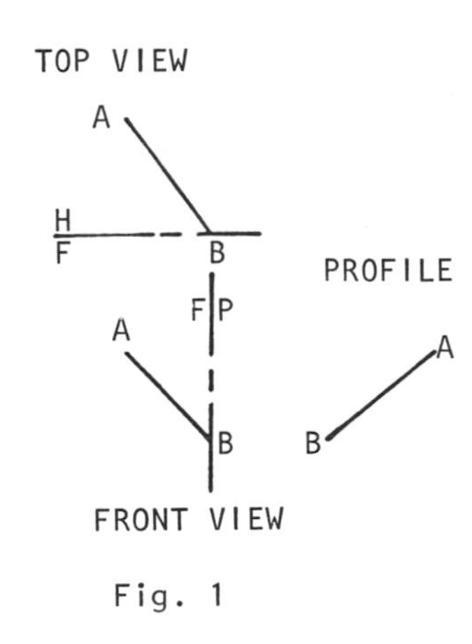

Fig. 1

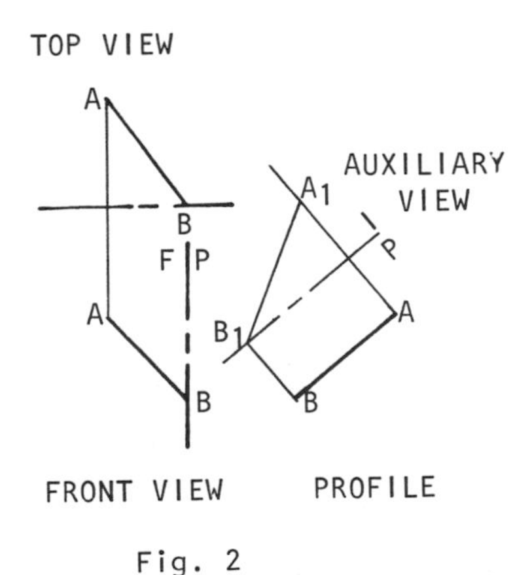

Fig. 2

Solution: Recall that the main purpose of a normal view is to show the true shape of the specific object, in this case the true length of the line. To accomplish this all that is necessary is to get an auxiliary view of the line by projecting any one of the views onto a parallel plane. In figure 2 this was done from the side view. Notice that to construct this auxiliary view the distances of A and B were measured from reference line F-P to the front view and were laid out from the reference line P-1 to the corresponding projection line on the auxiliary view. Keep in mind that this auxiliary view might have been constructed on any of the other views, as long as the auxiliary plane is parallel to the line. Also notice that the auxiliary plane was conveniently chosen to go through point B, to avoid measuring the distance of B from the reference lines to the auxiliary view.

• PROBLEM 4-18

Construct the normal view of a given machine part having a skew surface.

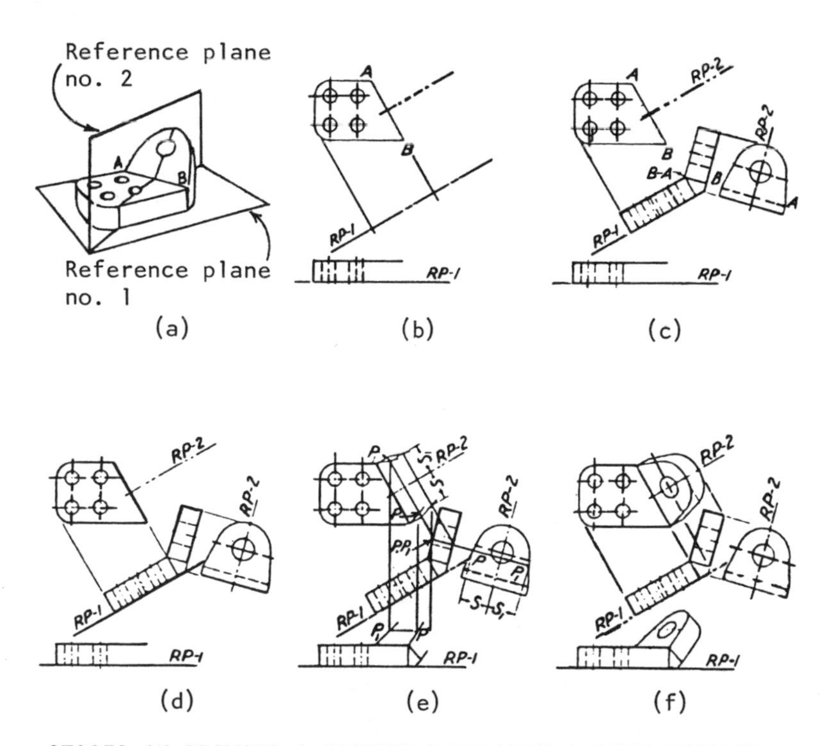

STAGES IN DRAWING A MACHINE PART WITH A SKEW SURFACE.

Solution: The accompanying figure illustrates the successive steps in drawing a normal view. The pictorial illustration, frame (a), shows a typical object with a skew surface. The line of intersection between the skew portion and the horizontal base is line AB. In order to get an edge view of the skew surface, a view may be taken looking in the direction of line AB, thus giving an end view of AB. Because AB is a line of the skew surface, the edge view will result. The reference plane (RP No. 1) for this view will be horizontal. In other words, lines of sight for the end view are parallel to A-B. The direction of observation for the normal view will be perpendicular to the skew surface. The reference plane (RP No. 2) for the normal view will be perpendicular to the edge-view direction and thus perpendicular to edge AB, as shown in frame (a).

At (b) partial top and front views are shown. The projectors and reference plane for the required edge view are also shown. The projectors, or lines of sight, are parallel to A-B. The reference plane is perpendicular to the projectors.

At (c), the edge view has been drawn. Note that line AB appears as a point in the edge view. The angle that the skew surface makes with the base is laid out in this view from specifications.

At (d), the normal view is added. The projectors for the view are perpendicular to the edge view. The reference plane, RP-2, is drawn perpendicular to the projectors for the normal view and at a convenient distance from the edge view. The reference plane in the top view is drawn midway between points A and B in the top view because the skew surface is symmetrical about this reference plane. The normal view is drawn from specifications of the shape. The projection back to the edge view can then be made.

The views thus completed at (d) describe the object, but the top and front views may be completed for illustrative purposes or as an exercise in projection. The method is illustrated at (e) and (f). Any point, say P, may be selected, in the normal view, and projected back to the edge view. From this view a projector is drawn back to the top view. Then the distance S from the normal view is transferred to the reference plane in the top view. A number of points so located will complete the top view of the circular portion, and the straight-line portion can be projected in a similar manner.The front view is found by drawing projectors to the front view, from the top view, for the points needed. Measure the heights from the reference plane in the edge view, RP-1, and transfer these distances to the front view. Note that this procedure for completing the top and front views is the same as for drawing the views originally but in reverse order.

• PROBLEM 4-19

Construct the normal view of a given machine part having a surface inclined to the top and front.

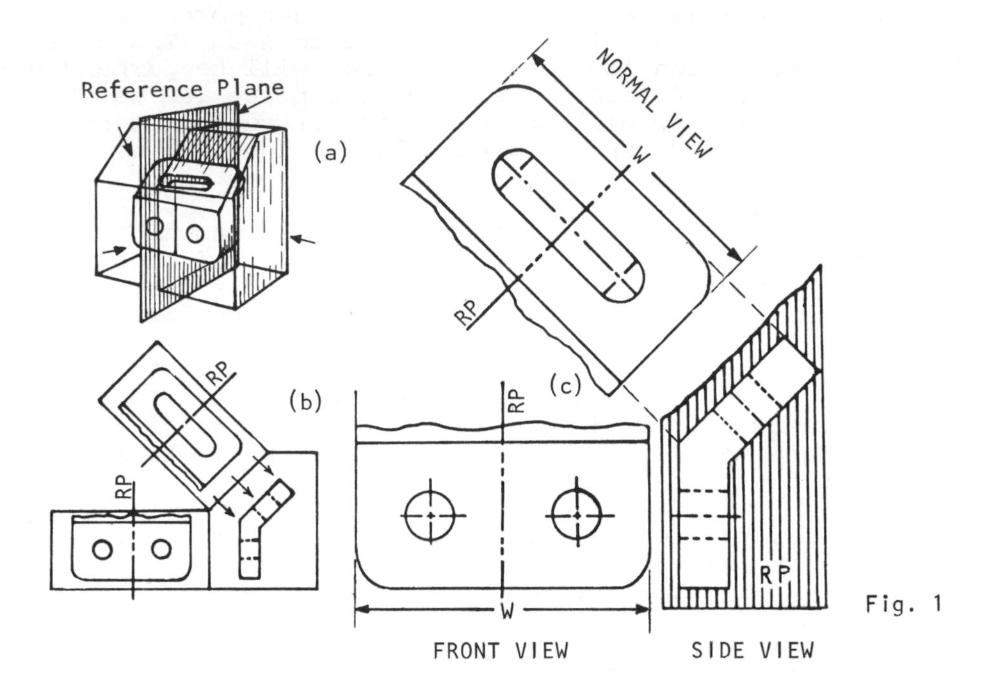

Fig. 1

Solution: A machine part may also be drawn with the inclined surface on the lower or upper front or rear. Orientation of this type is shown in Fig. 1, where the inclined surface is on the upper front. Illustration (a) shows the machine part in a "glass box". Recall that this visualization technique assumes a plane, either coinciding with or parallel to the surfaces of the given object, and that portion of the object projected onto this plane will appear in the drawing.

At (b), the planes of projection in (a) are opened up, properly oriented in the plane of the page. At (c), the front, side, and normal views of the inclined surface are represented. The steps in drawing an object of this type are given in Fig. 2 as follows:

Draw the partial front, top, and right-side views, as at (b). Locate the normal view direction by drawing projectors perpendicular to the inclined surface, as shown.

Locate the reference plane in the front view, RP-1, as shown in frame (c). This reference plane is taken at the left side of the object, because both the vertical and inclined portions have a left surface in the same profile plane. The reference plane in the normal view, RP-2, will be perpendicular to the projectors already drawn, and is located at a convenient distance from the right-side view, as shown in this same frame.

As shown at (d), measure the distances (widths) from

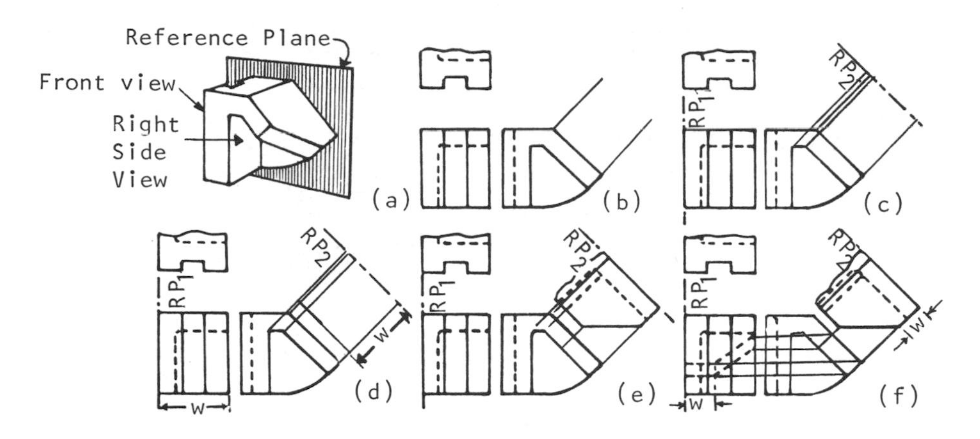

STAGES IN DRAWING A MACHINE PART WITH A SURFACE INCLINED TO TOP AND FRONT. Fig. 2

the reference plane, RP-1,(of various points needed), and transfer these measurements with dividers or scale to the normal view, measuring from the reference plane, RP-2, in the normal view. Note that points to the right of the reference plane in the front view will be measured in a direction toward the right-side view in the normal view.

From specifications of the surface contour complete the normal view (e).

Complete the right-side and front views by projecting and measuring from the normal view. As an example, one intersection of the cut corner is projected to the right-side view and from there to the front view; the other intersection is measured (distance W) from the normal view and then laid off in the front view.

• PROBLEM 4-20

Construct the normal view of a given machine part having a surface inclined to the top and side.

Solution: The accompanying figure illustrates the steps in drawing the normal view of a face inclined to the top and side:

Frame (a) shows a pictorial representation of the given machine part.

Draw the partial top and front views, as in frame

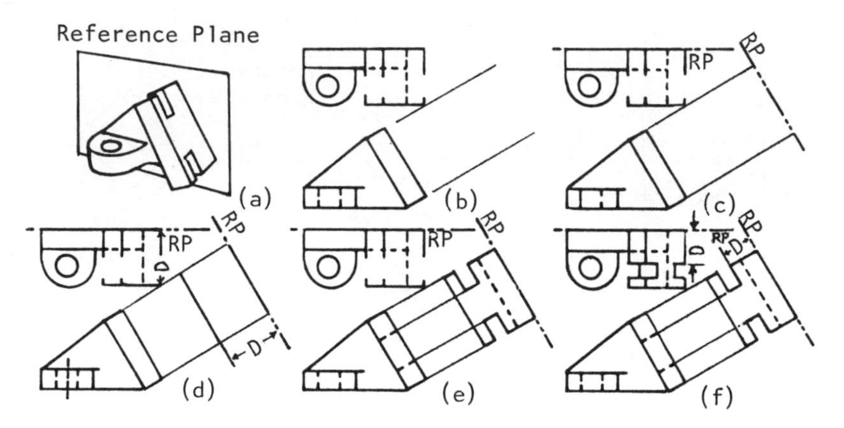

STAGES IN DRAWING A MACHINE PART WITH A SURFACE INCLINED TO TOP AND SIDE. Fig. 1

(b). Locate the view direction by drawing projectors perpendicular to the inclined surface, as shown.

Locate the reference plane (i.e., RP), in the top view, as shown in frame (c). The reference plane may be taken in front of, through, or to the rear of the view but is here located at the rear flat surface of the object because of convenience in measuring. The reference plane in the normal view will be perpendicular to the projectors already drawn, and is located at a convenient distance from the front view.

As shown in frame (d), measure the distance (depths) from the reference plane (of various points needed), and transfer these measurements with dividers or scale to the normal view, measuring from the reference plane in the normal view. Recall that the auxiliary plane is perpendicular to the frontal plane; consequently, the observer will see the frontal plane as an edge when his line of sight is perpendicular to the auxiliary plane. Distances that are perpendicular to the frontal plane will appear true length when the frontal plane appears as an edge. Therefore, line 1-2 can be located in the auxiliary (normal) view by measuring its distance, D, from the front reference plane in the top view and transferring this distance to the auxiliary plane, where the frontal plane appears as an edge also. Note that the points are in front of the reference plane in the top view and are therefore measured toward the front in the normal view.

From specifications in the pictorial drawing shown in frame (a), complete the normal view, as shown in frame (e).

Complete the drawing, as shown in frame (f). In this case the top view could have been completed before the normal view was drawn. However, it is considered better practice to lay out the normal view before completing the view that will show the surface foreshortened.

• **PROBLEM 4-21**

Given the front view of a hexagonal pyramid intersected by a cutting plane, C-P, construct the top and auxiliary views of the pyramid, and the true shape view of the plane of intersection.

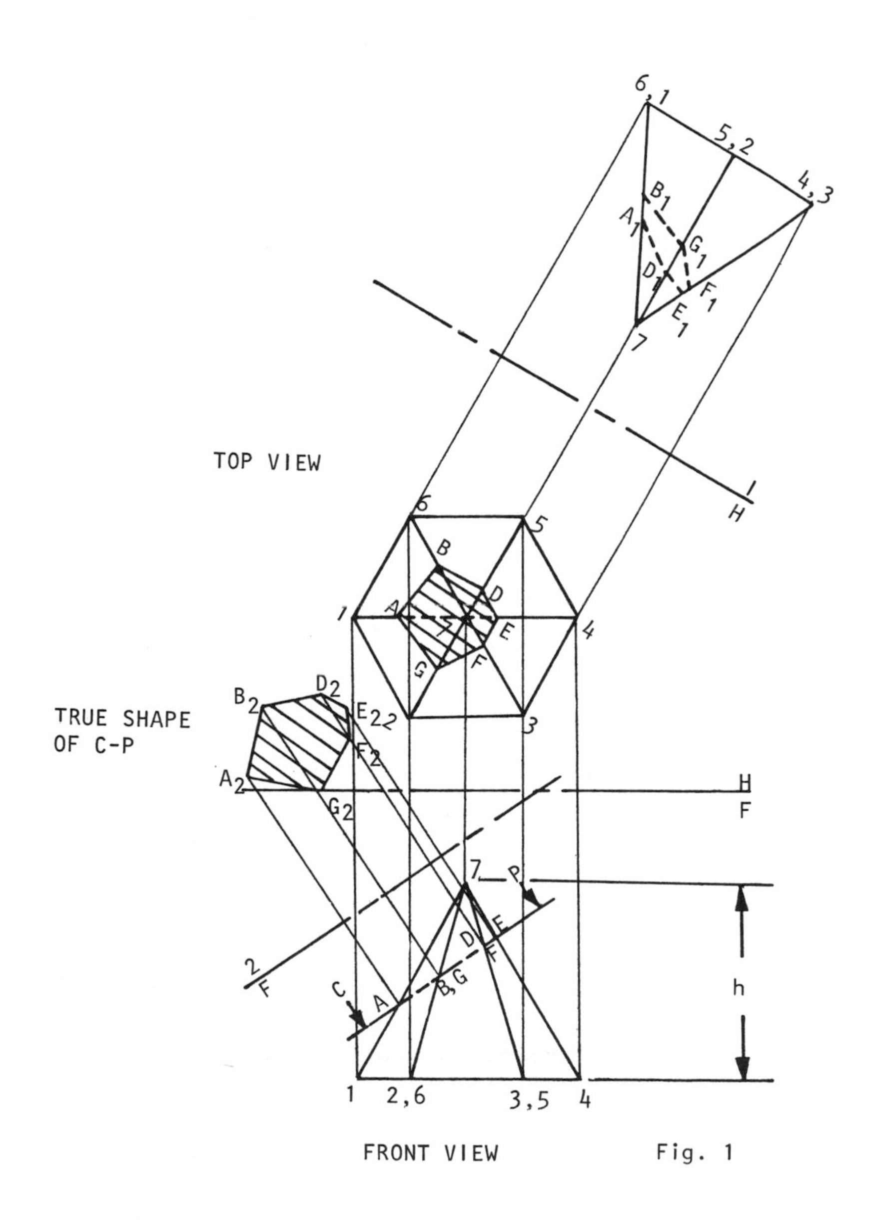

Fig. 1

Solution: The accompanying figure shows the given front view, as well as the required constructions. Construct the top view by establishing the horizontal rotation line, F-H. Draw the base by projecting points 1,2,3,4,5, and 6, the corner points; the length of each side of the base being equal to the true length side 2-3 or 5-6 in the elevation. Project the vertex point, point 7, and join the corner points to 7 by dashed lines. Project the points along C-P to the top view to intersect the corresponding edge line. For example, point A lies at the intersection of C-P and 1-7 in the elevation; and, is projected to A_1, along 1-7 in the plan. The plane formed by these points (i.e., A,B,D,E,F,G) is the plane of intersection, shown shaded. The edge lines from the base of the pyramid to the intersection are shown as solid lines; those from C-P to the vertex are shown as dashed lines. The auxiliary view is projected from the top view, having lines of sight perpendicular to H-1. Project point 7, the vertex, a distance "h" above H-1 (the same height as in the front view). Draw edge lines 6-7 and 4-7. Project points along the intersection to points A_1, B_1, D_1, E_1, F_1, and G_1 to arrive at the plane shown by the broken line in the figure. Notice that there are no other hidden lines in the pyramid because the rest of the lines are directly behind solid lines.

The true shape view of the intersection is found by establishing rotation line 2, F-2, parallel to C-P. This view has lines of sight perpendicular to C-P and thus yields the required true shape (i.e., normal) view. Project points A,B,D,E,F, and G. The plane formed by A_2, B_2, D_2, E_2, F_2,and G_2 is the required true shape view of the intersection. Recall that the distances from reference plane F-2 to each point on the true shape view are measured from reference plane F-H to the corresponding points on the top view.

• PROBLEM 4-22

Given the edge view of a circular disc of diameter A-B, making contact with a horizontal ground line, GL, and inclined at 45° to GL, construct a plan, front elevation, and auxiliary elevation which is projected from the plan onto a new ground line inclined at 45° to the horizontal.

Solution: The accompanying diagram shows the given edge view, A-B, making contact with the ground line, G-L, and inclined to G-L at 45°. Locate the center of the disc, (i.e., bisect A-B); label it O. With O as center, draw a semicircle of radius O-A or O-B. Divide this semicircle (which is a true plan or normal view of the disc) into any convenient number of equal arc lengths. The illustration employs four arc lengths which are labeled B-C, C-D, D-E, E-G. Project points C,D, and E from the circumference, downwards, perpendicularly, to the edge view of the

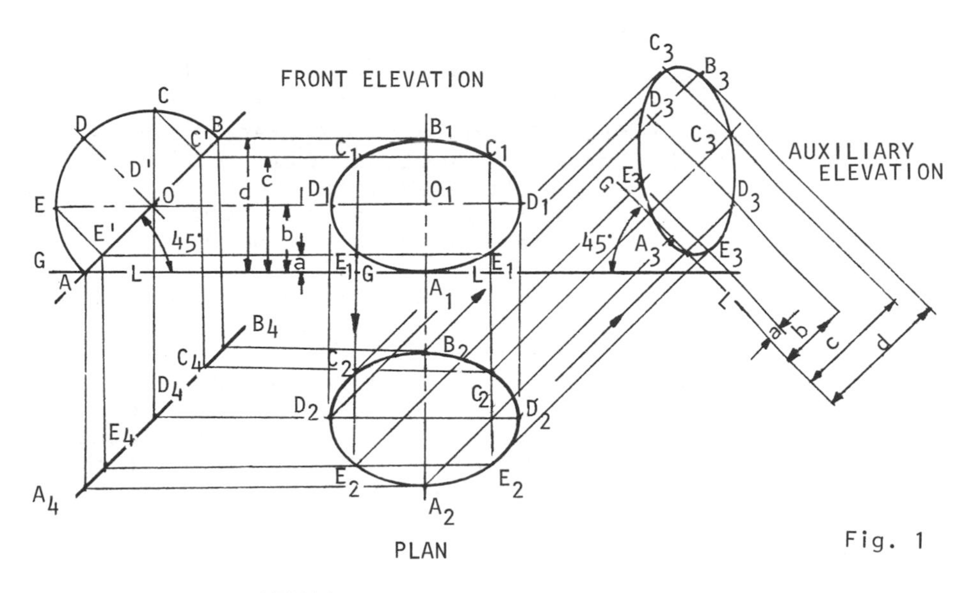

AUXILIARY ELEVATION OF A CIRCULAR DISC.

disc. These projections divide the diameter into four parts: A-E, E-O, O-C, and C-B. Project these points on A_1-B_1 horizontally.

Locate center O_1 at a convenient distance from the edge view of the disc along the horizontal projection from point O on A-B. Mark off distance O_1-D_1 on either side of O_1,using compasses or dividers, which is equal to distance O-D, O-A, or O-B on the semicircle. This is the major diameter and it always projects true length in any view. To locate points C_1 and E_1 notice that since angles COB and CC'O are 45 and 90 degrees respectively, angle OCC' is therefore 45 degrees also. These relations make triangle OCC' an isosceles triangle with sides OC' and C'C equal. Since segment C'C is actually perpendicular to the frontal plane, it projects in the same way onto the front elevation view. Therefore, to locate point C_1, take the length of OC' or CC' and measure that distance on both sides of segment A_1B_1. To locate point E_1, use the length of segement OE' or EE' and measure it on both sides of A'B'. Erect perpendiculars along D_1-D_1, at said distances from O_1. The points at which these perpendiculars intersect the horizontal projectors from C and E are labeled C_1 and E_1, respectively, as shown. Locate points B_1 and A_1 by erecting a perpendicular through O_1 along D_1-D_1. Points B_1 and A_1 are located at the respective points of intersection of this perpendicular and the horizontal projectors from points A and B. The smooth curve drawn through points A_1, E_1, D_1, C_1, and B_1 produces the required front elevation.

Construct the plan view of the disc by dropping vertical lines from the points along A-B to intersect a line inclined at 45°, since the horizontal as well as vertical lines of sight form an angle of 45° with the edge view of the disc. Project these latter intersections horizontally, to intersect with verticals dropped from corresponding points in the front elevation. As an example, the vertical projectors from E_1 intersect the horizontal projector from E_4 to locate E_2, as shown. A curve drawn through these intersections will produce a plan identical in shape to the elevation (once again, since the horizontal and vertical lines of sight form equal angles of 45° with the edge view of the disc).

In order to construct the auxiliary elevation view, such a view must be projected from the plan. Draw the new ground line G-L, making an angle of 45° with the horizontal ground line, as stipulated in the stated problem for the auxiliary elevation. Project the points produced in the plan perpendicularly to this new ground line. Mark off heights a, b, c and d from the horizontal ground line, as shown in the front elevation. These heights (or perpendicular distances from the edge view of the horizontal plane) are preserved in the auxiliary elevation since, here too, the horizontal plane appears as an edge. Mark off these heights from the ground line in the auxiliary view. Establish projectors, parallel to G-L in the auxiliary view, at the heights a, b, c, d. The intersections of these latter projectors with those from the plan locate the points on the auxiliary view of the disc. Draw a smooth curve through these points.

CHAPTER 5

APPLICATIONS OF LINES AND PLANES

PIERCING POINTS, VISIBILITY, INTERSECTIONS, STRIKES, AND DIPS

• PROBLEM 5-1

Find the piercing point of line A-B on plane 1-2-3 and the visibility in both the horizontal and frontal views by the auxiliary-view method (edge-view method).

Solution: Figure 1 shows the top and front views of given plane 1-2-3 and line A-B. Problem analysis reveals that

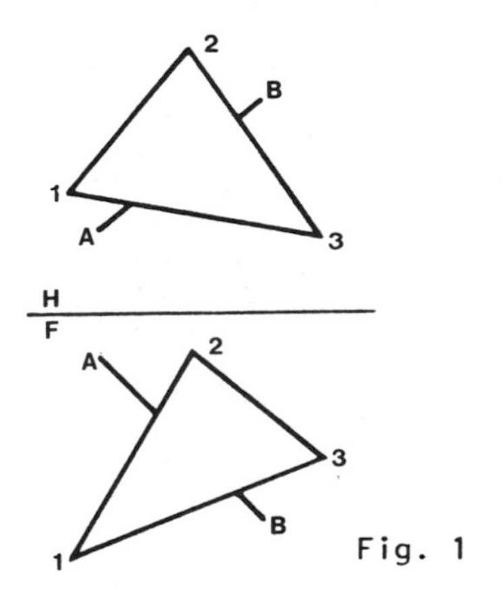

Fig. 1

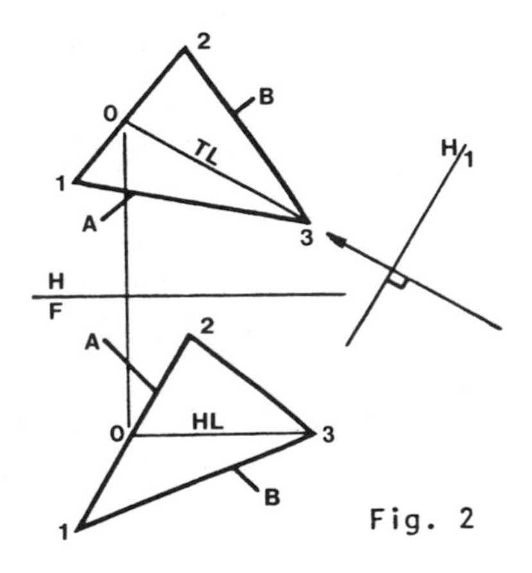

Fig. 2

the point on a line at which the line pierces a plane is apparent in any view that shows the plane as an edge. The position of the piercing point on the plane may then be found by projection. As illustrated in Fig. 2, draw the horizontal line, 0-3, in the front view and project it to the top view. Line 0-3 projects true length in the top view, since it is parallel to the plane of projection in the front view. Establish the line of sight for the primary auxiliary view parallel to 0-3 in the top view.

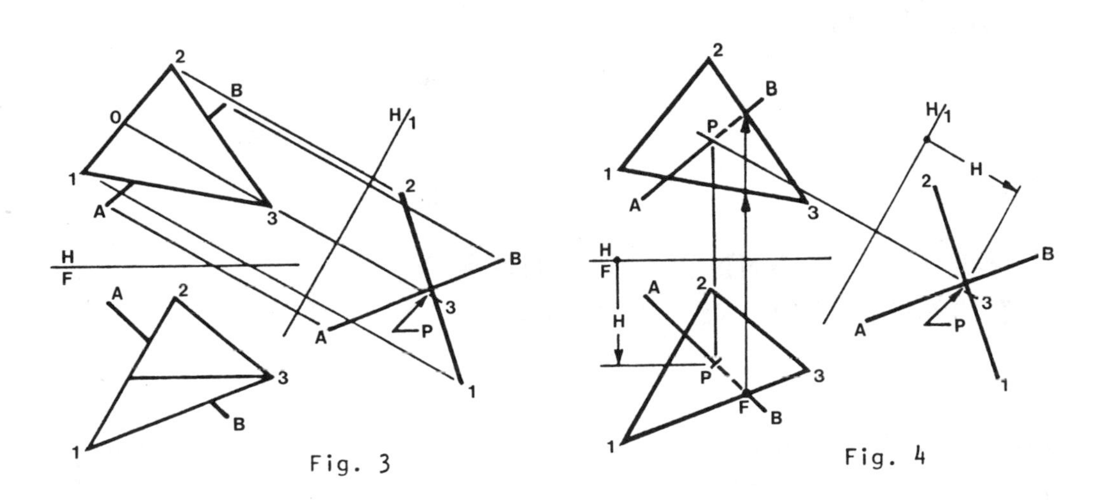

Fig. 3

Fig. 4

Reference plane H-1 is drawn perpendicular to the line of sight and 0-3. Referring to figure 3, find the edge view of plane 1-2-3 by obtaining the point view of line 0-3. Project line A-B into the auxiliary view, as well. Point P, the intersection of the projection of line A-B and line 1-2 in the auxiliary view, is the piercing point of line A-B on plane 1-2-3. Fig. 4 shows the projection of point P to the top and front views, in sequence. The piercing point could have been found just as easily by projecting an inclined view from the front view to show the plane as an edge. The front view of P can be verified by transferring distance H from the auxiliary view to the front view. By the definitions of the auxiliary and front views, this distance is constant. Point A is closer than line 1-3 to the H-1 plane in the auxiliary view; therefore, line A-P is higher than the plane and visible in the top view. Visibility of A-B in the front view is determined by analyzing point F where P-B and 1-3 cross. In projecting this point to the top view, line 1-3 is crossed before reaching line A-B; therefore 1-3 is visible in the front view, and P-F is hidden.

Find the piercing point of line X-Y on plane A-B-C by the cutting plane method.

Solution: (1) Fig. 1 shows the given plane, A-B-C, the given line, X-Y, and vertical cutting plane 1. (2) Problem analysis reveals that when a line in one plane pierces another plane, it must do so on the line of intersection of the two planes. The line of intersection of the two planes is most easily found in a view that shows one of the planes as an edge. This principle is used to find piercing points by introducing an additional plane, called the cutting plane, which contains the given line. (3) As shown in Fig. 1, vertical cutting plane 1 contains line X-Y and

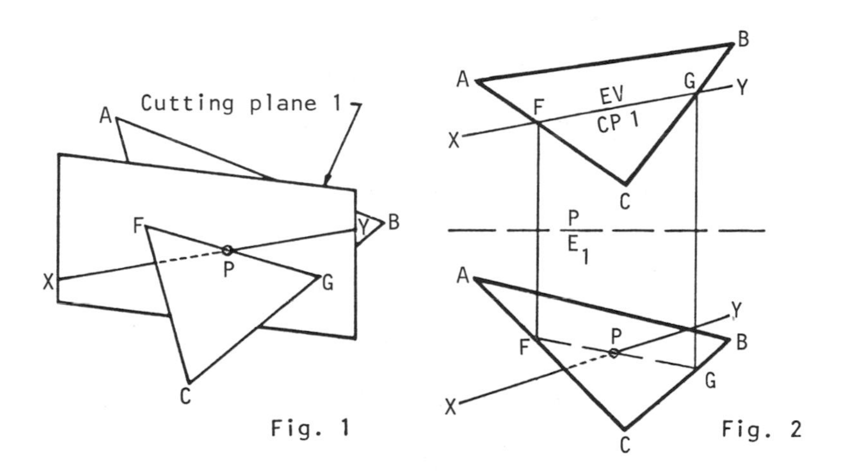

Fig. 1

Fig. 2

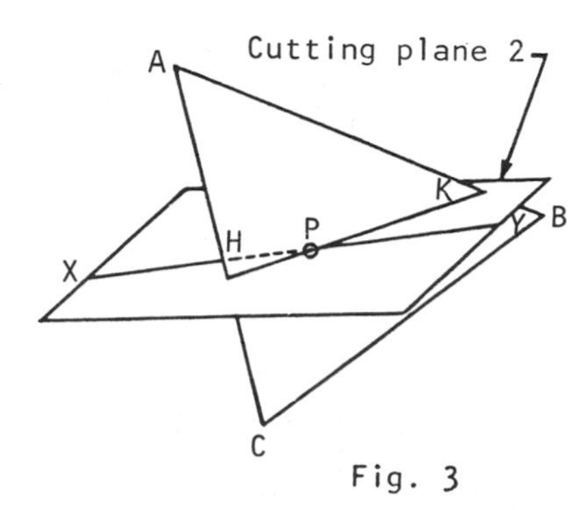

Fig. 3

intersects plane A-B-C along line F-G. The desired piercing point is the point at which F-G intersects X-Y. (4) Since a vertical plane must appear as an edge in the plan view, cutting plane 1 appears to coincide with line X-Y in the P-view of Fig. 2. The cutting plane cuts across line A-C in the P-view at F and B-C at G to yield two points on its line of intersection with plane A-B-C. (5) Points F and G are projected into view E_1, the primary elevation or front view, and connected by a line that crosses X-Y at piercing point P. (6) The visibility of

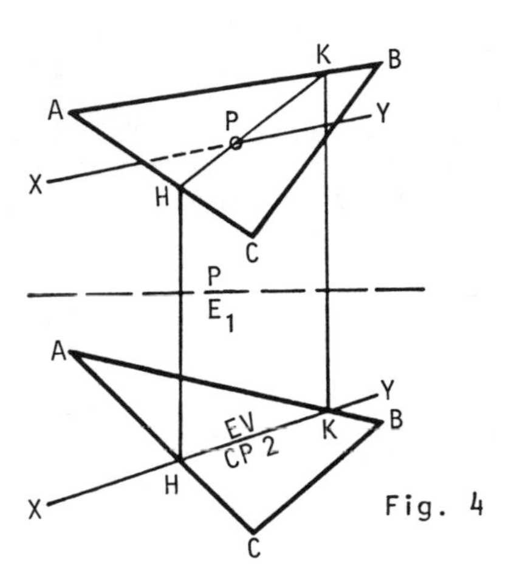

Fig. 4

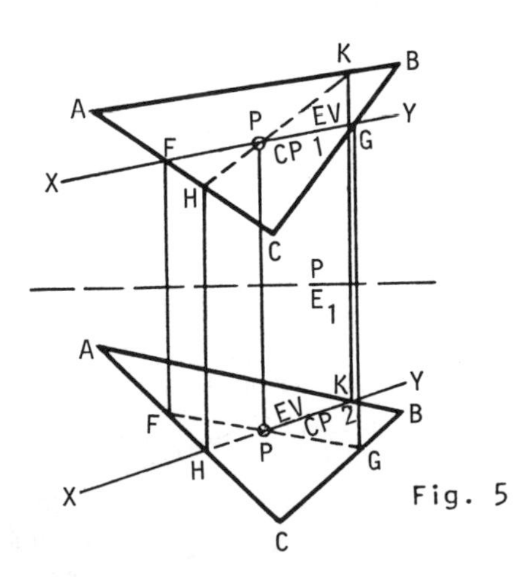

Fig. 5

line X-Y is then determined. (7) Fig. 3 illustrates the use of a cutting plane that appears as an edge in an elevation, or front, view to find the piercing point. Cutting plane 2 is passed through X-Y so that its edge view coincides with X-Y in view E_1 in Fig. 4. (8) This cutting plane cuts line A-C at H and A-B at K. H and K are projected into the plan view and connected by a line to locate the piercing point P. (9) Fig. 5 shows line X-Y and plane A-B-C with the piercing point located by both of the equivalent cutting plane procedures described. Note the location of the piercing point can be checked by projection between the two views.

• PROBLEM 5-3

Find the piercing point of line 1-2 with plane A-B-C, and the visibility of both views, by the projection method.

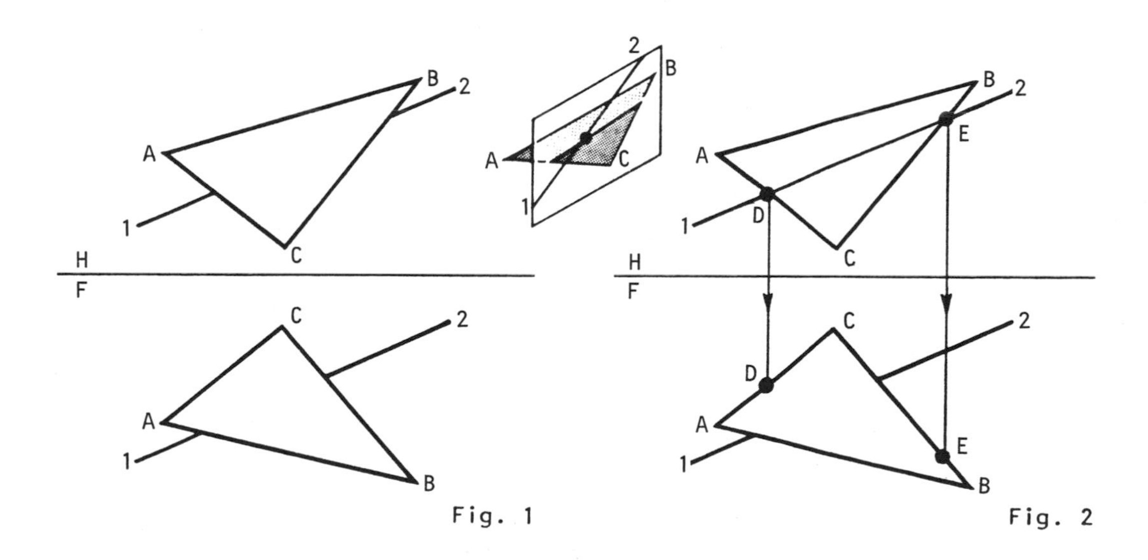

Fig. 1 Fig. 2

Solution: Figure 1 shows the given plane A-B-C and line 1-2 in the horizontal and frontal planes of projection. In figure 2, assume that a vertical cutting plane is passed through line 1-2 in the top view. The plane intersects A-C at point D, and B-C at point E. Project points D and E to the front view. Line D-E in figure 3 represents the trace of the line of intersection between the imaginary vertical cutting plane and plane A-B-C. Any line that lies in the cutting plane and intersects plane A-B-C will intersect along line D-E. Line 1-2 lies in the plane, therefore it intersects A-B-C at point P in the front view. Project point P to the top view. Figure 4 shows a method

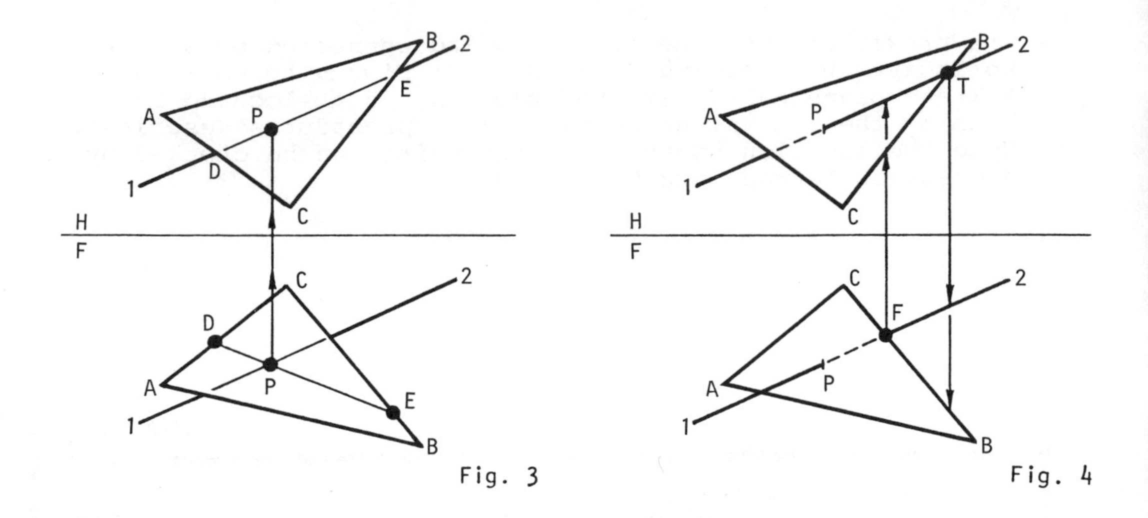

for determining the visibility of both views. The visibility of line 1-2 in the front view is determined by analyzing point F, the point at which P-2 crosses B-C. In projecting this point to the top view, line B-C is encountered (i.e., crossed) before reaching P-2. This illustrates that B-C is in front of P-2 in the top view; therefore, B-C is visible in the front view. The top view visibility is determined by analyzing point T in the same manner. In projecting this point to the front view, line P-2 is crossed before reaching the corresponding position on line B-C. This illustrates that P-2 is higher than B-C in the front view and is, therefore, visible in the top view.

• PROBLEM 5-4

Determine the visibility of a line which pierces a plane.

Solution: The concept of visibility is derived from the inherent notion that all planes are considered to be opaque. When the projection of a line, or any object, is below (as viewed in H), behind (as viewed in F), or on the far side of a plane relative to any viewing direction, the outline of any portion of the line so located will be represented as a hidden outline or dash line.

In the accompanying figure, line A-B pierces plane C-D-E at point X. It follows, then, that in any view

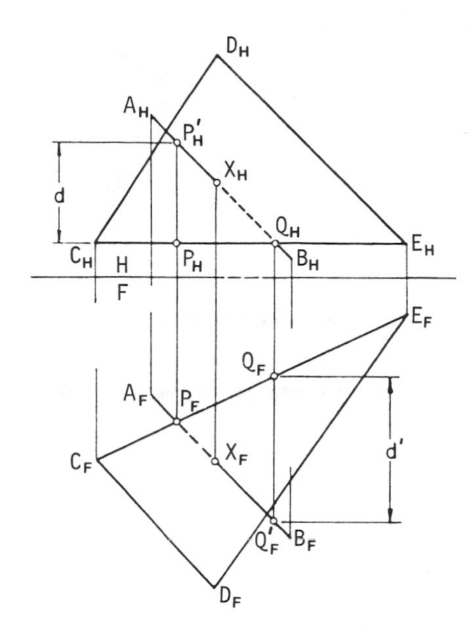

where the line is crossing the plane area, either that portion of the line from the piercing point X toward A or from X toward B must be hidden. Procedure for determining hidden lines is set forth in the following steps. (The subscript F refers to the frontal plane of projection, and H, to the horizontal plane of projection.)

(1) If A_F-X_F passes behind C_FE_F, the distance A_FX_F is behind C_FE_F and will be seen in the H view. To verify that A_F-X_F passes behind C_F-E_F, project point P_F, where line A_F-X_F appears to cross C_F-E_F, to the H view as P'_H. C_H-E_H is crossed before reaching A_H-X_H; therefore, C_F-E_F is in front of A_F-X_F. P'_H on A_HX_H is the distance "d" behind P_H on C_H-E_H.

(2) If line X toward A is behind plane C-D-E, it will be shown as hidden outline in the F projection until it passes beyond the edge of the plane.

(3) If X_F toward A_F is hidden, X_F toward B_F will be visible, and since A_HX_H is visible, X_H toward B will be hidden until A_HB_H passes the edge of the plane. The distance that X_H toward B is behind C_HE_H can be seen in the front view.

(4) Using the method employed in step (1), X_H toward B_H is proven to be below C_HE_H by the distance d' (as seen in F view).

(5) B_H-X_H is, therefore, hidden from the point Q_H where it appears to cross C_HE_H to X_H, and visible from X_H toward A_H.

• PROBLEM 5-5

Project line MN on plane ABC.

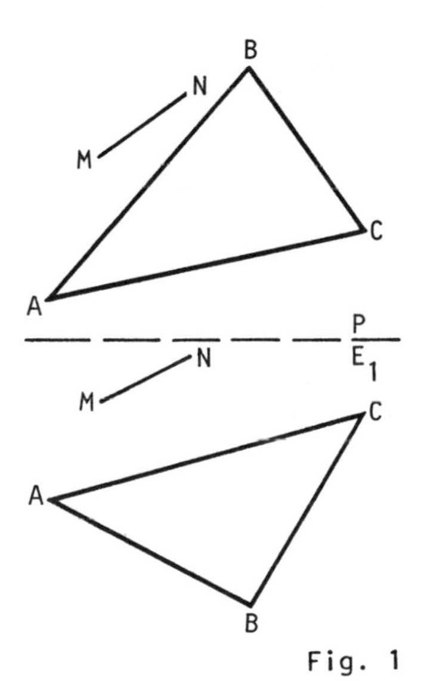

Fig. 1

Solution: Fig. 1 shows the line M-N to be projected on plane A-B-C, in both the plan, or horizontal, and primary elevation, or frontal plane of projection. Problem analysis reveals that when projectors are drawn perpendicular from a line to a plane, the line connecting their piercing points forms the projection of the line on the plane. Thus it follows that any orthographic view of a line is the projection of that line on an image plane. Any line may be projected on a plane that is not perpendicular to the line of sight by means of the edge-view. As in fig. 2, draw a horizontal line, AD, in the front view which projects to true length in the top view. Draw a horizontal line, CE, in the top view which projects to true length in the front view. In the top view, draw

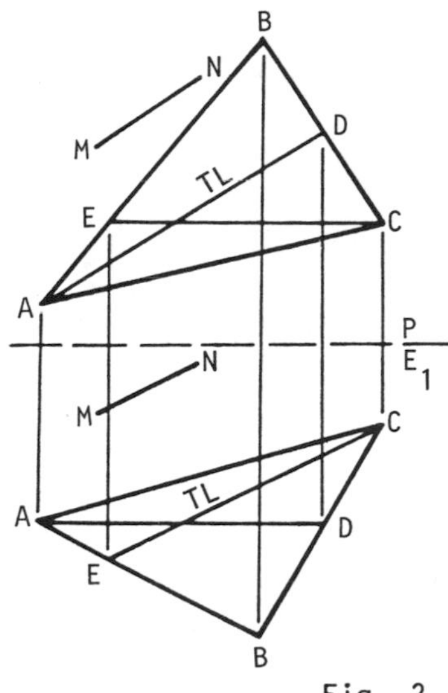

Fig. 2

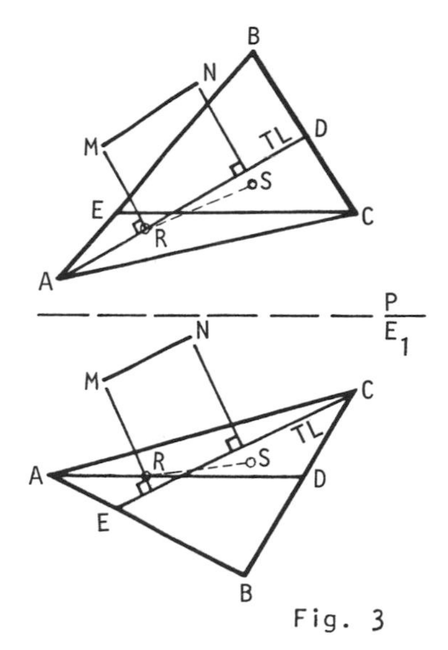

Fig. 3

lines from M and N that are perpendicular to line AD, and in the front view, draw lines from M and N that are perpendicular to line CE, as in fig. 3. The piercing points, R and S, are found by the cutting-plane method. The line R-S is the projection of the line M-N on plane A-B-C. The projection of M-N can also be found by an edge view of plane A-B-C and perpendiculars from M-N to the plane.

• PROBLEM 5-6

Locate the line of intersection of planes A-B-C and D-E-F, using the piercing point method.

Solution: An arbitrary line in one plane pierces another plane on the line of intersection of these two planes. Therefore, any two such piercing points will determine the location of the line of intersection of the two planes.

In the accompanying figure, planes A-B-C and D-E-F are given in the P, plan, and E_1, primary elevation, views and their line of intersection is desired. A vertical cutting plane is passed through line B-C in the plan view, cutting line D-E at G and line D-F at H. In view E_1,

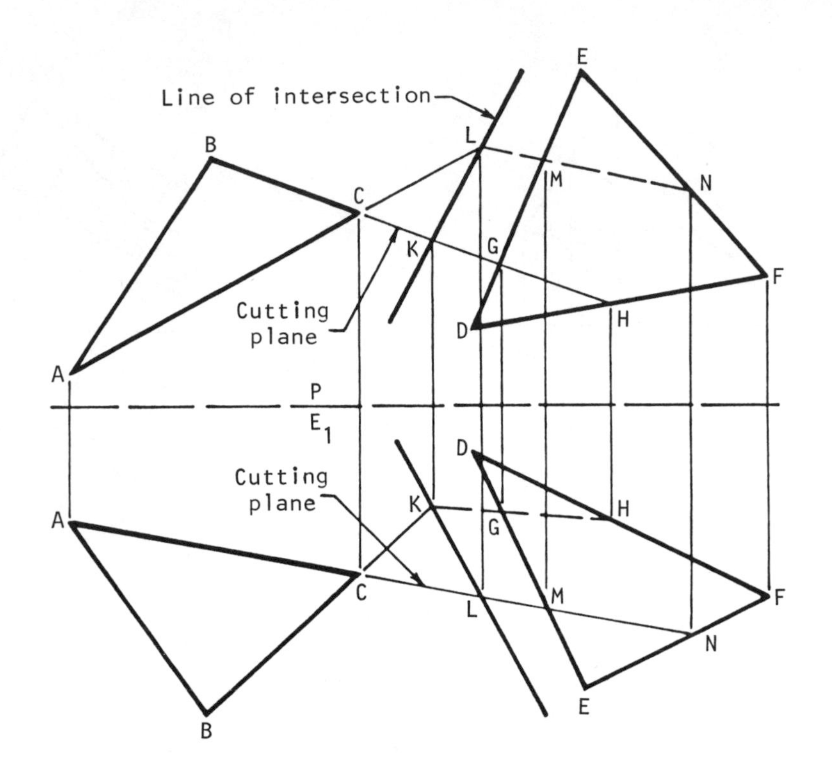

line G-H intersects the extension of line B-C at K, the point at which B-C pierces plane D-E-F. Point K is then projected into the plan view.

Likewise, a cutting plane appearing as an edge in view E_1 is passed through line A-C, cutting line D-E at M and E-F at N. In the plan view, line M-N intersects the extension of line A-C at L. Point L is then projected into view E_1. A line drawn through the piercing points K and L in both views represents the line of intersection. Note that the piercing point of any other line on either plane could have been used with the same result.

• PROBLEM 5-7

Locate the line of intersection of given planes MNOP and QRST, using the auxiliary view (edge view) method.

Solution: Fig. 1 shows the two given intersecting planes, MNOP and QRST, in the horizontal and frontal planes of projection. The line of intersection of the two planes will be apparent in a view in which one of the planes is seen as an edge (i.e., an auxiliary view). With the two given views, the necessary view can be drawn and the required line located.

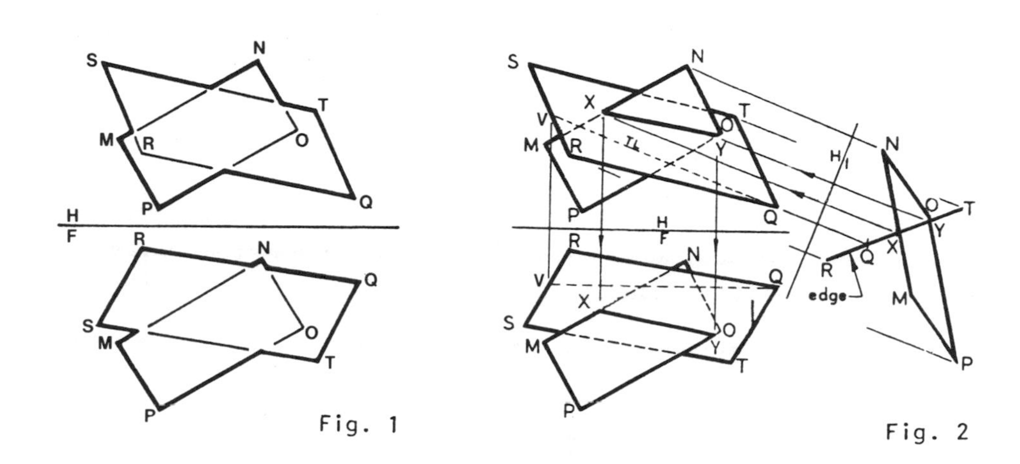

Fig. 1

Fig. 2

Draw the horizontal line QV, in plane QRST, in the front view, which will project to true length in the top view. With the line of sight parallel to line QV, draw the reference plane H-1 perpendicular to line QV to project plane QRST to an edge view. Also project plane MNOP to the auxiliary view. Line XY in this view is the line of intersection between plane MNOP and QRST. Project this line back to the top and front views in sequence, as shown in fig. 2. The front view of XY can be verified by transferring distances d_1 and d_2 to the front view.

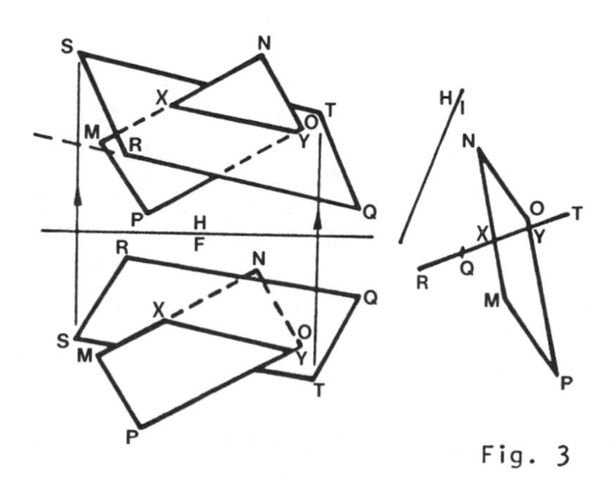

Fig. 3

Edge RQ is closer than edge MP to the H-1 plane in the auxiliary view; therefore RQ is higher than MP and visible in the top view. Visibility in the front view is verified by projecting line ST from the front view up to ST in the top view. Line RQ is crossed before reaching ST; therefore RQ is in front of ST and visible in the front view. (See fig. 3.)

• **PROBLEM 5-8**

Locate the line of intersection of planes M-N-O-P and Q-R-S-T by using the cutting-plane method.

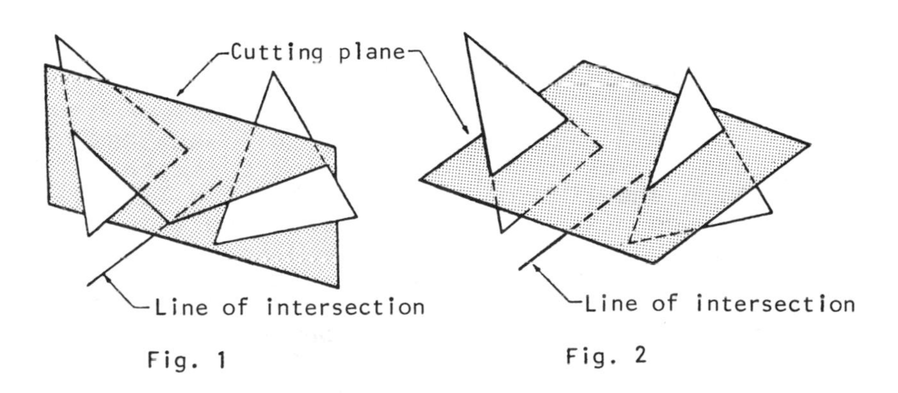

Fig. 1 Fig. 2

Solution: General problem analysis reveals that when any two given planes are intersected by a third plane, the two lines of intersection thus formed must be either intersecting or parallel. If the lines intersect, they must meet on the line of intersection of the two given planes, as shown in figure 1. If the lines are parallel, either they must be parallel to the line of intersection of the two given planes, as illustrated in figure 2, or the two given planes must be parallel to each other. Any two such cutting planes can be used to determine the line of intersection of two given planes.

Figure 3 gives a spacial analysis of the method of solution for the given problem. Pass an auxiliary plane, cutting a line from each of the two given planes. The point of intersection of these two lines is common to the three planes and, therefore, lies on the line of intersection of the given planes. Any second convenient cutting plane is used to find another point common to the two given planes. The line through the two points thus found will be the line of intersection of the two given planes.

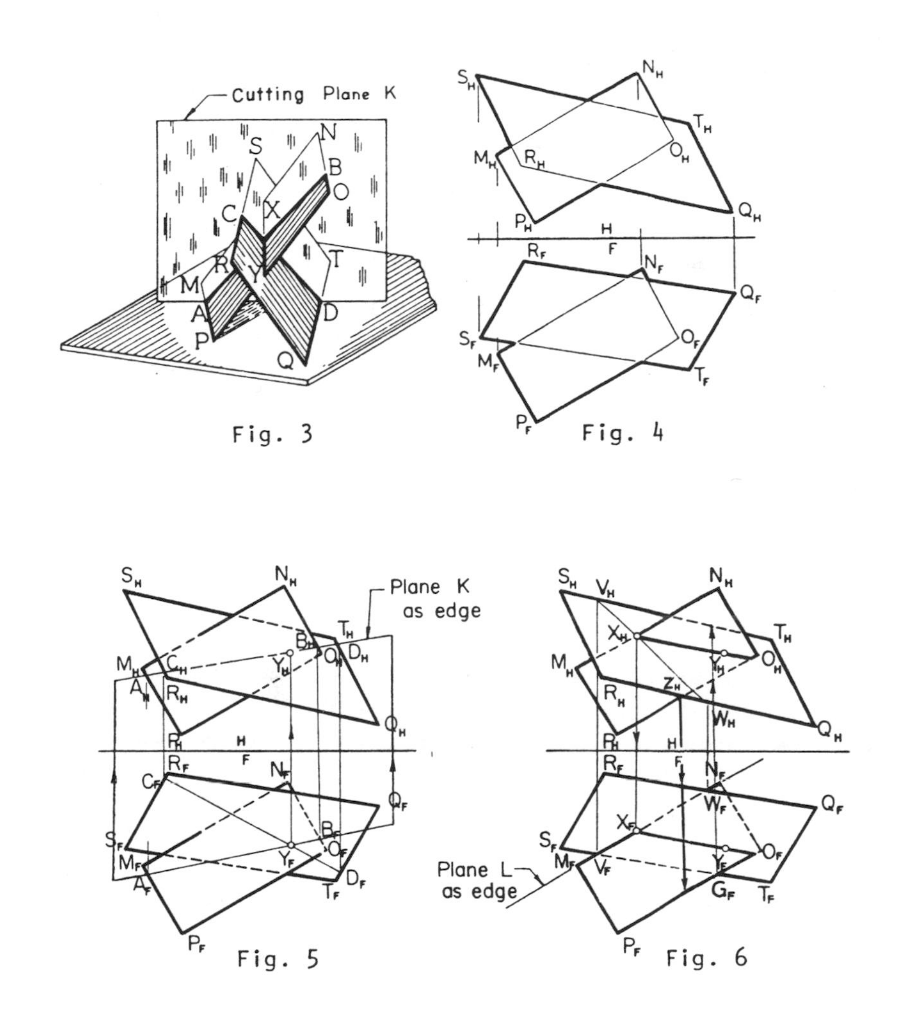

Fig. 3 Fig. 4

Fig. 5 Fig. 6

Fig. 4 shows the two given planes in the horizontal and frontal planes of projection (H and F, respectively). As shown in Fig. 5, pass the vertical frontal cutting plane K (which appears as an edge in the horizontal plane of projection), through the two given planes. In the H-view of Fig. 5, locate line A_H-B_H, where plane K cuts plane M-N-O-P. Similarly, locate line C_H-D_H, where plane K cuts plane Q-R-S-T. Project lines A_H-B_H and C_H-D_H to the F-view as A_F-B_F and C_F-D_F. In the F-view of fig. 5, locate point Y_F, where A_F-B_F and C_F-D_F intersect. Project Y_F to Y_H. As shown in fig. 6, pass the horizontal cutting plane L (which appears as an edge in the frontal plane of projection), through the two given planes. Following the same procedure as with plane K, locate line V_F-W_F where plane L cuts plane Q-R-S-T in the F-view. Plane L is coincident with the edge M_F-N_F of plane M-N-O-P in the F-view. Project lines V_F-W_F and M_F-N_F to the H-view as V_H-W_H and M_H-N_H. Locate, in the H-view, point X_H, where V_H-W_H and M_H-N_H intersect. Project X_H to X_F. The line through points X and Y is the required line of intersection of the two planes. Determine the visibility of the boundary lines of the given planes. For example, the intersection of R_H-Q_H and P_H-O_H is Z_H. Projecting Z_H to the F-view, R_FQ_F is crossed before P_H-O_H; therefore R-Q is higher than P-O in the front view, and visible in the top view. The intersection of O_F-P_F and S_F-T_F is G_F. Projecting G_F to the H-view, O_H-P_H is crossed before S_HT_H; therefore, O_F-P_F is in front and visible in the F-view.

• PROBLEM 5-9

Define and determine the strike and dip of a plane.

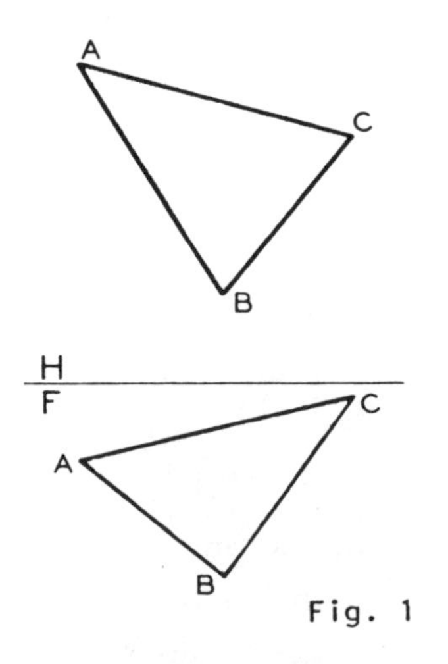

Fig. 1

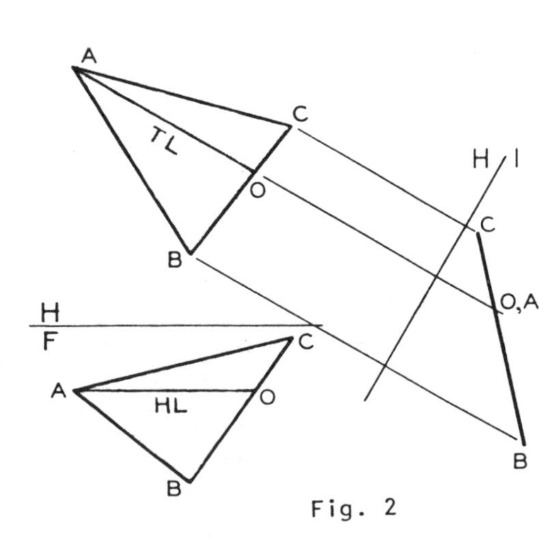

Fig. 2

Solution: Fig. 1 shows the given plane, A-B-C, in the H-, horizontal, and F-, frontal, planes of projection. The term "strike" refers to the compass direction of a level

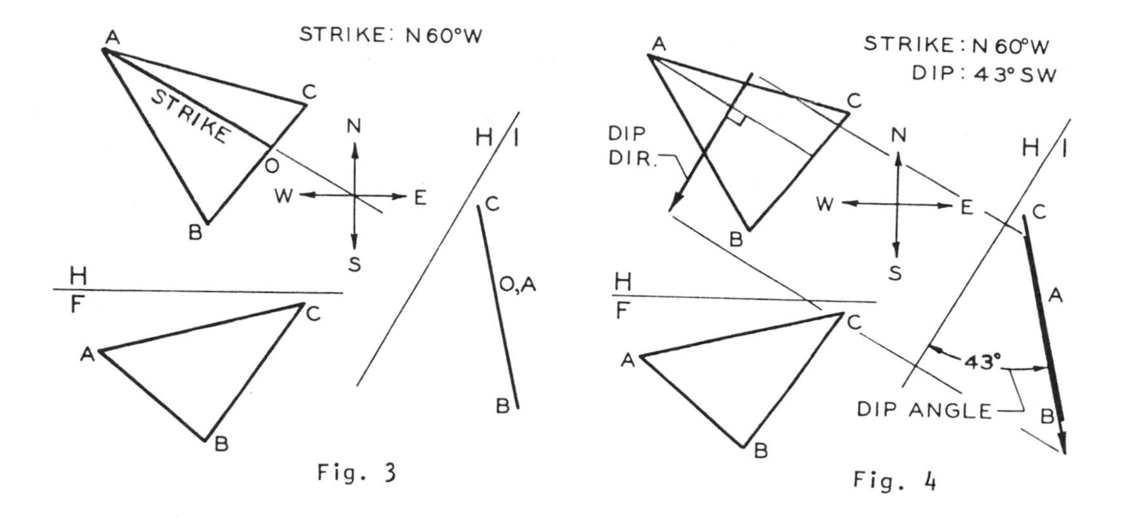

Fig. 3

Fig. 4

(i.e., horizontal) line in the top view of a plane. All level lines in a plane are parallel and have the same compass bearing.The term "dip" refers to the angle that the edge view of a plane makes with the horizontal plane, plus its general compass direction, such as NW or SW. The dip angle is found in the primary auxiliary view that is projected from the top view (i.e., a vertical plane at right angles to the strike), and its general direction is measured in the top view. Dip direction is measured perpendicular to a level line in a plane in the top view toward the low inside.

Fig. 2 illustrates the first step in determining the strike and dip of a plane. Draw a horizontal line, A-O, in the front view of plane A-B-C. This line projects true length in the top view. Project plane A-B-C as an edge in the auxiliary view, having lines of sight parallel to A-O, where A-O appears as a point. Project only from the top view in order to find the edge view of the horizontal plane.

Since the strike of a plane is the compass direction of a horizontal line in the plane, line A-O can be used to measure the strike of plane A-B-C, as shown in fig. 3. The compass direction is measured in the top view as either N60°W or S60°E. The line has no slope; therefore, either compass direction is correct.

Referring to Fig. 4, the dip angle is measured in the auxiliary view, and as mentioned above, its direction is perpendicular to the strike line in the top view. The dip of A-B-C is 43°SW. The strike and dip establish the plane. One application of the strike and dip of a plane is in the determination of ore vein locations and intersections.

• PROBLEM 5-10

Determine the intersection between ore veins by the auxiliary view method.

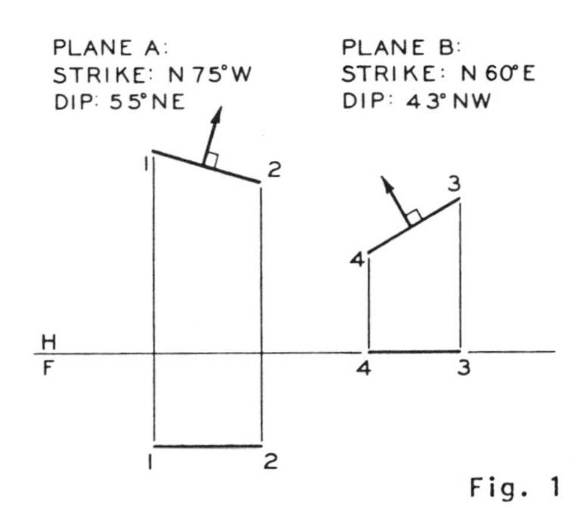

Fig. 1

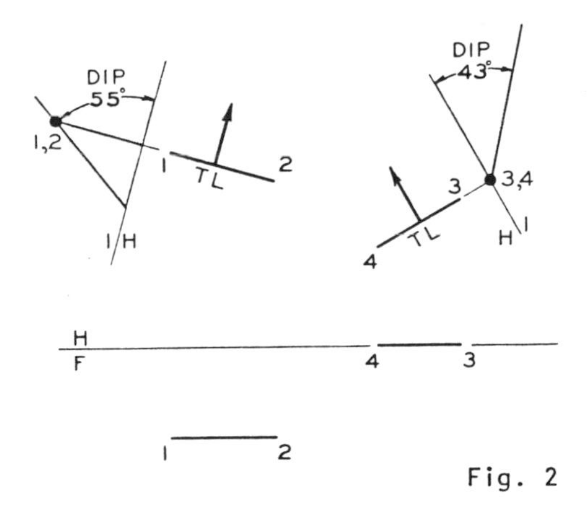

Fig. 2

Solution: Fig. 1 shows the strike and dip of two given ore veins, plane A and plane B, in the H-, horizontal, and F-, frontal planes of projection. Lines 1-2 and 3-4 are strike lines and are true length in the top view, since they appear as horizontal lines in the front view. The point view of each strike line is found in fig. 2 by an auxiliary view, using a common reference plane. The edge view of the ore veins can be found by constructing the dip angles with the H-1 plane through the point views. The low side is the side of the dip arrow.

A supplementary horizontal plane, H'-F', is constructed in fig. 3, at a convenient location in the front view, parallel to the H-F plane, at a distance H. This supplementary plane is shown in both auxiliary views located H distance from the H-1 reference plane. The H'-1' plane cuts through each ore vein edge in the auxiliary views.

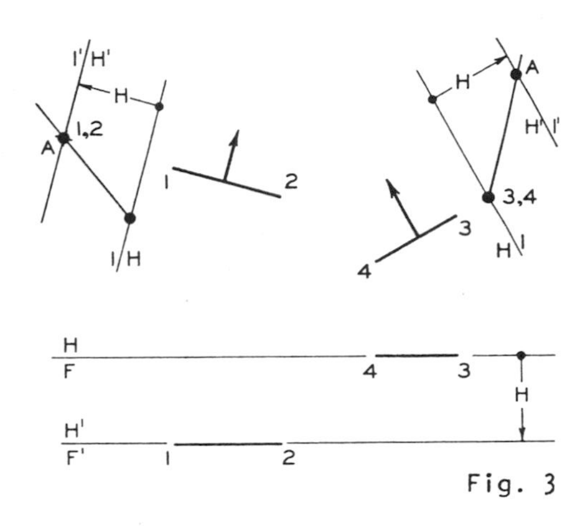

Fig. 3

Fig. 4 illustrates that points A, which were established on each auxiliary view by the H'-1' plane, are projected to the top view, where they intersect at point A. Points B on the H-1 plane are projected to their intersection in the top view at point B. Points A and B are projected to their respective planes in the front view. Line A-B is the required line of intersection between the two planes.

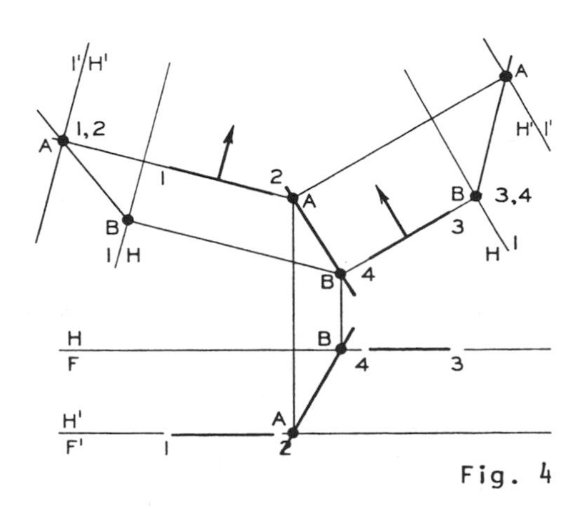

Fig. 4

SHORTEST DISTANCES BETWEEN POINTS, LINES, AND PLANES

• **PROBLEM** 5-11

Find the shortest connection from a given point C, to a given line, AB, using the plane method.

Solution: Fig. 1 shows the given point, C, and the given line, AB, in the plan view, P, and the primary elevation view, E_1. Preliminary problem analysis reveals that the true length of the perpendicular distance between a point and a line can be found in the view showing the true size of the plane formed by the given point and the given line.

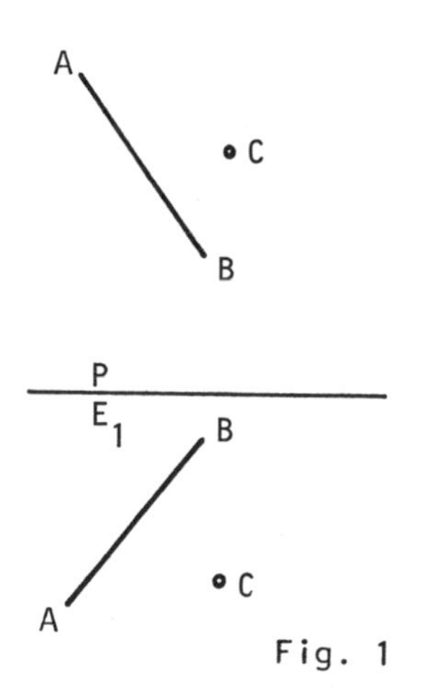

Fig. 1

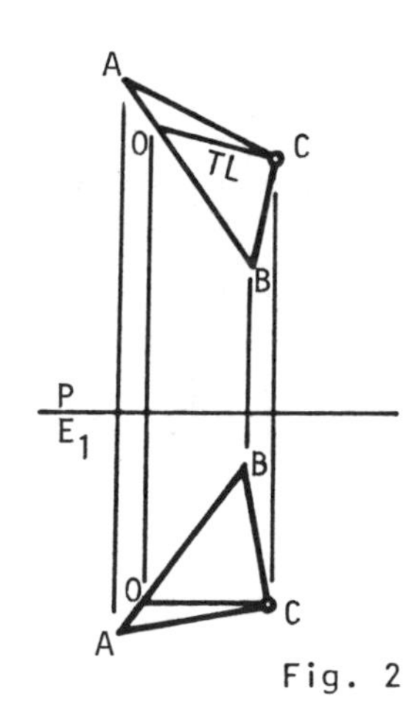

Fig. 2

Draw lines AC and BC to form plane ABC in both views. Draw in the primary elevation view a horizontal line, OC, which will project true length in the plan view, as in fig. 2. With the reference plane, P-E_2, drawn perpendicularly to the line OC, drawn an edge view of plane ABC by projection, as in fig. 3. Now make reference plane E_2-I_1 parallel to the edge view of ABC in order to project plane ABC to true size in view I_1. Once a true size view is obtained, a perpendicular to AB drawn from C will be true length and can be measured correctly. The length of CX is the required connection, as is seen in fig. 4.

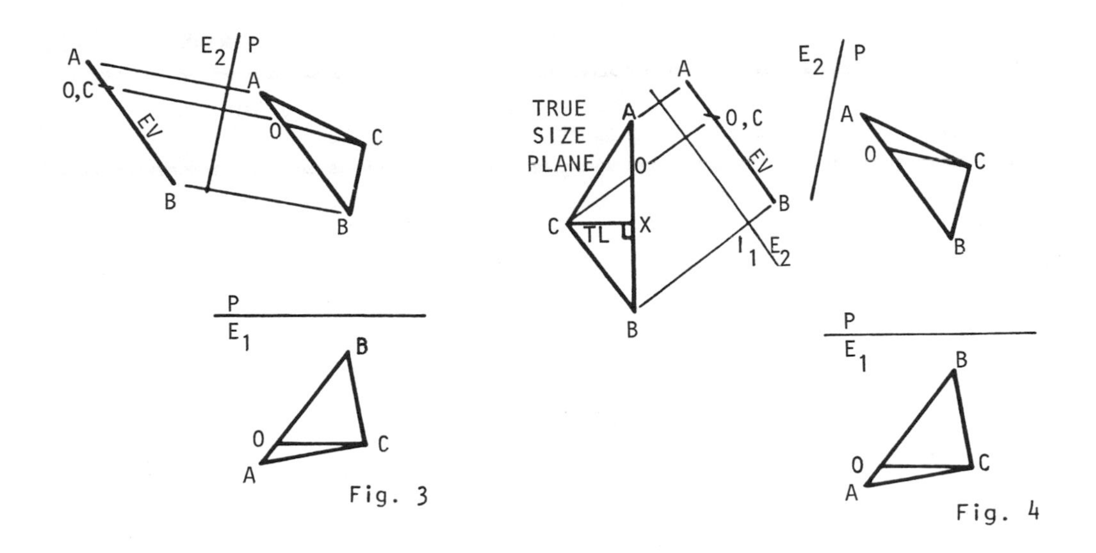

Fig. 3

Fig. 4

To find the slope of CX, project CX back to view E_2, the plan view, and the primary elevation view. Now a reference plane, P-E_3, drawn parallel to CX in the top view will enable CX to be projected to true length in view E_3. A line drawn parallel to plane P-E_3 through C will be the reference horizontal from which the slope can be measured. The slope of a line is the angle the line makes with the horizontal, and it can be seen only when the line appears in true length in an elevation view, as in fig. 5.

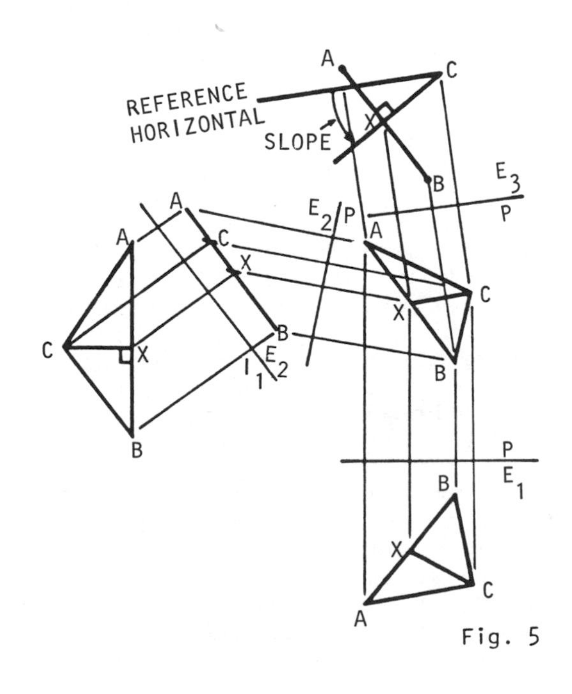

Fig. 5

Find the shortest distance between a given straight line and an arbitrary point P on a circle.

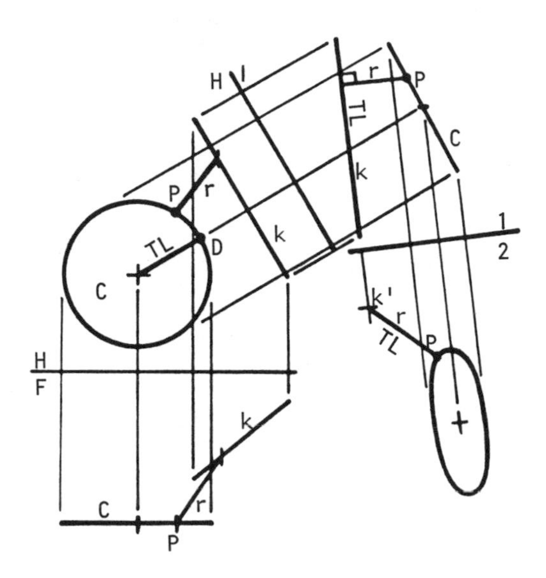

Solution: The accompanying figure shows the given line K and circle C in the H-, or horizontal, and the F-, or frontal, planes of projection. Problem analysis calls for the construction of a view of the line and circle in which the line projects as a point. The shortest line between the point view of the line and the curve is the required distance. The circle C projects true size and shape in the H-view, since the reference plane H-F is parallel to the edge view of the circle. Consequently, the radial line CD in the top view is true length. The lines of sight for auxiliary projection plane H-1 are parallel to CD; therefore, an edge-view of C is located in view 1. Furthermore, since H-1 is parallel to K in the top view, K appears true length in view 1. Line r is the shortest connection (but not true length distance) from point P on the circle to the straight line, K, in view 1. This is the case since the shortest connection from a point (i.e., any arbitrary point P on the circle) to a line is perpendicular to the line, and appears at right angles to the line in any view showing the line in true length. The lines of sight for view 2 are parallel to K in view 1, which is true length; therefore, K' is a point view of line K. In addition, since reference plane 1-2 is parallel to r (i.e., lines of sight for view 2 are perpendicular to r), r appears true length in view 2 and is the required distance. In other words, the radius of the arc centered on K', and tangent to the curve C at P, is the shortest distance between line K and point P on circle C.

Find the shortest level, or horizontal, connection between two given skewed lines, AB and CD, using the two-view method.

Solution: Fig. 1 shows the top and front views of the given lines AB and CD. The shortest horizontal connection between two lines is perpendicular to the plane which contains one of the lines and which is also parallel to the other line. This shortest horizontal connection has the same direction, length, and slope as a horizontal perpendicular to the plane from any other point on the line.

Fig. 2 shows plane ABE parallel to CD. ABE was constructed by first drawing BE parallel to CD in both views, drawing AE horizontally in the front view, and then projecting AE to the top view. (AE appears true length in the top view because it is parallel to the rotation line in the front view.)

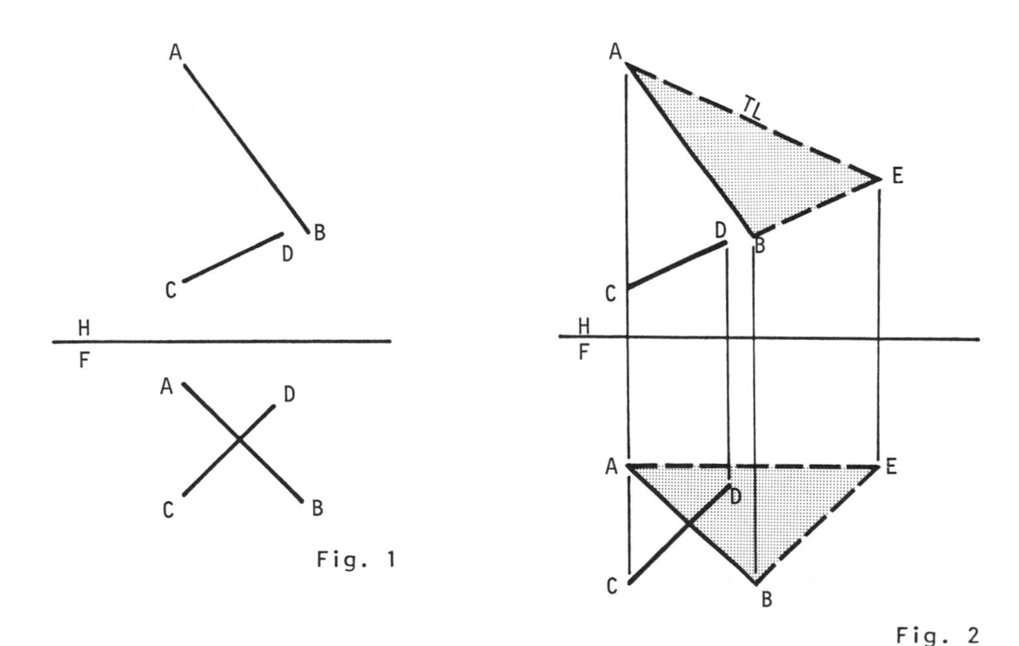

Fig. 1

Fig. 2

The distance from CD to plane ABE can be represented in the F-view by drawing a horizontal line DG, as shown in fig. 3. DG pierces the plane ABE at point F, which can be found by the cutting plane method of finding piercing points. Now a horizontal line on plane ABE is drawn in the F-view through point F, intersecting AB and BE at S and T. These points are projected to the H-view, where line DF can now be drawn, and it will be true length because it is parallel to the rotation line in the F-view. Projecting point G and connecting line DG shows that DG is perpendicular to AE. Line DF represents the shortest distance between line CD and plane ABE because: (1) DF is perpendicular to plane ABE; and (2) DF is true length in the H-view. The shortest distance between a line and a plane is the perpendicular distance, and only this distance will be true length in the view in which it is perpendicular to a line of the plane.

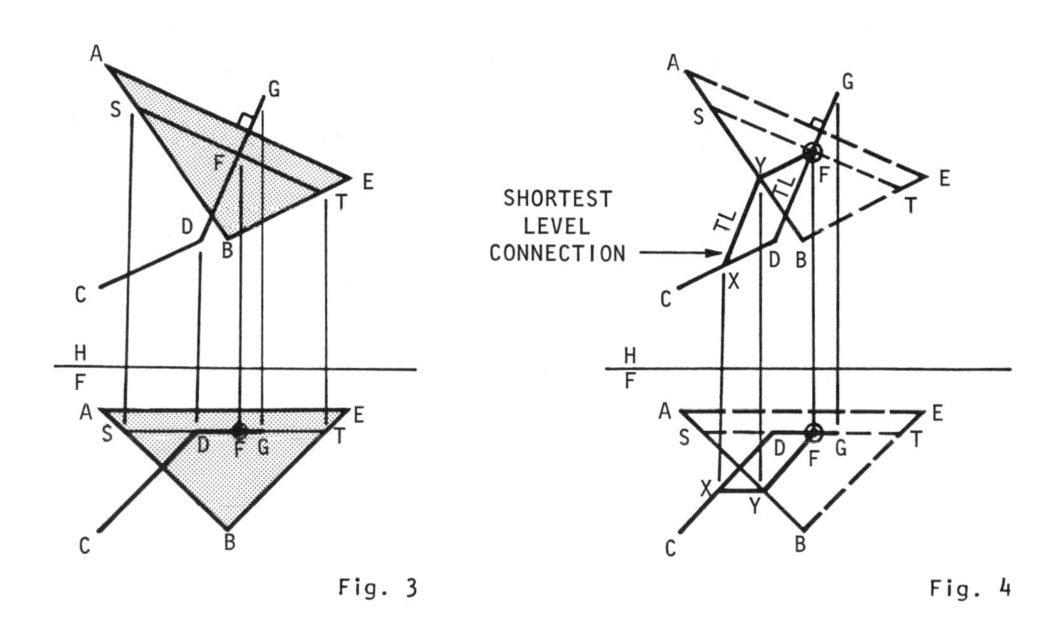

Fig. 3 Fig. 4

A line through point F parallel to CD in both views is drawn. Point Y is where this line intersects AB, as shown in fig. 4. Since FY is parallel to CD, all perpendicular distances between FY and CD will be equal in length and parallel. DF is one such distance, but DF does not go from AB to CD, as the problem calls for. A line from point Y, parallel to DF, going to CD will be equal in length to DF and will be the desired shortest distance from AB to CD. This line intersects CD at X, and XY is the required distance. Its length can be measured from the H-view, where it appears true length because it is parallel to the rotation line in the F-view.

Find the shortest horizontal (or level) distance between two skewed lines, A-B and C-D, using the plane method.

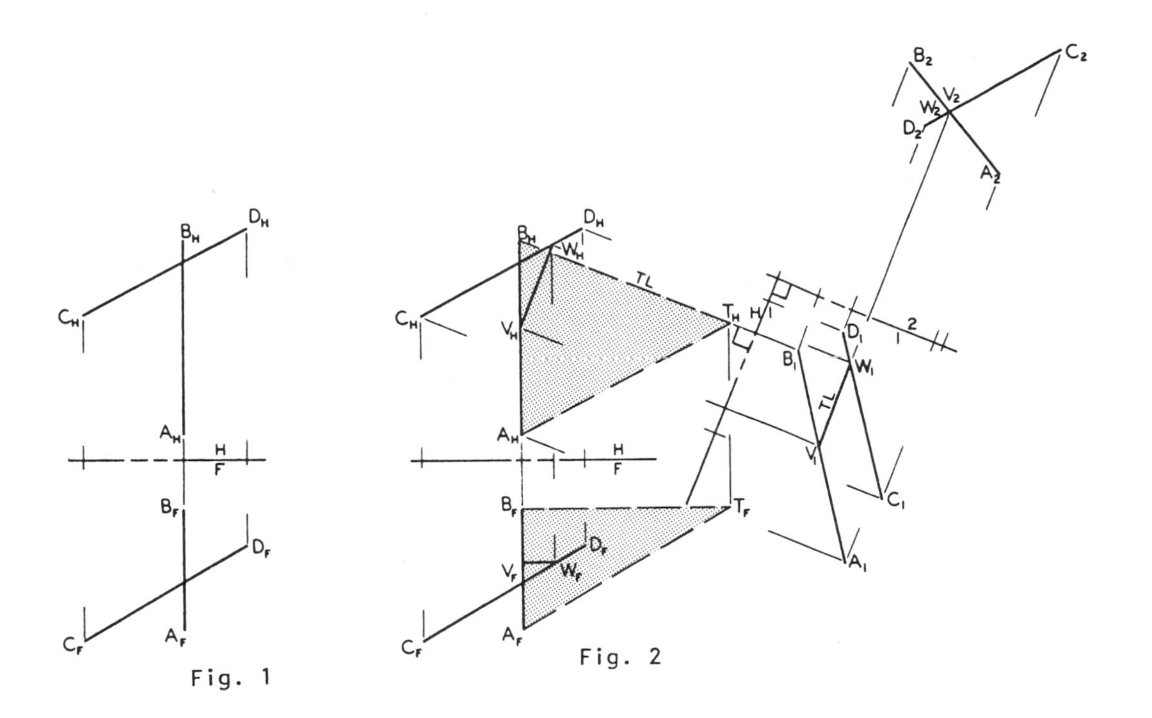

Fig. 1

Fig. 2

Solution: Fig. 1 shows the given lines in the H-, horizontal, and F-, frontal, planes of projection. Problem analysis reveals that the true length of the shortest horizontal connecting line can be found in an elevation view where the two given oblique, or skew lines appear to be parallel. Its location is determined from an adjacent projection on a plane at right angles to this elevation, which shows the horizontal connector as a point. Fig. 2 shows the method of solution. Construct a plane A-B-T parallel to line C-D by drawing line A_F-T_F parallel to C_F-D_F. This construction is based on the principle that a plane is parallel to a line if any line on the plane is parallel to the line. Project plane A_F-B_F-T_F to the H-view. The horizontal line B_F-T_F projects true length in the H-view. Obtain an edge-view of the plane A-B-T in an auxiliary elevation view, view 1, having lines of sight parallel to B_H-T_H. We are using the fact that the edge view of a plane can only be obtained in a view having lines of sight parallel to a true length line on the plane. Project line C-D to view 1 where it appears parallel to the edge view of plane A-B-T, the latter having been drawn parallel to the former. The direction of any horizontal connector is known in view 1 since a horizontal line is parallel to the $\frac{H}{1}$ rotation line. In view 1, any number of horizontal lines could be drawn, at different elevations, between the two given lines. Since the apparent length of all these horizontal connections is the same in this view, the one that appears in its true length must be the shortest. Its position is determined by drawing view 2 with horizontal lines of sight (i.e., parallel to the $\frac{H}{1}$ rotation line and to the true-length shortest horizontal connector) and thus obtaining a point view of the required line. Draw a projection at right angles to plane 1 (i.e., view 2) showing the shortest horizontal connector or distance as the point V_2W_2 (the point where the two oblique lines appear to intersect). Locate the shortest horizontal distance, V-W, in views 1, H, and F. Note that V-W is true length in view 1 since the point view of a line can be projected only from a true-length view of the line. Furthermore, since V-W is a horizontal line it is parallel to the $\frac{H}{1}$ rotation line as previously stated, and thus projects true length in the H-view. As a result, the V-W projections in the 1- and H-views are parallel. The F-projection also verifies V-W as a horizontal line, since here it must be parallel to the $\frac{H}{F}$ rotation line.

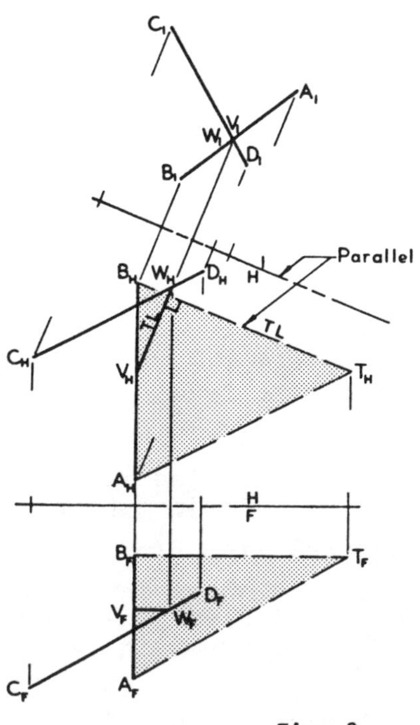

Fig. 3

Once the foregoing method of solution is understood it is possible to employ a short-cut version of this same method. This short-cut is illustrated in Fig. 3. A point view of the shortest horizontal connector can be obtained directly from the plan view because its direction is known to be perpendicular to a true length line in the plane of A-B (The plane of A-B is parallel to C-D). The two fundamental principles behind this analysis are: (a) a line perpendicular to a plane is perpendicular to all lines on the plane; (b) in any view a line perpendicular to a plane appears at right angles to any true-length line on the plane. The first step in solution requires that a plane be passed through line A-B, parallel to line C-D. View the two lines, A-B and C-D, on an elevation plane, 1, parallel to the true-length line B_H-T_H. This view shows the shortest horizontal connector as a point V_1W_1, where A_1-B_1 and C_1-D_1 appear to intersect. Locate V-W in the H projection where it will appear in true length, and in the F projection where it will check as a horizontal line.

• PROBLEM 5-15

Find the shortest distance between two nonintersecting, nonparallel (skew) lines, A-B and C-D, using the two-view method.

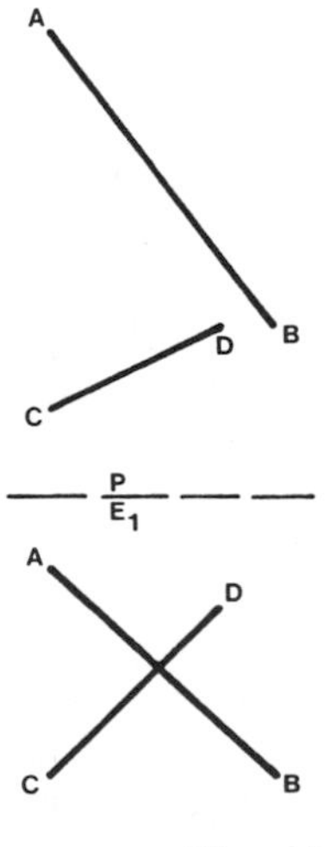

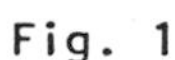

Fig. 1

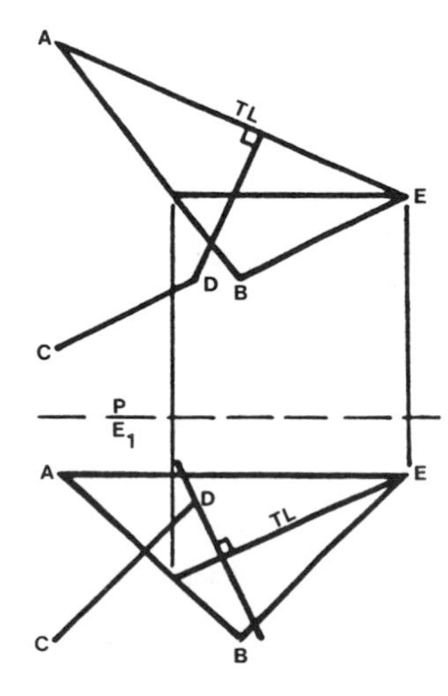

Fig. 2

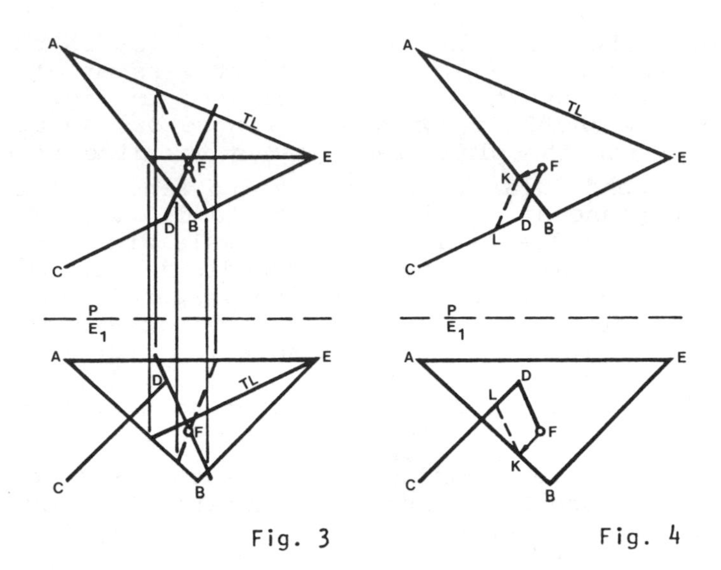

Fig. 3 Fig. 4

Solution: Fig. 1 shows the given lines in the P-, plan, and E_1-, primary elevation, views. The shortest connection between two lines is one drawn perpendicular to both lines. Therefore, the connecting line is perpendicular to a plane which contains one of the lines and is parallel to the other. This shortest connection has the same direction, length, and slope as a perpendicular to this plane from any other point on the line. First, refer to fig. 2. Plane A-B-E is constructed so that it contains line A-B and is parallel to line C-D because line BE is parallel to line C-D. Then, a line perpendicular to the plane through point D is constructed by drawing it at right angles to true-length lines on the plane in both P- and E_1-views. The reasoning behind this is that in any view, a line perpendicular to a plane appears perpendicular to any line on the plane that appears true length in that view. In fig. 3, the point F at which the perpendicular pierces the plane is found by the cutting plane method. Since the shortest connection from A-B to C-D is also perpendicular to plane A-B-E, it must be parallel to D-F and have the same length. To locate its position, as shown in fig. 4, a line is drawn parallel to C-D from F to its intersection with A-B at point K, in both views. From point K, a line parallel to D-F intersects line C-D at point L, thus establishing the shortest connection, K-L. The true length and slope of K-L can be determined by triangulation or by an additional elevation view.

• PROBLEM 5-16

Find the shortest distance between two nonintersecting, nonparallel (skew) lines, 1-2 and 3-4, using the plane method.

Solution: Fig. 1 shows the given lines in the H-, horizontal, and F-, frontal, planes of projection. The plane method of solution is based on the fact that if a plane contains one line and is parallel to another line, the shortest connection between the lines, being perpendicular to both of them, must be perpendicular to the plane. Therefore, any view showing this plane as an edge shows the true length of the shortest connection. In other words, the true length of the common perpendicular to two oblique lines can be seen in a view where the two lines appear to be parallel. The location of this connection is fixed in an adjacent view that shows both lines

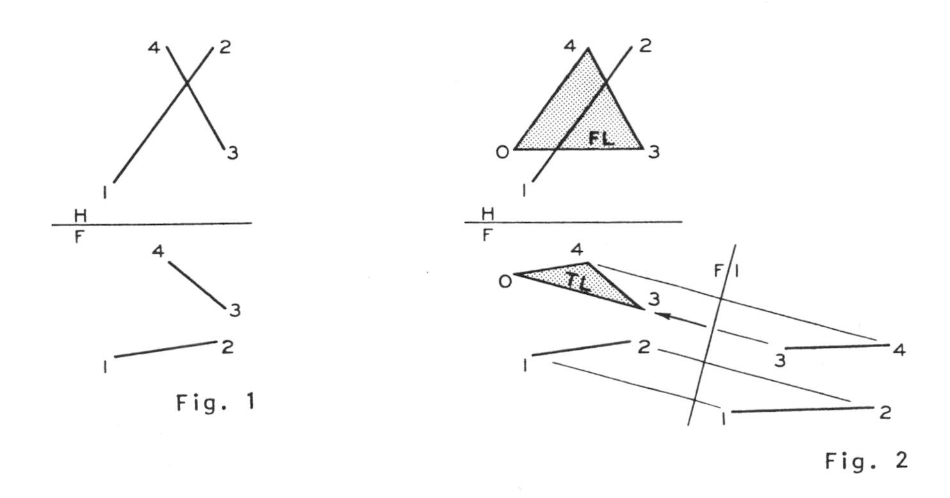

Fig. 1

Fig. 2

in true length. As shown in fig. 2, construct a plane through 3-4 that is parallel to line 1-2 by first drawing in both views a line 4-0 parallel to 1-2. Then draw the horizontal line 3-0 in the top view which projects to its true length in the front view. Since plane 3-4-0 contains a line parallel to line 1-2, the plane is parallel to the line. Since it is desirable to produce a view showing plane 3-4-0 as an edge, and since a true length line, 3-0, exists in this plane in the F- view, making the line of sight of the auxiliary view parallel to the line 3-0 will yield an edge view of plane 3-4-0, line 3-4. Note that 3-4 is parallel to 1-2 in this view. This follows logically from the fact that plane 3-4-0 is parallel to line 1-2. Now refer to fig. 3. The shortest distance will appear true length in the primary auxiliary view, where it will be perpendicular to both lines (i.e., the shortest distance is the perpendicular distance from line 1-2 to plane 3-4-0). In order to determine the location of this shortest connection, draw a secondary auxiliary view by projecting perpendicularly from the lines in the primary auxiliary view. Lines 1-2 and 3-4 cross in this view, and appear true length. As shown in fig. 4, the

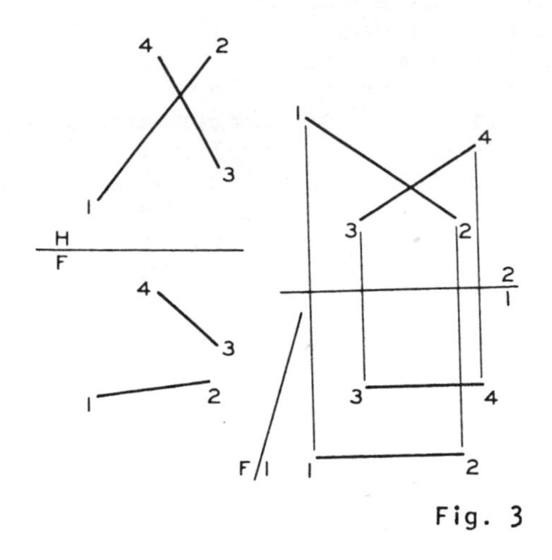

Fig. 3

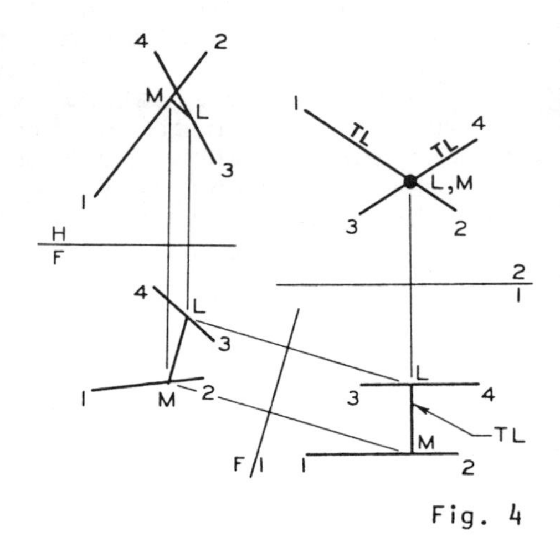

Fig. 4

crossing point of lines 1-2 and 3-4 establishes the point view of the perpendicular distance, L-M, between the true length lines. This distance is projected to the primary auxiliary view, where it is true length. The line is found in the front and top views by projecting points L and M to their respective lines in these views.

• PROBLEM 5-17

> Find the shortest grade distance, or distance of specified slope, between two given skewed lines using the two-view method.

Solution: Fig. 1 shows the given lines A-B and C-D in the P-, plan, and E_1-, primary elevation, views. The solution which follows applies to the specific downward slope of 30° from C-D to A-B but the procedure employed is applicable to any specified slope. The shortest specified grade distance between two lines is perpendicular to the plane that contains one of the lines and which is also parallel to the other line. This shortest specified grade distance has the same direction, length, and slope as a perpendicular to the plane having the same specified slope from any other point on the line. Now refer to Fig. 2. Plane A-B-E is constructed parallel to C-D by first drawing line BE parallel to CD in both views,

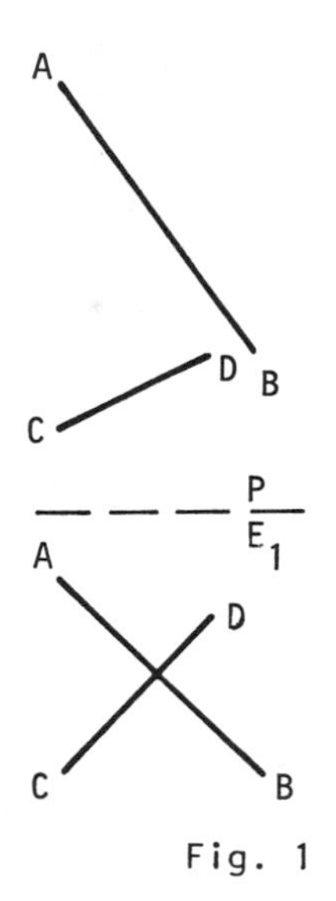

Fig. 1

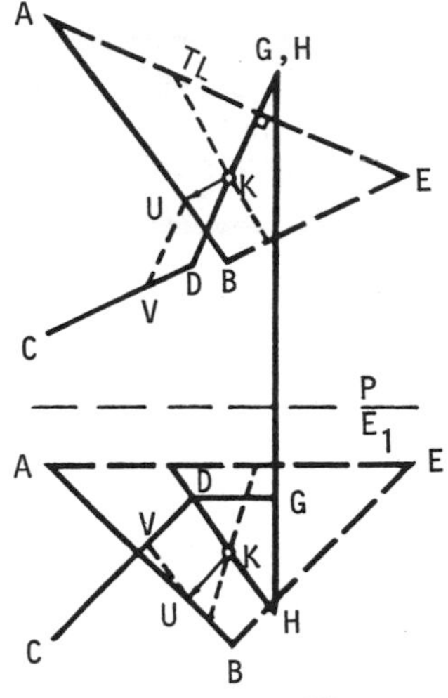

Fig. 2

drawing a horizontal line AE in the front view parallel to the line of rotation, and projecting AE to the top view. AE projects as true length to the top view.

As illustrated in Fig. 3, a line of any convenient length, D-H, is laid out at a slope of 30°. The plan length, D-G, and difference in elevation, G-H, are found. From D in the plan view of Fig. 2 line D-G is drawn perpendicular to the true-length line A-E, thus forming a line perpendicular to plane A-B-E. The length of D-G in the P- view corresponds to the plan length as shown in Fig. 3. Point G is projected to the E_1 view and point H is located using the difference in elevation, G-H (as shown in Fig. 3). The line D-H is drawn in both views, and the point K at which it pierces the plane A-B-E is found by the cutting plane method. From K, a line is drawn parallel to C-D in both views, to locate point U on line A-B The desired connection, V-U, is drawn parallel to line K-D,

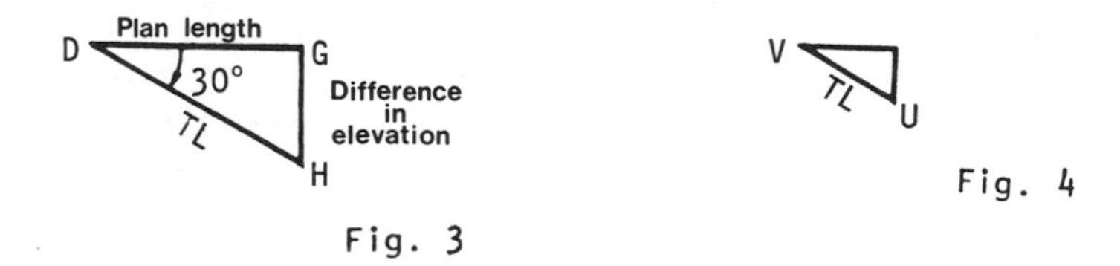

Fig. 3

Fig. 4

and therefore has a downward slope of 30° from C-D to A-B and is perpendicular to plane A-B-E. The true length of V-U can be found by triangulation, as shown in Fig. 4, or by an additional elevation view having lines of sight perpendicular to one of the established views of V-U.

• PROBLEM 5-18

Find the shortest grade distance, or distance of specified slope, between two given skewed lines using the plane method.

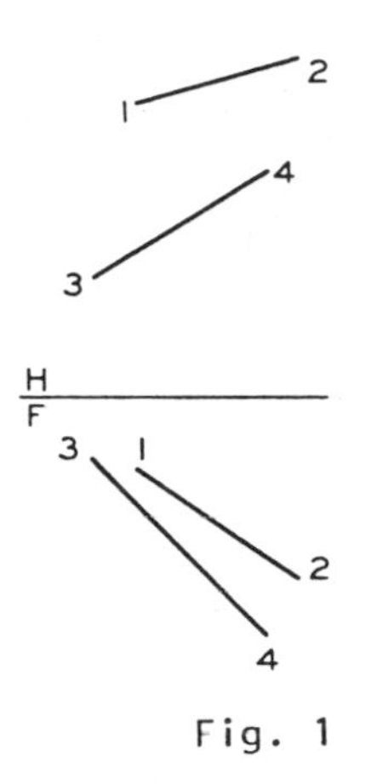

Fig. 1

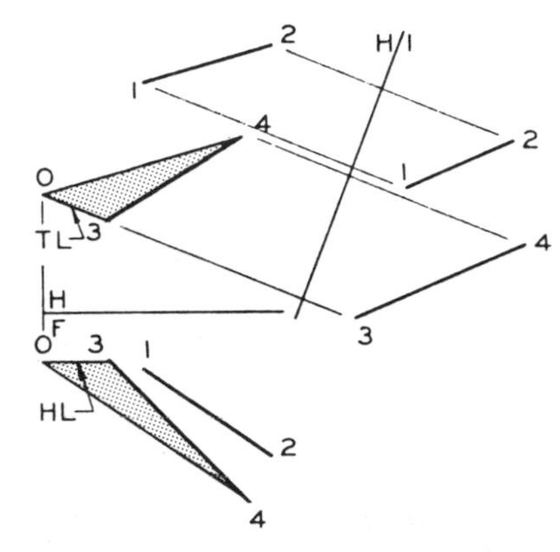

Fig. 2

Solution: Percent grade is an expression of the slope of a line, and is defined as the number of units rise in 100 units run. The plane method, a general approach to solving all skewed-line problems, is employed as the method of solution. The line-method, an alternative method of solving skewed line problems, is only applicable to the perpendicular distance between two skewed lines. Fig. 1 shows the given skewed lines, 1-2 and 3-4, in the H-, horizontal, and F-, frontal, planes of projection. The stated problem will be solved for a particular grade, 50 percent, by establishing the desired slope in the elevation view where the given lines appear parallel. Any other grade, either positive or negative, can be solved in the same manner. Referring to Fig. 2, draw plane 3-4-0 parallel to line 1-2 by drawing line 4-0 parallel to line 1-2 in both views. The horizontal line, 0-3, labeled HL in the F- view, projects true length in the H- view. An auxiliary elevation view, 1, having lines of sight parallel to this true length line, shows plane 3-4-0 as an edge, 3-4. Consequently, in this same view both lines project as parallel. Note that the primary auxiliary must be projected from the top view in order that the horizontal plane may appear as an edge in the primary view. It is from this edge view that the percent grade of a line can be drawn. Refer to fig. 3, and construct a 50 percent grade line with respect to the edge view of the H-1 plane (this edge is the run of the line) in the primary auxiliary view. Draw this line in the direction nearest to that which a line perpendicular to the two lines would assume. The grade could be drawn in two directions with respect to the H-1 reference plane. The shortest distance, however, between two lines will be the line drawn in a direction that is most nearly perpendicular to the lines. The shortest grade distance will appear true length in the primary auxiliary view. Project a secondary auxiliary view having lines of sight parallel to the direction of the grade line (i.e., having a specified slope with respect

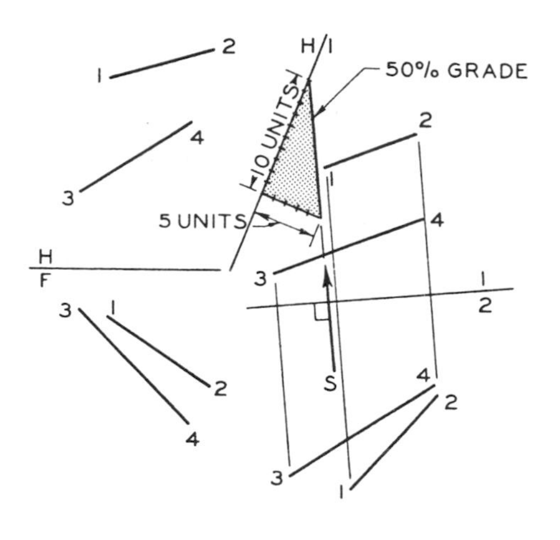

Fig. 3

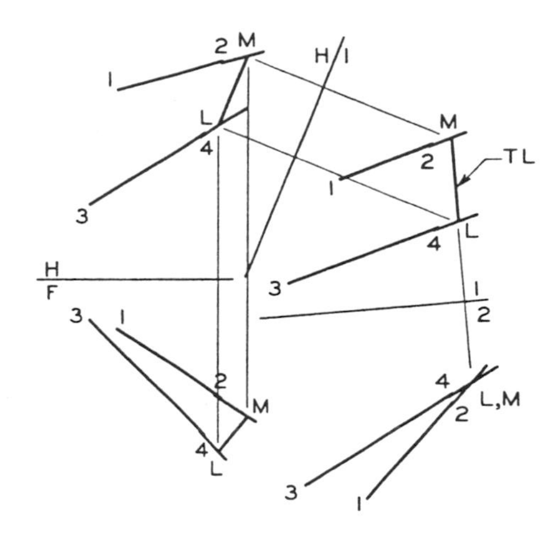

Fig. 4

to the horizontal plane). The lines in the secondary auxiliary view in fig. 4 are extended, since they are merely segments of longer, continuous lines, to establish their point of intersection. This intersection also locates the point view of L-M, the shortest line that can be drawn at 50 percent grade. Line L-M will appear true length at a 50 percent grade in the primary auxiliary view since it is parallel to the previously constructed 50 percent grade (i.e., the previously constructed 50 percent grade also projects as a point in view 2 since lines of sight are parallel to this true-length grade). Line L-M is projected back to the top and front views. Lines 1-2 and 3-4 must be extended in each view.

Note that the shortest distance of specified grade and the shortest horizontal connection must be parallel in the plan or horizontal view because they are both true length in view 1, the primary inclined or primary auxiliary view. For the same reason, they are also parallel to the shortest connection between the two skewed lines. Thus, it can be seen that all "shortest" connections between two given lines appear parallel in the plan or horizontal view. In other words, they all have the same bearing.

• PROBLEM 5-19

Find the shortest connection of specified bearing between two skew lines.

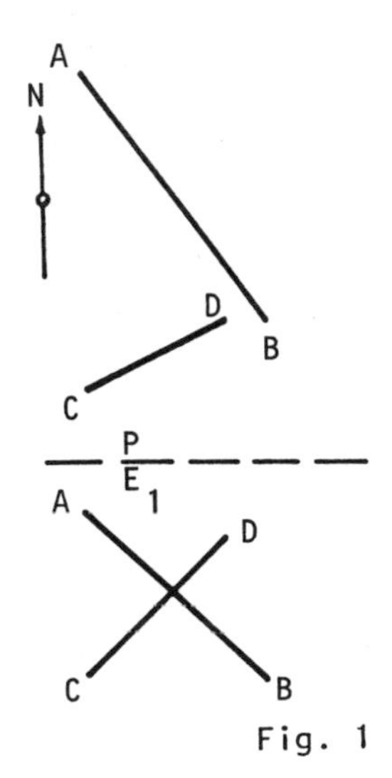

Fig. 1

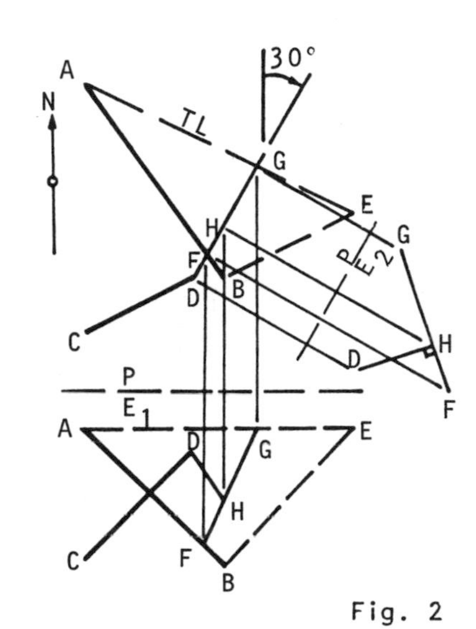

Fig. 2

Solution: In Fig. 1 lines A-B and C-D are given in the P-, plan, and E_1-, primary elevation, views. The following solution is based on the shortest connection with a bearing of N30°E from C-D to A-B; however, the general method is applicable to any specified bearing. Construct plane A-B-E parallel to C-D by drawing B-E parallel to C-D in both views, as shown in fig. 2. The shortest connection of the specified bearing from C-D to A-B has the same length and slope as the shortest connection of the specified bearing from any point on C-D to the plane A-B-E. Note that by its very definition, the shortest connection is perpendicular to the plane, A-B-E. Therefore, a line with a bearing of N30°E is drawn through point D in the plan view. A vertical plane containing this line intersects plane A-B-E along line F-G. The shortest connection with a bearing of N30°E from D to A-B-E lies in this vertical plane and is at right angles to F-G. View E_2 is drawn to show the true size and shape of the vertical plane. D-H is drawn perpendicular to F-G in this same view. View E_2 shows both the true length and slope of D-H, which is the shortest connection with a bearing of N30°E from point D to the plane A-B-E. However, it is not the desired connection from C-D to A-B because point H does not lie on A-B when projected into the given views.

To obtain the desired connection (fig. 3), a line is drawn parallel to C-D from H to locate point L on line A-B. A line is then drawn through L parallel to D-H to contact C-D at point K. The line K-L is the shortest connection with a bearing of N30°E from C-D to A-B. It has the same true length and bearing as line D-H. This problem was solved in the same manner as those requiring the shortest grade distance or the shortest level distance between two skewed lines.

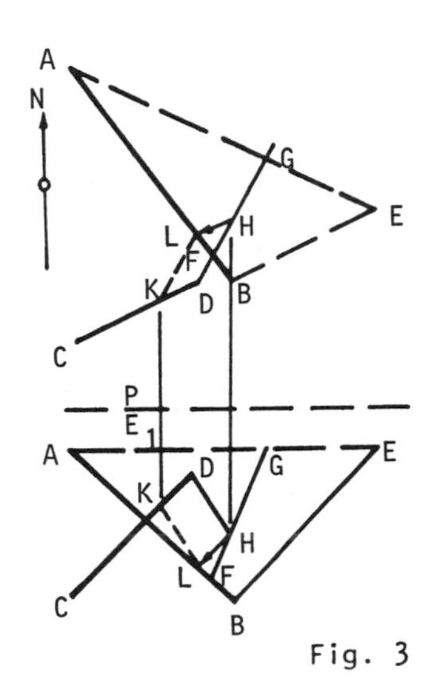

Fig. 3

The shortest connection of specified bearing and slope can be found by a variation of the foregoing procedure. The same steps are used, with the exception that the line D-H is drawn with the specified slope instead of perpendicular to F-G.

• PROBLEM 5-20

Find the shortest connection from a given point X to a given plane A-B-C, using the two-view method.

Solution: Fig. 1 shows the given plane and the given point in the P-view (plan or horizontal) and E_1-view (elevation or frontal). In the preliminary problem analysis, one must recognize that in any view, a line perpendicular to a plane appears at right angles to any line on the plane that is shown in its true length in that view. Thus the direction of the shortest connection in any view is determined. The piercing point can then be determined by the projection or the cutting-plane, or the edge-view methods. The horizontal line A-Y is drawn in view E_1 of

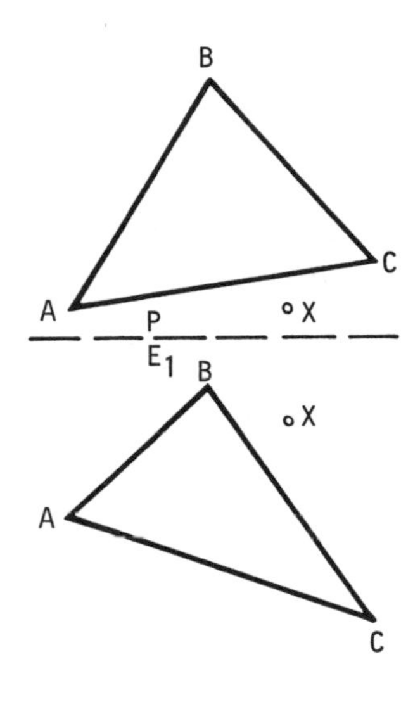

Fig. 1

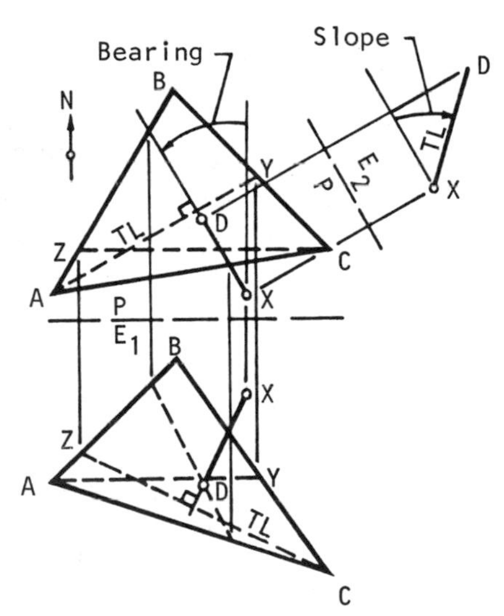

Fig. 2

fig. 2, and projected to the P-view where it appears in true length. The shortest connection is drawn from X, perpendicular to this line in the P-view. Similarly, the frontal line, C-Z, is drawn in the plan parallel to the E_1 reference plane so that it will appear true length in the E_1-view. Here, the shortest connection is drawn from X perpendicular to C-Z. The direction of the shortest connection has now been determined in both views. Point D, where the connection pierces the plane, is then determined by the cutting-plane method. If the true length and slope of the shortest connection are desired, they may be found by drawing an additional elevation view, E_2, as shown. Reference plane E_2 is parallel to D-X in the P-view, and therefore D-X is projected true length in E_2.

• PROBLEM 5-21

Find the shortest connection from a given point, X, to a given plane, A-B-C, using the edge view method.

Solution: Fig. 1 shows the given point and the given plane in the P- view (plan or horizontal) and E_1- view (elevation or frontal). The basic principle employed in solution is that the shortest connection from a point to a plane is a line from the point perpendicular to the plane. Further, any line that is perpendicular to a plane will appear at right angles to the plane in any view showing the plane as an edge. This view will show the true length of the perpendicular, or shortest, connection. If the edgeview is obtained in an elevation view, it will also show the slope of the shortest connection. Obtain view E_2, as illustrated in Fig. 2, showing the plane as an edge. The horizontal line A_E-G_E in view

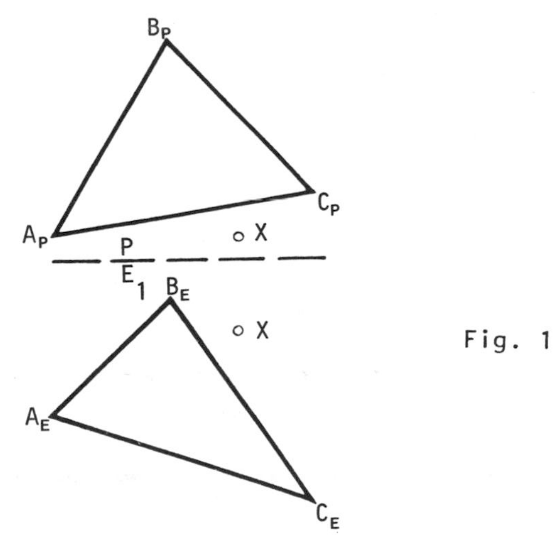

Fig. 1

E_1 is projected true length in the P- view. Lines of sight, for the auxiliary plane, are parallel to A_P-G_P in the P- view. Reference plane P-E_2 is perpendicular to these lines of sight. Find the edge view of plane A-B-C by obtaining the point view of line A_P-D_P. Project point X into view E_2. In E_2, line X-D is drawn from X perpendicular to the plane, and its true length and slope can be measured. Since X-D is true length in view E_2, in

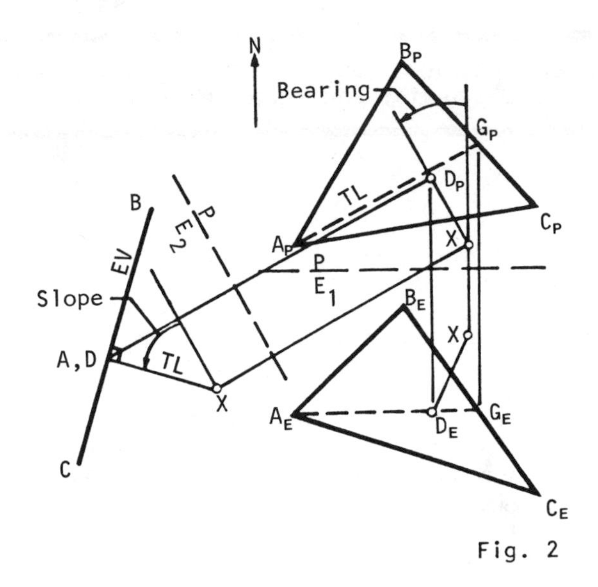

Fig. 2

the plan view it must be parallel to the reference plane P-E_2. The piercing point D can now be located in the P- and E_1- views by projection. The bearing is measured in the plan view, where A-D is true length.

• PROBLEM 5-22

Find the shortest connection from a given point, X, to a given plane, ABC, for a specified slope.

Solution: Fig. 1 shows the given plane, A-B-C, and point, X. Problem analysis reveals that the shortest connection of any specified slope from a given point to a plane lies in a vertical plane that is perpendicular to the given plane and contains the given point. In the plan (or horizontal) view, this vertical plane appears as an edge at right angles to any horizontal line on the given plane. An auxiliary elevation view showing the given plane as an edge shows this vertical plane in its true size. The vertical plane is true length since the lines of sight for this auxiliary view are parallel to true length projection of the horizontal line in the

P- view and perpendicular to the edge view of vertical plane in P- view. In this same auxiliary elevation view, a line of any specified slope drawn from the given point to the given plane determines the piercing point.

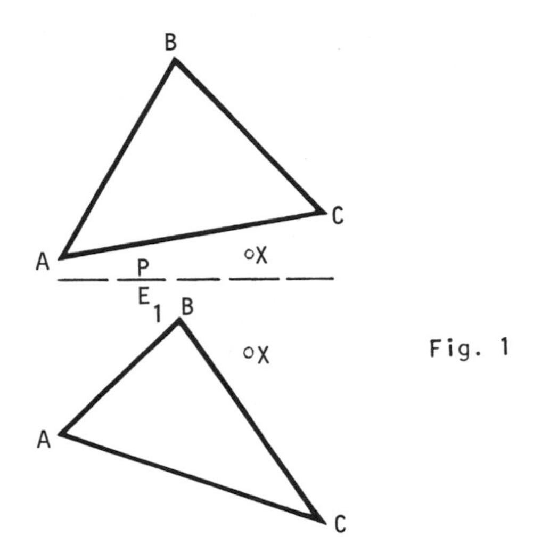

Fig. 1

Figs. 2-4 illustrate the solution to the stated problem, where the shortest connections with slopes of zero and 30° are desired. A horizontal line, AY, is drawn in view E_1 and projected to true length in the plan view. The plane ABC is then drawn as an edge in view E_2 by making the line of sight of the plan view parallel to line AY and the reference plane P-E_2 perpendicular to line AY, as in fig. 2. Point X is projected to view E_2, also. In

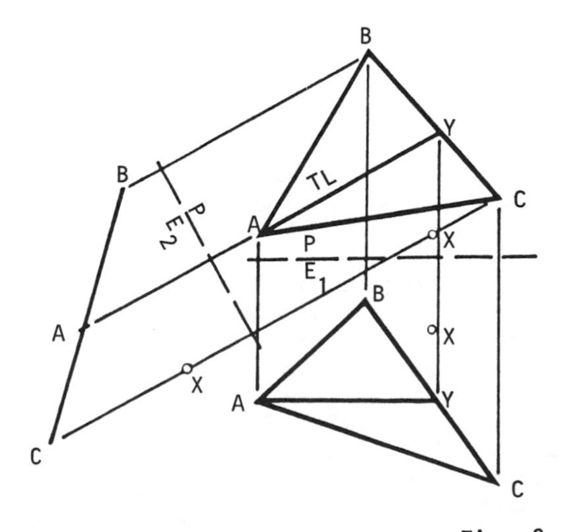

Fig. 2

fig. 3, a line parallel to reference plane $P\text{-}E_2$ is drawn through point X, intersecting plane ABC at point E. This is the shortest horizontal connection between X and ABC. A line with a slope of 30° from the horizontal is drawn through point X, intersecting plane ABC at point F. This is the shortest connection of 30° slope between X and ABC. In the plan view, a line is drawn through point X perpendicular to the true length line AY. Points E and F are projected to this view from view E_2, and where their projections intersect this line locates E and F in the plan view. These points are then projected again to view E_1,

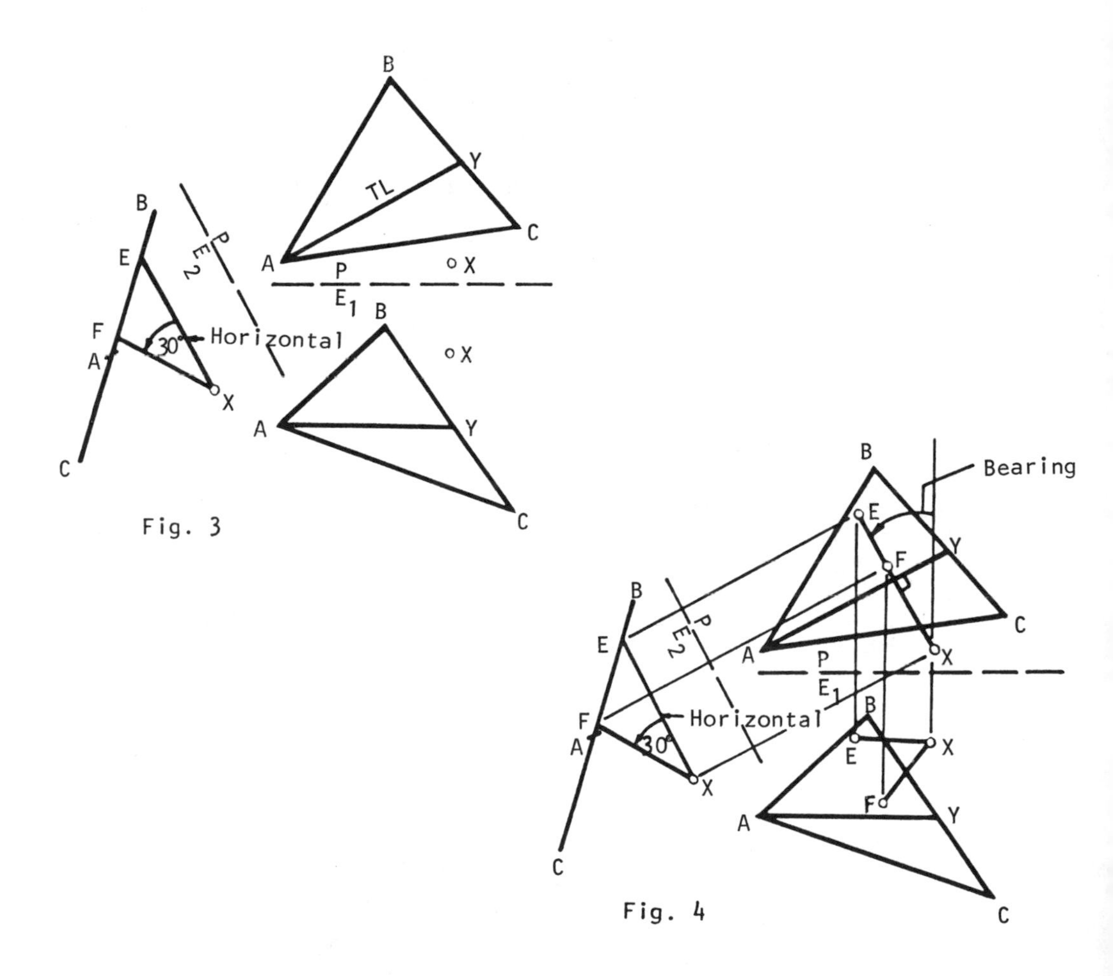

Fig. 3

Fig. 4

as in fig. 4. In the plan view, both X-E and X-F appear at 90° to the horizontal line A-Y on plane A-B-C. Consequently, both of these lines have the same bearing as the shortest possible connection. Keeping in mind that the bearing direction and bearing length are laid out in horizontal plane, the logic of this result is self-evident: all shortest connections, from a given point to a given plane, whatever their slope, must have the same bearing, since they all lie in the same vertical plane.

• PROBLEM 5-23

Find the shortest connection from a given point, X, to a given plane, A-B-C, for a specified bearing.

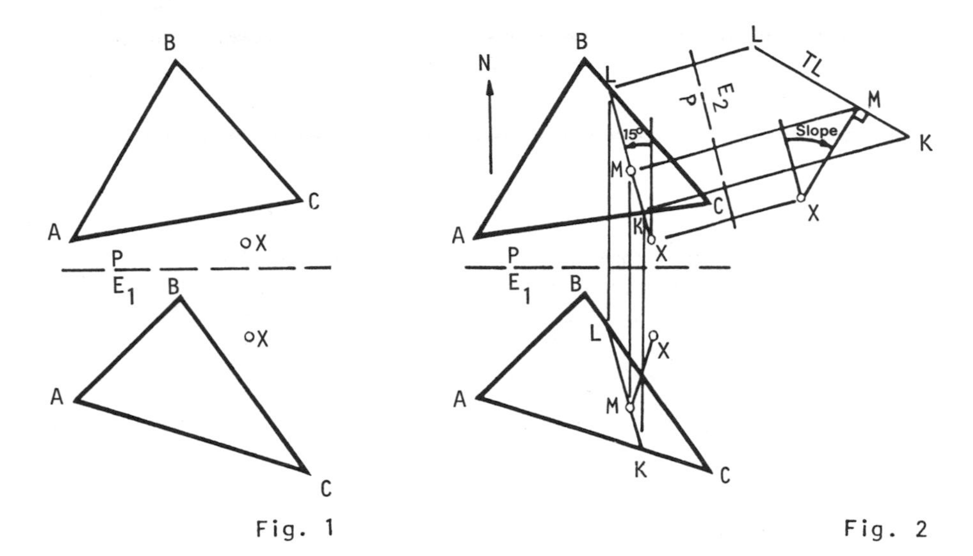

Solution: Fig. 1 shows the given plane, A-B-C, and given point, X, in the plan and primary elevation views. Problem analysis and the previous problem reveal that all lines of a specified bearing through a given point lie in the same vertical plane. The line of intersection between this vertical plane and the given plane contains the piercing points of all lines with the specified bearing. The shortest connection of a specified bearing is at right angles to this line of intersection. Fig. 2 illustrates the procedure for finding the shortest connection with a bearing of N15°W from point X to plane A-B-C. In the plan view, a line is drawn at an angle of 15° toward the west from the north. A vertical plane containing this line cuts across plane A-B-C along line K-L. Project K-L into view E_1. The edge view of said vertical plane is projected into view E_2, which has lines of sight perpendicular to the edge view. Consequently, E_2 shows the true size and shape of the vertical plane containing points X, K, and L. Since all lines on the vertical plane appear true length in view E_2, the shortest connection, X-M, is drawn perpendicular to K-L and its true length and slope are measured. Line X-M is then projected into the given views. Applying the principles from this problem and the previous one, the shortest connection of specified bearing and slope can be found by constructing a line with the specified slope in view E_2, and finding the point at which it strikes the line of intersection, K-L.

• PROBLEM 5-24

Find a point on a given line, K, which is equidistant from two given exterior points, A and B.

Solution: Geometrical loci are often useful in devising a solution for a problem. A locus is a geometrical element which contains every point that satisfies a given condition, and no other points. Thus, the locus of points equally distant from two given points is a plane which passes through the midpoint of the line segment joining the given points, and is also perpendicular to the line.

When the element required is a point, as in the stated problem, it may be determined by the intersection of loci. That is, the locus of points equidistant from points A and B is a plane, Q, which is perpendicular to the line A-B at its mid-point and the locus of points on the given line K, is the line K. The point P, in which line K cuts plane Q, is the intersection of the two loci and therefore is the required point .

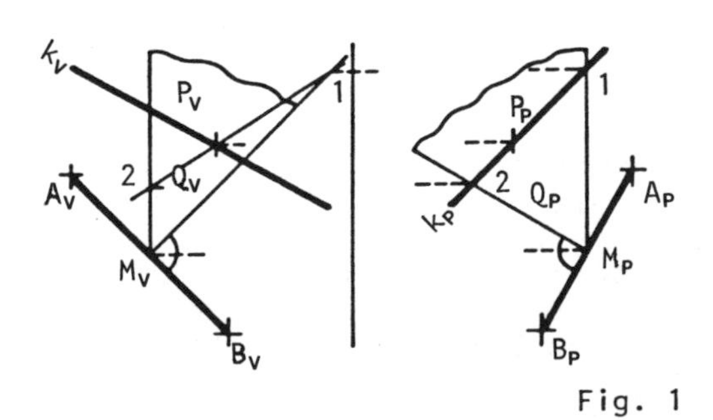

Fig. 1

Fig. 1 illustrates one approach to the problem, utilizing only two views, the horizontal and the profile views. Draw line A-B, connecting the two given points. Locate point M, the midpoint of A-B. Pass plane Q through point M, perpendicular to A-B. Locate the piercing point, P (the required point) in which line K pierces plane Q.

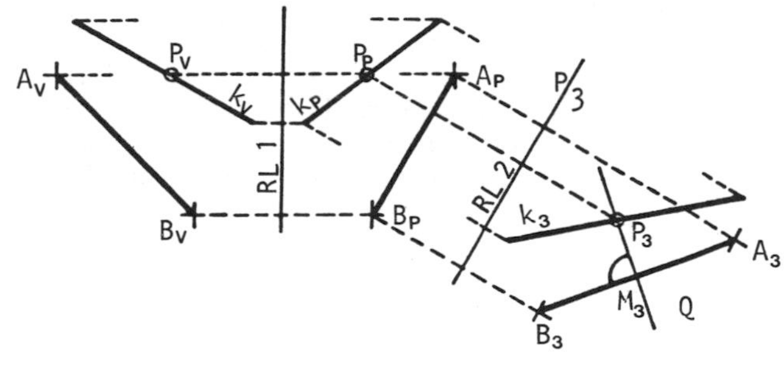

Fig. 2

Fig. 2 illustrates a second approach to this same problem, introducing an auxiliary projection plane. The auxiliary projection plane, P-3, is constructed parallel to line A-B. Consequently, line A-B projects true length in view 3. The same plane, Q, is constructed perpendicular to A-B at the midpoint M. Plane Q projects as an edge in view 3 since the projection plane is perpendicular to line 1-M in view P (see fig. 1). Line K intersects plane Q at P_3, the piercing point. Project point P (the required point) to all views.

• PROBLEM 5-25

Draw a plane that contains the given point X and is perpendicular to the given line AB.

Solution: Fig. 1 shows the given point, X, and line, A-B, in the P (plan or horizontal) and E_1 (elevation or frontal) planes of projection. Problem analysis calls for a recognition of the fact that a plane that is perpendicular to a line will appear as an edge in any view showing the true length of the line. Therefore, a plane may be constructed perpendicular to a line by obtaining a true-length view of the line and drawing, at right angles to it, a line representing an edge view of the plane. A plane may be drawn perpendicular to a line in two given views by applying the reverse of the principle employed to draw a line perpendicular to a plane. That is, when a plane is perpendicular to a given line, any line on the plane that appears true length in any view must be at right angles to the given line in that view.

Draw the horizontal line X-Y parallel to the rotation line in the E_1- view, as shown in Fig. 2. Project to the P- view. Since X-Y appears true length in the P- view, it must appear perpendicular to A-B if it is on a plane perpendicular to A-B. Likewise, frontal line X-Z is drawn parallel to the rotation line in the P- view, as shown in fig. 3. Consequently, the E_1- view projection of X-Z is true length and perpendicular to A-B. The

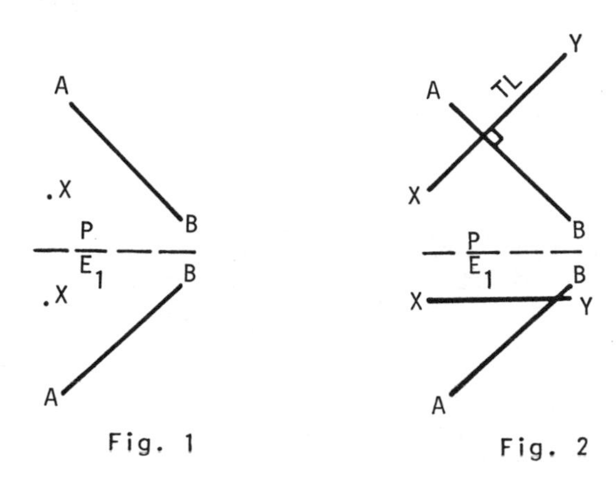

Fig. 1 Fig. 2

plane X-Y-Z thus formed in fig. 4 is the required plane, perpendicular to A-B. Note that the horizontal and frontal lines of the X-Y-Z plane are also known as the horizontal and vertical traces of the plane, respectively. The visibility of A-B and the piercing point are then determined.

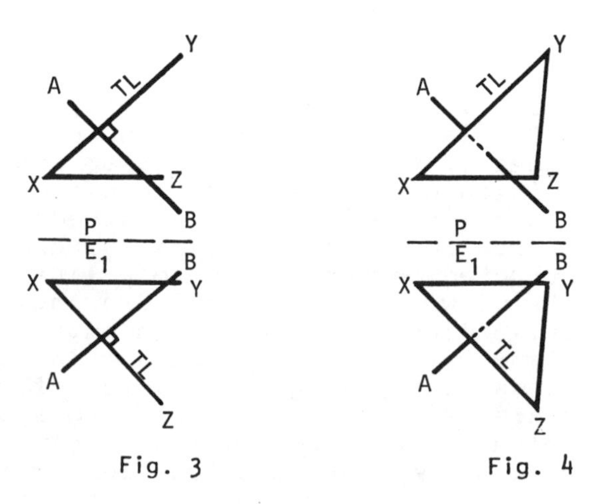

Fig. 3 Fig. 4

> Draw a plane parallel to a given line A-B through a given point C.

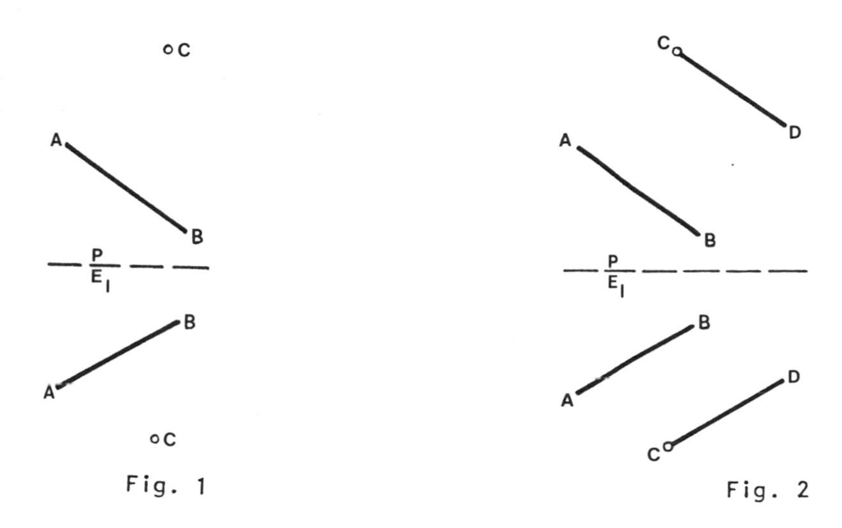

Fig. 1

Fig. 2

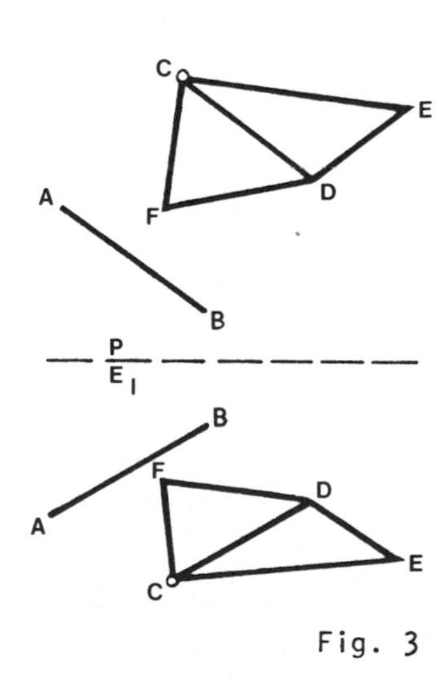

Fig. 3

Solution: Fig. 1 shows a given point C through which planes are to be drawn parallel to the given line A-B. In fig. 2, line C-D is drawn through C parallel to line A-B in both P and E_1 views (plan and elevation views, respectively). Consequently, C-D and A-B are parallel in space. Lines C-E and D-F are then added, as shown in fig. 3, to form planes C-D-E and C-D-F, both parallel to A-B. Note that any other line intersecting C-D could have been used to form a plane parallel to A-B.

Pass a plane through line A-B, perpendicular to plane 1-2-3.

Solution: Basic to the understanding of the perpendicularity of planes is the fact that a plane is perpendicular to another plane if a line in one plane is perpendicular to the other plane.

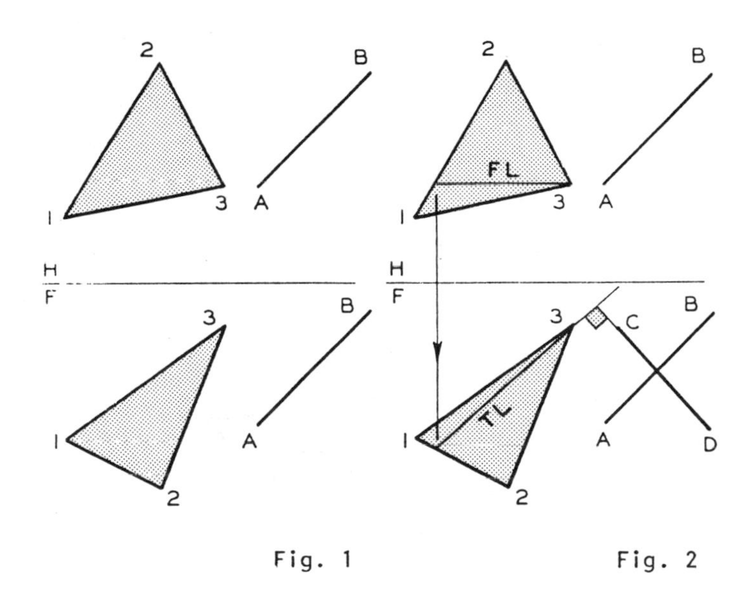

Fig. 1 Fig. 2

Fig. 1 shows the given line A-B and plane 1-2-3 in the horizontal and frontal planes of projection. A plane can be passed through a line if the line lies in the plane that is constructed; therefore, an infinite number of planes can be established through line A-B by intersecting it with another line. Since two intersecting lines form a plane, if the line drawn to intersect line A-B were drawn perpendicular to plane 1-2-3, the plane formed would be perpendicular to plane 1-2-3.

A true-length line is found in the front view of plane 1-2-3 in fig. 2 by constructing a frontal line in the top view and projecting it to the front view. Line C-D is drawn through a convenient point on line A-B, perpendicular to the extension of the true-length line in the front view. Note that the true size of a right angle formed by two intersecting lines will appear in any view showing either line in true length.

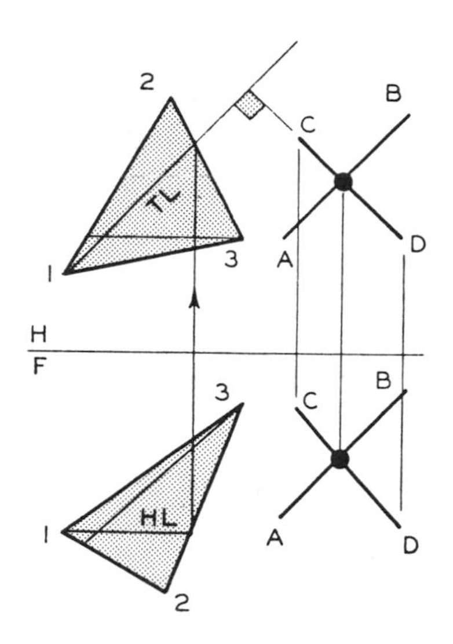

Fig. 3

A horizontal line is constructed in the front view of Fig. 3, and projected to the top view, where it is true length. The top-view projection of line C-D is drawn perpendicular to this true-length line which goes through the top view of the point on line C-D.

Line C-D has been constructed perpendicular to two intersecting lines on plane 1-2-3 and is known to be perpendicular to the plane. This is a direct result of the fact that a line perpendicular to a plane is perpendicular to all lines on the plane. Lines A-B and C-D form the required plane which passes through A-B and is perpendicular to plane 1-2-3.

• PROBLEM 5-28

> Draw a line which is parallel to a given plane formed by lines a and b, and intersects two given lines, d and e.

Solution: Fig. 1 shows the given lines in the horizontal and frontal planes of projection. The solution readily follows after we recognize that the required line lies in a plane parallel to plane a-b and is determined by the points at which this plane cuts lines d and e. Naturally, an infinite number of lines can be drawn which will satisfy the stated conditions; therefore, any point may be assumed on d and a line drawn through it parallel to

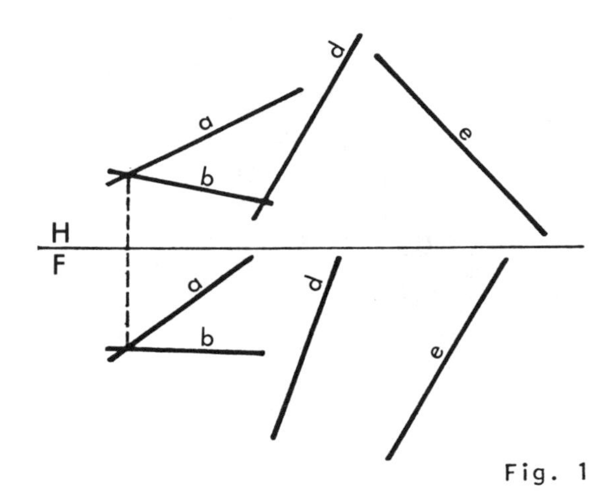

Fig. 1

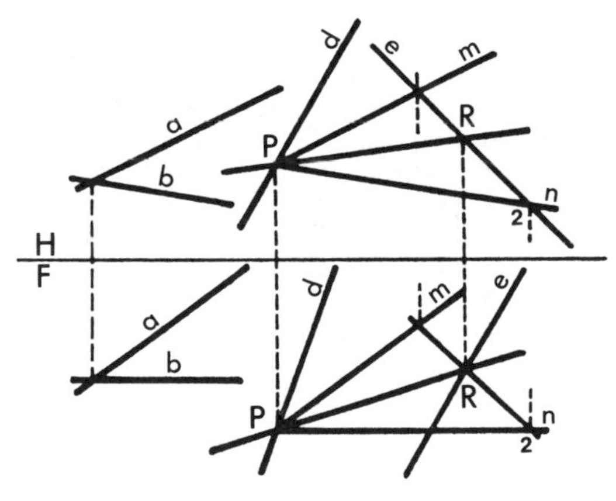

Fig. 2

plane a-b and cutting line e. As shown in Fig. 2, assume point P on line d. Through point P, draw lines m and n parallel to lines a and b, respectively. Show these lines in both views. Lines M and N and point P form a plane, Q. Where this plane cuts the line e, at point R in fig. 2, determines the second point needed to draw the required line intersecting lines d and e. Line PR is this required line. Note that the aggregate of all lines intersecting d and e and parallel to plane a-b is a curved surface called the hyperbolic paraboloid.

• PROBLEM 5-29

Construct a line perpendicular to a plane, given its piercing point in orthographic projection.

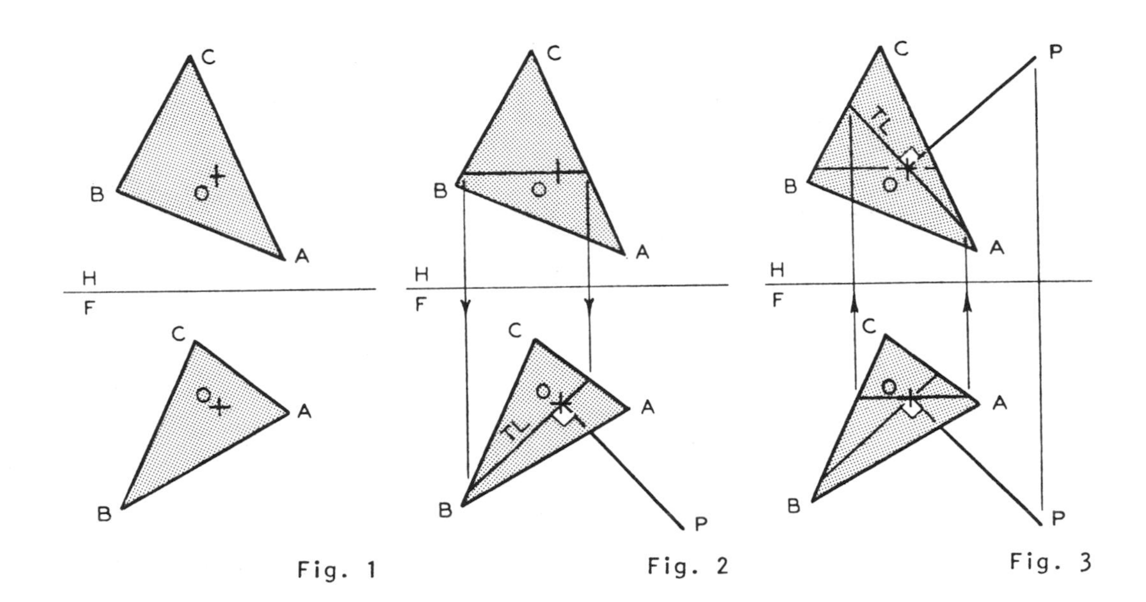

Solution: Plane A-B-C is given in fig. 1 with piercing point O located on the plane. Problem analysis reveals that a line will be perpendicular to a plane if it is perpendicular to two intersecting lines on the plane. In other words, if a line is perpendicular to two intersecting lines in a plane, it will be perpendicular to all lines in the plane. Furthermore, in any view, a line perpendicular to a plane appears perpendicular to any line on the plane that appears true length in that view. Locate a true-length line in the front view by constructing a frontal line through point O in the top view. Project the frontal line to the front view, as shown in fig. 2. The line marked TL is true length and is a line on the plane; consequently, line O-P can be drawn through point O perpendicular to the TL line in the F- view. Since line O-P is to be perpendicular to another line as well, draw a horizontal line on the plane in the front view, as shown in fig. 3. Project this line to the top view, where it appears TL. The top view of O-P is constructed perpendicular to this line. Line O-P is the required perpendicular to the plane because it is perpendicular to two intersecting lines on the plane.

• **PROBLEM 5-30**

Construct a line through a given point, O, making a given angle, 45°, with a given line, 1-2.

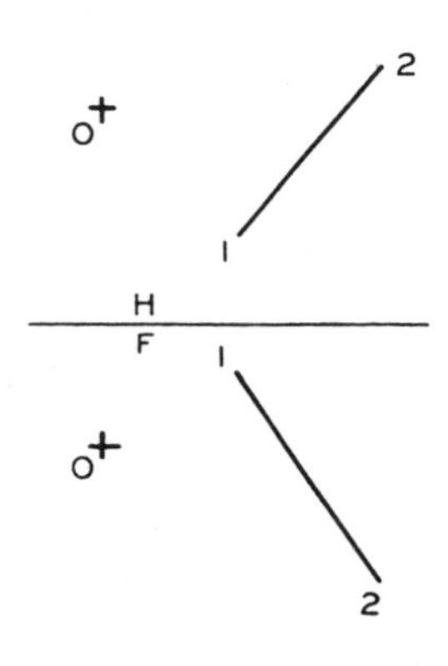

Fig. 1

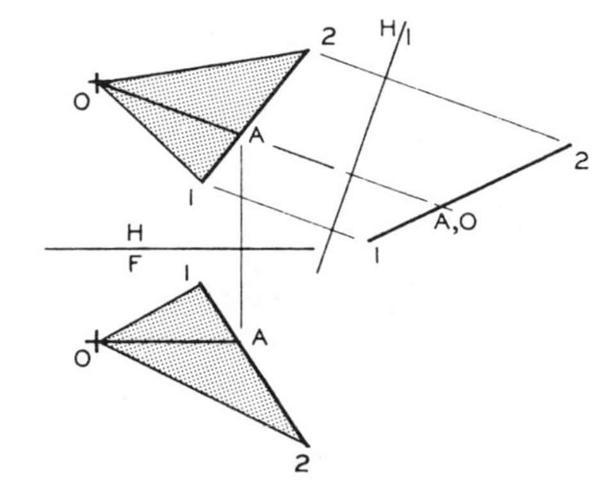

Fig. 2

Solution: Fig. 1 shows the given point, O, and line, 1-2, in the H-, horizontal, and F-, frontal, planes of projection. Referring to fig. 2, connect point O to each end of the line to form plane 1-2-O in both views. Draw a horizontal line in the front view of the plane and project it to the top view, where it is true length. Determine the edge view of the plane by finding the point view of line O-A; project the plane into auxiliary view 1, which has lines of sight parallel to the true length line O-A. Determine the true size of plane 1-2-O,

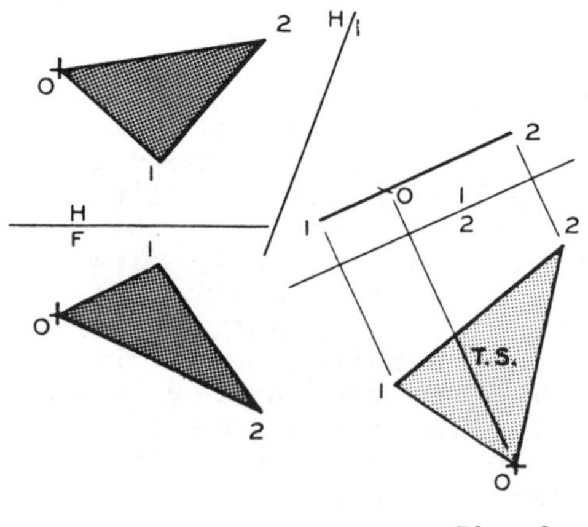

Fig. 3

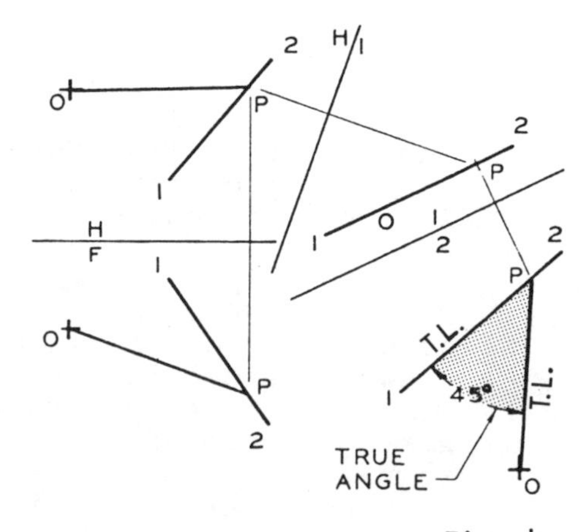

Fig. 4

as shown in Fig. 3 by establishing an auxiliary view having lines of sight perpendicular to the edge view of plane 1-2-O. The plane projects true shape-and-size in this secondary auxiliary view. Referring to fig. 4, line O-P can be constructed at the specified angle, in this case 45°, with the line 1-2 in the secondary auxiliary view, 2. This construction is based on the fact that both 1-2 and O-2 are true length in view 2; both lines lie in the same true-size-and-shape plane, 1-2-O. Furthermore, the true size of any angle, except a right angle, between two intersecting lines will appear only in a view showing both lines in true length. Project point P back to the primary auxiliary, top, and front views, and connect it with point O. Note that this problem could have been solved by projecting from the front view as well.

• PROBLEM 5-31

Find the locus of a line making given angles with:

(A) Intersecting lines

(B) Parallel lines

(C) Nonintersecting, nonparallel lines.

Solution: (A) The two intersecting lines define a plane in which the required line must lie. To make a solution possible, the sum of the two given angles must be equal to the supplement of the angle between the two given lines.

(B) The required line must lie in the plane defined by the parallel lines. A solution is possible only if the two given angles are equal (or supplementary) angles.

(C) In any plane, a given line makes equal angles with all parallel lines. Therefore, any line that makes a given angle with one of the given lines will make the same angle with a parallel line through the other given line. This principle makes it possible to use two cones of revolution with the same vertex when the two given lines do not intersect.

In the figure, lines A-B and C-D, given in the P (plan) and E_1 (primary elevation) views, are to be connected by a line that makes angles of 51° with A-B and 40° with C-D. The cones that show the possible positions of lines making these angles with the given lines can be drawn only in a view that shows both lines in their true

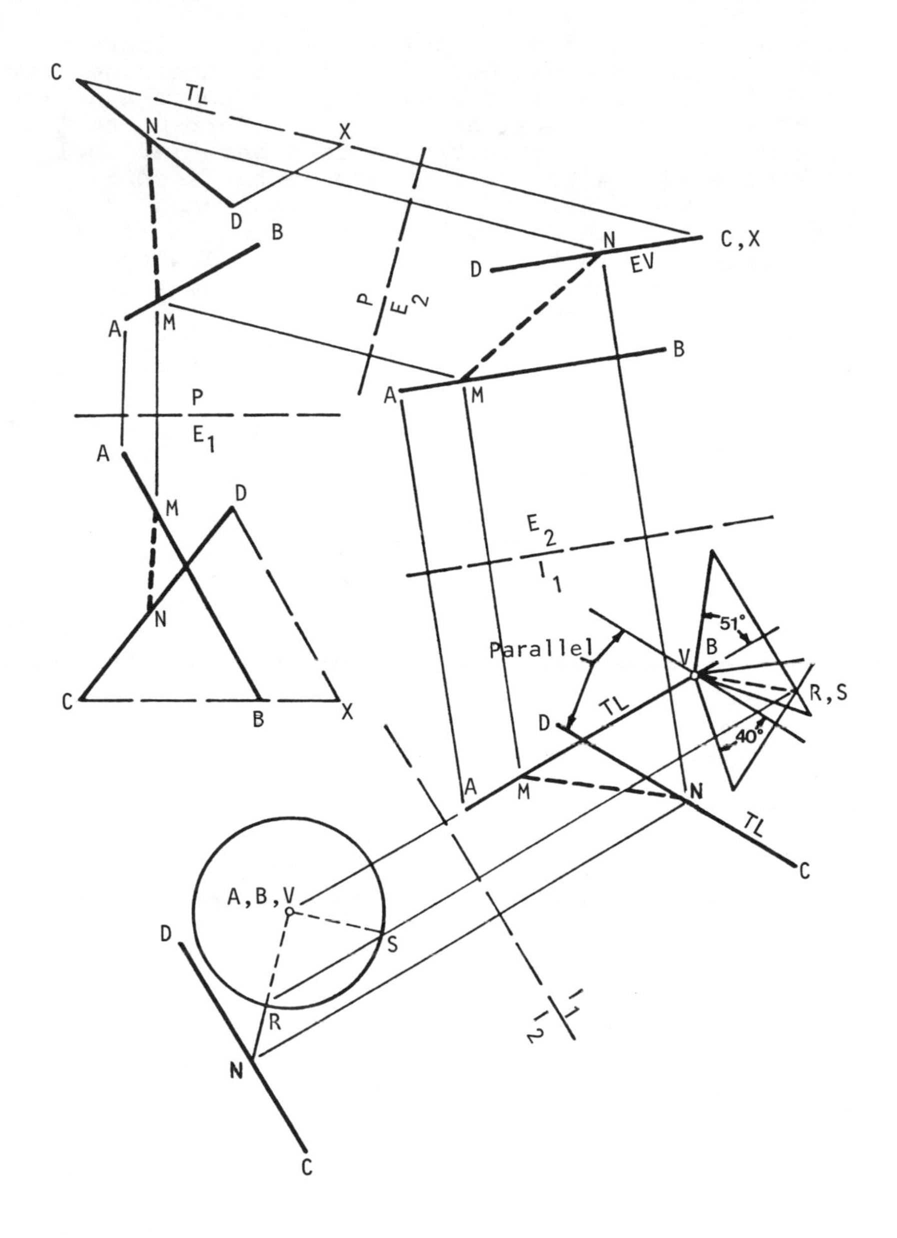

length. Therefore, plane C-D-X is constructed parallel to A-B by drawing D-X parallel to A-B. C-D-X appears as an edge in view E_2, the secondary elevation view, since the lines of sight for E_2 are parallel to the true length line C-X. View I_1, the primary inclined view, is then drawn having lines of sight perpendicular to the edge view of C-D-X and line A-B, to show both A-B and C-D in true length. Point V is selected at any point on line A-B, and a 51° cone is constructed with A-B as an axis and V as a vertex. A line parallel to C-D is drawn through V, and a 40° cone is constructed with this line as an axis and V as a vertex. View I_2, the secondary inclined view, is drawn to show A-B as a point (i.e., lines of sight for I_2 are parallel to the true-length line A-B) and the base of the 51° cone as a circle. Note: The diameter of the circle is equal to the distance across the base and not to the length of the elements. The bases of the two cones are found to intersect at R and S, thus determining the elements V-R and V-S common to both cones.

It can now be seen that neither V-R nor V-S actually intersects line C-D. This was to be expected, since it is unlikely that point V would be accidentally selected at the proper point. However, the direction of lines making the proper angles with A-B and C-D has been determined by the intersection of the cones. The correct position can now be determined by sliding the cones along A-B until a common element intersects C-D. Therefore, a plane that appears as an edge in view I_2 is passed through line V-R, and C-D is seen to pierce this plane at N.

Point N is projected into view I_1, and a line parallel to V-R is drawn in this view, intersecting A-B at M. This procedure is equivalent to sliding the cones along A-B until the vertex is at M, although the cones are not actually drawn in this position. The line M-N is the required line, which can now be located in the given views by projection. Note that if C-D were extended, an alternate solution could be obtained by passing a plane through line V-S in view I_2.

• PROBLEM 5-32

Find the locus of a line making given angles with two perpendicular planes.

Solution: As shown in the accompanying figure, a line 2 feet long is to be drawn in a southerly direction from M so that it makes an angle of 45° with the given vertical plane and an angle of 30° with the horizontal plane. Problem analysis calls for the following discussion. The locus of a point or line is the assemblage of all possible positions of the point or line that satisfy a given set of geometric conditions. The locus of the point or line might also be defined as the line or surface formed when the point or line moves in a specified manner. The locus of a line that makes a specified angle with a given plane is a cone of revolution with its axis perpendicular to the plane. Each of the possible positions of the line is an element of the cone. If the line is to make a specified angle with some other plane, the locus of the line will again be a cone of revolution. In such a situation, the two cones must have the same vertex and

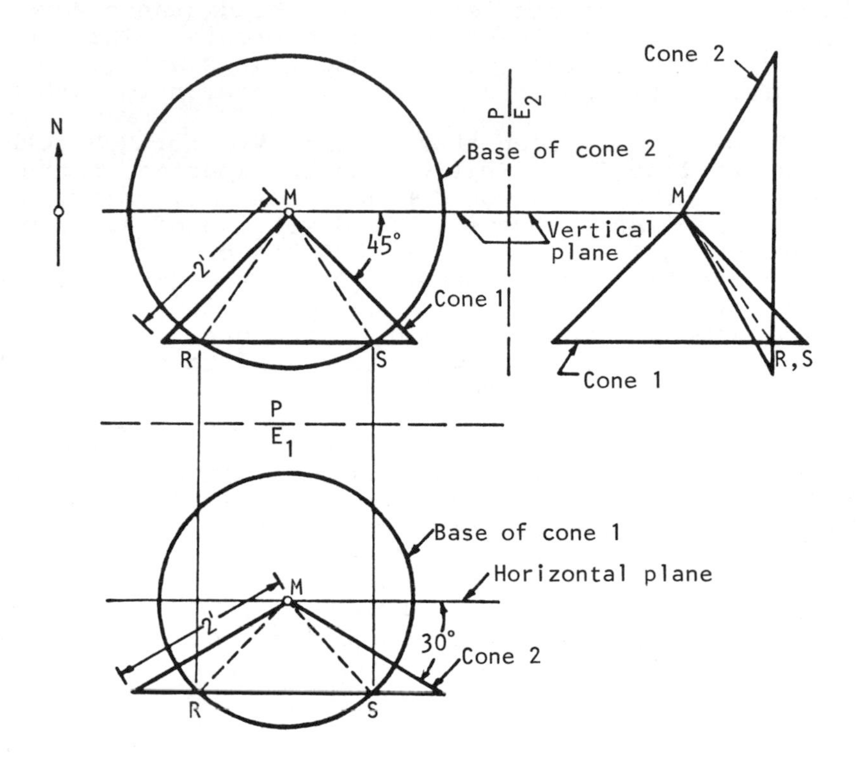

elements of equal length, since both cones are formed by the same line. The elements common to both cones satisfy both given conditions, or, in other words, these elements are the only possible positions of the required line. In the P, or plan, view, cone 1 is constructed so that its extreme elements are 2 feet long and make an angle of 45° with the given vertical plane. The base of this cone appears as an edge parallel to the vertical plane, and therefore appears as a circle in the elevation view, E_1. Similarly, cone 2 is constructed in the elevation view so that its extreme elements make an angle of 30° with a horizontal plane. This cone also has elements 2 feet long. The base of cone 2 is horizontal and appears as a circle in the plan view. From the figure, it is easy to recognize that the diameter of each circle is equal to the distance across the base of the corresponding cone, and not to the length of the elements. Points R and S, at which each circle crosses the base of the other cone, are the points at which the two bases intersect. These points should check by projection between views. Lines M-R and M-S are elements of both cones and are, therefore, the two possible positions of lines that satisfy all the given conditions. If the conditions had specified that the line extend in a southeasterly direction, line M-S would be the only possible solution. The line could also be located by drawing view E_2 showing the vertical plane as an edge. In this view, the bases of both cones appear as edges and their intersection is apparent.

• **PROBLEM 5-33**

Find the locus of a line from a given point making given angles with two nonperpendicular planes.

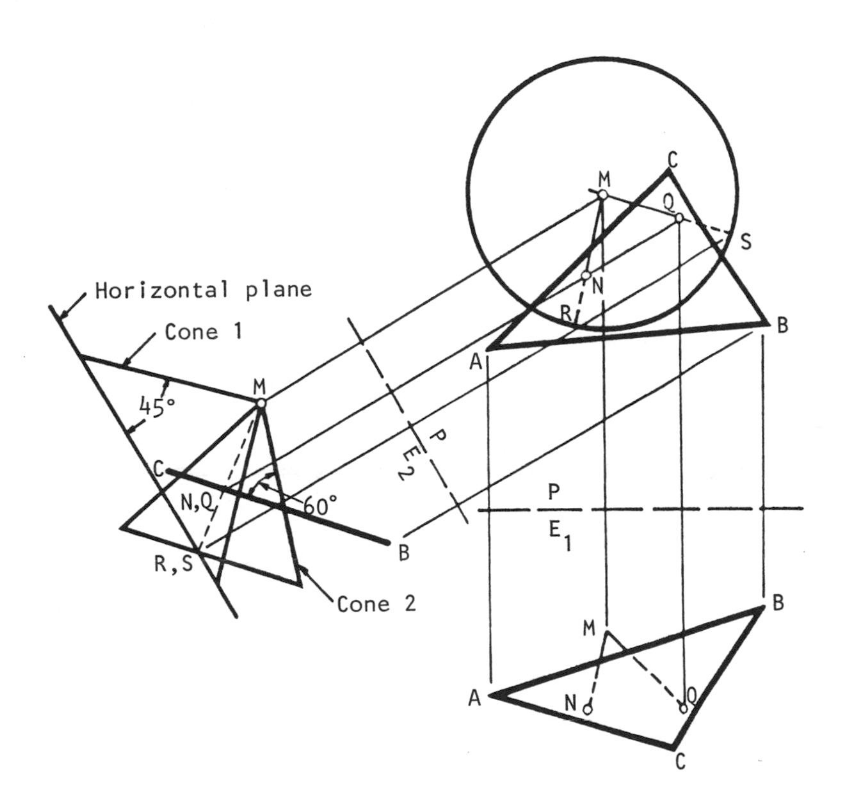

<u>Solution</u>: The accompanying figure shows the point M and inclined plane A-B-C in the P, or plan, and E_1, or primary elevation, views. Employ the cone of revolution method to draw a line from M so that it has a slope of 45° with respect to the horizontal, and makes an angle of 60° with the plane A-B-C. When the two given planes are not perpendicular to each other, a view should be drawn to show both planes as edges. View E_2 is drawn to show both plane A-B-C and a horizontal plane as edges. Point M is located in this view by projection. In view E_2, the two cones are drawn with elements of any convenient length. Note that these elements can be of any length as long as they are the same length on both cones, since both cones are formed by the same line. Cone 1 has elements that make an angle of 45° with the horizontal, and cone 2 has elements that make an angle of 60° with plane A-B-C. Since the bases of both cones appear as edges in this view, the points R and S at which they

intersect are apparent. The base of cone 1 is drawn in the plan view as a circle of diameter equal to the distance across the base of the cone. The points of intersection are projected to this circle. Lines M-R and M-S are the two possible lines that satisfy the given conditions. The points N and Q at which these lines pierce plane A-B-C are clearly seen in view E_2, where A-B-C appears as an edge. If in the figure the line had been required to make the 45° angle with some plane other than a horizontal plane, an additional view would have been necessary to show both given planes as edges in the same view.

ANGLES BETWEEN LINES AND PLANES

• PROBLEM 5-34

Determine the angle between a line and a plane, using the line method.

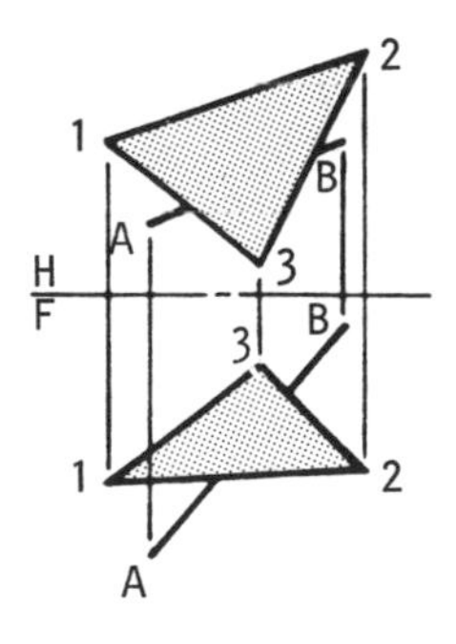

Fig. 1

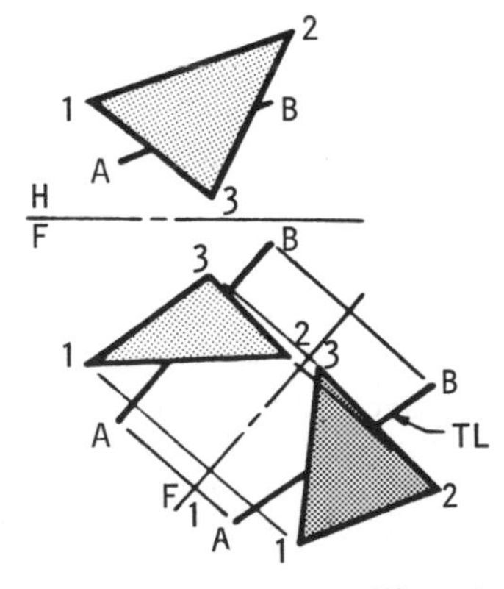

Fig. 2

Solution: Fig. 1 shows the given line, A-B, and plane, 1-2-3, in the horizontal and frontal planes of projection. The key principle involved in the solution is that the angle a line makes with a plane can be measured only when the line appears in its true length and the plane appears as an edge in the same view. The line method of solution is based on the fact that all views projected from the point view of a line will show the line in true length.

Two of these views will also show the given plane as an edge. As shown in fig. 2, determine the true length of line A-B in a primary auxiliary view by projecting from either principal view. The lines of sight for this auxiliary view are perpendicular to line A-B. Plane 1-2-3 is projected also; however, it does not appear true size-and-shape in this auxiliary view, except in a special case. A special case would imply that this view also shows the plane as perpendicularly projected from the

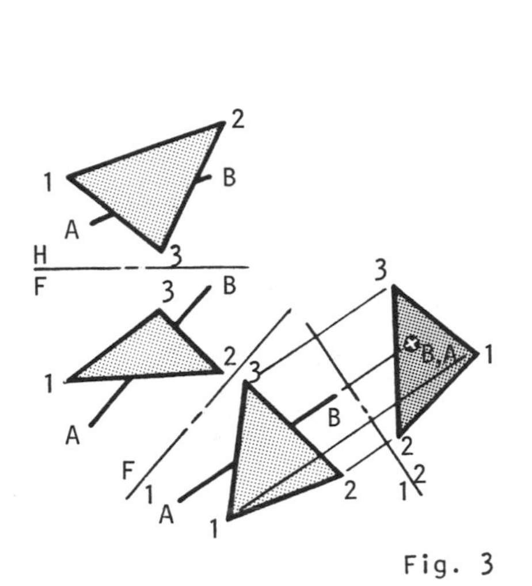

Fig. 3

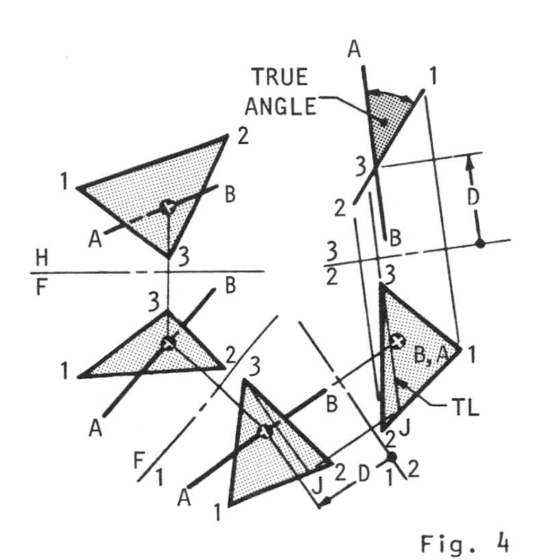

Fig. 4

edge-view of the plane. Fig. 3 illustrates the construction of the point view of line A-B in the secondary auxiliary view, 2; lines of sight are parallel to line A-B in view 1. Plane 1-2-3 does not appear true size in view 2 unless line A-B is perpendicular to the plane. Note also that the point view of the line in view 2 is also the piercing point on the plane. In order to locate the edge view of the plane in the third auxiliary view, 3, construct a true length line, J-3, on plane 1-2-3 in the secondary auxiliary view. The lines of sight for view 3 are parallel to this line. Line A-B will be true length in this view, since it appeared as a point in the secondary auxiliary view. Measure the angle in the third auxiliary view and determine the piercing point and visibility in the previous views.

• PROBLEM 5-35

Determine the angle between a line and a plane, using the plane method.

Solution: Fig. 1 shows the given plane, 1-2-3, and line A-B in the horizontal and frontal planes of projection. The angle a line makes with a plane is defined as the smallest angle between the line and the plane. Consequently, this angle must be measured in a plane that is perpendicular to the given plane. In other words, the angle a line makes with a plane can be measured only when the line appears in its true length and the plane appears as an edge in the same view. Although edge views of the plane can be obtained from any given view, it is unlikely that the line will appear in its true length in these views. However, all views projected from the true-shape-and-size view of a plane will show the plane as an edge, and two of these views will also show the true length of the line. Determine the edge view of plane 1-2-3 by projecting from

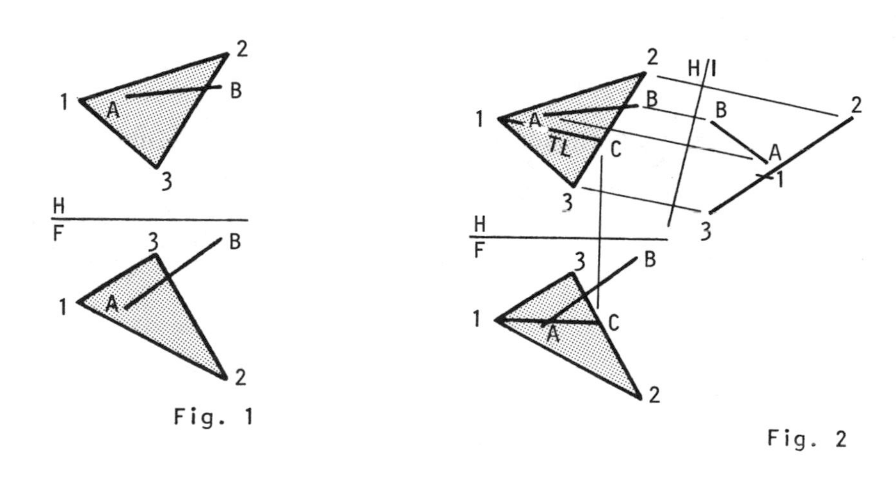

Fig. 1

Fig. 2

either the F- or H-view. Fig. 2 shows the edge view by projecting from the H-view to the first auxiliary view, 1. The horizontal line 1-C in the F-view is true length in the H-view. Establish lines of sight parallel to 1-C in the H-view. The first auxiliary plane is perpendicular to these lines of sight and shows plane 1-2-3 as an edge. Project line A-B from the H-view to the auxiliary view.

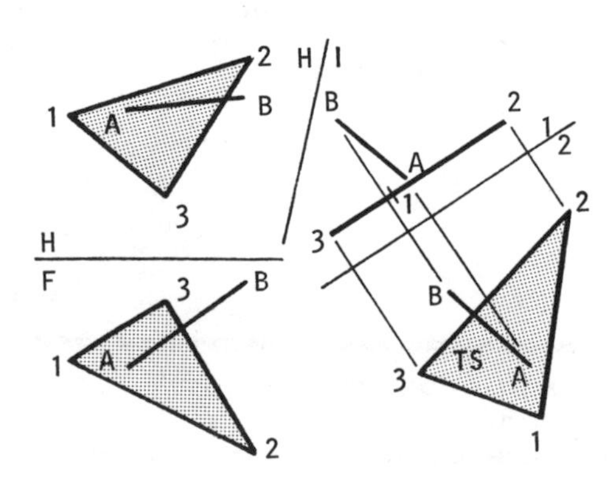

Fig. 3

The angle cannot be measured in this view since the line is not true length. As illustrated in fig. 3, determine the true size of plane 1-2-3 in a secondary auxiliary view, 2, projected perpendicularly from the edge view of the plane. Again, line A-B is not true length in this view. Recalling that a view projected in any direction from a true-size-and-shape view of a plane will result in an edge view of the plane, and that line A-B must appear

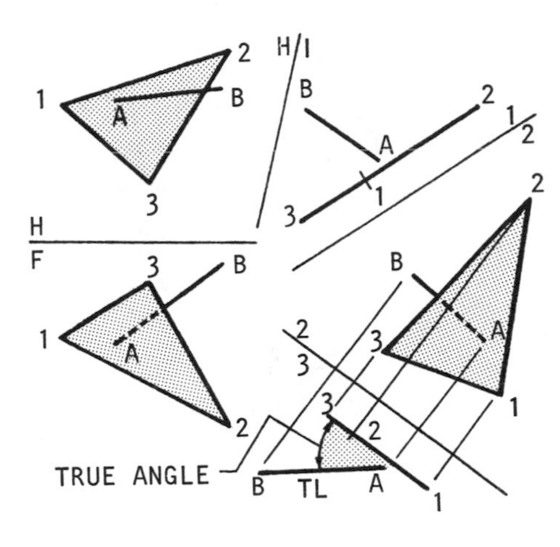

Fig. 4

true length in order to measure the required angle, project, as in Fig. 4, a third auxiliary view perpendicularly from line A-B in the second auxiliary view. The line appears true length (TL) and the plane appears as an edge in view 3. Measure the true angle in view 3. Visibility is shown in all views.

• PROBLEM 5-36

Determine the angle between a line and a plane, using the complementary angle method.

Solution: Fig. 1 shows the given plane, A-B-C, and line, M-N, in the horizontal and frontal planes of projection. The key principle involved in the solution is that the angle a line makes with a plane can be measured only when the line appears in its true length and the plane appears as an edge in the same view. The complementary angle method of solution is based on the fact that the angle between a line and a plane is defined as the angle between

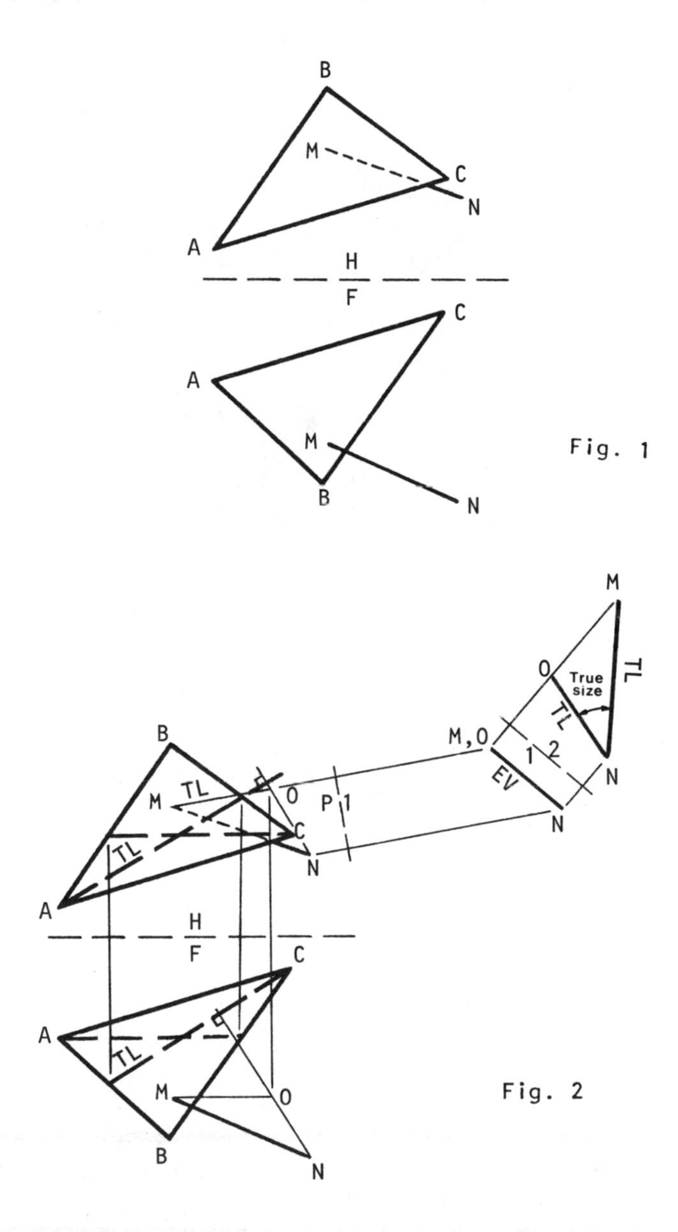

Fig. 1

Fig. 2

the line and its projection on the plane. Consequently, the angle between a line and a perpendicular to a plane is the complement of the angle the line makes with the plane. As shown in fig. 2, a perpendicular through point N to the given plane is constructed in the F- and H-views by means of the cutting plane method. The line M-O, shown as horizontal in the F-view, is drawn on the plane formed by the perpendicular and line M-N. As a result, M-O appears true length in the H-view. Auxiliary view 1 is drawn to show plane M-N-O as an edge. Auxiliary view 2, having lines of sight perpendicular to the edge view of M-N-O, shows plane M-N-O in true size-and-shape. The complementary angle, M-N-O, can then be measured and its value subtracted from 90 degrees to obtain the angle between line M-N and plane A-B-C. The true size of the angle a line makes with a plane can also be found by projecting the line onto the plane and then constructing the true size of the plane formed by the line and its projection on the plane.

• PROBLEM 5-37

Determine the dihedral angle between two planes using the true-size-of-plane method.

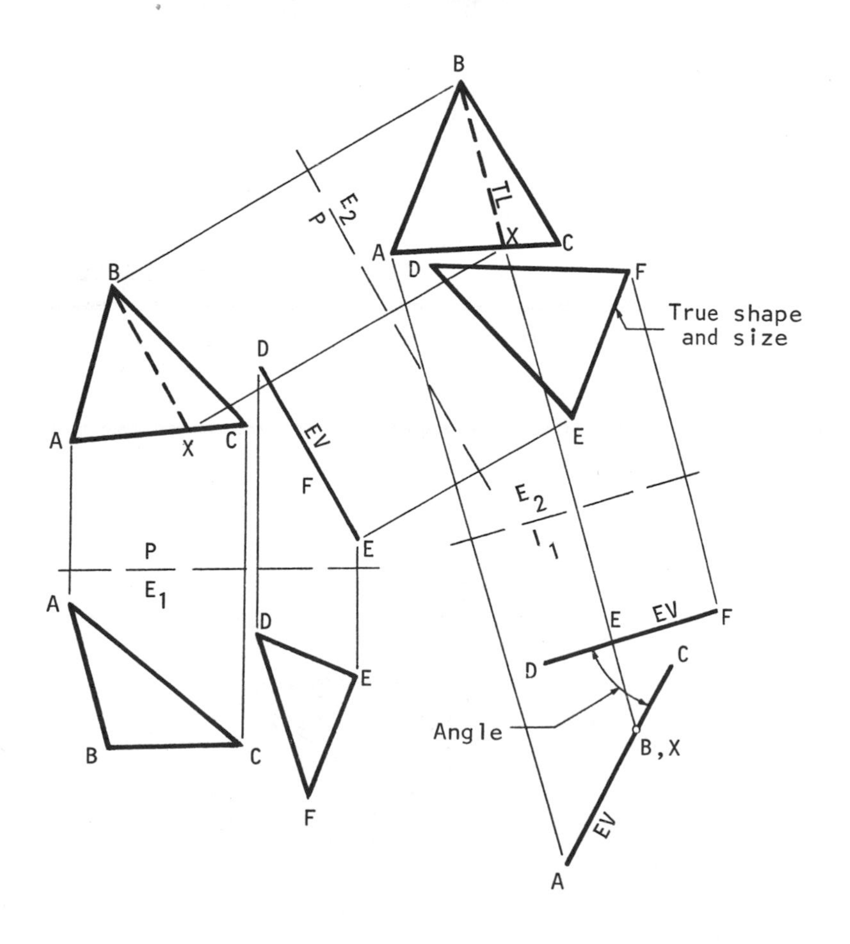

Solution: Problem analysis reveals that any view projected from the true-shape-and-size view of a plane will show the plane as an edge. Furthermore, an edge view of a plane may be projected from any view of the plane. Consequently, when a view shows one of the planes in true shape and size, another view may be projected from it to show both planes as edges.

In the accompanying figure, planes A-B-C and D-E-F are given in the P, plan, and E_1, primary elevation, views and the angle between them is desired. Since D-E-F appears as an edge in the plan view, view E_2, the secondary elevation, is drawn having lines of sight perpendicular to this edge view and thus showing D-E-F in true shape and size. A line is drawn from B to X in the plan view, parallel to the rotation line $\frac{P}{E_2}$, so that B-X appears true length in view E_2. View I_1, the primary inclined view drawn to show B-X as a point, shows both planes as edges. The true size of the angle between two intersecting planes can be seen only in a view that shows both planes as edges. View I satisfies this requirement by showing the two planes in edge view and here the required dihedral angle can easily be measured.

If plane D-E-F had not been given as an edge, an additional view showing either of the planes as an edge would have been required before the foregoing procedure could be followed.

• PROBLEM 5-38

Determine the dihedral angle between planes G-K-L and M-N-O, using the edge-view (or line of intersection) method.

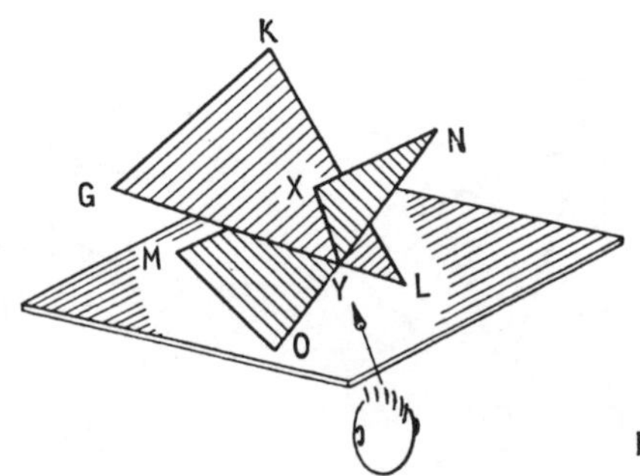

Fig. 1

Solution: Fig. 1 shows a spacial analysis of the stated problem. The line of intersection of two planes is a line common to both planes. When this line is viewed as a point, both planes appear as lines or edges. The dihedral angle between the planes is the angle between these edge lines. Fig. 2 shows the two given planes in the horizontal and frontal planes of projection, the H- and F-views, respectively. Fig. 3 shows the location of the line of intersection, X-Y, by means of the cutting plane method (the edge-view method would be equally valid). Fig. 4

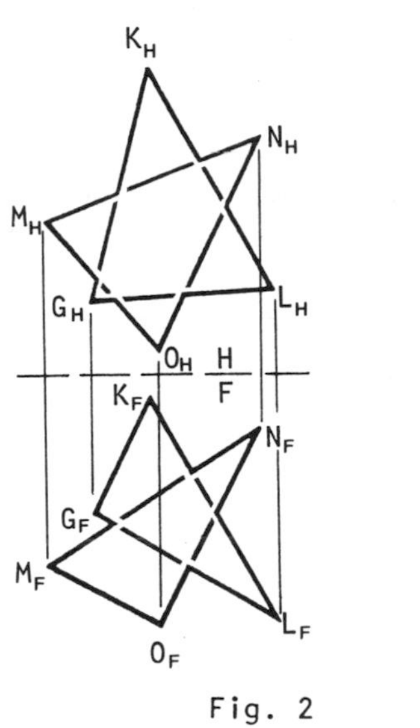

Fig. 2

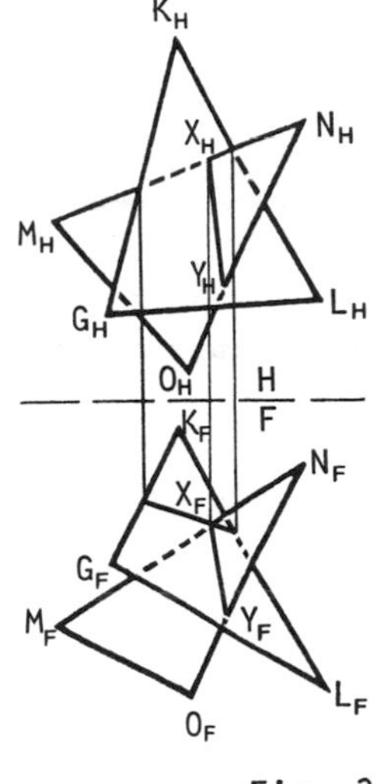

Fig. 3

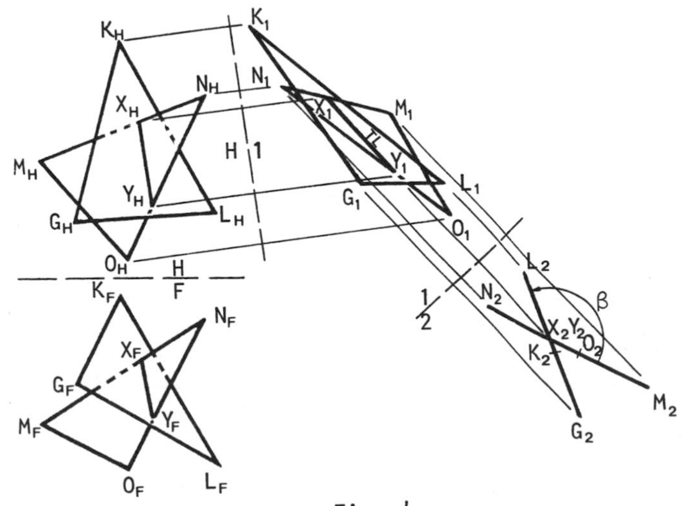

Fig. 4

shows a view of X-Y as a point, (X_2, Y_2). To show a line as a point, it is necessary to have the line of sight parallel to the true length view of the line and a reference plane perpendicular to the line of sight. View 1 shows X-Y as true length. View 2, perpendicular to view 1 and X_1-Y_1, is the required view showing X-Y as a point, (X_2,Y_2). Both given planes show as edges in view 2. Measure the angle β, the dihedral angle, between the edge views.

• PROBLEM 5-39

Determine the dihedral angle between two given planes using the line method.

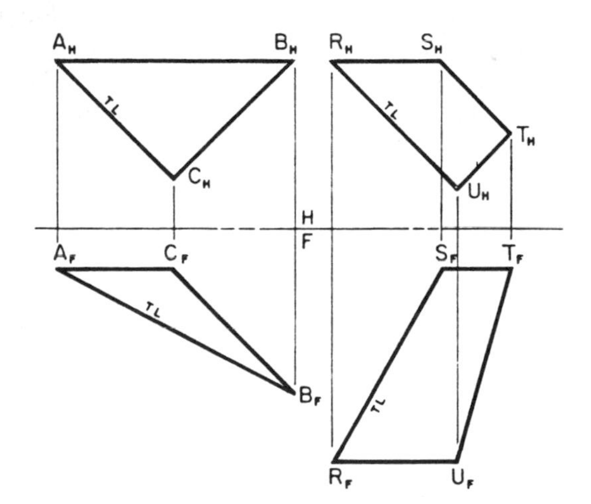

Fig. 1

Solution: Fig. 1 shows the two given planes, A-B-C and R-S-T-U, in the horizontal and frontal planes of projection. The key principle involved in the solution is that the true angle between two planes is the supplement of the angle between two lines respectively perpendicular to the planes from any point in space. Select a convenient point, P, as shown in fig. 2. Drop perpendiculars, using either the cutting-plane or edge view method (the former is shown in fig. 2), from P to each plane, P-X and P-Y. Measure angle θ in auxiliary view 1,

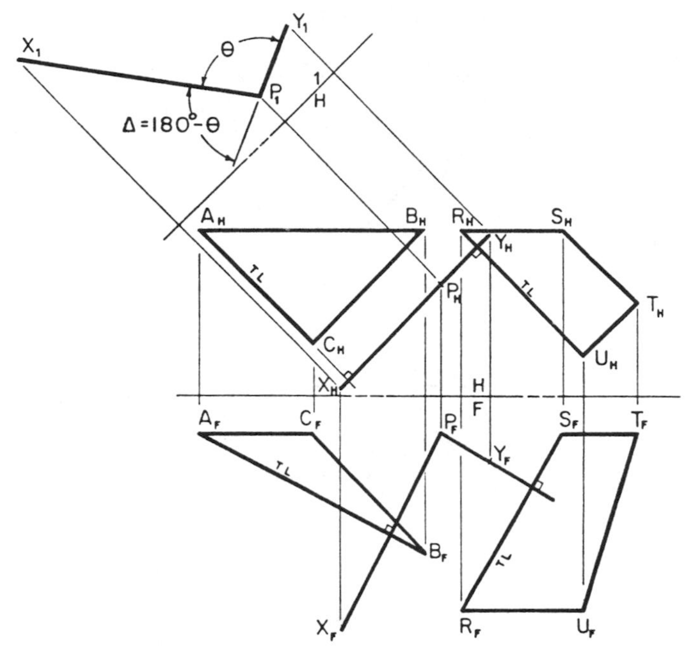

Fig. 2

where P_1-Y_1 and P_1-X_1 are true length. The latter are true length since the lines of sight for view 1 are perpendicular to the edge view of plane X_H-P_H-Y_H. Subtract θ from 180 degrees to give the dihedral angle, Δ.

• PROBLEM 5-40

Locate the following details upon a plane:

(A) A hole for a 24 inch circular, vertical stack centered at point P, is to be cut in roof A-B. Find the H and F projections and a scale size template of the area to be cut on the roof's surface.

(B) A given frustum of a right square pyramid is to be erected upon plane R-S-T. The base is to be on the upper side of the plane with its center point equidistant from R, S, and T, and with one diagonal of the base parallel to R-S.

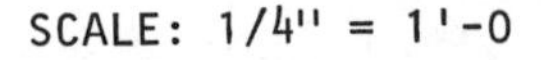

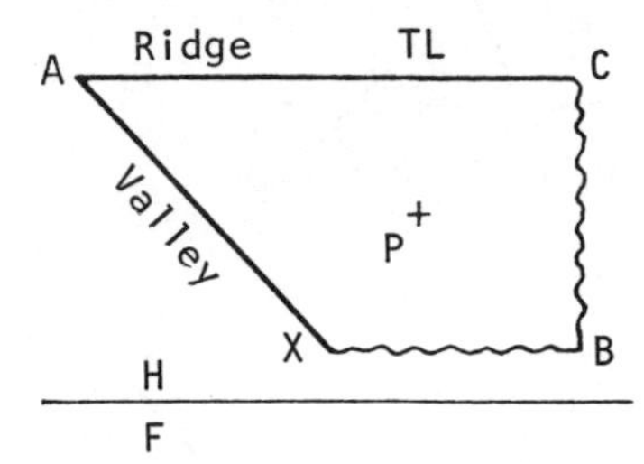

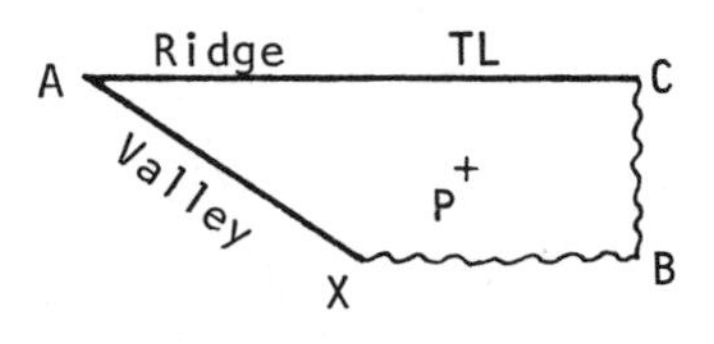

Fig. 1

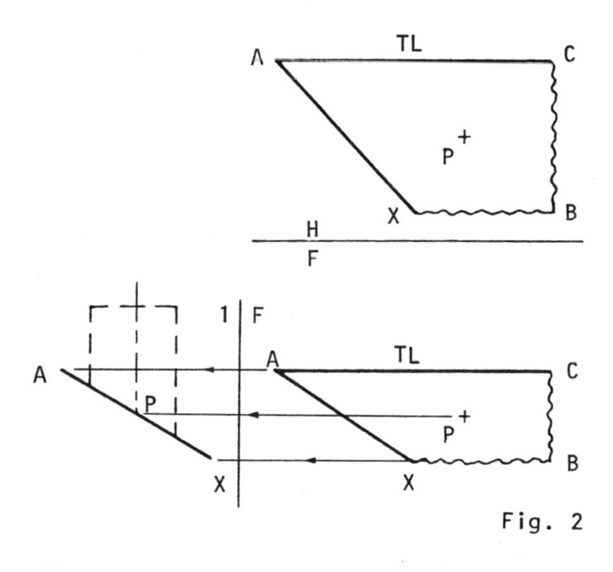

Fig. 2

Solution: (A) Fig. 1 shows the front and top views of the plane of the roof, on which a hole is to be cut for a circular pipe. The required template is the line of intersection between a plane (the roof plane) and a cylinder (the pipe). Because the roof plane is not perpendicular to the axis of the pipe, the line of intersection which outlines the hole to be cut will be an ellipse. Line AC appears parallel to the H-F reference plane in both views and therefore is true length in both views. Point P is to be the center of the hole. The template must be made from a true shape and size view of the plane of the roof, and this view must be projected from an edge view. As in fig. 2, make the F-1 reference

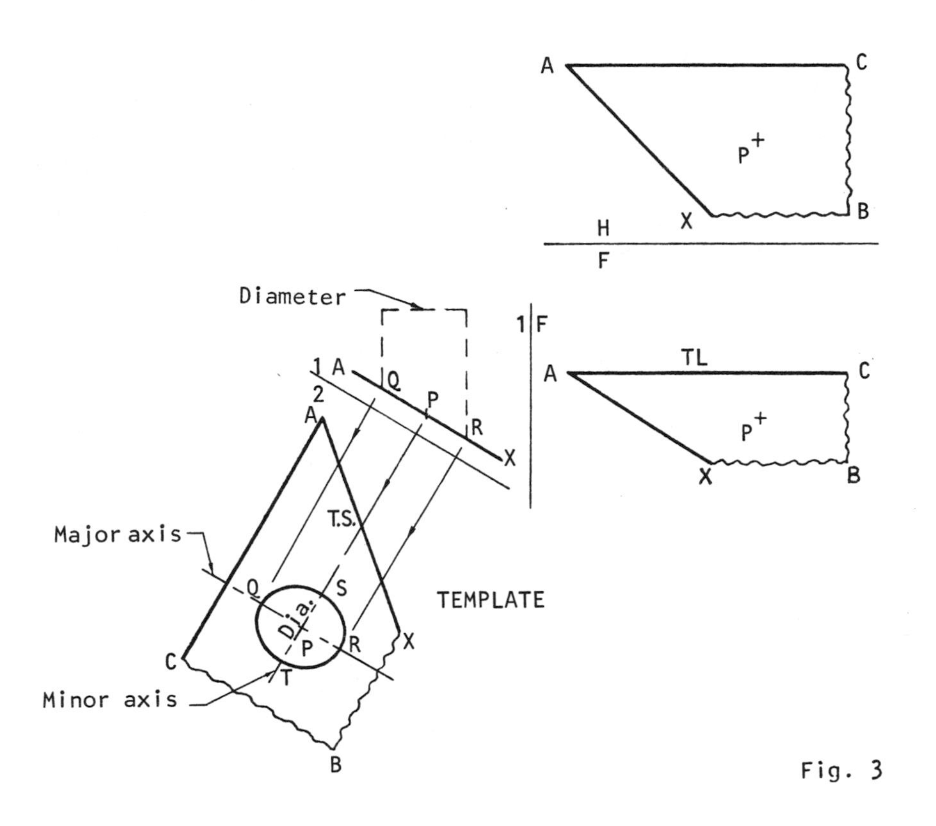
A
C
P+
X
B
H
F
Diameter
1
F
1 A
2
Q
P
R
X
A
TL
C
P+
X
B
A
T.S.
Major axis
Q
S
Dia.
P
R
X
C
T
Minor axis
B
TEMPLATE

Fig. 3

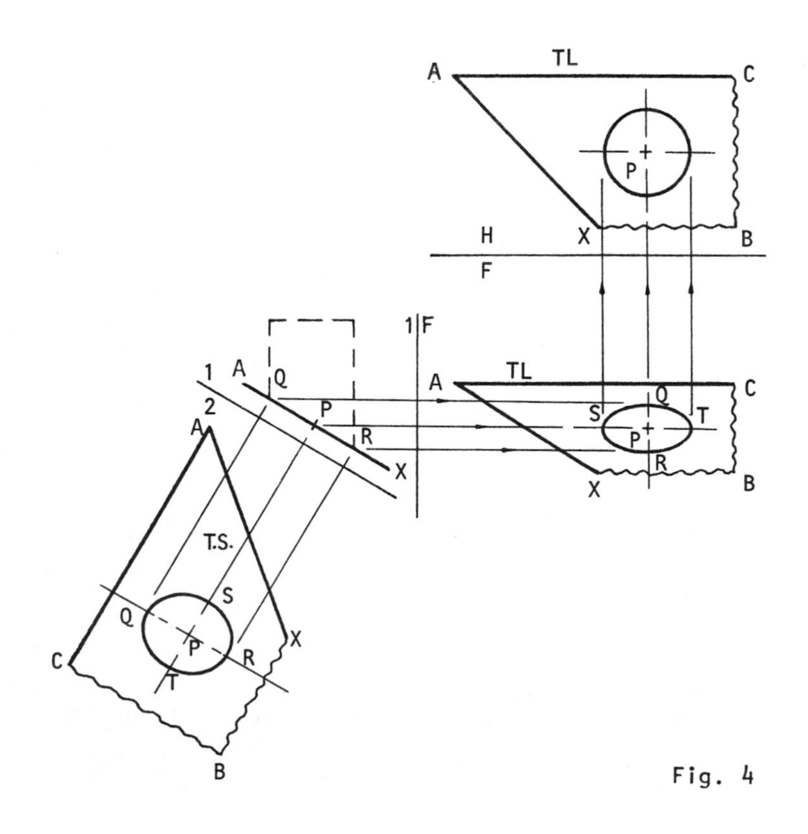
A
TL
C
P
X
B
H
F
1
F
1 A
Q
P
R
X
2
A
A
TL
C
S
Q
T
P
R
X
B
T.S.
Q
S
P
R
X
C
T
B

Fig. 4

plane perpendicular to the true length line AC in either view (this solution uses the F-view). Project the roof plane to view 1 as an edge, and draw the elevation view of the circular pipe. From this edge view, the true shape and size view of the plane, and therefore the true shape and size view of the hole can be drawn. As in fig. 3, make the 1-2 reference plane parallel to the edge view of view 1 and project the roof plane to view 2. Where the pipe cuts the roof plane in view 1 (points Q and R) locates the major axis of the elliptical hole. The minor axis will be the diameter of the pipe. Construct the required ellipse template in view 2 with axes QR and ST. Project the ellipse back to the front and top views, as in fig. 4.

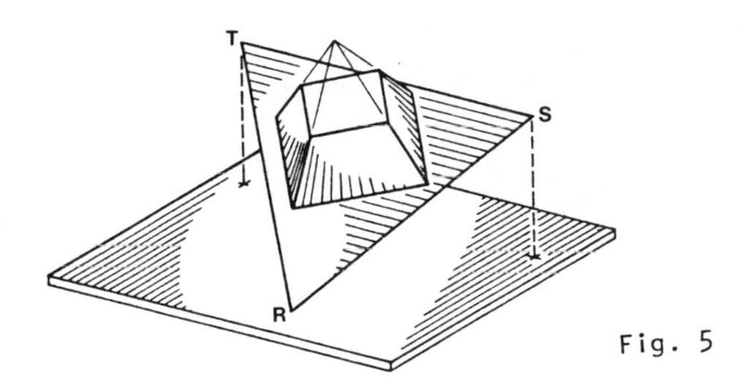

Fig. 5

(B) Fig. 5 illustrates a space analysis of the stated problem. Fig. 6 shows the given plane R-S-T in the H- and F-planes of projection. Fig. 7 shows the given frustum in the H- and F-planes of projection. Fig. 8 illustrates

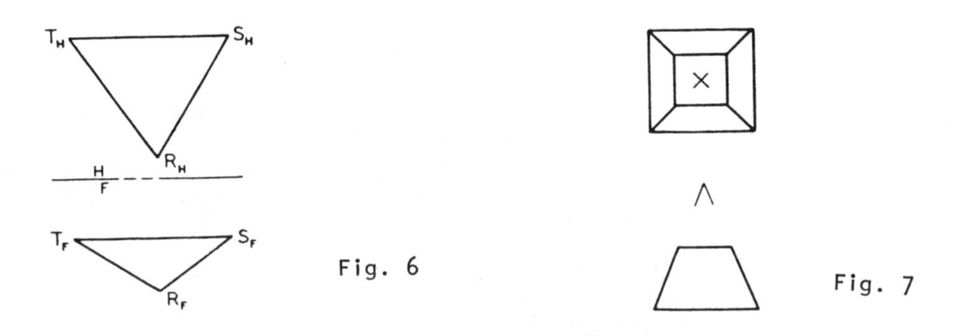

Fig. 6

Fig. 7

the method of solution. Since the horizontal line, T_F-S_F, is parallel to the H-F rotation line and to the frontal line, T_H-S_H, both T_F-S_F and T_H-S_H are true length.

The lines of sight for auxiliary view P are parallel to T_F-S_F; therefore the plane appears as an edge in view P.

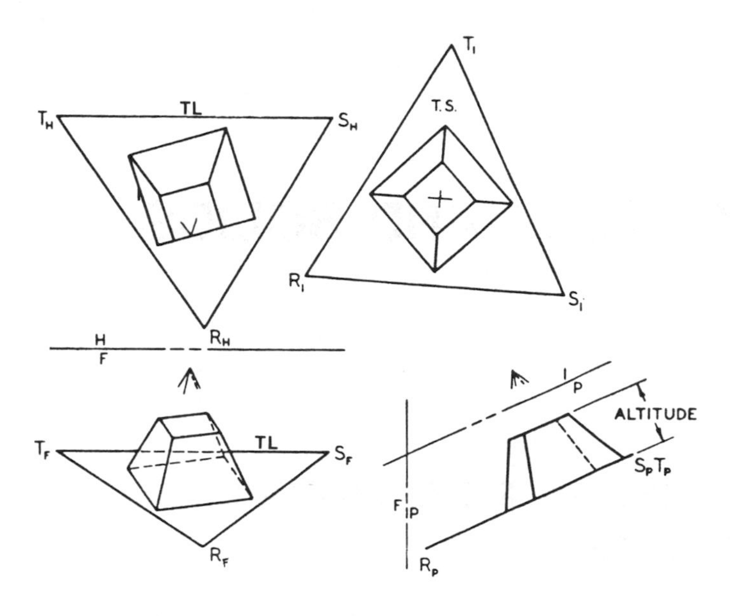

Fig. 8

Rotation line 1-P is parallel to the edge-view of plane R-S-T; consequently, the plane appears true size-and-shape in view 1. Locate the base area of the frustum on the true-size view. Referring to view P, erect the altitude of the frustum perpendicular to the edge view of the plane. Visualize which is the upper side of the plane. Locate all projections in the given views, and show proper visibility in all views.

CHAPTER 6

SURFACES AND SOLIDS

DEFINITION OF TERMS CONCERNING SURFACES

• PROBLEM 6-1

STATE THE DEFINITIONS OF THE TERMS USED IN THE DISCUSSION OF CURVED AND WARPED SURFACES.

Solution: (1) A surface is generated by a moving line, called the generatrix, which may be either straight or curved.

(2) Any one of the infinite number of positions of the generatrix is called an element of the surface.

(3) A directrix is a straight or curved line that controls or directs the motion of the generatrix in forming a surface. A surface may have one or two directrixes.

(4) In forming certain surfaces, the generatrix must make a constant angle with a given plane called a plane director.

(5) A developable surface is one that can be rolled out into a plane. In other words, it is a surface that can be cut from a plane and then rolled or bent into the desired shape without deforming the material.

(6) A ruled surface is any surface, such as a cylinder, that can be formed by a straight-line generatrix, or, in other words, a surface on which straight lines can be drawn.

(7) A double-ruled surface is a surface, such as a plane on which two intersecting straight lines may be drawn through the same point.

(8) A single-curved surface is a surface, such as a cylinder or cone, that can be generated by a straight line moving along a curved directrix, so that any two adjacent straight-line elements are either parallel or intersecting. It is, therefore, a develop-

able ruled surface.

(9) A double-curved surface is a surface, such as a sphere, on which no straight lines can be drawn, that can be generated by a curved line moving in a curved path. Double-curved surfaces cannot be developed, although in practice approximate developments may be used to form surfaces that are nearly correct.

(10) A warped surface is a ruled surface that cannot be developed. Any two adjacent elements on a warped surface are nonintersecting and nonparallel.

(11) An axis of a surface is a straight line about which the surface is symmetrical. Some surfaces have more than one axis and some have none.

CONES, CYLINDERS, PRISMS, SPHERES, AND SPHERICAL TRIANGLES

• PROBLEM 6-2

DEFINE A CONE AND DESCRIBE THE DIFFERENT TYPES OF CONES.

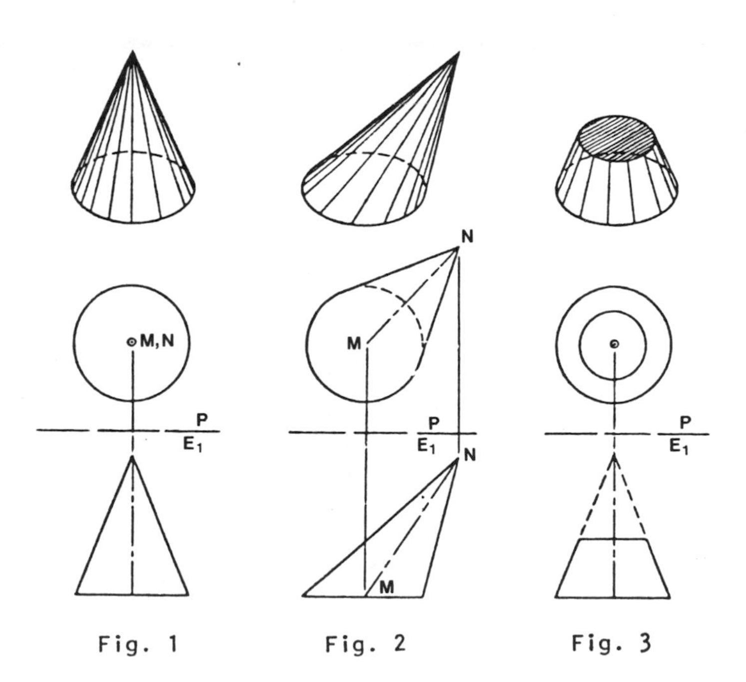

Fig. 1 Fig. 2 Fig. 3

Solution: (1) A cone is defined as a surface formed by a straight line that passes through a fixed point called the vertex and follows a curved-line directrix.

(2) A cone is also the surface generated by a straight line moving in a curved path about a straight-line axis that it intersects at the vertex.
(3) Each of the infinite positions of the straight line generating the cone is an element of the cone.
(4) If all the elements make equal angles with the axis, a section perpendicular to the axis is a circle, and the cone is called a cone of revolution or a right circular cone, as illustrated in Fig. 1.
(5) If the elements do not make equal angles with the axis, the right section may have many different shapes, the most common being the ellipse. If the right section is an ellipse, the cone is called an elliptical cone, as shown in Fig. 2.
(6) When the vertex has been cut off, the cone is called a truncated cone, as shown in Fig. 3.
(7) If the generatrix of a cone extends through the vertex, it will generate two equivalent surfaces called nappes. In most cases, only one nappe of a cone is considered, and it is usually limited by a plane section called the base.
(8) The vertex, the base and two extreme elements are usually shown to represent a cone, as illustrated in Figs. 1, 2 and 3.

• PROBLEM 6-3

Given a point P on the surface of a cone in the F-, frontal, plane of projection, determine its projection in the H-, horizontal, plane of projection.

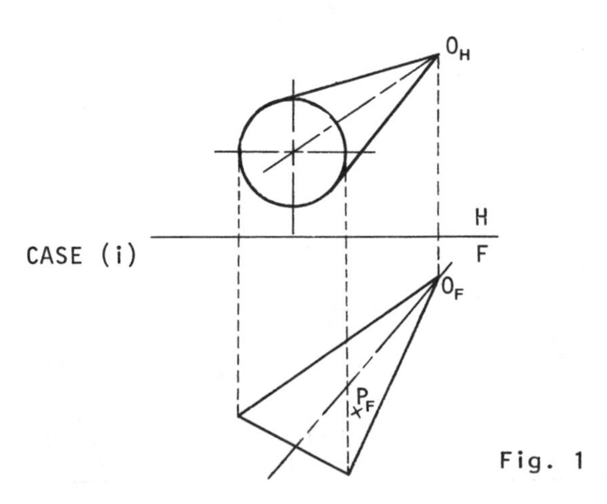

Fig. 1

Solution: (1) Fig. 1 shows the given point, P_F, on the surface of the cone. This is case (i).(2) Fig. 2 shows the given point P_F on the surface of the cone whose H- and F- projections are different from those in Fig. 1. This is case (ii). (3) the method of solution for both cases is the same, and is outlined in the following steps. The solution to case (i) is shown in Fig. 3, and that for case (ii) is shown in Fig. 4. (4) Draw element $O_F - P_F$, cutting the base in point 1 or 2. This element has two possible positions since P may lie on either the front or rear side of the cone. (5) Draw the projection $O_H - 1$ and $O_H - 2$.

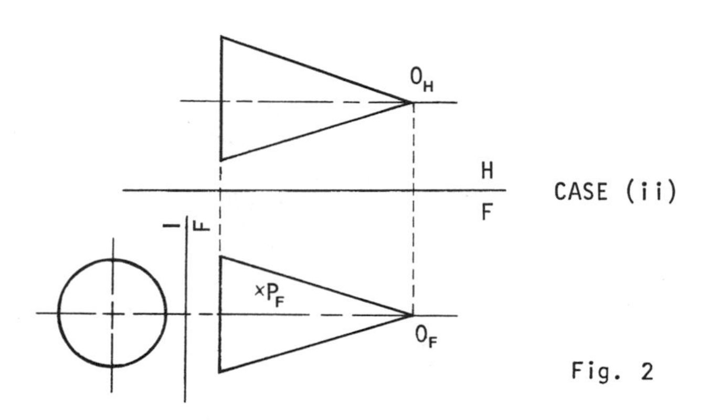

Fig. 2

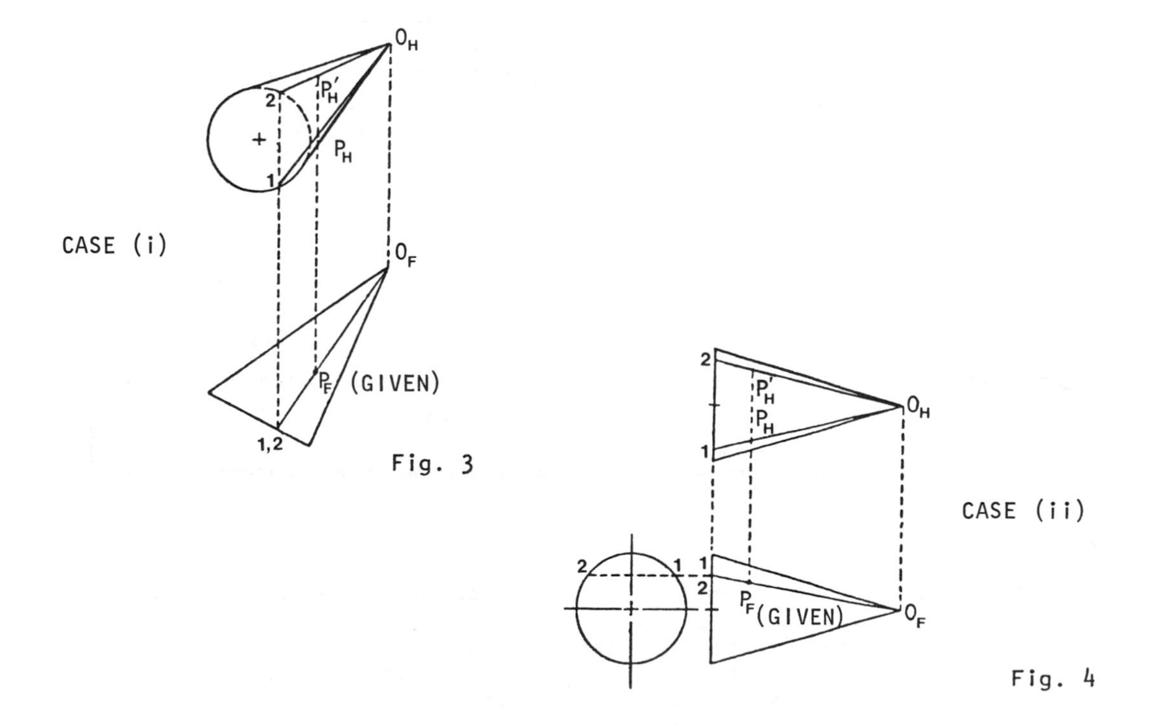

Fig. 3

Fig. 4

Project point P to the H-view. P_H may lie on either line. (6) Note: a similar construction applies when the point is given on the surface of a cylinder.

• **PROBLEM 6-4**

Construct a plane tangent to a cone (or cylinder) at a given point, P, on the surface.

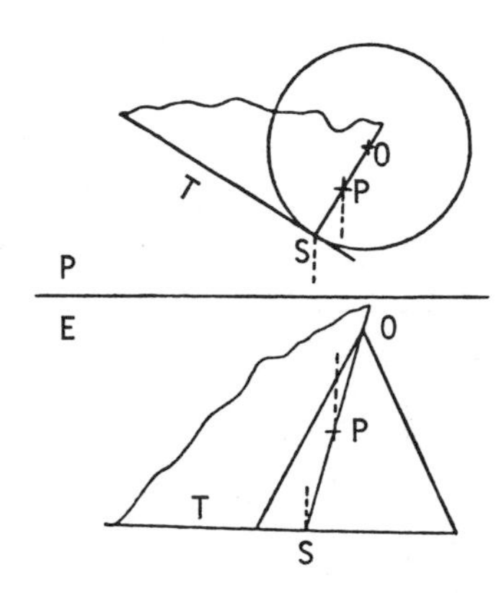

Solution: (1) The accompanying figure shows the given cone and the location of point P, in the P-plan , and E-elevation views.
(2) Problem analysis reveals that the tangent plane will contain the element of the surface through the given point and a line tangent to the base at the point in which the element intersects the base.
(3) The required plane is determined by the element O-S, which contains the given point, and the line T, which is tangent to the base at point S.

• **PROBLEM 6-5**

Construct a plane tangent to a cone (or cylinder) through a given exterior point P.

Solution: (1) The accompanying figure contains the given cone and point P in the front and top views.
(2) Problem analysis reveals that the required tangent plane must contain the vertex of the cone. As a result, the required plane must contain the line passing through the vertex and the given point P. The second line determining the plane is a tangent to the base, drawn through the point Q where the line joining the given point and the vertex pierces the plane of the base. Note: If the given surface is a cylinder, the auxiliary line PQ is drawn parallel to the axis of the cylinder.

(3) As illustrated in the accompanying figure, draw line OP in the F-view, intersecting the plane of the base of the cone in point Q. Project point Q to the H-view.

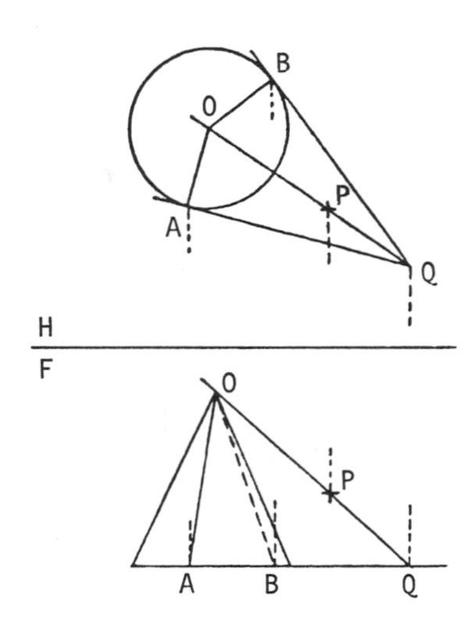

(4) Draw lines QA and QB tangent to the base of the cone in the H-view. Planes O-Q-A and O-Q-B are each tangent to the cone.
(5) The elements of tangency are O-A and O-B, respectively. Project points A and B to the F-view and draw the elements.

• PROBLEM 6-6

Define a cylinder. Explain and sketch the three classes of cylinders: circular, elliptical, and parabolic.

Solution: (1) By definition, a cylinder is a surface formed by a straight-line generatrix that is perpendicular to a plane and follows a curved-line directrix in that plane.
(2) By definition, a cylinder is also a surface formed by a straight line that moves in a curved path, while remaining parallel to a straight line directrix.
(3) From the above definitions, it can be seen that a cylinder is not necessarily round and does not have to be a closed surface.
(4) A plane that is perpendicular to the axis of a cylinder is perpendicular to all of its elements, and cuts a right section of the cylinder. Cylinders are classified according to the shapes of their right sections.
(5) If the right section is a circle, the cylinder is a cylinder of revolution, or a right circular cylinder.

(6) Other examples are the elliptical cylinder and the parabolic cylinder, where the right sections are an ellipse and a parabola, respectively.

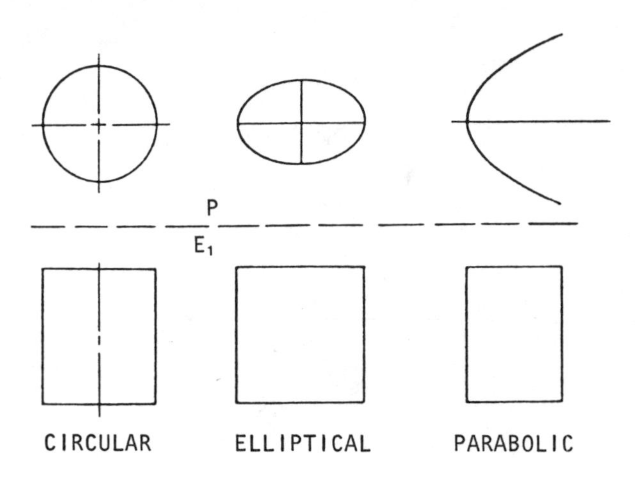

(7) The accompanying figure shows the P, plan, and E_1, primary elevation, views of these three classes of cylinders.
(8) Any view of a cylinder that shows the true shape and size of its right section also shows the axis and all the elements as points. Such a view is an edge view of the cylinder.
(9) In any view, the extreme outside elements are called the limiting elements of the cylinder.
(10) The limiting elements are the only elements that need to be drawn in any view (except the edge view) to represent a cylinder. Since, in practice, a cylinder has a definite length, one or both bases are usually shown.

• PROBLEM 6-7

Transfer the extreme elements of a cylinder of revolution from one view to another.

Solution: Fig. 1 shows the given cylinder of revolution in the P, plane, and E_1, primary elevation, views. Problem analysis reveals that the diameter of a cylinder of revolution appears true length in every view; consequently, extreme elements in one view can be transferred to another by employing this fact.

Before proceeding with the required solution to the stated problem, it is necessary to first understand, spacially, the concept of the limiting elements of the cylinder. Fig. 2 shows four views of a cylinder of revolution with axis A-B and vertical bases. It is important to realize that the shortest distance between extreme elements in every view is equal to the diameter. Since the axis, A-B, and the elements of the cylinder are parallel to the rotation line, P-E_2, as shown in view P, the axis and the elements of the cylinder appear true length in view E_2, the secondary elevation view. View I_1, the primary inclined view, has lines of sight parallel to the true length lines in E_2; consequently, I_1, shows the axis and all elements as points. In other words, I_1, gives an edge view of the cylinder and the true shape and size of its right section, a circle. Observe that the limiting element in any one view is not the limiting element in any of the other views. For example, element 1,which is the limiting element in the plan, appears directly in front of axis A-B in view E_2 because it is the element closest to the rotation line in view I_1. In view E_1, the primary elevation view, element 1 appears higher than the axis A-B,rather than directly in front of it, as can be verified by measuring from the rotation line. Element 2, which is a limiting element in view E_2, appears directly above the axis A-B in the plan view, since both must be the same distance from the rotation line in this view. This latter point is further illustrated in view I_1, where the point views of A-B and element 2 lie on a line which is parallel to the rotation line $\frac{I_1}{E_2}$. Element 2 is located in view E_1 by measurement from the rotation line.

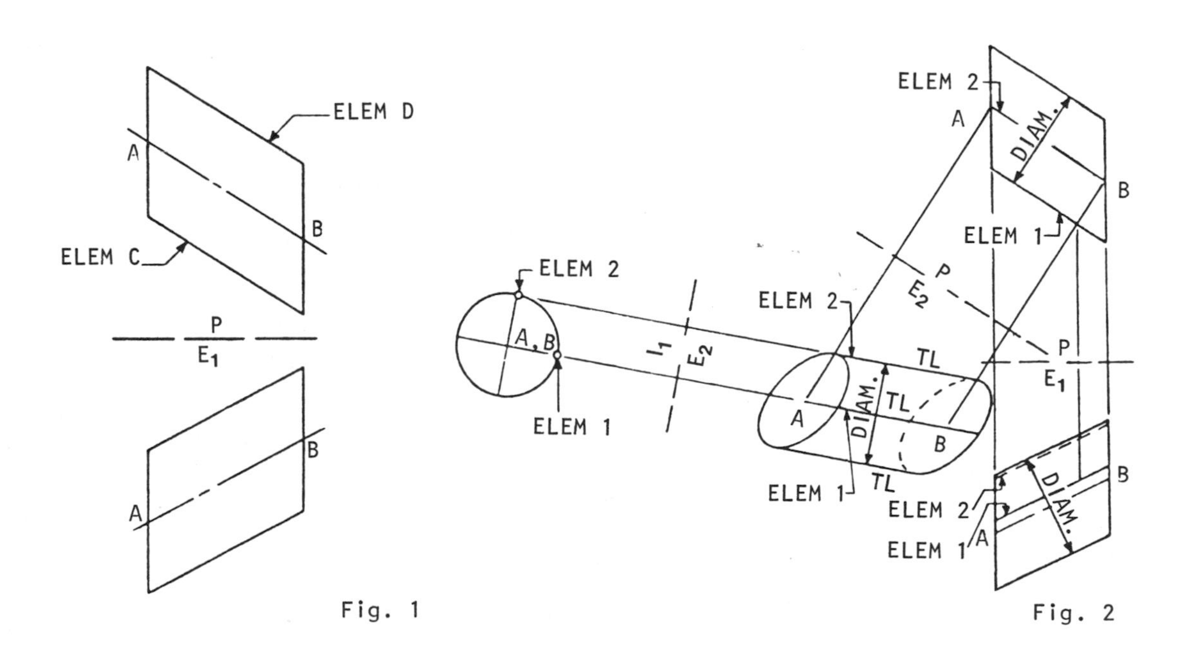

To locate the limiting elements, C and D, in the elevation view, a diameter is drawn through any point P on the axis in the plan view. Because X-Y is true length in the plan, it must be drawn parallel to the rotation line in view E_1. Note: Since all elements of a cylinder are parallel to axis, only one point on an element need be projected to transfer it from one view to another. As a result, elements C and D are drawn through X and Y, respectively, parallel to the axis A-B, in view E_1. Following this same line of reasoning, the extreme elements of view E_1 could be transferred to the plan by a similar procedure.

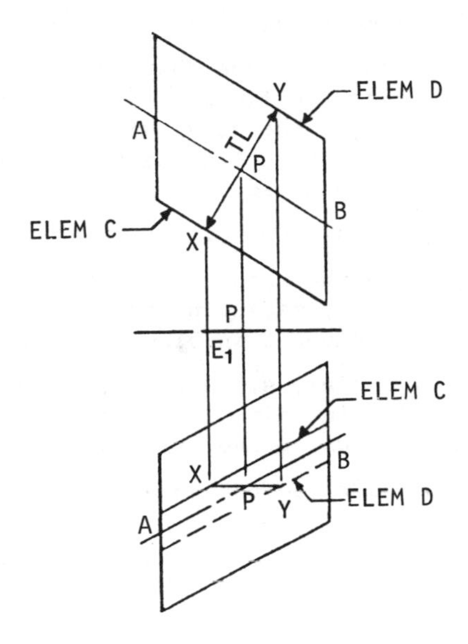

Fig. 3

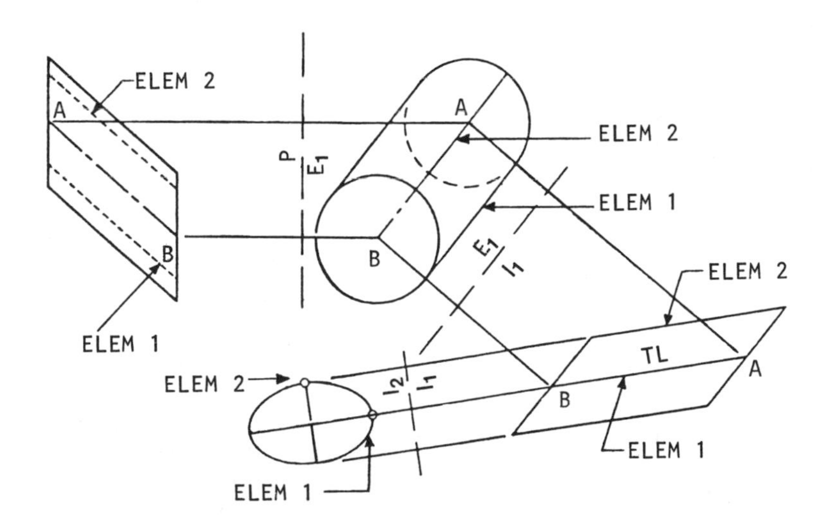

Fig. 4

Four views of an elliptical cylinder are shown in Fig. 4. The axis AB and all elements appear true length in view I_1; view I_2 is an edge view of the cylinder showing the true shape and size of its right section. The extreme elements in one view lie along the same line as the axis in an adjacent view when, and only when, one of the views is a true-length view. Thus, element 1 appears to coincide with the axis in view I_1, but is at some distance from the axis in the plan view.

Construct a plane tangent to a given cylinder, containing a given external point, using the two-view method.

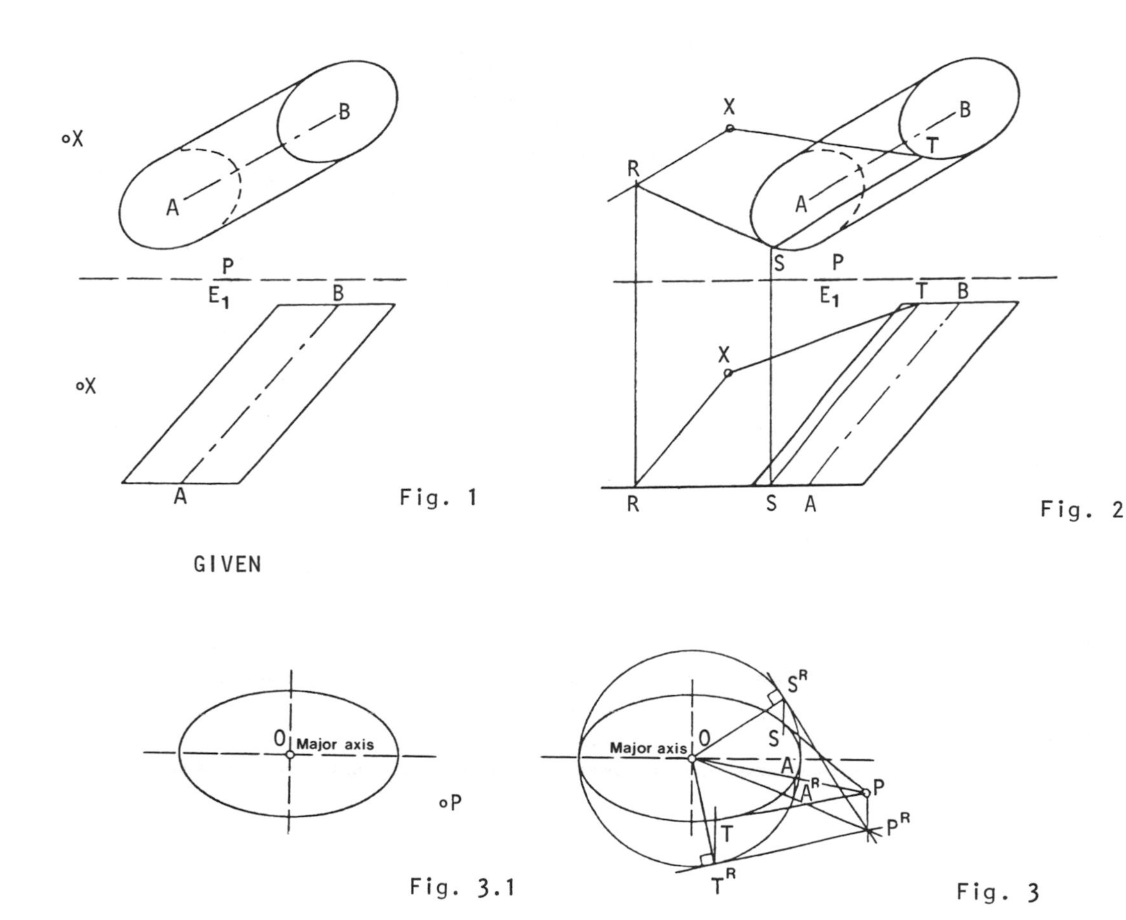

Solution: (1) Fig. 1 shows the given cylinder, with axis A-B, and external point, X, in the P, plan, and E_1, primary elevation, views.

(2) Problem analysis reveals that any plane that is tangent to a cylinder contains a single element of the cylinder. In addition, a plane tangent to a cylinder intersects the plane of the base of the cylinder along a line that is tangent to the base.

(3) As shown in Fig. 2, draw a line through point X, in both views, parallel to the elements of the cylinder.

(4) In view E_1, this line pierces the plane of the base extended at point R. Project point R to the plan view.

(5) From R in the P-view, draw a line tangent to the base at point S. Note: The exact point of tangency, S, is located by the method illustrated in Fig. 3, and described in step (9), below.

(6) Project S into view E_1.

(7) Plane X-R-S, thus formed, is tangent to the cylinder along element S-T.
(8) Note: One other tangent to the base could be drawn to locate the plane V-R-W, shown in the preceding problem.
(9) When it is necessary to draw tangents to an ellipse, the exact points of tangency can be determined by the method illustrated in Fig. 3. Here, it is desired to draw tangents from point P to the ellipse, as shown in Fig. 3.1.
(10) Problem analysis reveals that an ellipse can be considered as a view of a circle that is inclined with respect to the lines of sight. The diameter of this circle is equal to the major axis of the ellipse.
(11) Fig. 3 illustrates the revolution of the ellipse about its major axis until it appears as a circle.
(12) In this same figure, draw a line from O to P, crossing the ellipse at point A. As the ellipse is revolved, point A moves to A^R.
(13) Revolve point P about this same axis, thereby causing it to move at right angles to the axis, until it intersects line $O\text{-}A^R$ at P^R.
(14) Draw lines from P^R, tangent to the circle at points S^R and T^R.
(15) Revolve the circle back to its original position, moving points S^R and T^R to S and T, respectively. S and T are the exact points of tangency of the tangents P-S and P-T.

• PROBLEM 6-9

Construct a plane tangent to a given cylinder, containing a given external point, using the edge-view method.

Solution: (1) Fig. 1 shows the given cylinder with axis A-B, and external point, X, in the P, plan, and E_1, primary elevation, views.
(2) Problem analysis reveals that any plane that is tangent to a cylinder contains a single element of the cylinder. Furthermore, a plane tangent to a cylinder will appear as an edge in a view showing the edge view of the cylinder.
(3) As shown in Fig. 2, construct view E_2, the secondary elevation view, showing the cylinder in true length, by drawing the rotation line $\frac{E_2}{P}$ parallel to the cylinder in the P-view.
(4) Draw line VW parallel to AB in view E_1 and project it to the plan and E_2-views. Line VW appears true length in view E_2. Construct view I_1, the primary inclined view, showing the cylinder as an edge, and having lines of sight parallel to the true-length elements, AB and VW, in view E_2.
(5) In view I_1, draw a line through point X, tangent to the cylinder:

the edge view of the required plane.
(6) Determine the exact point of tangency, the point view of element V-W, by drawing a radius of the circle perpendicular to the tangent.
(7) Project element V-W into the other views to locate the required plane, X-V-W.

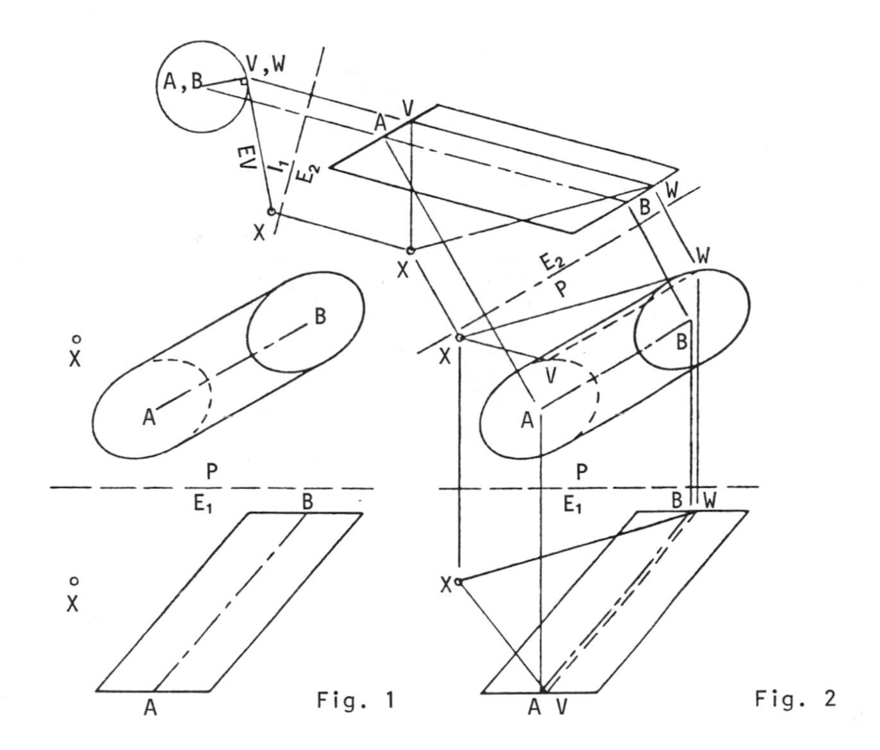

(8) Note: A second line could be drawn tangent to the circle in view I_1, also.

• PROBLEM 6-10

Construct a plane tangent to a given cylinder and parallel to a line, R-S, which is located outside the cylinder.

Solution: (1) Fig. 1 gives a spacial analysis of the stated problem and required solution.
(2) Problem analysis reveals that in order to determine a plane that is tangent to a given cylinder and, at the same time, parallel to a line located outside the cylinder, a plane must be created that contains the line and that is parallel to the cylinder axis (and, therefore, parallel to all elements of the cylinder). Any other plane traces parallel to the trace of the constructed plane on the cylinder

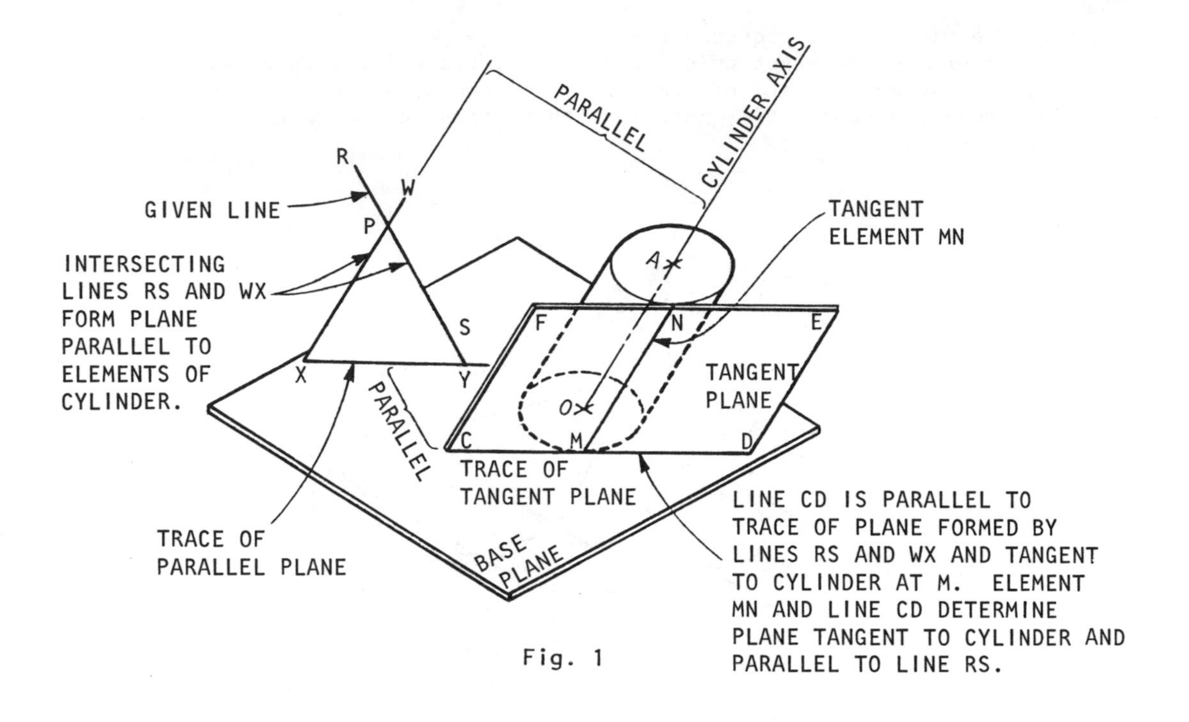
PARALLEL
CYLINDER AXIS
R
W
GIVEN LINE
P
TANGENT
ELEMENT MN
A
INTERSECTING
LINES RS AND WX
FORM PLANE
PARALLEL TO
ELEMENTS OF
CYLINDER.
S
F
N
E
X
Y
TANGENT
PLANE
O
PARALLEL
C
M
D
TRACE OF
TANGENT PLANE
TRACE OF
PARALLEL PLANE
BASE
PLANE
LINE CD IS PARALLEL TO
TRACE OF PLANE FORMED BY
LINES RS AND WX AND TANGENT
TO CYLINDER AT M. ELEMENT
MN AND LINE CD DETERMINE
PLANE TANGENT TO CYLINDER AND
PARALLEL TO LINE RS.

Fig. 1

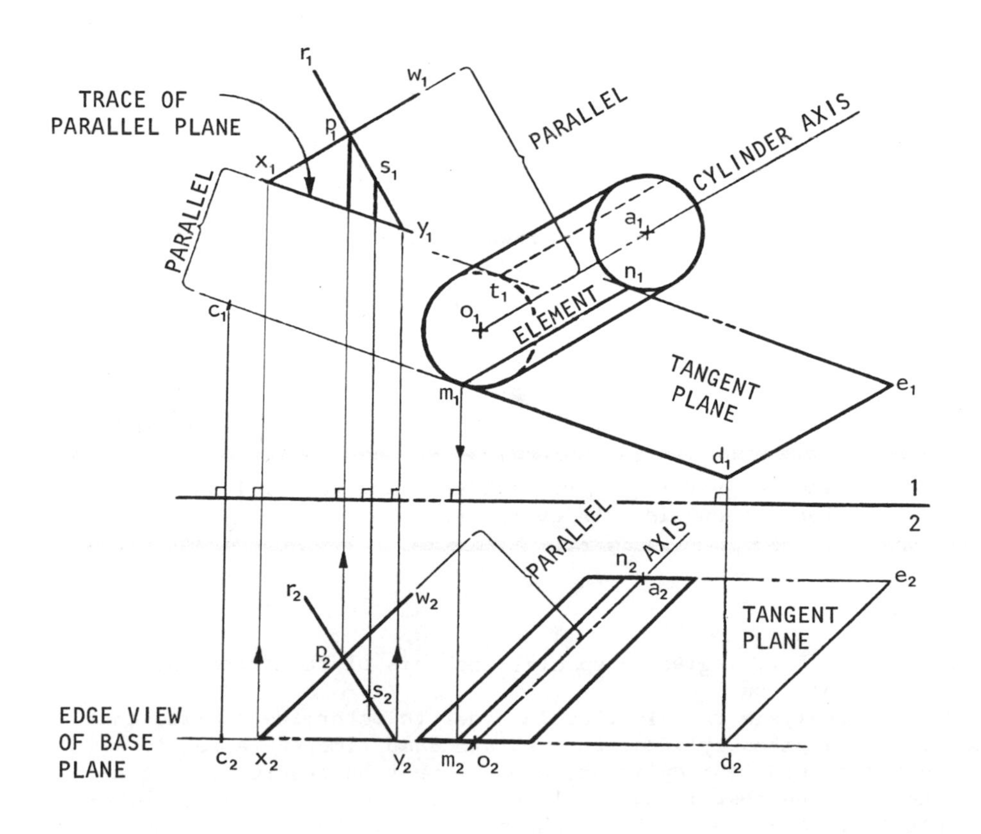
TRACE OF
PARALLEL PLANE
PARALLEL
CYLINDER AXIS
PARALLEL
ELEMENT
TANGENT
PLANE
1
2
PARALLEL
AXIS
TANGENT
PLANE
EDGE VIEW
OF BASE
PLANE

Fig. 2

base plane represent traces of planes parallel to the cylinder axis. For example, as illustrated in Fig. 1, the trace of the tangent plane C-D-E-F, C-D is parallel to the trace X-Y of the plane formed by intersecting lines R-S and W-X. Note: W-X is parallel to the cylinder axis A-O. The tangent plane, C-D-E-F, is, therefore, parallel to the given line R-S, which is located outside the cylinder.
(3) The construction of the required plane, tangent to a cylinder and parallel to a line outside the cylinder surface is outlined in the following steps. All steps are illustrated in Fig. 2.
(4) Starting in view #2, assume point p_2 on line r_2s_2. Construct a line w_2x_2 that passes through p_2, that is parallel to the cylinder axis a_2o_2, and that intersects the base plane at point x_2.
(5) Extend r_2s_2 until it intersects the base plane at point y_2. Line x_2y_2 is the frontal projection of the trace of the plane formed by the intersecting lines r_2s_2 (extended) and w_2x_2 (parallel to the cylinder axis a_2o_2).
(6) Project p_2 upward to view #1 in order to determine p_1 on r_1s_1. Through p_1 construct a line w_1x_1 parallel to the horizontal projection of the cylinder axis a_1o_1.
(7) Project x_2 and y_2 upward to view #1 in order to determine x_1 and y_1. Line x_1y_1 is the trace of the plane, formed by intersecting lines w_1x_1 and r_1s_1, which is parallel to the cylinder axis.
(8) Draw a line parallel to trace x_1y_1 and tangent to the base of the cylinder at point m_1.
(9) In view #1, construct at m_1 the element m_1n_1 parallel to the cylinder axis. Element m_1n_1 and the parallel trace c_1d_1 (which determines m_1) determine a plane that is tangent to the cylinder and parallel to the line r_1s_1.
(10) Project m_1 downward to view #2 and locate m_2. Draw element m_2n_2 and completely determine the tangent plane with line c_2d_2. (The tangent plane has been limited as $m_2n_2d_2e_2$.)
(NOTE: In view #1, a tangent trace has been constructed at point t_1 parallel to the trace x_1y_1. In other words, there are two planes tangent to the cylinder and parallel to the given line RS outside the cylinder.)

• PROBLEM 6-11

Construct a parallelepiped having been given the directions of three concurrent edges and the body diagonal passing through the given corner.

Solution: (1) In fig. 1, lines r,s, and t, intersecting at point A, and the body diagonal A-K, are given in the H-,horizontal, and

F-,frontal, planes of projection.

(2) The construction consists of the following steps:

(a) Pass a plane Q through lines s and t.

(b) Through point K, draw line r' parallel to line r.

(c) Find point D in which r' intersects plane Q.

(d) Draw lines from D parallel to s and t, establishing face A-B-C-D, as in fig. 2.

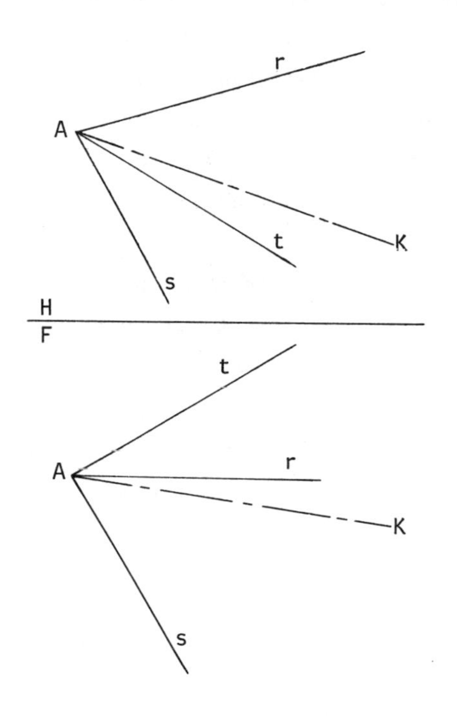

Fig. 1

Fig. 2

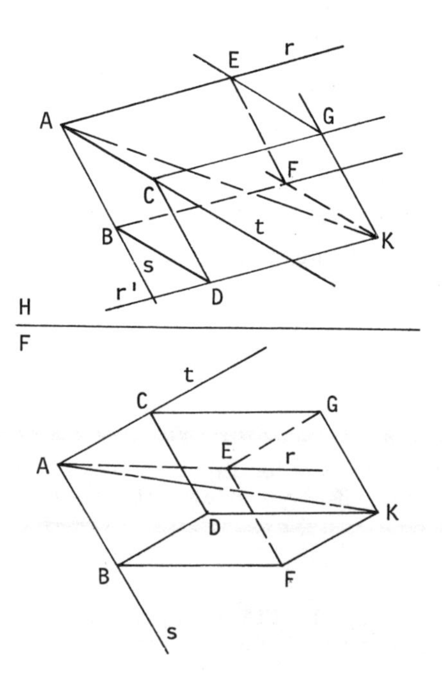

Fig. 3

(e) Locate point G, the point of intersection of the line from C, drawn parallel to r, with the line from K, drawn parallel to s.
(f) Locate point F, the point of intersection of the line from K drawn parallel to t with the line from B drawn parallel to r.
(g) Locate point E, the point of intersection of the line from G drawn parallel to t with r, as in fig. 3.
(h) Draw E-F completing the parallelepiped, A-B-C-D-E-F-K-G.
(i) Project to the other principal view.

• PROBLEM 6-12

Find the true shape and size views of a given rectangular prism, having a right section with face widths equal to "a" and "b" and a given axis, AB.

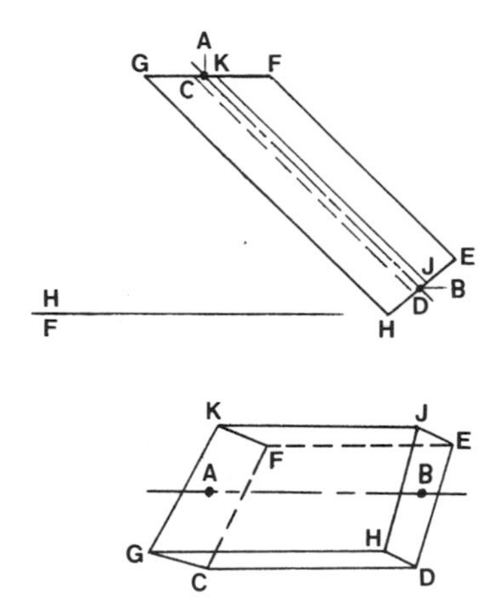

Fig. 1

Solution: Fig. 1 shows the front and top views of the given rectangular prism, C-D-E-F-G-H-J-K, with a right section having face widths equal to "a" and "b". Line AB is the axis of the prism. Base C-G-K-F has been cut on an angle and base D-H-J-E is a right section.

To draw the end view (right section) of the prism, pass a cutting plane through line AB parallel to lines KJ, FE, CD, and GH. The trace of this plane on base DHJE is line XY, passing through point B, parallel to JH and ED. Take reference plane H-1 perpendicular to line AB in the top view, which appears true length because it is parallel to H-F in the front view. Line AB now projects as a point to view 1. Project line XY to view 1, also, and construct the end view with the given dimensions, with the sides of length "b" parallel to line XY, as seen in fig. 2.

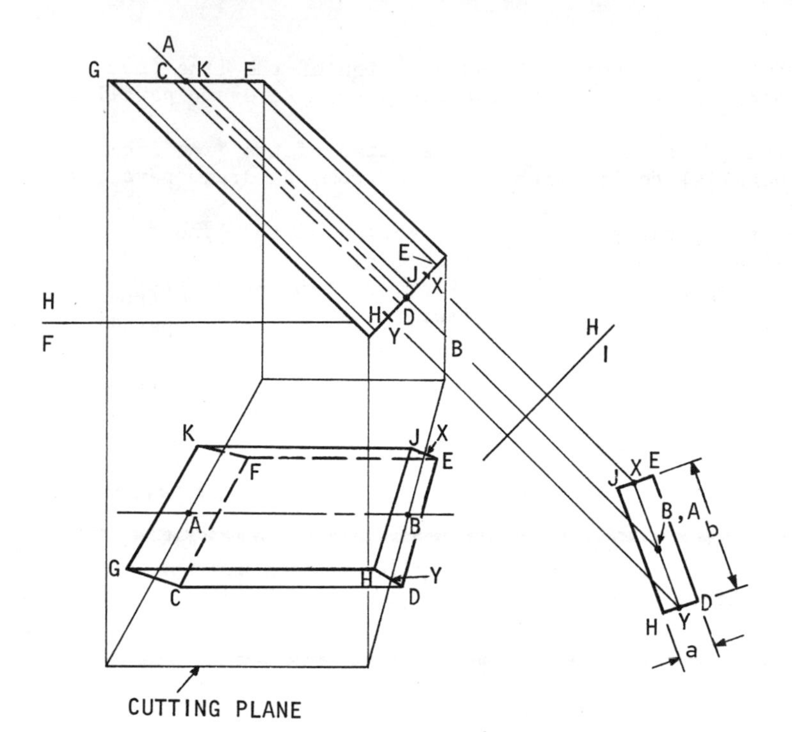
A
G C K F
E
J
X
H
F
H D
Y
B
H
I
K J X
F E
A B
G Y
H
C D
J X E
B, A
b
H Y D
a
CUTTING PLANE

Fig. 2

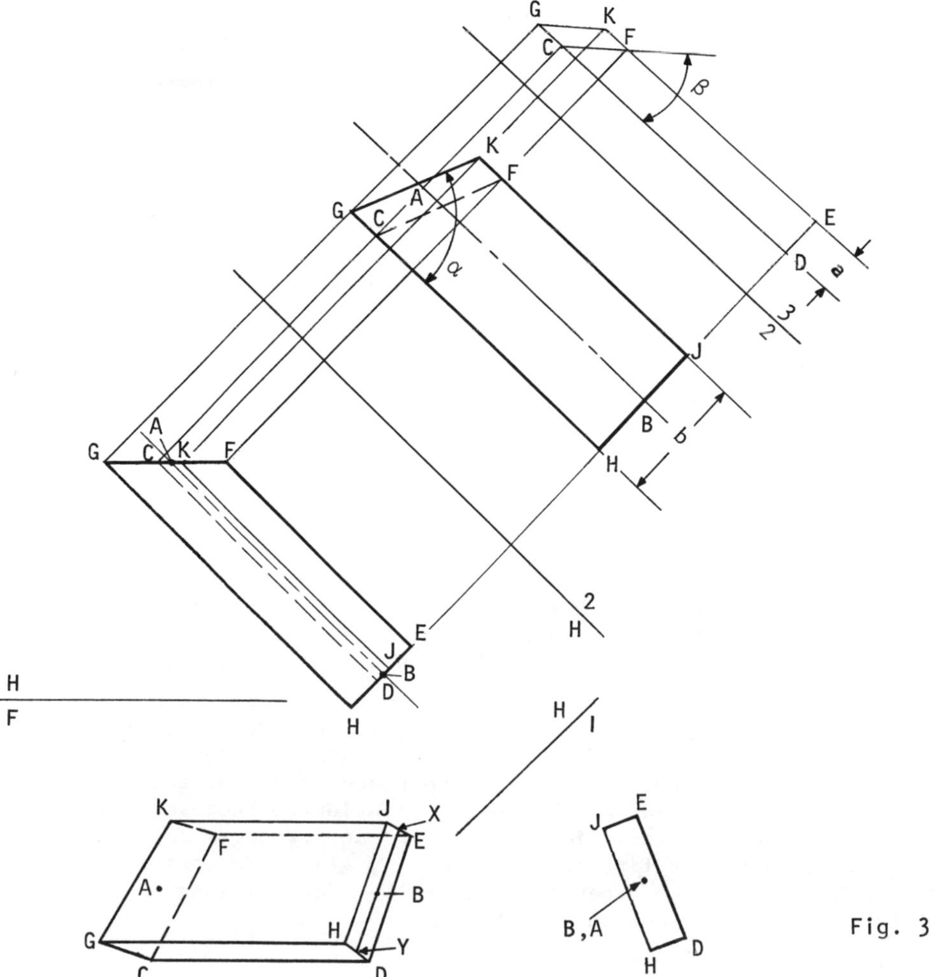
G K
C F
β
K
F
G A
C
E
α
D
a
3
2
J
B
b
H
G A
C K F
2
H
E
J
B
D
H
H
F
H
I
K J X
F E
A B
H Y
G
C D
J E
B, A
H D

Fig. 3

Project the true size views of the prism from the top view, in which line AB appears true length. Take reference plane H-2 parallel to line AB, and project to view 1, making the width of the face equal to "b". The true angle of the bevel cut on the wide face is equal to α. Project from view 2 in the same manner, with reference plane 2-3 parallel to lines KFJ and GCH, making the face width equal to "a" and the true angle of the bevel equal to β, as in fig. 3. Views 1, 2, and 3 are the required true shape and size views of the prism.

• PROBLEM 6-13

Given the right section and axis of the surface, draw the views of a prism.

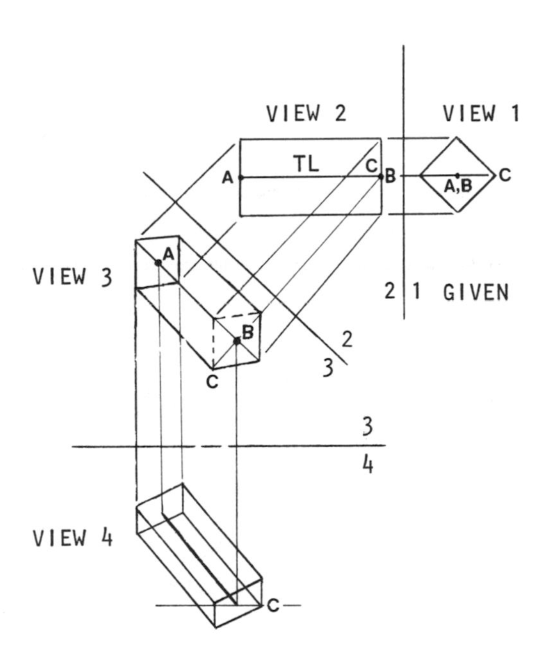

Solution: As shown in the accompanying diagram, draw the normal (true length: TL) and end (point) views of the axis of the prism. The true size of the right section appears in the view in which the axis projects as a point: the given view. The right section appears edgewise in the view in which the axis projects in its true length. This is in view 2. Having been given the right section on the end view of the given axis, complete the required views by projecting from this view.

Construct view 2 by extending the horizontal diagonal shown on the end view across rotation line 1-2, which is perpendicular to said horizontal diagonal. Construct, on this horizontal line, the given true length (TL) axis A-B. The basis for this step is the fact that a line appears as a point in a view having lines of sight parallel to the true length projection of the line in an adjacent view. As constructed, the lines of sight for the given view are parallel to the true length line in view 2. Complete view 2 by projecting the extreme elements from the given view.

Establish rotation line 2-3 between view 2 and view 3. The axis of the prism in view 3, the top view, must be parallel to line 2-3 since A-B projects true length in view 2. Complete view 3 by projecting the extreme elements and right section diagonals from view 2, as shown.

Since one diagonal of each base is horizontal (i.e., the diagonal labeled C), as shown in the given view, this same line must appear as a horizontal line in view 4 - the front view. Establish the horizontal rotation line 3-4 between view 3 and view 4. Project A-B to the front view. Draw the horizontal line C through point B. Said line C is perpendicular to A-B and parallel to 3-4. One diagonal of the base centered on B will lie along line C. Determine the endpoints by projection from view 3, as shown. Similarly, construct a horizontal line through point A, and check the endpoints by projection from the H-view. Complete the front view by constructing the extreme elements parallel to axis A-B, and drawing the right sections.

• PROBLEM 6-14

Discuss spherical triangles and their elements.

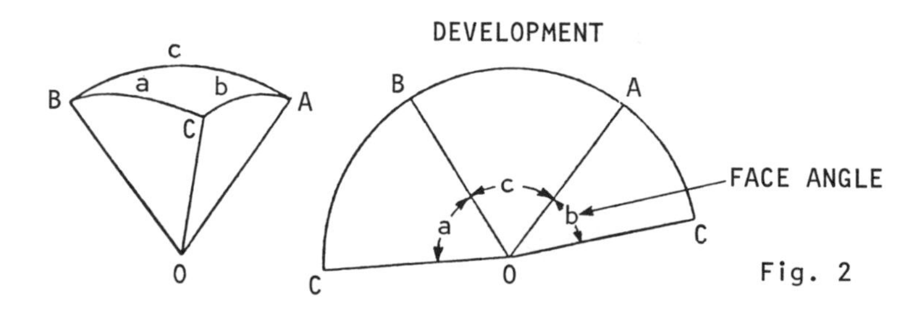

Fig. 2

Fig. 1

Solution: (1) The portion of a spherical surface bounded by three arcs of great circles is a spherical triangle. If each vertex of a spherical triangle ABC is joined to the center O of the sphere by a straight line, a corresponding trihedral angle, O-ABC, is formed (Figure 1). The sides of the spherical triangle opposite the vertices A, B, and C are denoted by the letters a, b, and c respectively. These sides are measured by the face angles of the trihedral angle. The angles of the spherical triangle are measured

by the dihedral angles of the trihedral angle. In order that a trihedral angle may be constructed, any one face angle must be less than the sum of the remaining face angles. The development of the faces is a circular sector, its radius being equal to the radius of the sphere (Figure 2).
(2) The spherical triangle or trihedral angle is composed of six elements: three dihedral angles and three face angles. It can be constructed if any three elements are given. The remaining elements can be deduced from those given. In certain cases, more than one solution may be possible.

• PROBLEM 6-15

Given a point P on the surface of a sphere in the F-frontal, plane of projection, determine its projection in the H-,horizontal, plane of projection.

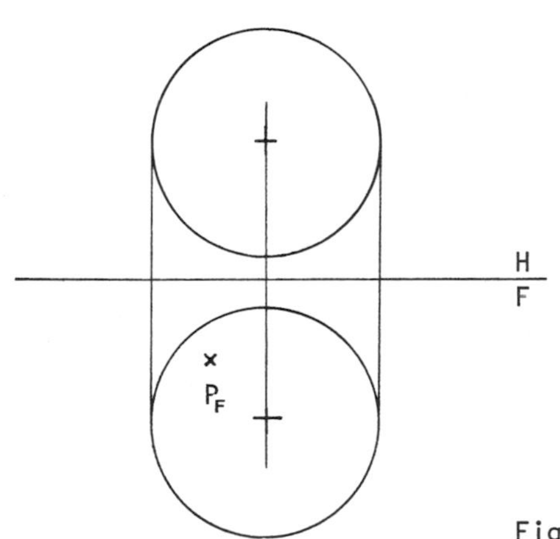

Fig. 1

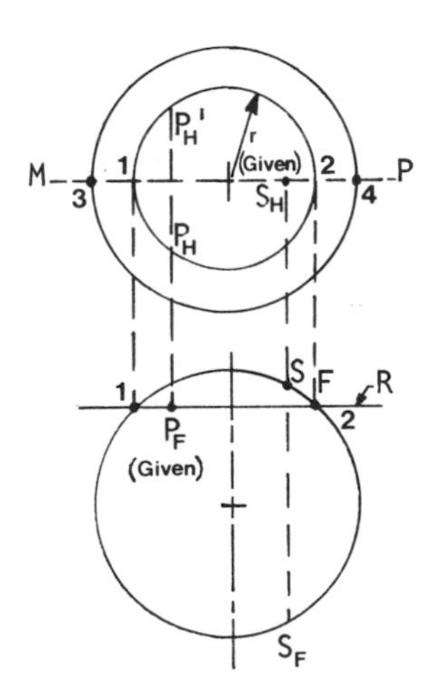

Fig. 2

Solution: (1) Fig. 1 shows the given sphere in the H- and F- planes of projection. Point P is given in the front view and denoted P_F.
(2) As shown in Fig. 2, a horizontal plane, R, through P_F cuts the surface of the sphere in the circle of radius r.
(3) Project this circle of radius r, whose diameter is equal to the distance along the edge view of plane R between points 1 and 2, to the H-view.
(4) The possible locations of P_H, the point P in the H-view, are at the intersections of the projector from the F-view and circle of radius r.
(5) A second case, superimposed on the diagram for the first, involves the given point S_H. The solution is similar to that of point P.
(6) Pass a horizontal cutting plane P through S_H. Plane P cuts the surface of the sphere in a circle whose diameter is equal to the

distance along plane P between points 3 and 4. This circle is a great circle of the sphere since the cutting plane passes through the center of the circle.
(7) The possible locations of S_F, the point S in the F-view, are at the intersections of the projector from the H-view and the great circle. In other words, S_F lies on the frontal great circle of the sphere .

• PROBLEM 6-16

Construct a plane tangent to a sphere (or surface of revolution) at a given point, P, on the surface.

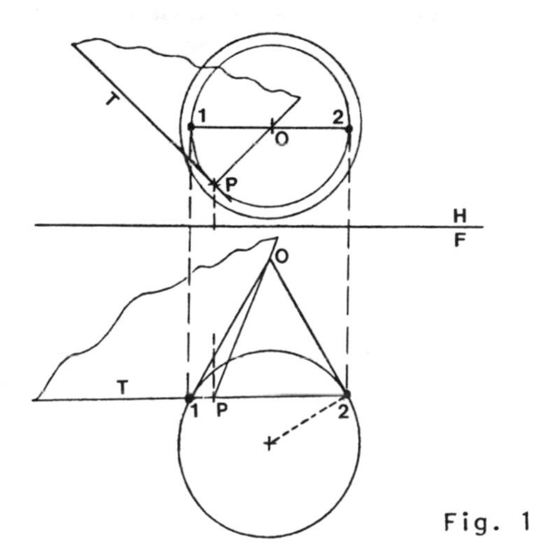

Fig. 1

Solution: (1) The accompanying figure contains the given sphere and point P in the H-,horizontal, and F-,frontal, planes of projection.
(2) Problem analysis reveals that one must first construct a cone of revolution tangent to the sphere and containing the given point. The required plane is a plane tangent to the cone at the given point.
(3) Draw the horizontal circle of the sphere passing through point P, as shown in the H-view.
(4) Using this circle as the base, construct the cone which is tangent to the sphere in the F-view.
(5) The axis of the cone is vertical with vertex at point O.
(6) Draw a line through points O and P in the H-view; and, project this element to the F-view.
(7) The frontal line, 1-2, in the H-view represents the diameter of the base of the cone. Similarly, draw a horizontal line through point P in the F-view, whose distance is equal to the distance 1-2 from the H-view. This is the base of the cone of revolution.
(8) As in a previous problem, the plane tangent to the cone is de-

termined by the element O-P of the surface through the given point, and a line, T, tangent to the base circle at the point in which the element intersects the base. Plane O-P-T is the required plane.

ANGLES OF A TRIHEDRAL ANGLE

• PROBLEM 6-17

Determine the three dihedral angles of a trihedral angle, ABC, when the three face angles, a,b, and c, are given.

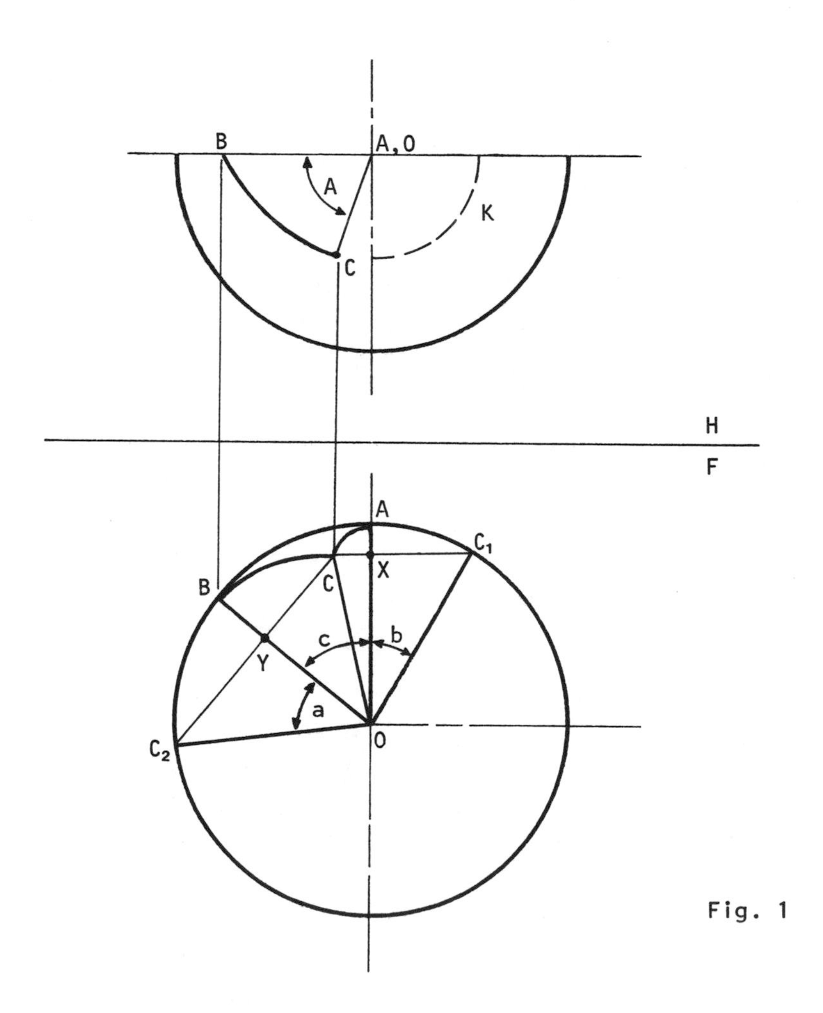

Fig. 1

Solution: As in fig. 1, construct the front and top views of the development of the trihedral angle, making edge OA vertical and using any radius of convenient length and the given angles a, b, and c. To form the trihedral, or solid, angle, fold the faces AOB, AOC_1, and BOC_2, in space into their correct positions by holding face AOB fixed and rotating faces AOC, and BOC, about the edges OA and OB respectively until the edges OC_1 and OC_2 meet.

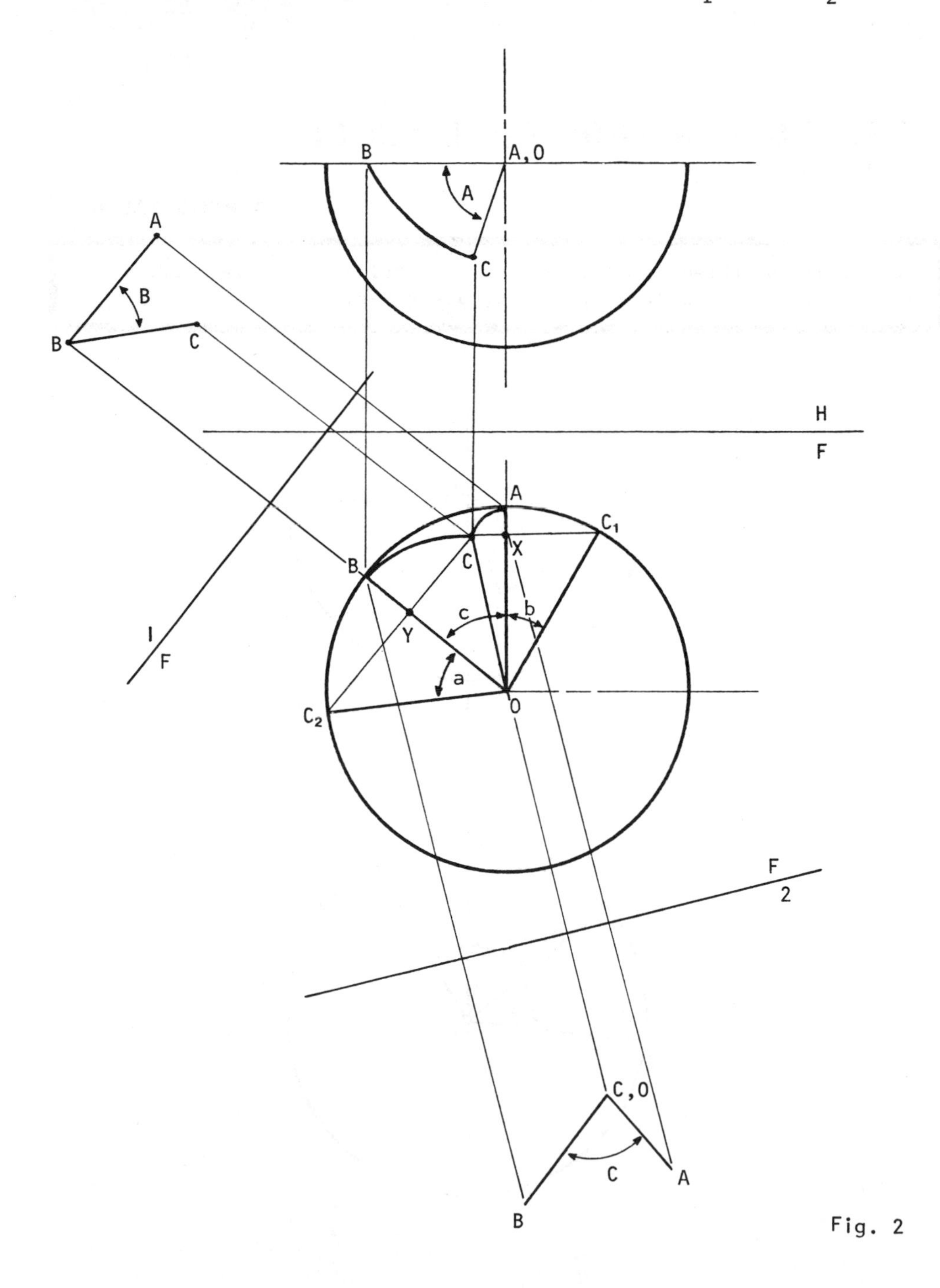

Fig. 2

The line C_1X through C_1 and perpendicular to OA represents the path of rotation in the front view for point C_1. In the top view, this path is the circular arc K. The line C_2Y through C_2 and perpendicular to OB represents the path of rotation of point C_2. Lines C_1X and C_2Y extended intersect at point C, one vertex of the spherical triangle. The other vertices are A and B.

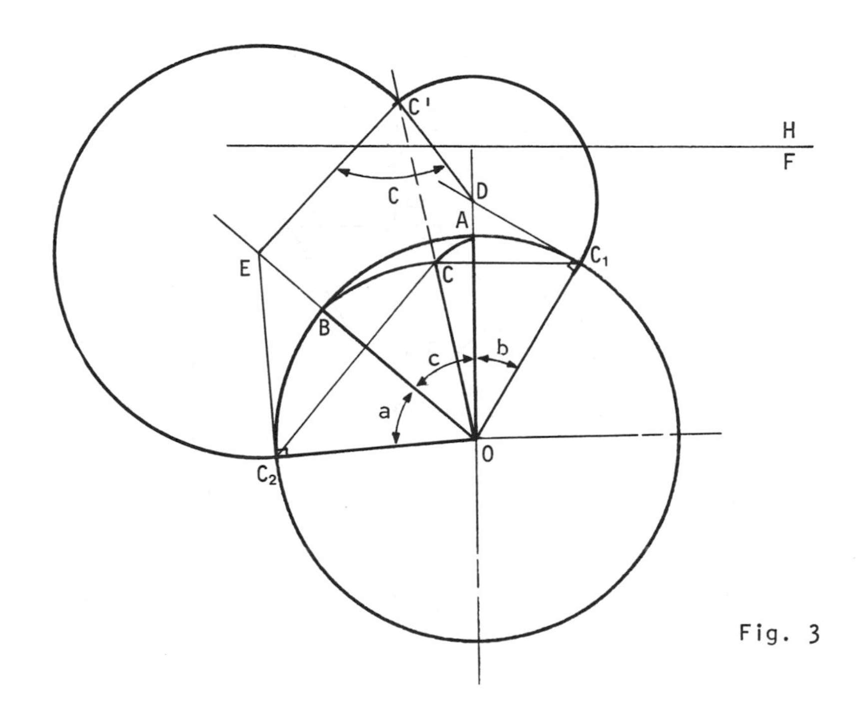

Fig. 3

Locate point C in the top view by projecting point C from the front view to where it intersects arc K. Now all the determining points of the trihedral angle have been obtained. Draw BC and AC in the front view and BC in the top view to complete the drawing.

Problem analysis reveals that the true size of the angle between two intersecting planes, the dihedral angle, can be seen only in a view that shows both planes as edges. The two intersecting planes will both appear as edges in a view that shows their line of intersection as a point. Consequently, since edge OA is a point in the top view, the dihedral angle A is given by angle BAC in this view.

As shown in fig. 2, apply the same reasoning as above to find the dihedral angle B. Draw the reference plane F-1 perpendicular to the line OB in the front view to project OB to the auxiliary view as a point and faces AOB and BOC as edges. Angle B is the angle ABC in view 1. Angle C can be found by the same method. Draw the reference plane F-2 perpendicular to the line OC in the front view to project OC as a point and faces BOC and AOC as edges. Angle C is the angle BCA in view 2.

Another method of determining the dihedral angle C is more precise and is shown in fig. 3. Draw a line through point C_1 in the front view perpendicular to line OC_1, and extend line OA. These lines intersect at point D. Draw a line through point C_2 perpendicular to line OC_2, and extend line OB. These lines intersect at point E.

Extend line OC in the front view. Construct an arc of radius C_1D centered at D and another arc of radius C_2E centered at E. These arcs intersect at point C', located on the extension of line OC.

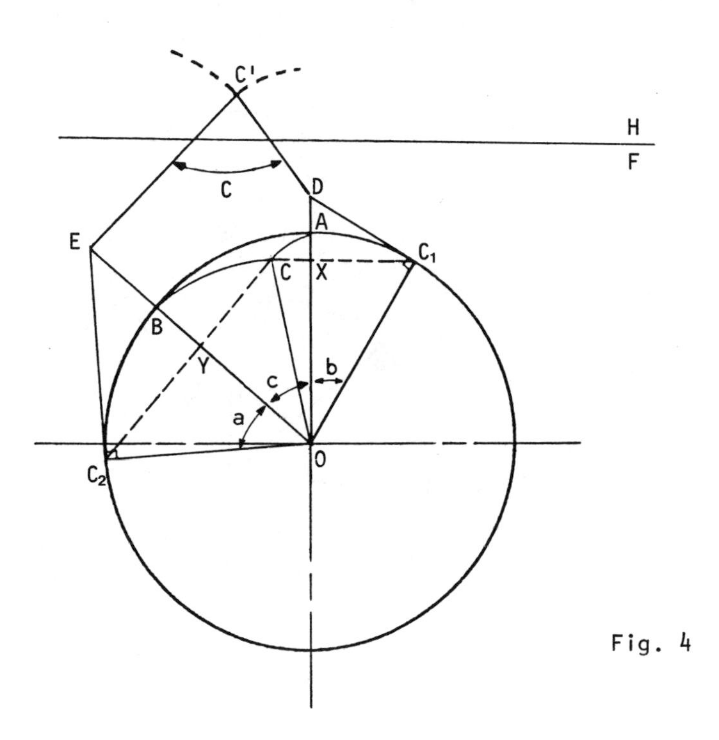

Fig. 4

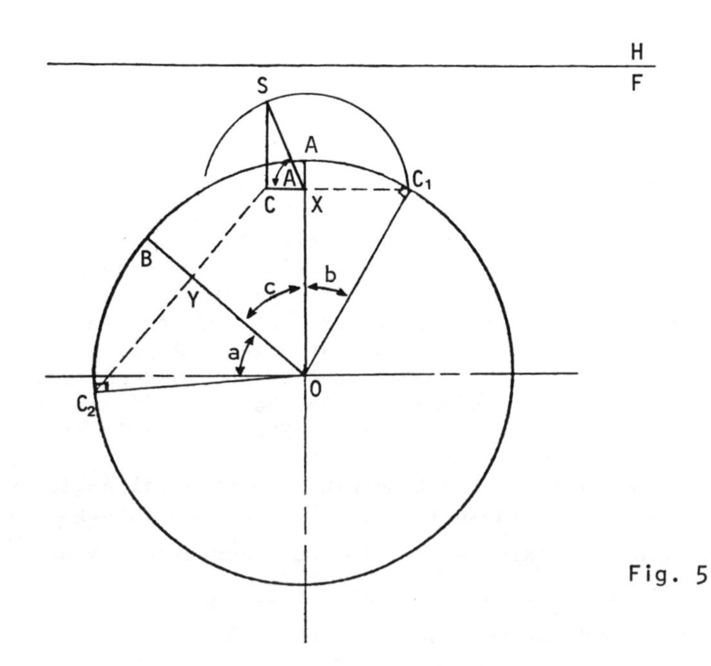

Fig. 5

The dihedral angle C is the angle EC'D. This angle would be formed by the paths of lines C_1D and C_2E when the faces BOC and AOC were rotated to their true positions in space.

A simpler, more concise method of solution of the problem is shown in figs. 4, 5, and 6, using only the front view of the development. Construct the front view as before and locate the true size view of the dihedral angle C, as in fig. 4. (Omit projection lines for clarity.) Draw an arc of radius C_1X centered at X. Draw a line through point C perpendicular to line C_1X, intersecting the arc at point S. Draw a line from X to S. The angle CXS is the dihedral angle A, as shown in fig. 5. To locate angle B, draw an arc of radius CS centered at C. Draw a line through C perpendicular to C_2Y, intersecting the arc at point T. Draw a line from Y to T

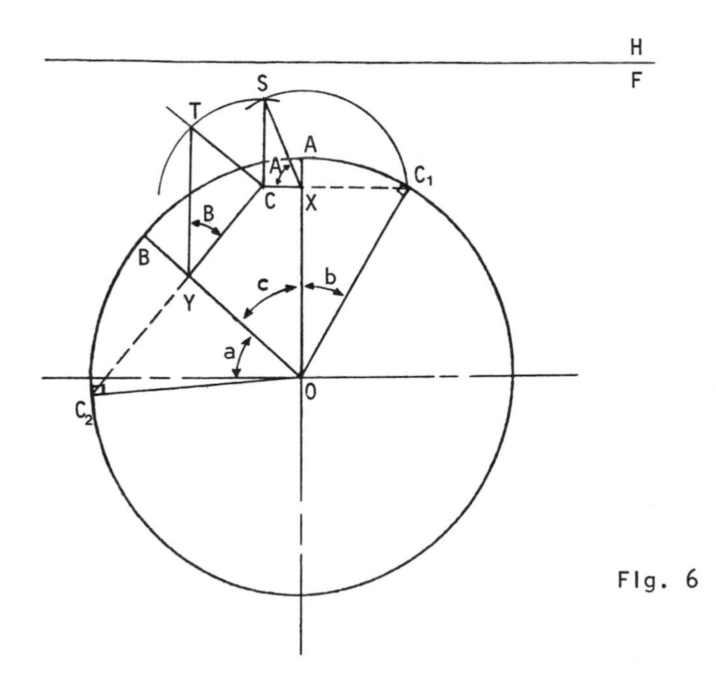

FIg. 6

The angle CYT is the dihedral angle B, as in fig. 6. These views of angles A and B are the partial top and auxiliary views of the same angles inverted (to avoid overlapping). Note that in the construction illustrated in figs. 5 and 6, regarding the views of angles A and B as right sections of the dihedral angles rotated into the vertical projection plane will aid in comprehension.

Determine one face angle and two dihedral angles of a trihedral angle, given two face angles and the included dihedral angle.

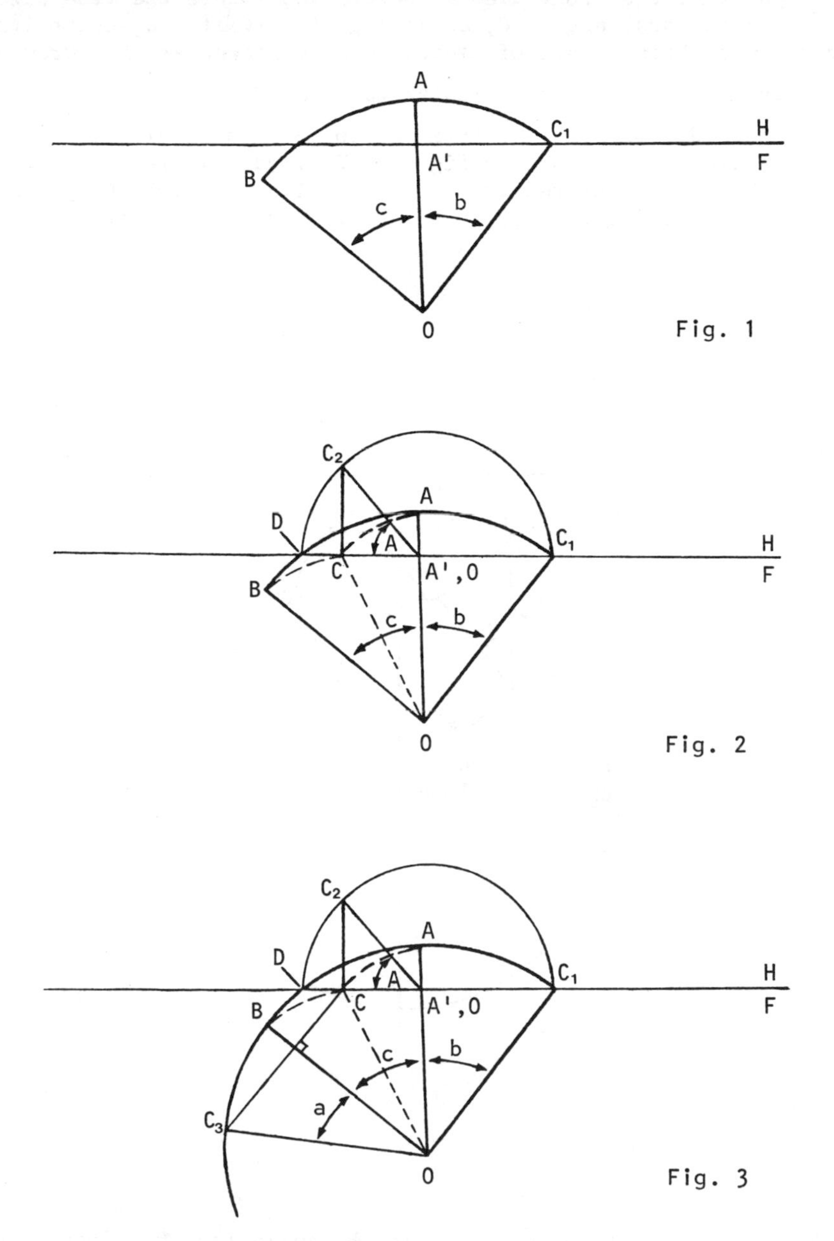

Solution: As shown in fig. 1, draw a vertical line, AO, of any convenient length in the front view. From this line, construct the development of the two given faces, AOC, and AOB, of the trihedral angle, with given face angles b and c respectively. Locate the horizontal reference plane H-F through point C_1 with H-F intersecting line AO at point A'. This point is the top or point view of line AO of the front view. As shown in fig. 2, construct a semicircle of radius AC_1 centered at A', which intersects plane H-F at point D. With a vertex at A' and a terminal side A'D, construct the given dihedral angle A, which locates point C_2 on the semicircle. The arc from C_1 to C_2 represents the top view of the path of the rotation of face AOC, about line AO into its true position in space. Project from point C_2 to the front view and across from point C_1 to locate point C, the third vertex of the trihedral angle.

To find the third face angle, a, draw a line through point C perpendicular to line OB in the front view to intersect the arc C_1AB extended at point C_3, as shown in fig. 3. The angle BOC_3 is the face angle a. (The line C-C_3 represents the path of the rotation of face BOC about line OB until BOC lies in the frontal plane.) This completes the development of the three faces of the trihedral angle. With the three face angles known, the remaining dihedral angles, B and C, can be found as described in a previous problem.

• PROBLEM 6-19

Find the two dihedral angles and one face angle of a trihedral angle, given two face angles and an opposite dihedral angle.

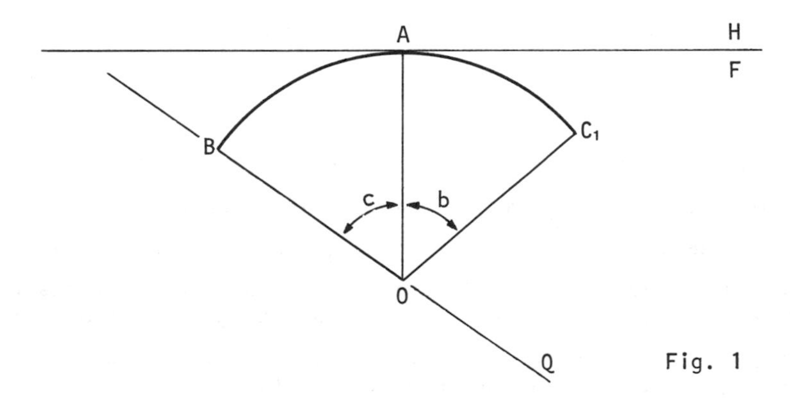

Fig. 1

Solution: As in fig. 1, draw a vertical line AO of any convenient length in the front view. Construct the development of the given faces, AOB and AOC, of the trihedral angle, with given face angles b and c, respectively. Take the horizontal reference plane H-F through point A. Through line OB in the front view, pass a plane Q which makes the given dihedral angle, B, with the frontal projection plane. The plane Q will contain the face BOC of the trihedral angle.

As in fig. 2, draw a perpendicular line through any convenient point X_1 on the extension of OB in the front view, intersecting plane H-F at point Y_1. Construct an arc of radius X_1Y_1 centered at Y_1 to intersect plane H-F at point X_2. With X_2Y_1 as a terminal side and vertex at X_2, construct the given dihedral angle B in the top view, intersecting a vertical line through point Y_1 at point Y_2. Extend line OB in the front view to intersect H-F at point T. The trace of the plane Q in the horizontal projection plane is the line joining T and Y_2.

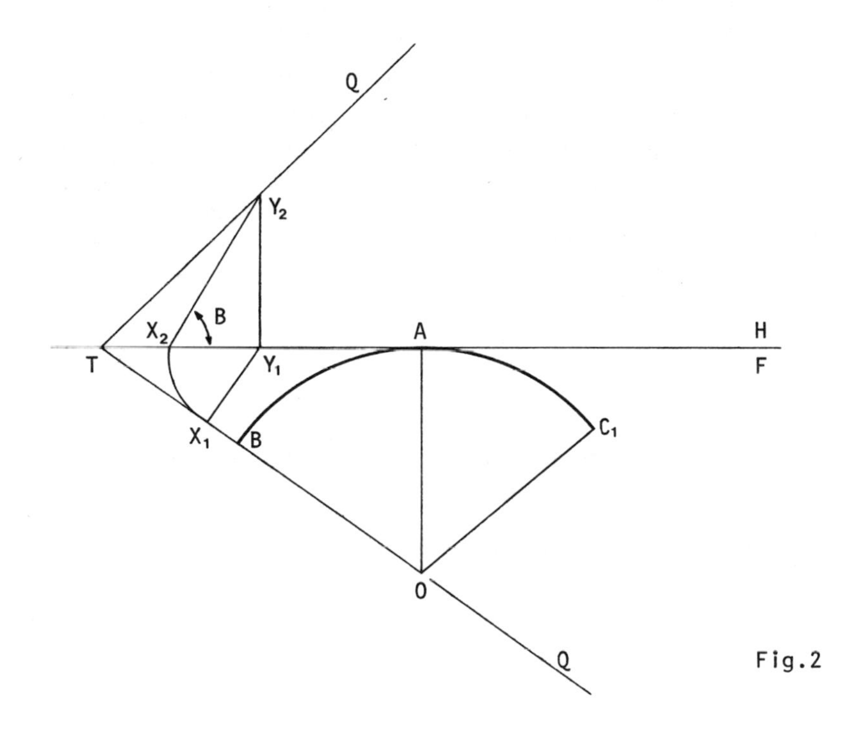

Fig.2

As in fig.3, extend line OC_1 in the front view to intersect plane H-F at point K_1. Draw an arc of radius AK_1 centered at A to intersect the H-trace of Q at points K_2 and K_3. Note that obtaining two points of intersection indicates two possible solutions to the stated problem. If the arc through K_1 were tangent to the H-trace of Q, only one solution would exist, and if it did not intersect the H-trace of Q at all, no solution would exist.

To complete one solution, draw AK_2, which represents the top view of plane AOK containing face AOC_1 in its true position in space. Project K_2 to plane H-F, intersecting at point K, and draw OK in the front view. Point C, the third vertex of the trihedral angle, lies on line OK where a horizontal projection line from point C_1 intersects line OK. Line $C\text{-}C_1$ represents the front view of the path of rotation of plane AOC_1 about line OA as an axis.

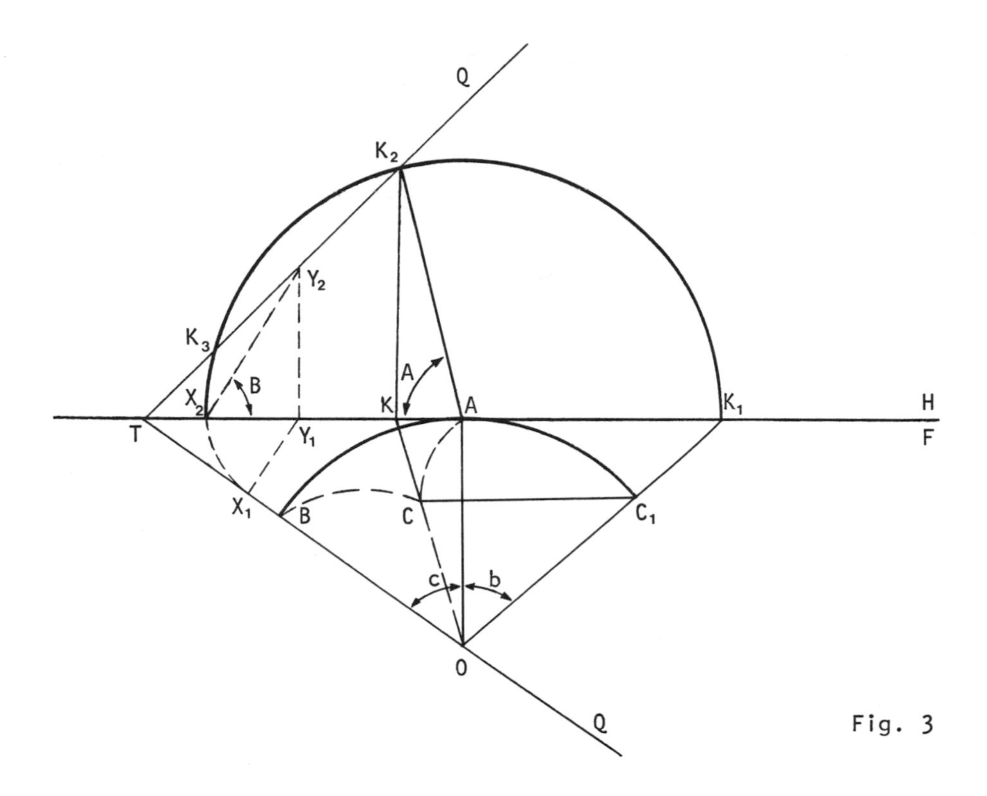

Fig. 3

The dihedral angle A is seen in the top view in fig.3 as the angle TAK_2, as both faces AOB and AOC are seen as edges and their line of intersection,AO,appears as a point. Knowing the two given face angles and the included dihedral angle, along with a given opposite dihedral angle, the remaining angle, can be found as described in a previous problem.

ORTHOGRAPHIC DRAWINGS OF SOLIDS

• PROBLEM 6-20

Complete the front view of a valve handwheel, given the top and partial front views.

Solution: Fig. 1 shows the given top and partial front views of the valve handwheel. The connections from points F to A, from points B to C, and from points D to E are required to complete the front view. As shown in fig. 2, pass horizontal cutting planes, 1 through 5, through the front view of the handwheel, where the planes appear as edges. Project the points where each plane intersects the edges of the handwheel to the horizontal center line of the top view, and construct the circles with the corresponding diameters which are cut from the handwheel by the planes. As shown in fig. 3, project the points, 6 through 19, where the circles of the top view intersect the semi-elliptical connections from F to A, B to C and

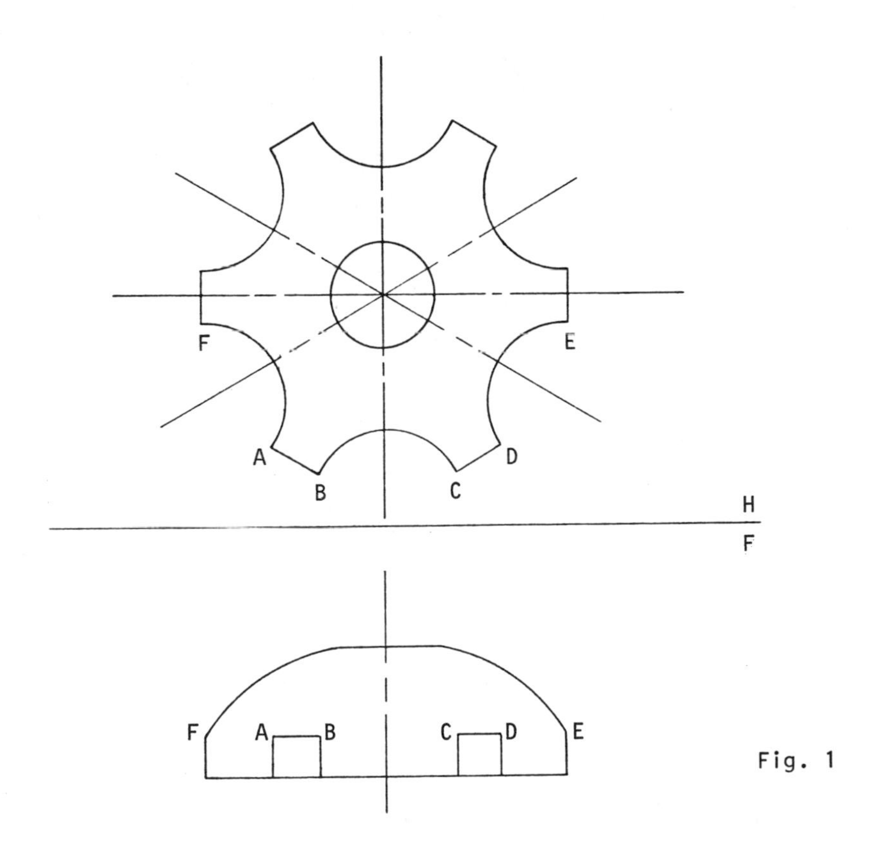

Fig. 1

D to E to the front view to where they lie on the edge views of their respective cutting planes. Connect these points in the front view with a smooth curve for the required completed front view of the valve handwheel, as seen in fig. 3.

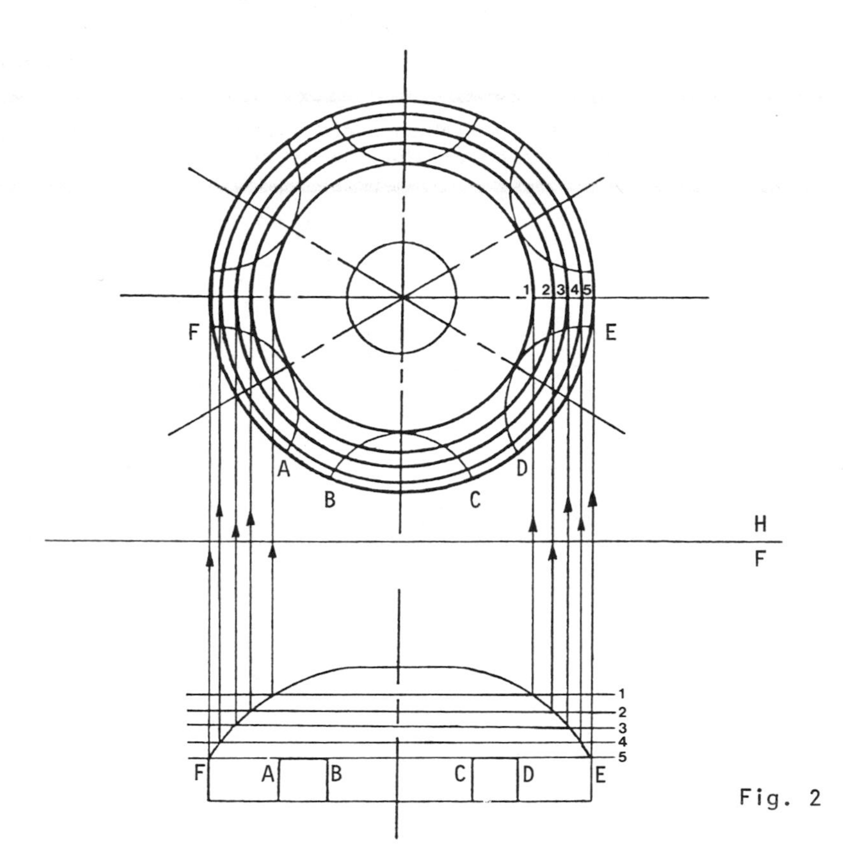
1 2 3 4 5
F
E
A
D
B
C
H
F
1
2
3
4
5
F A B C D E

Fig. 2

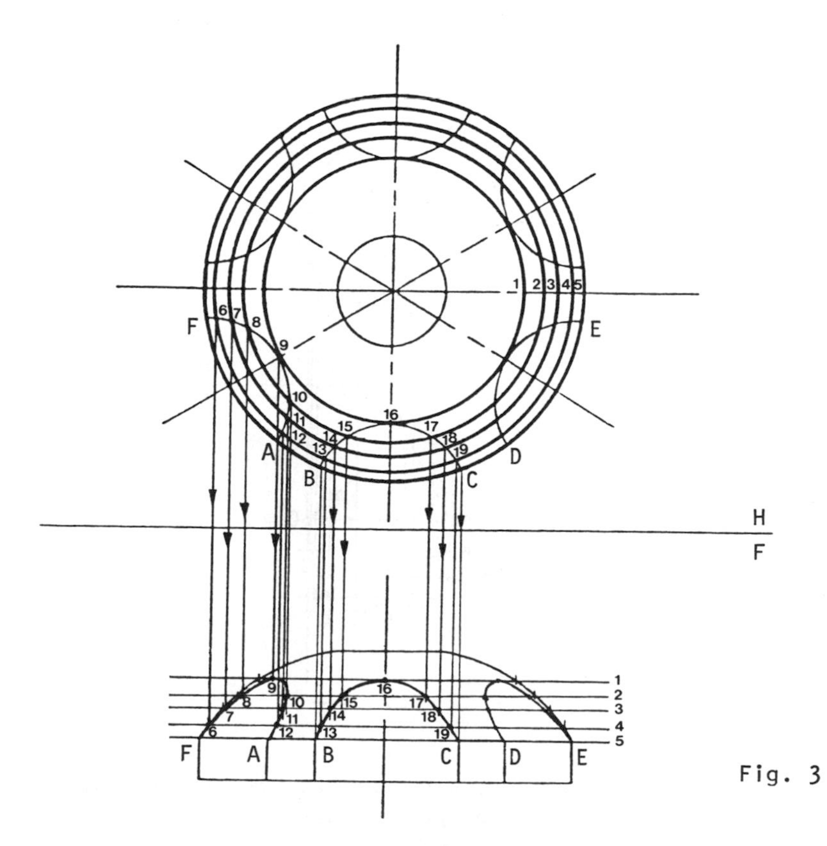
1 2 3 4 5
F
E
6 7 8 9
10 11 12 13 14 15 16 17 18 19
A
B
C
D
H
F
1
2
3
4
5
F A B C D E

Fig. 3

Complete the front view of a nail-ended rod, given the top and partial front views. The oblique faces of the end of the rod are flat surfaces.

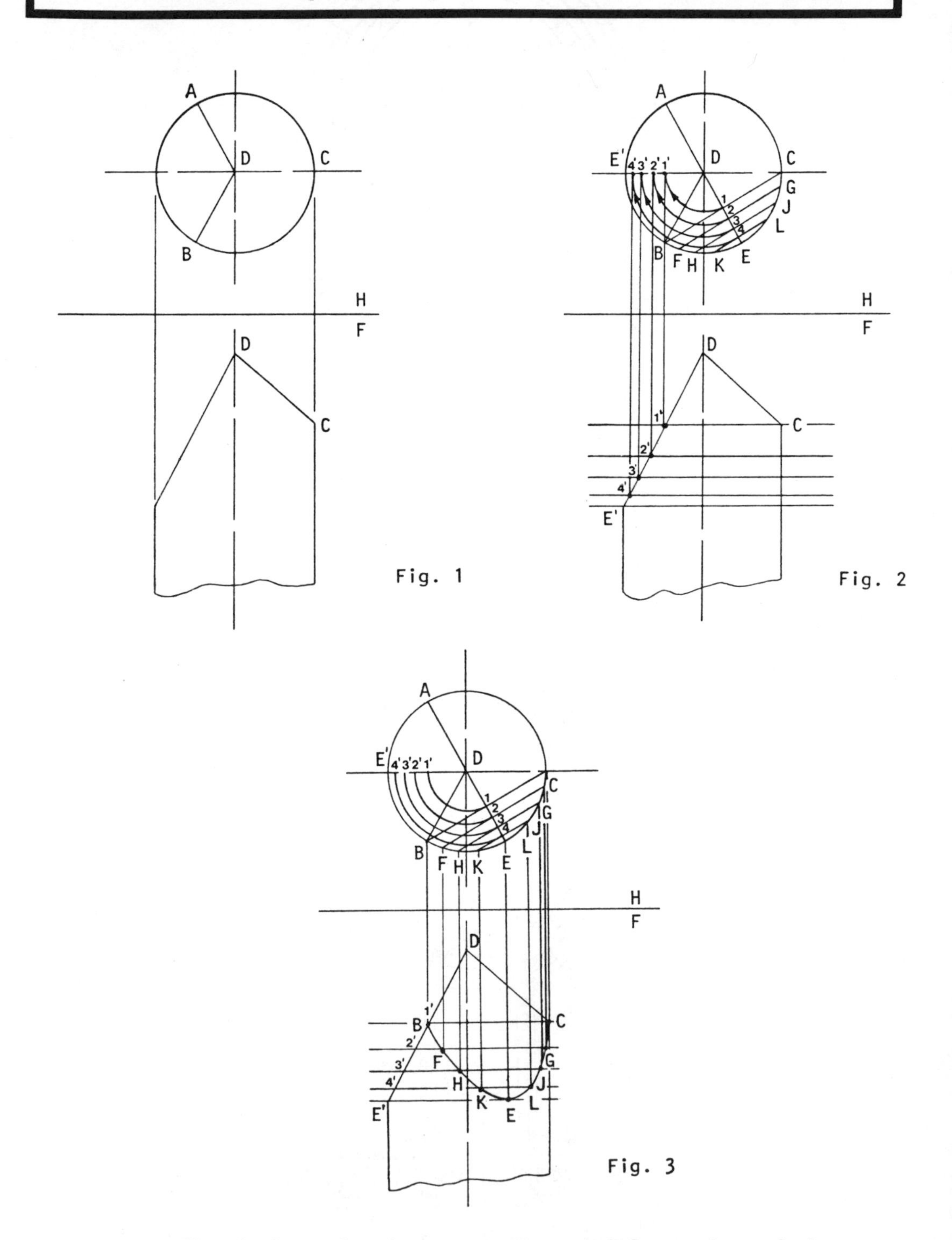

Fig. 1

Fig. 2

Fig. 3

Solution: Fig. 1 shows the given top and partial front views of the nail-ended rod. Analysis of the problem reveals that the completed front view will show the line of intersection formed between an oblique cutting plane and a right cylinder. The face visible in the front view will be DBC. As in fig. 2, connect points B and C in the top view, and extend line AD to point E on the circle. Draw lines FG, HJ, and KL parallel to BC in the top view, intersecting line DE at points 2 through 4, respectively. Draw the arcs centered at D in the top view with radii D1, D2, D3, and D4 from line DE to the center line DE', as in fig. 2. Project points 1' through 4' to the front view to the edge DE', and draw horizontal lines through these points. As in fig. 3, project points B and C to the front view to the horizontal line through point 1',and follow the same procedure for points F, G, H, J, K, and L. Connect these points with a smooth curve for the completed front view of the nail-ended rod.

• PROBLEM 6-22

Complete the front view of a palm-ended rod, given the top and partial front views.

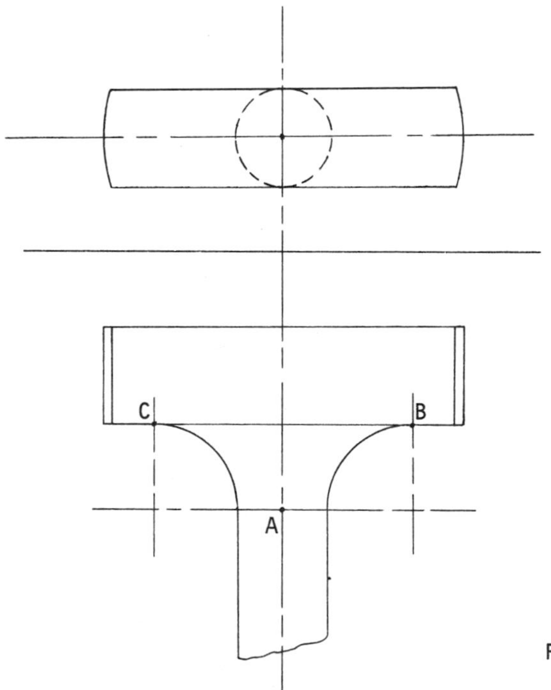

Fig. 1

Solution: Fig. 1 shows the given top and partial front views of the palm-ended rod. As in fig. 2, pass horizontal cutting planes through the arcs AB and AC of the rod handle in the front view. These planes cut circles 1, 2, 3, and 4 from the handle, as seen in the top view. Label the points where circles 1, 2, 3, and 4 intersect the edge of the rod in the top view points 1', 2', 3' and 4', correspondingly. As in fig. 3, project points 1', 2', 3', and 4' to the front view to where they intersect their respective cutting planes. These points of intersection in the front view determine the points through which a smooth curve may be drawn to complete the front view of the palm-ended rod.

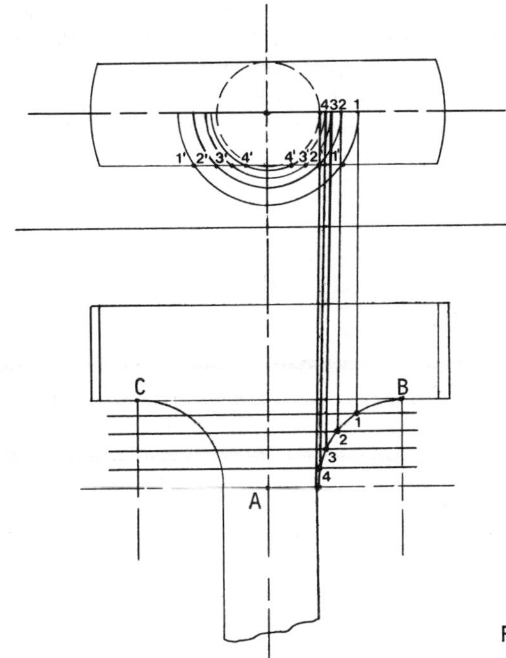

Fig. 2

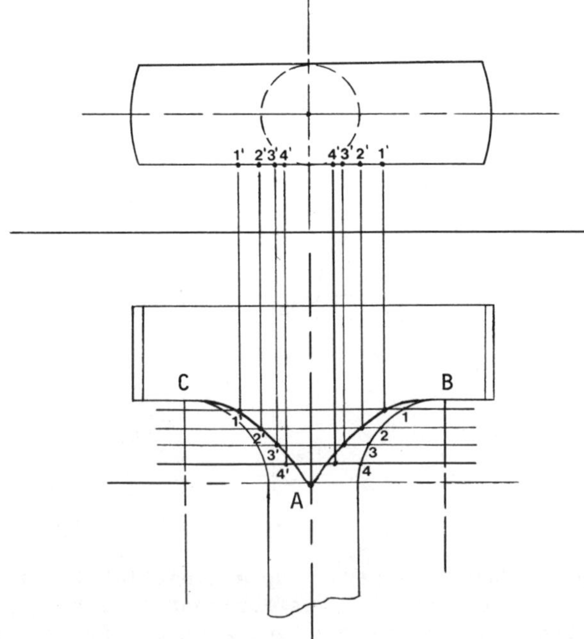

Fig. 3

Complete the bottom view of a lathe half-center, given the front, right side, and partial bottom views.

Solution: Fig. 1 shows the given front, right side, and partial bottom views of the lathe half-center. As in fig. 2, pass vertical cutting planes a through j through the front view of the object.

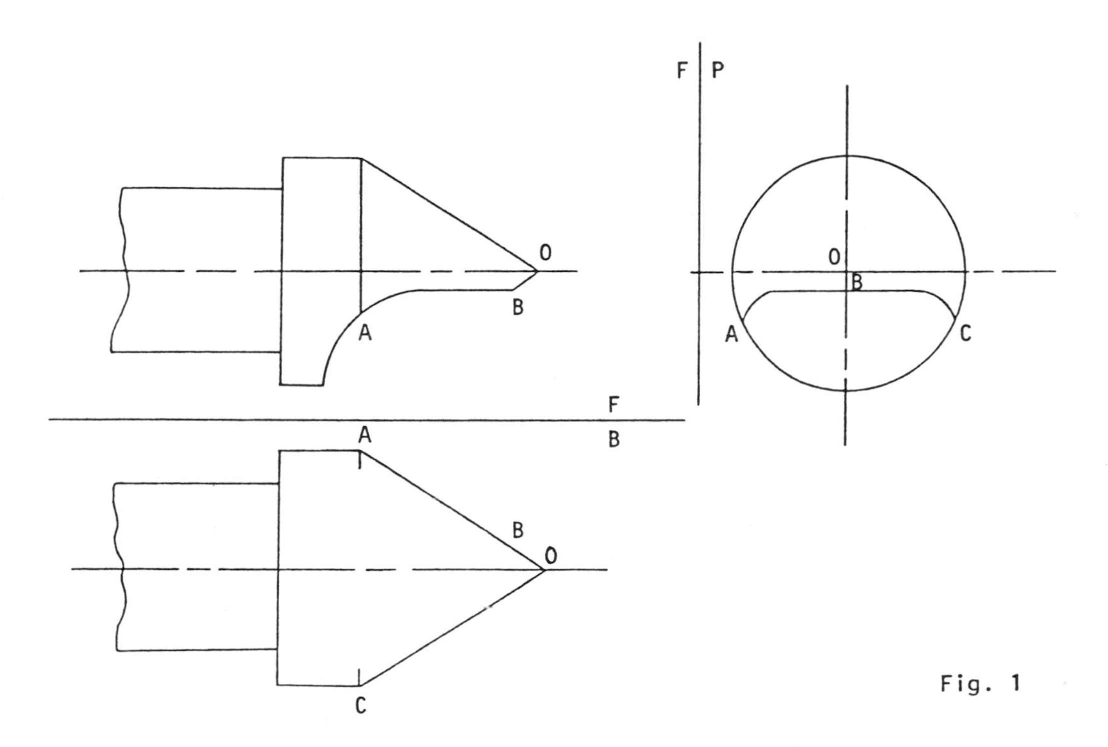

Fig. 1

Project the points where the planes cut the inclined edge of the lathe to the right side view, where the circles cut by the planes appear. Revolve points d through j in the right side view about the apex O to points d' through j' on the line through A, B, and C. As in fig. 3, project the cutting planes to the bottom view from the front view, and project the points d' through j' to the bottom view from the right side view. Where these projections intersect locates points on the bottom view of the curved section of the lathe. Connect these points with a smooth curve for the required completed bottom view of the lathe half-center, seen in fig. 3.

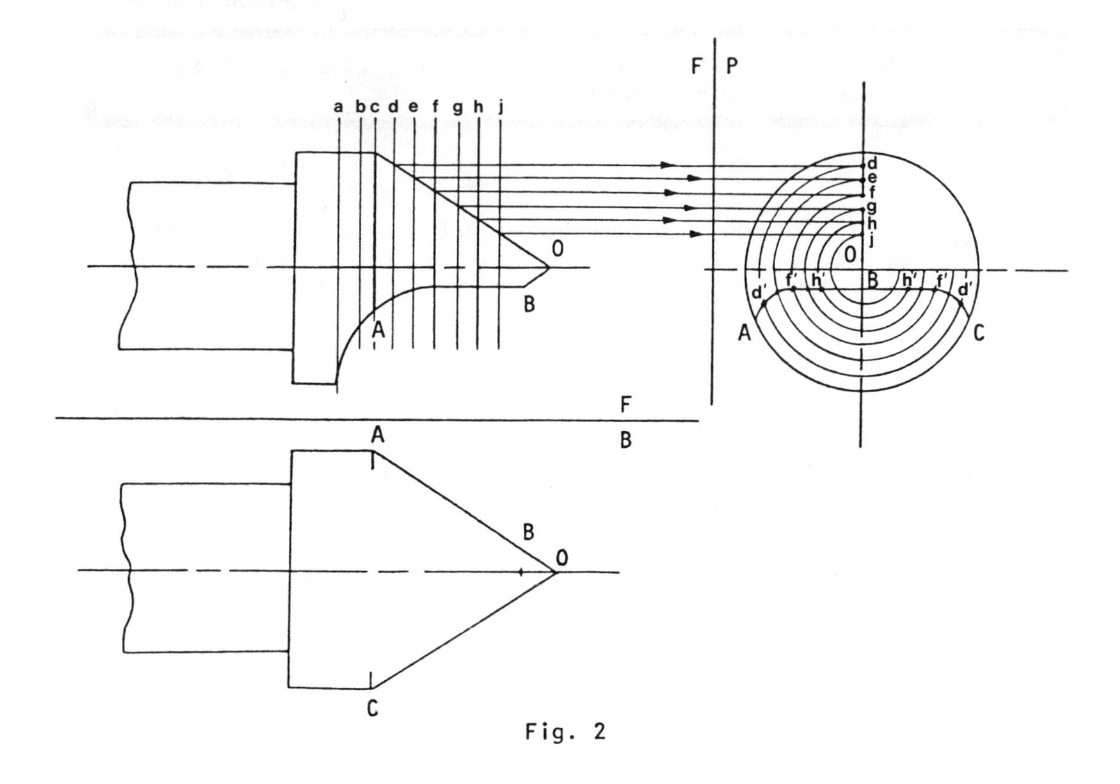
F
P
a
b
c
d
e
f
g
h
j
O
B
A
d
e
f
g
h
j
O
B
d'
f'
h'
h'
f'
d'
A
C
F
B
A
B
O
C

Fig. 2

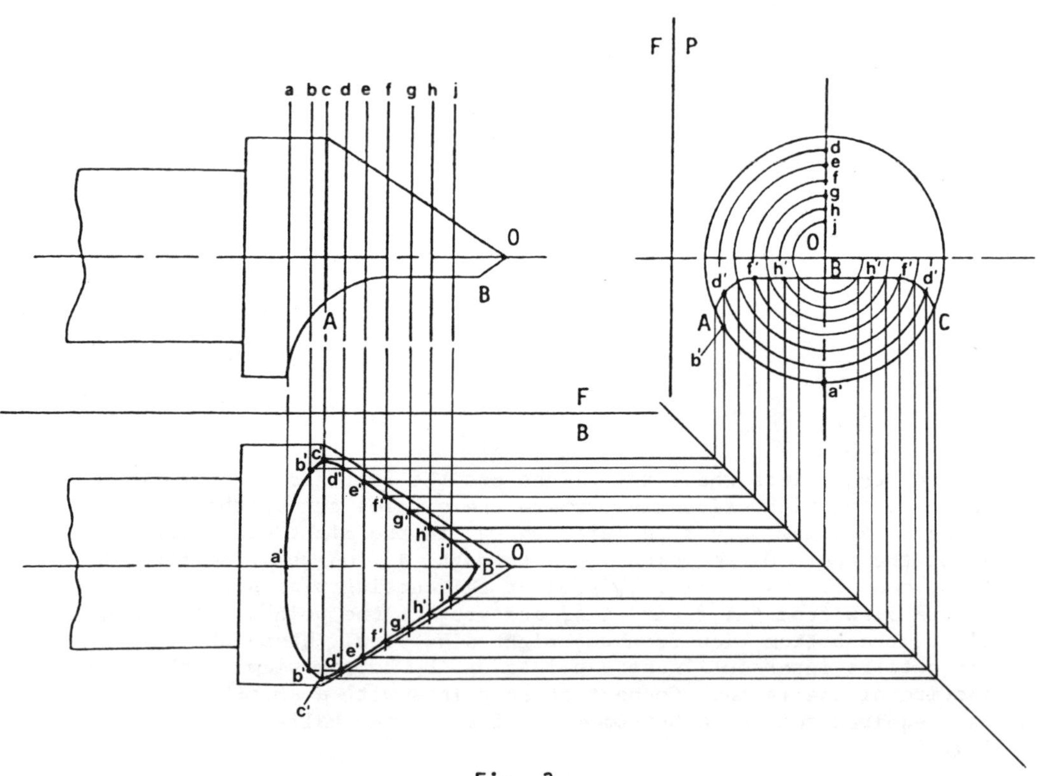
F
P
a
b
c
d
e
f
g
h
j
O
B
A
d
e
f
g
h
j
O
B
d'
f'
h'
h'
f'
d'
A
C
b'
a'
F
B
c'
b'
d'
e'
f'
g'
h'
j'
a'
B
O
j'
h'
g'
f'
e'
d'
b'
c'

Fig. 3

• PROBLEM 6-24

Construct the primary views of a helix.

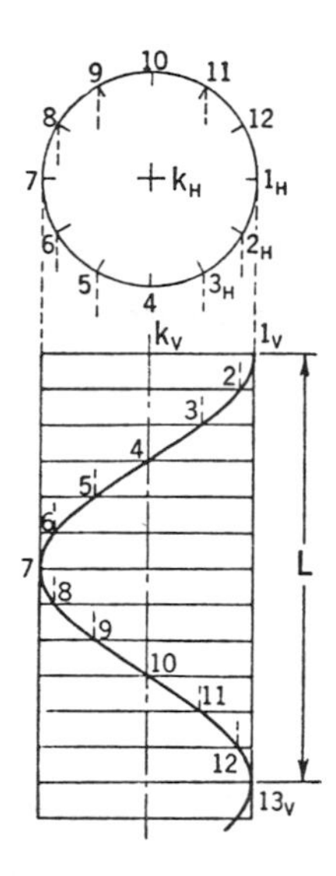

Solution: One of the most important warped surfaces has for one of its directrices a space curve called the helix. In its usual form, the helix may be defined as the path of a point which remains at a constant distance from a straight line, rotates uniformly around this line as an axis, and at the same time moves uniformly in a direction parallel to the line. Any distance traversed by the generating point in its motion parallel to the axis is the axial advance. The axial advance for one complete revolution of the generating point is the lead or pitch of the helix.

The projection of the helix on any plane perpendicular to the axis is a circle. As shown in the accompanying figure, let the vertical line k be the axis, L the lead, and P the generating point. Let the circle centered on k_H represent the path of rotation. Since the respective motions of point P around and parallel to k are uniform, the distances traversed by P in these two motions will be proportional. Divide the circle and the lead into the same number of equal parts, as 12, and draw horizontals through the points that divide the lead. Starting from 1_H, at the end of 1/12 of a complete revolution 2_H will be the H-projection of the generating point. At the same time, point P will have advanced 1/12 of the lead. Therefore, 2_V will be located by projecting from 2_H to the second horizontal line from the top. The V-projection for the position given by 3_H will lie on the third horizontal from the top. The remaining positions are located in a similar manner. A smooth curve joining the points 1_V to 13_V is the V-projection of the helix.

Since the axial advance of the point P is uniform, every portion of the helix is equally inclined to the H-plane. When the direction of motion of the generating point (in the circular view) is clockwise as it moves away from the observer, the helix is right-handed. The opposite direction of rotation gives a left-handed curve. The helix in the figure is right-handed.

• PROBLEM 6-25

Plot the path of a point moving horizontally along the surface of a rotating cylinder, the longitudinal center line of which is horizontal.

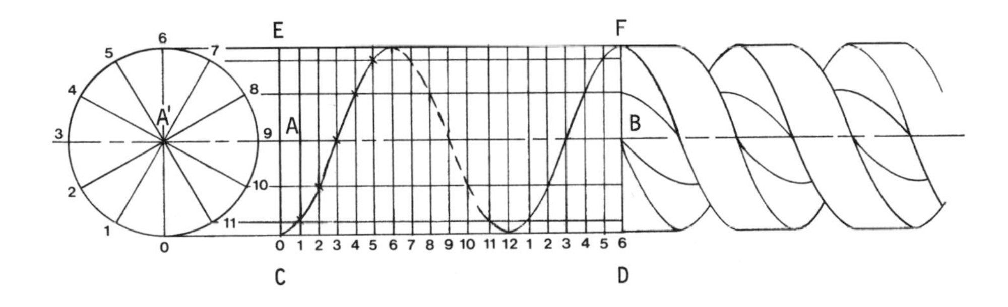

Solution: Assume clockwise rotation of the cylinder and the point moving from A to B. Draw the center line AB any convenient length. With A', located on the line B-A extended, as center, radius equal to the radius of the cylinder, draw a circle and divide it into twelve equal, numbered parts, as shown. Project the diameter of the semicircle towards A and complete a rectangle, EFDC, to represent the cylinder. Divide the rectangle into eighteen equal parts as 0, 1, 2, 3,... etc., and erect perpendiculars. Transfer the divisions of the circle horizontally across the rectangle. Commencing with 0, the intersections of the numbered divisions 0, 1, 2, 3,..., etc., vertically and horizontally, will give points through which a smooth helical curve, a helix, may be drawn. It will be seen that in moving from 0 to 6 the point will have completed one-half a revolution of the cylinder; therefore, from 6 to 12 the line will be hidden and is shown dotted. The distance 0 to 12 is one complete revolution of the cylinder and is called a pitch. Screw threads, coil springs and twist drills are examples of the application of helix construction.

Plot the path of a point traversing the surface of a right circular cone while moving at a constant rate both around and along the axis.

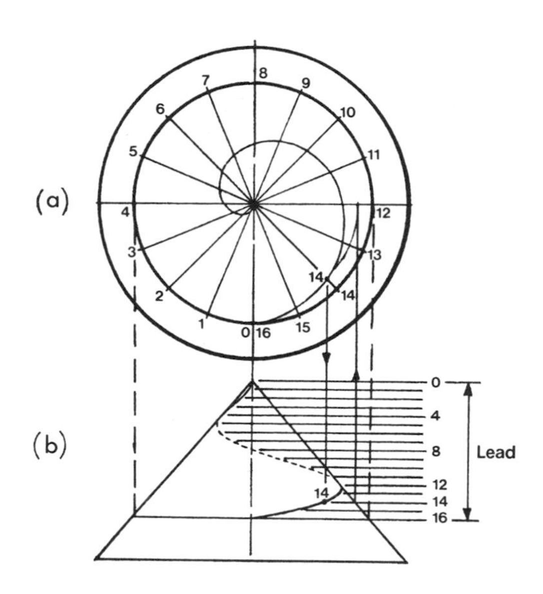

Solution: Make two views of the right circular cone on which the helix will be generated. (a) the top view and (b) the front view. Lay out uniform angular divisions in the top view, which show the end view of the axis. Divide the lead into the same number of parts. Points can now be plotted on the curve. Each plotted point will lie on a circle cut from the cone by a plane dividing the lead, as well as on the angular division line. Thus, for example, to plot point 14, measure the radius of the circle cut from the cone by plane 14 dividing the lead (b). Draw the circle of said radius on view (a). The intersection of this circle and radial division line 14 locates point 14. To locate this point in view (b), project from the top view to plane 14 in the front view. Similarly, locate all other points. A smooth curve drawn through these points yields (a) a top view and (b) a front view of a conic helix.

Assuming a clockwise rotation and that the point originates at the vertex, angular division lines 0 to 4 and 12 to 16 constitute the front view, and angular division lines 4 to 12 constitute the back view. Consequently, in moving from 0 to 4 and 12 to 16, the first and fourth quarter revolutions, respectively, the path of the point is seen and, therefore, shown by a solid curve. In moving from 4 to 12, the second and third quarter revolutions, the path of the point is hidden and, therefore, shown by a dotted curve.

Construct a tangent to a helix at a given point.

Solution: The helix may be regarded as drawn on the surface of a right circular cylinder having the same axis. As seen in fig. 1 which shows the front and top views of the cylinder, the base of this cylinder is the circle lying in plane H. When the cylinder is developed, this base will develop into the straight line 1, 2, 3, ... 1. Since every portion of the helix is equally inclined to the plane of the base, this curve will develop into the straight line 1-13, which makes the angle θ with the developments of the base. Angle θ is the helix angle.

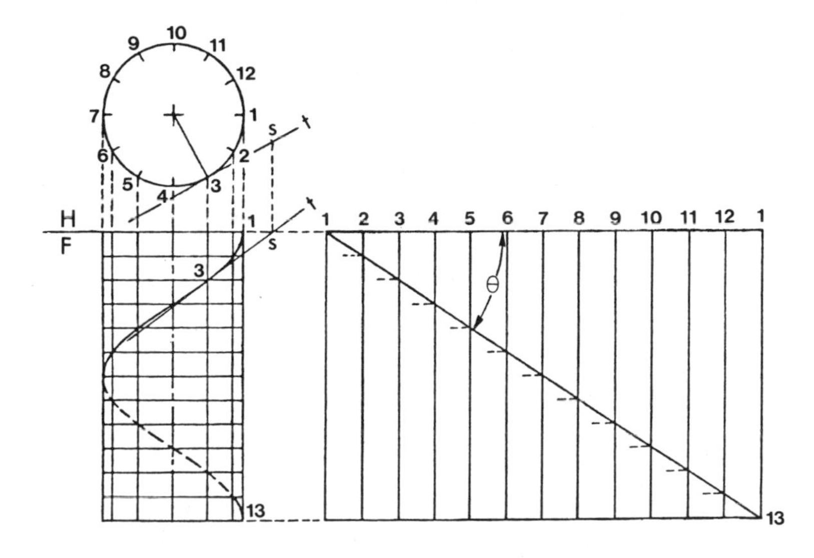

Fig. 1

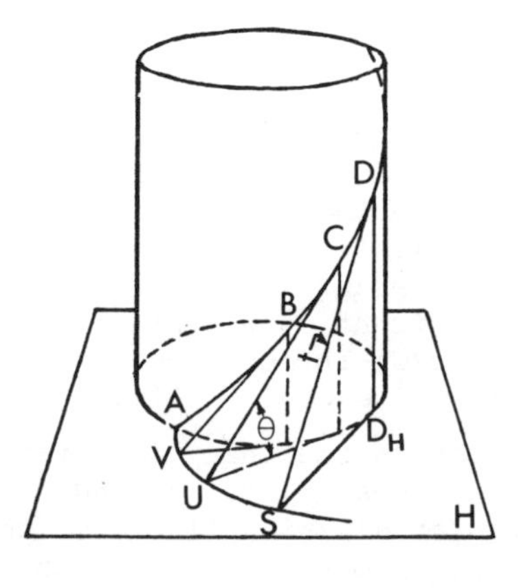

Fig. 2

Since every element of the helix makes the same angle θ with plane H, every line tangent to the helix will also make the angle θ with H. The length of the tangent line between the point of tangency and H will be equal to the length of the helix between this point and H.

Let line t be tangent to the helix at D and pierce H at point S. The length of t is ABCD on the helix, and the length of the H-projection of line t, $S\text{-}D_H$, is equal to the arc $A\text{-}D_H$. See fig. 2.

Let point 3 be the given point, as in fig. 1, and draw a line t tangent to the circle at this point. Mark off the distance 3-S, in the top view, equal to the arc length 1-3. (The point S is the piercing point of line t in plane H.) Project S to the front view to where it intersects the H-F reference plane. (Point S lies in plane H, so it must lie on H-F.) Draw line t in the front view through 3 and S. Note that any plane perpendicular to the axis may be used as the base of the curve.

• PROBLEM 6-28

Define the developable helicoid, or helical convolute.

(a) Construct the convolute.
(b) Develop a given portion of a helical convolute.
(c) Find the radius of the development of the helix.
(d) Construct the exact development of the surface: the surface between the circular arc and its involute.

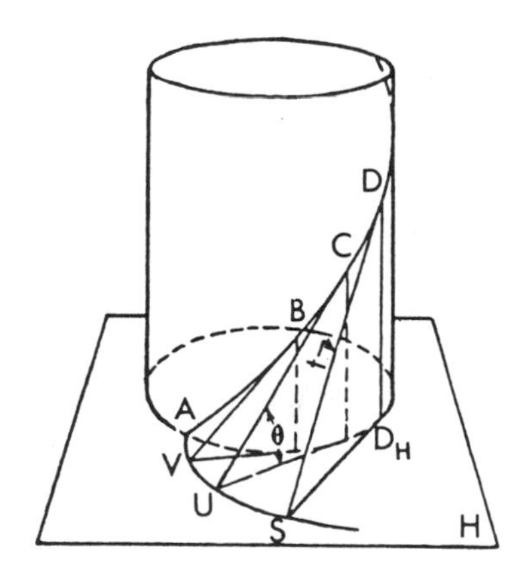

Fig. 1

Solution: The developable helicoid or helical convolute is generated by a moving straight line which remains always tangent to a helix. In the pictorial illustration, Fig. 1, ABCD ... is the helical director; BV , CU, DS positions of the generating line; and AVUS the line in which the surface intersects the plane H taken perpendicular to the axis of the helix. If the elements are produced indefinitely on either side, two nappes of the surface are formed. The surface is single curved and is therefore capable of development.

(a) Construct the convolute. Let the helix a, Fig. 2, be given. Draw tangents to the helix at points 0, 1,... 8, and find their traces on the H-plane. A smooth curve b drawn through these traces represents the intersection of the surface with H, and may be regarded as the base of the convolute. It will be seen that b_H is the involute of the circle a_H.

Since the convolute is a single curved surface, a plane tangent to the surface at a given point is tangent along the entire length of the element passing through the point. The tangent plane will also contain a line tangent to the base at the point in which the element intersects the base.

(b) Develop a given portion of a helical convolute. Let it be required to develop a portion of the helical convolute represented in Fig. 3. Divide the helix into a number of equal parts and draw the elements A1, B2, and so forth. Assume that the elements intersects as follows: B2 and C3 meet in W; C3 and D4 meet in X; D4 and E5 meet in Y; and so forth. This assumption is nearly exact, becoming entirely true when the divisions of the helix are infinitely small. The surface can now be developed by triangulation, a portion being shown in Fig. 4.

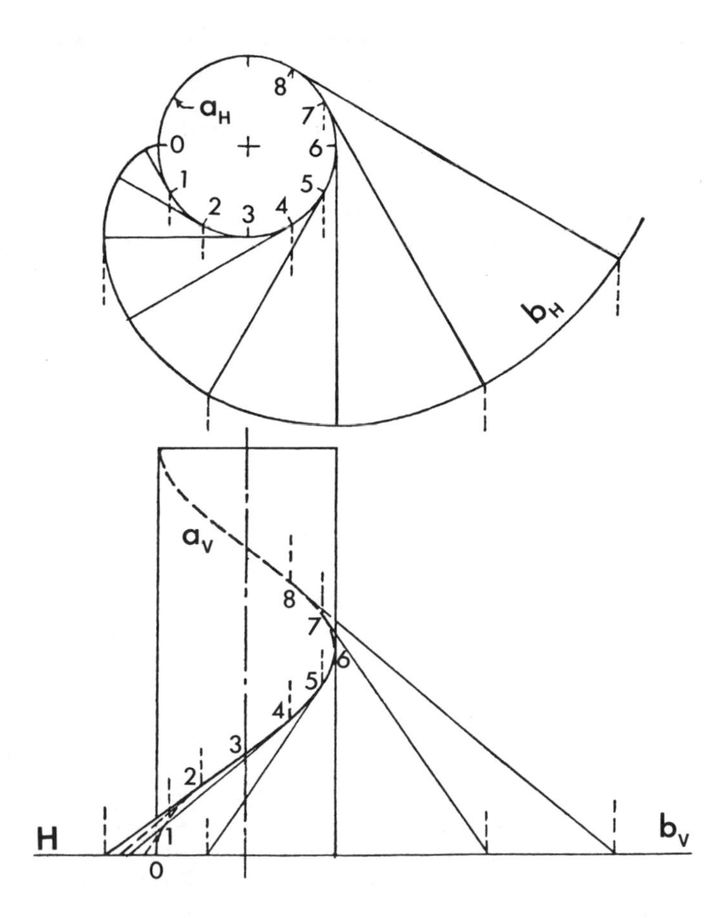

Fig. 2

As an illustration of development by triangulation, consider the development of the lateral faces of a right pyramid: Fig. 5. By definition, a development of a polyhedron is a drawing which shows the true size and relative position of each face of the solid. It represents the surface cut open along certain edges and folded out into a single plane. When a development is cut from sheet material and properly bent, it reproduces the surface of the solid rather closely. In practical work, the development usually shows the inside of the surface, since the working dimensions of sheet-metal structures are often the inside dimensions. In addition, certain allowances must be made for seams and "crowding" due to the thickness of the material. In this example, only the purely geometrical aspects of the development will be treated.

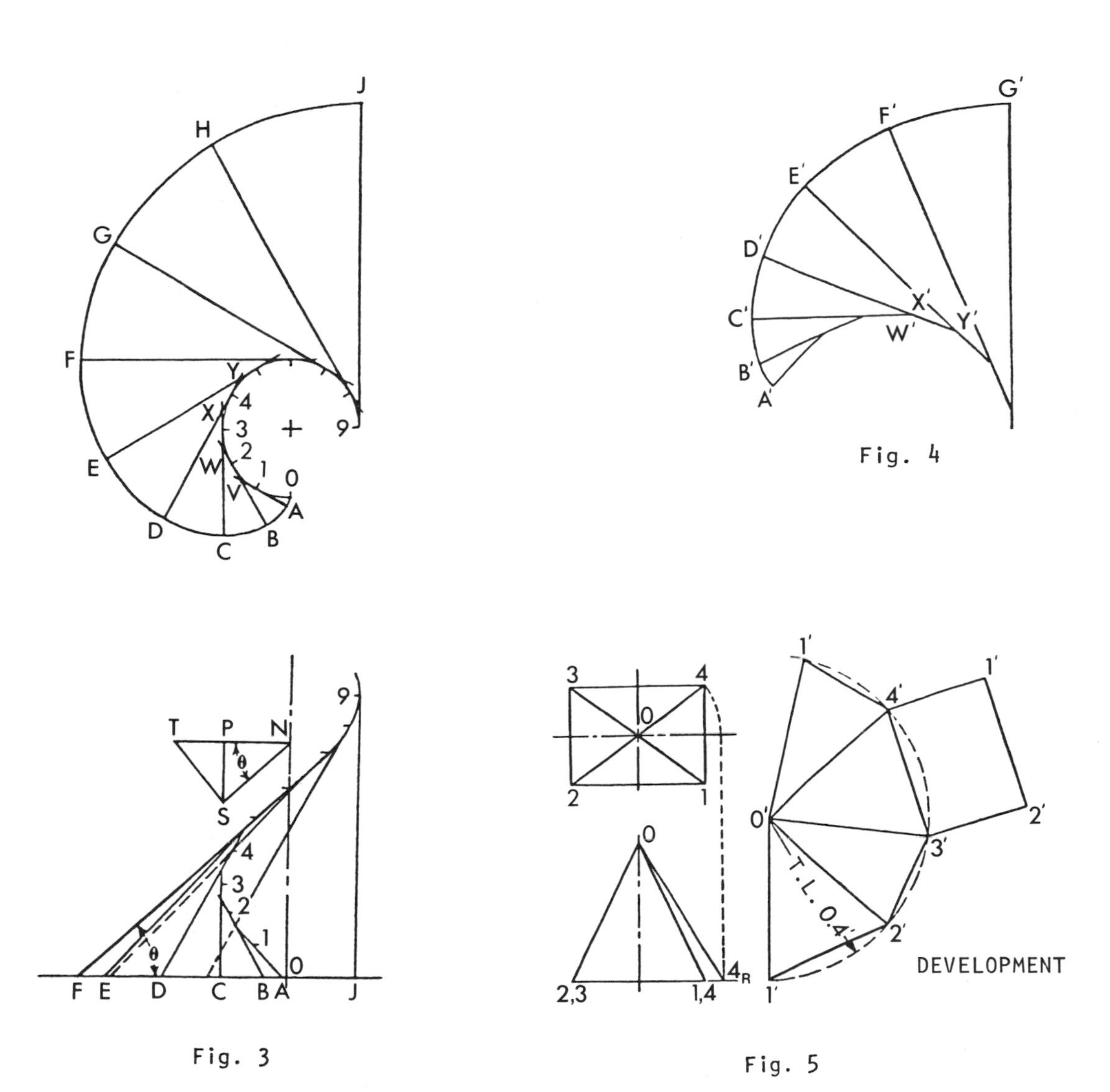

Fig. 4

Fig. 3

Fig. 5

The development of the lateral faces in Fig. 5 consists of four triangles. The surface is opened along the edge 0-1, and the faces are laid out in sequence. The lateral edges are equal in length. Assume the point 0' as the vertex of the development. Using the true length of 0-4 is a radius and point 0' as the center, strike an arc. Starting at point 1', set off the true lengths of the basal edges, 1'-2', 2'-3', 3'-4', 4'-1' along this arc. Connect these points to form the faces.

The rectangular base may be attached to any basal edge of the development.

Examining the development, Figure 4, it can be seen that the angles W', X', Y',... are equal since they are the true sizes of the angles W_H, X_H, Y_H... of the H-projection which are equal. Also, the lengths W'X', X'Y'... are equal. Therefore, a circle drawn tangent to these segments will represent the development of a portion of the helix. When the divisions of the helix are infinitely small, the tangent circle becomes the exact development of the helix, and the curve BCDEFG becomes its involute. Hence, the true development of a helical convolute is bounded by a circle and its involute. The radius of the circle is equal to the radius of curvature of the helix.

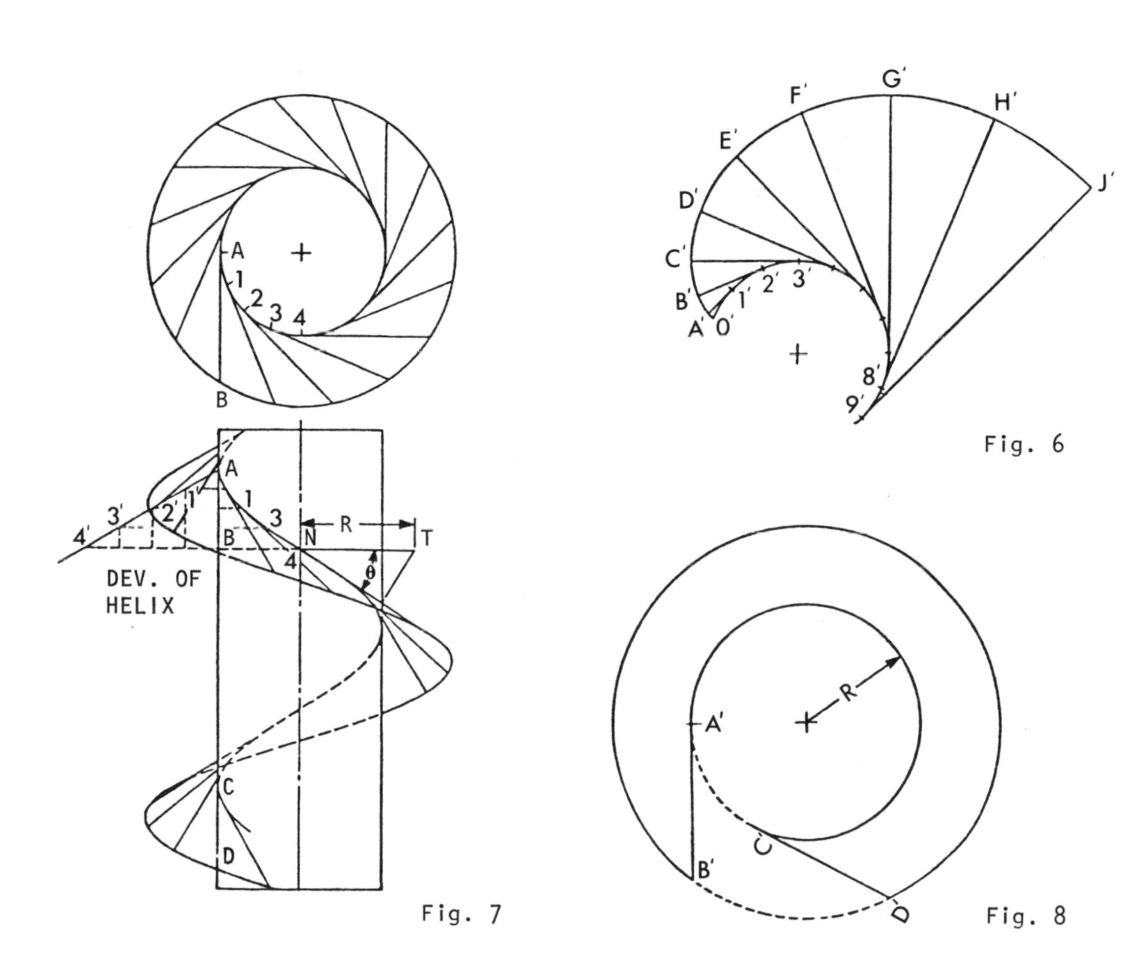

Fig. 7

Fig. 6

Fig. 8

(c) Find the radius of the development of the helix. The radius of curvature of a helix is constant, and is equal to $R \sec^2\theta$ (by Differential Calculus), where R is the radius of the cylinder containing the helix, and θ is the helix angle. This radius can be found graphically as follows: (Figure 3) . In any convenient position in the V-projection, lay off NP = R; draw NS making the angle θ with NP; also draw PS perpendicular to NP. Then $NS = R \sec \theta$. Draw ST perpendicular to NS; then $NT = NS \sec \theta = R \sec^2\theta$. Therefore, the radius of curvature of the helix is equal to NT.

(d) Construct the exact development of the surface. With radius NT, draw a circular arc (Figure 6), Make the distances 0'-1', 1'-2',...8'-9' equal to the true lengths of the segments 0-1, 1-2, and so forth, of the helix. These lengths may be taken from the development of the helix. Construct the involute of this arc. The required development is the surface between the circular arc and its involute.

The portion of a helical convolute lying between the helical directrix, and a circular cylinder (Figure 7) having the same axis has been used in practice as the working surface of a screw conveyor. The surface between elements AB and CD represents one turn or flight of the convolute.

In order to develop the surface, it is necessary to determine the radius of the development R = NT and the development of a portion of the helix A-4'.

The development (Figure 8) is constructed by striking a circle of radius R and laying off along the arc the developed length of the helix A'C'. At point A', the element AB is drawn in true length tangent to the arc, locating point B'. A circular arc passing through point B' represents the developed edge of the convolute. The element C'D' drawn tangent to the arc A'C' at point C' completes the development.

• PROBLEM 6-29

Define a helicoid. Construct a plane tangent to the helicoid at a given point P.

Solution: In its most common form, the screw surface or helicoid is generated by a straight line which slides uniformly along a fixed straight line or axis with which it makes a constant angle and at the same time rotates uniformly around that axis. Note: Any given point of the generating line describes a helix. The surface thus obtained is a warped surface and, therefore, not developable.

A second form of the helicoid is obtained when the generating line does not intersect the axis but remains at a constant distance form it. When the constant angle between the generating line and the axis is 90°, the surface generated is a right helicoid. When the constant angle is other than 90°, the surface is an oblique helicoid. Due to the fact that the generatrix is of indefinite length, the helicoid is composed of two nappes, separated by the axis of the helix. In this problem (see the accompanying figure) only one nappe is shown. The surface is usually represented by its axis, its helical directrix, and several positions of the generating line.

In constructing the tangent plane, recall that a plane tangent to a warped surface at a given point will contain the element of the surface which passes through the point. In general, the plane will be tangent to the surface at only one point and elsewhere it will be secant to the surface. Any plane containing an element of a warped surface will be, in general, tangent to the surface at some point. For the case of a doubly ruled surface, a plane tangent to the surface at a given point will contain the two straight-line elements of the surface which pass through the given point. For a singly ruled surface a plane tangent at a given point may be determined by one straight-line element of the surface drawn through the point and the tangent to any curve lying in the surface and containing the given point. For the singly ruled surface in the accompanying figure, let the given helix, a, be the directrix and the given line, k, be the axis of a right helicoid.

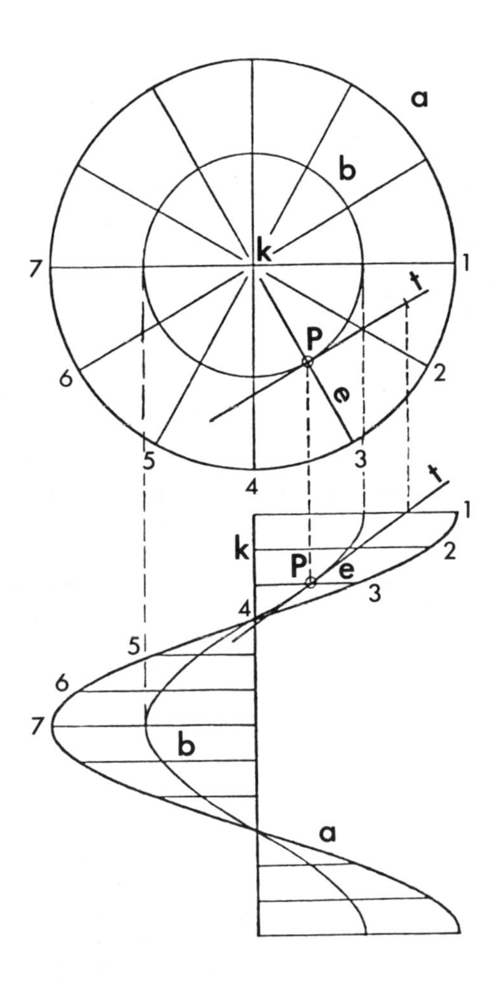

A plane tangent at the given point P is determined by the element e (the one straight-line element of the surface, drawn through the point) and the line t (the tangent to any curve lying in the surface and containing the given point). The line t is tangent at P to the helix b passing through P. The helix b is determined by the intersection of the cylinder of radius kP, passing through point P, and the helicoid.

HYPERBOLIC PARABOLOID, AND UNPARTED HYPERBOLOID

• PROBLEM 6-30

Describe a hyperbolic paraboloid.

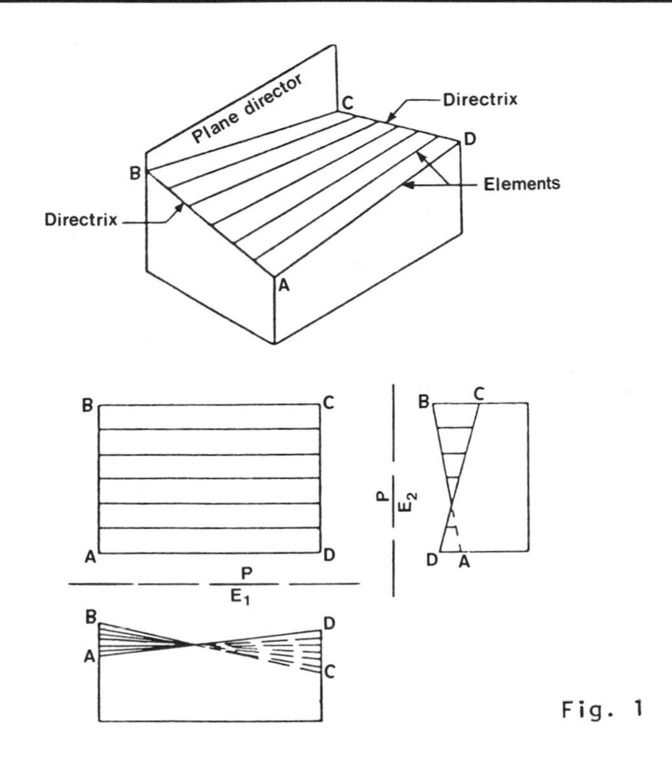

Fig. 1

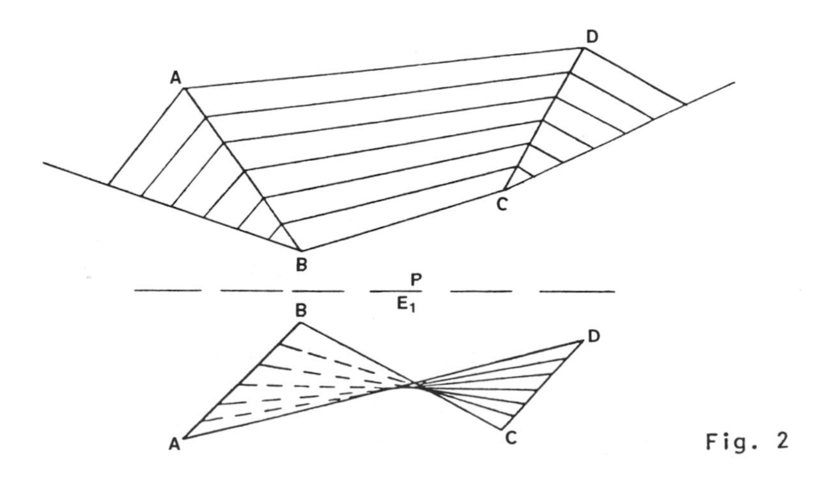

Fig. 2

Solution: The surface generated by a straight line that moves along two nonintersecting, nonparallel lines and at the same time remains parallel to a plane is called a hyperbolic paraboloid (Figure 1).

It has a straight-line generatrix, two straight-line directrixes, and a plane director, and is a double-curved warped surface.
The position of the plane director need not be known in order to draw elements of a hyperbolic paraboloid when the positions of the straight-line directrixes are given. Each of the directrixes can be divided into equal segments, using the same number of segments on each directrix (Figure 2). The points thus established can then be connected by straight-line elements.

• PROBLEM 6-31

Discuss the properties of the unparted hyperboloid in detail.

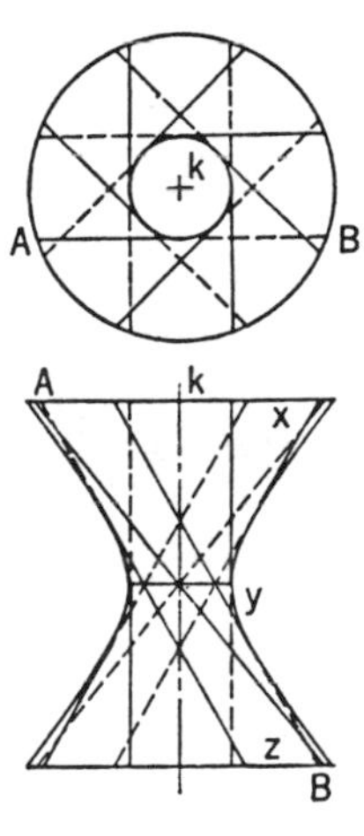

Fig. 1

Solution: The unparted hyperboloid is a doubly ruled surface in which the motion of the generating line may be controlled by three straight-line directrices so selected that the surface has no plane director. This surface is generally obtained when three straight-line directrices are taken at random. The surface is symmetrical, having one principal axis. Any plane containing the axis cuts the surface in a hyperbola. A plane perpendicular to the axis will, in general, cut the surface in an ellipse. For this reason, the surface is sometimes known as the elliptic hyperboloid. It is possible to choose the directrices so that the elliptical cross section becomes a circle. The hyperboloid then becomes a surface of revolution.
The hyperboloid of revolution is most easily constructed by rotating one straight line about another not in the same plane.
Let the line AB, in Fig. 1, rotate about the vertical axis k. Each point of line AB will describe a circle perpendicular to k. Points A and B describe the circles x and z respectively. The point

of AB nearest the axis will describe the circle y, which is called the circle of the gorge. Eight successive positions of the generating element AB are represented.

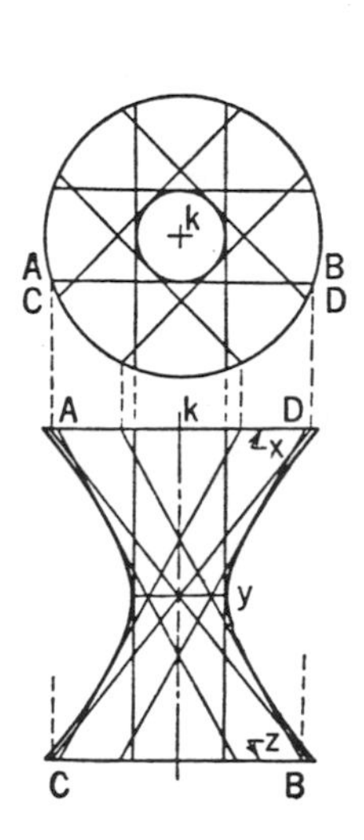

Fig. 2

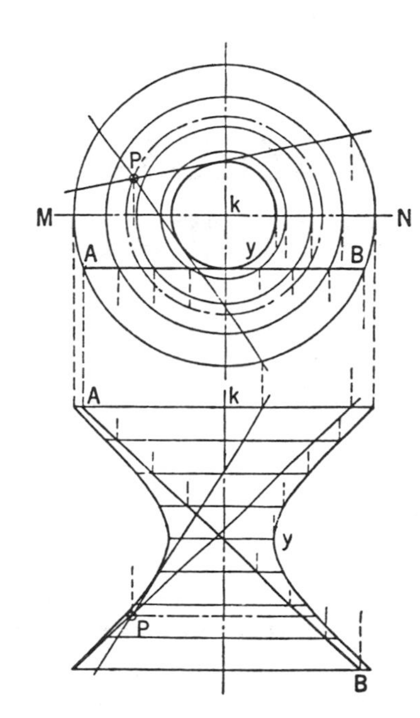

Fig. 3

Since this surface is symmetrical with respect to the axis k, it is evident that the line CD (Figure 2), which is placed the same as AB except that it has the opposite inclination, can be rotated to generate the same surface. The hyperboloid is thus doubly ruled. In Figure 2 eight elements of each ruling are drawn. It will be noted that each projection of an element of one system is also a projection of an element of the second system.

The hyperboloid of revolution can be represented by a series of parallel circles (Figure 3). Let k be the axis and AB the generating line. Each point of AB will describe a circle having its plane at right angles to line k. The hyperbola which represents the apparent contour of the front view is determined by the points in which the frontal plane MN, containing the axis k, cuts the parallel circles of the surface.

A plane tangent to the surface at point P contains one element of each generation drawn through the point.

CHAPTER 7

INTERSECTIONS OF SOLIDS AND SURFACES

CONES

• PROBLEM 7-1

Describe the conic sections which result from the intersections of a plane with a cone.

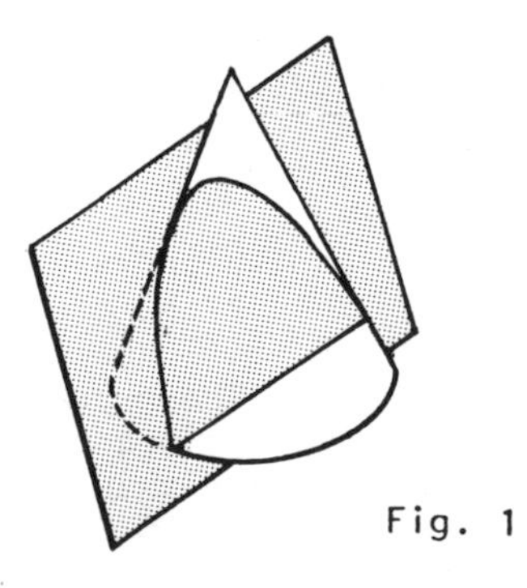

Fig. 1

Solution: As shown in fig. 1, the line of intersection formed by a cutting plane that passes through the cone parallel to an element of the cone is a parabola. (The cutting plane intersects only one nappe of the cone.) This figure can be defined geometrically as the locus of coplanar points equidistant from a straight line, known as the directrix, and a point, known as the focus. Its equation is in the form

$$Y = AX^2 + BX + C ,$$

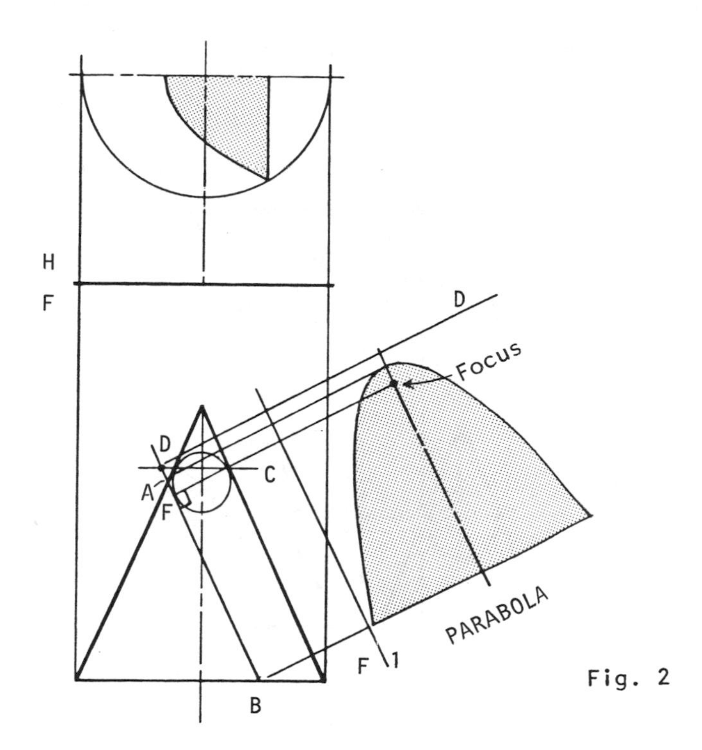

Fig. 2

where A does not equal zero.

The cutting plane AB is seen as an edge in the front view of fig. 2. The top view is drawn by projecting plane AB using a series of horizontal planes. Now the true size and shape view of the parabola can be drawn in a primary auxiliary view, projected perpendicularly from the edge view of plane AB.

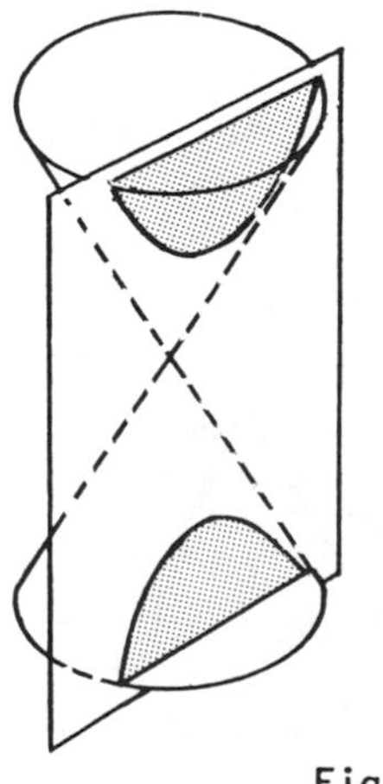
Fig. 3

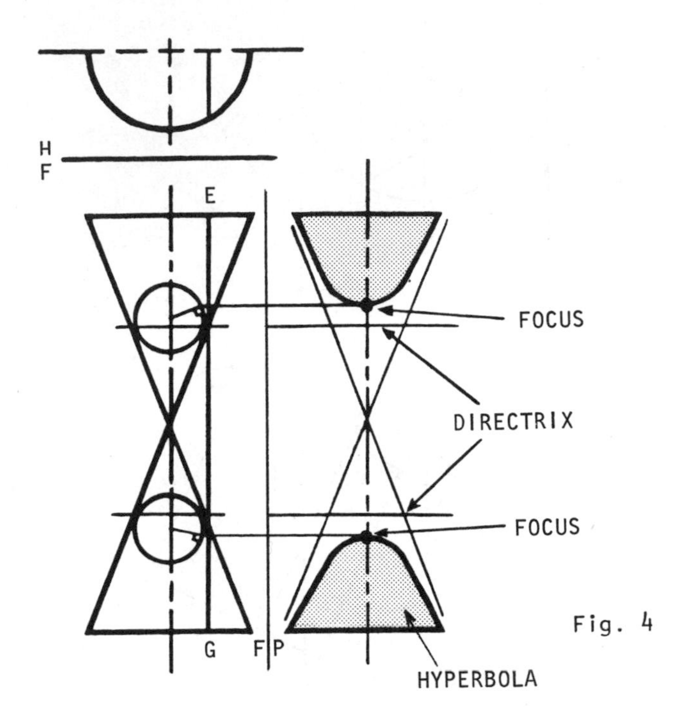

Fig. 4

To locate the foci and directrices of any conic section, spheres are inscribed in the cone, tangent to the plane. This is an application of the geometrical theorem stating that (1) the points of tangency between such spheres and the cutting plane are the foci of the conic section, and (2) the straight lines in which the cutting plane is intersected by the planes of the circles of contact of the spheres and the cone (i.e., the planes which pass through the points of tangency between the spheres and the cone) are the directrices of the conic section.

As seen in fig. 2, point F in the front view projects to the auxiliary view as the focus of the parabola. Plane C intersects the edge view extension of plane AB at point D, which is the point view of the directrix of the parabola, and which is projected to the auxiliary view as a line. All lines entering the open end of the parabola parallel to its axis of symmetry will be reflected to a common point, the focus.

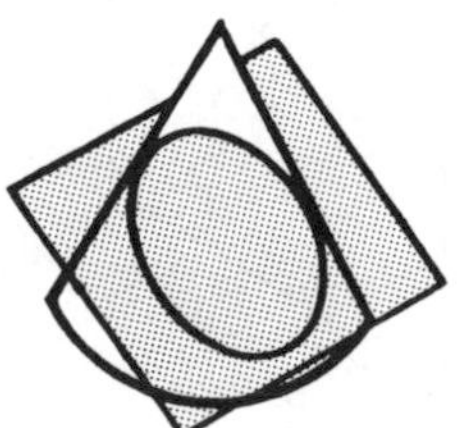
Fig. 5

The line of intersection that is formed by a cutting plane which is passed through a cone in a manner such that it intersects both nappes of the cone is a hyperbola, as seen in fig. 3. A hyperbola can be geometrically defined as the locus of coplanar points, the difference of whose distances to two fixed points, called foci, is constant. Its equation is in the form

$$\frac{X^2}{A^2} - \frac{Y^2}{B^2} = 1 ,$$

where A and B do not equal zero.

The cutting plane EG is seen as an edge in the front view in fig. 4. The line of intersection of this cutting plane is found in the top view by using horizontal cutting planes. An auxiliary view, a right side view in this case, shows the true size and shape of the hyperbola. The circles constructed in fig. 4 tangent to the limiting elements of the cone and to plane EG locate the foci and directrices of the hyperbola.

As shown in fig. 5, the line of intersection of a plane that cuts all elements of a cone but is not perpendicular to the axis of the cone is an ellipse. An ellipse can be defined geometrically as the locus of coplanar points the sum of whose distances to the foci is constant.

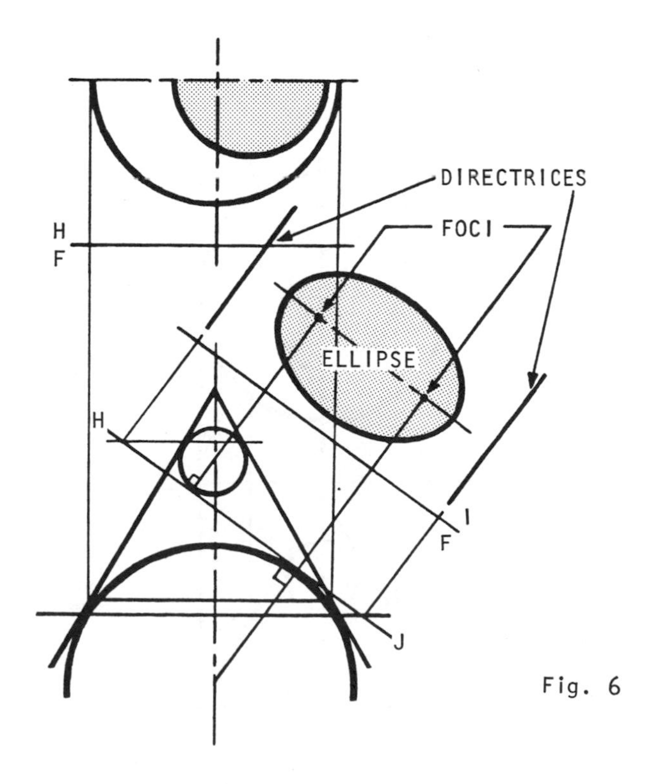

Fig. 6

The equation of an ellipse is in the form

$$\frac{X^2}{A^2} + \frac{Y^2}{B^2} = 1 ,$$

where A and B do not equal zero.

The edge view of the cutting plane HJ is seen in fig. 6, in the front view. The top view of the line of intersection is found by using horizontal cutting planes. The auxiliary view results in a true size view of the ellipse. The foci and directrices are located in the manner described above and shown in fig. 6.

The line of intersection of a plane that is passed through a cone perpendicular to the axis of the cone is a circle, as shown in fig. 7. A circle can be defined

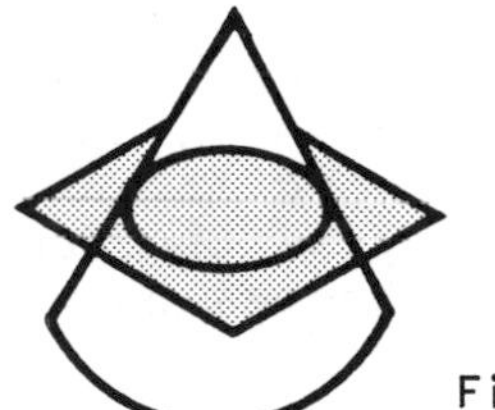

Fig. 7

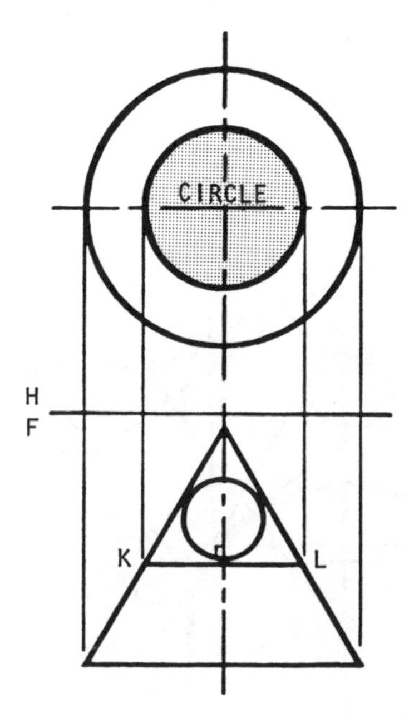

Fig. 8

geometrically as the locus of coplanar points equidistant from a fixed point, called the center. Its equation is in the form of the Pythagorean theorem, $X^2 + Y^2 = R^2$, where R is the length of the hypotenuse, or, in this case, the radius of the circle.

The edge view of the cutting plane, KL, is seen in the front view in fig. 8. The top view is drawn by projecting the edge view and constructing the circle whose diameter equals the distance between the projections of plane KL. This view is also the true size and shape view of the circle. The center of the circle is located on the axis of the cone, and is projected to the top view.

There are three limiting or degenerate conic sections that can be formed by passing cutting planes through the vertex in various manners. If in fig. 2 plane AB is moved parallel to itself until it contains the vertex of the cone, the section is a straight line. If in fig. 4 plane EG is moved parallel to itself until it passes through the vertex of the cone, the section is two intersecting straight lines. If in fig. 6 plane HJ is moved parallel to itself until it contains the vertex of the cone, the section is a single point.

• PROBLEM 7-2

> Determine the line of intersection between a plane and a cone.

Solution: (1) Fig. 1 shows the top and front views of

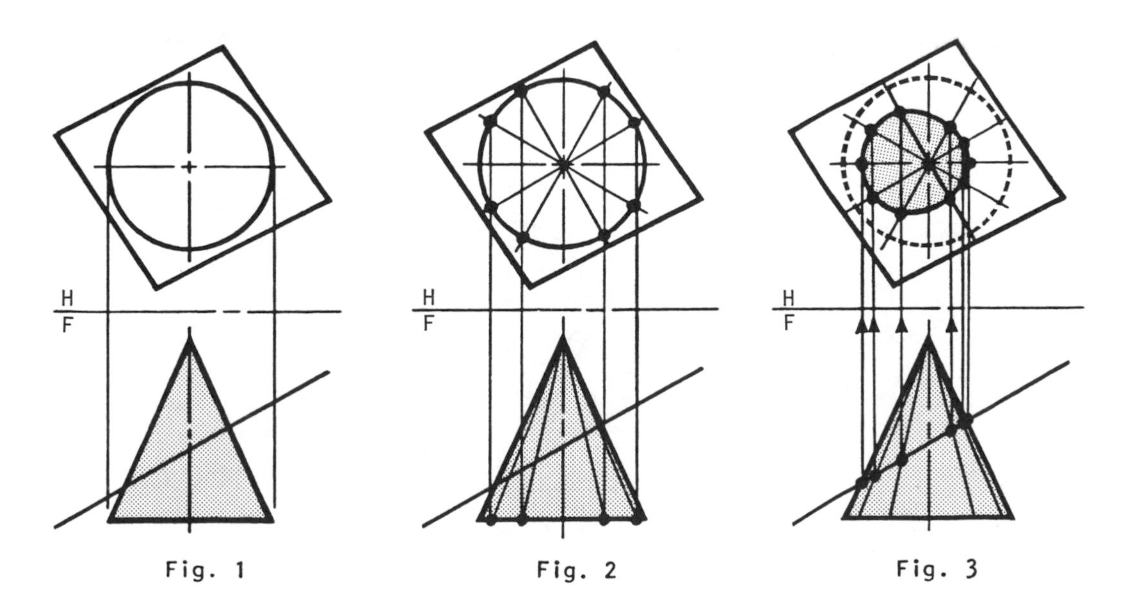

Fig. 1 Fig. 2 Fig. 3

a cone and a plane, with the plane appearing as an edge in the front view. (2) In order to determine the intersection between the plane and cone, a series of elements is drawn in the top view on the surface of the cone, as shown in Fig. 2. (3) These elements are projected from the top view of the base to the front view of the base. (4) The projected points are connected with the apex of the cone in the front view.

(5) The piercing points of the elements of the cone are located in the front view along the projection lines which connect the elements along the base with the apex of the cone, and intersect the edge view of the plane. (6) The top view of the line of intersection is found by connecting in sequential order the piercing points of the elements of the cone, which are projected from the front view up to the corresponding element in the top view. (7) The front view of the line of intersection coincides with the edge view of the plane. Fig. 3 shows the completed line of intersection between the plane and the cone. Note that if the edge view of the plane is not given, an auxiliary view must be drawn to show the plane as an edge. This method of solution is exemplified in the following problem.

• PROBLEM 7-3

Show the plane section in the given views for a cone of revolution cut by the given plane K-L-M-N, using

(a) the two view method,

(b) the edge view method.

Solution: (a) The Two View Method:

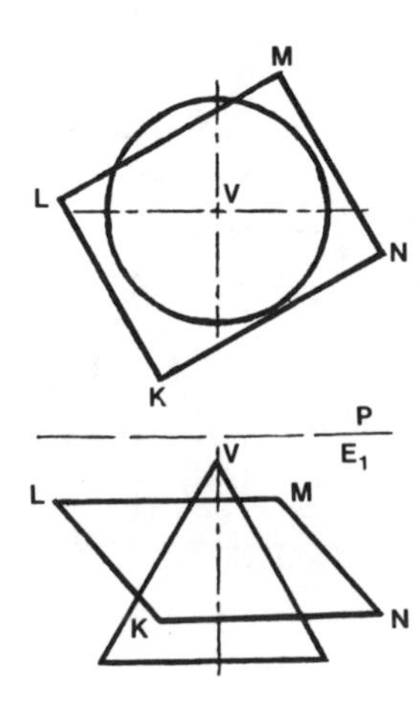

Fig. 1

(1) Fig. 1 shows the given plane and cone of revolution in the P, plan, and E_1, primary elevation, views. (2) In Fig. 2, the base of the cone in the P view is divided into 12 equal parts and 12 straight-line elements of the cone are drawn. (3) These elements are projected to the E_1 view. (4) As shown in the P view, a vertical cutting plane through the vertex containing elements 1 and 7 intersects the plane K-L-M-N along line A B. (5)

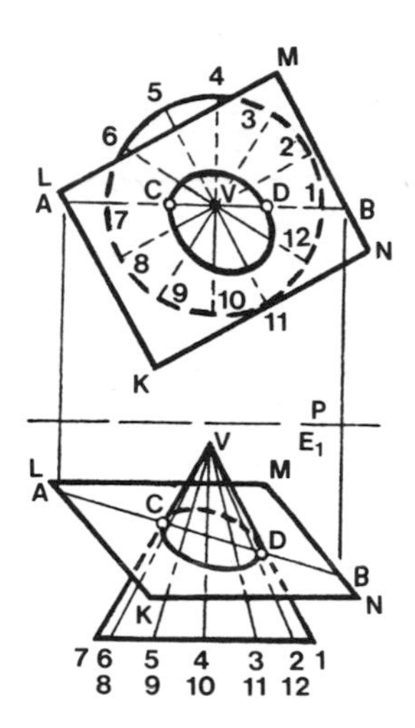

Fig. 2

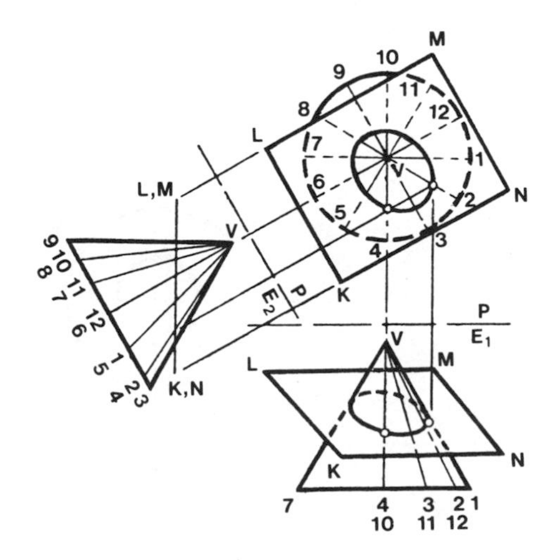

Fig. 3

Projecting line A B to view E_1, these elements are found to pierce the plane at points C and D. C and D are then projected into the plan view. (6) Likewise, cutting planes are used to locate the remaining piercing points (i.e., cutting planes through the vertex and elements 8 and 2, 9 and 3, etc.). These points are then connected by a smooth curve. For clarity, only the cutting plane through elements 1 and 7 is shown.

(b) The Edge View Method:

(1) This method of solution also refers to the cone and plane, as given in Fig. 1. (2) In Fig. 3, View E_2, the secondary elevation is drawn to show plane K L M N as an edge. This results from the fact that the lines of sight for view E_2 are parallel to the true length lines L M and KN in the P-view. Lines L M and K N are true length in the P-view since they are horizontal lines in E_1, perpendicular to the lines of sight for view P. (3) Twelve equally spaced elements of the cone are drawn in the P-view, and projected to views E_1 and E_2. (4) The points at which these elements cross the edge view of the plane in view E_2 are projected into views P and E_1 as shown for elements 2 and 4. (5) After all points have been located, they are connected by a smooth curve.

• PROBLEM 7-4

Obtain a true size and shape view of the parabolic intersection of a plane A-B and cone of revolution.

Solution: (1) Fig. 1 shows the given intersecting plane, A B, and cone of revolution in the front view. The plane

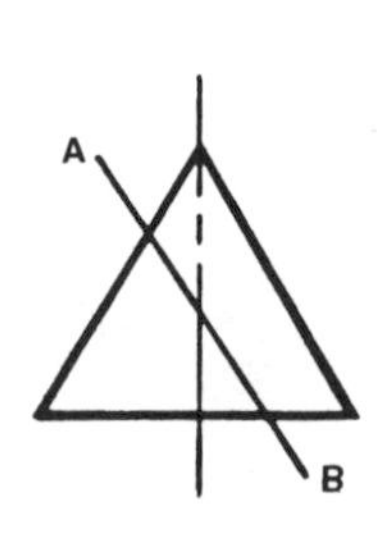

Fig. 1

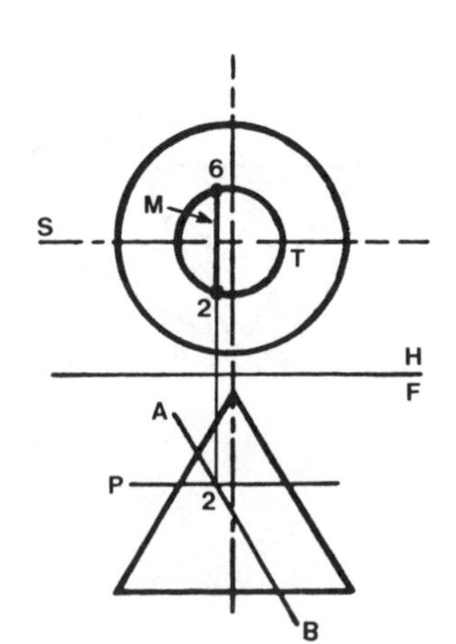

Fig. 2

A B appears edgewise in the front view. (2) Pass a horizontal cutting plane, for example, P, through the cone in the F-view, as shown in fig. 2. The horizontal plane P cuts the cone in the circle T. The diameter of circle T is equal to the distance along plane P between the two extreme elements of the cone. Project this distance to the plan (or H) view, and construct circle T. (3) Project the point of intersection of plane A B with the horizontal plane P, up to the plan view. This projector intersects circle T at two points, 2 and 6. The line M through these points represents the line cut from plane A B by the horizontal cutting plane, P. Points thus obtained are points on the intersection of plane A B and the cone. As in fig. 3, pass additional horizontal cutting planes, Q and

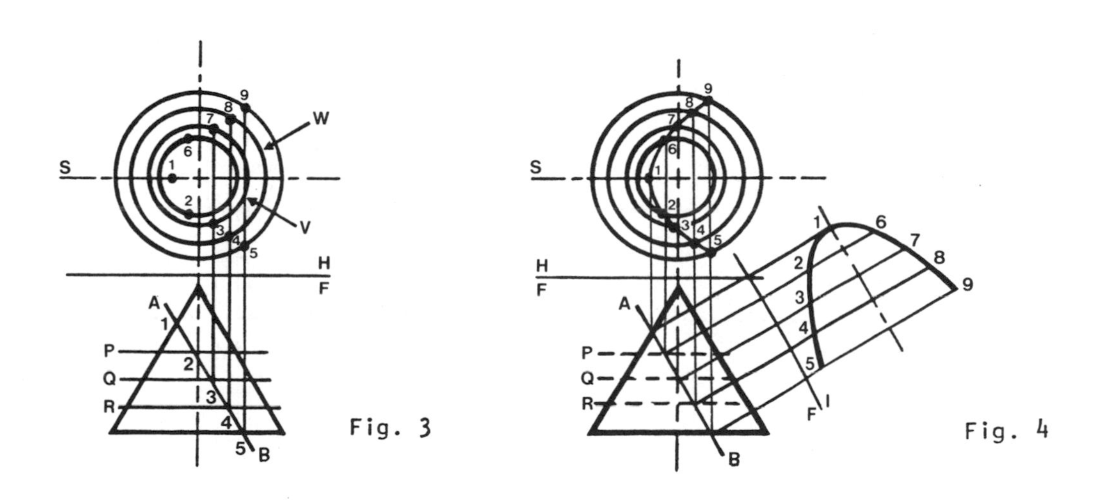

Fig. 3

Fig. 4

R, through the cone in the F-view, cutting the cone in circles V and W, respectively. Project the points of intersection of plane A B with planes Q and R to the plan view, locating points 3, 4, 7, and 8. Also project the points where plane A B cuts the limiting elements of the cone in the F-view, points 1, 5, and 9 in the plan view.

(4) Note: Point 1 is the highest point on the curve of intersection and lies, therefore, on the contour element. (5) Since the angle between plane A B and the axis of the cone is equal to that between the elements and the axis (plane A B is parallel to the right limiting element in the F-view), the intersection is a parabola. The top view of the parabola is symmetrical about the center line S. (6) The true shape of the parabola is obtained by drawing an auxiliary view, view 1, on a projection plane taken parallel to A B. This results from the fact that projection from a view which shows the edge view of a

plane parallel to the rotation line, will produce a true shape and size view of the plane. Fig. 4 shows the completed true size and shape view of the parabola, in the auxiliary view.

• PROBLEM 7-5

Obtain the true view of the hyperbolic conic section cut from a cone by a cutting plane, ST.

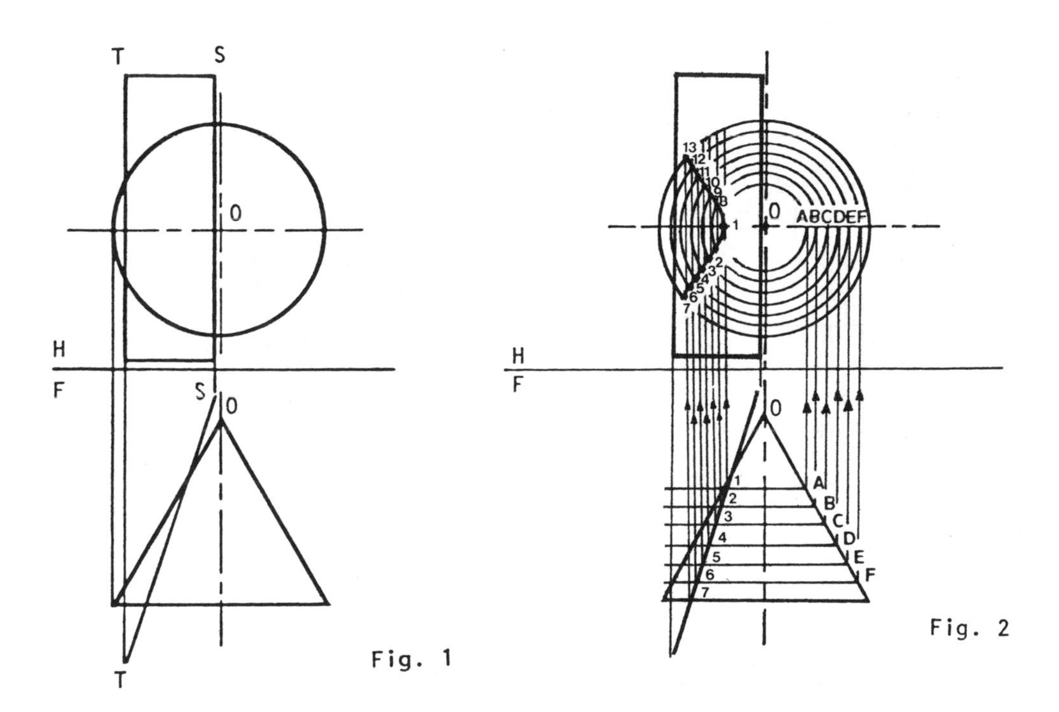

Fig. 1

Fig. 2

Solution: Fig. 1 shows the top and front views of the cone with vertex O and the cutting plane, ST. As shown in fig. 2, pass several horizontal planes through the cone in the front view. The points where the planes intersect the right limiting element of the cone are labeled A,B,C,D,E, and F, and the points where the planes intersect the cutting plane are labeled 1 through 7. Project points A through F to the top view of the cone to where they intersect the horizontal center line, and construct the circles with radii OA, OB, OC, ... , OF, centered at O. These circles represent the lines of intersection between the cone and the horizontal planes. Project points 1 through 7 to the top view to where they intersect the

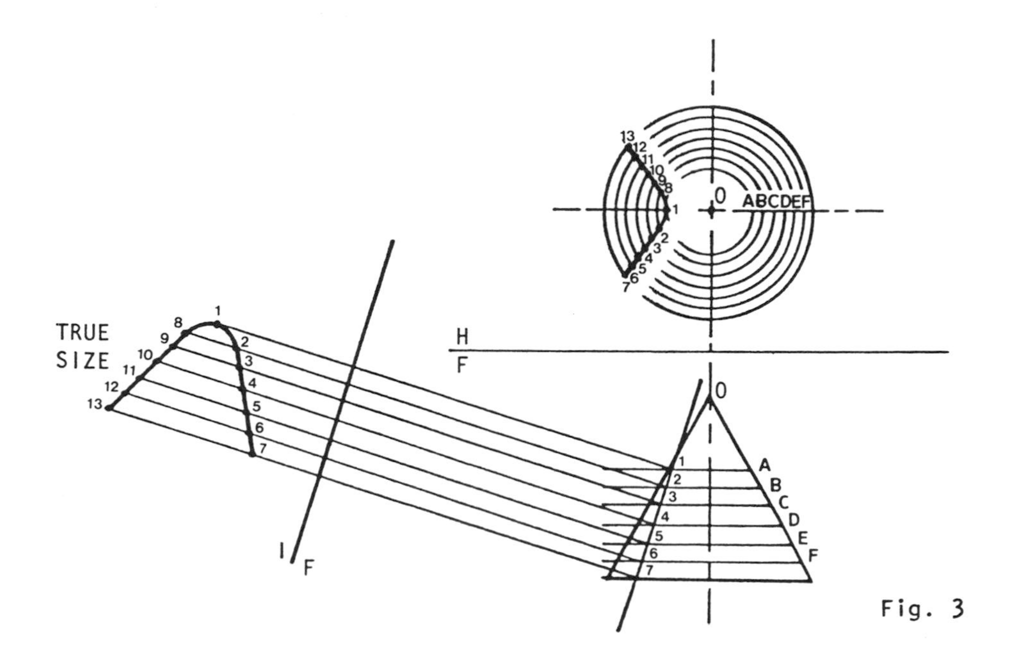

Fig. 3

circles of radii OA through OF, respectively. Connect these intersection points, labeled 1 through 13, with a smooth curve for the top view of the section cut by plane ST. For the true shape and size view of the section, draw an auxiliary view from the front view, with lines of sight perpendicular to the edge view of cutting plane ST and the reference plane F-1 parallel to plane ST. Project points 1 through 13 to view 1, as in fig. 3. This view is the required true view of the hyperbolic section.

• PROBLEM 7-6

Determine the intersection between two cones.

Solution: (1) Fig. 1 shows the two intersecting cones in the H-, horizontal, and F-, frontal, planes of projection. (2) Problem analysis reveals that a series of cutting planes is needed to cut elements on each cone. These elements, in turn, are used to find piercing points on the lines of intersection. A plane drawn to contain a line that passes through both apexes of the cones and a line passing through both bases of the cones will cut elements in each cone.

(3) Fig. 2 illustrates the construction of this plane.

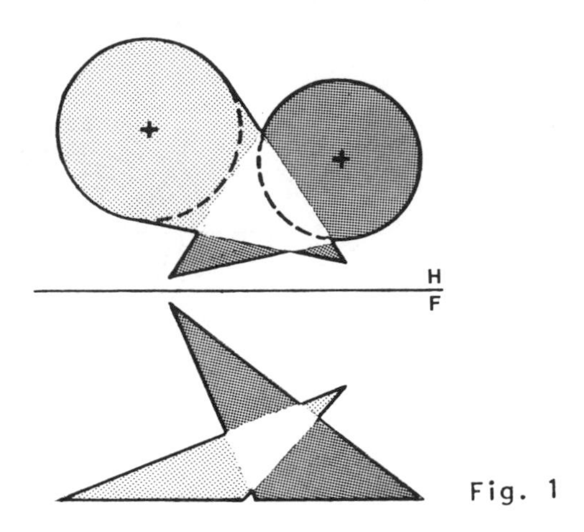

Fig. 1

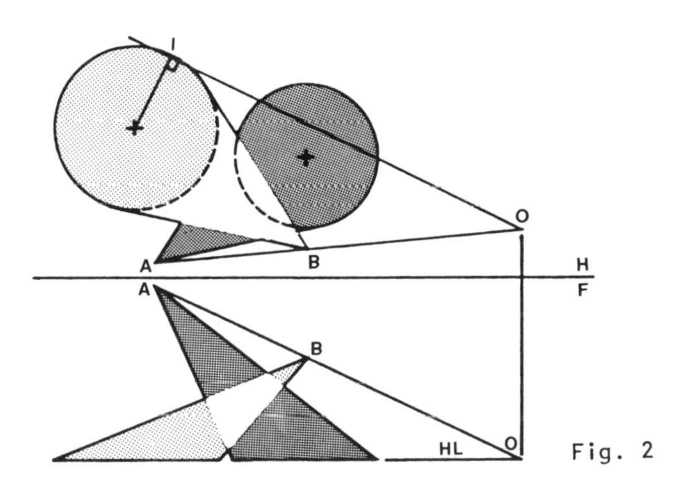

Fig. 2

Draw a line through apexes A and B, and extend it to point O on the plane of the bases of the two cones in the front view. Project line ABO to the top view, and draw line O1 to intersect the two bases. This plane will cut elements on each cone, since the apexes lie on a common line. Note: Line O1 is tangent to cone B.

Pass a cutting plane through line OA, as shown in fig. 3 which joins both vertices, and intersects the bases of the cones. Line O3 is the trace of the cutting plane in the bases. Draw the elements of cone A that lie on this cutting plane by connecting vertex A to the points where line O3 intersects base A, R and S. Draw the elements of cone B on the cutting plane by connecting vertex B to the points where line O3 intersects base B, T and 3. Where these elements intersect each other, points E,F,G, and H, determines the locations of four points on the lines of intersection, as seen in fig. 3. Pass additional cutting planes through the vertices and bases in a similar manner

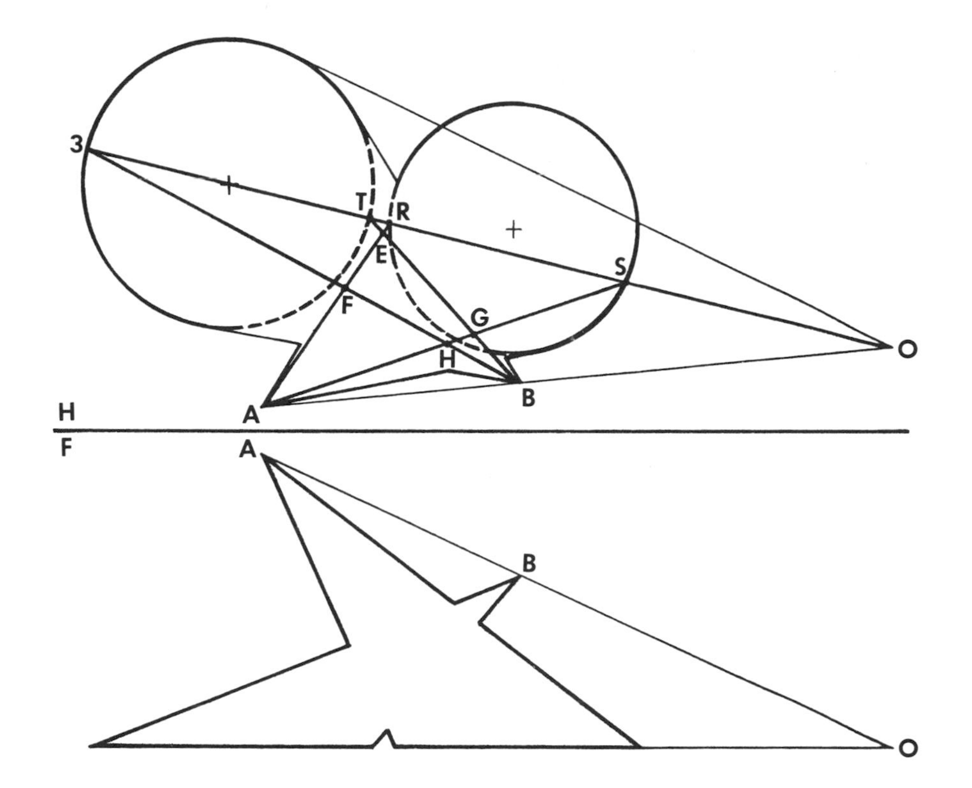

Fig. 3

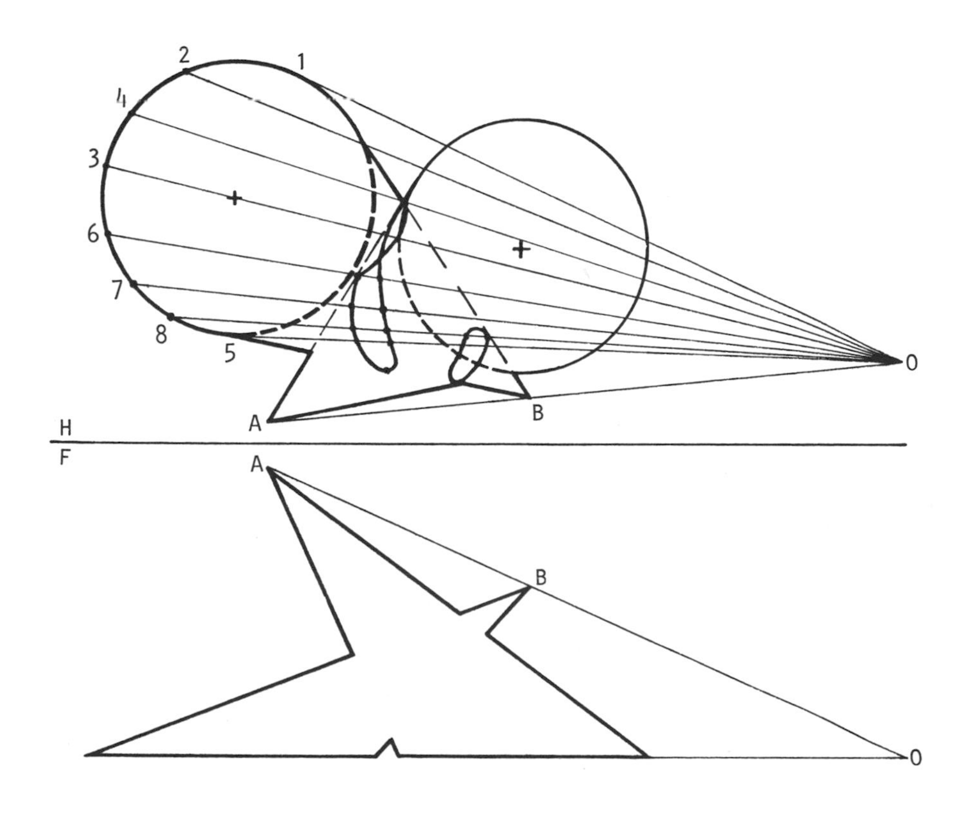

Fig. 4

to determine more points on the lines of intersection. Connect the points to form the two intersection lines, as shown in fig. 4.

(4) Project the elements used in the top view to the front view, as illustrated in Fig. 5. Project also the

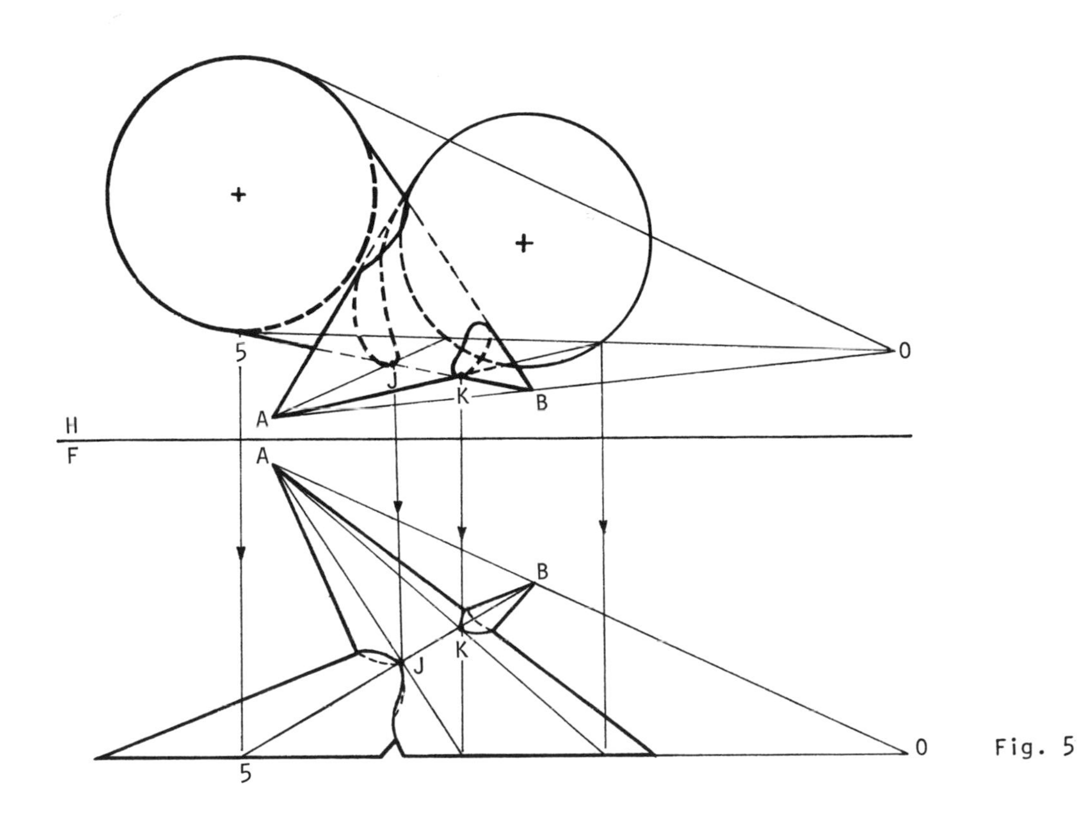

Fig. 5

points lying on them in the top view to the front views of these elements. (5) Points J and K on plane A-O-5 are projected to illustrate this technique of finding the line of intersection in the front view. The line joining point B with the point view of the element cut from cone B by cutting plane O-5, contains points J and K in the top view. As a result, this same line contains J and K in the front view. The intersections of the lines from point A to the point views of the elements cut from cone A by cutting plane O-5, with line B-5, locate J and K in both views. (6) Determine the visibility.

(7) Note: Cutting plane A-O-5 establishes only two points since it is tangent to one of the cones.

• **PROBLEM 7-7**

Determine the line of intersection between two cones whose axes intersect.

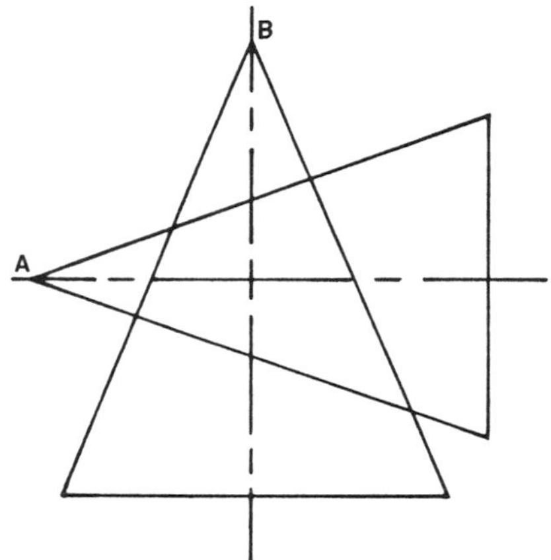

Fig. 1

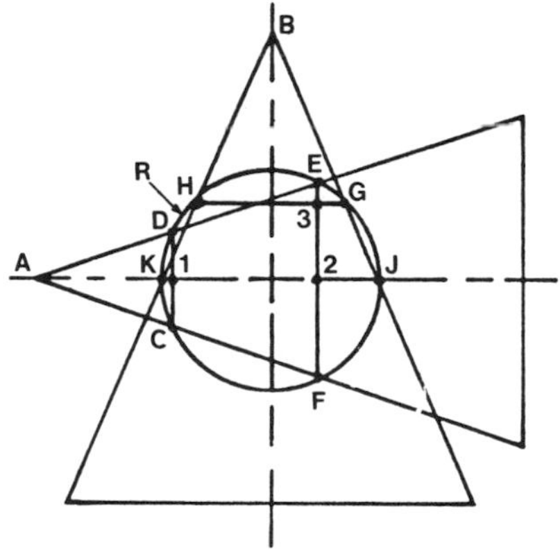

Fig. 2

Solution: Fig. 1 shows the two given intersecting cones, A and B, whose axes intersect. As in fig. 2, construct the circle R, centered at the point where the axes intersect with any convenient radius that will cut the surfaces of both cones. Circle R cuts cone A at points C,D,E, and F, and it cuts cone B at points G and H. Connect CD and EF, as shown in fig. 2, which are both perpendicular to the axis of cone A. Connect GH, perpendicular to the axis

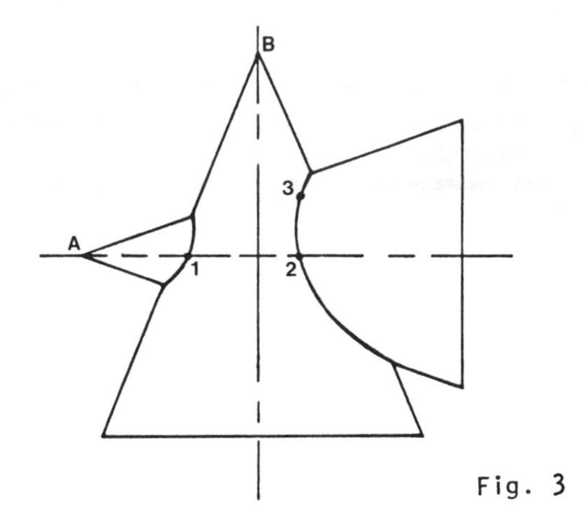

Fig. 3

of cone B. The points where these lines intersect, points 1, 2, and 3 are points located on the line of intersection between the two cones. Other points of intersection are located similarly. Connect these points with a smooth curve for the required line of intersection, as seen in fig. 3.

• PROBLEM 7-8

Determine the points at which the given line, AB, pierces the given cone of revolution.

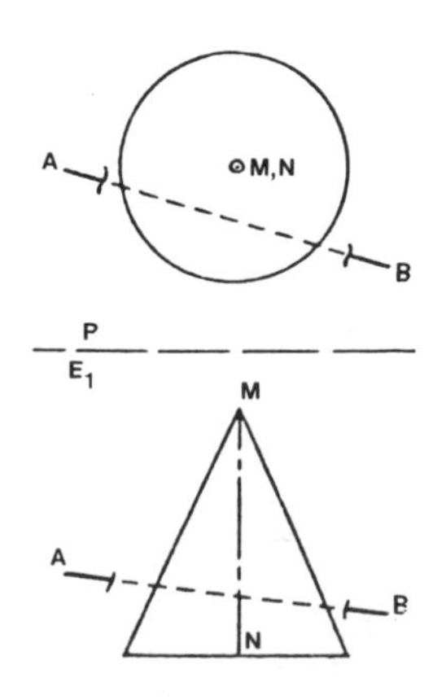

Fig. 1

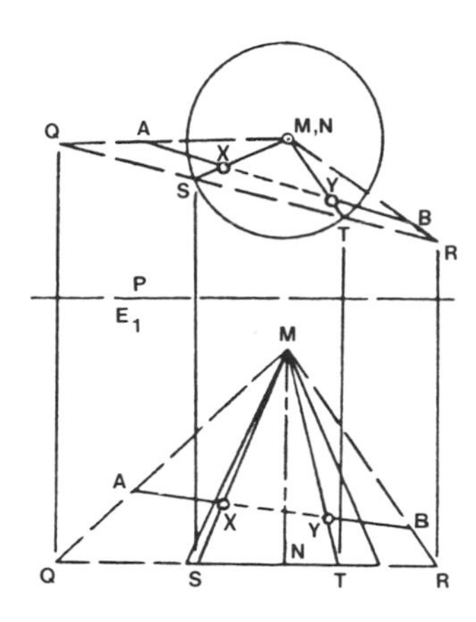

Fig. 2

Solution: (1) Fig. 1 shows the given cone, with axis M-N, and the piercing line, A-B, in the P, plan, and E_1, primary elevation, views. (2) Problem analysis reveals that any given line that pierces a cone must intersect two straight-line elements of the cone. A plane that contains these two elements must also contain the given line and must pass through both the vertex and the base of the cone.

(3) In both views, construct a plane that contains both the vertex of the cone and the line A-B by drawing lines from M through A and B as shown in Fig. 2. (4) In view E_1, the lines from M through A and B, strike the plane of the base at points Q and R. Project Q and R into the plan view. (5) In the plan view, the line Q-R crosses the base at points S and T, thereby locating the two elements, M-S and M-T, which the plane cuts from the cone. (6) Since both these elements and the line A-B lie in the plane M-Q-R, the points X and Y at which they intersect are the required piercing points. (7) Project points S and T into view E_1 drawing elements M-S and M-T to locate the piercing points X and Y. Check the location of these piercing points by a projection between views.

CYLINDERS

• PROBLEM 7-9

Find the line of intersection between a given cylinder and a given oblique plane.

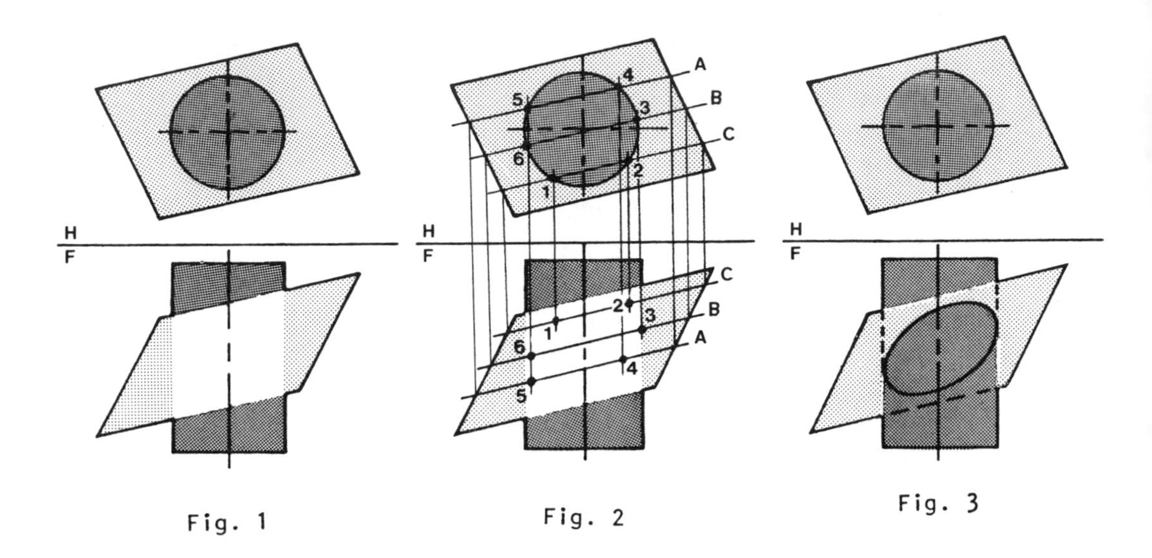

Fig. 1 Fig. 2 Fig. 3

Solution: (1) Fig. 1 shows the given cylinder and oblique plane in the horizontal and frontal planes of projection. (2) The line of intersection is located in the front view by determining the points at which lines on the surface of the cylinder intersect lines on the oblique plane. (3) In order to establish lines on the surface of the cylinder and lines on the intersecting plane, pass vertical cutting planes through the top view of the cylinder as shown in Fig. 2. (4) Points 1 and 2 are selected for purposes of illustration of the forementioned technique. Points 1 and 2 are the points of intersection of cutting plane C and the surface of the cylinder in the top view. (5) Project these points to the front view, where they lie on the line of intersection of cutting plane C and the oblique plane. (6) Similarly, locate additional piercing points in the two given views, in order to obtain a more accurate line of intersection. The points thus located in the front view are connected by a smooth curve to form an elliptical line of intersection. (7) To complete the solution of the problem, visibility is shown in Fig. 3. Since points 1 and 2 are closer than points 4 and 5 to the F-H rotation line, points 1 and 2 are visible in the front view.

• PROBLEM 7-10

Determine the line of intersection between a given oblique plane, A-B-C-D, and a given oblique cylinder.

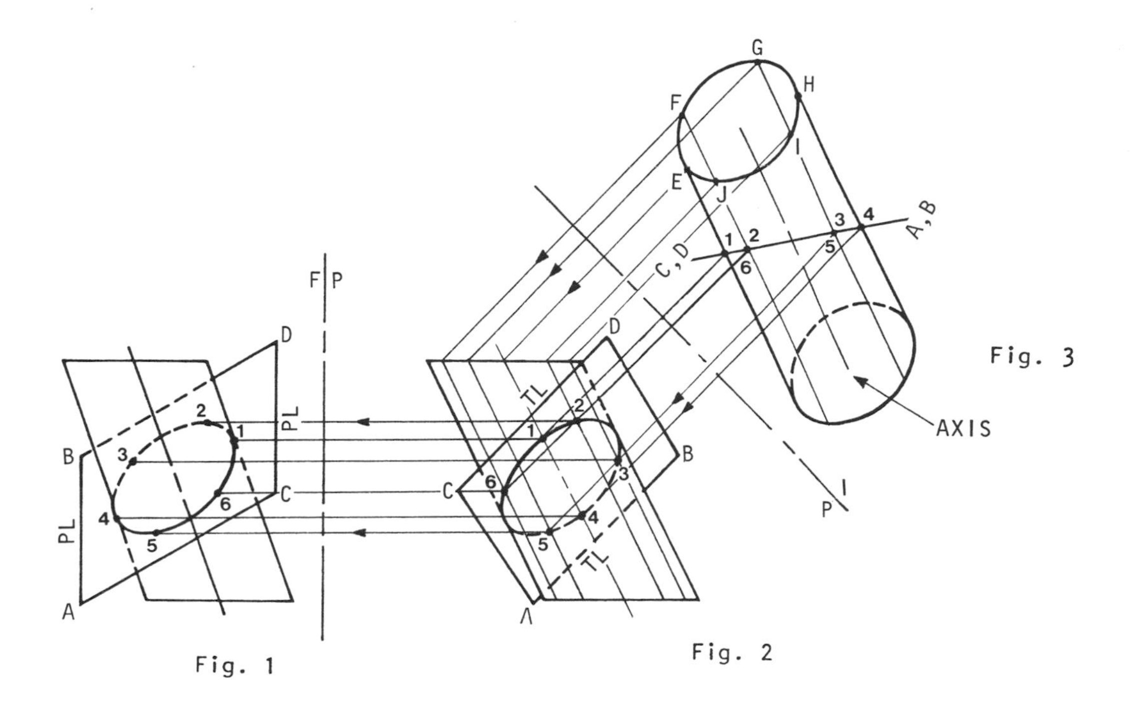

Fig. 1

Fig. 2

Solution: (1) Figs. 1 and 2 show the given plane and cylinder in the F, front, and P, profile, views. (2) This problem illustrates the general case for the intersection between an oblique plane and an oblique cylinder, in which the cylinder does not appear true length, and the plane does not appear as an edge in the principal views. (3) The line of intersection between these forms is determined by finding an edge view of the plane in a primary auxiliary view, view 1 (shown in Fig. 3), where the projection of the cylinder is foreshortened. (4) In establishing this auxiliary view which shows the oblique plane as an edge, note that the sides A-B and C-D of said plane are parallel to the F-P rotation line in the F-view. Consequently, these lines project true length in the P-view. Establish the 1-view having lines of sight parallel to the true length lines A-B and C-D in the P-view. Thus, an edge view of plane A-B-C-D is obtained since the lines A-B and C-D appear as points. (5) Cutting planes parallel to the axis of the cylinder are used to establish lines on the surface of the cylinder in the primary auxiliary view. (6) These lines are found to intersect the circumference of the cylinder's top surface at points E,F,G,H,I, and J. (7) These same lines on the surface of the cylinder intersect the edge view of A-B-C-D at the points 1,2,3,4,5, and 6. Note that points 1 through 6 are the respective projections of points E through J, from the top of the cylinder to the oblique intersecting plane. (8) The lines on the surface of the cylinder are projected from the auxiliary view to the profile view. Each of the points of intersection found in the auxiliary view is projected to its respective line in the profile view. These points are connected to establish the elliptical line of intersection. (9) Visibility in the P-view is determined by the fact that points 1, 2 and 6 are closer than points 3, 4, and 5 to the 1-P rotation line in view 1. As a consequence, the curve containing points 1, 2, and 6 is visible in the P-view, as in Fig. 2.(10) the line of intersection is found in the front view by transferring measurements of the points from the primary auxiliary view to the front view using dividers. Visibility in the F-view is determined by applying the same reasoning as that used to determine the visibility in the P-view.

(11) The above method of determining the line of intersection between a cylinder and a plane is known as cross projection, since the two projection lines used to locate each point form a cross.

• PROBLEM 7-11

Find the line of intersection between a given cylinder and a given plane, whose side view projection is an edge.

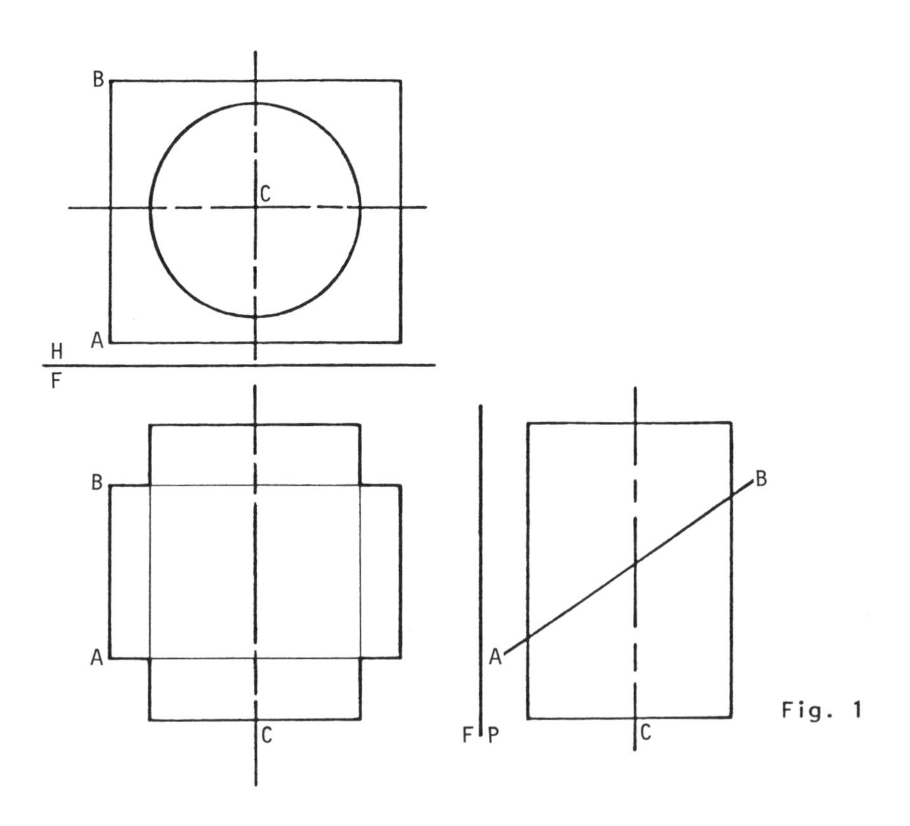

Fig. 1

Solution: Fig. 1 shows the principal views of the given cylinder and plane: the top, front, and side views. To determine the line of intersection between the plane and the cylinder, first project to both the top and front views (1) the points where plane AB intersects the limiting

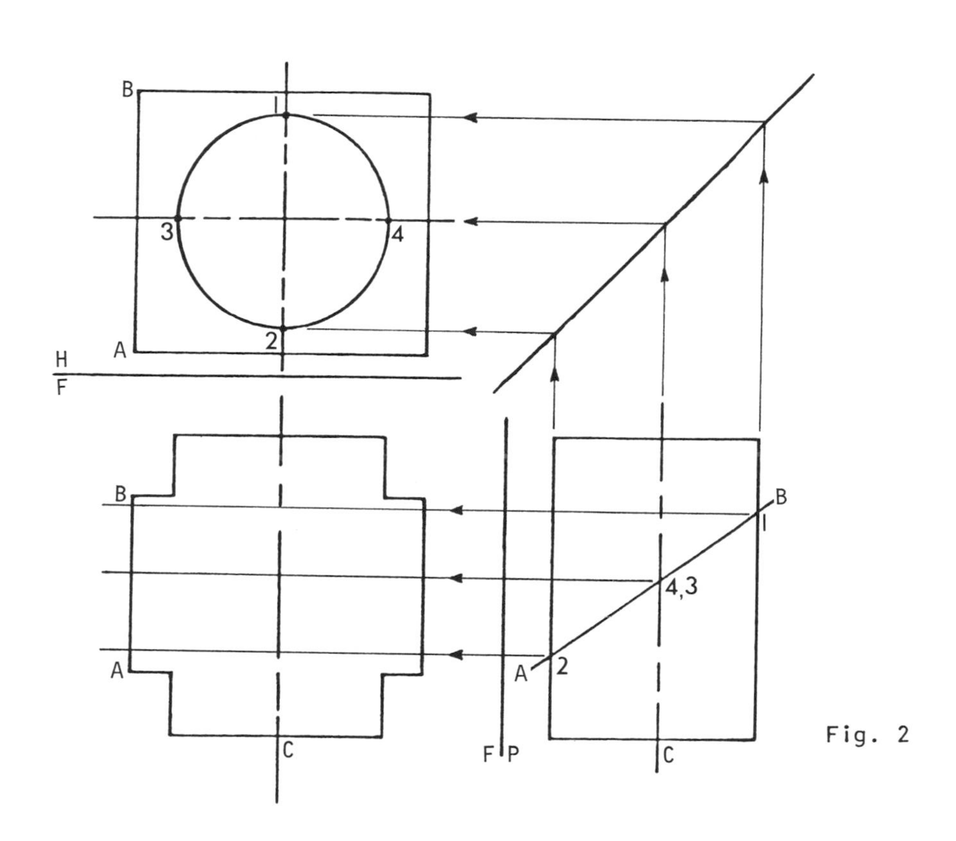

Fig. 2

elements of cylinder C in the side view, and (2) the points where plane AB intersects the axis of cylinder C in the side view. These are points 1,2,3, and 4, as shown in fig. 2. Now project to the front view the points in the top view where the previous projections intersect the circle, as in fig. 3. Where these projections intersect in the front view locates four key points of the line of intersection, as labeled. For a more complete drawing of the intersection line, continue to project points on the plane from the side view to the top and front views and then project the respective points from the top view to the front view. These are points 5 through 12, as in fig. 4.

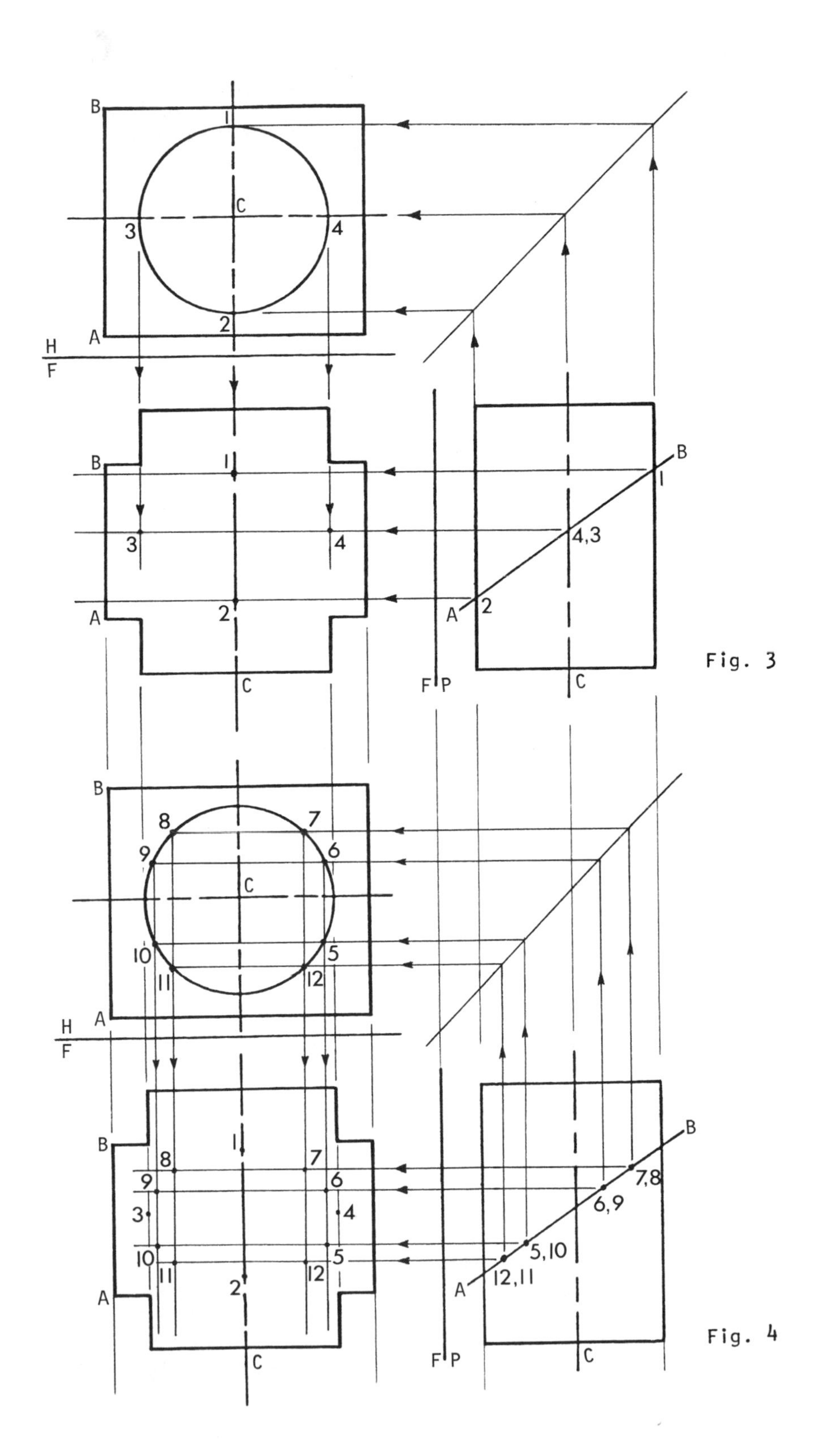
B
1
C
3
4
2
A
H
F
B
1
3
4
A
2
C
B
1
4,3
A
2
F
P
C
B
8
7
9
6
C
10
5
11
12
A
H
F
B
1
8
7
9
6
3
4
10
5
11
12
A
2
C
B
7,8
6,9
5,10
A
12,11
F
P
C

Fig. 3

Fig. 4

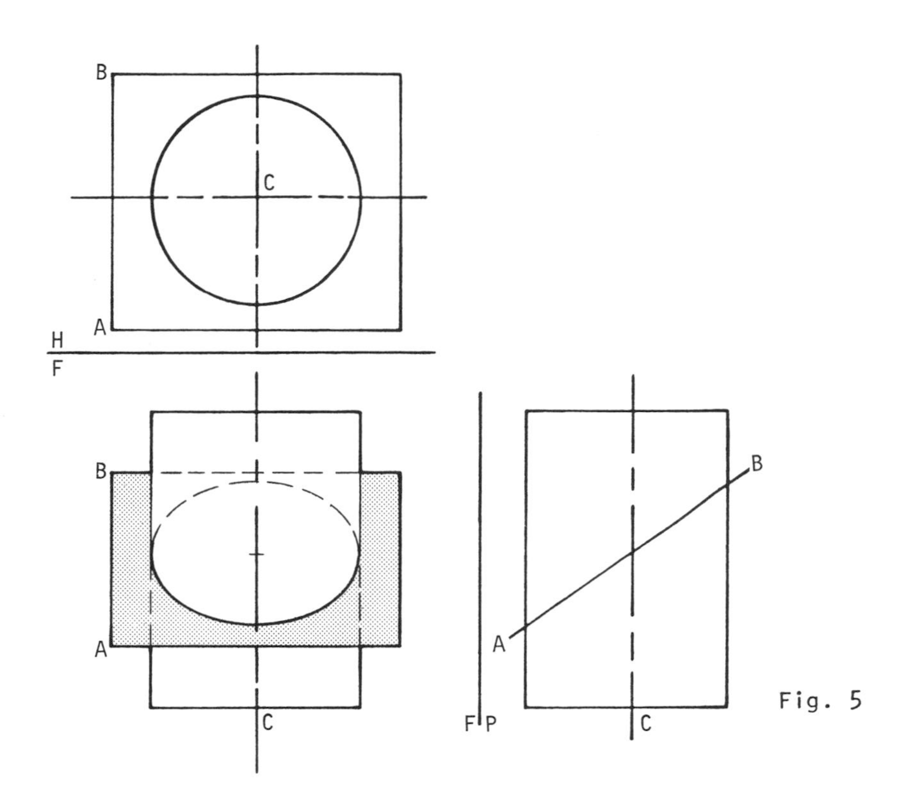

Fig. 5

Draw the line of intersection as a smooth curve and determine the visibility, shown in fig. 5. Since the lower point A on the edge view of plane AB is closer to the rotation line, F-P, than the upper point B, the lower edge of the plane is visible in the front view.

• PROBLEM 7-12

Find the true shape and size of the plane section of an elliptical cylinder cut by an inclined plane.

Solution: (1) Fig. 1 shows the given elliptical cylinder and inclined plane in the P, plan, and E_1, primary elevation, views. (2) Problem analysis reveals that the true shape and size of any plane section of a cylinder can be seen only in a view for which the lines of sight are perpendicular to the plane section (i.e., the lines of sight appear perpendicular to the edge view of the plane section). (3) As shown in Fig. 2, draw a rotation line parallel to

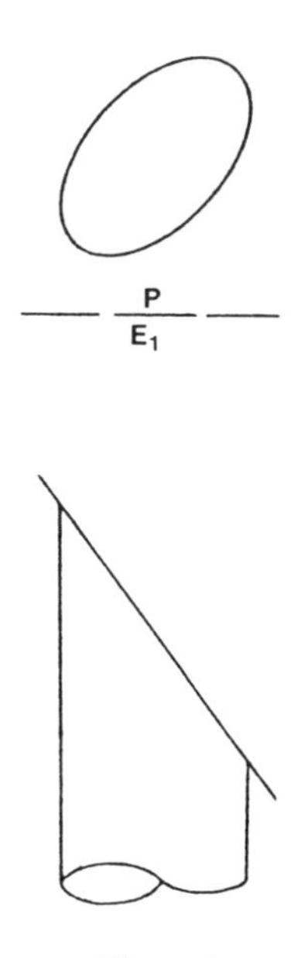

Fig. 1

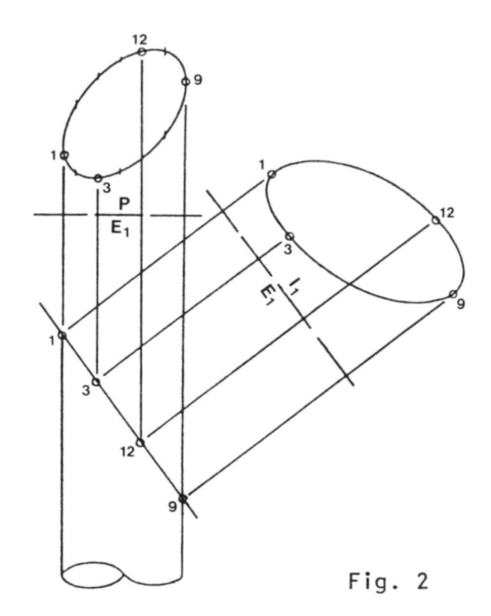

Fig. 2

the edge view of the inclined plane to obtain a true-shape view, which is labeled I_1, the primary inclined view. (4) Select a number of elements on the edge view of the cylinder in the plan view, as illustrated in Fig. 2. The extreme elements corresponding to those numbered 1,3,9, and 12 should always be included. (5) Project these elements to view E_1 to find the points at which they pierce the plane. (6) The piercing points are then projected into view I_1, where they are connected to give the true shape and size of the section. Note: The selection of additional elements on the edge view of the cylinder in the P-view will allow for the construction of a more accurate line of intersection.

(7) If the views of a cylinder corresponding to views E_1 and I_1 of Fig. 2 were given, the true shape of the right section could be found by the same procedure, the elements being selected in view I_1.

• PROBLEM 7-13

Find the true shape and size of the plane section of a cylinder of revolution cut by a vertical plane.

Solution: Although the method illustrated in a previous

problem for obtaining the true shape and size of the plane section of cylinder works for any cylinder, the following method is easier when the cylinder is a cylinder of revolution and its axis is given. Any plane section of a cylinder of revolution not perpendicular or parallel to the elements will be an ellipse with a minor axis equal to the diameter.

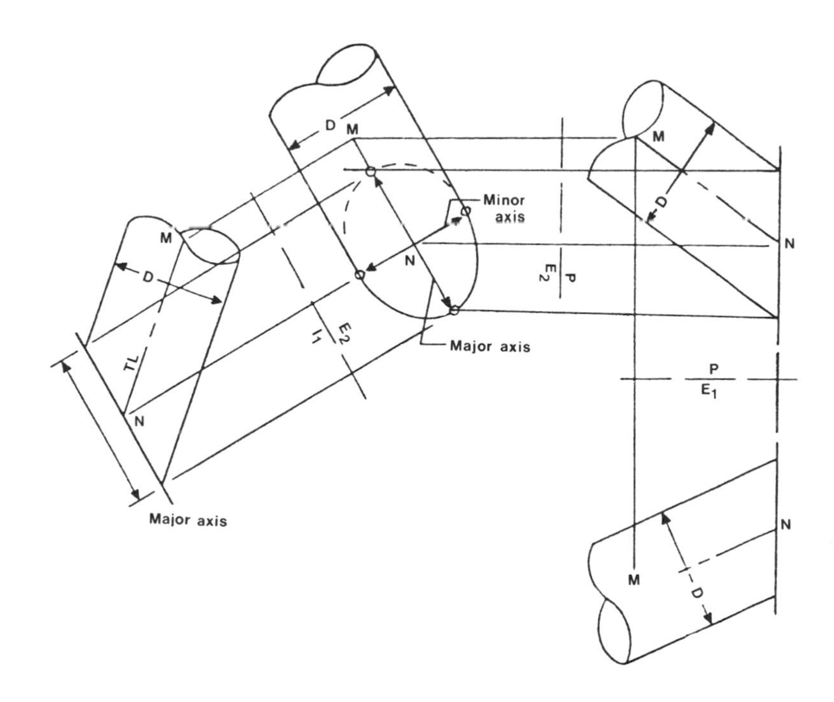

In the accompanying figure, a cylinder of revolution with axis MN is given in the P, plan, and E_1, primary elevation, views. The true shape and size of the section of the cylinder cut by the vertical plane is wanted. The true shape and size of the plane section will appear in view E_2, secondary elevation, where the lines of sight are perpendicular to the edge view of the plane in the plan. View I_1, the primary inclined view, is drawn to obtain the true length of the axis MN, since the lines of sight for this view are perpendicular to the axis in E_2. In this view, the plane section appears as an edge due to the fact that all views projected from a true shape and size view are edge views. The distance between the points at which the extreme elements of the cylinder strike this plane is the length of the major axis of the ellipse. The major axis is then projected into view E_2, where it coincides with the axis of the cylinder. The minor axis is a diameter perpendicular to the elements. With these axes, the ellipse can be drawn.

• **PROBLEM 7-14**

Determine the intersection between two given oblique cylinders, using the cutting plane method.

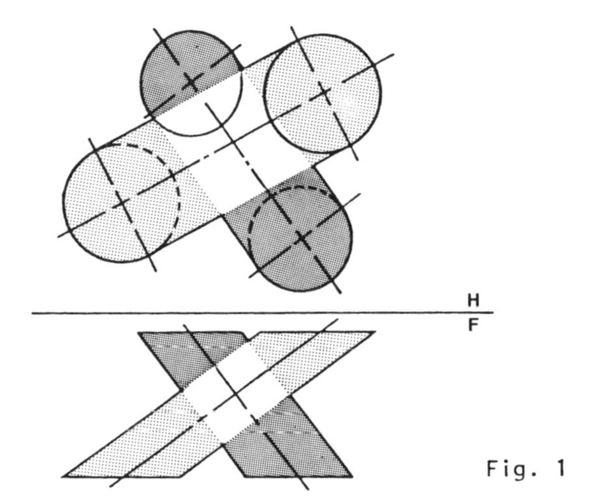

Fig. 1

Solution: (1) Fig. 1 shows the intersecting cylinders in the H-, horizontal, and F-, frontal, planes of projection. (2) As shown in Fig. 2, in order to determine the line of intersection of the two cylinders, a plane must be constructed in space which is parallel to both cylinders. This condition will be satisfied if the plane con-

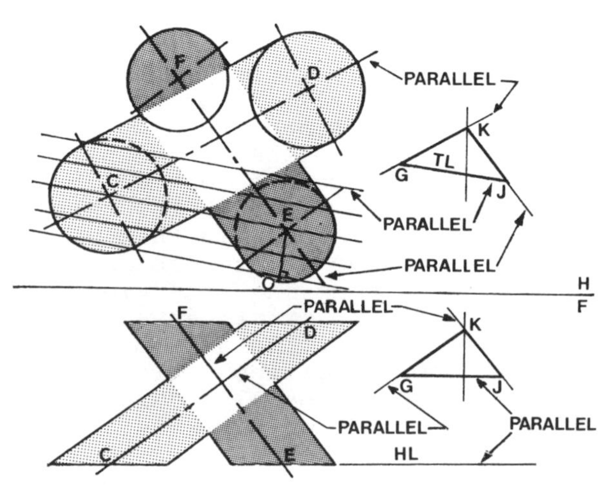

Fig. 2

tains lines that are parallel to the axes of each cylinder. Draw said plane. (3) It is necessary that the planes be passed through the top views of the intersecting cylinders as cutting planes. As a result of this requirement, construct a horizontal line in the triangular plane, which is parallel to the edge view of the base planes of the cylinders in the front view. (4) Project this line to the triangular plane in the top view, where it appears true length (TL) and labeled G-J. (5) The direction of line G-J is used as the direction for the cutting planes that will be drawn in the top view, to pass parallel to

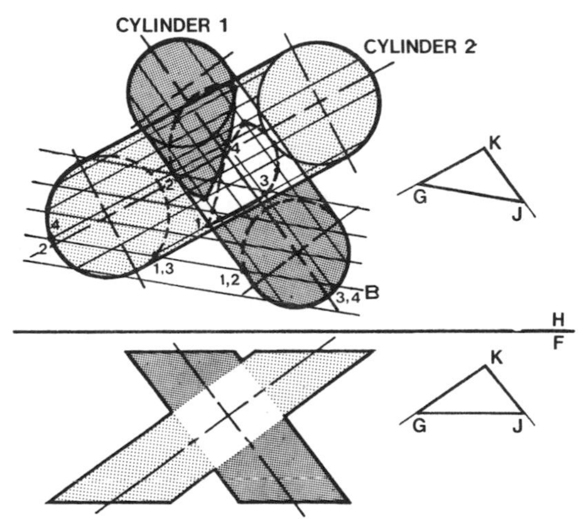

Fig. 3

the axes of the cylinders. In other words, the direction of line G-J represents the direction of the line of intersection of the cutting planes on the circular bases of the cylinders in the top view. Pass a series of cutting planes through the bases, parallel to line G-J.

(6) Fig. 3 illustrates that where the cutting planes intersect the bases of the cylinders in the top view, elements parallel to the axes of each cylinder are formed on the cylinders. Note that four elements are cut by each cutting plane. The points where these common elements intersect locate points on the lines of intersection. (These points should be labeled, as are the four example points shown.) (7) As an illustration of step (6), consider cutting plane B. This cutting plane cuts elements 1-2 and 3-4 from cylinder #1; and elements 2-4 and 1-3 from cylinder #2. Element 2-4 intersects 1-2 at point 2; and, intersects 3-4 at point 4. Element 1-3 intersects 1-2 at point 1; and, intersects 3-4 at point 3.

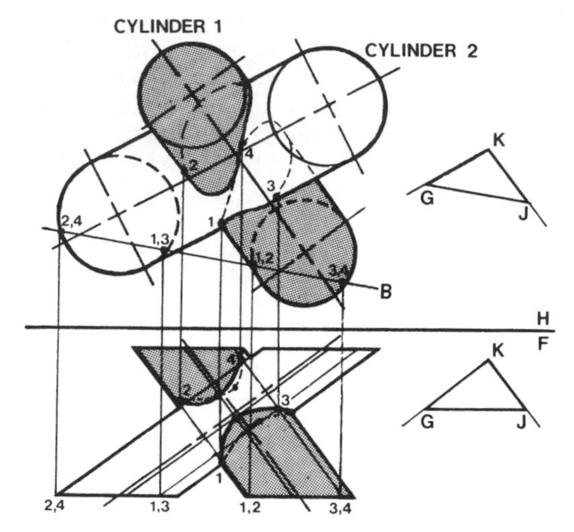

Fig. 4

(8) Project the elements on the cylinders from the top view to the front view. The points found to lie on specific elements in the top view are projected to their respective elements in the front view. Several points have been projected as examples. (9) Visibility in the front view is determined by analysis of the top view.

• PROBLEM 7-15

Determine the points at which a line pierces a cylinder using the two-view method. Show visibility.

Solution: (1) Fig. 1 shows the given cylinder and the exterior segments of line M-N in the P, plan, and E_1, primary elevation, views. (2) Referring to Fig. 2 of the previous problem, any plane appearing as an edge in view I_1 would be parallel to the elements of the cylinder. If this plane were drawn so that it contained the line M-N, it would also contain the elements S and T. Applying this principle, it is possible to locate the piercing points by using only two views.

(3) In each view in Fig. 2 of the accompanying diagram, a line is drawn through M, parallel to the elements of the cylinder. This line and M-N form a plane containing M-N which is parallel to all the elements of the cylinder. (4) In view E_1, this plane intersects the plane of a base of the cylinder along line Q-R. (5) In the plan view, points S and T, at which line Q-R crosses the base, are

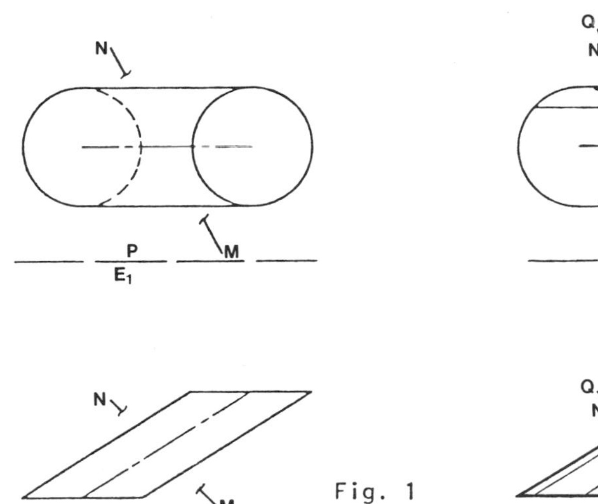

Fig. 1

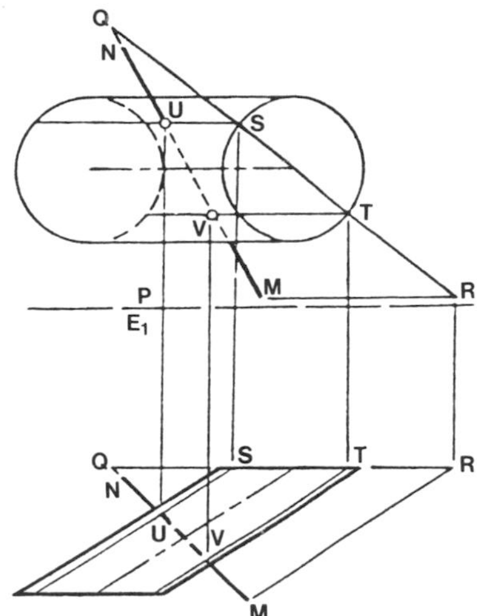

Fig. 2

the ends of the two elements cut from the cylinder by the plane M-N-Q-R. (6) Points S and T are projected into view E_1. The elements are drawn in both views to locate the designed piercing points, U and V. (7) To verify the location of these points, project between the views.

(8) The portion of the given line between the piercing points is invisible, as indicated by the dashed lines. (9) The portion of the given line between the piercing point and the limiting element is visible if the element through the piercing point is visible. (10) As an illustrative example, the given line in the plan view is visible from N to U because the element S is visible in this view. In this same view, the portion of the given line between the limiting element and V is hidden because the element T is hidden. The visibility in view E_1 is similarly determined.

• PROBLEM 7-16

Determine the points at which a line pierces a cylinder using the edge-view method. Show visibility.

Solution: (1) Fig. 1 shows the given cylinder and the exterior segments of line M-N in the P, plan, and E_1, primary elevation, views. (2) Problem analysis reveals that any line that passes through a cylinder must intersect at least one element on the cylinder. If the line is straight and the cylinder is a closed figure, the line must intersect two elements. When employing the edge-view method the points at which any line pierces a cylinder can be found by locating the line in a view showing the cylinder

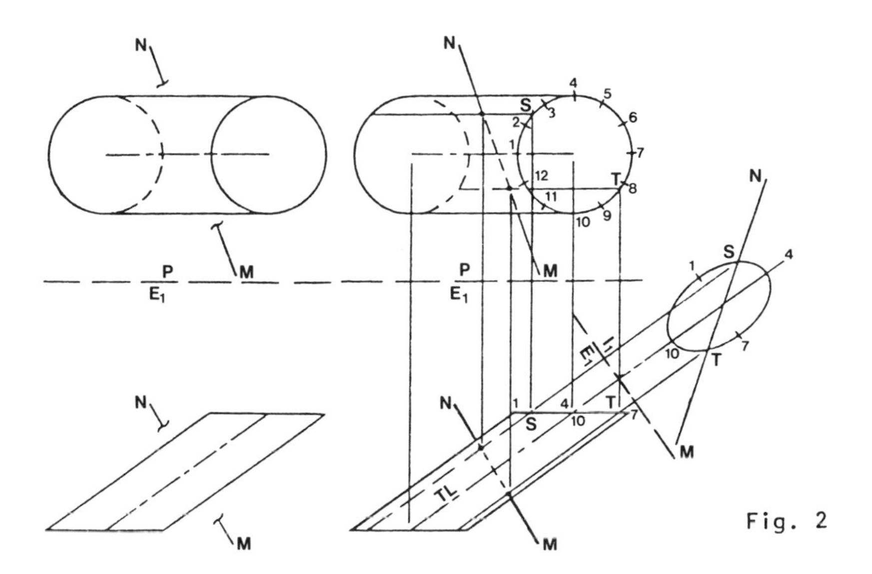

Fig. 1

as an edge. (3) The primary inclined view, I_1, shows the edge view of the cylinder, since view E_1 shows its true length (see Fig. 2). View E_1 shows the true length of the cylinder since the cylinder axis is parallel to the rotation line P-E_1 in the P-view; and consequently, appears true length in view E_1. (4) Points on a number of elements are selected on the base in the plan view, as shown in Fig. 2. These points are projected to the base in view E_1 and then into view I_1 where they are connected to form the edge-view. (5) In the inclined view, line M-N intersects the edge view at two points that determine the elements S and T. (6) Project elements S and T into view E_1, where their intersections with line M-N determine the required piercing points. Project these piercing points into the P-view.

• PROBLEM 7-17

Determine the intersection between a given cone and a given cylinder, using the cutting plane method.

Solution: (1) In the accompanying figure, the P, plan, and E_1, primary elevation, views of the cone and cylinder are given. (2) Problem analysis reveals that the intersection of a cone and a cylinder is found by using cutting planes that cut straight line elements from both surfaces. These cutting planes must contain the vertex of the cone and must be parallel to the axis of the cylinder. The points at which elements of the cone and cylinder intersect can be found most easily by drawing a view showing the cylinder as an edge. Since this view shows the point view of elements of the cylinder, the points at which the elements intersect the cone are readily apparent.

(3) As shown in the given diagram, line C-D, the horizontal line in view E_1, appears true length in the

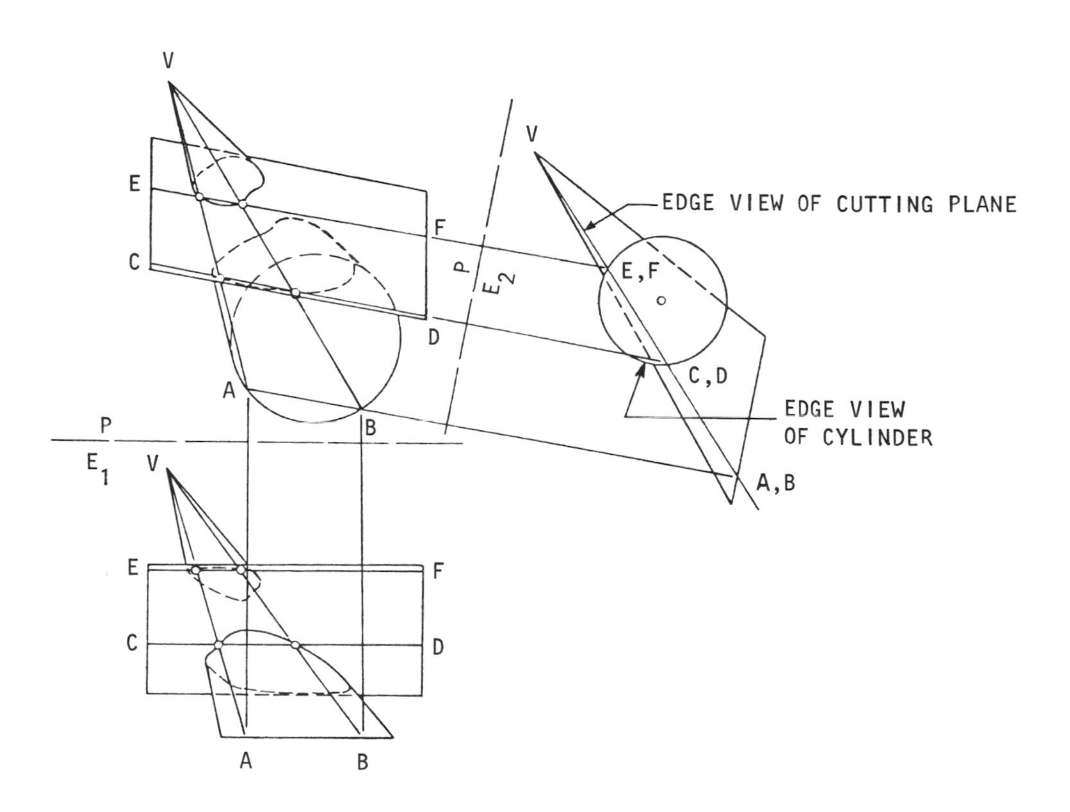

plan view. (4) Construct E_2, the secondary elevation view, showing the cylinder as an edge. To achieve this, the lines of sight for E_2 must be parallel to the true-length lines in the plan. (5) Draw a cutting plane, that appears as an edge, through the vertex of the cone in E_2. This plane cuts elements C-D and E-F from the cylinder and elements A-V and B-V from the cone. (6) Project these four elements into the given views, where they cross at points on the lines of intersection of the cone and cylinder.

(7) Note: The line of intersection between the cylinder and the cone is shown in the accompanying figure; but the cutting planes used to locate additional points, by application of the foregoing method, have been omitted for clarity.

(8) For the case when one base of the cylinder lies in the same plane as the base of the cone, the line of intersection can be located in the two given views. Cutting planes through the vertex of the cone and parallel to the axis of the cylinder are used to locate intersecting elements of the cone and the cylinder.

PRISMS

• PROBLEM 7-18

Find the projections and true shape and size of a right section of a triangular oblique prism.

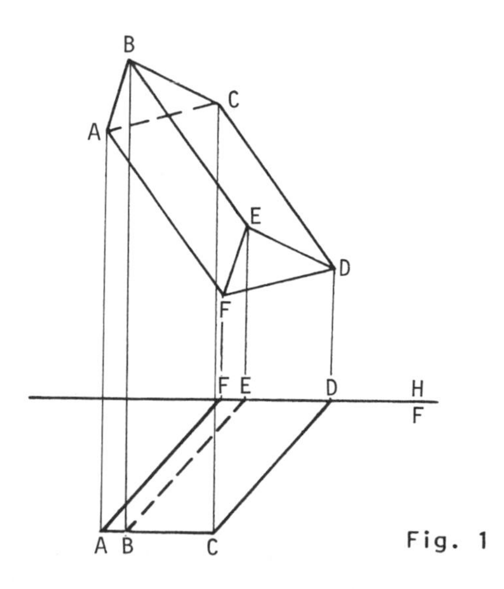

Fig. 1

Solution: By definition, a right section of a solid is a section cut by a plane perpendicular to the axis or center line of the solid. It is commonly called a cross

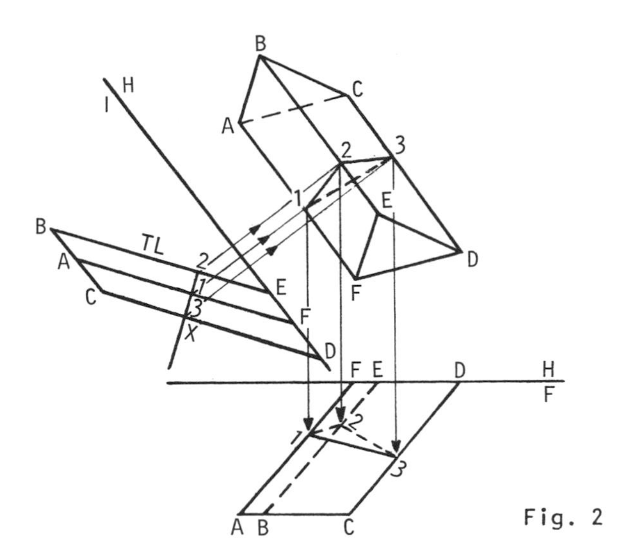

Fig. 2

section. The right section plane appears edgewise and perpendicular to the axis in a view showing the axis in true length. Points in which edges of the solid pierce the section plane are found in this view. The true size of a right section of a solid appears in the view in which the axis of the solid projects as a point.

Fig. 1 shows the triangular oblique prism in the front and top views. Since a view in which the axis appears true length is required first, draw reference plane H-1 parallel to line AF in the top view. The lateral edges of the prism now project true length to the first auxiliary

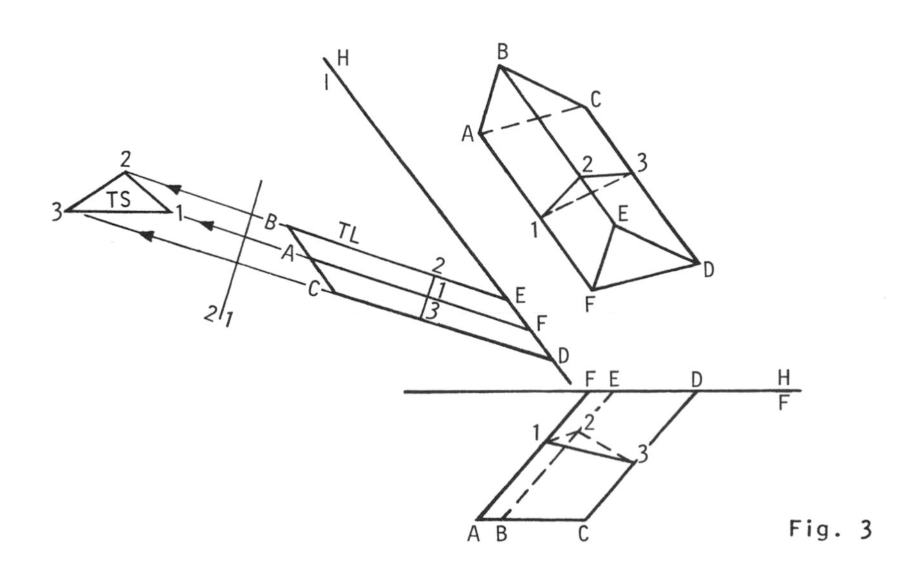

Fig. 3

view, as in fig. 2. A plane, X, passed perpendicular to these edges in view 1 will form a right section of the prism, cutting the edges in points 1, 2, and 3. Project these points back to the top and front views. For the true size and shape view of the right section, draw reference plane 1-2 perpendicular to the true length lines in view 1, projecting these lines into view 2 as points. View 2 is the required view of the prism, as shown in fig. 3.

• PROBLEM 7-19

Given the front and profile views of a plane, A-B-C, and a hexagonal prism, find the points in which the edges of the prism intersect the plane.

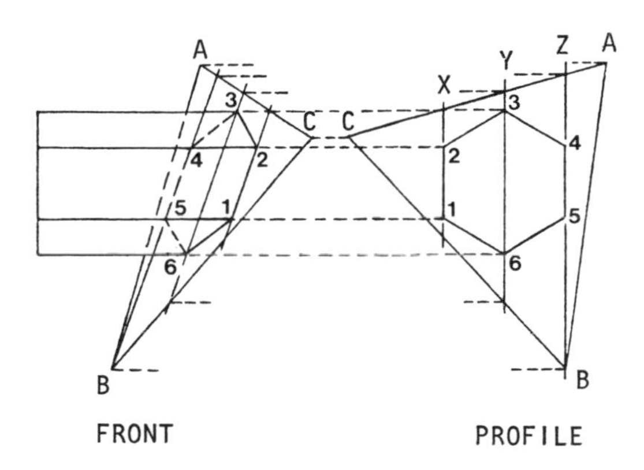

Solution: The most appropriate method of solution is the cutting plane method. The accompanying figure contains the given front and profile views of the plane and prism.

The point in which each edge of the prism pierces the plane is established by passing vertical cutting planes through the corners of the prism in the profile view. The polygon having these points as vertices represents the section cut from the prism by plane A-B-C. As illustrated, cutting plane X, which appears edgewise in the profile view, determines the points 1 and 2 in which the two front edges of the prism intersect plane A-B-C. Likewise, cutting planes Y and Z determine points 3,6 and 4,5, respectively. The hexagon 1-2-3-4-5-6 is the required intersection.

• PROBLEM 7-20

Determine the intersection of an oblique plane and an oblique prism.

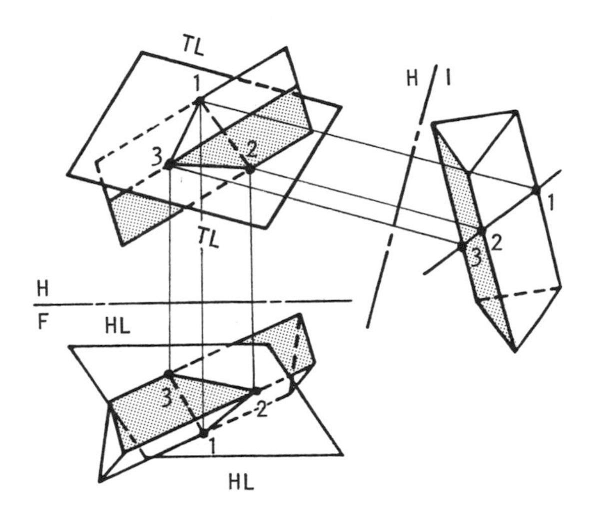

Solution: (1) The accompanying figure illustrates the general case of the intersection between an oblique plane and an oblique prism. (2) Problem analysis reveals that since both the plane and the prism are oblique in the H, horizontal, and F, frontal, planes of projection, neither the plane nor any of the planes of the prism appears as an edge. Once a view in which the plane projects as an edge is located, the solution follows from the principles applied in the foregoing problem: the cutting plane method.

(3) The edge view of the plane is found in a primary auxiliary view by taking the point view of a line on the plane by projection from either view. As illustrated, two edges of the plane appear as horizontal lines (HL) in the front view. Consequently, these same lines appear true length (TL) in the top view. An auxiliary view, view 1, having lines of sight parallel to the true length lines on the plane, shows the plane as an edge. (4) Note that the lateral edges of the prism do not appear true length in the auxiliary view; although this will not complicate the problem.

(5) In the auxiliary view, locate the points where the corner edges of the prism intersect the plane: points 1, 2, and 3. (6) Project these points to the top and front views, locating the intersection of the plane with the prism. (7) Determine the visibility in both views by inspection.

• **PROBLEM 7-21**

Determine the intersection of a given oblique plane and a given prism, using the cutting plane method.

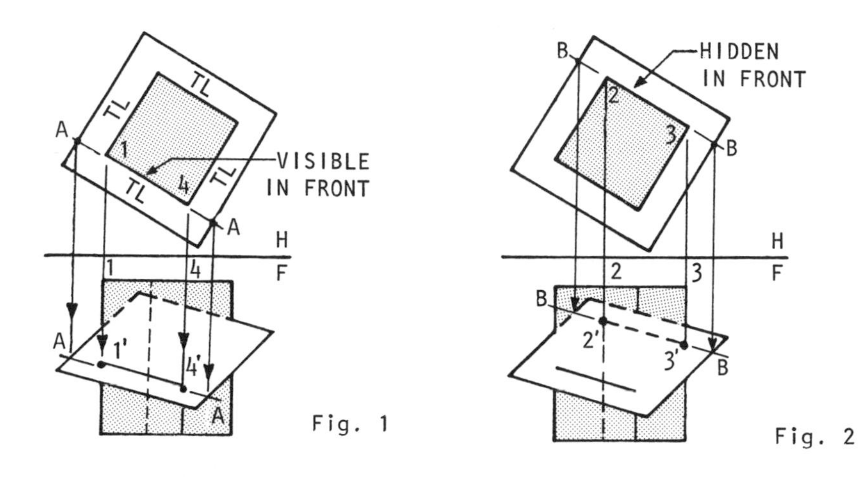

Fig. 1

Fig. 2

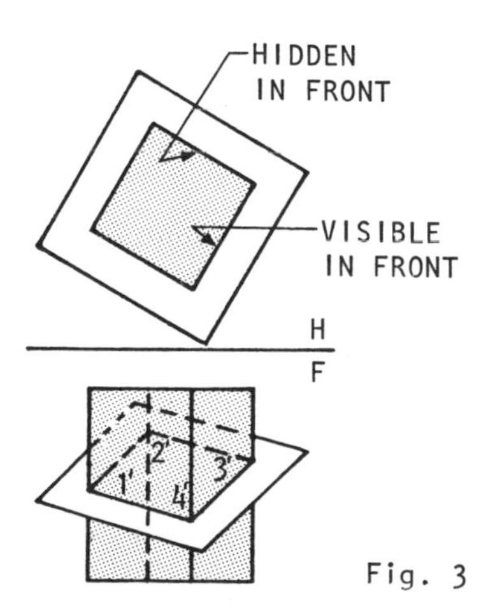

Fig. 3

Solution: (1) The accompanying figures illustrate the general case of intersection between a plane and a prism. (2) Analysis of Fig. 1 reveals that the vertical corners of the prism project true length in the front view and the plane appears foreshortened in both views. (3) The method of solution consists of passing vertical cutting planes through the planes of the prism in the top view in order to locate piercing points of the corners of these planes in the front view. Connecting the points locates the desired intersection.

(4) In Fig. 1, the vertical cutting plane A-A is passed through the vertical plane 1-4, in the top view, and is projected to the front view. The intersection of plane A-A with the projections of points 1 and 4 in the front view are the required piercing points 1' and 4'.

(5) In Fig. 2, the vertical plane B-B is passed through the top view of plane 2-3 and is projected to the front view. The intersection of plane B-B with the projections of points 2 and 3 in the front view are the required piercing points 2' and 3'.

(6) As shown in Fig. 3, the line of intersection is completed by connecting the four points, 1', 2', 3', and 4', in the front view. (7) Visibility in the front view is found by inspection of the top view.

• PROBLEM 7-22

Determine the intersection between two given prisms.

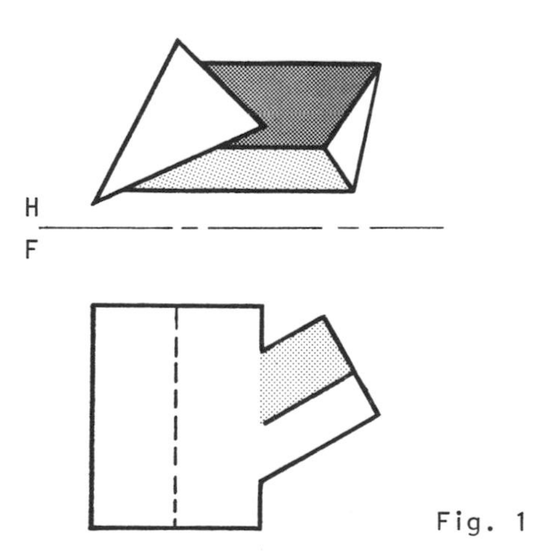

Fig. 1

Solution: (1) Fig. 1 shows the top and front views of the two intersecting prisms. (2) As shown in Fig. 2, construct the end view of the inclined prism by projecting an auxiliary view from the front view. The frontal line, 3-3', in the top view projects true length in the front view. The lines of sight for the auxiliary view are parallel to the true length (TL) lines in the front view, thereby establishing a true shape and size end view of the prism. (3) In reference to the vertical prism, show only line A-B, the corner

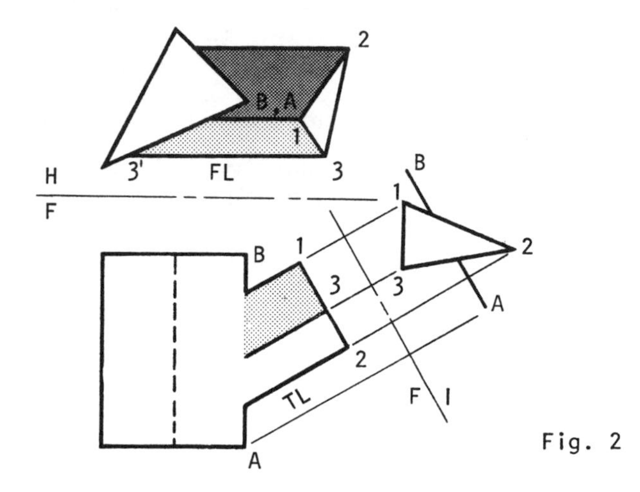

Fig. 2

line, in the auxiliary view because this is the only critical line (i.e., the line needed for determination of the piercing points in the front view). (4) Letter the points on the endview: points 1, 2, and 3.

(5) Locate the piercing points of lines 2-2' and 3-3' in the top view, as shown in Fig. 3. Project these points to the front view to the extension of the corresponding corner lines. For example, 3' from the top view is projected to 3 extended at 3' in the front view. Likewise, apply the same method to points 2' and 2, extended. (6) Spacial analysis reveals that a line connecting points 2' and 3' will not be a straight line, but will bend around the corner line A-B. The point where this line intersects the corner (i.e., line A-B) is found to be point X. Point X is initially located in the primary auxiliary view as the point of intersection of line 2-3 and A-B. (7) Project

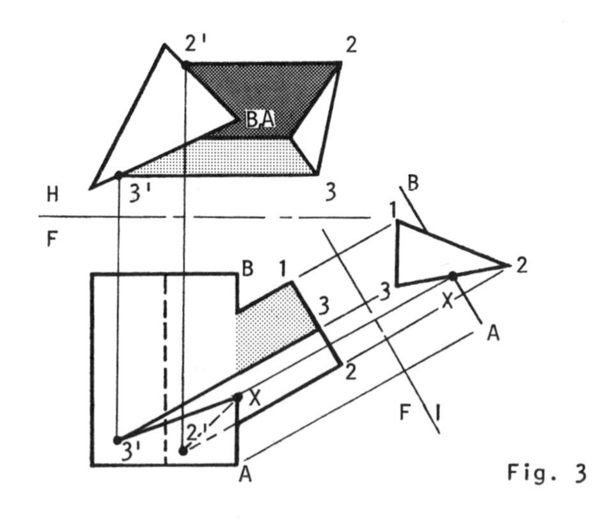

Fig. 3

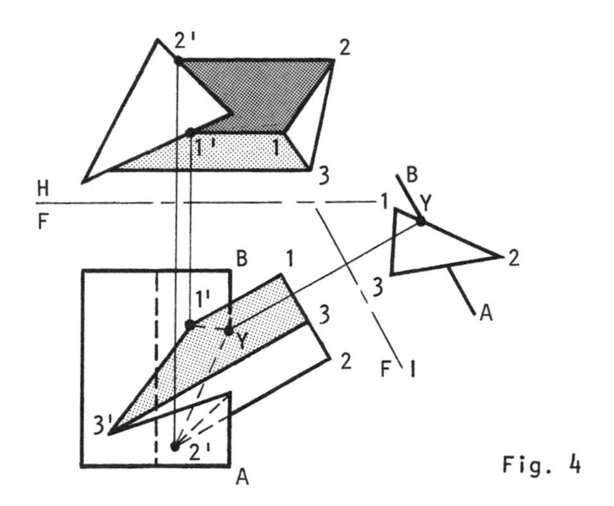

Fig. 4

X to the front view and construct lines 3'-X and 2'-X (further lines of intersection between the two prisms).

(8) With reference to Fig. 4, locate point 1' in the front view: the intersection of the projection of 1' from the top view with 1 extended in the front view. (9) Just as line 2'-3' was found to bend around line A-B (see auxiliary view), line 1'-2' bends around corner line A-B at point Y. Point Y is initially located in the auxiliary view as the point of intersection of line 1-2 and A-B. (10) Draw line 1'-Y-2' in the front view. (11) By means of inspection of the primary auxiliary view, line 1'-Y-2' is found to be invisible in the front view. (12) Draw line 1'-3' as a visible straight line since it does not bend around a corner (see the auxiliary view: 1-3 does not cross A-B).

• PROBLEM 7-23

Determine the intersection between a given cone and a given prism.

Solution: (1) Fig. 1 shows the intersecting cone and prism in the H-, horizontal, and F-, frontal, planes of projection. In addition, a spacial analysis is provided to allow for a clearer understanding of the problem and the method of solution.

(2) Referring to Fig. 2, construct a primary auxiliary view, view 1, to obtain the edge views of the lateral sur-

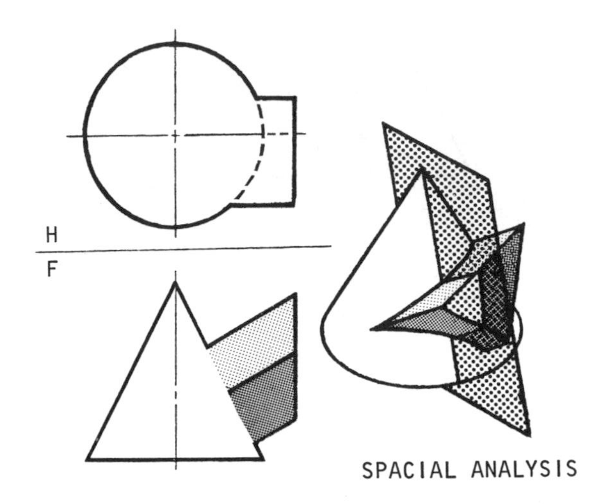

Fig. 1

faces of the prism. The lines of sight for view 1 are parallel to true length corner lines of the inclined prism in the front view. (3) In the auxiliary view, pass cutting planes through the cone which radiate from the apex. These planes establish elements on the cone and, therefore, lines lying on the surface of the cone and surface of the prism. (4) Project these elements to the principle views.

(5) Locate the piercing points of the cone's elements with the edge view of plane 1-3 in the auxiliary view, as shown in Fig. 3. Project these points to the front and top views. (6) As an illustration of step (5), consider

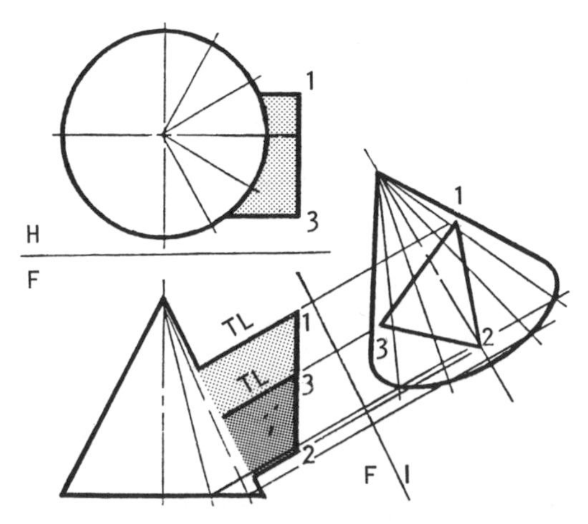

Fig. 2

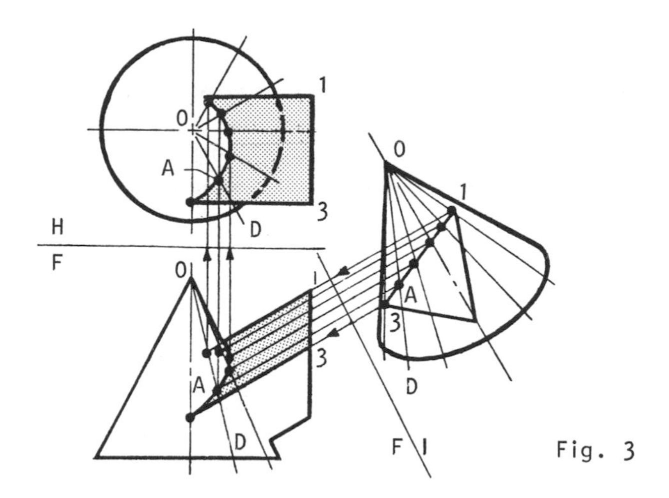

Fig. 3

point A which lies on element O-D in the primary auxiliary. It is projected to the front and top view of element O-D. (7) Similarly, locate other points along the line of intersection.

(8) The location of the remaining piercing points is illustrated in Fig. 4 as a repetition of step (4). Piercing points are established where the conical elements intersect the edge views of the other planes of the prism in the

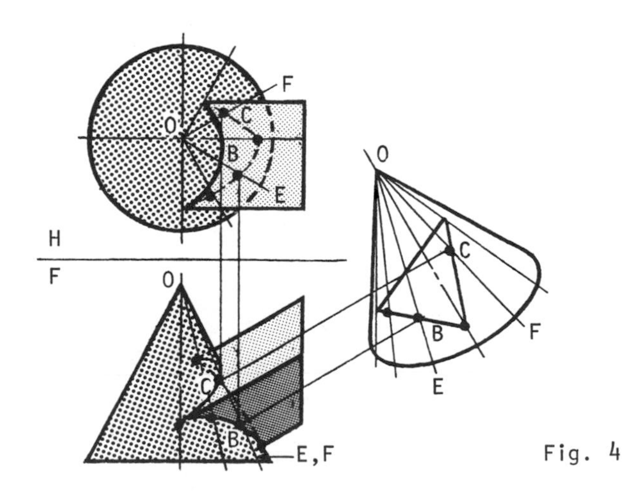

Fig. 4

primary auxiliary view. (9) Show visibility in each view after the location of a sufficient number of points.

Determine the intersection between a given cylinder and a given prism, using the cutting plane method.

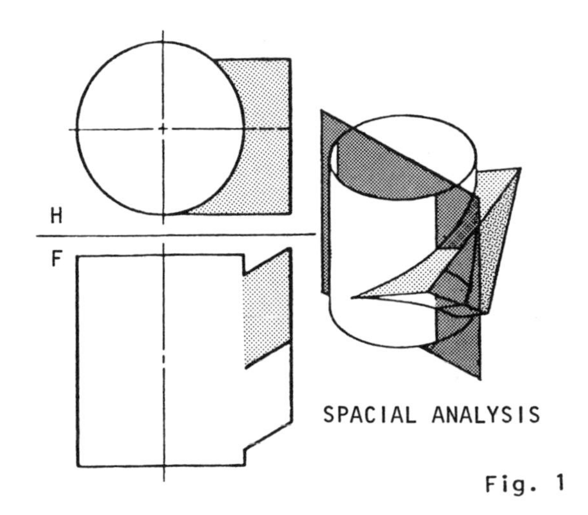

Fig. 1

Solution: (1) Fig. 1 shows the intersecting prism and cylinder in the H-, horizontal, and F-, frontal, planes of projection. In addition, a spacial analysis of the problem is provided in order to promote a clearer understanding of problem analysis and the chosen method of solution.

(2) As shown in Fig. 2, a series of cutting planes are used to establish lines that lie on the surfaces of the

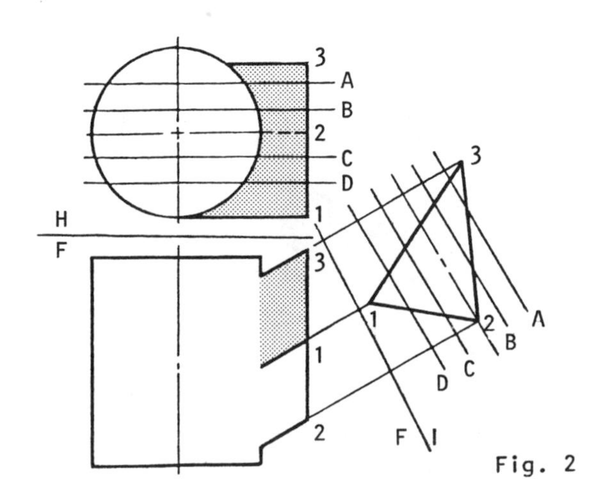

Fig. 2

cylinder and the prism. These lines intersect where they cross in the views of projection, since they lie in a common cutting plane. A primary auxiliary view is needed to locate the lines on the surface of the prism in the front view. Project an auxiliary view of the triangular prism from the front view to construct its end view and to show its surface as edges. This auxiliary view has lines of sight parallel to the true length corner lines of the prism in the front view. These lines are true length since they appear as frontal lines in the top view. (3) Pass frontal cutting planes through the top view of the cylinder and project them to the auxiliary view. Note: the spacing between the cutting planes is equal in both views.

(4) Fig. 3 illustrates the location of points along

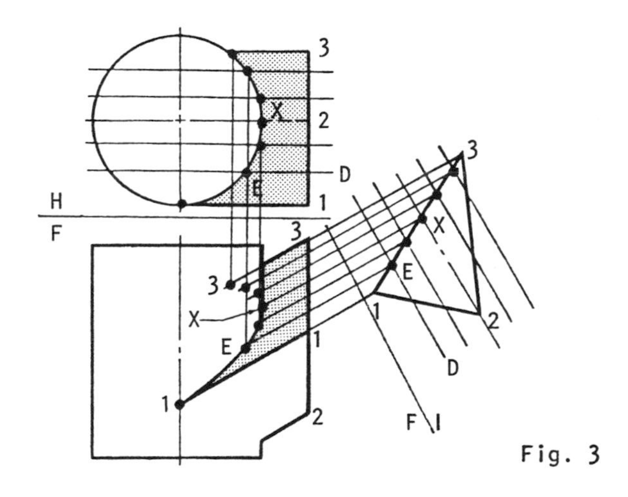

Fig. 3

the line of intersection of the cylinder and plane 1-3 in the top view. Project these points to the front view. The intersection of the projectors coming from the top view with those coming from the auxiliary view, locates points on the line of intersection in the front view. (5) As an example of step (4), consider point E on cutting plane D: Point E is found in the top and primary auxiliary views and projected to the front view where the projectors intersect. (6) Point X on the center line is the point where visibility changes from visible to hidden in the front views.

(7) The remaining points of intersection are determined by using the same cutting planes, as shown in Fig. 4. For

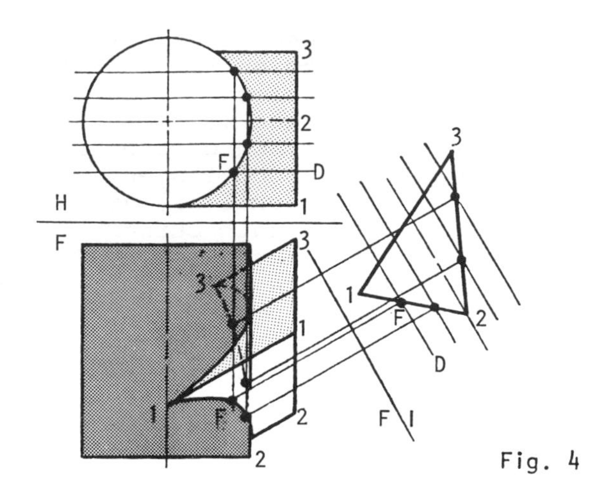

Fig. 4

example, point F is shown in the top and primary auxiliary views and is projected to the front view on the line of intersection 1-2. (8) Connect the points thus located and determine visibility. (9) Note: Sufficient cutting planes should be used in order to produce the most accurate representation of the line of intersection.

SPHERES

• PROBLEM 7-25

Determine the points at which a line pierces a sphere using (A) the small circle and (B) the great circle methods.

Solution: Problem analysis reveals that the points at which a line pierces a sphere lie on a circle that is cut from the sphere by any plane containing the line. This circle may be either a great circle or a small circle.

(A) Small Circle Method:

(1) In Fig. 1, the sphere with radius R and center X, and the line MN are given in the P, plan, and E_1, primary

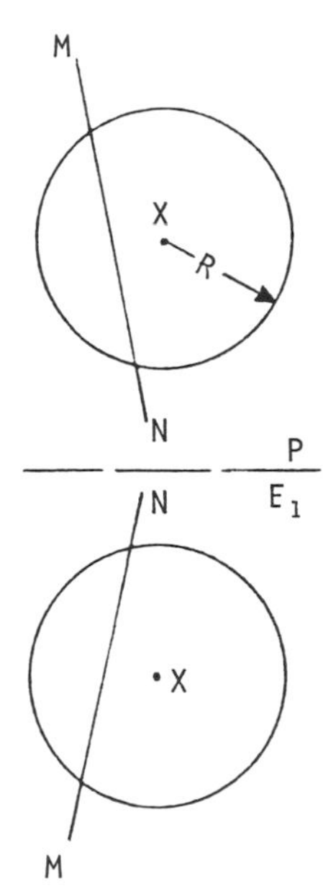

Fig. 1

elevation, views. (2) A vertical plane passed through MN cuts a small circle, since it does not contain the center of the sphere. (3) Referring to Fig. 2, the secondary elevation view, E_2, is drawn to show the true size of this small circle, with radius r. The rotation line P-E_2 is parallel to line MN ; consequently, MN and the small circle which represents the intersection of the vertical cutting plane and the sphere, appear in true size. (4) The points at

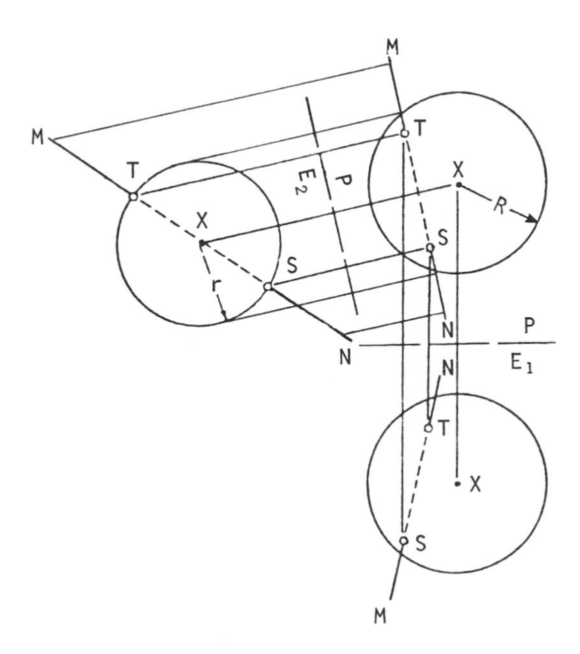

Fig. 2

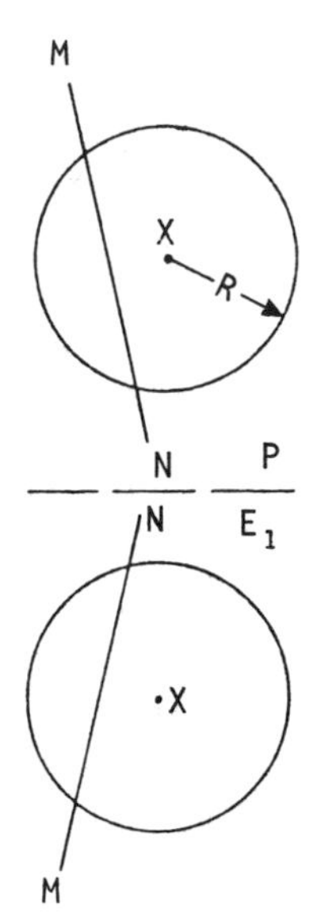

Fig. 3

which line MN intersects this true size small circle, S and T, are the required piercing points. (5) Project these points back into the given views.

(B) Great Circle Method:

(1) Fig. 3 represents the same data given in the preceding small circle problem. (2) As illustrated in Fig. 4, a plane is drawn that contains MN and the center of the sphere, X, and therefore cuts a great circle from the sphere. (3) The

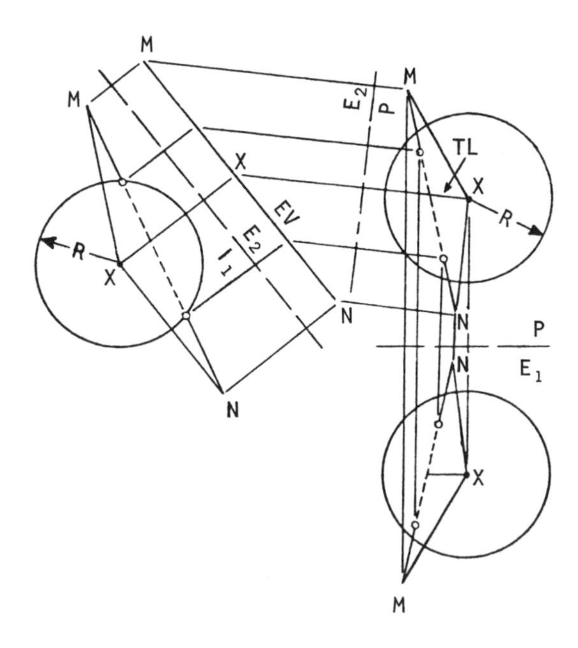

Fig. 4

piercing points will be located in the view showing the plane in true size. (4) In order to locate the view showing the plane in true size, obtain a view having lines of sight perpendicular to the edge view of the cutting plane. (5) To obtain the edge view of the cutting plane, draw a horizontal line in view E_1, which projects true length (TL) in the plan view. Construct view E_2 having lines of sight parallel to this true length line, and thus showing the plane which passes through MN as an edge. (6) View I_1, the primary inclined view, is drawn having lines of sight perpendicular to the edge view of the cutting plane; thus, said plane appears true size in I_1. (7) The piercing points appear at the intersection of the great circle and line MN in view I_1. (8) The piercing points are then located in the given views by projection.

• PROBLEM 7-26

Find the line of intersection between a given sphere and a given cylinder.

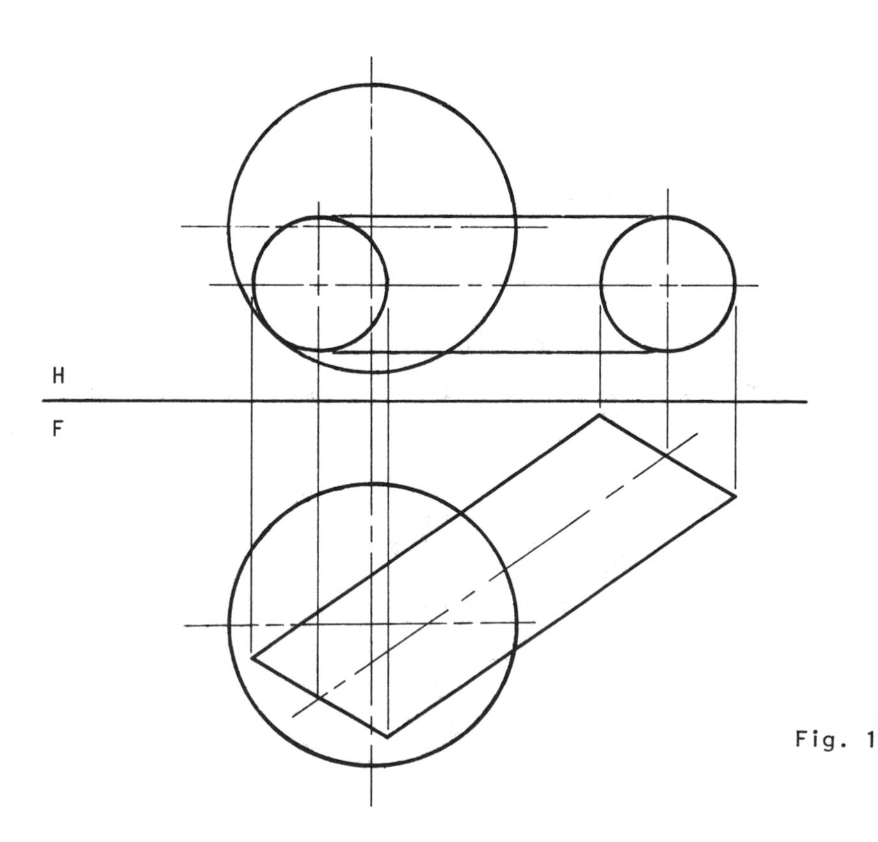

Fig. 1

Solution: Fig. 1 shows the top and front views of the given sphere and cylinder. The axis of the cylinder appears true length in the front view since it is parallel to the reference plane, H-F, in the top view. The line of intersection can be found by locating the points at which various elements of the cylinder pierce the sphere. As shown in fig. 2, pass a vertical cutting plane through the cylinder in the top view, where the plane appears as an edge. This plane intersects elements AC and BD of the

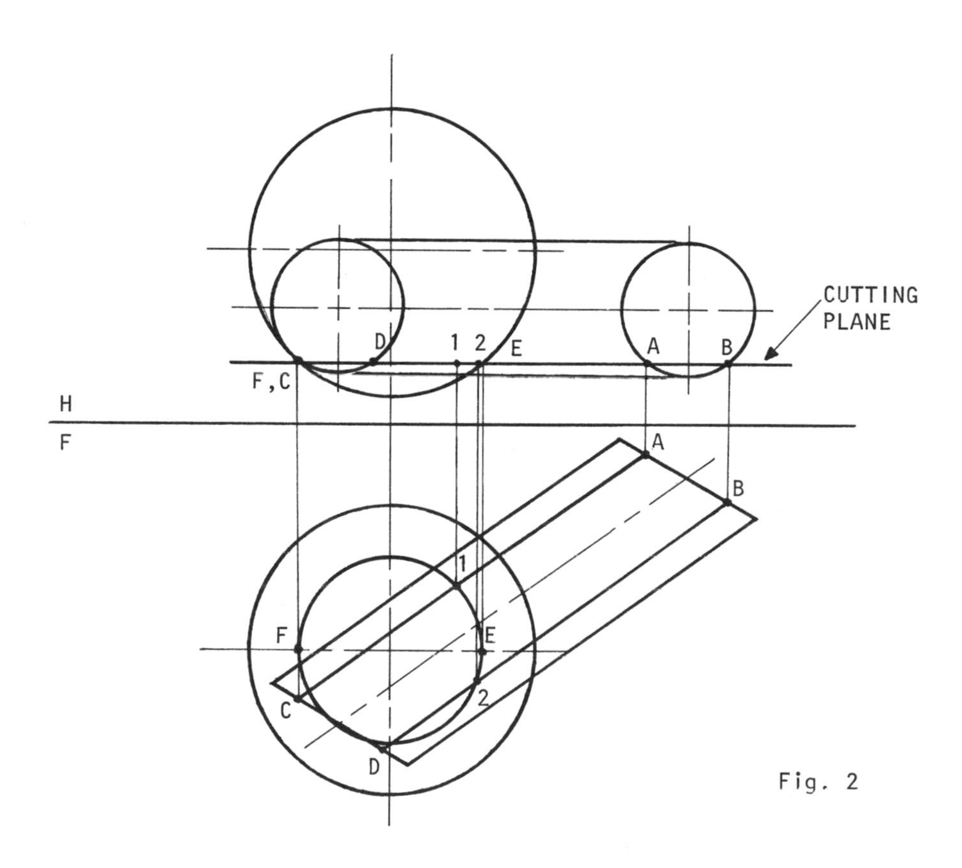

Fig. 2

cylinder, and it intersects the sphere in line EF, representing the edge view of the small circle cut from the sphere by the plane. Project elements AC and BD to the front view, and construct the small circle of the sphere with diameter EF in the front view, also. Elements AC and BD intersect the small circle at points 1 and 2, respectively. Project points 1 and 2 back to the top view to where they lie on the edge view of the cutting plane, thus locating two points on the line of intersection. Pass additional vertical cutting planes through the cylinder and sphere, as seen in fig. 3, and locate more points on the line of intersection by the method just described.

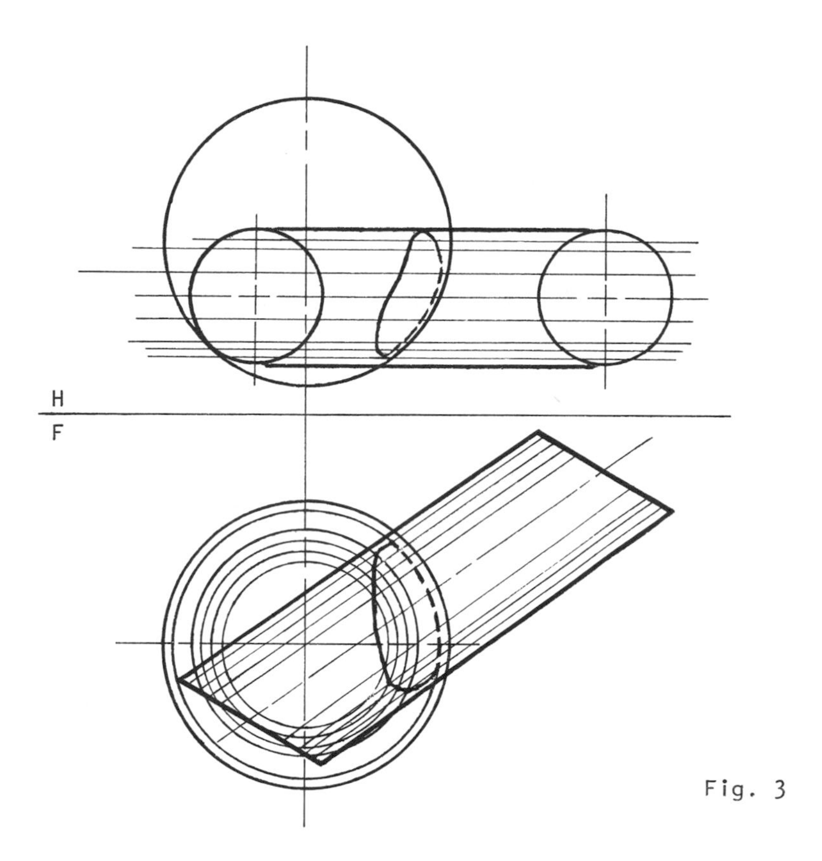

Fig. 3

Connect these points with a smooth curve and determine the visibility.

• **PROBLEM 7-27**

Find the line of intersection between a given sphere and a given prism.

Solution: Fig. 1 shows the top and front views of the given sphere and prism. Any cutting plane parallel to the H-F reference plane will intersect the sphere in a circle and the prism in a quadrilateral congruent to its bases. Where these two sections intersect (i.e., have common points) determines points on their line of intersection. As in fig. 2, pass a horizontal cutting plane, E, through the sphere and prism in the front view. Project the corresponding sections to the top view, where the circle cut from

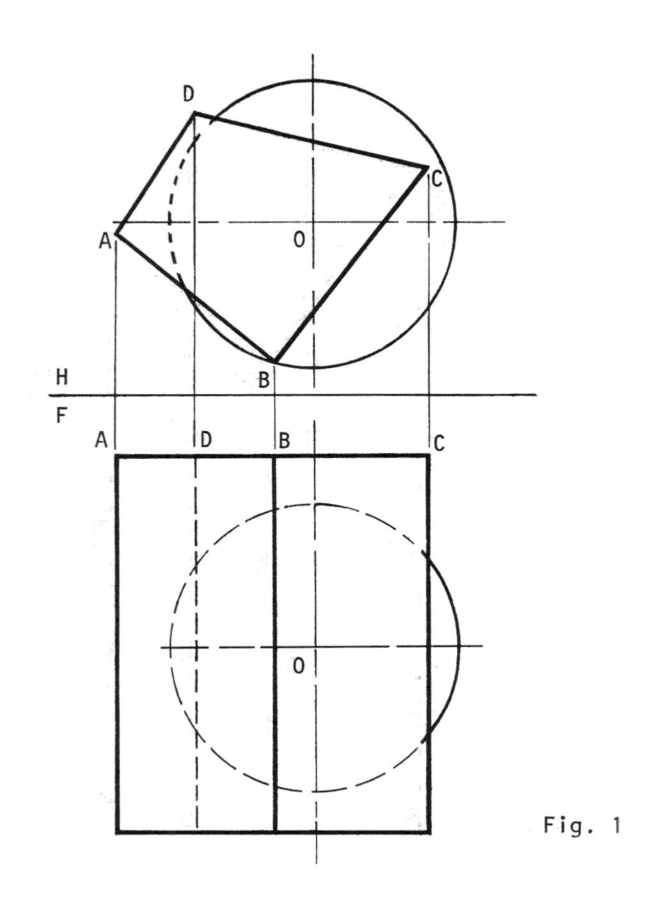

Fig. 1

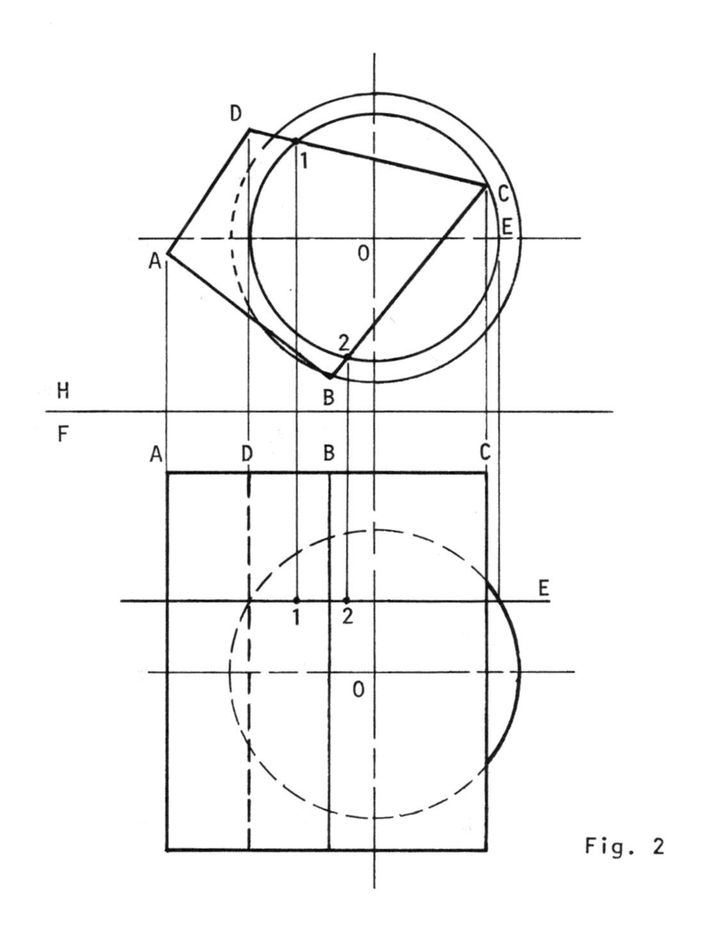

Fig. 2

the sphere appears true size and shape, and the quadrilateral cut from the prism coincides with the base ABCD. Points 1 and 2 where the two sections intersect are on the line of intersection between the sphere and the prism. Project 1 and 2 to the front view to where they lie on cutting plane

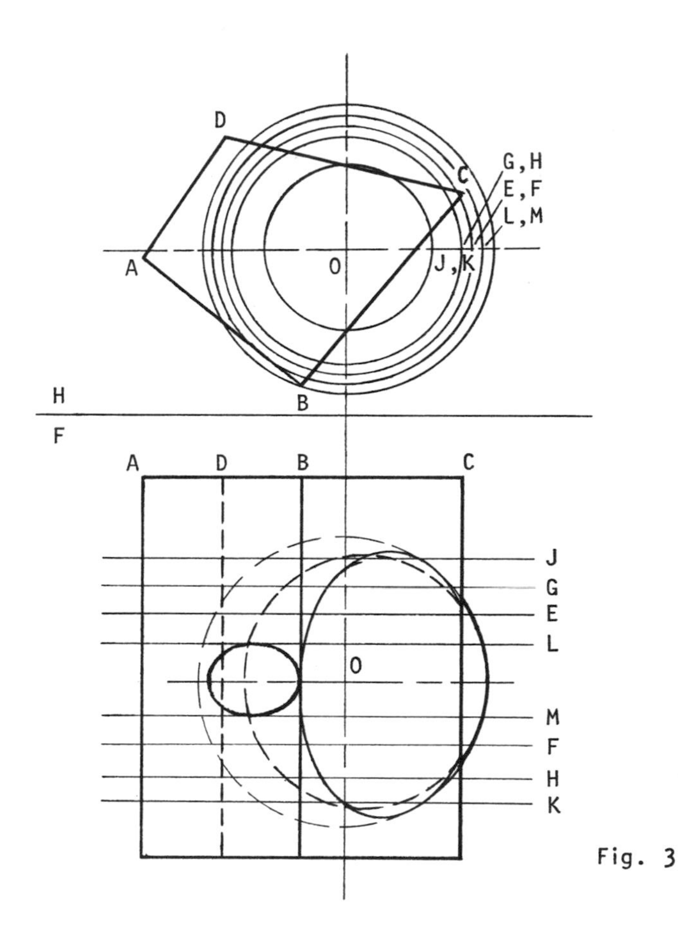

Fig. 3

E. As shown in fig. 3, pass additional horizontal planes through the sphere and prism in the front view, and locate other points on the line of intersection in the same manner in which points 1 and 2 were located. Connect these points with smooth curves and determine visibility for the required line of intersection, as seen in fig. 3.

TORUS AND ELLIPTICAL CYLINDER

• **PROBLEM** 7-28

Find the line of intersection between a torus and an elliptical cylinder.

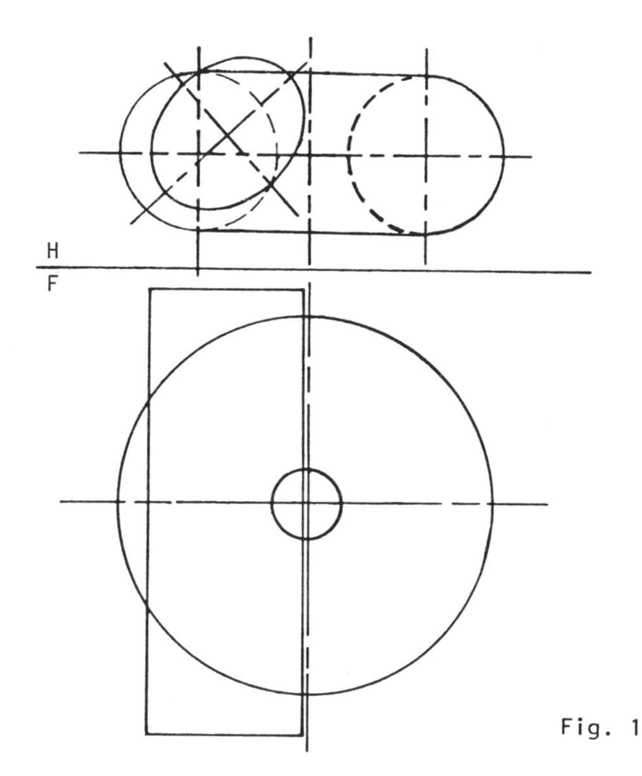

Fig. 1

Solution: Fig. 1 shows the top and front views of the given torus and elliptical cylinder. As shown in fig. 2, pass a vertical cutting plane through the torus and cylinder in the top view, where the plane appears as an edge. This plane cuts elements A and B from the cylinder and circles C and D from the torus, as seen in the front view in fig. 2. Elements A and B intersect circles C and D at points 1 through 6, points which lie on the line of intersection between the cylinder and the torus. Locate additional points in the same manner by passing more cutting planes

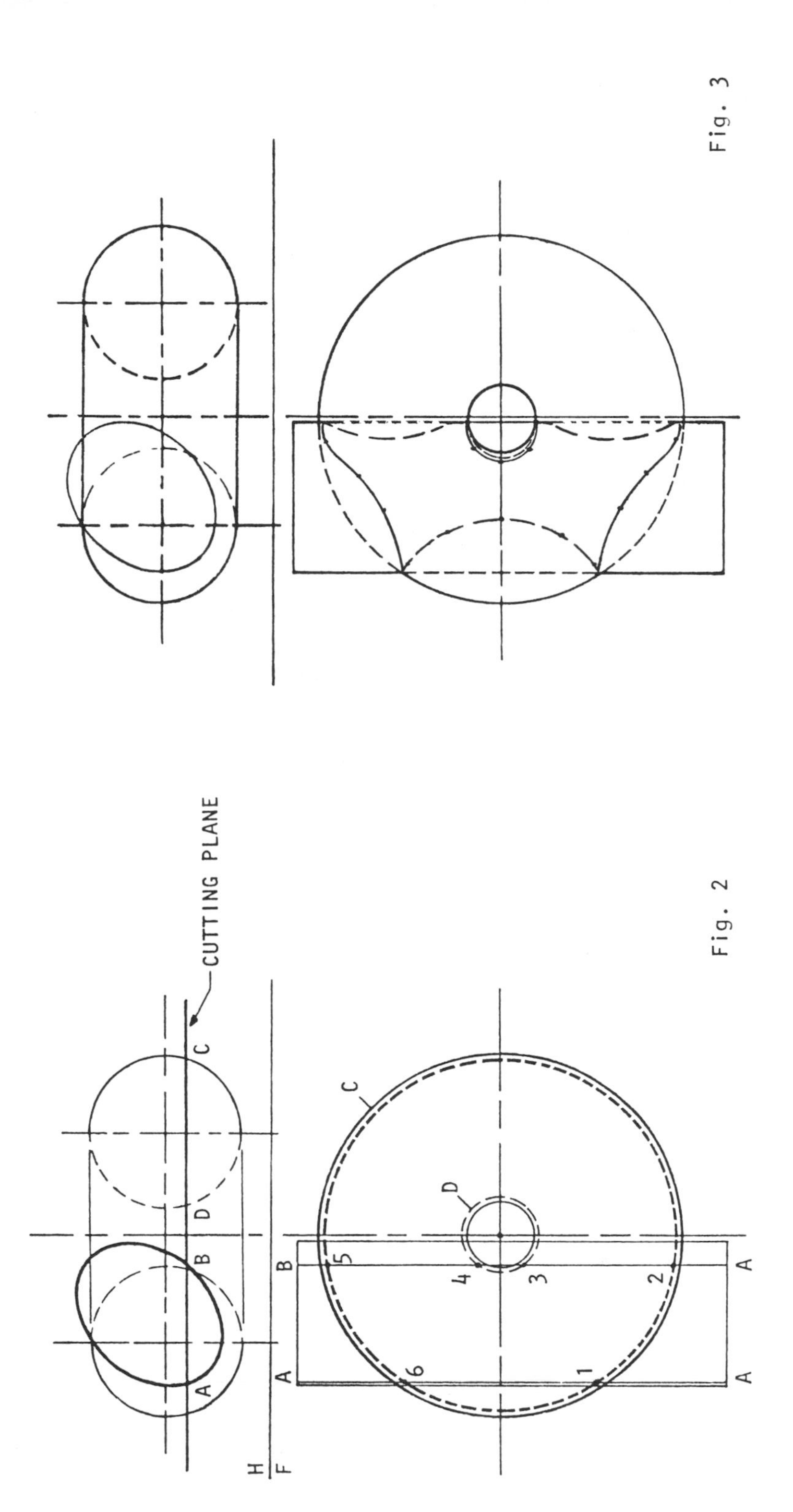

Fig. 2

Fig. 3

through the cylinder and the torus. Connect these points with a smooth curve for the required line of intersection, as seen in fig. 3.

PYRAMIDS

• PROBLEM 7-29

Determine the line of intersection between a given prism and a given pyramid, using the edge-view method.

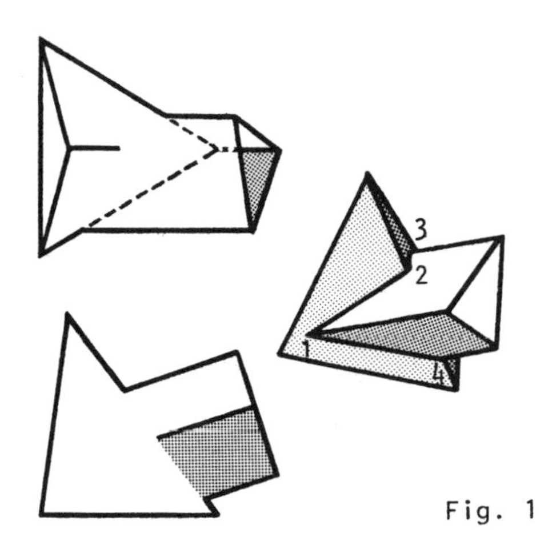

Fig. 1

Solution: Fig. 1 shows the top and front views of the given prism and pyramid, along with the spacial analysis diagram. The line of intersection can best be determined in a view in which the surfaces of one of the solids appear as edges. Line CD of the prism is true length in the front view since it appears parallel to reference plane H-F in the top view, as seen in fig. 2. The surfaces of the prism can therefore

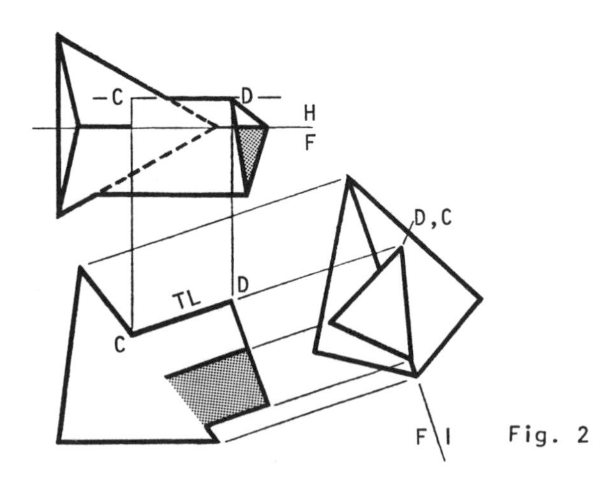

Fig. 2

be drawn easily as edges by constructing the primary auxiliary view from the front view. Make reference plane F-1 parallel to the edge view of the base of the prism in the front view and project both the prism and the pyramid to the auxiliary view. The surfaces of the prism appear as edges in view 1 since the true length line CD appears as a point in view 1. As shown in fig. 3, pass planes OA and OB through the apex

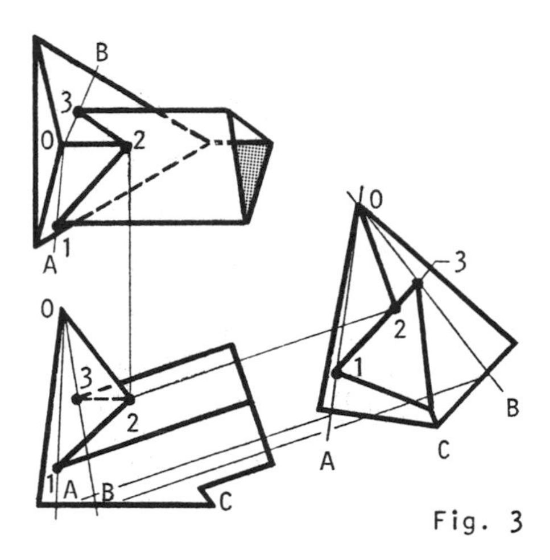

Fig. 3

of the pyramid, point O, in view 1 and through the points 1 and 3 of the prism. Project the points where planes OA and OB intersect the surfaces of the pyramid to the front and top views, and construct the respective elements of the pyramid. Project points 1 and 3 to elements OA and OB of the front and top views, correspondingly. Project point 2 from view 1 to element OC of the pyramid in the principal views, and connect points 1 and 2 and points 2 and 3 for the partial line of intersection, as shown in fig. 3.

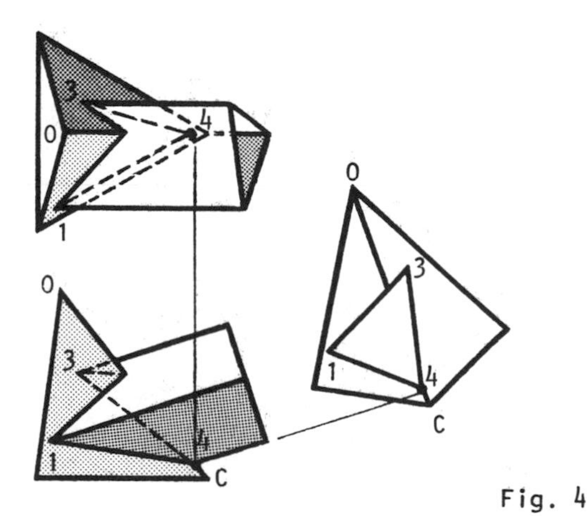

Fig. 4

Point 4 on the edge view of the surfaces of the prism lies on element OC of the pyramid in view 1, as labeled in fig. 4. Project point 4 to the front and top views, and connect points 1 and 4 and points 3 and 4 in each view to complete the line of intersection between the prism and the pyramid. Determine the visibility, as shown in fig. 4.

• PROBLEM 7-30

Find the line of intersection between a given prism, with base ABC, and a given pyramid, V-RSTU, using the two-view method.

Solution: Fig. 1 shows the top and front views of the given prism and pyramid. The line of intersection can be found by locating the points in which the edges of one of the solids pierce the plane surfaces of the other solid.

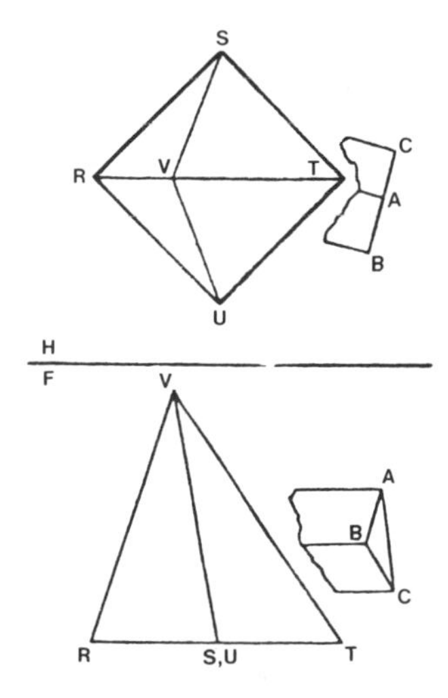

Fig. 1

As in fig. 2, pass a horizontal cutting plane through edge line A in the front view of the prism, which intersects edges VS and VT at points E and D, respectively. Project E and D to the top view, and extend edge line A to intersect line ED at point F. Project F, which is the piercing point of edge A through plane VST of the pyramid, to the front view to edge line A. As in fig. 3, pass a cutting plane through the edge line VT of the pyramid in the front view. This plane intersects the edges A, B, and C of the prism through points G, H, and J, respectively. Project points G, H, and J to the top view to their corresponding edges, and connect point G with point H and point J with point H, showing the cutting plane in the top view. This plane intersects the edge VT of the pyramid through points K and L. Project K and L to the front view. Locate additional piercing points in the same manner, and connect

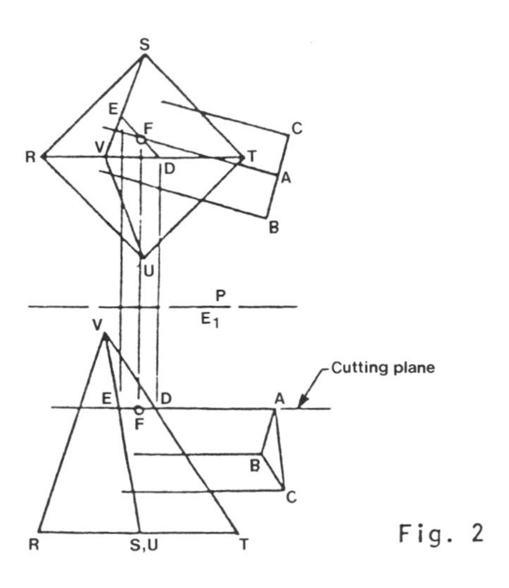

Fig. 2

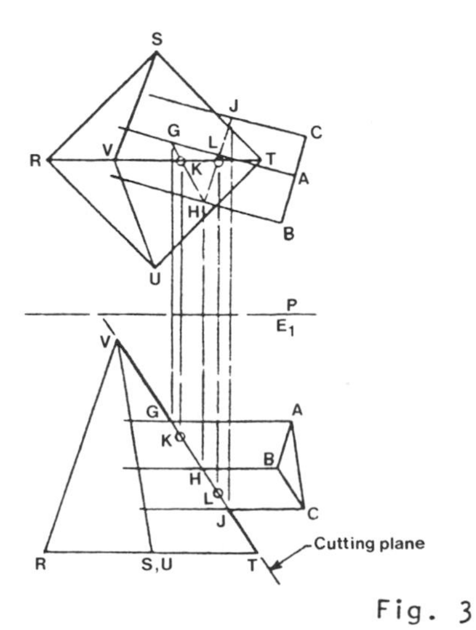

Fig. 3

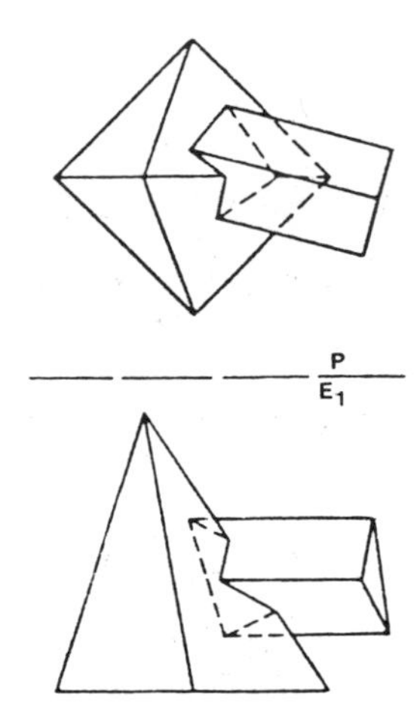

Fig. 4

them to form the line of intersection. Determine the visibility in both the top and front views, as shown in fig. 4.

• PROBLEM 7-31

Determine the line of intersection between a given hexagonal pyramid, O-ABDEFG, and an inclined cutting plane, CP, and construct its true size and shape view.

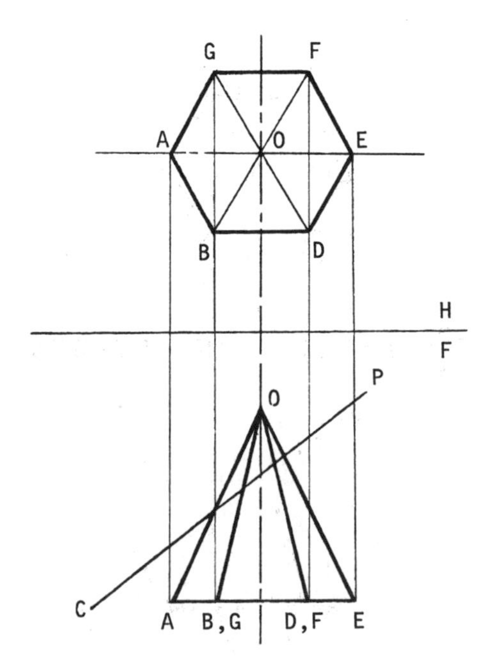

Fig. 1

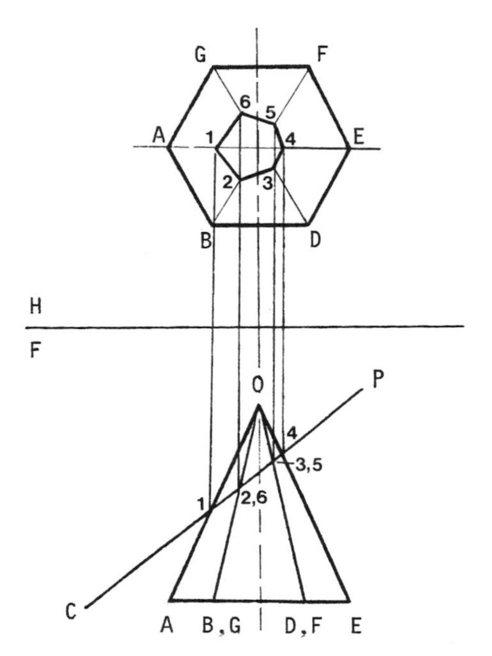
G
F
6
5
A
1
4
E
2
3
B
D
H
F
O
P
4
3,5
1
2,6
C
A
B,G
D,F
E

Fig. 2

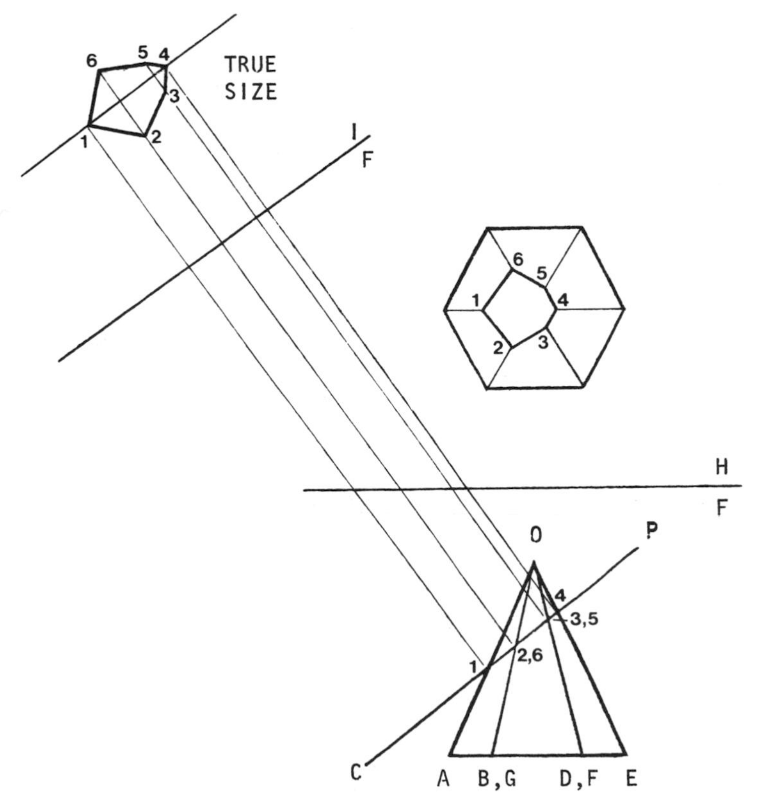
6
5 4
TRUE
SIZE
3
1
2
I
F
6
5
1
4
2
3
H
F
O
P
4
3,5
1
2,6
C
A
B,G
D,F
E

Fig. 3

Solution: Fig. 1 shows the top and front views of the given hexagonal pyramid, O-ABDEFG, cut by plane CP, which shows as an edge in the front view. To determine the line of intersection in the top view, project from the front view the points where plane CP intersects the elements OA, OB, OD, OE, OF, and OG, labeled 1 through 6, respectively, to their corresponding elements in the top view. Connect points 1 through 6 in order, in the top view, to complete the line of intersection, as shown in fig. 2. The true size and shape view is constructed by projection from the edge view of the cutting plane. With lines of sight perpendicular to plane CP in the front view, make reference plane F-1 parallel to plane CP and construct the primary auxiliary view. The true size and shape of the line of intersection between the plane and the hexagonal pyramid is seen in view 1, as in fig. 3.

• PROBLEM 7-32

Find the line of intersection between two given pyramids, G-DEF and O-ABC.

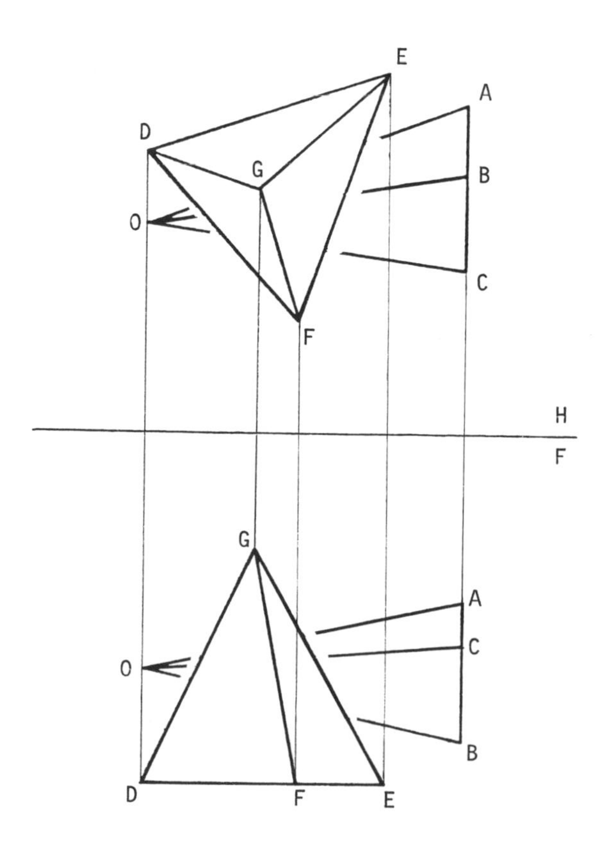

Fig.1

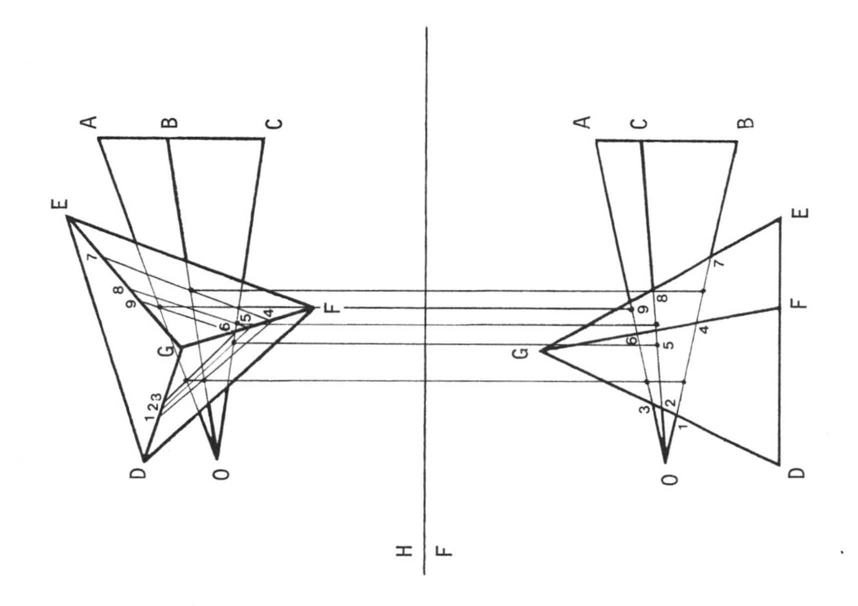

Fig. 3

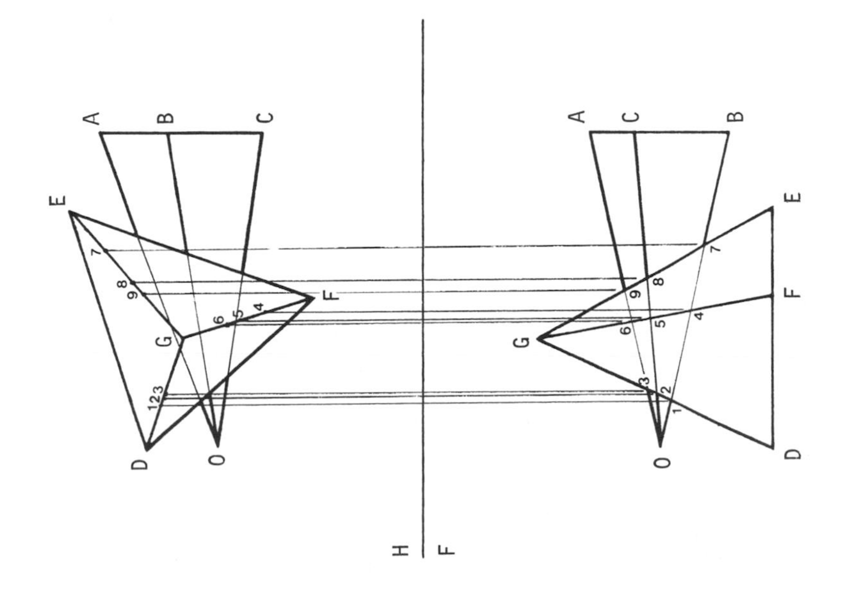

Fig. 2

Solution: Fig. 1 shows the top and front views of the two given pyramids, G-DEF and O-ABC. The line of intersection can best be found by locating the points at which the edges of one pyramid pierce the plane surfaces of the other pyramid. The piercing point of a line through a plane can be determined using two views. As shown in fig. 2, complete all the edge lines of pyramid O-ABC in both the front and top views. Edge OB crosses edge lines GD, GF, and GE of pyramid G-DEF in the front view at points 1, 4, and 7, respectively. Project these points to the top view to where they lie on the corresponding edge lines of pyramid G-DEF, as in fig. 2. Points 2, 5, and 8 of edge OC and points

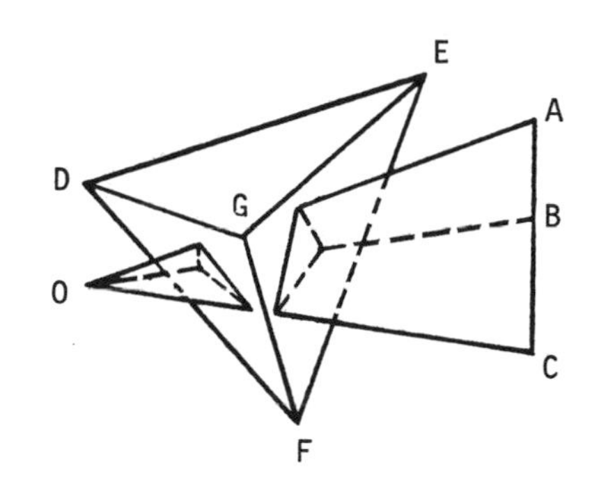

H

F

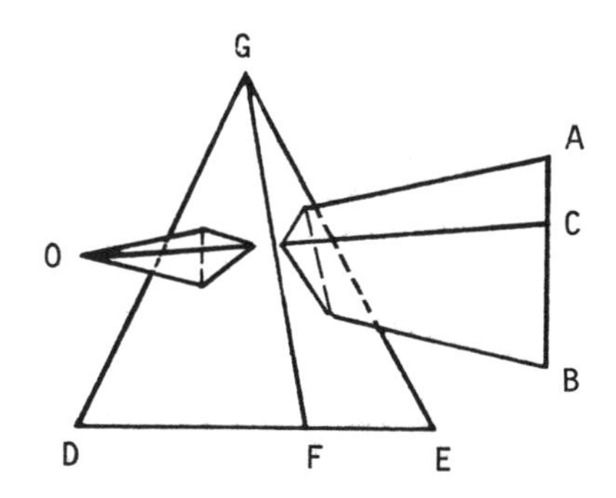

Fig. 4

3, 6, and 9 of edge OA are projected to the top view in the same manner. Since points 1 and 4 lie on the same line in the front view, connect them in the top view as in fig. 3. Likewise connect the remaining corresponding points in the top view. The point at which line 1-4 in the top view crosses edge OB is the piercing point of edge OB through plane GDF. Similarly, locate the other piercing points and project all of them to the front view to where they lie on the respective edges of pyramid O-ABC. Determine the visibility and complete the front and top views of the line of intersection, as shown in fig. 4.

• PROBLEM 7-33

Find the points in which a line intersects a polyhedron.

Solution: The accompanying figure shows the given line K and prismoid (polyhedron) A-B-C-D-E-F-G-H in the H, horizontal, and F, frontal, planes of projection. Problem analysis reveals that the cutting plane method will readily provide a solution.

A vertical cutting plane through line K cuts the pris-

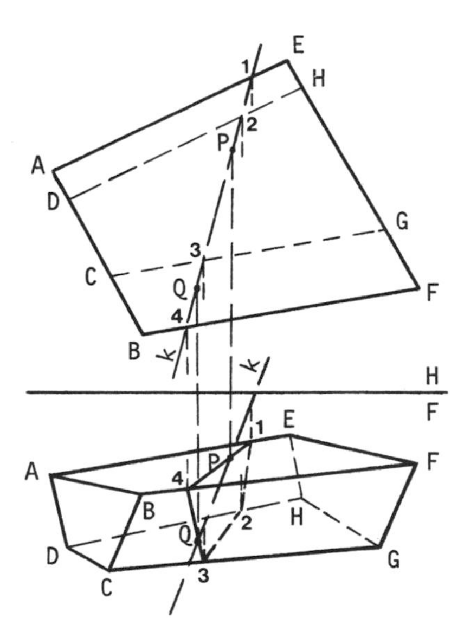

moid in the section 1-2-3-4, as is clearly seen in the H-view. Project these points to the F-view, and construct the section. Line K intersects the edges of this section at points P, on line 1-4, and Q, on line 3-4. Project these two points to the H-view. The points P and Q, where line K crosses the edges of the plane section, are the required points where line K intersects the solid.

• **PROBLEM 7-34**

Find the intersection of a given plane and a given polyhedron using the auxiliary view method and the cutting plane method.

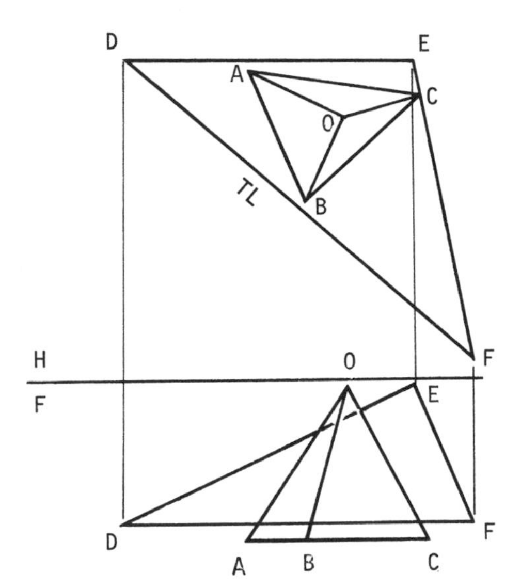

Fig. 1

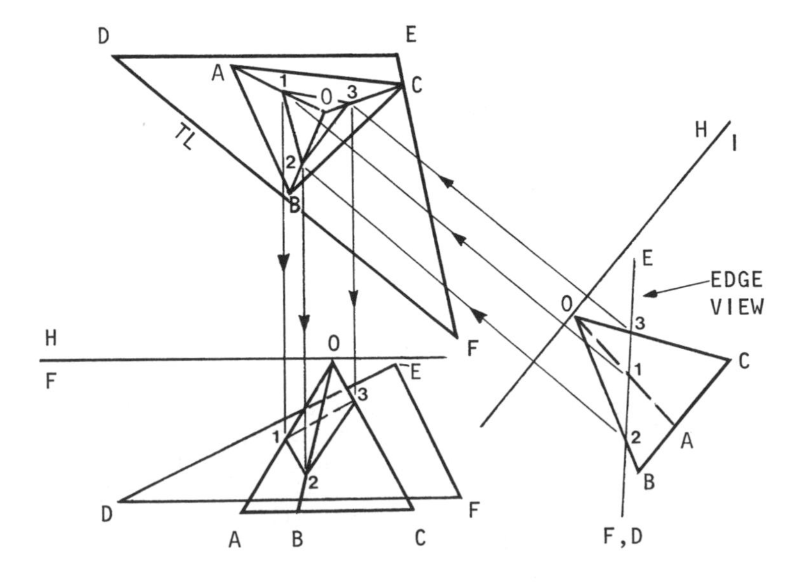

Fig. 2

Solution: For the general case of intersection of a solid and a plane, the line of intersection is determined by the points in which the edges of the solid pierce the given plane. The problem, in effect, reduces to one of finding the point in which a line pierces a plane.

Fig. 1 shows the front and top views of the given pyramid, O-ABC, and the given intersecting plane, DEF. With the auxiliary view method, an edge view of plane DEF is necessary to locate the points at which the edges of the pyramid pierce the plane. The edge view will appear in any

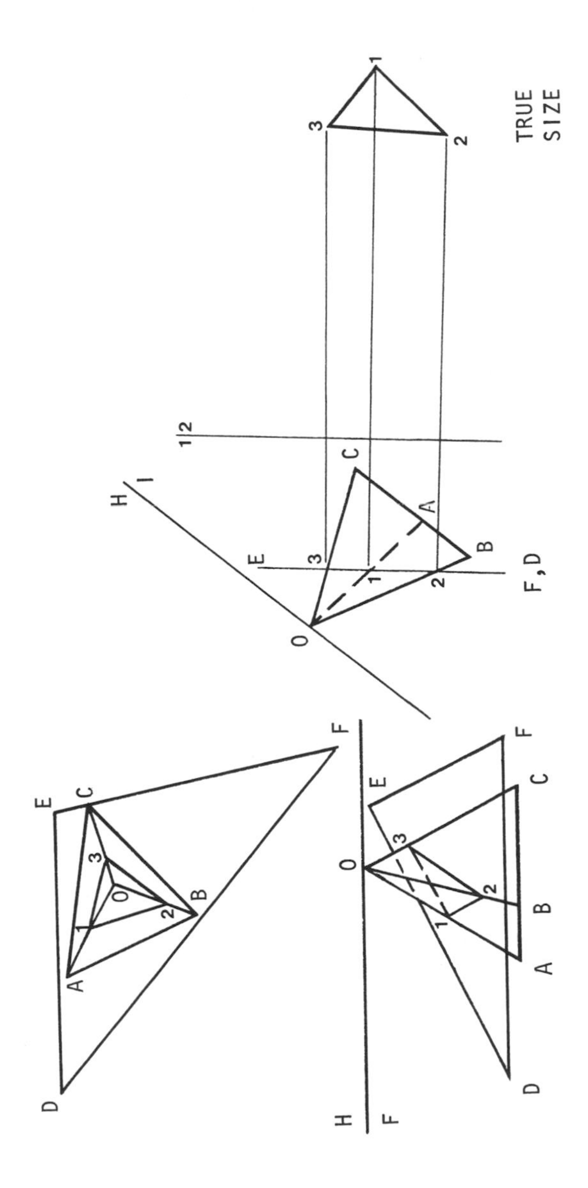

Fig. 3

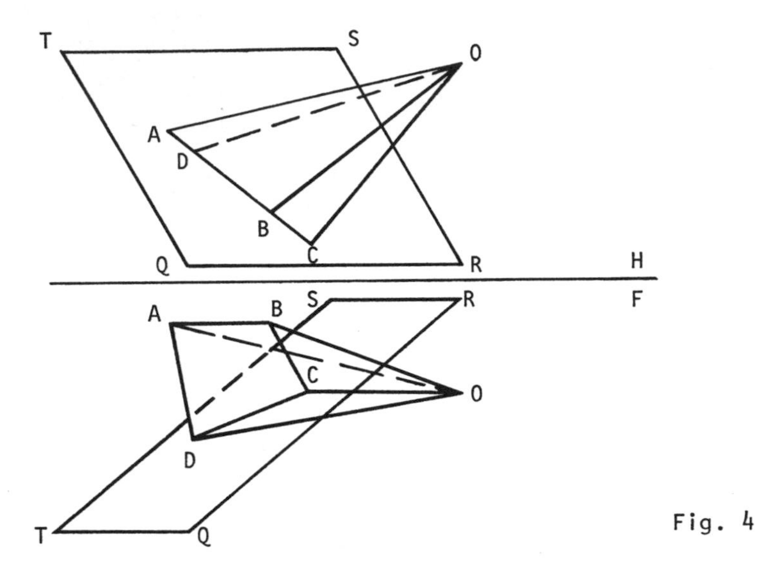

Fig. 4

view with lines of sight parallel to a true length line of the plane. Line DF in the front view is parallel to the reference plane H-F, and it projects to true length in the top view. As shown in fig. 2, make reference plane H-1 perpendicular to the true length line DF and construct auxiliary view 1, which shows plane DEF as an edge. The piercing points of the edges of the pyramid can now be located and projected back to the top and then the front views. The true shape and size view of the intersection

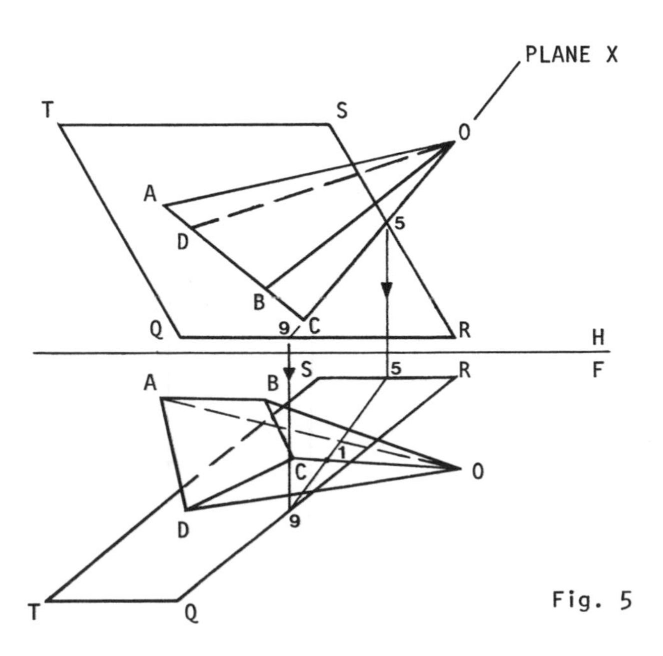

Fig. 5

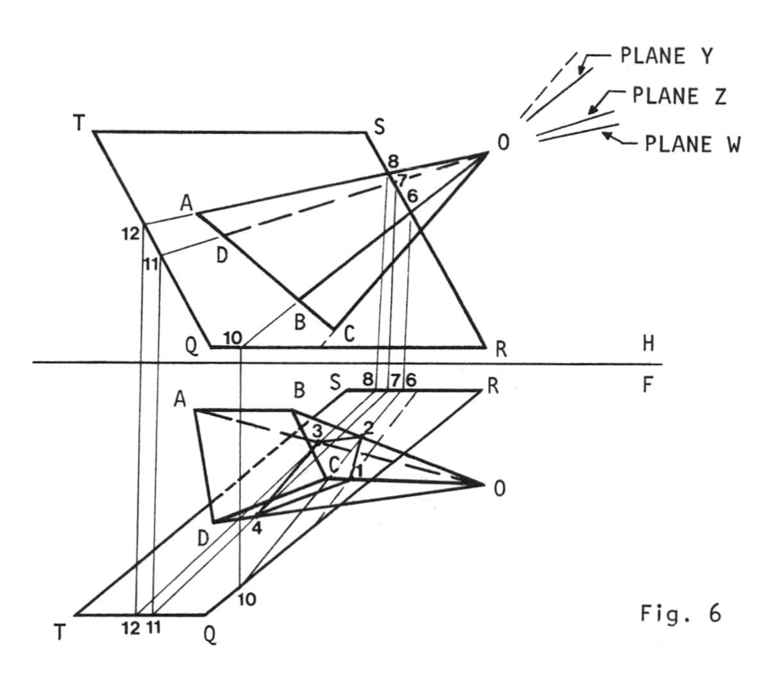

Fig. 6

can be obtained by a second auxiliary view with lines of sight perpendicular to the edge view of plane DEF in view 1. Reference plane 1-2 is parallel to DEF to yield the true size view 2, as in fig. 3.

Fig. 4 shows the front and top views of the given pyramid, O-ABCD, and the given intersecting plane, QRST. Using the cutting plane method, pass a vertical cutting plane X through edge OC in the top view of the pyramid, as shown in fig. 5. The trace of plane X in QRST is the line

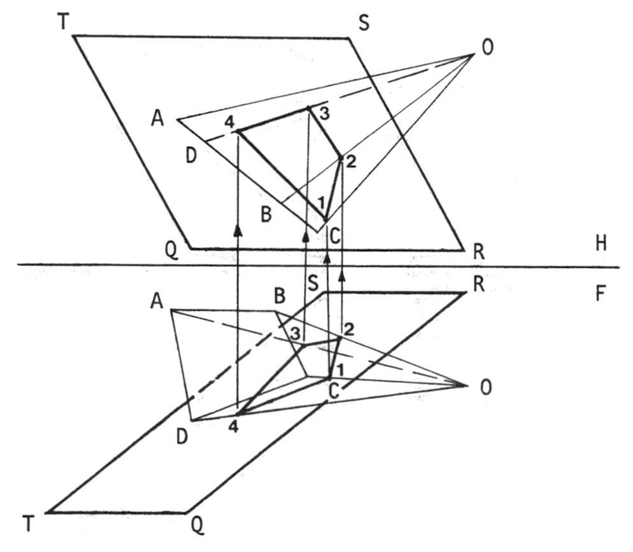

Fig. 7

5-9. Project line 5-9 to the front view, and where this line intersects the edge OC is point 1 on the line of intersection.

Pass vertical cutting planes Y, Z, and W through edges OB, OD, and OA of the pyramid in the top view, as shown in fig. 6. Project their traces on plane QRST to the front view and where they intersect the corresponding edges of the pyramid locates points 2, 3, and 4 on the intersection line. Connect points 1-4 sequentially for the view of the line of intersection. Project these points to the top view, as in fig. 7.

CHAPTER 8

SECTIONAL VIEWS AND CONVENTIONAL PRACTICES

OFFSET SECTIONS, SECTIONAL VIEWS

• PROBLEM 8-1

Explain the uses and construction of an offset section. Sketch the figure with the offset section and draw the front offset view.

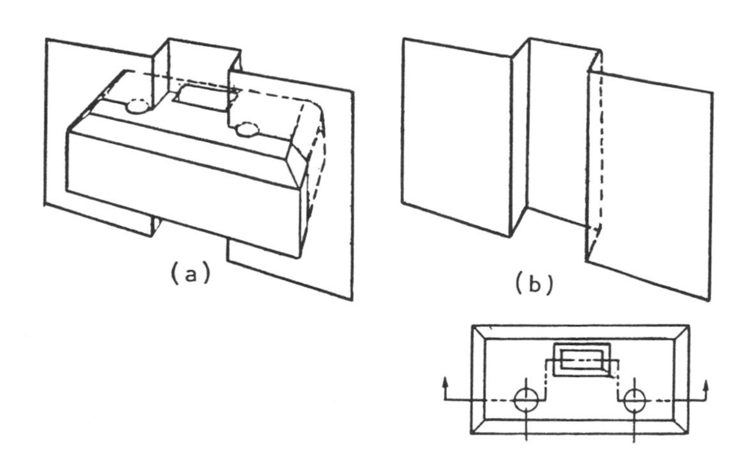

Solution: In the different types of section views the cutting plane has almost always passed continuously through the object as a straight line. Many objects are designed in such a way, however, that certain features which require sectioning do not lie in the same straight line. By changing the direction of the cutting plane, it is possible to show all the desired features in one section view.

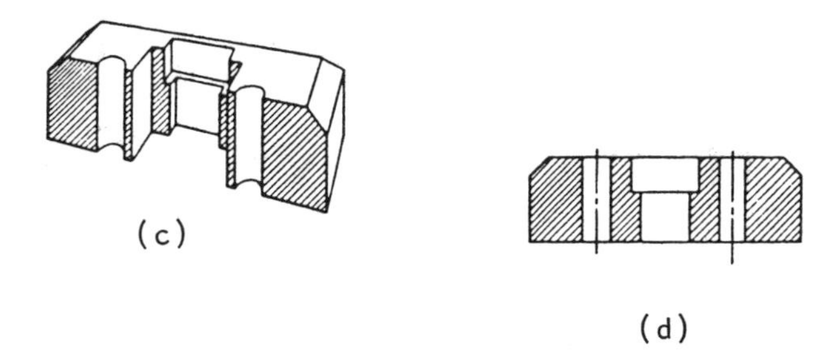
(c) (d)

It can be seen in Fig. A that the cutting plane passes through the object to include all the inaccessible features. Fig. B shows the cutting plane removed. Fig. C shows how the object appears with the portion in front of the cutting plane removed. The orthographic drawing of the object, with the portion of the object in front of the cutting plane removed for sectioning, is shown in Fig. D. It should be noticed that there are no lines in the section view to show that the cutting plane changes directions.

• PROBLEM 8-2

Draw the section views of figures 1(a) and 2(a).

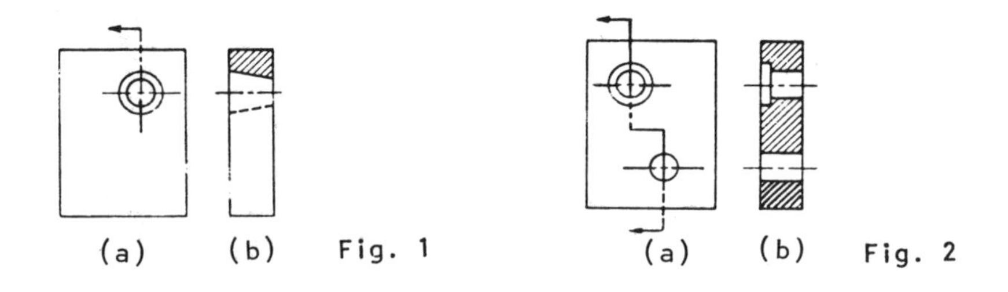
(a) (b) Fig. 1 (a) (b) Fig. 2

Solution: A cutting-plane line is used in the drawing of an object to show where the section has been removed. The cutting plane is indicated only in the view showing the plane as a line. The cutting-plane line shown in

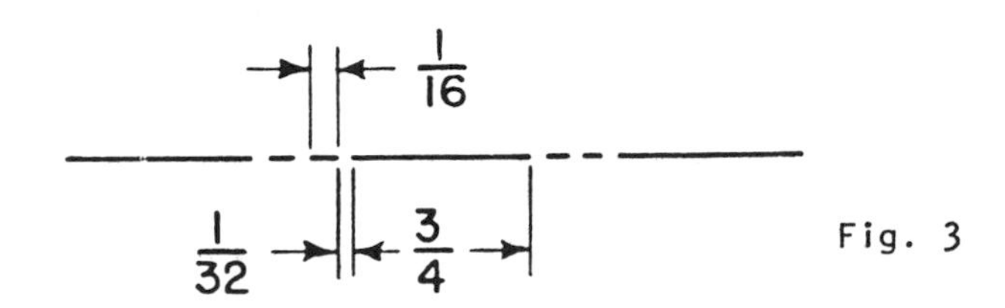

Fig. 3

Fig. 3 is illustrated in full scale or dimension. It is represented as a heavy line and is composed of a 3/4" line, a 1/32" space, a 1/16" line, a 1/32" space, a 1/16" line, and a 1/32" space, then repeating a 3/4" line, etc. Arrowheads on the cutting-plane line are used to indicate the direction in which the remaining portion of the object is to be viewed (Figs. 1(a) and 2(a)). The part of the object in front of the cutting-plane line is visualized as having been removed. Whenever it is obvious that the cutting plane passes through the center line, the cutting-plane line symbol may be omitted. Figures 1(b) and 2(b) are the section views of 1(a) and 2(a).

• **PROBLEM** 8-3

Describe the procedure for the construction of section views and sketch a section view of the given figure.

Solution: The following procedure is observed in the drawing of section views.

1. The object should be first studied and visualized pictorially (Fig. 1).

2. The type of section showing the clearest description of the object should be determined, and the cutting-plane line should then be indicated.

3. The portion of the object in front of the cutting-plane line should be imagined as having been removed (Fig. 2) and the student should then make an orthographic projection of the remaining portion for the desired section view.

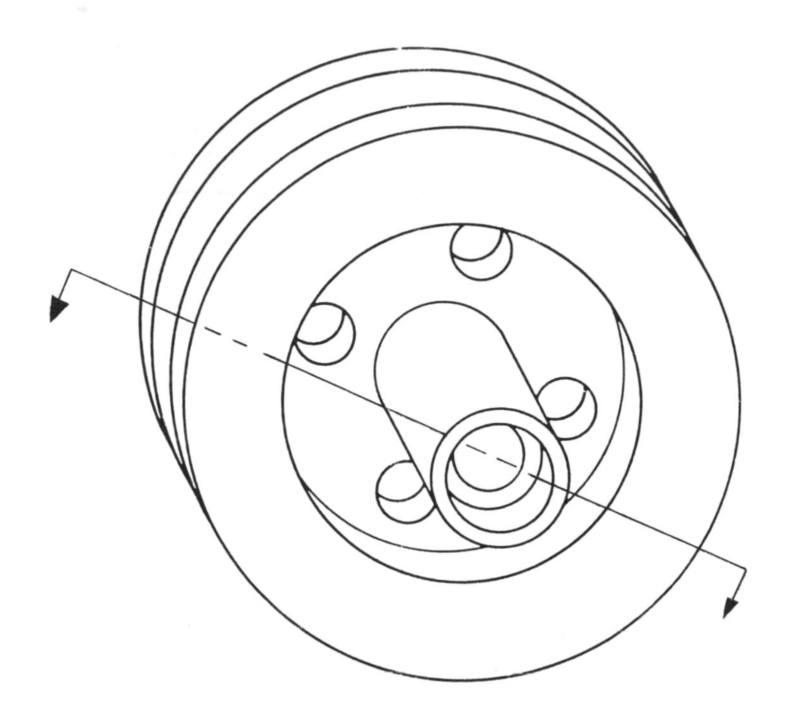

PICTORAL VIEW Fig. 1

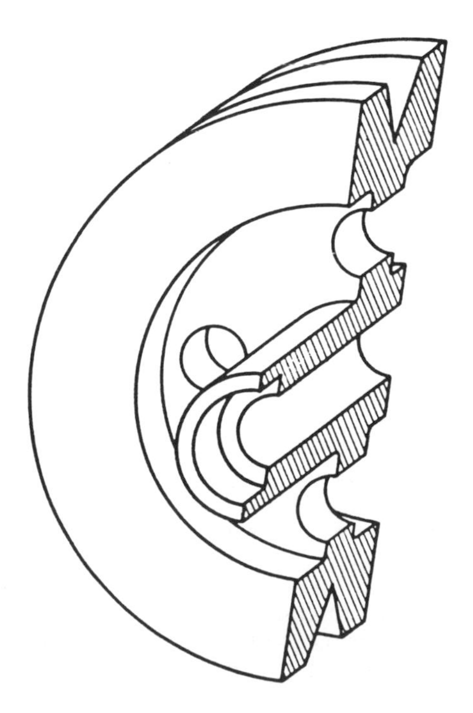

PORTION REMOVED IN FRONT OF CUTTING-PLANE LINE. Fig. 2

• PROBLEM 8-4

Compare the ordinary end elevation and section views of figure 1.

Solution: Often it is difficult to show clearly the true shape or construction of interiors by means of exterior views. This source of confusion is generally eliminated by the drawing of partial views or full-section views of the object. In these drawings, one must visualize a portion or section as having been removed or cut away; consequently, interior details and construction become clearly visible. Invisible lines become visible in all section views.

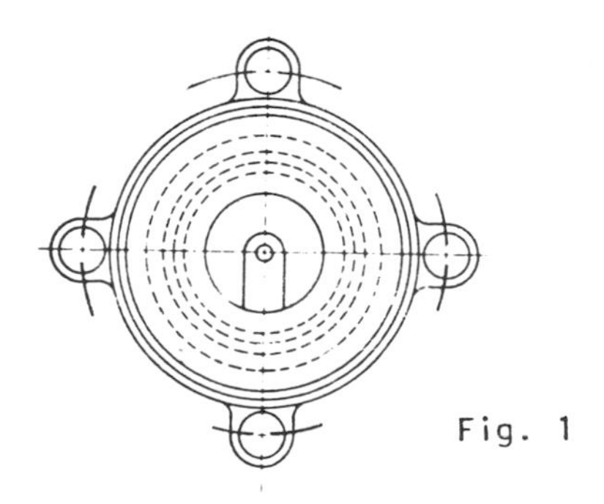

Fig. 1

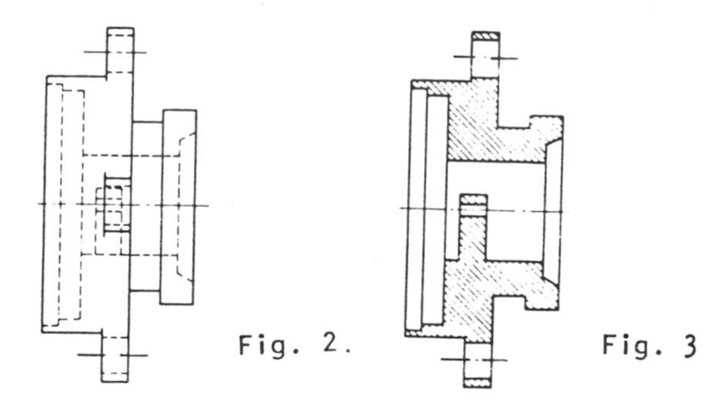

Fig. 2. Fig. 3

Figs. 1-3 show a comparison between an ordinary end elevation (Fig. 2) and a section view (Fig. 3), which simplifies the object and makes it more easily understood.

Section views must be shown in a way that is consistent with conventions. In engineering drawing a convention is any drawing procedure based upon approved practice and accepted generally in industry. One of the prinicipal functions of the American Standards Association has been to induce American industry to standardize drawing procedures and drafting room practices.

FULL, FRONT, AND HALF SECTION PROFILES

• PROBLEM 8-5

Draw a full section profile view of the given pulley.

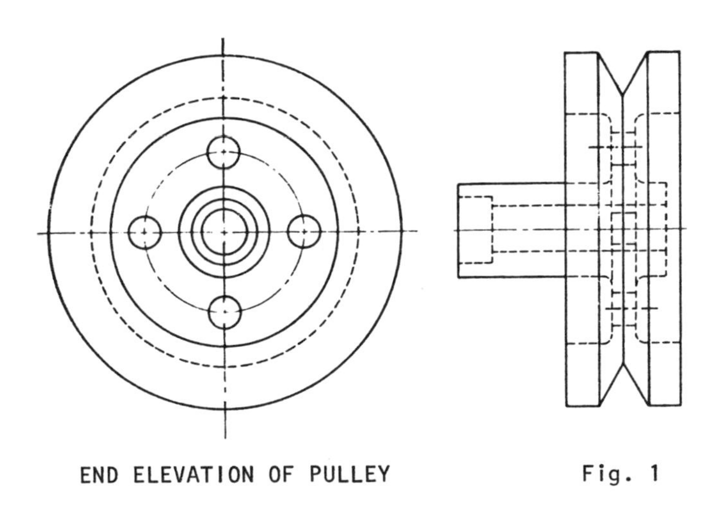

END ELEVATION OF PULLEY Fig. 1

Solution: A full section is the view obtained by passing a cutting plane entirely through an object.

The end elevation of a pulley is shown in Fig. 1. It can be seen from this drawing that considerable confusion is created by the number of invisible lines. A full section of the pulley is recommended to clarify its interior details and construction.

The pulley is first visualized pictorially as shown in Fig. 1 of problem #3. The cutting plane is now assumed to pass through the center, and the portion of the pulley in front of the cutting plane is imagined removed as shown in Fig. 2 of problem #3. An orthographic drawing of the remaining portion is then constructed as in Fig. 2, and section lines are drawn where the material has been cut by the cutting plane as indicated in the drawing.

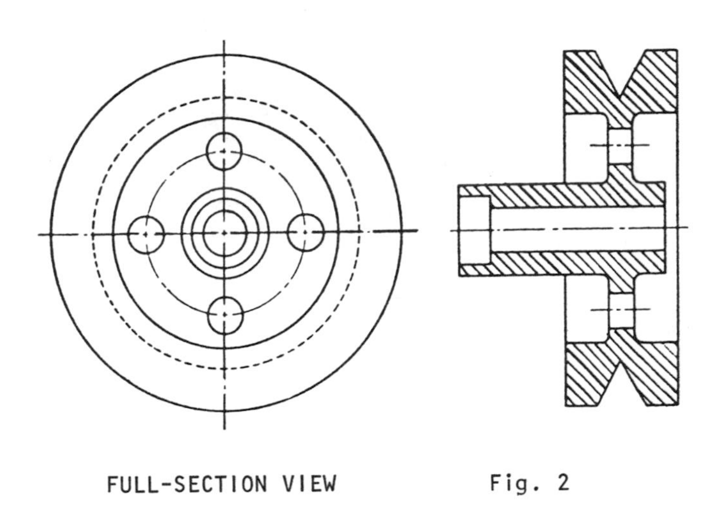

FULL-SECTION VIEW Fig. 2

It should be noticed that in this drawing the cutting-plane line has been omitted, as it obviously passes through the center line.

• PROBLEM 8-6

> Sketch a half section drawing of the pulley shown in Fig. 1 of problem 3. Also draw the front and half section profile views.

Solution: A half section is a view in which only one-quarter has been removed from the object (Fig. 1).

In half sections, the cutting-plane line ends on the center line of the object. This center line divides the sectioned from the unsectioned part of the object as shown in Fig. 2.

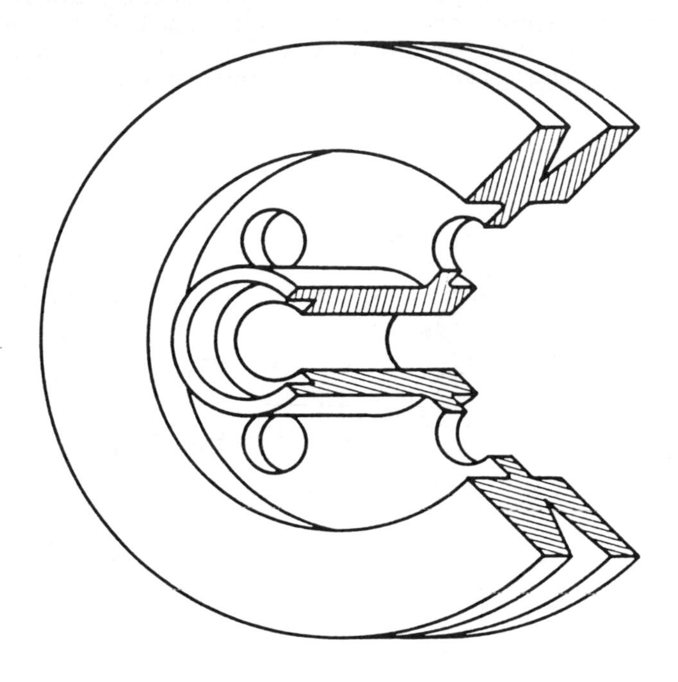

HALF-SECTION VIEW. Fig. 1

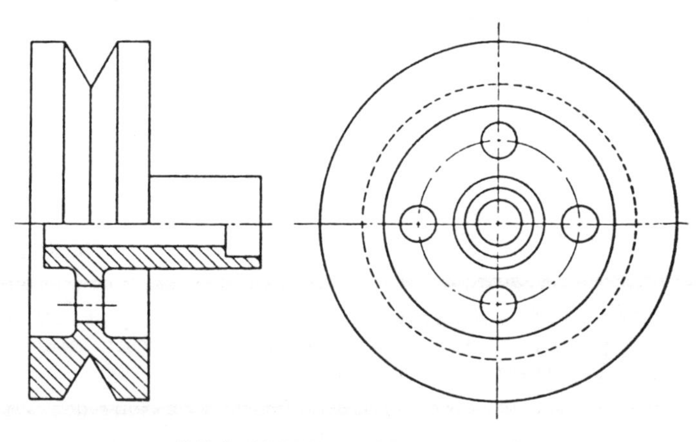

HALF-SECTION VIEW. Fig. 2

Fig. 2 shows the orthographic drawing of the remaining portion of the pulley after one-quarter has been removed. It should be noted that the invisible lines in the unsectioned part are permissible only when they are necessary for dimensioning.

The junctions of sectioned and unsectioned portions of an object are separated by a center line. Theoretically, a solid line should be shown; but since the portion of the object is only imagined to be removed, a center line is generally used.

REVOLVED AND DETAILED SECTIONS

• PROBLEM 8-7

Explain a revolved section using these illustrations:

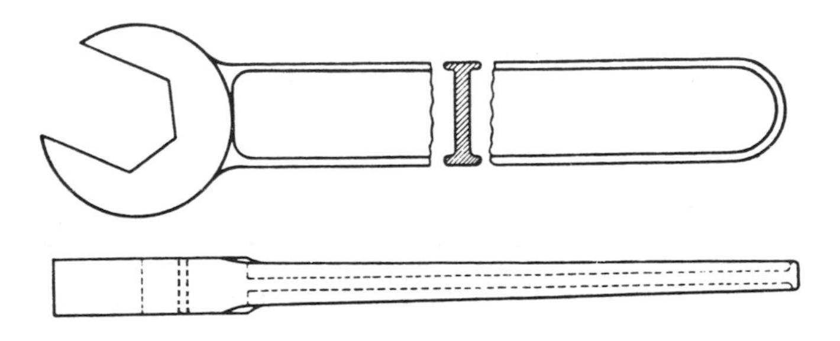

WRENCH, INCLUDING VIEW OF REVOLVED SECTION. Fig. 1

Solution: A revolved section is used to show the true shape of an object, such as a handle, spoke, etc. Fig. 1 shows two views of a wrench. Without the revolved section, the exact shape of the handle would be unknown. This revolved section is obtained by passing the cutting plane perpendicular to the axis of the piece to be sectioned. The portion of the object cut by the cutting plane is revolved until it coincides with the plane of the paper. Since this revolved section is superimposed on one of the original views, any line of the object interfering with the revolved section may be broken, as in the case of the wrench (Fig. 1), or omitted, as shown in the drawing of the C-clamp (Fig. 3).

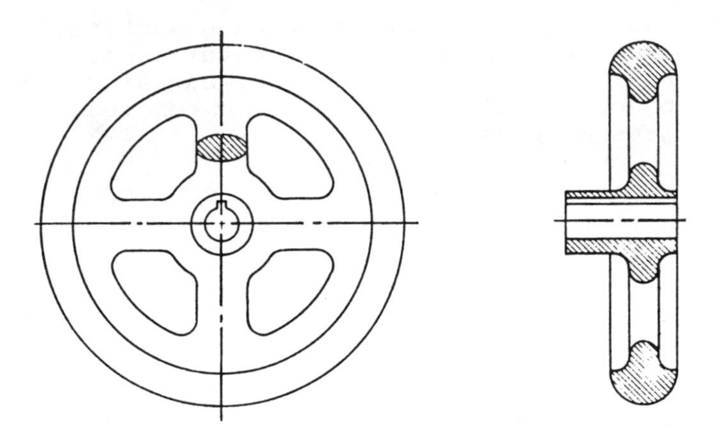

HANDWHEEL AND REVOLVED SECTION. Fig. 2

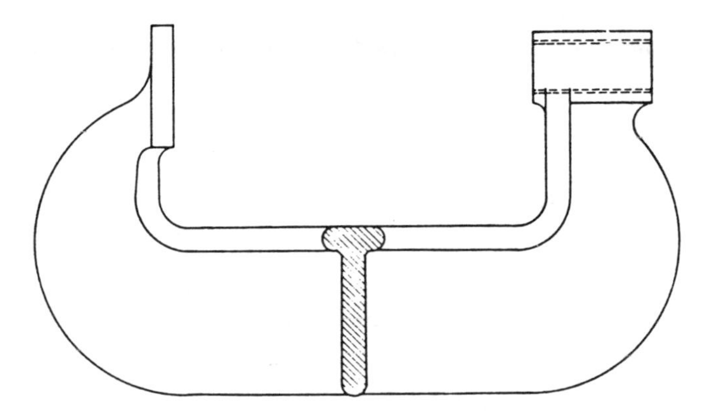

C-CLAMP, INCLUDING VIEW OF REVOLVED SECTION. Fig. 3

A drawing of a handwheel is shown in Fig. 2, together with a revolved section to indicate the true shape of the spoke. This information is given neither in the front elevation nor in the full-section view. Consequently, a revolved section is absolutely necessary to obtain a description of the spoke.

• **PROBLEM** 8-8

Explain detailed sections, or removed sections. Illustrate removed sections for an aircraft propeller blade.

Solution: Detailed sections, or removed sections, are used for the same purpose as revolved sections but, instead of being drawn on the view, they are set off, or shifted, to some adjacent place on the paper. The cutting plane, with reference letters, should always be indicated unless the place from which the section has been taken is obvious. Removed sections are used whenever restricted space for the section, or the dimensioning of it, prevents the use of an ordinary revolved section. When the shape of a piece changes gradually, or is not uniform, several sections may be required, as in Fig. 1.

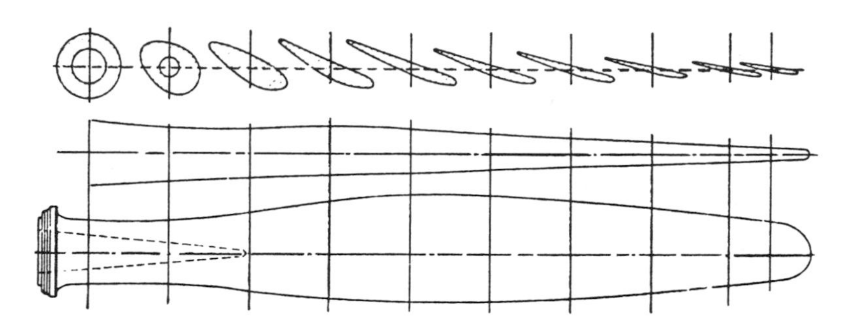

REMOVED SECTIONS. Fig. 1

It is often an advantage to draw them to larger scale than that of the main drawing, in order to show dimensions more clearly. Sometimes sections are removed to a separate drawing sheet. When this practice is employed, the section must be carefully shown on the main drawing with cutting plane and identifying letters. Often these identifying letters are made as a fraction in a circle, with the numerator a letter identifying the section and the denominator a number identifying the sheet. The sectional view is then marked with the same letters and numbers. Detail sections are sometimes called "separate sections," "shifted sections," or "sliced sections."

CHAPTER 9

ROTATION AND REVOLUTION

REVOLUTION OF POINTS AND LINES AROUND AN AXIS

• PROBLEM 9-1

State and illustrate the rule for measurement from rotation lines.

Solution: Rule: Any point on an object must appear the same distance from the rotation line in all views connected to a given view by projection lines.

This rule is illustrated in the figure. Point X is a distance d from the rotation lines in views E_1, E_2, and E_3, each of which is connected to view P by projection lines. The distance from the rotation line to point Y is s in the three views connected to view E_3, and m in the three views connected to view I_1.

It is essential that this rule be thoroughly understood, since measurement from rotation lines is the key to drawing any new view.

Alternate Rule: Any point is projected from an adjacent view, and its distance from the rotation line is obtained from an alternate view. "Project from adjacent; measure from alternate."

In the figure, views E_1, E_2, and E_3 are connected to view P by projection lines. In other words, view P is adjacent to each E view. However, the E views are not adjacent to each other since to get from one to another,

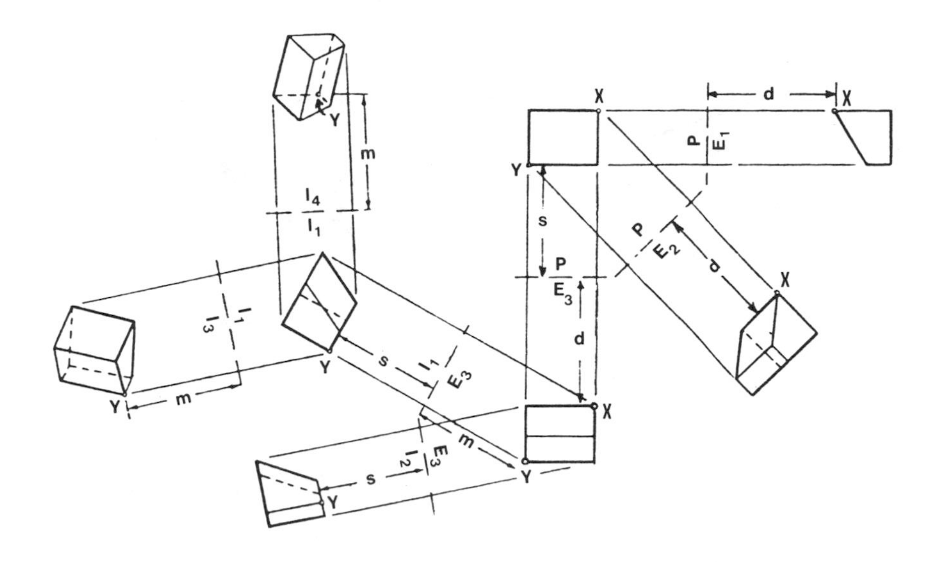

we must proceed through the P view. The E views are therefore called alternate views. Note that view E_2 is projected from the adjacent P view and the rotation-line measurements are taken from the alternate view E_1. Also note that E_3 is projected from P, and measurements can be taken from either E_1 or E_2.

Similarly, view E_3 is adjacent to views P, I_1, and I_2, and P, I_1, and I_2 are alternate views. Also, the alternate views I_3, I_4, and E_3 are adjacent to I_1.

It may also be helpful for the student to "count back" two rotation lines and two views. The measurements are then taken between the second rotation line and points in the second view.

• PROBLEM 9-2

Revolve a given point, O, about a given axis, 1-2, to the highest location of the point above the axis.

Solution: Fig. 1 shows the top and front views of the given axis, line 1-2, and the given point, O. The true view of the circular path of revolution can only be seen in a view which shows the axis as a point. With the line of sight perpendicular to the axis 1-2 in the front view, make the reference plane F-1 parallel to the axis and

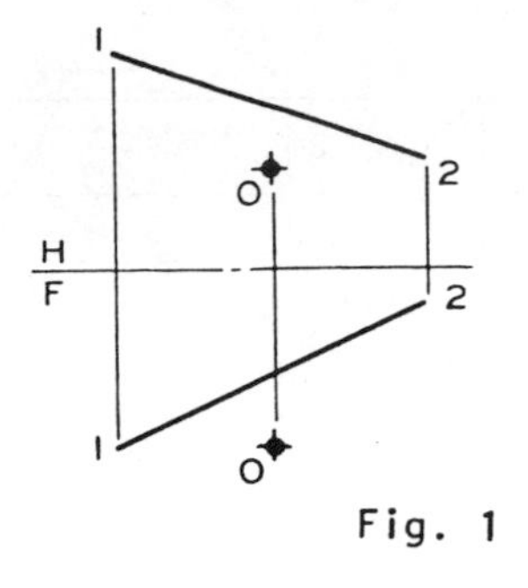

Fig. 1

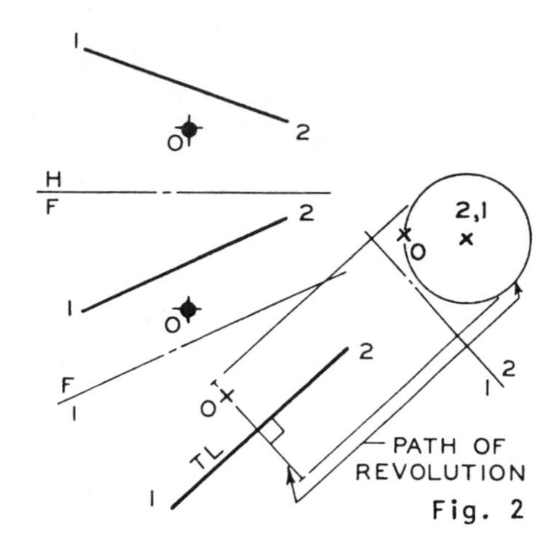

Fig. 2

construct the primary auxiliary view, as shown in fig. 2. Axis 1-2 projects true length in view 1. Draw the point view, the secondary auxiliary view, of the axis from view 1 by drawing the reference plane 1-2 perpendicular to the axis and projecting both the axis and point O to view 2. Draw the path of revolution in view 2 by constructing a circle of radius 2-0 centered at 2. This path appears as an edge in view 1. To determine the highest location of point O, construct the vertical line 1-3 to the reference plane H-F in the front view, as in

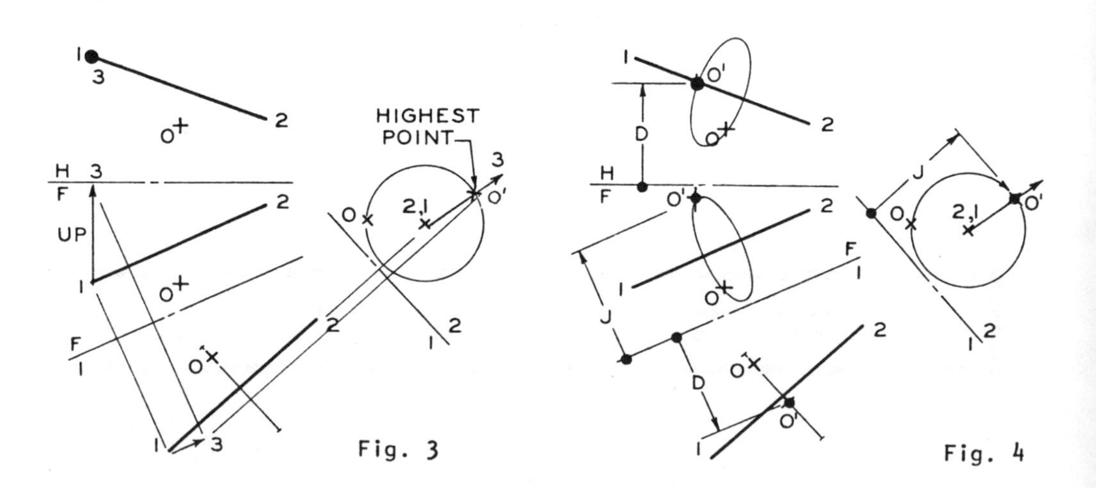

Fig. 3

Fig. 4

fig. 3. Project this directional line to the top view, where it appears as a point, and to views 1 and 2. The point at which line 1-3 intersects the circle of revolution in view 2, point O', is the highest location of point O as it revolves about the axis 1-2. Project point O' and its path of revolution back to view 1, the front, and the top views, as in fig. 4. Note that point O' lies on line 1-2 in the top view, which verifies that it is at its highest location. Projection from the top view would have also provided a solution to the problem.

• PROBLEM 9-3

Revolve a given line, 3-4, about a given axis, 1-2, a specified number of degrees, θ, and project the revolved position to all views.

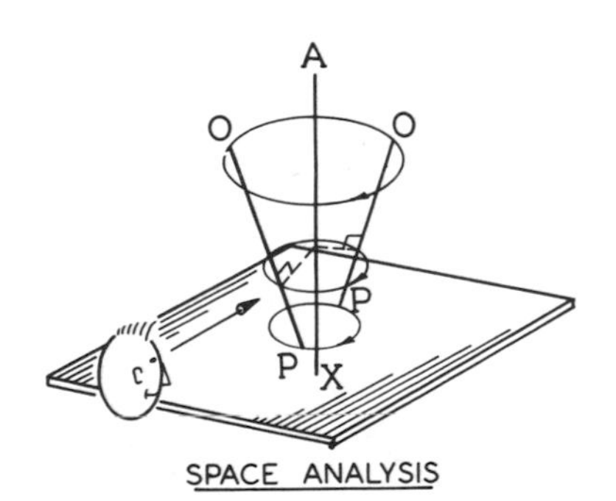

SPACE ANALYSIS

Solution: The space analysis diagram shows a three-dimensional view of the revolution of a line about an axis. Fig. 1 shows the front and top views of the given line, 3-4, and the given axis, 1-2. The true view of the circular path of revolution of a line about an axis can be seen only in a view in which the axis appears as a point, as is apparent in the space analysis diagram. The axis 1-2 appears true length in the top view since it is parallel to the reference plane H-F in the front view. Therefore, draw the point view of the axis by making the reference plane H-1 perpendicular to the true length axis 1-2 in the top view and projecting 1-2 as a point to view 1, as in fig. 2. project line 3-4 to view 1 also.

As shown in fig. 3, construct the circles of revolution for the endpoints 3 and 4 of line 3-4 in view 1. Draw the circle of revolution of point 3 with radius 2-3 centered at 2 and the circle of revolution of point 4

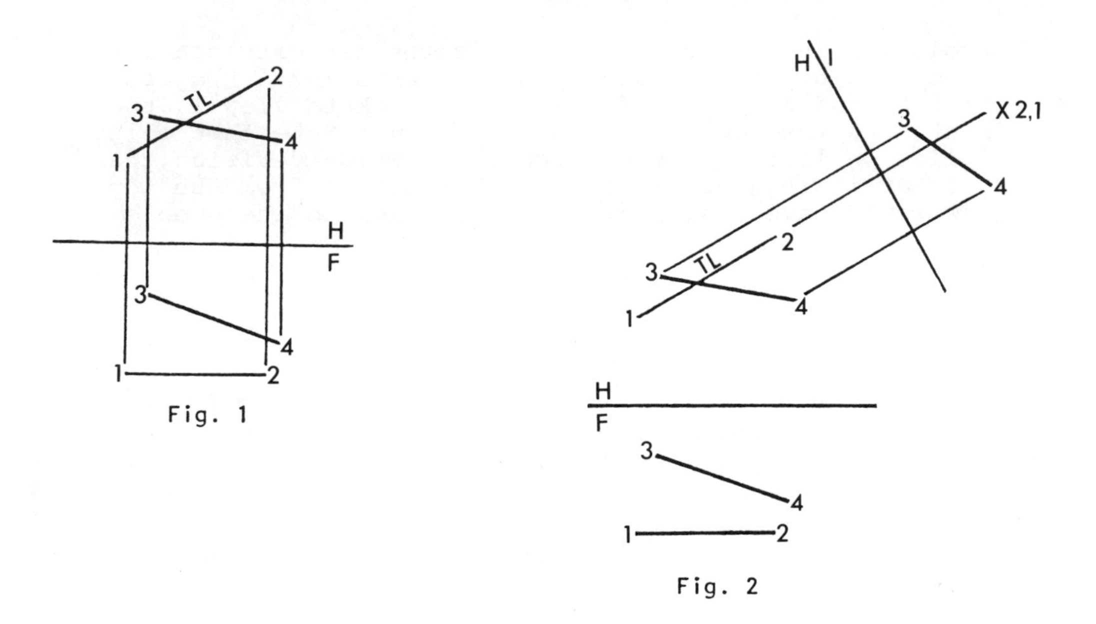

Fig. 1

Fig. 2

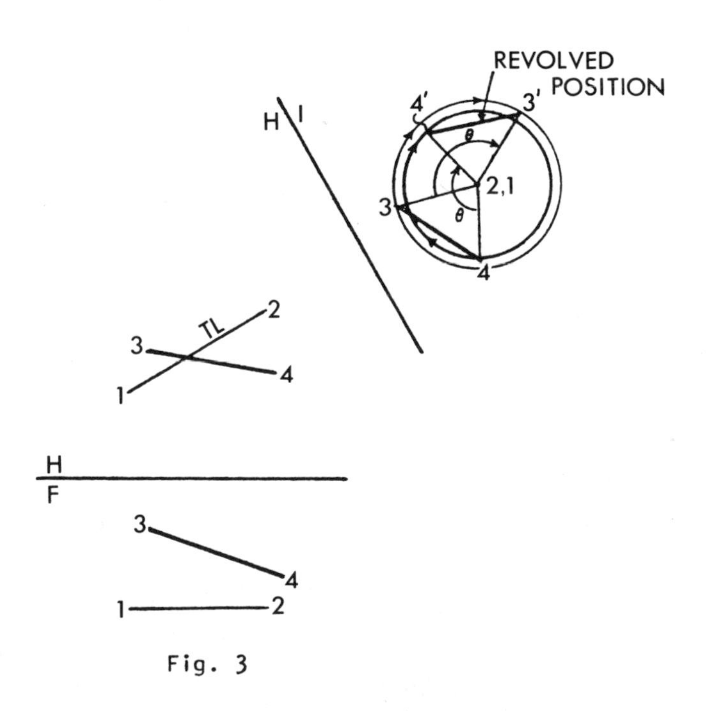

Fig. 3

with radius 2-4 centered at 2. Revolve points 3 and 4 the specified number of degrees, θ, to their new positions, 3' and 4'. Project the revolved points 3' and 4' back to the top view, as in fig. 4, to where their projections intersect projections parallel to plane H-1 from points 3 and 4 in the top view. Then project 3' and 4' to the front view, measuring the distances down from the plane H-F on the projection lines equal to their distances,

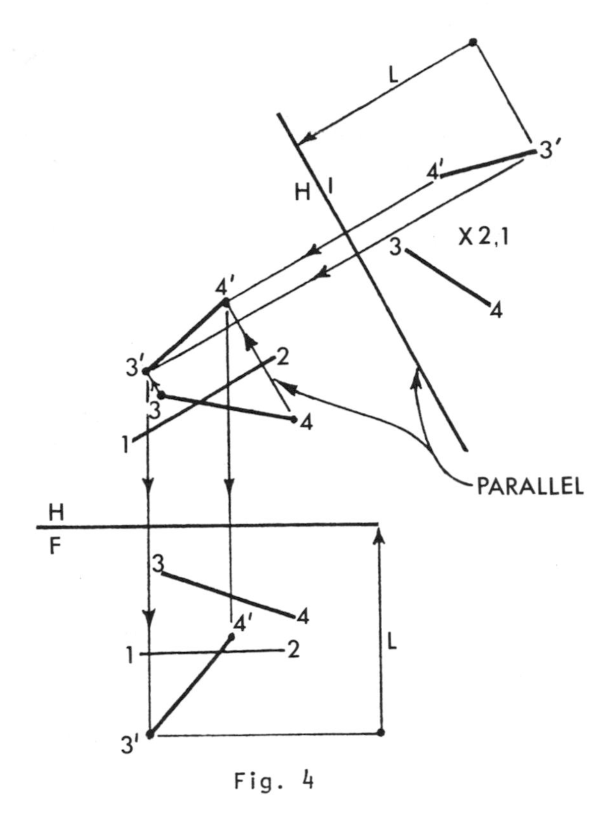

Fig. 4

such as L, from the plane H-1 in view 1. Connect 3' and 4' in the front and top views for the required revolved position of line 3-4, as shown in fig. 4.

TRUE LENGTH OF A LINE

• PROBLEM 9-4

Find the true length of a given line, AB, by the method of revolution.

Solution: A line projects true length into any view when it appears in an adjacent view parallel to the image plane of the first view. An oblique line can be revolved

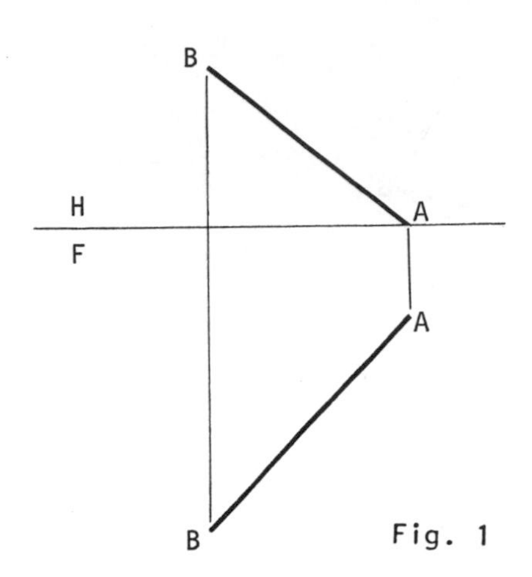

Fig. 1

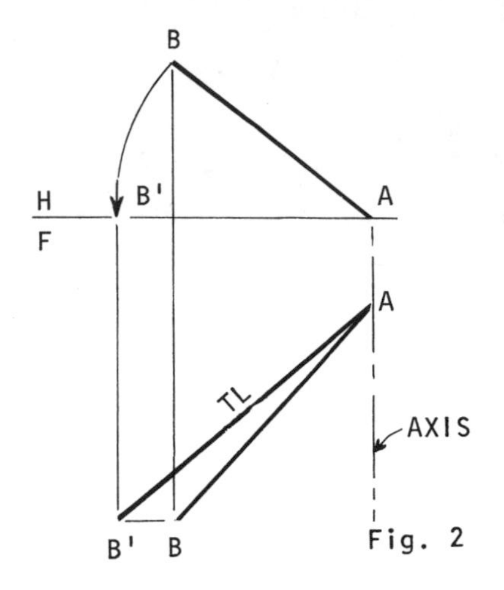

Fig. 2

about an axis through a point of the line until the line is parallel to an image plane only when the axis appears as a point. Fig. 1 shows the top and front views of the given line AB, whose true length is to be found. A vertical axis through point A in the front view will project as a point in the top view, as shown in fig. 2. Revolve point B in the top view about the axis through point A until line AB is parallel to the image plane H-F. Project the revolved position of point B, labeled B', to the front view to where it intersects a horizontal projection from point B in the front view. Connect A and B' to measure the true length of line AB.

The above construction can be simplified by omitting the revolution and merely laying out a distance from the axis at point O, along a horizontal projection line, equal to the length AB in the top view, as seen by the line OB' in fig. 3.

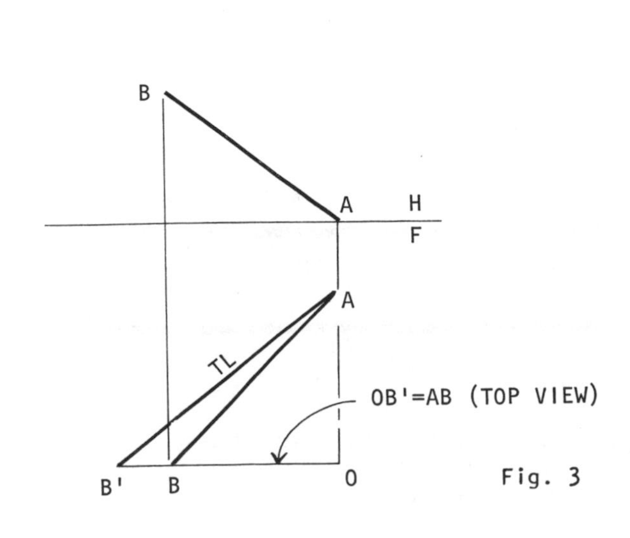

Fig. 3

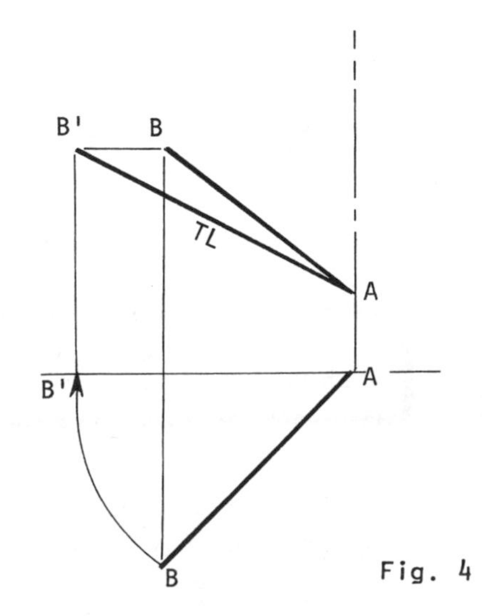

Fig. 4

The true length of line AB can also be found in the top view by revolving point B in the front view about the point view of a vertical axis through point A in the top view, as shown in fig. 4. The revolved position of B in the front view, B', is projected to the top view, where the line AB' shows the required true length of the line AB.

• PROBLEM 9-5

Find the true slope of a given line, AB, by the method of revolution.

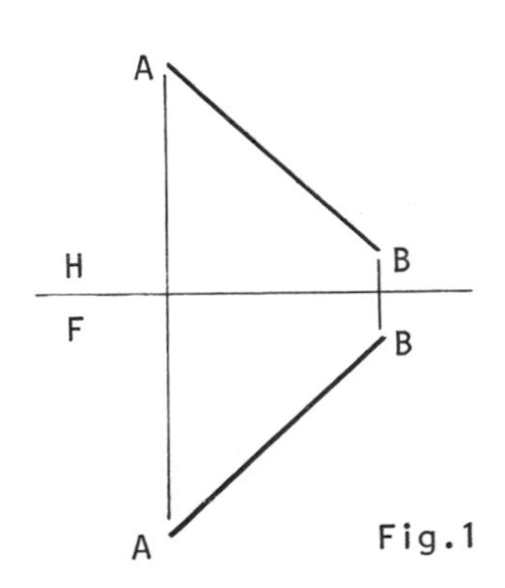

Fig.1

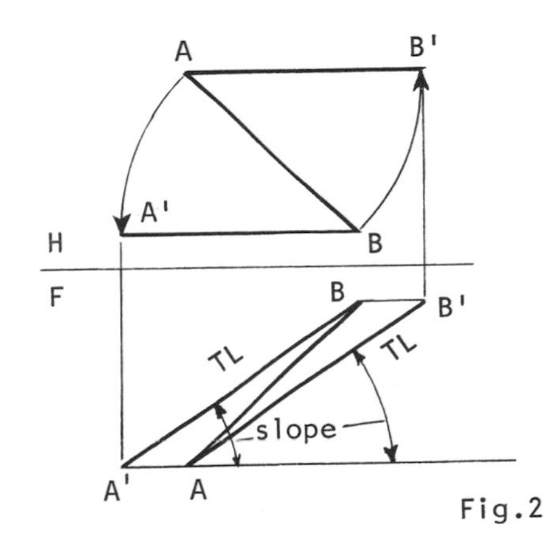

Fig.2

Solution: Fig. 1 shows the front and top views of the given line AB, whose true slope is to be found. The true slope of a line can be measured in a view in which the line appears true length. As shown in fig. 2, the axis about which line AB is to be revolved can be through either point A or point B. If the axis is chosen through point A, revolve point B to B' where line AB' is parallel to the reference plane H-F. Project B' to the front view, where line AB' appears true length. The required slope is the angle between line AB' and a horizontal line. If the axis is chosen through point B, revolve point A to A', where line A'B is parallel also to plane H-F. Project A' to the front view, where line A'B is true length. The required slope is then measured as the angle between line A'B and the horizontal line. Note that A'B and AB' are parallel to each other, verifying the fact that their slopes must be equal since they represent the same line.

Lay out a line, AB, with a given length, bearing, and slope.

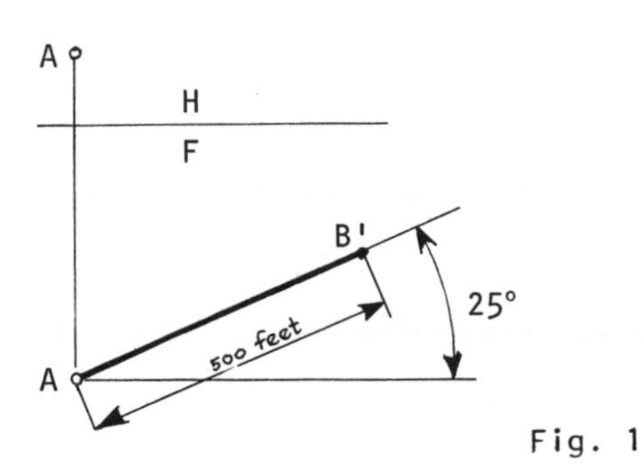

Fig. 1

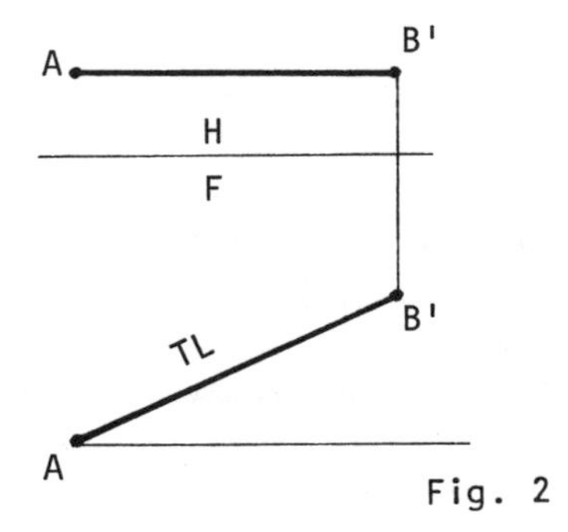

Fig. 2

Solution: Let the line AB have a length of 500 feet, a bearing of N40°E and a slope of +25 degrees. Locate point A conveniently in the front view and project it to the top view. In the front view, draw a line through point A with the given slope of +25°, since the slope of a line is the angle between the line and the horizontal. Mark off a distance on this line equal to the given length of 500 feet to scale, and label the endpoint B', as in fig. 1. Project B' to the top view to where it intersects a horizontal line through point A in the top view, as in fig. 2. Line AB' in the front view is true

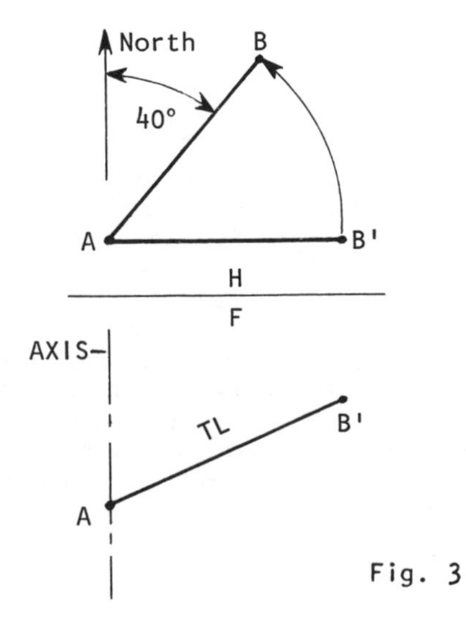

Fig. 3

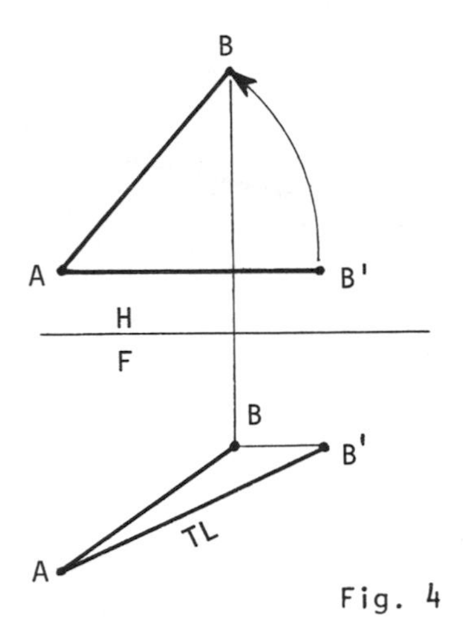

Fig. 4

length because AB' in the top view is parallel to the reference plane H-F. With the axis of revolution through point A, revolve B' in the top view to point B, as in fig. 3, where line AB makes an angle of 40 degrees with north, which is the given bearing. Locate point B in the front view by projecting from the top view to where the projection line intersects a horizontal projection from point B' in the front view, as in fig. 4. Line AB in the top and front views is the required line.

• PROBLEM 9-7

Find the true size of the angle between two given intersecting lines, AB and CD, using the method of revolution.

Solution: The true size of the angle between two intersecting lines can be measured only in a view in which both lines appear true length. Fig. 1 shows the front and top views of the given lines AB and CB. Reference plane H-F intersects line CB at point D and passes through point A. Therefore, line AD projects true length in the top view. Pass the axis of revolution through AD in the top view, about which the lines AB and CB will be revolved. The true paths of revolution will appear only in a view in which the axis projects as a point. Construct the point view of the axis AD by making the reference plane H-1 perpendicular to the true length line AD and projecting

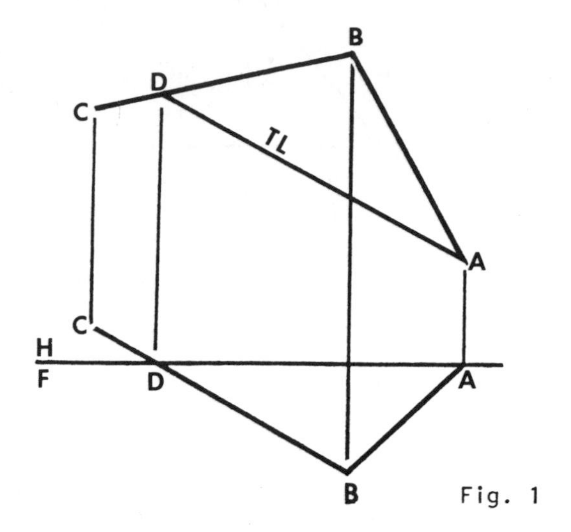

Fig. 1

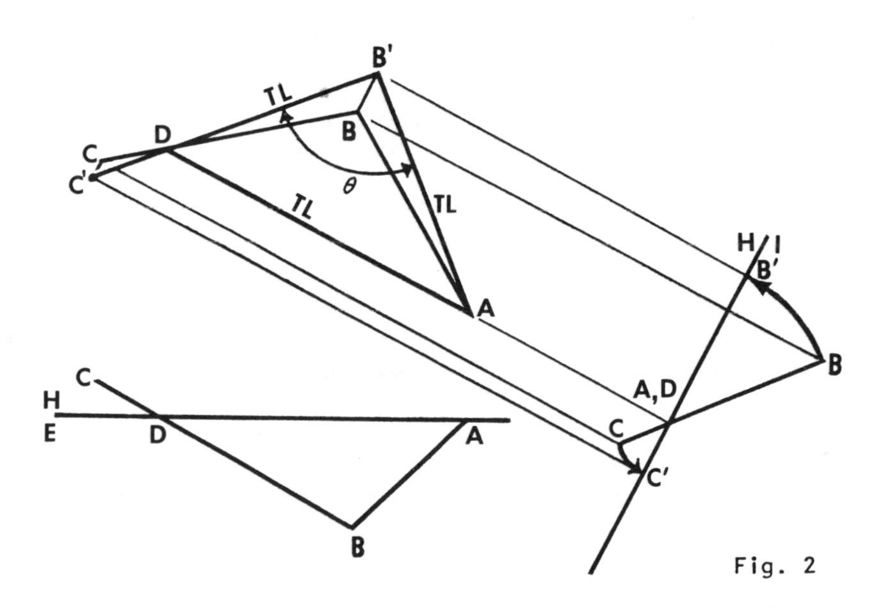

Fig. 2

AD as a point in the primary auxiliary view, as in fig. 2. Project points B and C to view 1 also, and draw BC. Revolve point B to B' and point C to C' about the axis until lines C'B' and AB' are parallel to plane H-1. Project points C' and B' back to the top view to where they intersect projection lines parallel to plane H-1 from C and B, respectively. Lines C'B' and AB' both appear true length in the top view. Angle C'B'A is the required angle between the intersecting lines.

• PROBLEM 9-8

Find the true length of a perpendicular from a given external point, P, to a given line, AB, using methods of revolution.

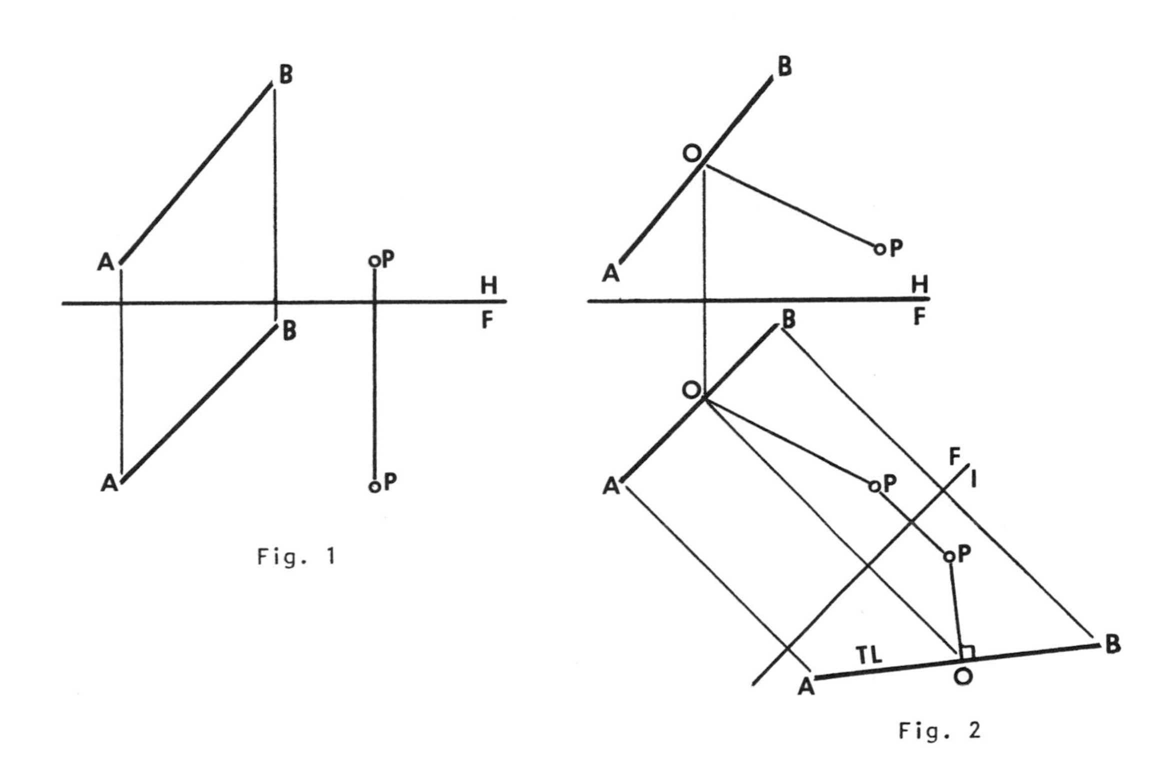

Fig. 1

Fig. 2

Solution: Fig. 1 shows the top and front views of the given point P and the given line AB. A true perpendicular to a line from an external point can only be seen in a view in which the line appears true length. To show the true length of line AB, construct the primary auxiliary view by making the reference plane F-1 parallel to AB in the front view, as in fig. 2. Line AB projects true length in view 1. Project point P to view 1, also, and draw a line through P perpendicular to AB at point O. Project OP to the front and top views. To find the true length of perpendicular OP, revolve OP in view 1 about an axis through O to OP', where OP' is parallel to plane F-1, as in fig. 3. Project line OP' to the front view, where it appears true length.

The same solution can be arrived at by using only two views, but three revolutions. Fig. 4 again shows the top and front views of point P and line AB, placed farther apart to avoid overlapping of revolutions. Find the true length of line AB by revolving AB about an axis through

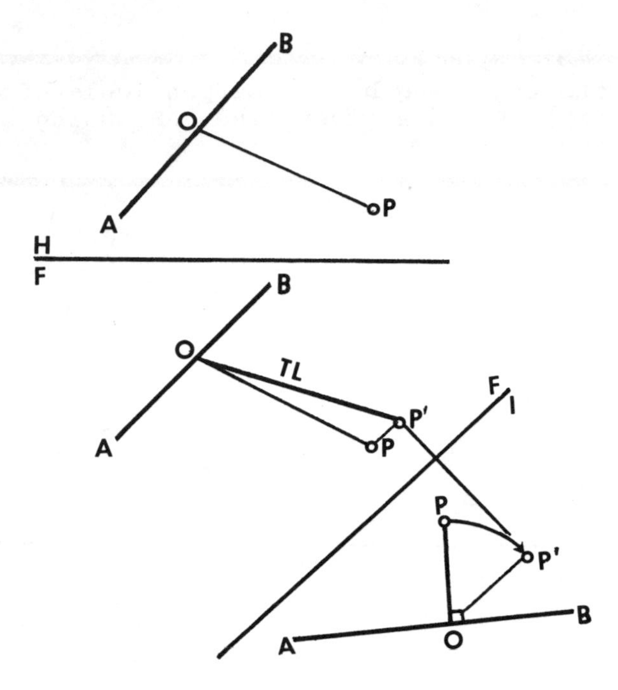

Fig. 3

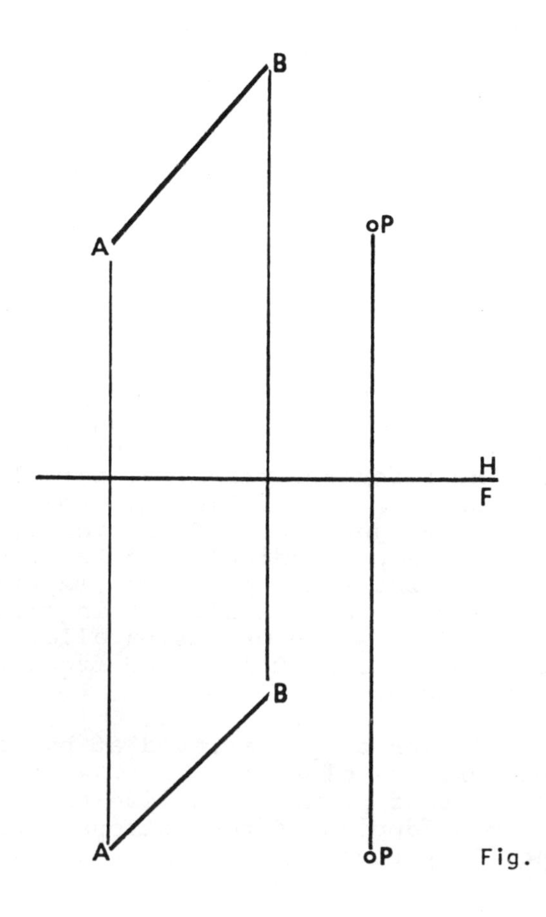

Fig. 4

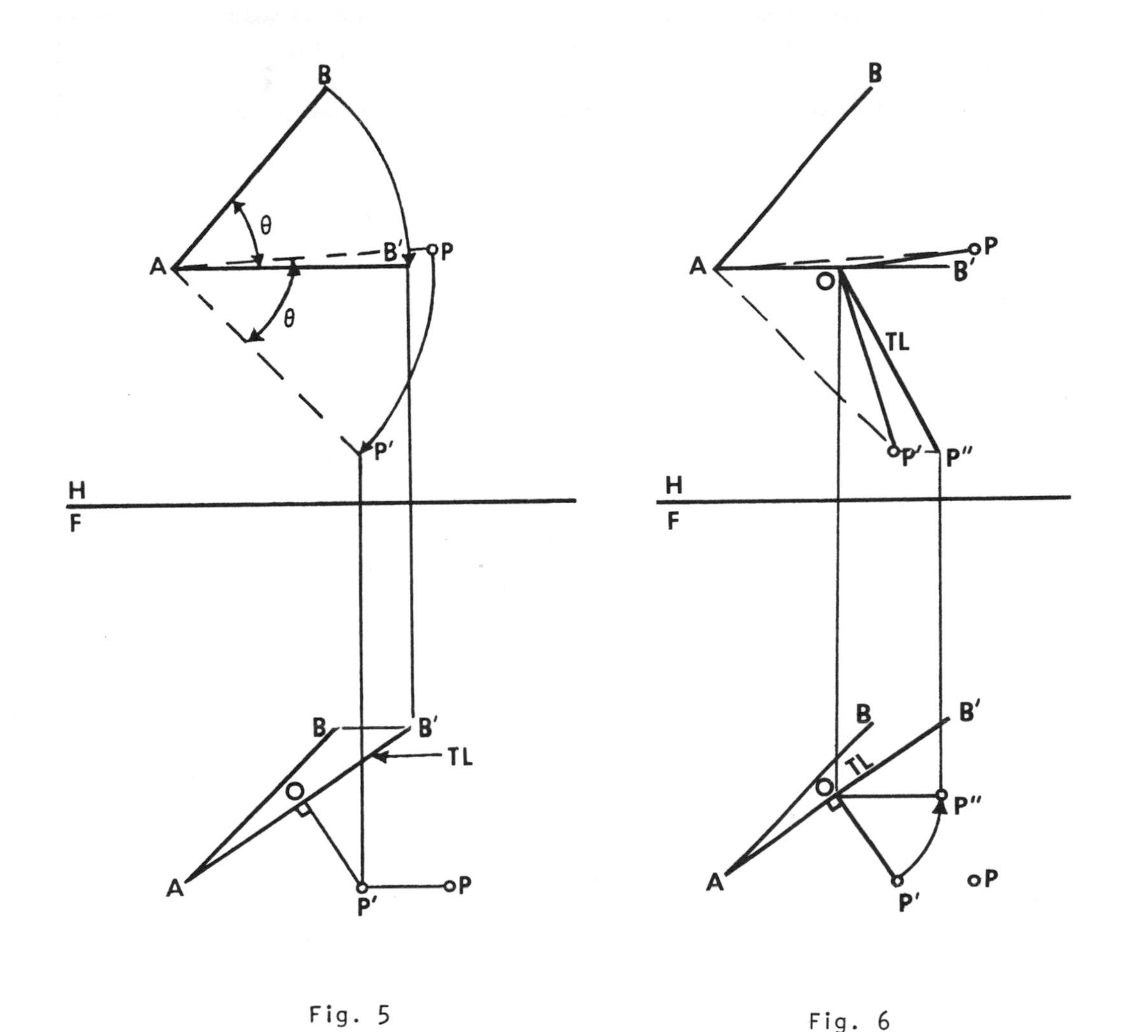

Fig. 5

Fig. 6

point A to AB' in the top view, where AB' is parallel to plane H-F, as in fig. 5. Project AB' to the front view, where AB' appears true length. The position of point P in the top view relative to line AB must also be revolved, now, to its new position relative to AB'. Since AB was revolved through θ degrees to AB' in the top view, revolve point P, about the same axis through A, through θ degrees to P'. Project P' to the front view to where it intersects the horizontal projection line from P. Construct the true perpendicular OP', as shown in fig. 5. To find the true length of OP', revolve OP' about an axis through point O in the front view to OP", which is parallel to plane H-F, as in fig. 6. Project OP" to the top view, where it appears true length.

Using only the two given views, find the shortest connection between two given skewed lines, 1-2 and 3-4.

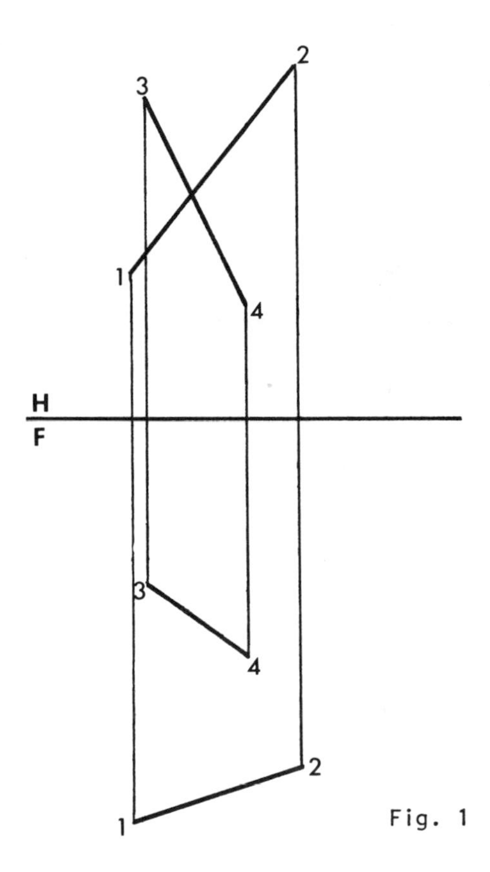

Fig. 1

Solution: Fig. 1 shows the two given skewed lines, 1-2 and 3-4, in the top and front views. The shortest connection, which is the perpendicular distance, between two skewed lines can be drawn in a view in which one line appears as a point. As shown in fig. 2, revolve point 3 in the top view about an axis through point 4 to point 3', where line 3'-4 is parallel to the reference plane H-F. Project point 3' to the front view to where it intersects a horizontal projection line from point 3, and draw line 3'-4, which is true length. As point 3 was revolved θ degrees about the axis, revolve points 1 and 2 θ degrees about the axis to 1' and 2', and project 1' and 2' to the front view to where they intersect horizontal projection lines from 1 and 2. The true length line 3'-4 will appear as a point in the top view if it is perpendicular to plane H-F in the front view. As in fig. 3, revolve line 3'-4 about an axis through point 4 to 3"-4 which is a vertical

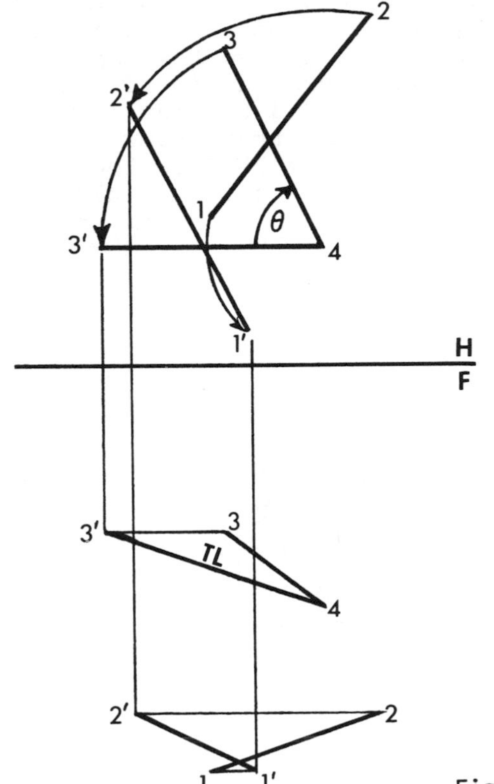

Fig. 2

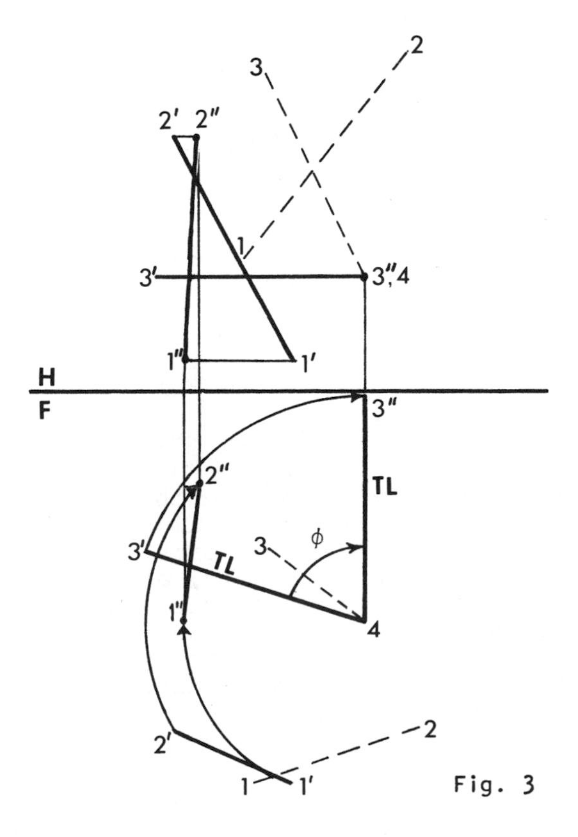

Fig. 3

line. As 3'-4 was revolved ϕ degrees about the axis, revolve line 1'-2' ϕ degrees about the axis to 1"-2". Project line 1"-2" and line 3"-4 to the top view, where 3"-4 appears as a point, as shown in fig. 3. Construct the perpendicular from point 3"-4 to line 1"-2" at point P", and project point P" to the front view to line 1"-2" as in fig. 4. Point O" on line 3"-4 is located by drawing a horizontal line through P" perpendicular to 3"-4. Line O"-P" is perpendicular to line 3"-4 in the front view because 3"-4 is true length.

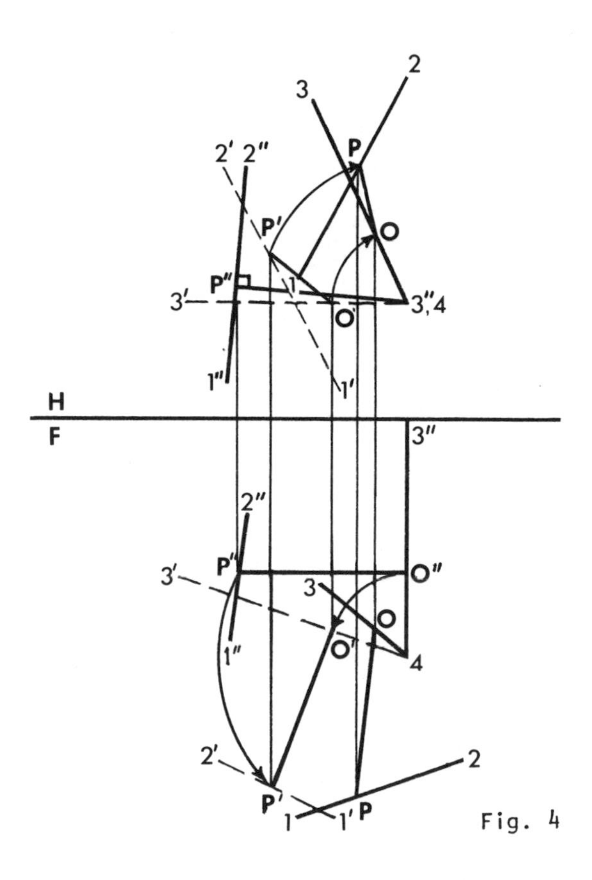

Fig. 4

Project the line O"-P" back to the other views following the reverse order of the revolutions. As in the front view of fig. 4, revolve P" and O" back to P' and O' on lines 1'-2' and 3'-4 , respectively, about the axis through point 4. Project P' and O' to the top view. Revolve P' and O' in the top view back to P and O on lines 1-2 and 3-4, respectively, about the axis through point 4. Project P and O to the front view to show the required perpendicular in all views, as illustrated in fig. 4.

Find the true shape and size of a given plane, ABC, using the method of revolution.

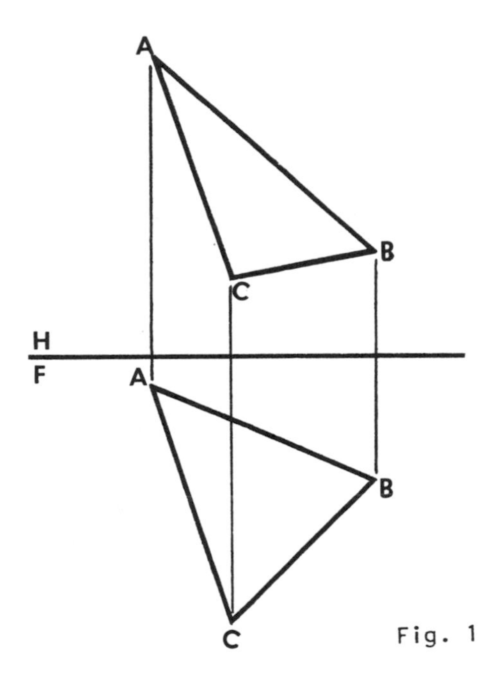

Fig. 1

Solution: The true shape and size of a plane can be determined only in a view in which the lines composing the plane appear true length. Fig. 1 shows the top and front views of the given plane ABC. In order to revolve a plane about an axis, the plane must appear in a view as an edge and the axis as a point. The true path of revolution can be seen in this view. Draw a horizontal line BD in plane ABC in the front view, as in fig. 2. Line BD projects true length in the top view, and it can then be used as an axis about which the plane can be revolved. Construct the primary auxiliary view, view 1, by making the reference plane H-1 perpendicular to the true length line BD, in the top view, and projecting BD as a point and the plane ABC as an edge. Revolve edges AC, BC, and AB about the axis to A'C', BC', and A'B, all parallel to plane H-1, as shown in fig. 2. Project these edges to the top view, where the plane A'BC' appears true size and shape, as in fig. 3.

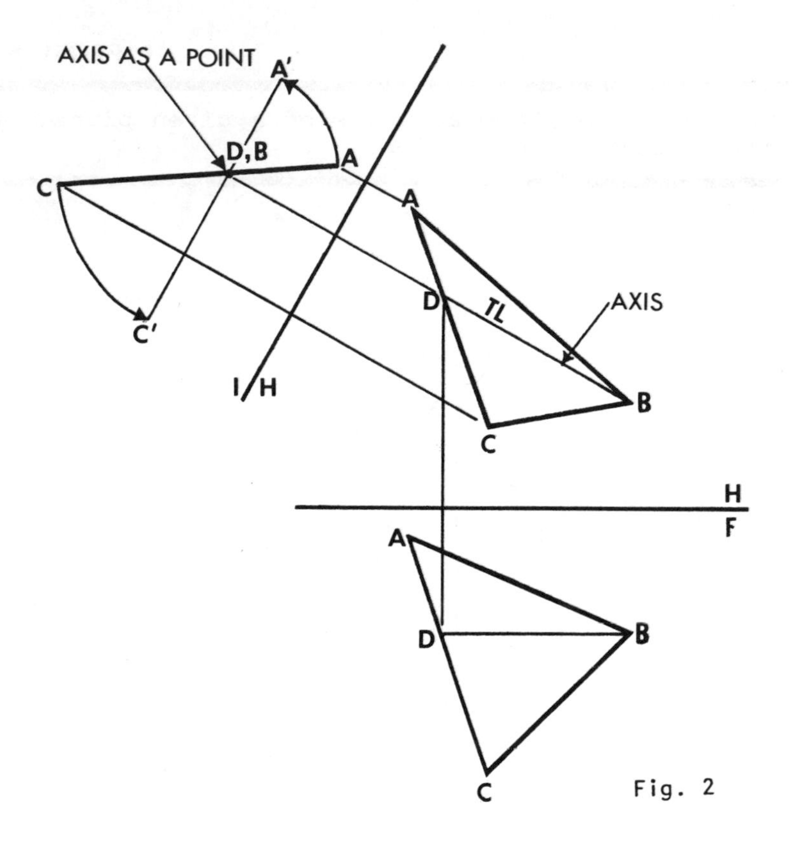
AXIS AS A POINT
A'
D,B
A
C
C'
I/H
A
D
TL
AXIS
B
C
H
F
A
D
B
C
Fig. 2

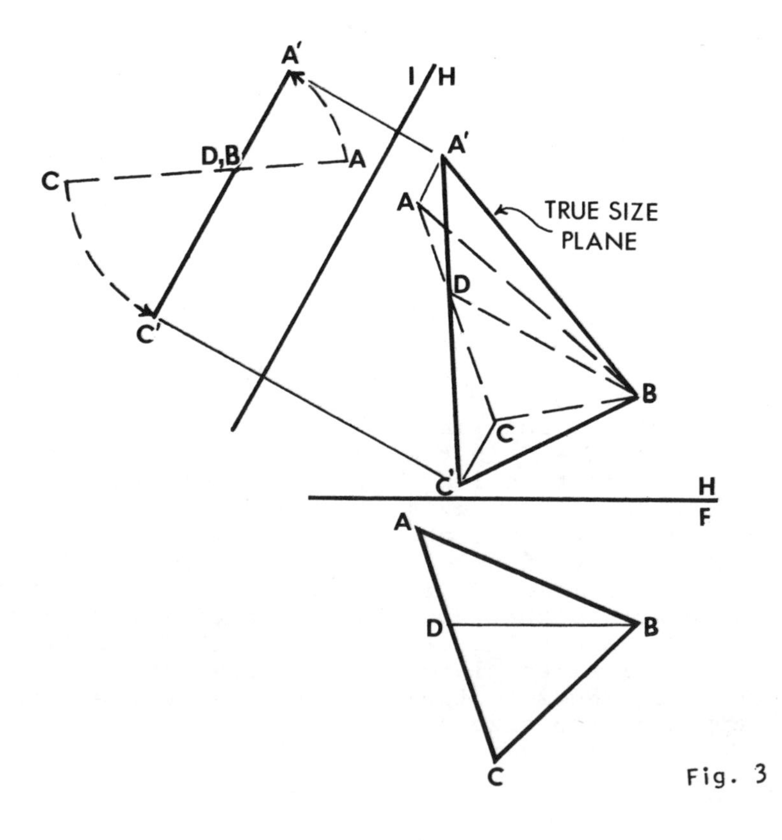
A'
I/H
D,B
A
C
C'
A'
A
TRUE SIZE
PLANE
D
B
C
C'
H
F
A
D
B
C
Fig. 3

Construct the edge view of a given plane, ABC, using the method of revolution, and indicate the dip of the plane.

Solution: Fig. 1 shows the top and front views of the given plane ABC. The edge view of a plane can be constructed if a true length line in the plane projects to a view as a point. As in fig. 2, draw a horizontal line

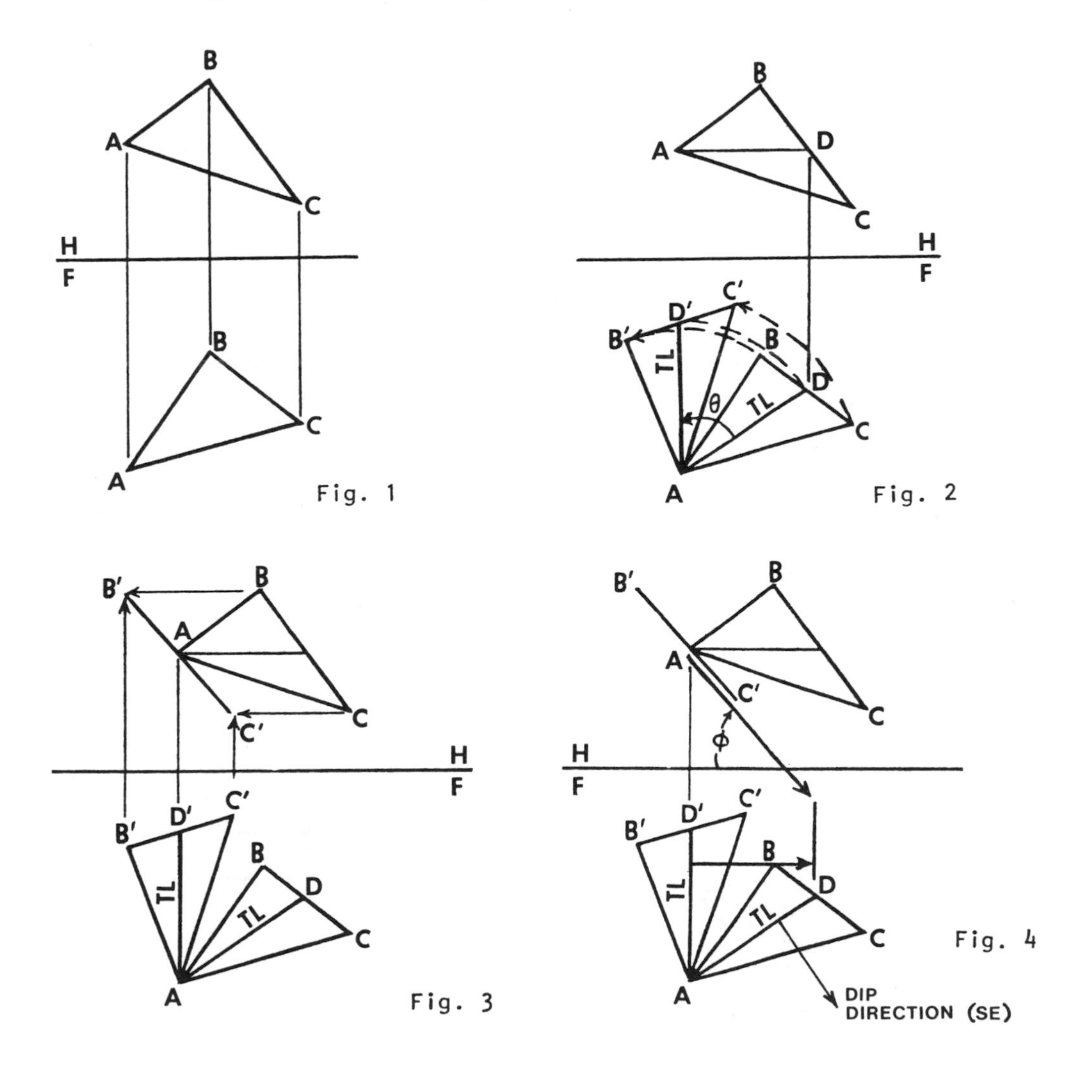

Fig. 1

Fig. 2

Fig. 3

Fig. 4

AD in plane ABC in the top view. Line AD projects true length in the front view. In order for the true length line AD to project to a view as a point, and therefore the plane ABC as an edge, AD must appear perpendicular to the reference plane for the view to which it is to be projected. With a vertical axis through point A in the top view that will project as a point in the front view, revolve line AD in the front view to AD' which is perpendicular to plane H-F. Since line AD was revolved θ degrees to AD' revolve lines AB and AC θ degrees, also, to AB' and AC', as shown in fig. 2. Project points B' and C' to the top view to where they intersect horizontal projection lines from B and C, respectively. Line B'AC' is the required edge view of plane ABC, as in fig. 3. Once the edge view of a plane is obtained, the dip, which is the angle and compass direction between the plane and the horizontal plane, can be measured. The angle ϕ of the edge view, as in fig. 4, is the dip angle. The arrow drawn perpendicular to the true length line AD in the front view indicates the dip direction of plane ABC.

TRUE ANGLE BETWEEN LINES AND PLANES

• PROBLEM 9-12

Determine the angle a given line makes with (a) the horizontal projection plane, (b) the frontal projection plane, and (c) the profile projection plane, using the method of revolution.

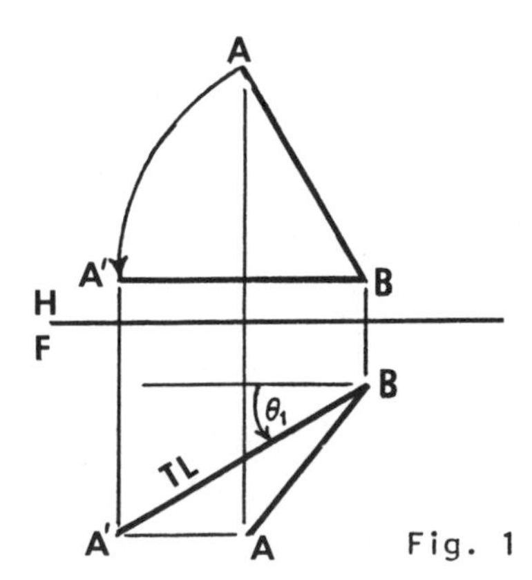

Fig. 1

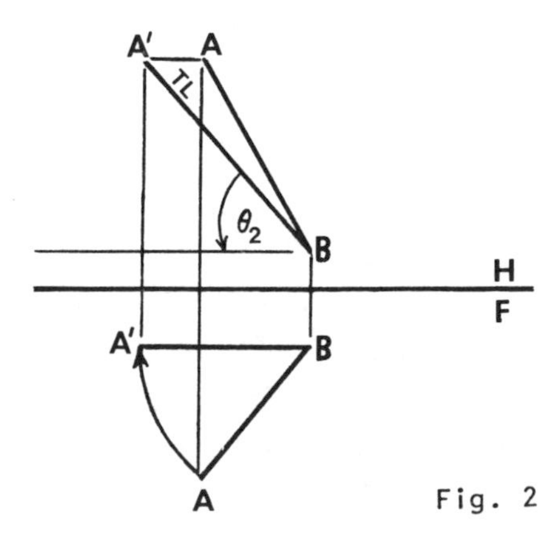

Fig. 2

Solution: The angle between a line and any projection plane can only be measured in a view in which the line appears true length. Fig. 1 shows the top and front views of the given line AB. Rotate AB in the top view about an axis through B to A'B, which is parallel to plane H-F. A'B projects true length in the front view, and the angle θ_1, which is the angle between the horizontal projection plane and line AB, can be measured. Fig. 2 again shows the top and front views of line AB, but the rotation was performed in the front view for the true length of AB in the top view. Angle θ_2 is the required angle between the frontal projection plane and line AB.

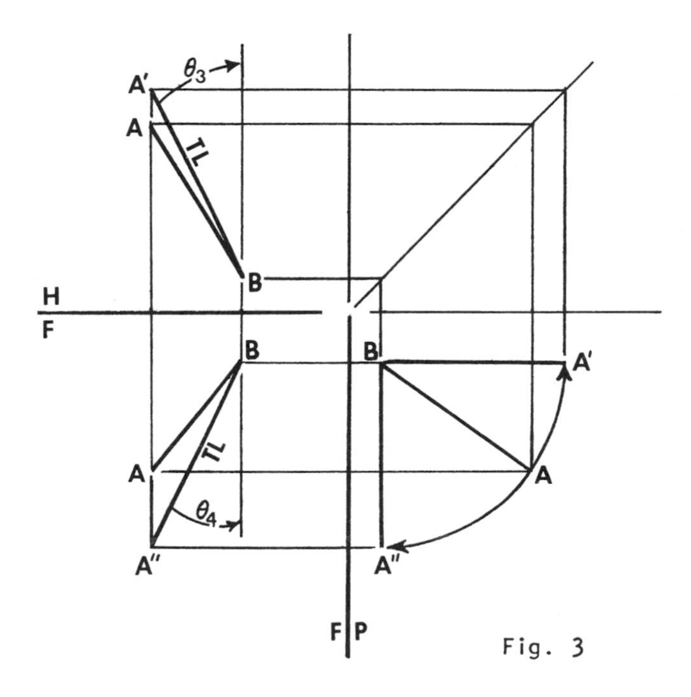

Fig. 3

Fig. 3 shows the top, front, and right side views of line AB. Rotate AB in the side view about an axis through B to A'B, which is parallel to plane H-F extended. A'B projects true length to the top view, and angle θ_3, the angle between the profile projection plane and line AB, can be measured. As confirmation of this angle, AB can also be rotated to A"B parallel to plane F-P. A"B projects true length to the front view, and the angle θ_4, equal to angle θ_3, is the required angle between the profile projection plane and line AB.

• **PROBLEM 9-13**

Determine the angle between a given line, AB, and a given plane, PQRS, using the method of rotation.

Solution: The true angle between a line and a plane can only be measured in a view in which the line appears true length and the plane appears as an edge. Fig. 1 shows the top and front views of the given line, AB, and the given plane, PQRS. Locate point C where AB pierces the plane PQRS, as shown in fig. 2. The required angle is the angle between AC and PQRS, so the true length of AC is necessary.

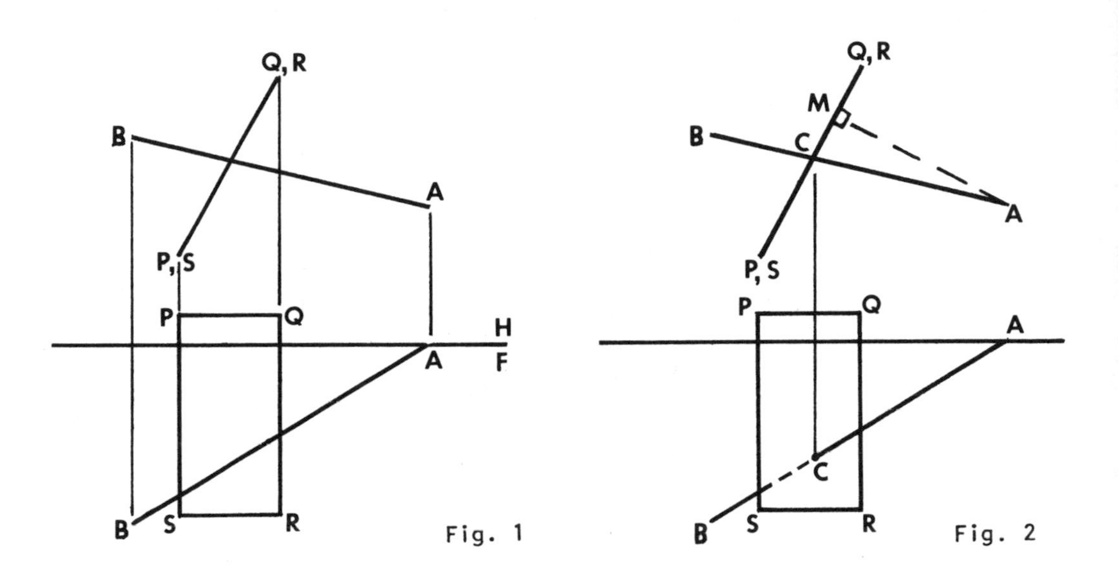

Fig. 1 Fig. 2

As in fig. 2, draw the line AM, through point A, perpendicular to plane PQRS. This is the axis about which line AB will be revolved. As in fig. 3, construct auxiliary view 1 with lines of sight parallel to AM and reference plane H-1 perpendicular to AM. Axis AM projects to view 1 as a point, so in this view revolve AC to AC', which is parallel to plane H-1. Line AC' projects true length in the top view, and the angle θ, the required angle, can be measured, as in fig. 3.

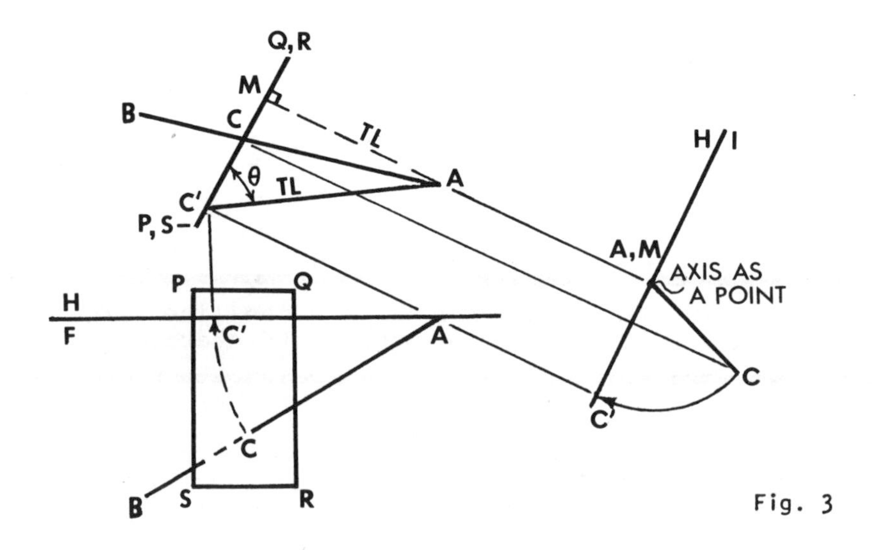

Fig. 3

Note that this problem could also have been solved by revolving AC in the front view, about an axis through A, to AC', parallel to plane H-F and projecting true length in the top view, as indicated in fig. 3 by the dashed rotation line in the front view.

• PROBLEM 9-14

Find the dihedral angle between two given planes, ABC and BCD, using the method of revolution.

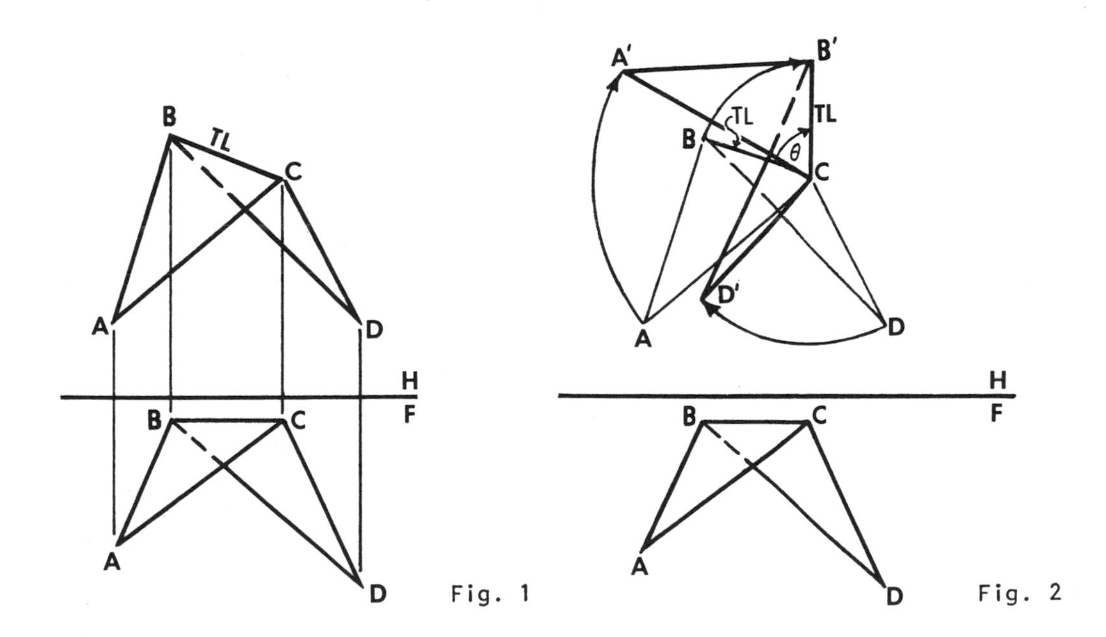

Fig. 1 Fig. 2

Solution: The true size of the dihedral angle between two planes can be seen in a view in which both planes appear as edges and their line of intersection appears as a point. Fig. 1 shows the top and front views of the given planes, ABC and BCD. Line BC, the intersection line between the planes, appears parallel to the H-F reference plane in the front view and true length in the top view. In order for the line of intersection, BC, to appear as a point in a view, it must be true length and perpendicular to the reference plane in an adjacent view. As in fig. 2, revolve BC in the top view about an axis through C to B'C, which is perpendicular to the H-F reference plane. Since BC was revolved θ degrees to B'C, revolve AC and DC θ degrees to A'C and D'C. Connect A'B' and B'D', as in fig. 2. Pro-

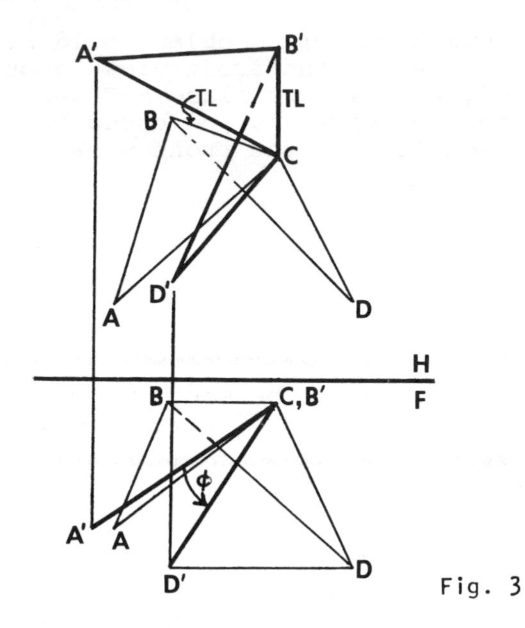

Fig. 3

ject A', B', and D' to the front view to where they intersect horizontal projection lines from A, B, and D, respectively. Connect A'C and D'C, as in fig. 3, to complete the edge view of the two planes. The angle ϕ, the required angle, can now be measured.

• PROBLEM 9-15

Determine the dihedral angle between two given planes, using the plane angle method and revolution.

Solution: (1) The dihedral angle between two planes can also be measured by the plane angle cut from the two planes by a third plane passed perpendicular to their line of intersection.

(2) Figure 1 shows the given planes, V-A-B and V-A-C, in the horizontal and vertical planes of projection. (3) Assume any point P on the line V-A, the line of intersection of the two planes, as illustrated in Fig. 2. (4) Pass plane X through P, perpendicular to the frontal line V-A in the front view. Here plane X appears as an edge. (5) Draw lines P-Q and P-R in which plane X cuts planes

V-A-B and V-A-C respectively, in the front view. PQ will coincide with PR in this view. The line RQ appears as a point, in the front view, since the lines of sight are parallel to it. (6) Project lines P-Q and P-R to the top view.

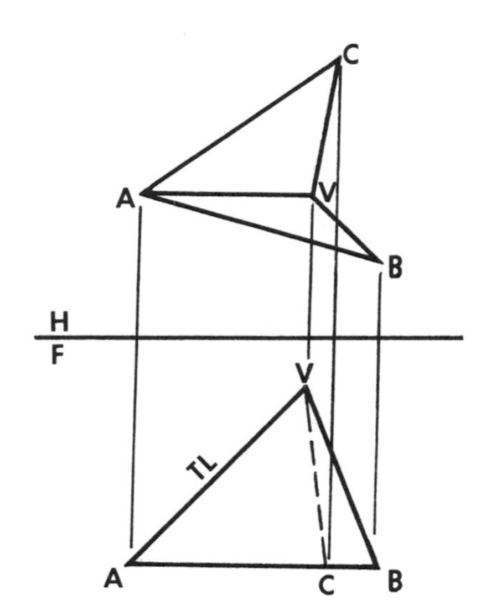

Fig. 1

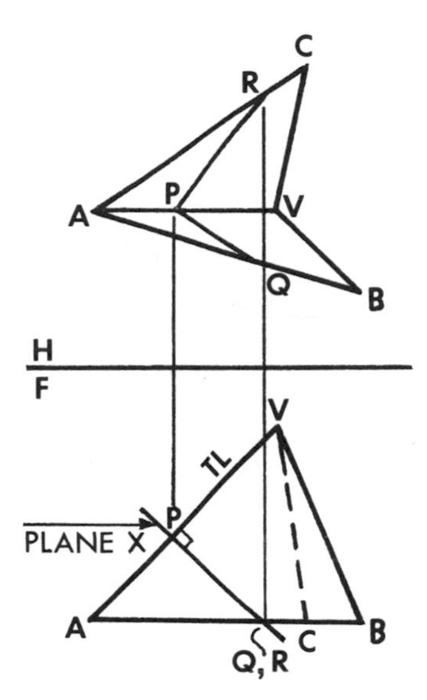

Fig. 2

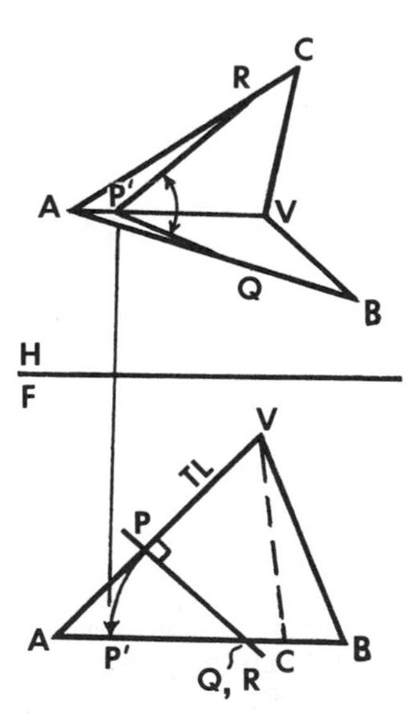

Fig. 3

(7) Find the true size of angle Q-P-R. This is accomplished by rotating the plane angle Q-P-R about the point view of axis Q-R in the front view. This rotation brings point P to P' in the horizontal plane A-B-C. Projecting P' up to the top view allows for the measurement of the required dihedral angle, by measuring angle QP'R, as in fig. 3.

• PROBLEM 9-16

Find the dihedral angle between a given plane, shown by its H- and F- traces, and the horizontal and frontal projection planes.

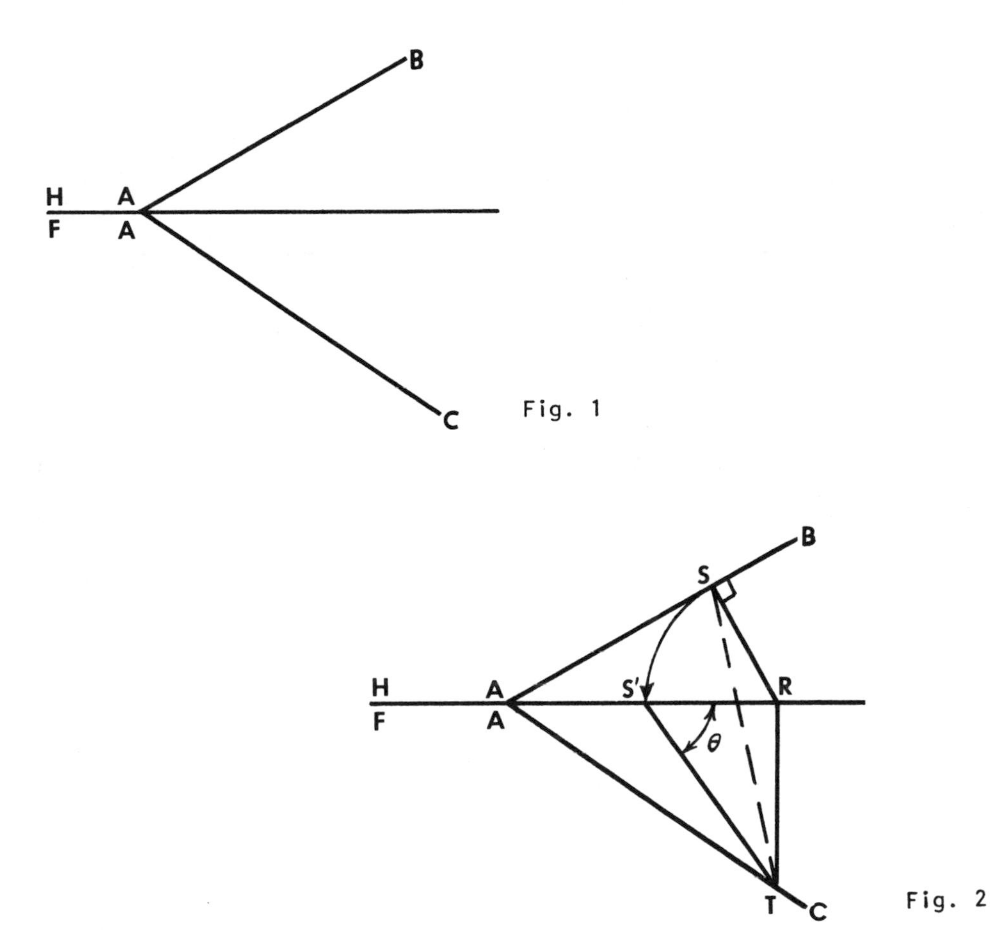

Fig. 1

Fig. 2

Solution: As previously shown, the angle between any two planes can be measured in a plane perpendicular to their line of intersection. Fig. 1 shows the top and front views of the trace of the given plane in the horizontal and frontal projection planes. Through any point S on the H- trace, AB, of the given plane, ABC, pass a plane SRT perpendicular to AB. The true size of angle RST is the required angle between plane ABC and the horizontal projection plane. As seen in fig. 2, revolve SR about an axis through R to S'R parallel to the H-F reference plane. The angle RS'T, angle θ, is the true, required angle since S'R is true length because it is parallel to plane H-F and RT is true length because it appears as a point in the top view.

The angle between the plane ABC and the frontal projection plane is found in a similar manner. As in fig. 3, pass the plane LMN through any point L on the F-trace, AC, of plane ABC such that LMN is perpendicular to AC. Revolve LM about an axis through M to L'M parallel to the H-F plane. Line L'M and MN are both true length, and the angle ML'N, angle ϕ, is the true, required angle between plane ABC and the frontal projection plane.

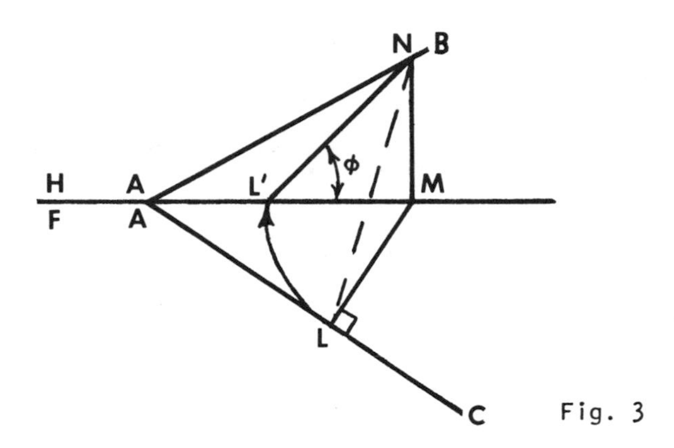

Fig. 3

TRUE SIZE AND SHAPE OF A PLANE

• PROBLEM 9-17

> Find the H- trace of a plane, given its F- trace and the angle between the plane and the frontal projection plane.

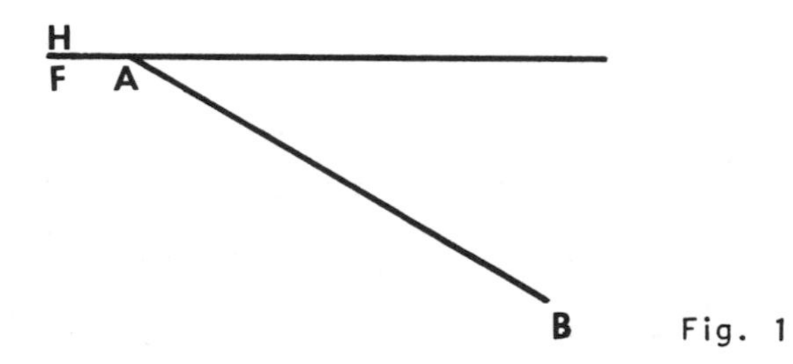

Fig. 1

Solution: Fig. 1 shows the F- trace, AB, of the given plane which makes an angle θ with the frontal projection plane. The angle θ will be seen in a view which shows the true size and shape of a plane perpendicular to AB.

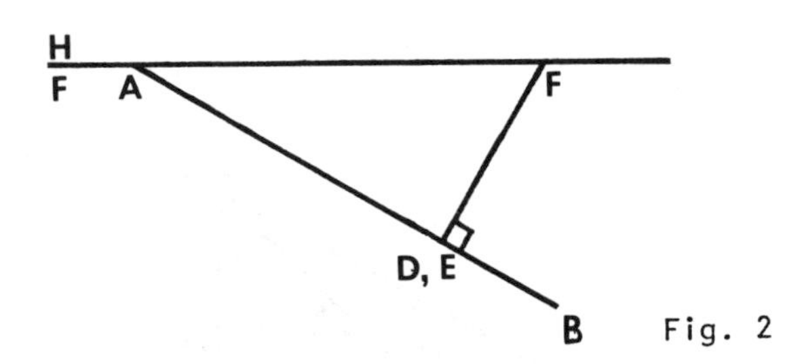

Fig. 2

Through any point D on AB, pass plane DEF perpendicular to AB, as shown in fig. 2. Line DF is the edge view of this plane. In order to construct the true size and shape view of plane DEF, the edge view must be parallel to the reference plane of an adjacent view. Revolve DF about an axis

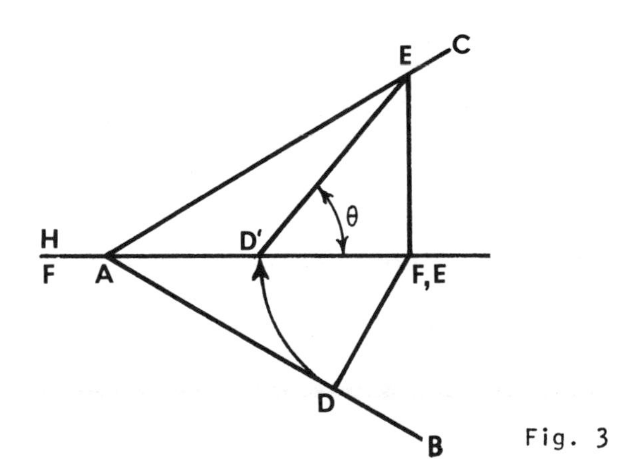

Fig. 3

through point F to D'F, parallel to the H-F reference plane. Draw a line through point D' which makes the given angle θ with the plane H-F. Project vertically from point F to intersect the line through D' at point E. Construct the H- trace, AC, through point E.

• PROBLEM 9-18

Find the line through point O which makes the angle θ with the horizontal plane, and the angle ϕ with the vertical plane.

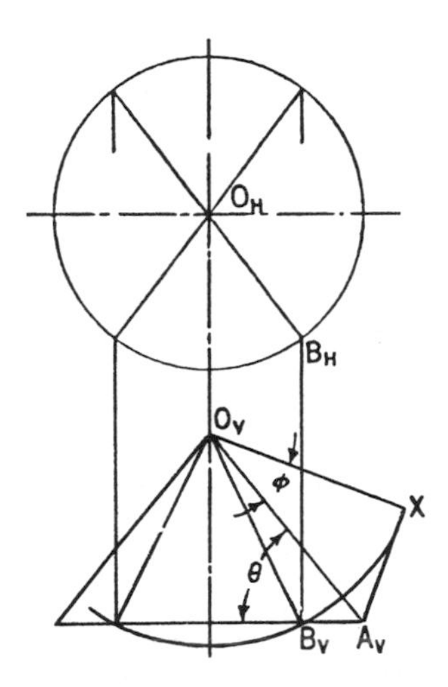

Solution: (1) Draw the horizontal and vertical projections of a right circular cone whose elements make the angle θ with the horizontal plane, assuming any convenient length, O-A, for the elements. (2) The remaining problem is to select the particular element, or elements, of this cone that makes the angle ϕ with the vertical plane. (3) From the geometry of the given figure, it is evident that the length of the vertical projection of the required element must equal O-A $\cos\phi$. This is found graphically by constructing the right triangle A-O-X, having the angle A-O-X = ϕ. (4) Distance O-X is the length of the vertical projection of the required element. This length is transferred to the cone by describing arc X-B. Determine the projection $O_V B_V$, which equals O_V-X. The required line is O-B. (5) Note that there are four possible solutions to this problem when $(\theta+\phi)$ is less than 90°. If $(\theta+\phi) = 90°$, there are two solutions. There is no solution if $(\theta+\phi)$ is greater than 90°.

• PROBLEM 9-19

Find the true size and shape view of the inclined face of a given solid, using the method of single revolution.

Solution: Fig. 1 shows the top, front, and right side views of the given solid, with the indicated inclined face. The true shape and size view of this face can be seen in a view adjacent to one in which the plane of the face is parallel to the reference plane between the two views. Thus the solid can be revolved in the front view, where the inclined face appears as an edge, to a position where the inclined face is parallel to the reference plane

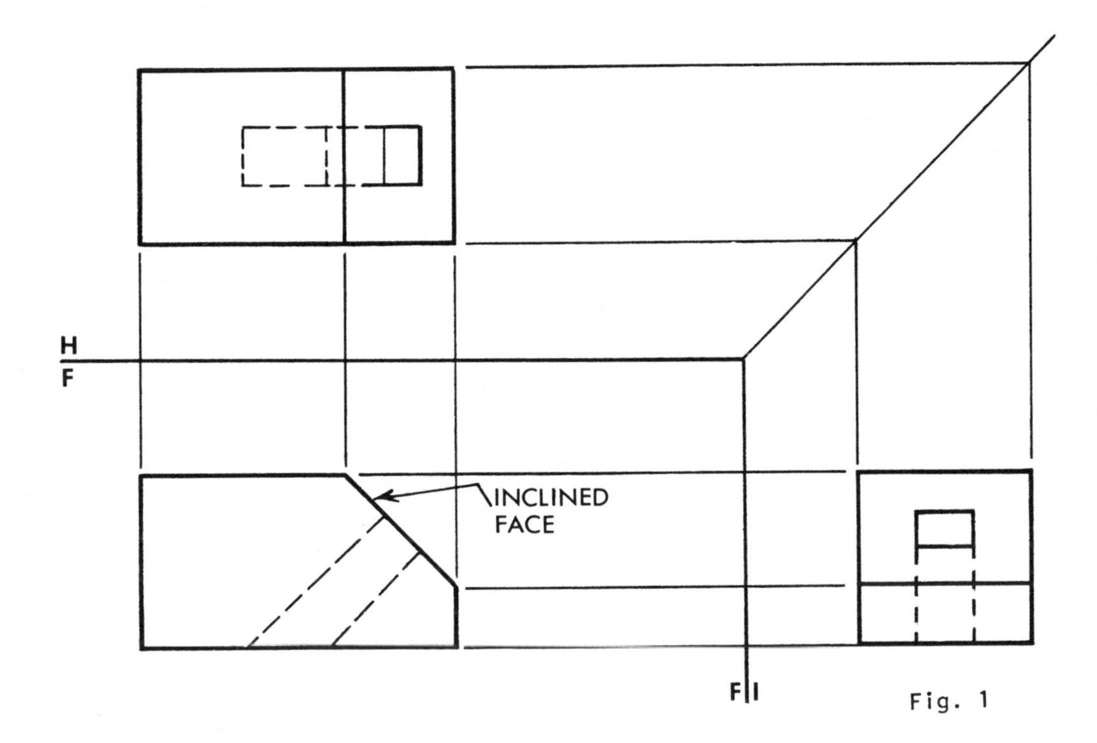

Fig. 1

H-F or F-P. As illustrated in fig. 2, in the front view revolve line 2-3, the edge view of the inclined face, about an axis through point 3 to 2'-3 which is parallel to reference plane F-P. Note that, as stated above, line 2-3 could have been revolved about the same axis until it was parallel to plane H-F. Revolve the remaining points of the solid through the same number of degrees, θ, as point 2 was revolved, and connect the corresponding points for the completed front view of the new position of the solid. Project these revolved points to the top view to where they intersect horizontal projection lines from their respective, original points. Complete the top view of the revolved solid, seen in fig. 2, by connecting corresponding points. Project from both the top and front views of the revolved solid to the right side view for the true size and shape view of the inclined face. Note that in the revolved top and right side views of fig. 2, the original views have been omitted for clarity.

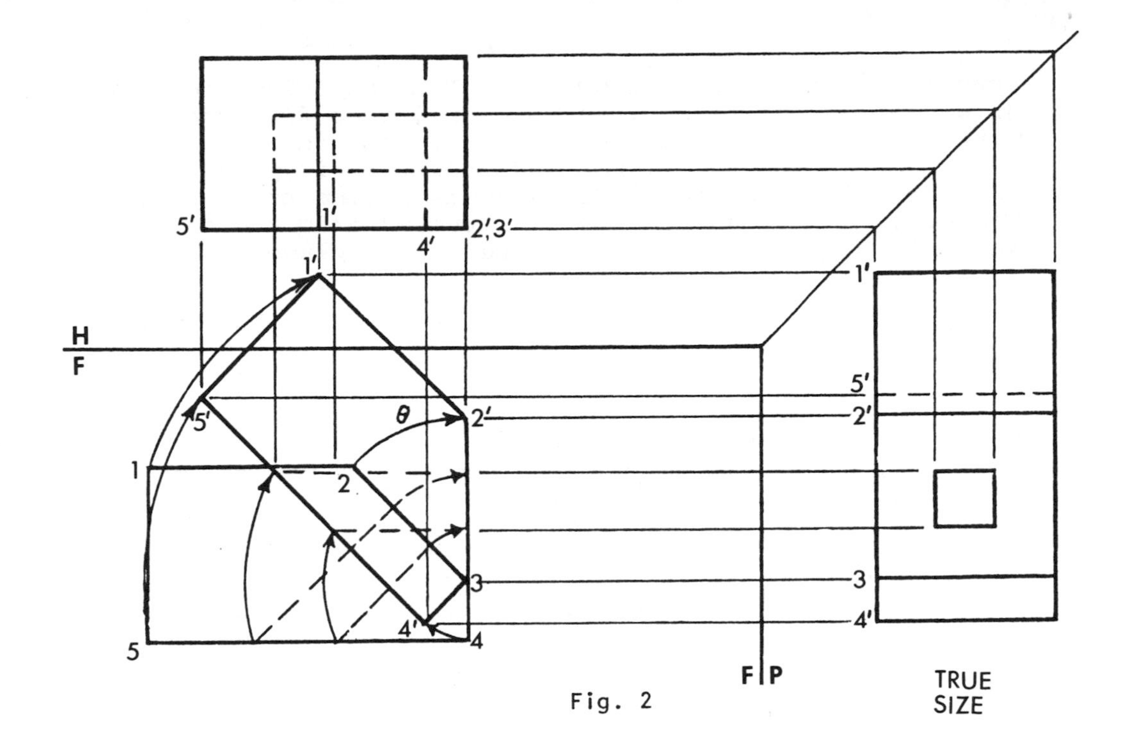

Fig. 2

• PROBLEM 9-20

> Find the true size and shape of an inclined plane face of a solid, using double revolution.

Solution: Fig. 1 shows the top, front, and side views of solid 1-2-3-5-A-B-4-C, with the inclined face ABC, whose

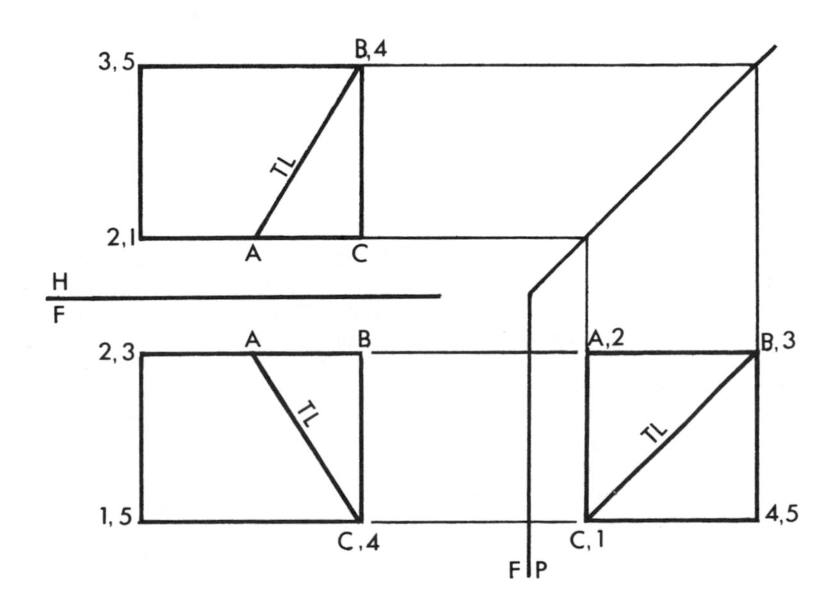

Fig. 1

true size and shape is required. The true size and shape of a plane can be seen in a view projected from the edge view of the plane. Line AB in the front view is parallel to the H-F reference plane, and therefore it projects true length in the top view. Since AB is a line of plane ABC, the edge view of ABC can be constructed from a view in which AB is perpendicular to a reference plane. As in

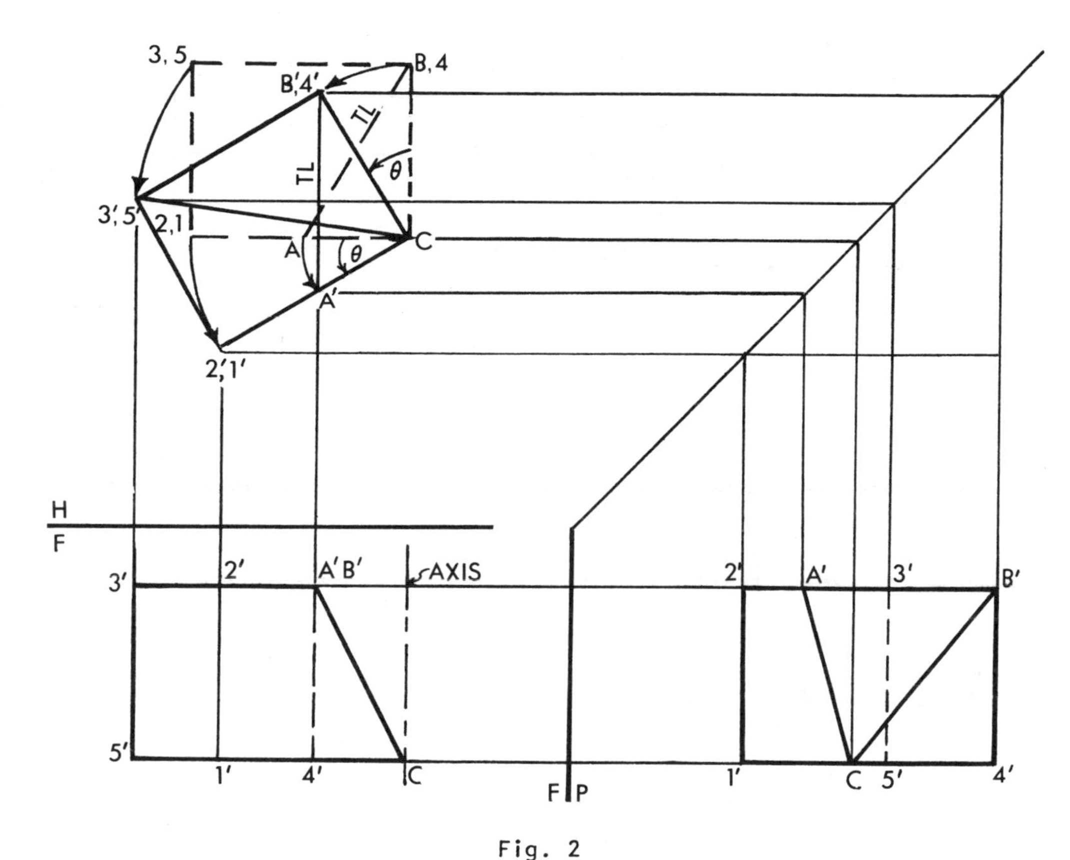

Fig. 2

fig. 2, revolve plane ABC in the top view about an axis through point C to A'B'C, where the true length line A'B' is perpendicular to plane H-F. As AC and BC were both revolved θ degrees, revolve the remaining points of the solid θ degrees also, and project these revolved points to the front and side views to where they intersect horizontal projection lines from their corresponding original points. (Note that only the revolved front and side views have been shown in fig. 2 to avoid overlapping.) Plane A'B'C appears as an edge in the front view. As illustrated in fig. 3, revolve the edge view of ABC about an axis through point C to A"B"C which is parallel to the F-P reference plane. Revolve the remaining points of the solid through the same number of degrees, and project all double-revolved points to the side view to where they intersect vertical projection lines from their corresponding single-revolved points. Plane A"B"C appears true size and shape in the double-revolved side view. Project from both the front and the side views to construct the double-revolved top view.

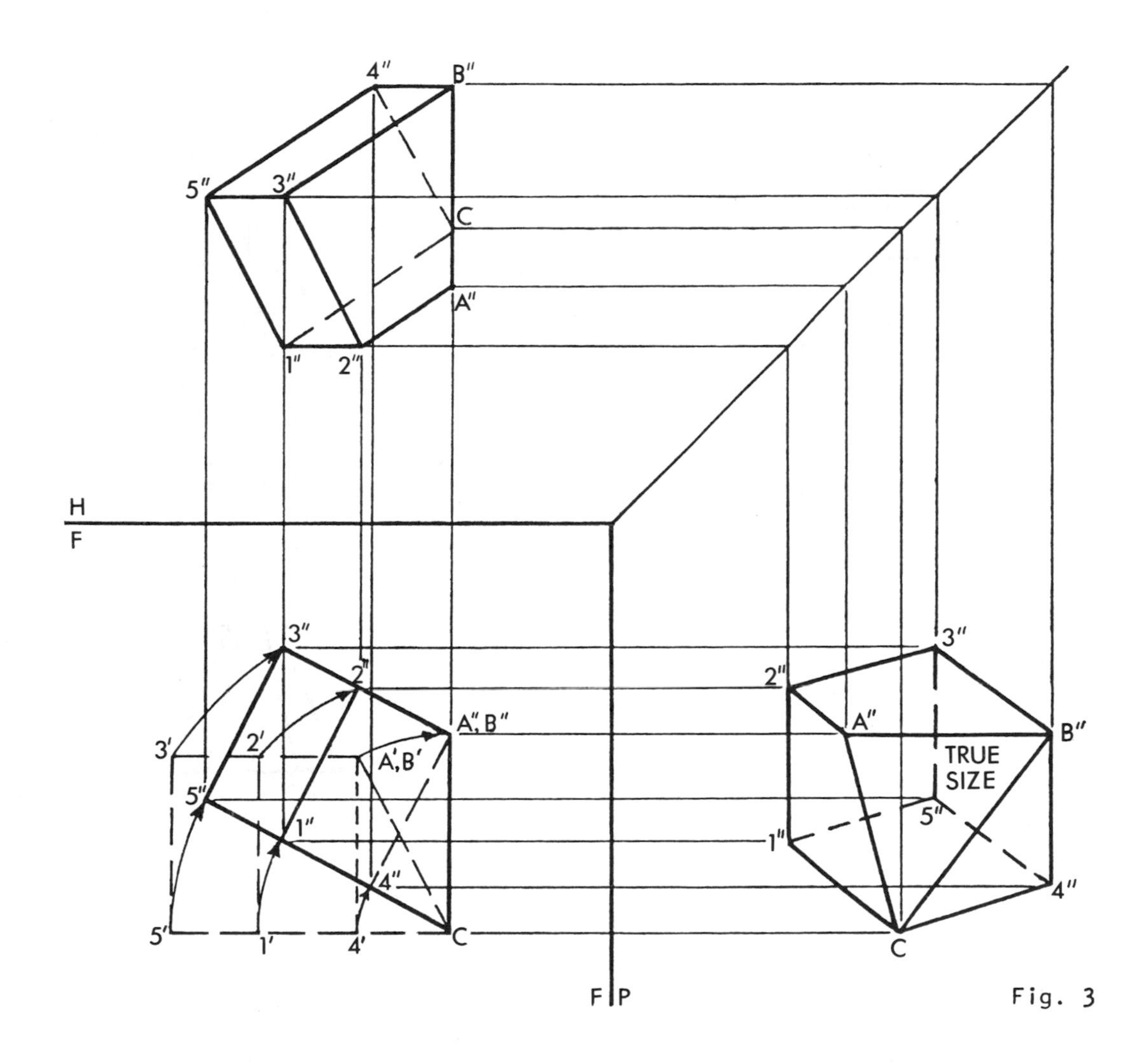

Fig. 3

• **PROBLEM 9-21**

> Find the true shape and size view of the section cut from a given hexagonal pyramid, O-ABCDEF, by the cutting plane, JG, using the method of revolution.

Solution: Fig. 1 shows the top and front views of the

given hexagonal pyramid, O-ABCDEF, and the cutting plane, JG. The points where the cutting plane, which appears as an edge in the front view, intersects the edges of the pyramid are labeled 1 through 6 in the front view, corres-

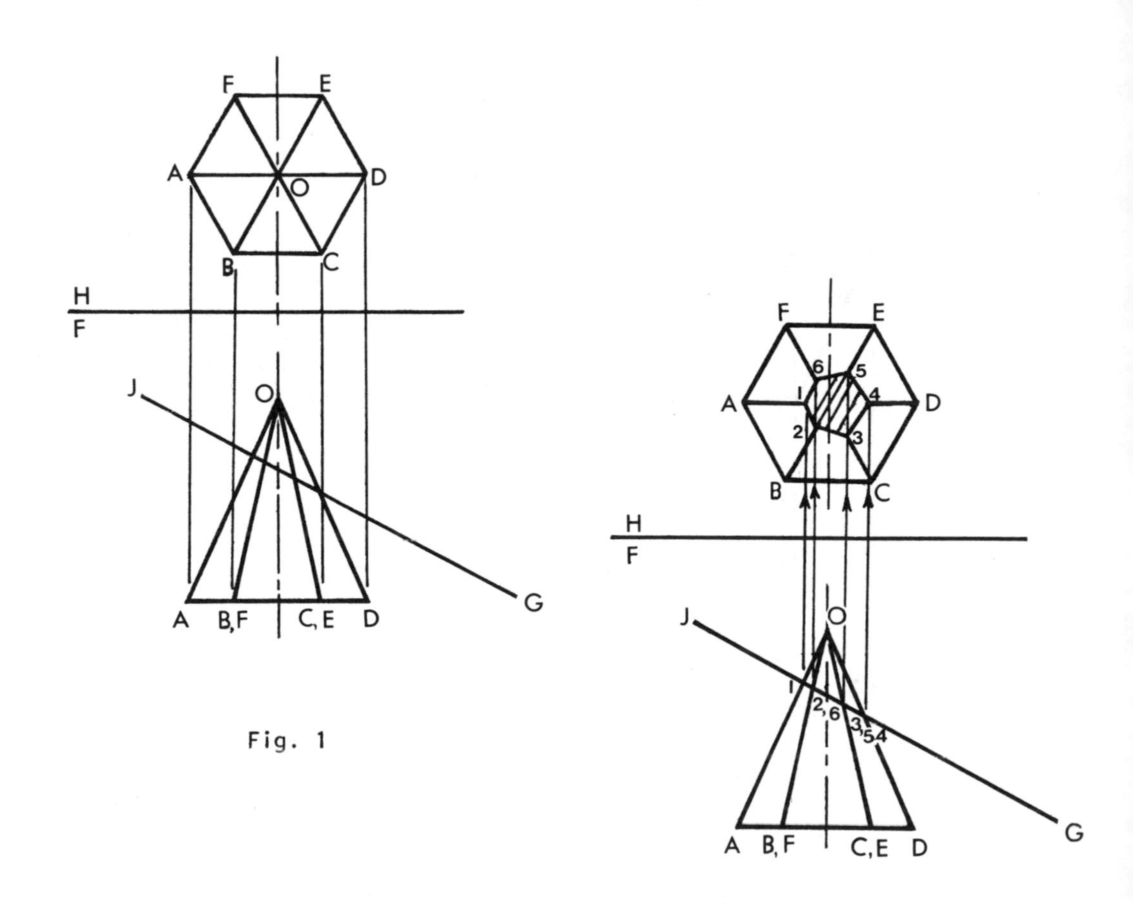

Fig. 1

Fig. 2

ponding to edges OA, OB, OC, OD, OE, and OF of the pyramid. Locate these points in the top view by projecting them to where they intersect their respective cone edges in the top view. Connect points 1 through 6 in this view to complete the line of intersection between the pyramid and the plane. The true shape and size view of the section can be seen in a view adjacent to one where the edge view of the section is parallel to the reference plane between the two views. In order for the edge view of this section to appear parallel to reference plane H-F, revolve the edge view about an axis through point G to its new position. Project points 1' through 6' to the top view to where they intersect horizontal projection lines from points 1 through

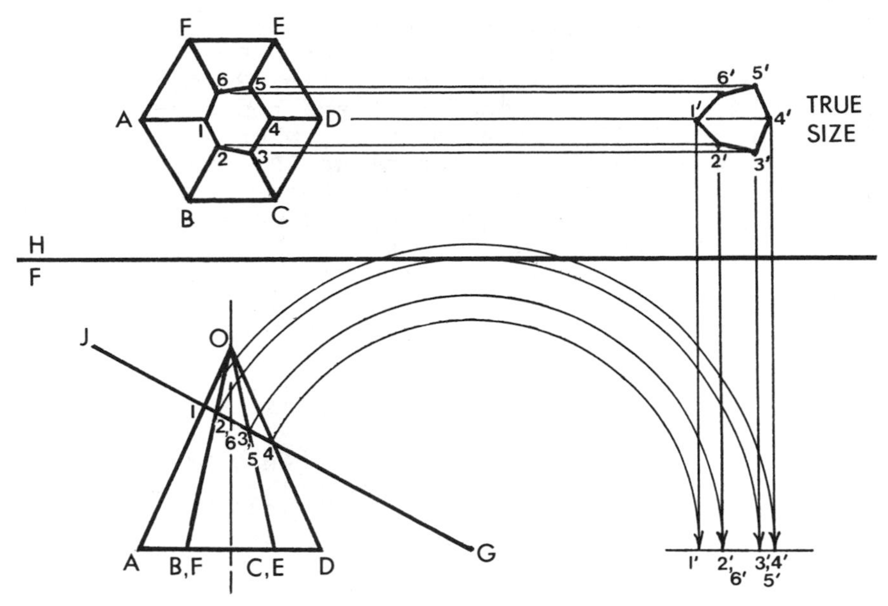

Fig. 3

6, respectively, as shown in fig. 3. Connect these points in order for the true shape and size view of the section.

• PROBLEM 9-22

> Find the true shape and size view of an elliptical conic section cut from a cone by a given plane, OP, using the method of revolution.

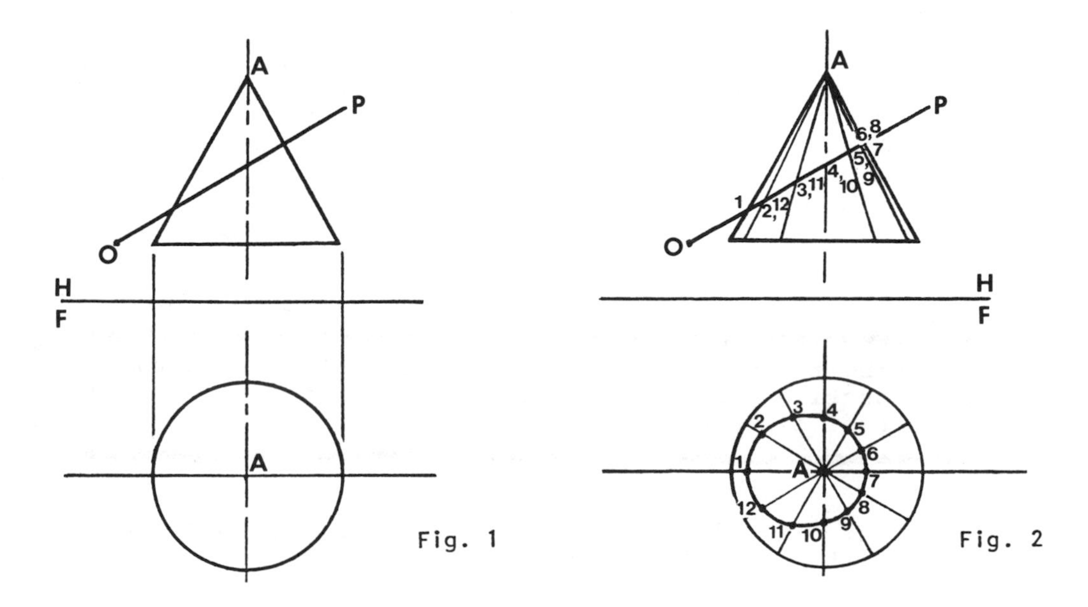

Fig. 1

Fig. 2

Solution: Fig. 1 shows the top and front views of the cone with vertex A and the given plane OP, which appears as an edge in the top view. As in fig. 2, draw several elements of the cone in the front and top views. The points 1 through 12 in the top view, where these elements are intersected by the cutting plane OP, are located on the line of intersection. Project these points to the front view to their corresponding elements, and connect them with a smooth curve to complete the line of intersection. The true shape and size of the section can be seen in a view adjacent to one in which the edge view of the section appears parallel to the reference plane be-

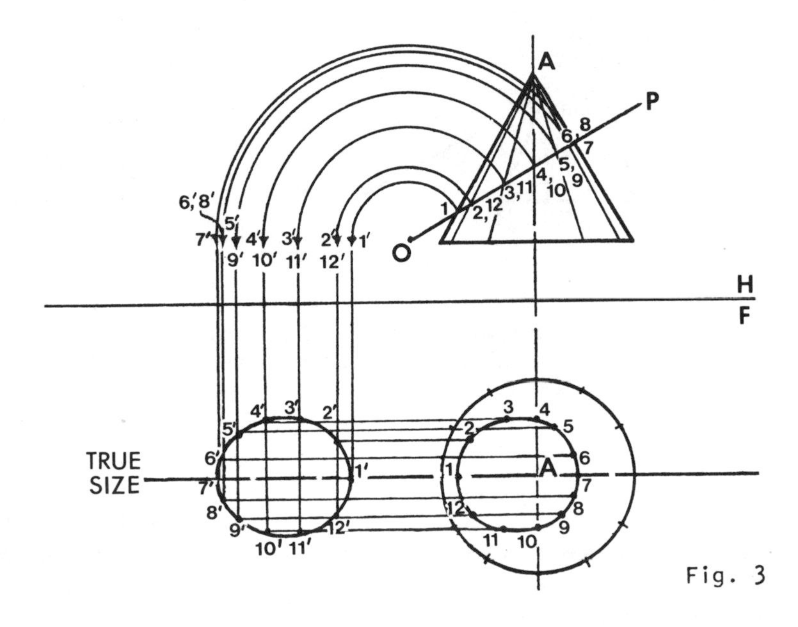

Fig. 3

tween the two views. As in fig. 3, revolve the edge view of the section about an axis through point O until it is parallel to the reference plane H-F. Project points 1' through 12' to the front view where they intersect horizontal projection lines from points 1 through 12, respectively. The ellipse through points 1' to 12' is the required true shape and size view of the section cut by plane OP.

• PROBLEM 9-23

> Find the true size and shape view of the hyperbolic conic section cut from a cone by a given plane, P, using the method of revolution.

Solution: Fig. 1 shows the top and front views of the cone, with vertex O, and the vertical plane P, which ap-

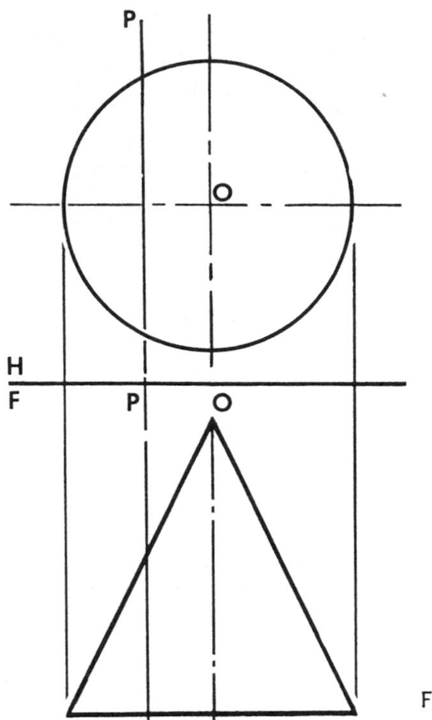

Fig. 1

pears as an edge in both views. As in fig. 2, pass several horizontal planes through the cone in the front view, intersecting the right limiting element in the

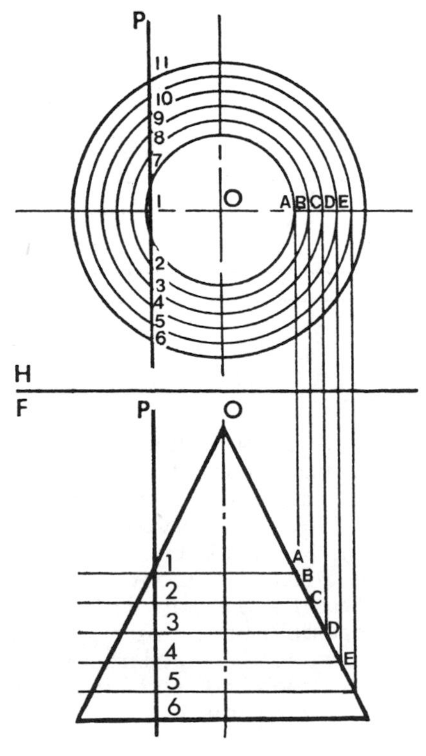

Fig. 2

points labeled A, B, C, D, and E, and the plane P in the points labeled 1, 2, 3, 4, and 5. Project points A through E to the top view to the center line, and construct the circles of radii OA through OE centered at point O. These circles represent the sections cut from the cone by the horizontal planes, and they intersect the plane P in points 1 through 11, forming the edge view of the section. The true size and shape of the section can be seen in a view adjacent to one in which the edge view of the section is parallel to the reference plane between the two views. (Note that in this case the edge view of the section appears in both the top and front views, so the true view can be projected from

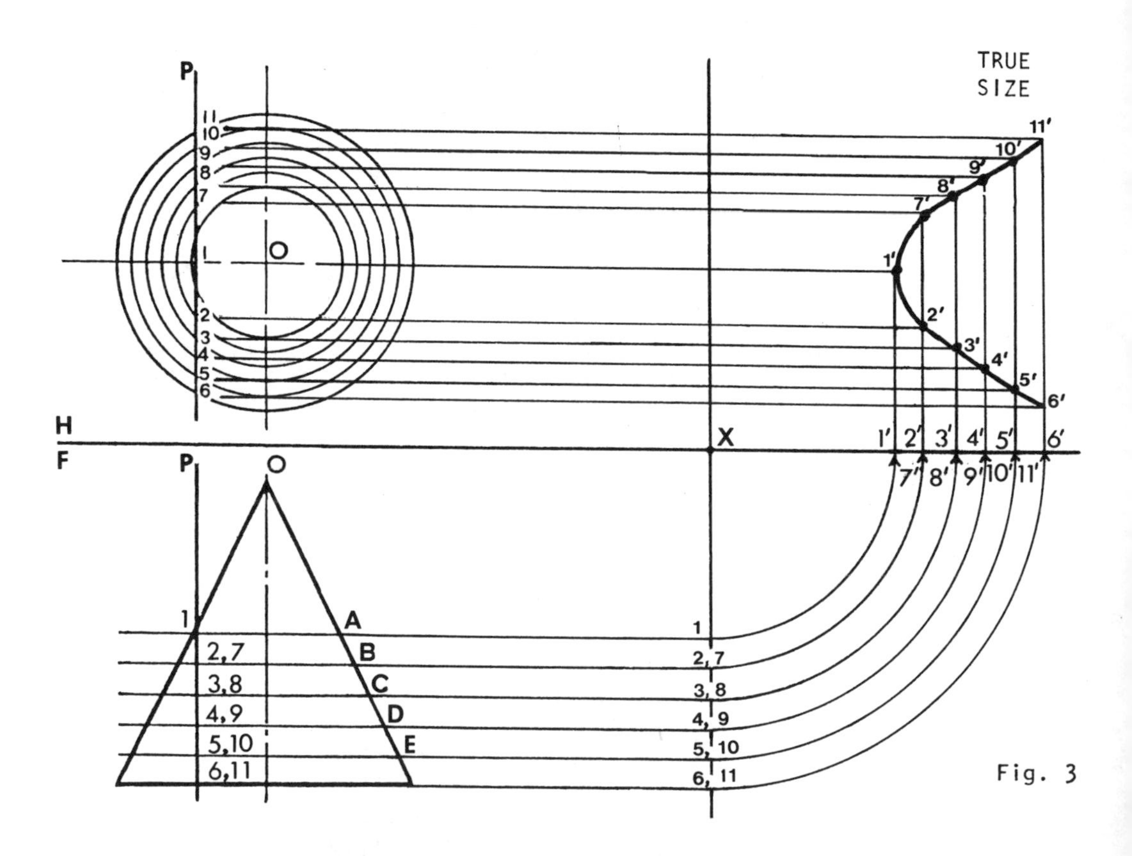

Fig. 3

either one. Fig. 3 shows the section revolved in the front view.) Project points 1 through 11 horizontally to a vertical line through a convenient point X on the plane H-F, as in fig. 3. Revolve these points about an axis through point X until the edge view is parallel to the H-F reference plane. Project points 1' through 11' to the top view to where they intersect horizontal projection lines from points 1 through 11, respectively. Connect these intersection points with a smooth curve for the required true size and shape view of the hyperbolic section cut by plane P.

• **PROBLEM** 9-24

Find the true size and shape view of the parabolic section cut from a cone by a given plane, OP, using the method of revolution.

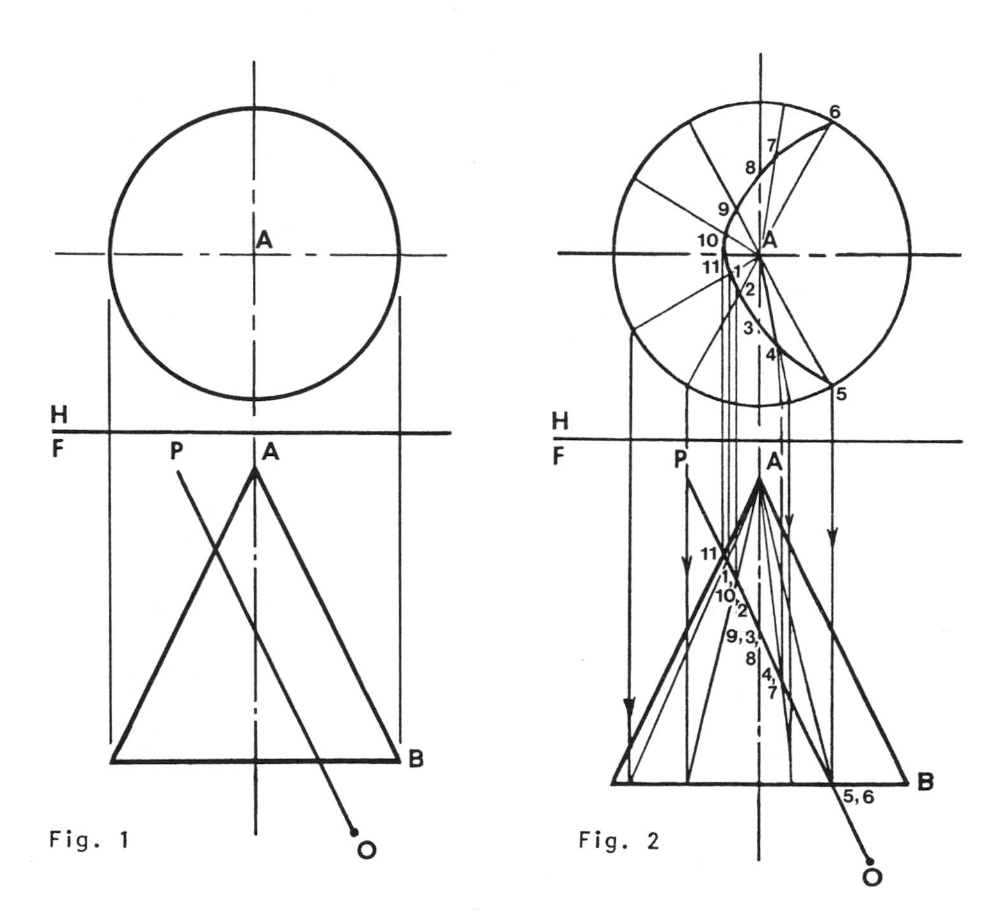

Solution: Fig. 1 shows the top and front views of the cone, with vertex A, and the given plane, OP, which appears as an edge parallel to element AB in the front view. As in fig. 2, draw several elements of the cone in the top and front views. The points 1 through 11 in the front view, where these elements cross the edge view of plane OP, lie on the line of intersection between the cone and the plane. Project these points to the top view to their corresponding elements, and connect them with a smooth curve to complete the line of intersection. The true size and shape of the section will appear in a view with lines of sight perpendicular to the edge view of the section. Revolve the edge view of plane OP containing the section about an axis through point O until the edge view is parallel to the reference plane H-F. Project points 1' through 11' to the top view to intersect horizontal projection lines from points 1 through 11, respectively. Connect these intersection points with a smooth curve for the required true size and shape view of the parabolic section cut by plane OP.

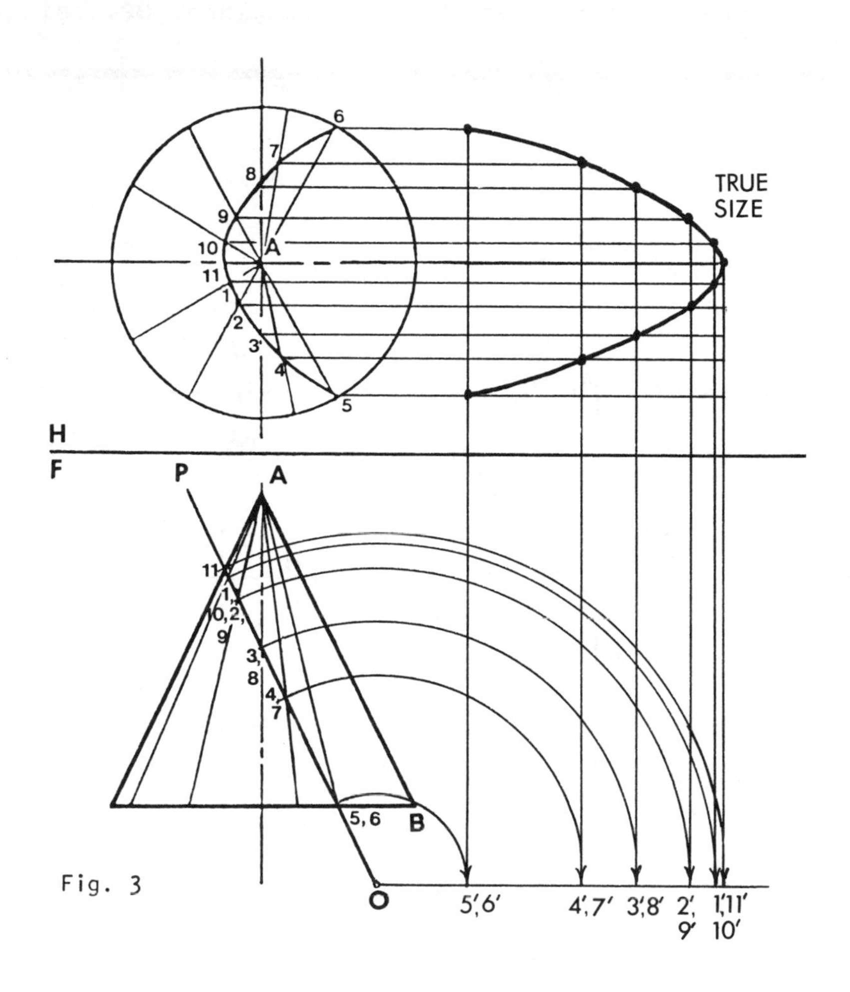

Fig. 3

REVOLUTIONS

• PROBLEM 9-25

> Using only the two given views, construct a horizontal connection of a specified length between two given lines, AB and BC.

Solution: Fig. 1 shows the front and top views of the two given lines, AB and BC. The required horizontal connection can be drawn only in a view in which the lines AB and BC appear true length. This view can be drawn using revolutions and the two given views. As in fig. 2, draw

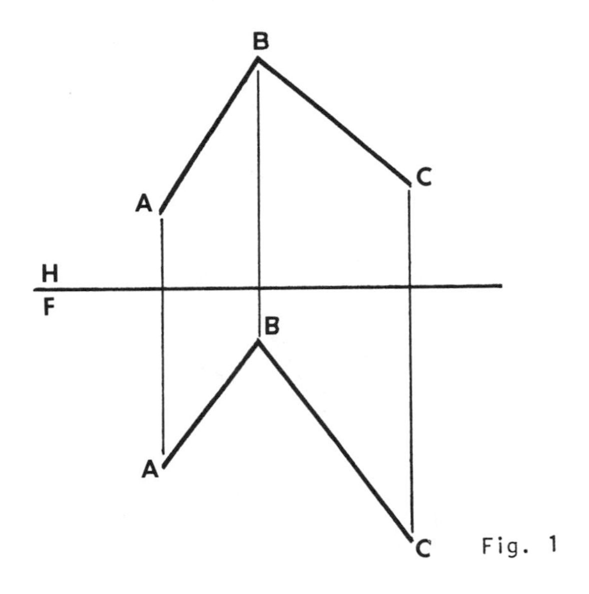

Fig. 1

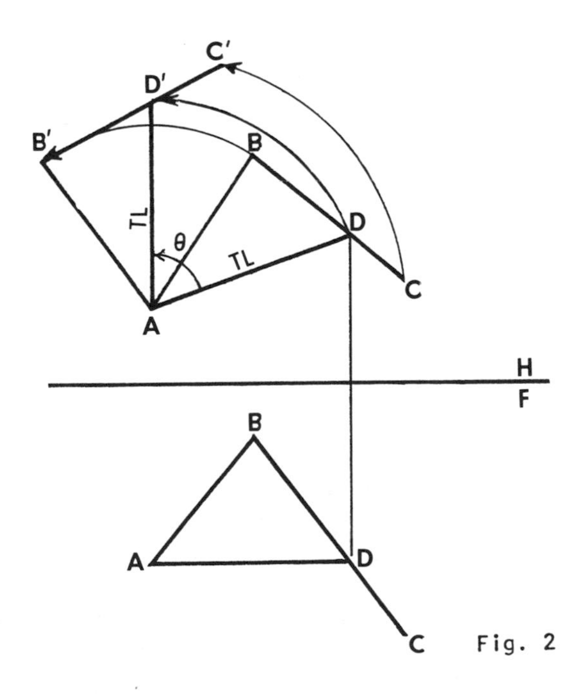

Fig. 2

a horizontal line AD in the front view which projects to true length in the top view. The edge view of plane ABC can be drawn in the front view if the true length line AD appears perpendicular to the reference plane H-F in the top view. As illustrated in fig. 2, revolve line AD about an axis through point A to AD', which is perpendicular to plane H-F. As AD was revolved through θ degrees, revolve line AB and point C through θ degrees, also, to AB' and C', respectively. Project AB' and B'C' to the front view, as in fig. 3, to where B' and C' intersect horizontal

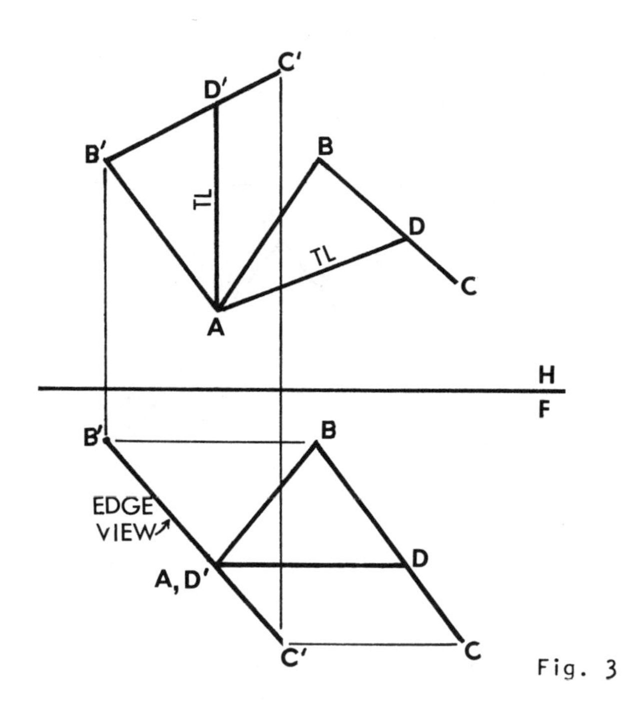

Fig. 3

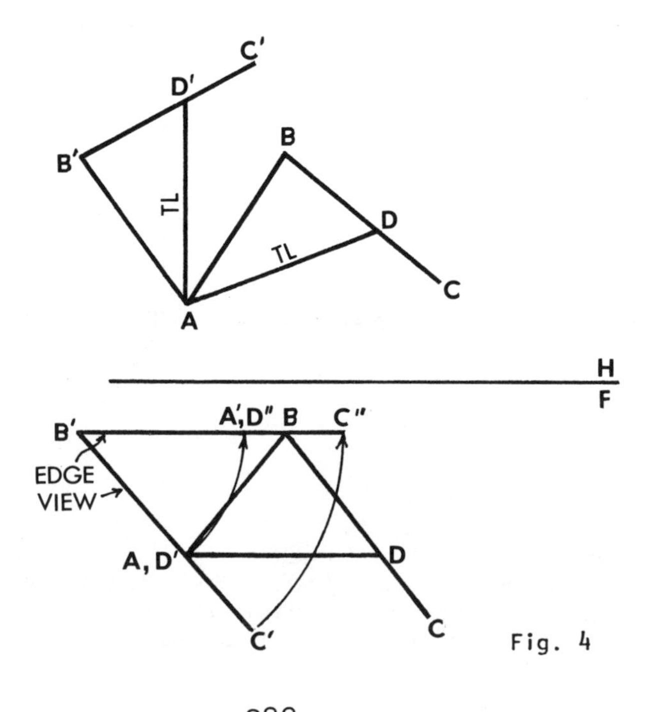

Fig. 4

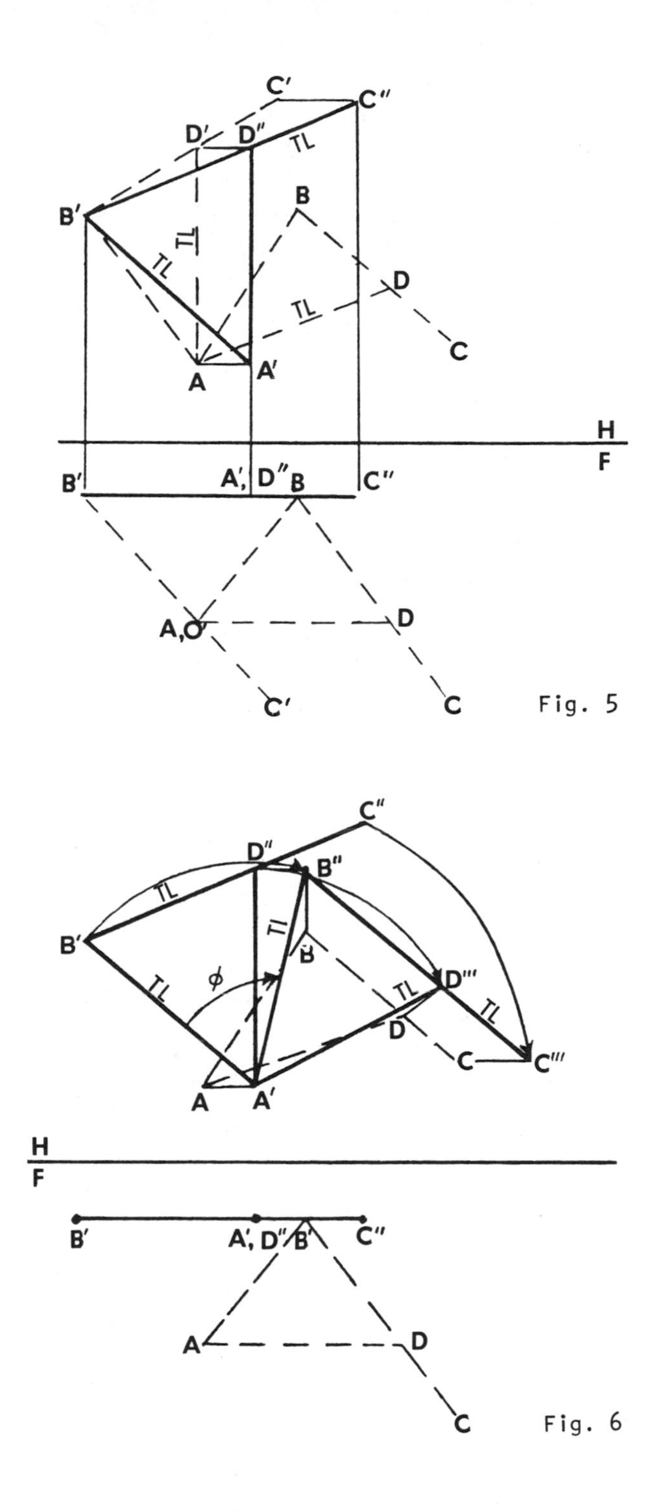

Fig. 5

Fig. 6

projection lines from B and C, accordingly. Line B'AC' is the edge view of plane ABC. The true size and shape of plane ABC, and thus the true lengths of AB and BC, can be drawn in the top view if the edge view of plane ABC appears parallel to plane H-F in the front view. As shown in fig. 4, revolve the edge view, B'AC', about an axis through point B' to B'A'C", which appears parallel to

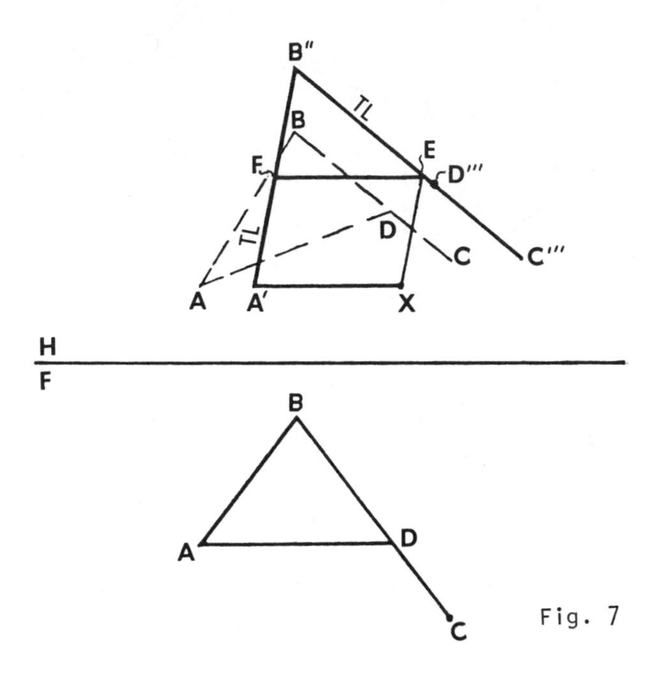

Fig. 7

plane H-F. Project points A', D", and C" to the top view to where they intersect horizontal projection lines from A, D', and C', respectively, as in fig. 5. All views except the edge view and the true size and shape view appear in dashed lines in fig. 5 for clarity. Because the required connection was a horizontal one, the true size and shape view must be revolved again. To show the final position of lines AB and BC, as in fig. 6, revolve point B' about an axis through A' to B", intersecting a vertical projection line from the originally given point B. Revolve point C" through the same number of degrees as B' was revolved, ϕ, to C''', intersecting a horizontal projection line from the originally given point C. The required horizontal connection can now be drawn from B"C''' to A'B". As in fig. 7, construct a horizontal line, A'X, of the specified length. Through point X, construct a line parallel to A'B", intersecting B"C''' at point E. The horizontal line EF, through point E and intersecting A'B" at point F, is the required horizontal connection of specified length.

• PROBLEM 9-26

Using only the two given views, construct the bisector of angle DEF from point E to line DF.

Solution: Fig. 1 shows the given top and front views of

plane DEF. The angle DEF can only be bisected by a line constructed in a view in which the plane appears true size and shape. Using only these two views and methods of revolution, the appropriate view of the plane can be drawn. As in fig. 2, draw the horizontal line AF in the

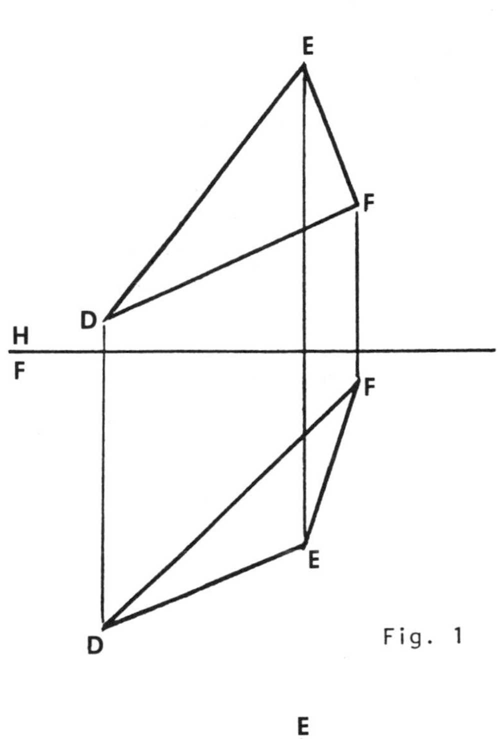

Fig. 1

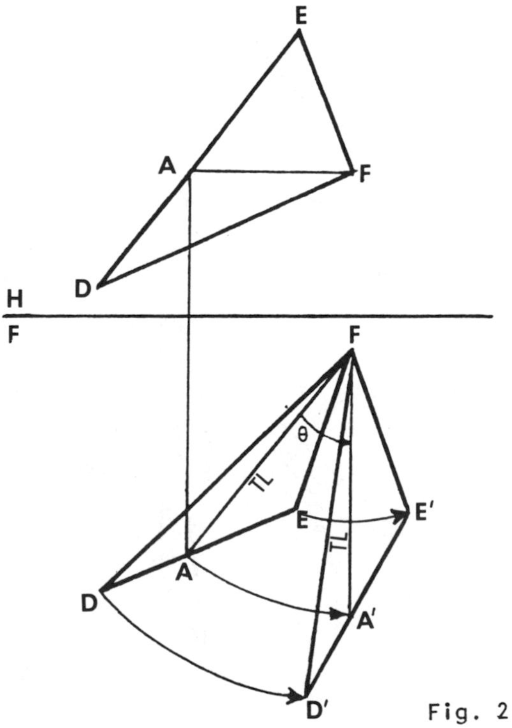

Fig. 2

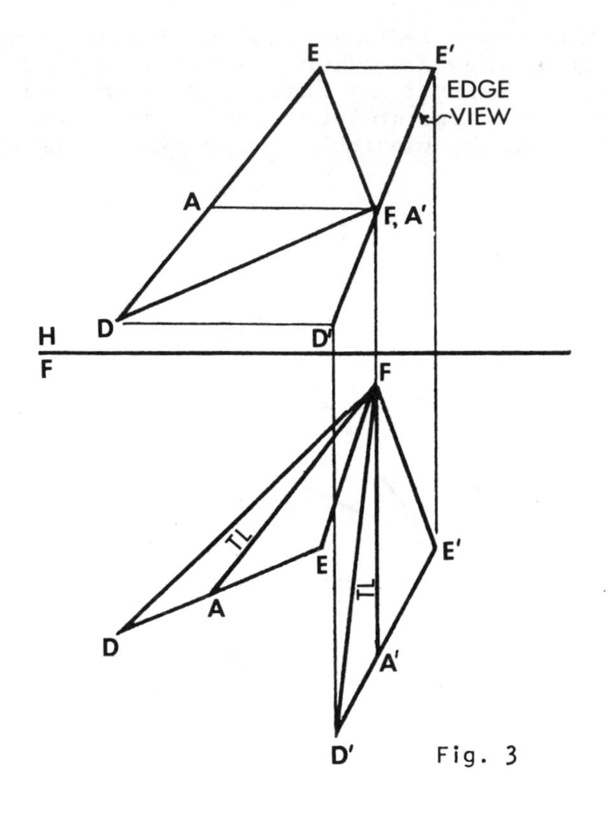
E
E′
EDGE
VIEW
A
F, A′
D
D′
H
F
F
TL
E
E′
TL
A
D
A′
D′
Fig. 3

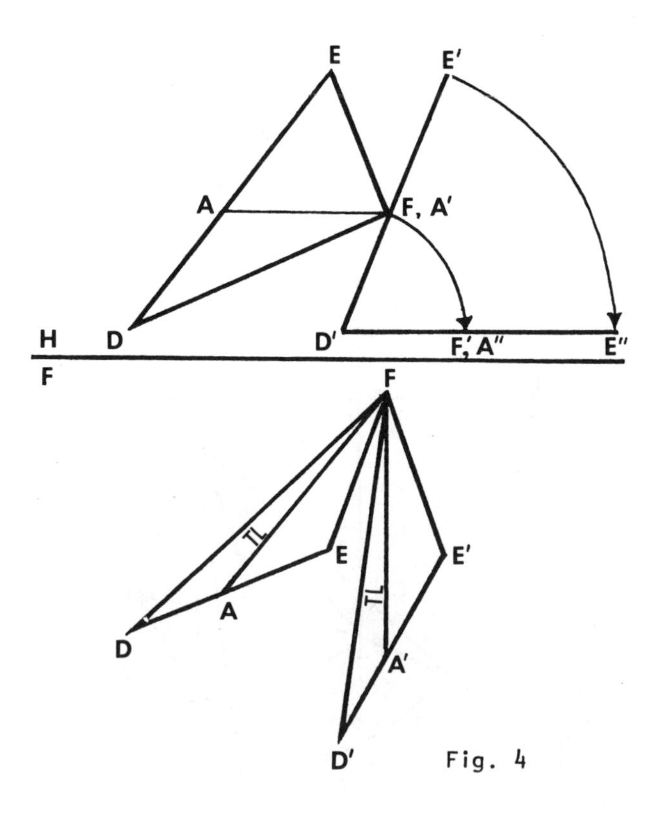
E
E′
A
F, A′
H
D
D′
F,′ A″
E″
F
F
TL
E
E′
TL
A
D
A′
D′
Fig. 4

top view, which projects true length in the front view. With a true length line in the plane DEF, the edge view can be constructed, from which the true size and shape view can be drawn. The edge view will appear in the top view if the true length line in the front view is perpendicular to the reference plane, H-F. Revolve line AF about an axis through point F in the front view to A'F, which is vertical. As AF was revolved θ degrees, revolve EF and DF θ degrees also, to E'F and D'F, and complete the new view of plane DEF. Project points D' and E' to the top view where they intersect horizontal projection lines from D and E, respectively. Construct line D'FE', which is the edge view of plane DEF, as seen in fig. 3. The true size and shape view of the plane will appear in the front view if the edge view appears parallel to plane H-F in the top view. As in fig. 4, revolve the edge view

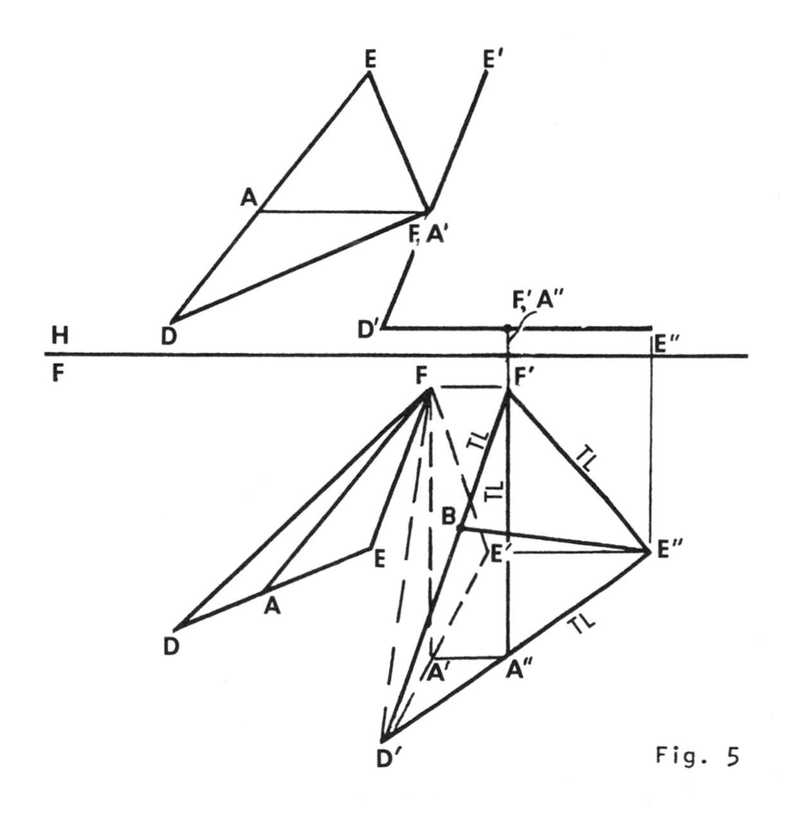

Fig. 5

of plane DEF about an axis through point D' to D'F'E", which is parallel to plane H-F. Project points F' and E" to the front view where they intersect horizontal projection lines from F and E', respectively. Point D' remains in the same location in the front view, so draw D'F', D'E", and F'E" to complete the true size and shape view, as seen in fig. 5. Construct the bisector of angle DE"F' from E" to D'F'. Line BE" is the required angle bisector, as shown in fig. 5.

Draw the views of the given lavatory handle after it has been revolved 100° counterclockwise.

Solution: Fig. 1 shows the top and front view of the

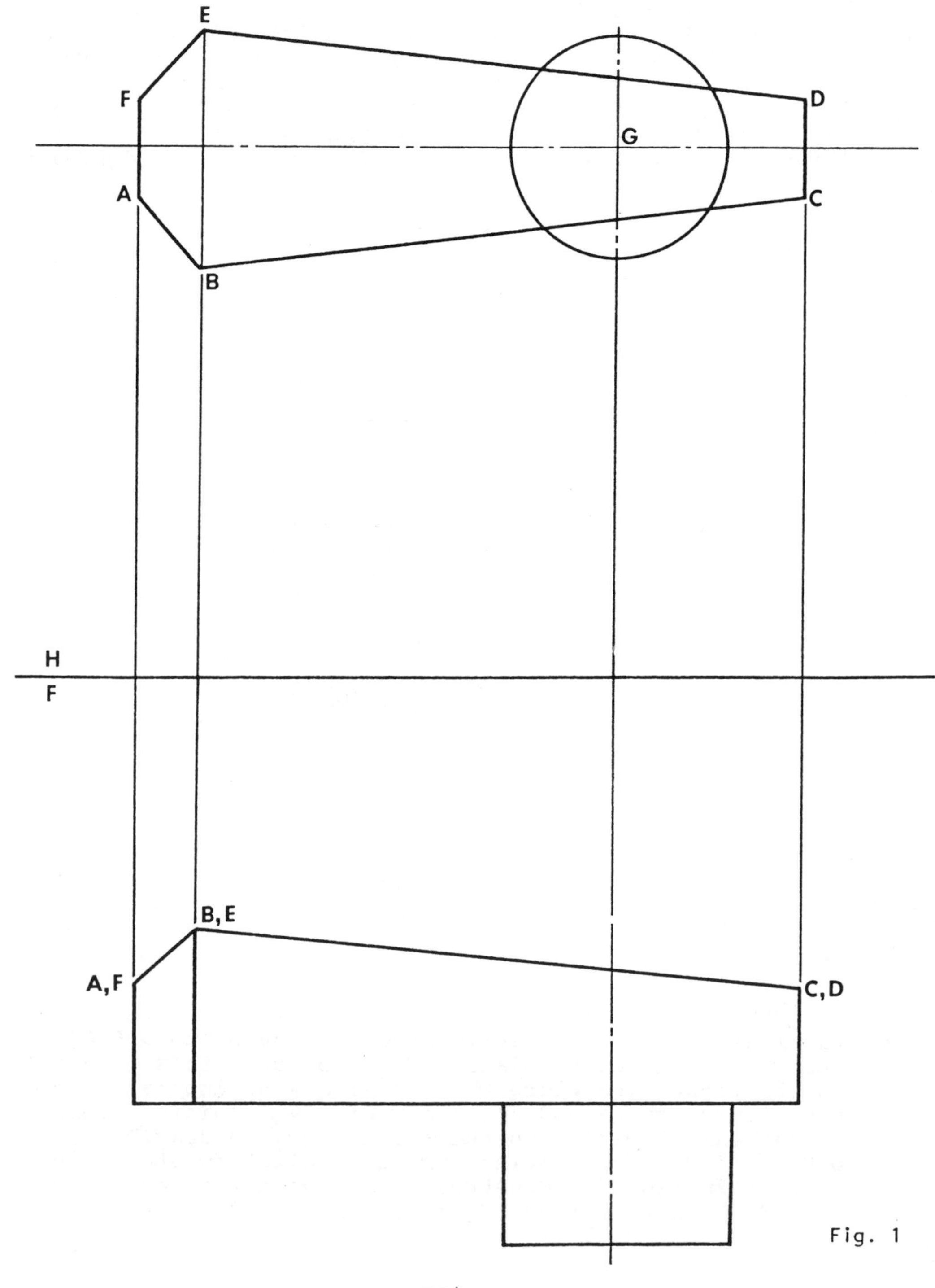

Fig. 1

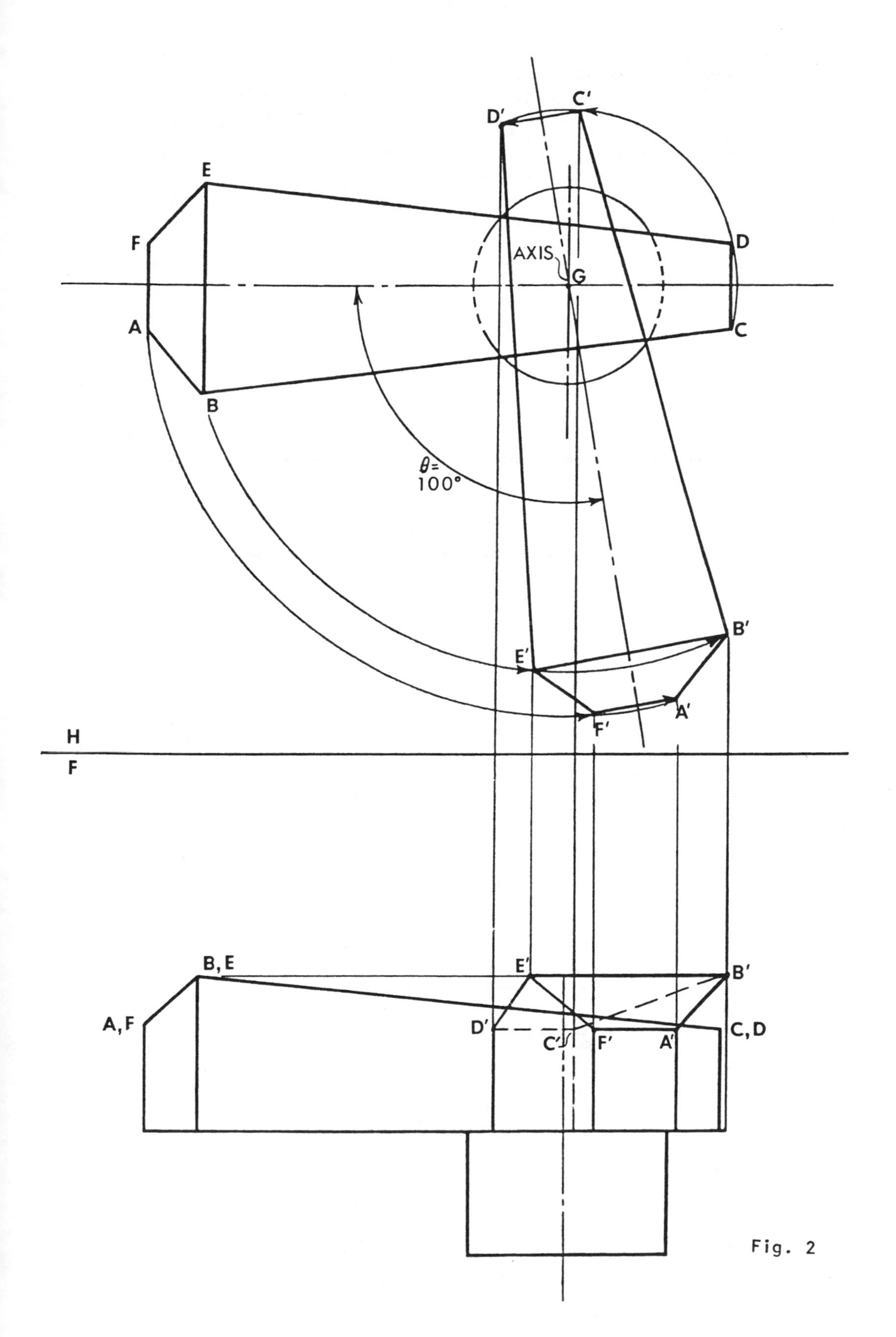

Fig. 2

given lavatory handle, which pivots about an axis through the center of the stationary base. As shown in fig. 2, revolve the limiting points of the handle, points A, B, C, D, E, and F, about an axis through point G, 100° in a counterclockwise direction to points A' through F', respectively. Connect the corresponding points to complete the top view of the revolved handle. Project points A' through F' to the front view to where they intersect horizontal projection lines from points A through F, respectively, and connect corresponding points and determine visibility for the completed front view of the revolved handle.

• PROBLEM 9-28

A crack in a given cylinder appears as a straight line in the top view. Construct the front view of the crack after the cylinder has been revolved 90° clockwise.

Solution: Fig. 1 shows the given cylinder in the top and front views. The line 1-9 in the top view represents

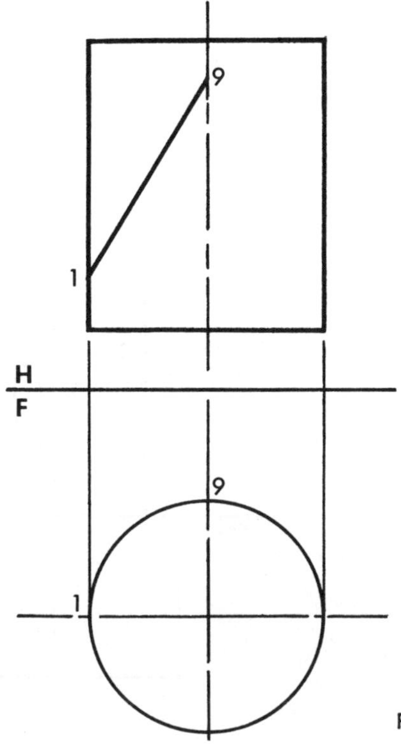

Fig. 1

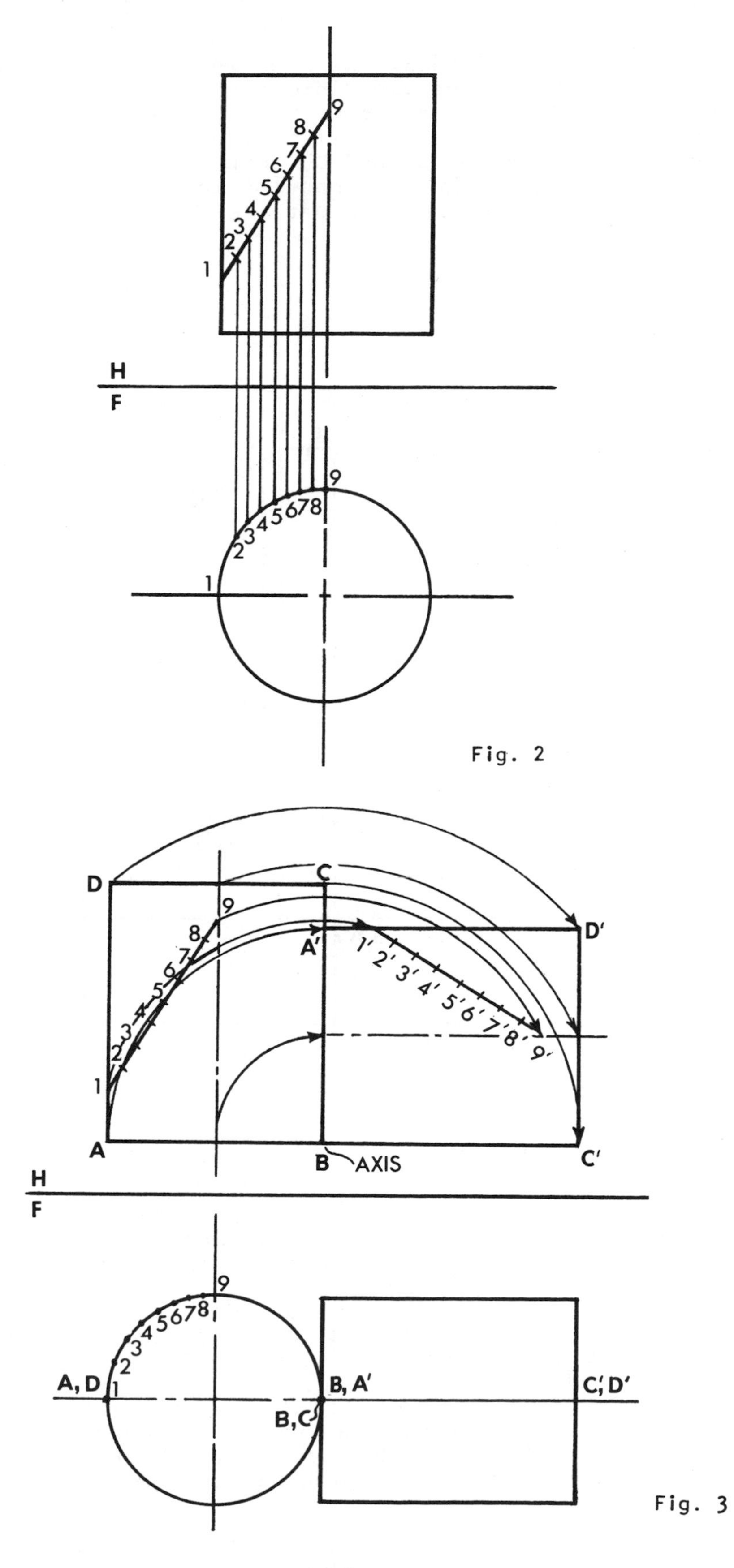

Fig. 2

Fig. 3

the crack in the cylinder. As seen in fig. 2, divide the line 1-9 in the top view into eight portions, and project points 2-8 to the front view to where they lie on the circle. Revolve the limiting points of the cylinder, points A, B, C, and D, about an axis through point B, 90° in a clockwise direction in the top view. Edge BC, which is now BC', is parallel to reference

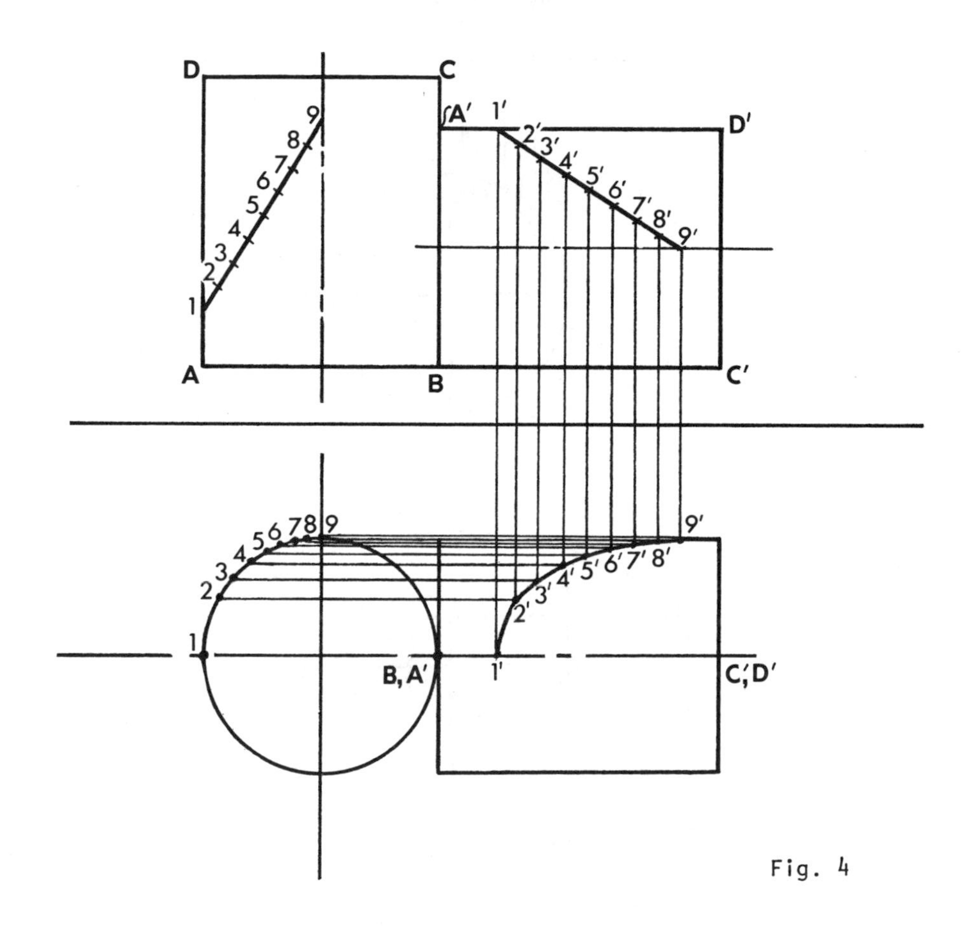

Fig. 4

plane H-F as seen in fig. 3. Revolve points 1 through 9 on the crack 90° to their respective positions on the revolved cylinder. Project points A', B, C', and D' to the front view to where they intersect horizontal projection lines from points A, B, C, and D, correspondingly, and construct the revolved front view of the cylinder. As illustrated in fig. 4, project points 1' through 9' to the front view to where they intersect horizontal projection lines from points 1 through 9, respectively. Connect points 1' through 9' with a smooth curve for the required front view of the crack on the revolved cylinder.

• PROBLEM 9-29

Two fire doors opening from a corridor into a stairwell are shown in the given plan and partial elevation views. If the door labeled ABCD is opened 110 degrees, will the door labeled EFGH clear the closer of door ABCD? If not, find the number of degrees door EFGH can swing through before striking the closer.

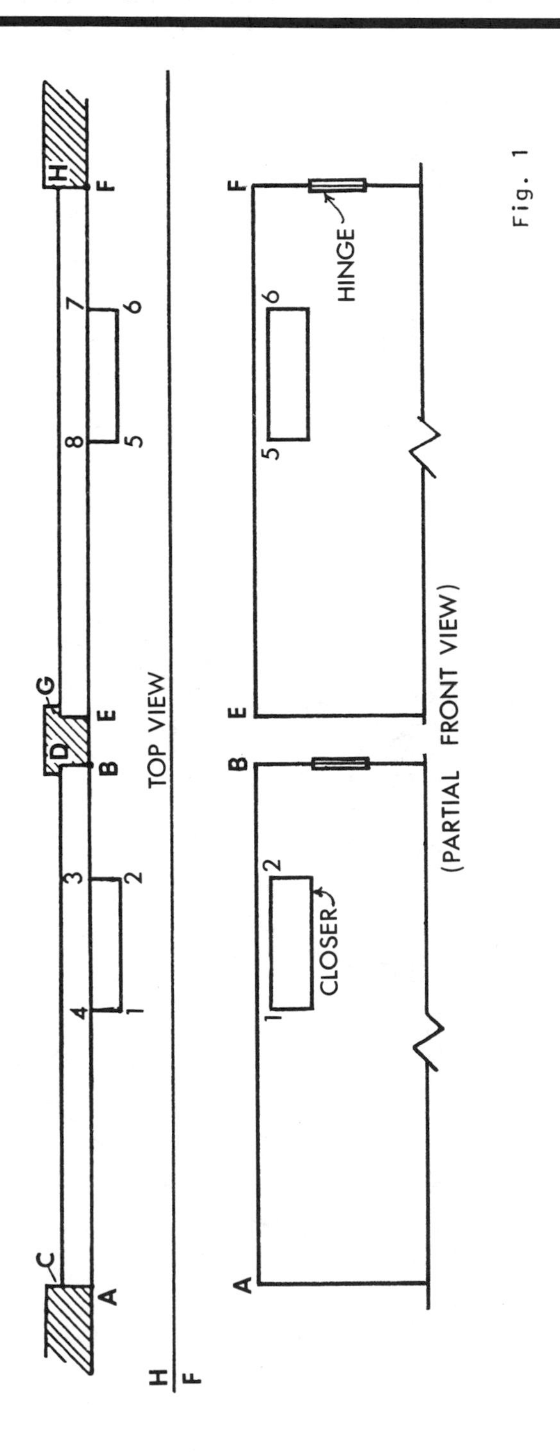

Solution: Fig. 1 shows the two given doors in the plan and partial elevation views. Note that only the plan view is needed to solve the problem. As in fig. 2, revolve point A 110 degrees about an axis through B, which in this case is the hinge of the door, to point A'. Revolve points C and D and points 1 through 4 110 degrees about the same axis, and connect appropriately to show the new position of door ABCD. As in fig. 3, revolve line EF about an axis through point F to determine whether or not door EFGH will clear the closer of door

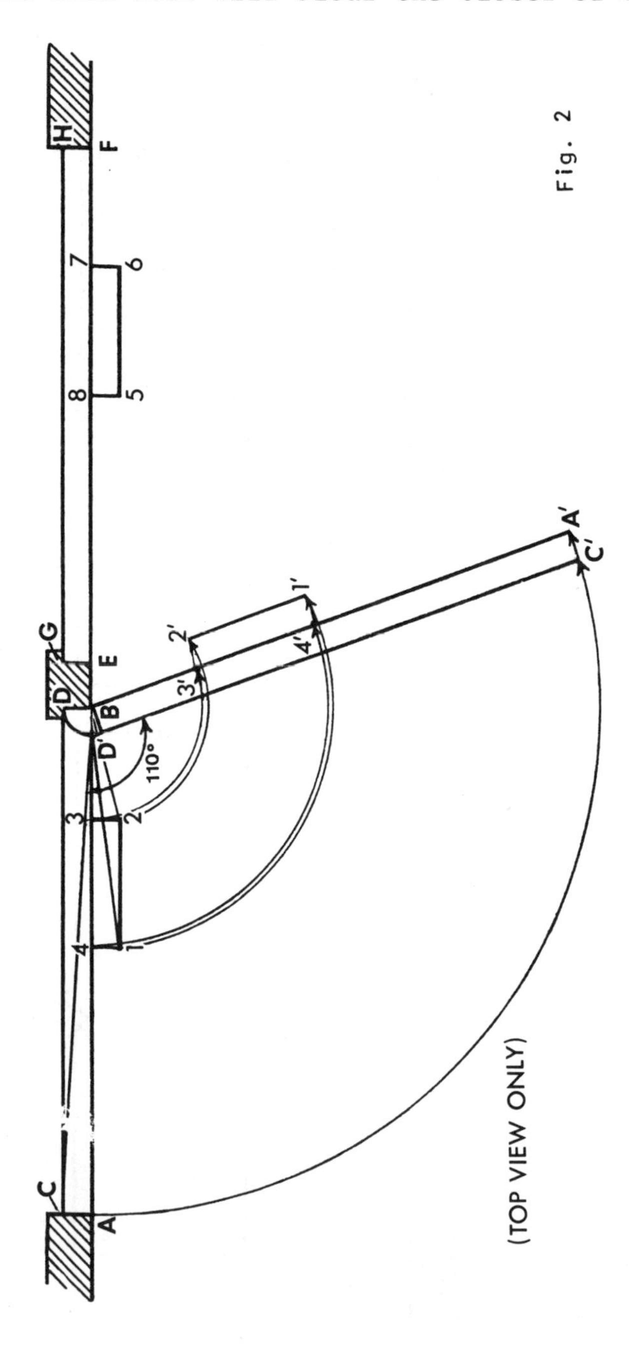

Fig. 2

ABCD. As seen in fig. 3, line EF intersects line 2'3' of the closer of door ABCD, and it is concluded that door EFGH does not clear the closer of door ABCD. Measure the angle formed by line EF and line E'F, labeled θ in fig. 3 and which is 10.5 degrees in this case, and revolve the remaining lines of door EFGH through the same number of degrees and about the same axis to show the new position of door EFGH.

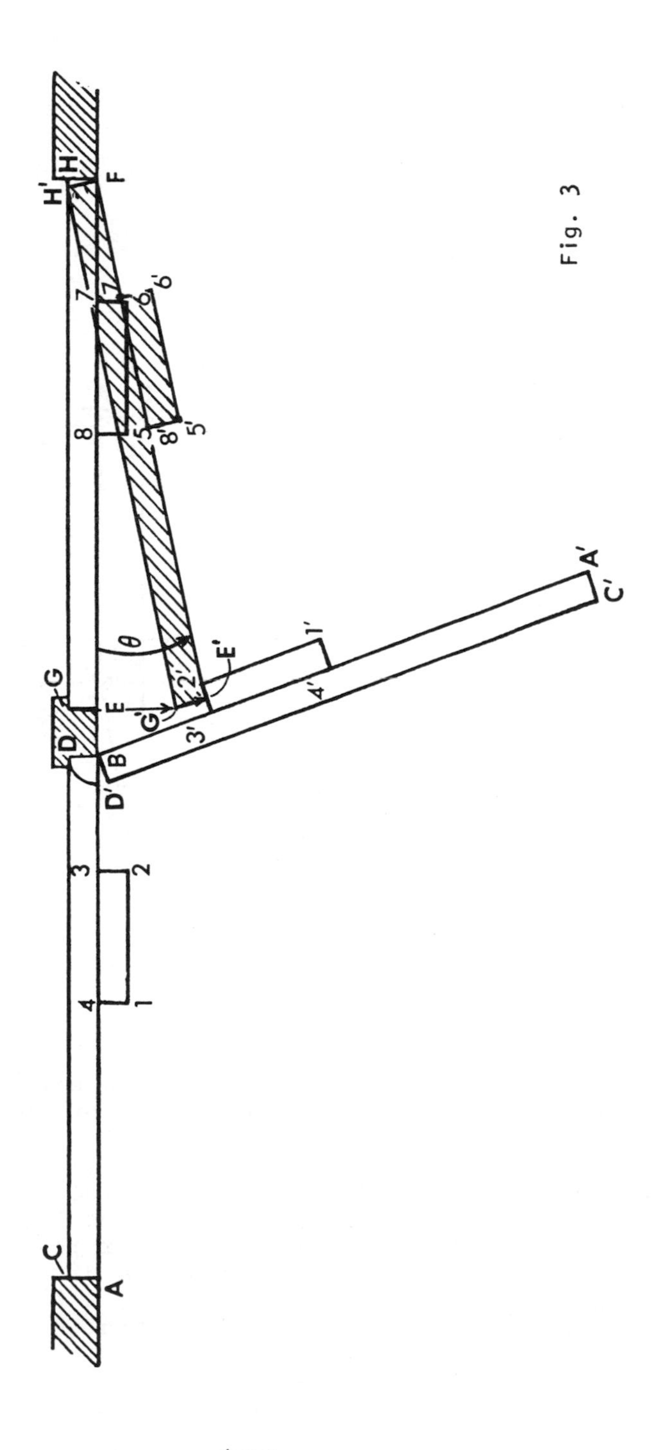

Fig. 3

Rod BC is attached to a post at C by a ball-and-socket joint. If the cable AB is kept taut while the rod is rotated southward, find the bearing of BC when B is a specified distance, L, lower than A.

Solution: Fig. 1 shows the top and front views of rod BC and cable AB. Rod BC and cable AB are true length in the front view as both appear parallel to reference plane H-F in the top view. The directional arrow, D, indicates north.

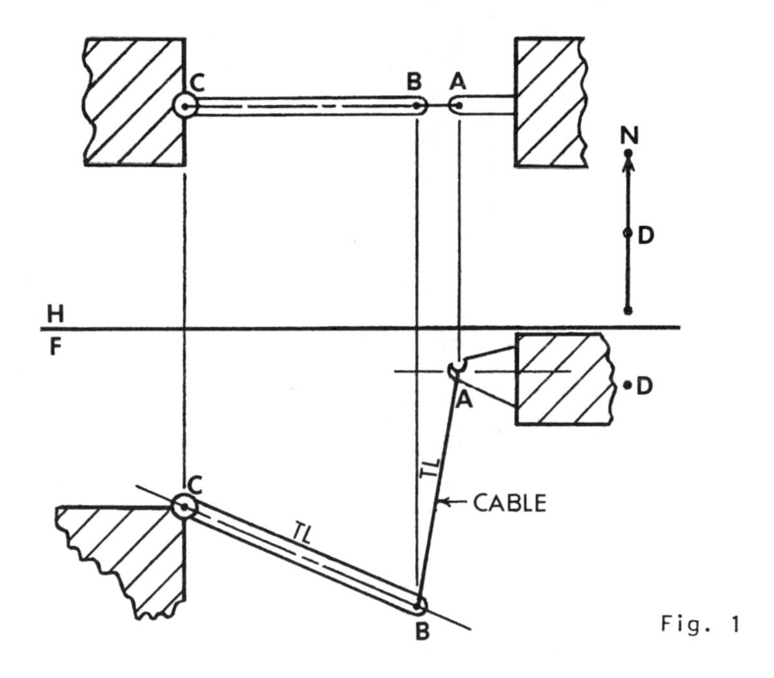

Fig. 1

Since both the rod and the cable are to remain the same length, they will be rotated about an axis through points A and C, as seen in fig. 2. The path of revolution of BC and AB can be seen in a view in which the axis appears as a point. With the line of sight parallel to the axis AC and the reference plane F-1 perpendicular to AC, construct the circular path of revolution in the primary auxiliary view. Project the directional arrow to view 1, also. As in fig. 3, draw a horizontal line the specified distance, L, below point A in the front view. Project to the front view from view 1 the path of revolution, which appears as an edge in the front view. The point where the edge view of the circle of revolution intersects the horizontal line a distance L below point A, B', locates the revolved position of point

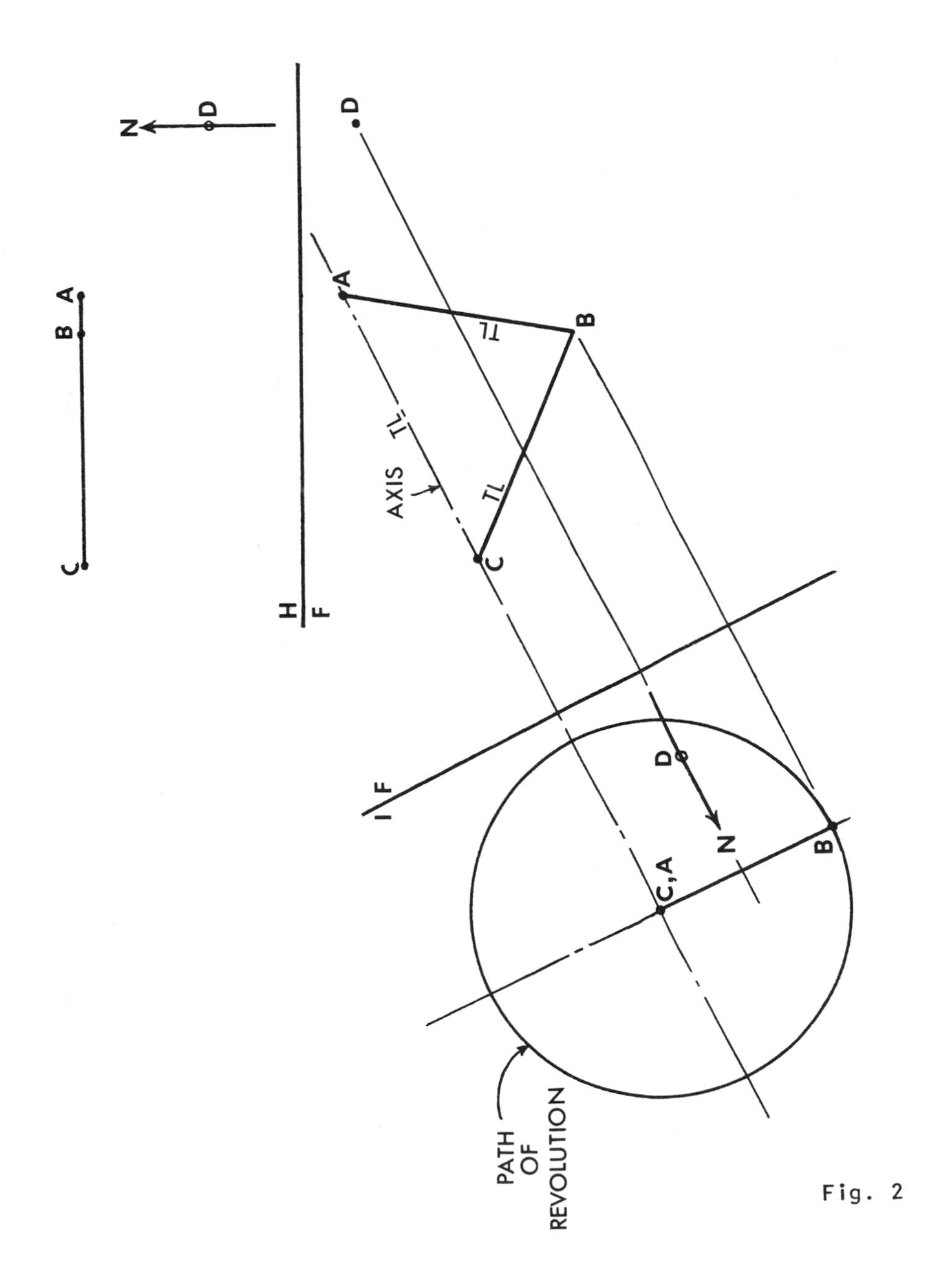

Fig. 2

B in a southward direction. Draw CB' and AB', as shown in fig. 3. Project point B' to the top view at a distance "S" from the plane H-F, since B' is a distance "S" from the plane F-1 in view 1, as illustrated in fig. 4. Construct CB' and AB' in the top view and an arrow on the

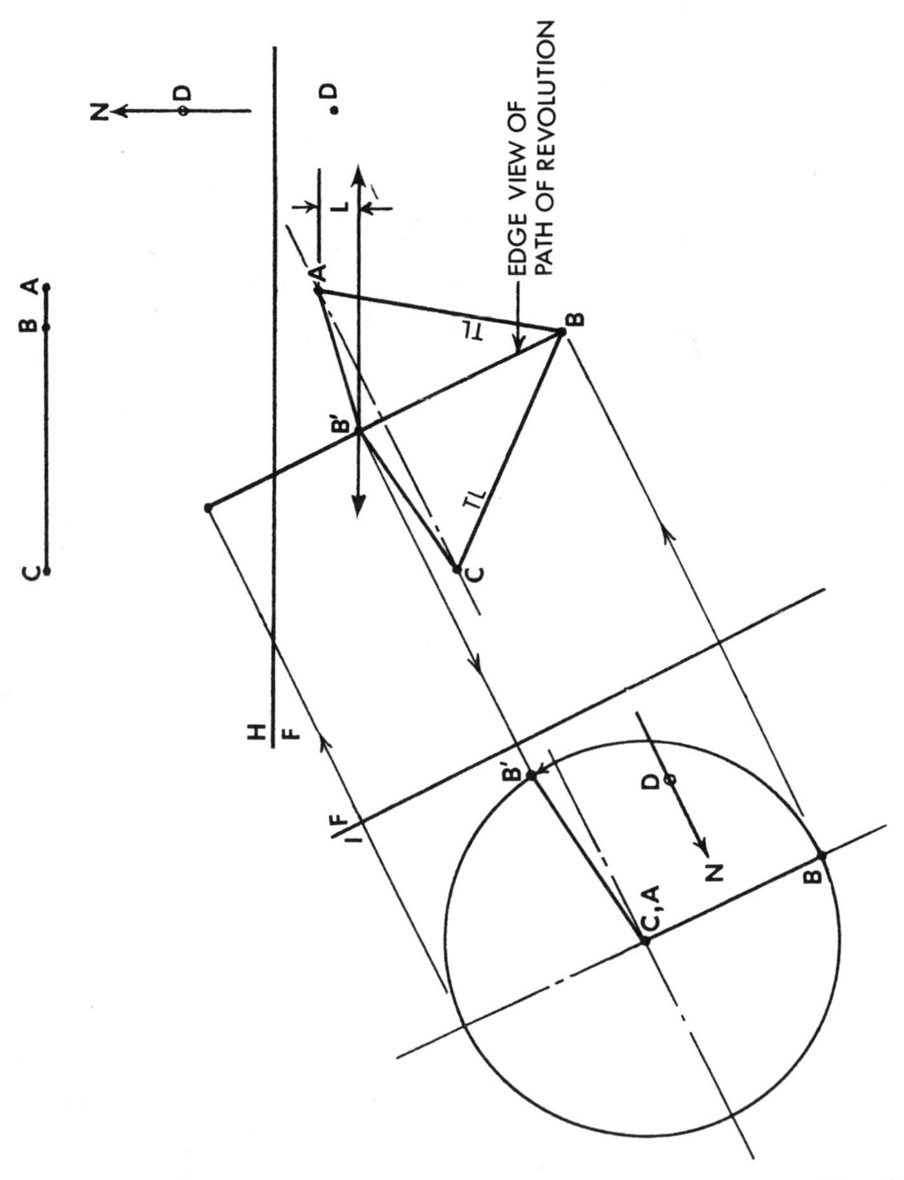

Fig. 3

other side of CB' parallel to the directional arrow D. This second arrow is simply drawn for convenience and to avoid overlapping. Extend CB' to the second arrow. The

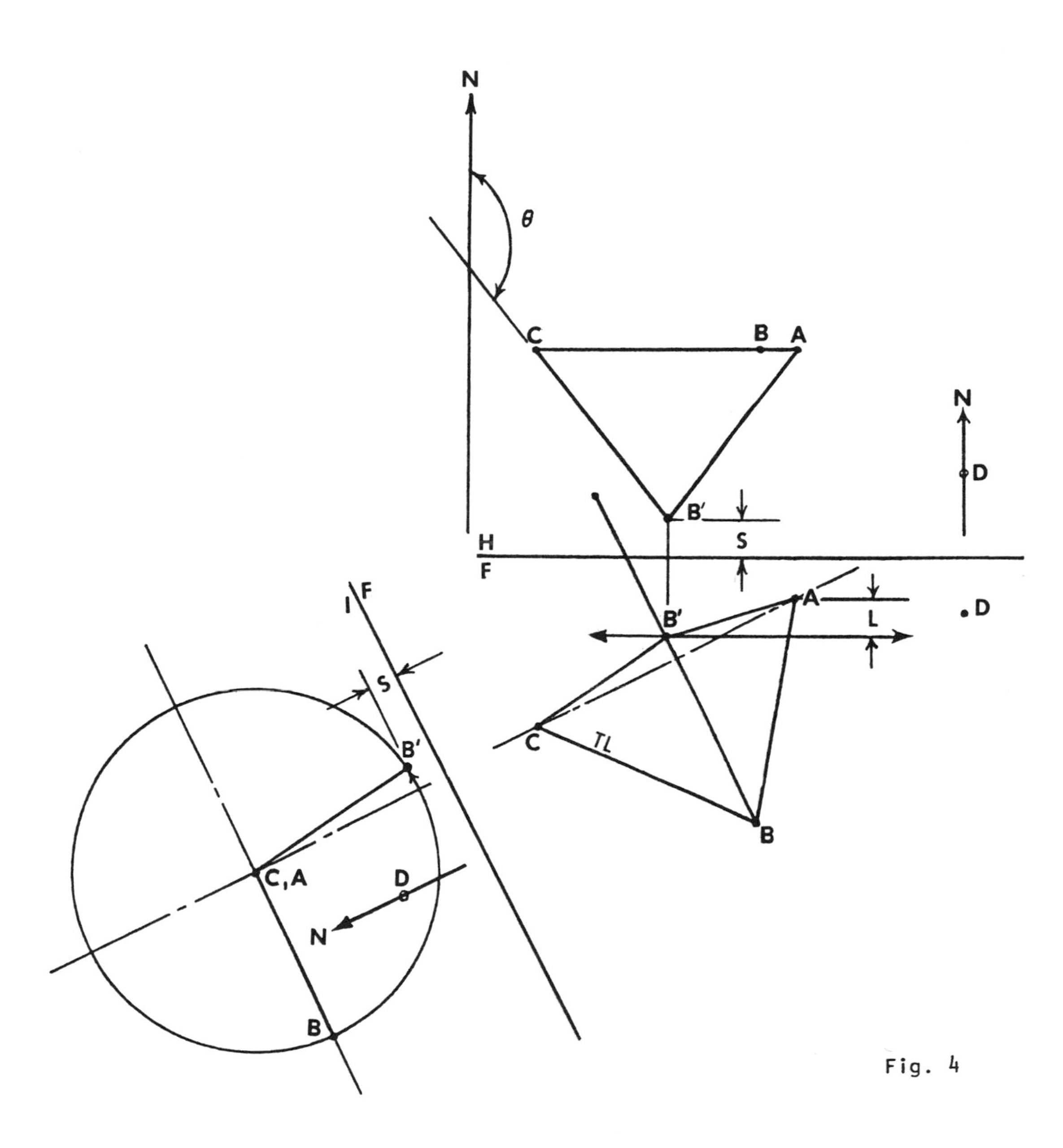

Fig. 4

bearing of CB', which must be measured in the top view, is the angle θ between north and the line CB' extended, as seen in fig. 4.

A lever of length "r" is perpendicular to the cable AC at point B. Find the maximum number of degrees, ϕ, through which the lever can revolve between the floor and the wall, and show the points at which the lever strikes the floor and the wall in both views.

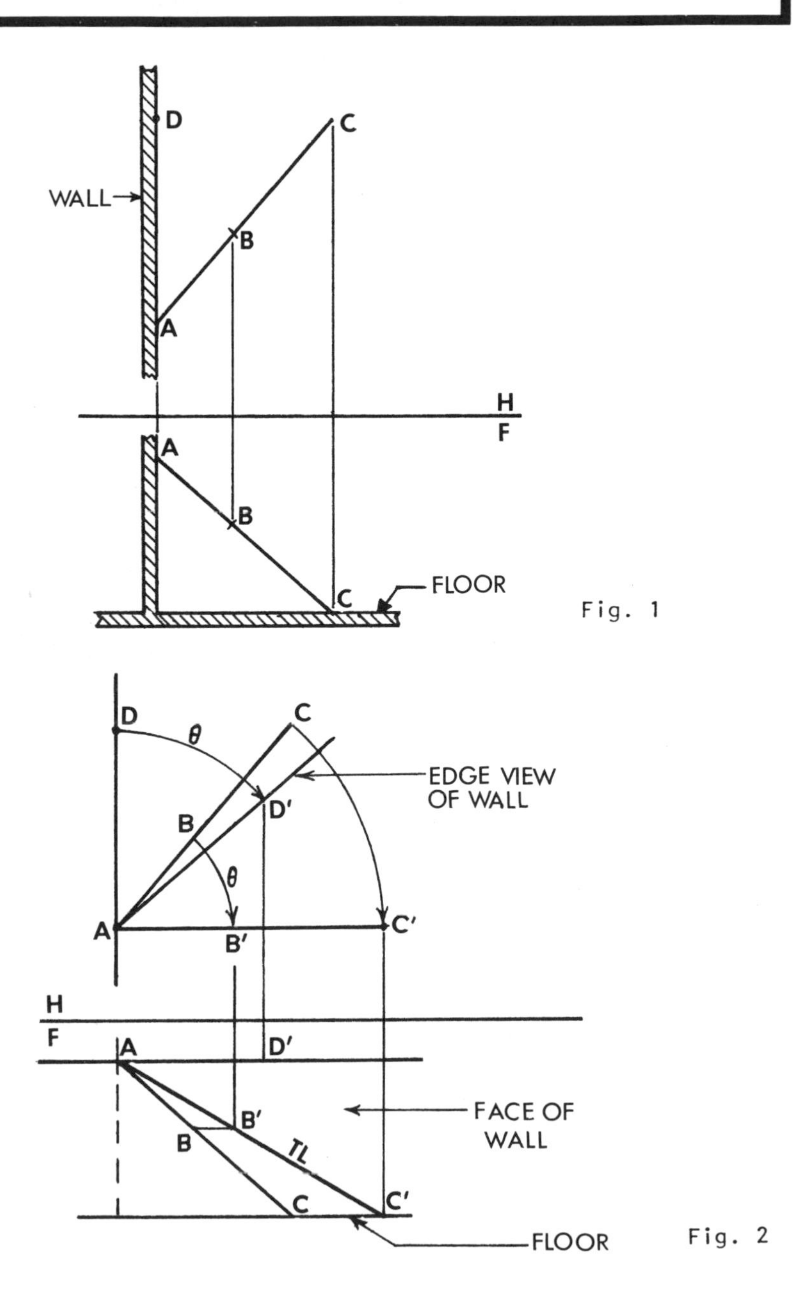

Solution: Fig. 1 shows the given cable, AC, between the wall and the floor, in the top and front views. In order to construct the lever perpendicular to the cable, the true length view of the cable is needed. As in fig. 2, revolve the cable, line AC, about an axis through point A to AC', which is parallel to the reference plane H-F in the top view. As line AC was revolved through θ degrees, revolve line AD, which represents the wall, through θ degrees, also, to D'. Project AC' and AD' to the front view where AC' appears true length. The face of the wall now appears in the front view, just as the face of the floor appears in the top view. Since the lever can revolve about the cable, construct the primary auxiliary

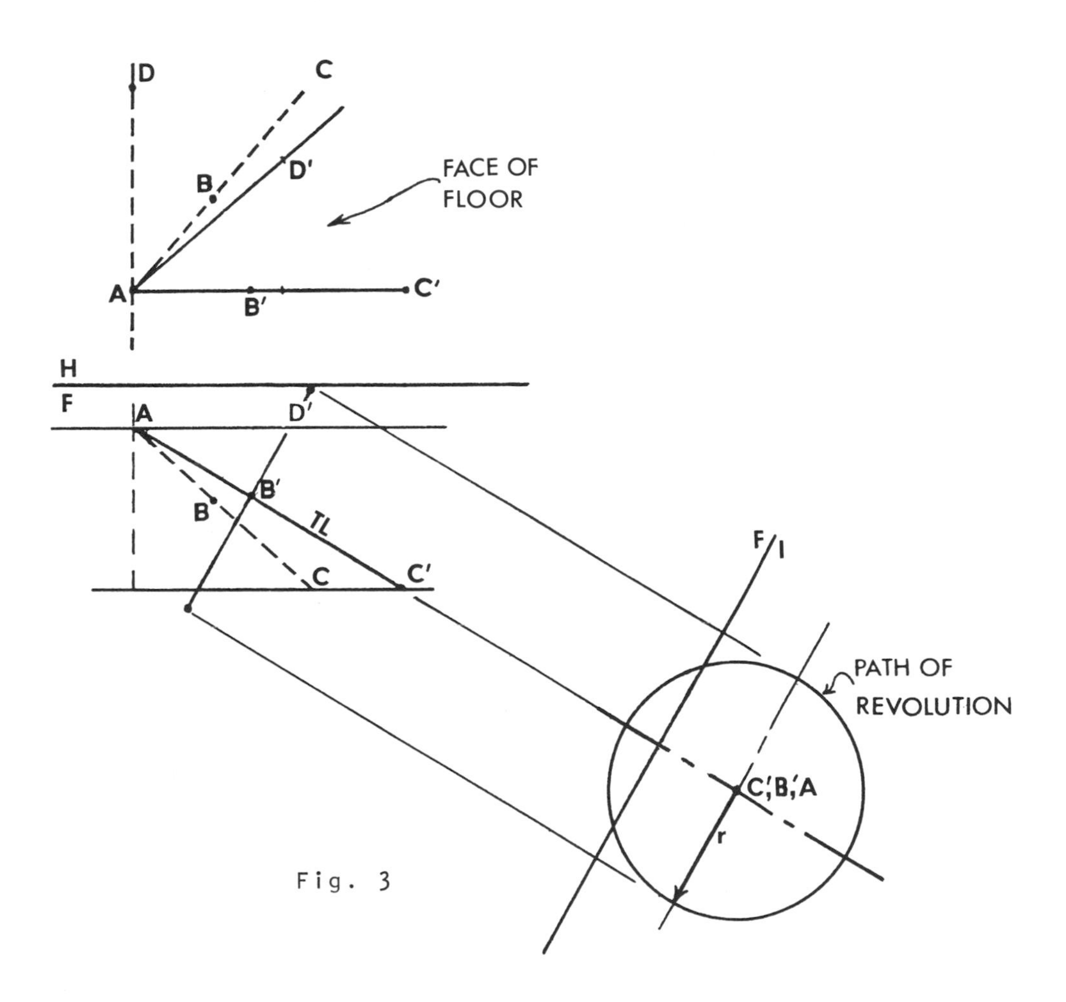

Fig. 3

view, view 1, to show the path of revolution. With the line of sight parallel to the true length line AC' in the front view, make the reference plane H-F parallel to AC', and project AC' to view 1 as a point, as in fig. 3. Construct a circle of radius "r", equal to the length of the lever, centered at the point view of the cable AC'. Project the circle back to the front view, where it appears as an edge. As seen in fig. 4 in the front view, the lever strikes the floor at point E on the path of revolution. Project E to view 1 on the circle, where it is located at a distance "p" from the plane F-1. Project point E to the top view at a distance "p" from the plane H-F, and draw B'E, as shown in fig. 4. To locate the point at which the lever strikes the wall, construct the

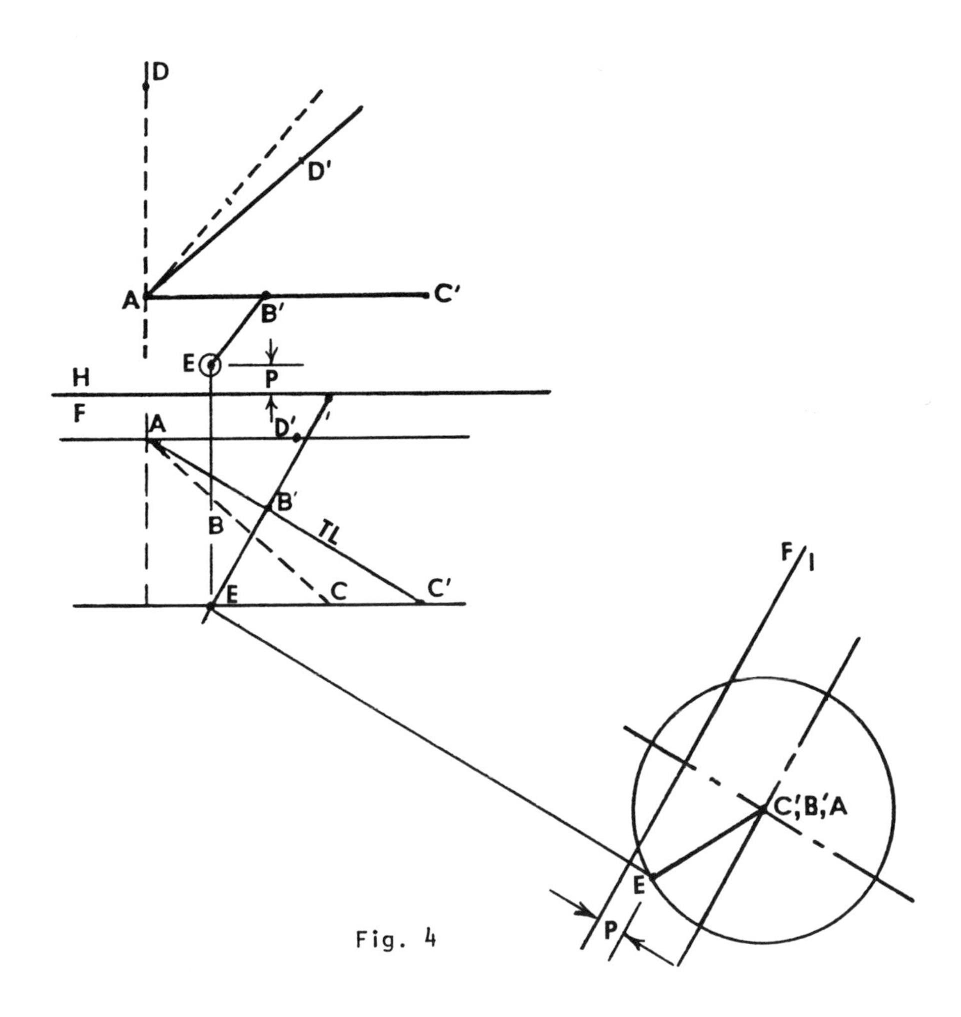

Fig. 4

partial view of the circle of revolution in the top view, where it appears as an ellipse, as in fig. 5. The ellipse has a major diameter through B' equal to the diameter of the circle, and it has a minor diameter through B' projected from the edge view of the circle in the front view. The ellipse intersects the edge view of the wall in the top view at points F and G. Project both points to the front view to where they lie on the edge view of the circle of revolution. Point F cannot be the desired striking point of the lever because it lies below the floor in the front view. Point G, on the face of the wall in the front view is therefore the correct striking point. Project point G to the auxiliary view, as in fig. 5. The maximum angle through which the lever can turn between the wall and the floor is labeled ϕ in view 1 of fig. 5. Note that the angle cannot be $360°-\phi$ because that angle contains the portion of the path of revolution which goes behind the wall.

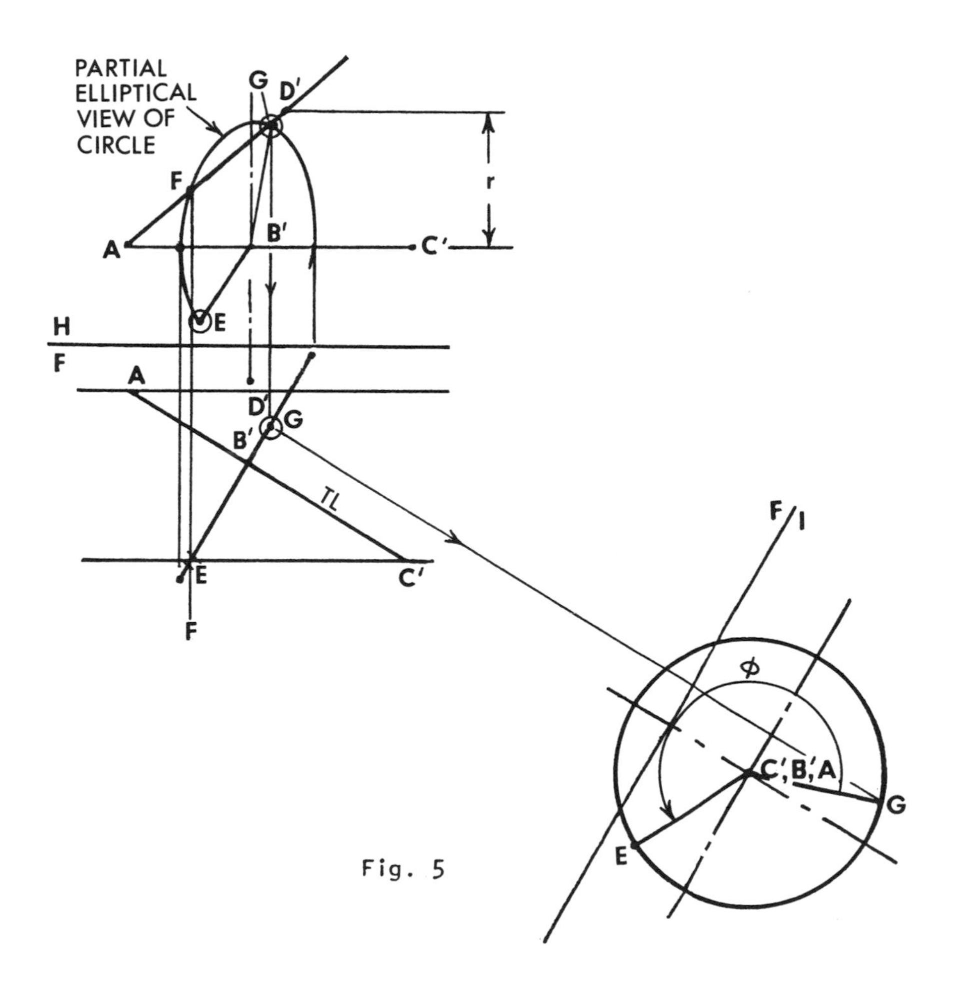

Fig. 5

CHAPTER 10

PICTORIAL VIEWS

ISOMETRIC DRAWINGS

• PROBLEM 10-1

Construct the isometric drawing of the jig body, given the top, front, and left side views.

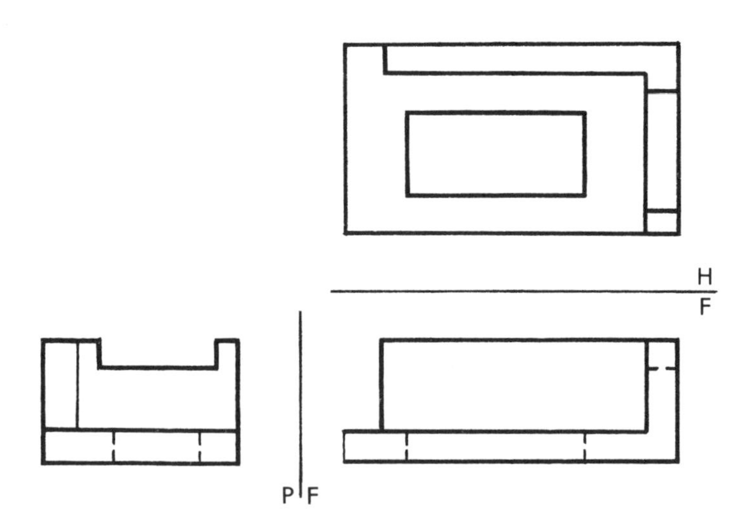

Fig. 1

Solution: Fig. 1 shows the given views of the jig body. To draw an isometric view of an object, the orthographic views are first enclosed in rectangles to obtain the overall dimensions for length, width, and height. As the top view is already in a complete rectangle, enclose the front and left side views in rectangles also, as shown in fig. 2. Now the isometric box in which the jig body will be drawn can be constructed. In all isometric drawings, the height remains vertical, the width recedes on an axis which makes a 30° angle with the horizontal, and the depth recedes on another axis which also makes a 30° angle with the horizontal. As illustrated in fig. 2, construct the isometric box whose overall dimensions are the total height, width, and depth of the jig body.

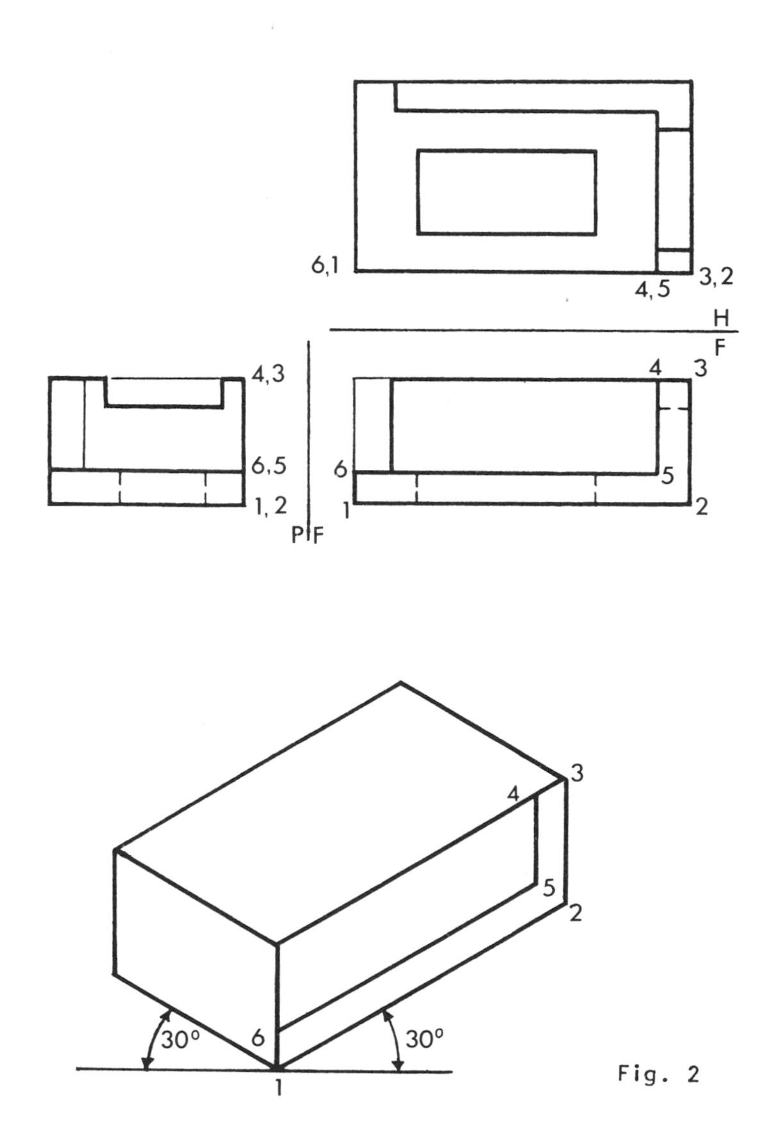

Fig. 2

These dimensions can be transferred directly onto the isometric axes from the orthographic views of the jig body using dividers.

Each line in an orthographic view represents a plane that composes a portion of an object. The plane labeled 1-2-3-4-5-6 in the orthographic views in fig. 2 is drawn in the box, transferring measurements with dividers. Measure heights 1-6 and 2-3, and place them on the corresponding vertical axes of the isometric box.

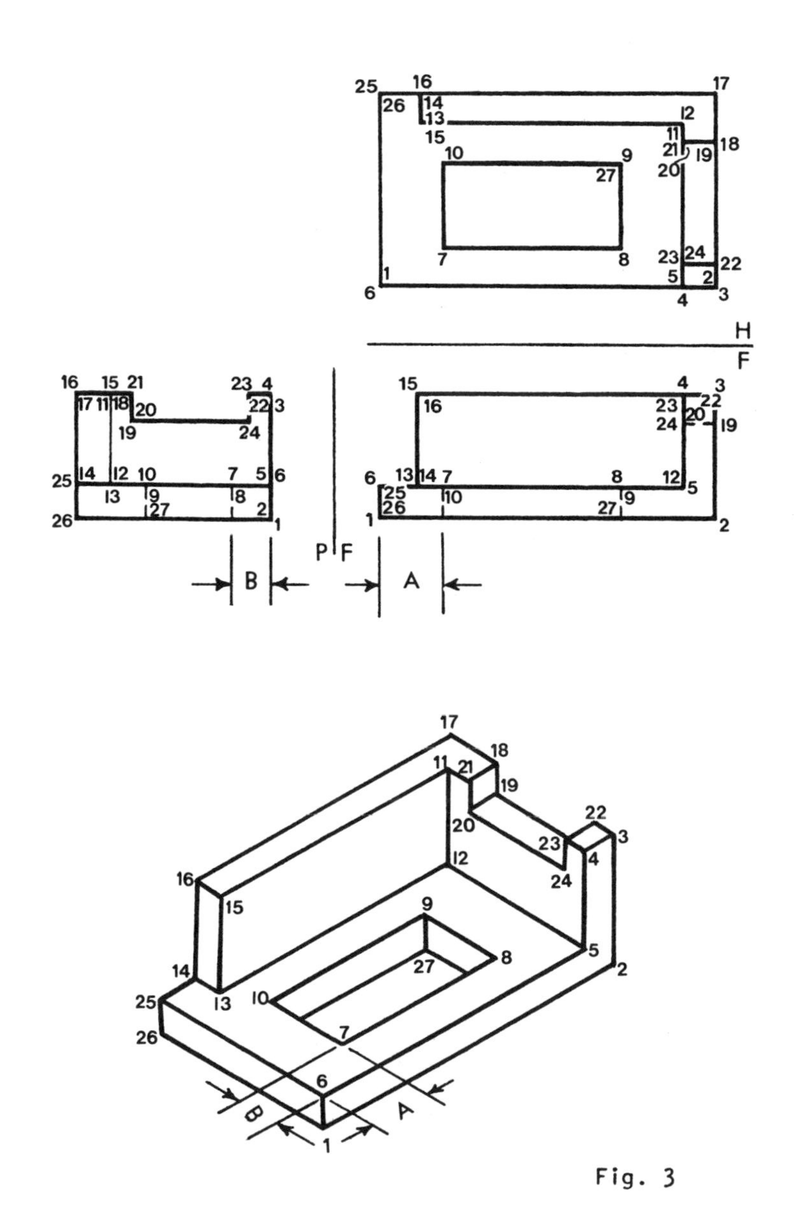

Fig. 3

Measure depths 6-5, 1-2, and 4-3, and place them on the corresponding receding 30° axis. Continue to locate the remaining planes in the same manner until all the visible planes have been transferred to the isometric box. Darken the principal lines and erase the guidelines used to complete the required isometric drawing of the jig body, as seen in fig. 3. Note that in isometric drawings hidden lines are not usually shown.

• PROBLEM 10-2

Construct the isometric drawing of the miter box, given the top, front, and right side views.

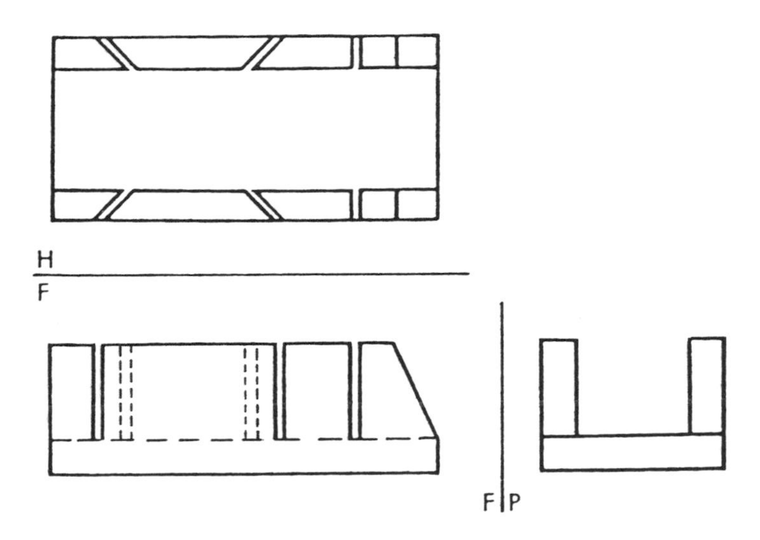

Fig. 1

Solution: Fig. 1 shows the given views of the miter box. Construct the isometric box by enclosing each of the orthographic views in rectangles and transferring overall height, width, and depth dimensions to the isometric axes with dividers, as in fig. 2. Straight lines that are not parallel to any of the isometric axes are located simply by finding their endpoints and connecting them with lines. These non-isometric lines will not be true length in the isometric view. As shown in fig. 2, point D is found in the isometric box by transferring the distance F, the distance that D is from the front vertical edge of the box, with dividers. Plane A-B-C-D-E is then drawn after height BC has been transferred. Continue to transfer the planes composing the miter box to the isometric view until all but the angled cuts in the box have been drawn. As can be seen in fig. 3, once the basic skeleton of the miter box has been drawn, the angled cuts can be easily located by transferring distances from the orthographic views. Complete the required isometric view of the miter box, omitting hidden lines.

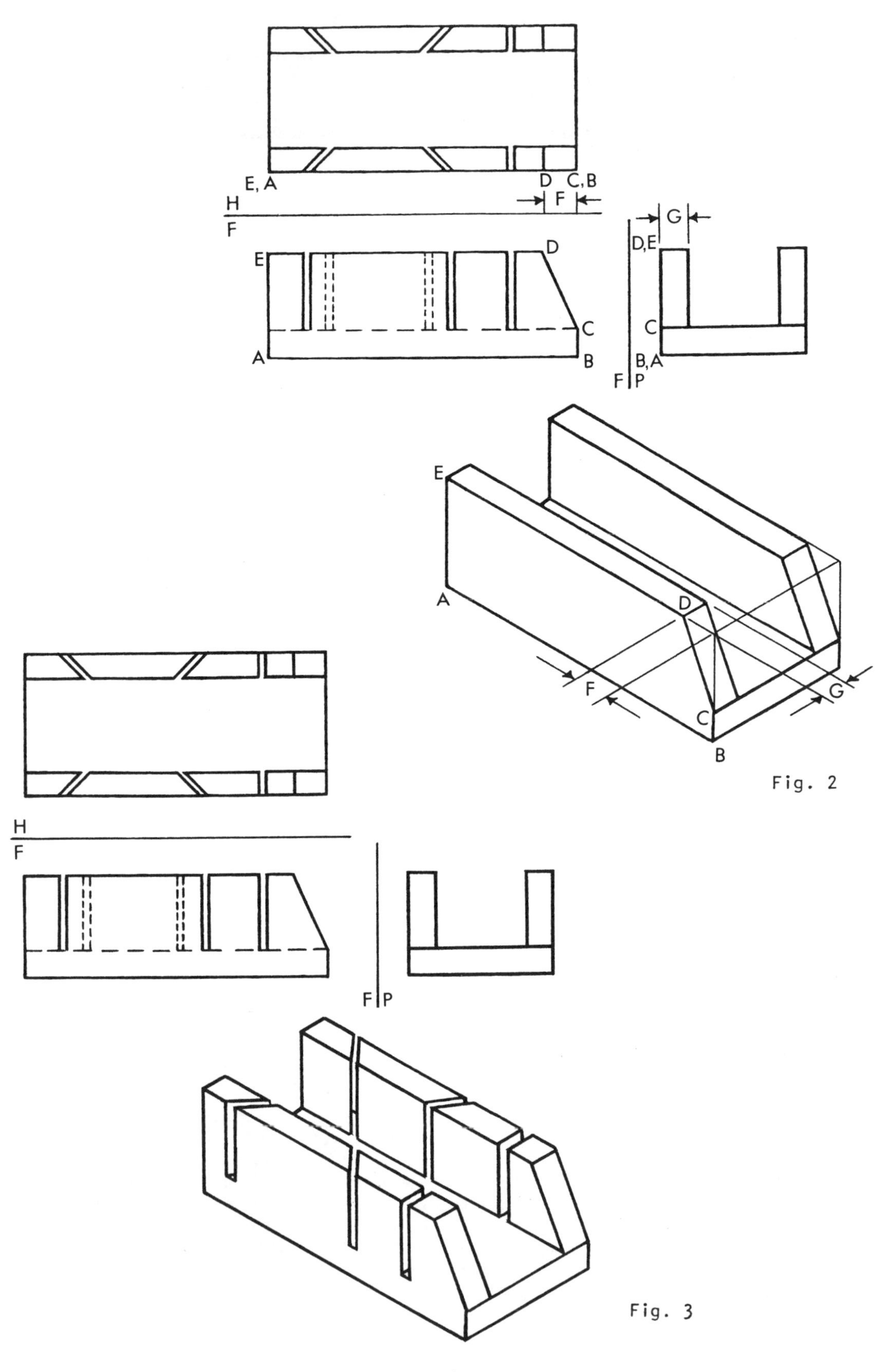

Fig. 2

Fig. 3

Construct the isometric drawing of the inclined guide, given the top, front, and right side views.

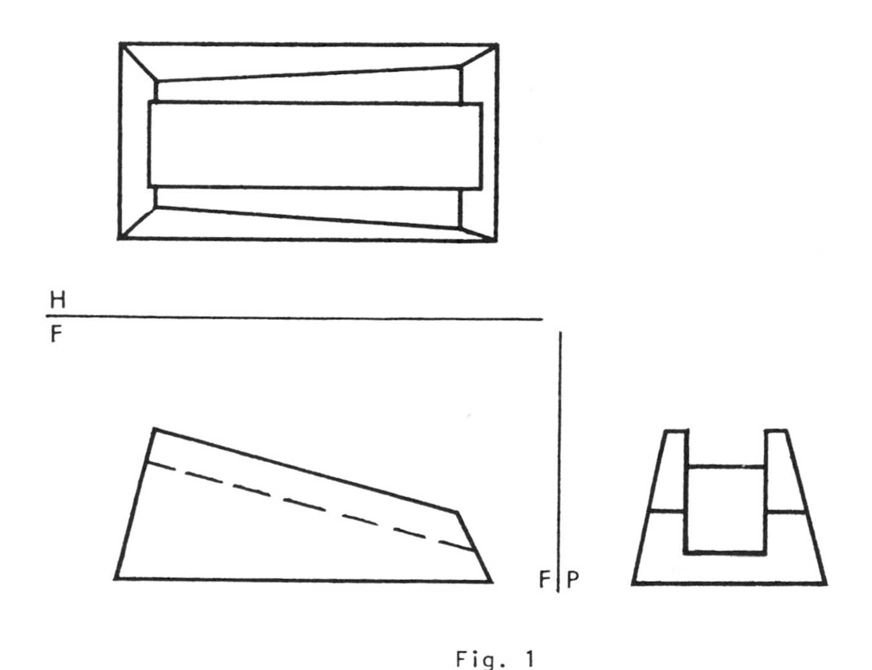

Fig. 1

Solution: Fig. 1 shows the given views of the inclined guide, which has corners that are not located on any isometric lines. The method used to find these corners in the isometric view is called the co-ordinate construction method. Essentially locating these non-isometric corners is like locating coordinates on the X, Y, and Z planes in mathematics. As shown in fig. 2, point 1 is found in the completed isometric box by measuring the distances it lies away from the axes and transferring the measurements to the isometric view. For point 1, these distances are labeled a, e, and h in the orthographic views and also in the isometric view. Points 2, 3, and 4 are also found in this manner. Connect these points to the corner points of the base to form the oblique plane on the surface of the inclined guide. Continue to locate remaining corner and end points and connect them with straight lines to complete the required isometric view of the guide, as shown in fig. 3.

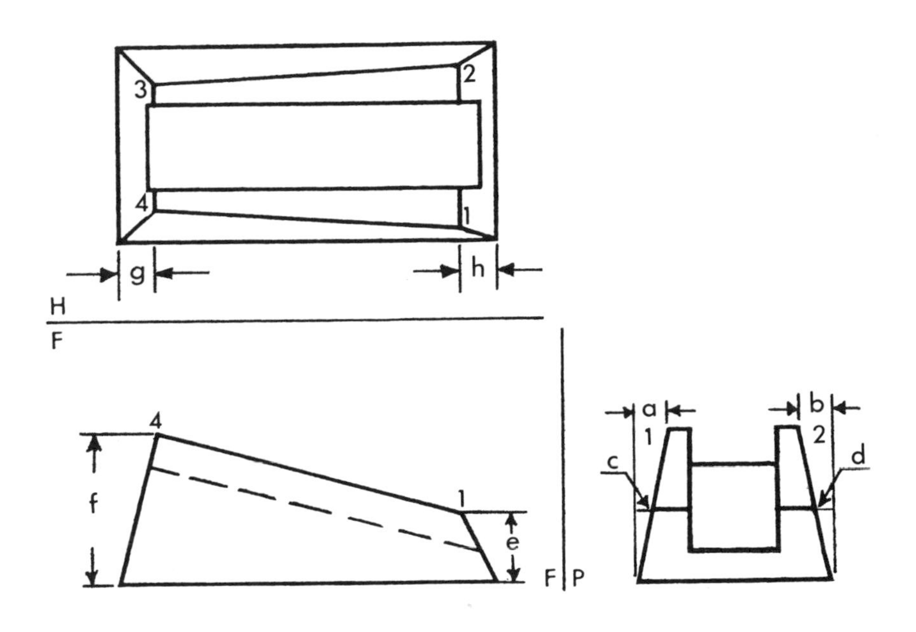

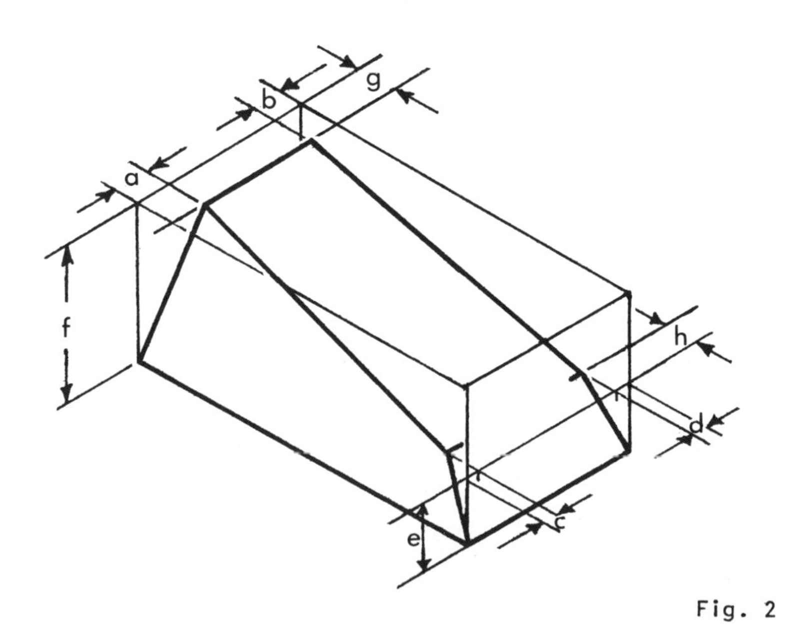

Fig. 2

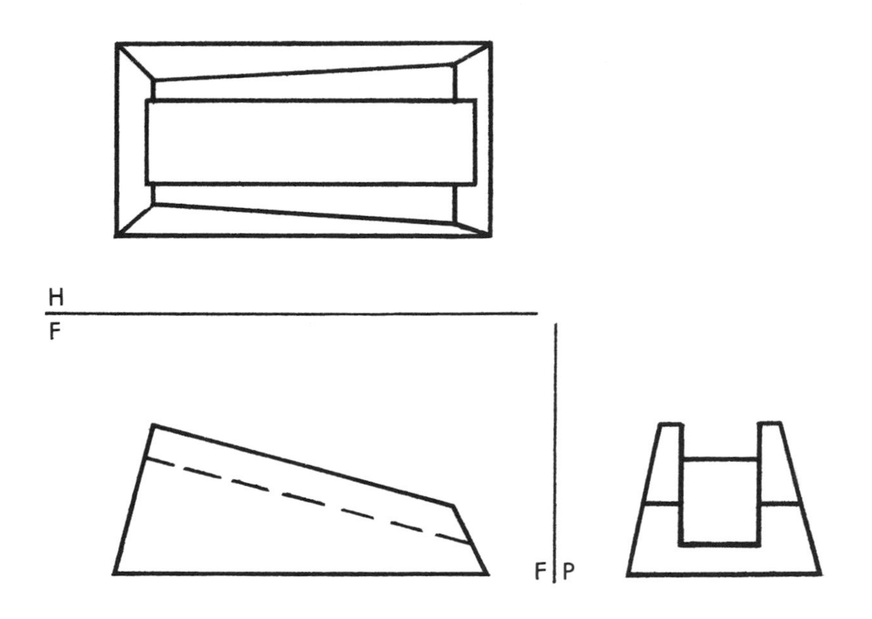

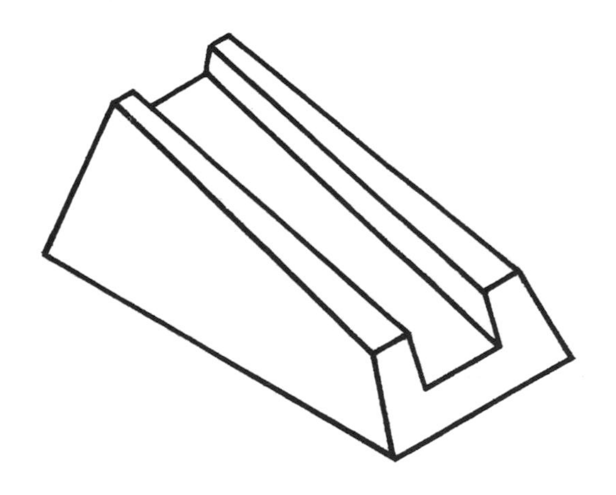

Fig. 3

• PROBLEM 10-4

Construct the isometric view of the support, given the top, front, and right side views.

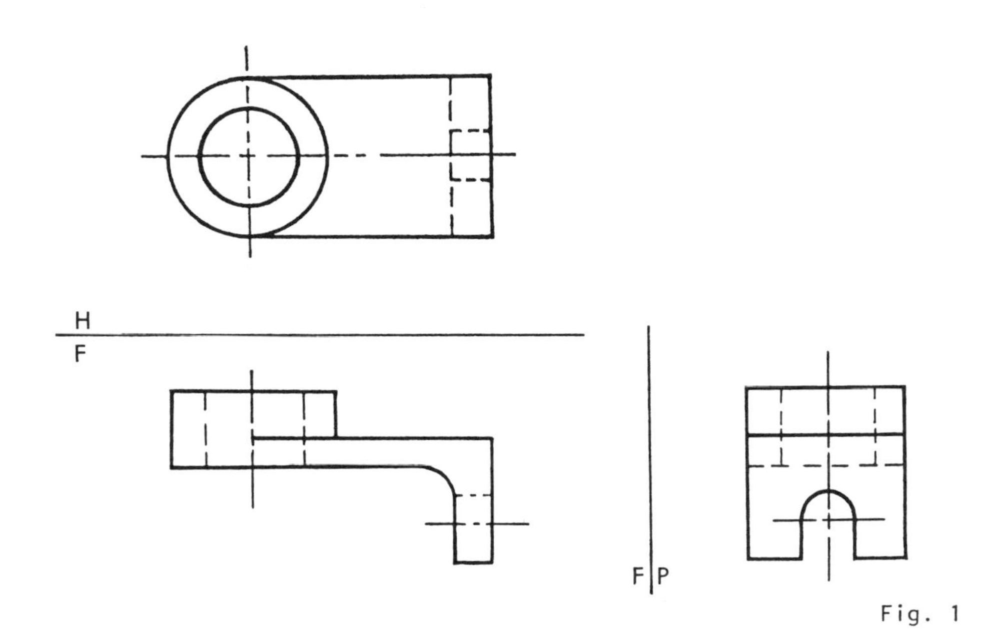

Fig. 1

Solution: Fig. 1 shows the given views of the support, which has circular planes. Circles appear in isometric views as ellipses, because the principal planes of projection are inclined. As in fig. 2, construct the isometric box, in which the support will be drawn, with overall dimensions equal to the total height, width, and depth of the support. In the orthographic views, enclose all circles and circular arcs in squares and construct these squares in the isometric box to determine the location of the ellipses to be drawn. Construct also the basic outline of each plane of the support. Using the four-center method, construct the ellipses and elliptical arcs in the designated boxes in the isometric view, as in fig. 3. Note that the longer diagonal of these boxes is the major diameter of the ellipse and the shorter diagonal is the minor diameter of the ellipse. Darken in the principal lines for the complete required view of the support, and omit any hidden lines.

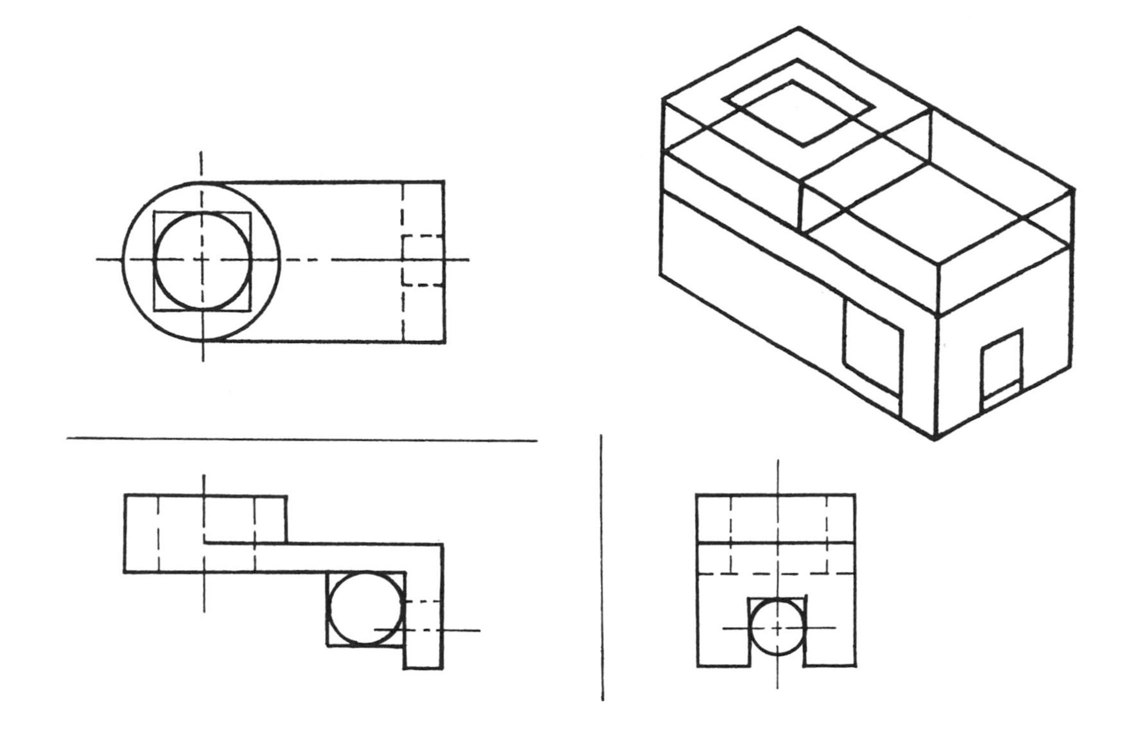

Fig. 2

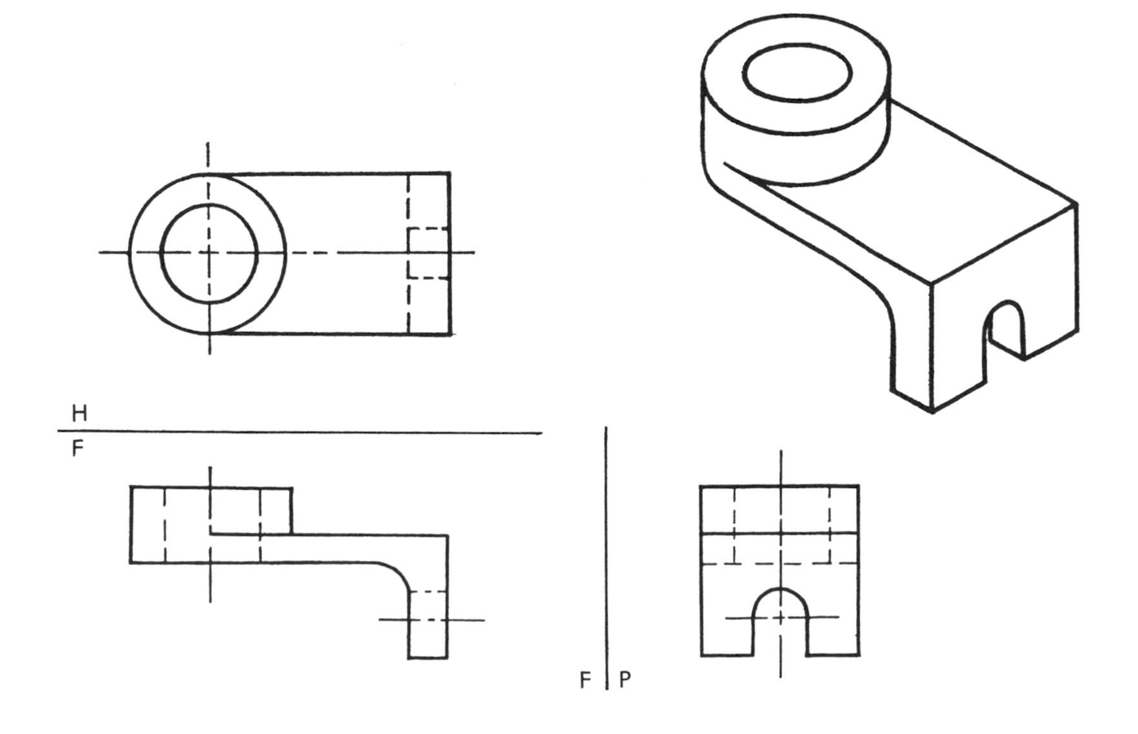

Fig. 3

• PROBLEM 10-5

Construct the isometric view of the rod guide, given the top and front views.

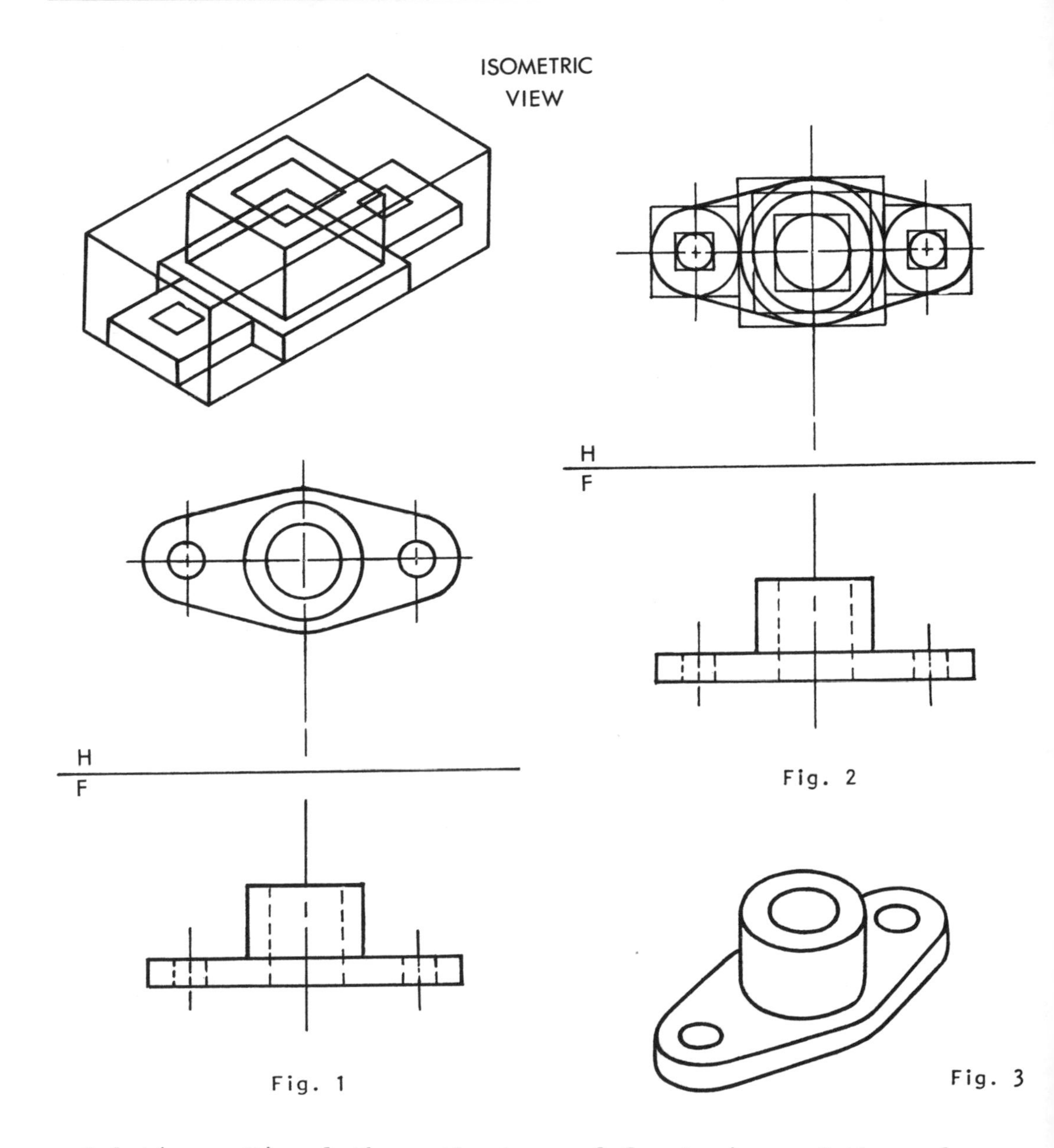

Fig. 1

Fig. 2

Fig. 3

Solution: Fig. 1 shows the top and front views of the rod guide, which is composed primarily of circles and arcs. As in fig. 2, enclose each circle in a square so that they can be transferred to the isometric view. Construct the isometric box with overall dimensions equal to the total height, width, and depth of the rod guide. Draw the boxes which enclose the circles in the isometric view to determine the location of the ellipses and elliptical arcs. As in fig. 3, construct the ellipses and arcs using the four-center method.

The arcs are connected with tangents that are approximated using a triangle. Darken the principal lines and erase the construction lines used. Omit hidden lines. The complete required isometric view of the rod guide is shown in fig. 3.

• PROBLEM 10-6

Construct the isometric view of the object whose top and front views are given in fig. 1.

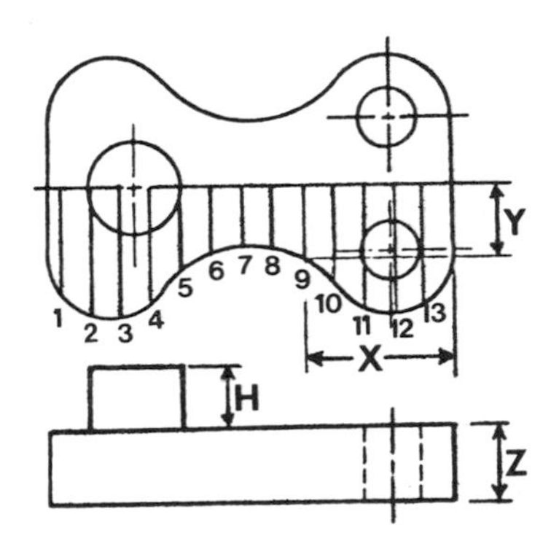

Fig. 1

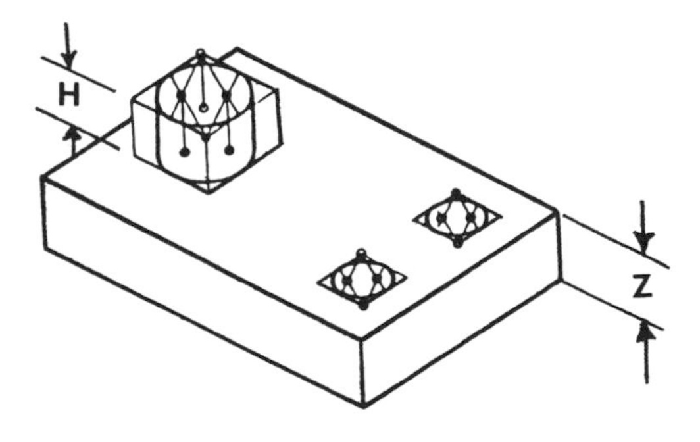

Fig. 2

Solution: The object whose top and front views are shown in fig. 1 contains many curved surfaces. These surfaces will appear irregular in the isometric view to be constructed. Establish the isometric box in which the object is to be drawn by constructing one box, of height "Z", to contain the base of the object and a second box, of height "H", located on top of the first one to contain the cylinder on the base. As in fig. 2, use the four-center method to construct the ellipses of the cylinder and the two holes in the base.

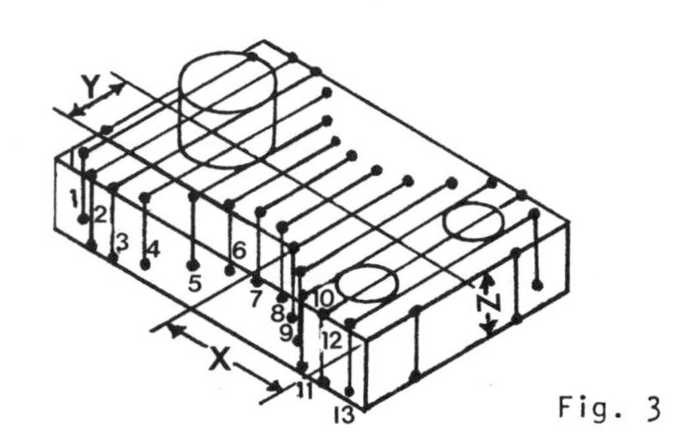

Fig. 3

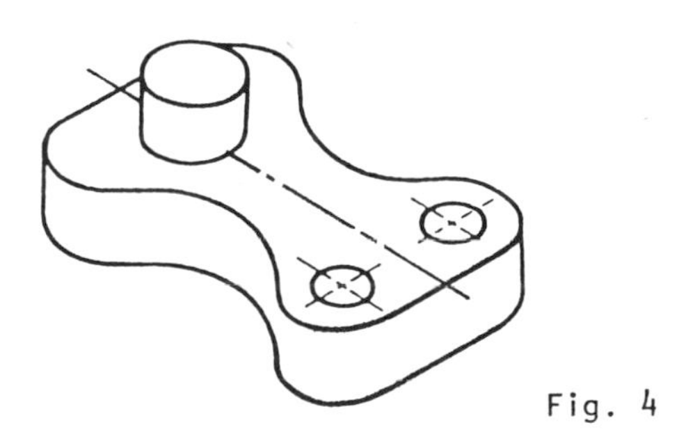

Fig. 4

The curved surfaces of the base of the object cannot be drawn in the isometric view using ellipses because they are irregular. Therefore, various points on the edges of the surface must be plotted using the coordinate method. As seen in fig. 1, pass a series of vertical cutting planes through the top view of the base of the object, intersecting the edges at points 1 through 13. Each edge point thus located appears at a certain distance from the center line and a certain distance from one end. Point 9, for example, is at a distance Y from the center line and a distance X from the right end. Transfer these width and depth distances from the top view of the object to the isometric box, as illustrated in fig. 3. Project each point downward a distance Z to locate points on the bottom edge of the curves. Connect these points with a smooth curve for the required isometric view, as shown in fig. 4. Darken principal lines and erase guide lines used.

• **PROBLEM 10-7**

Construct the isometric view of the block, ABCD, whose top and front views are given in fig. 1.

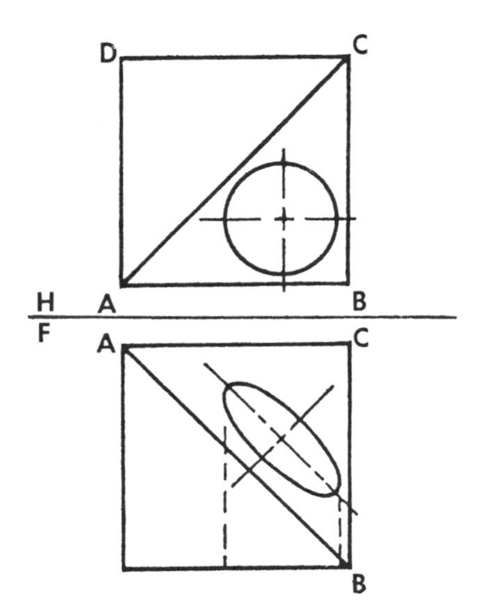

Fig. 1

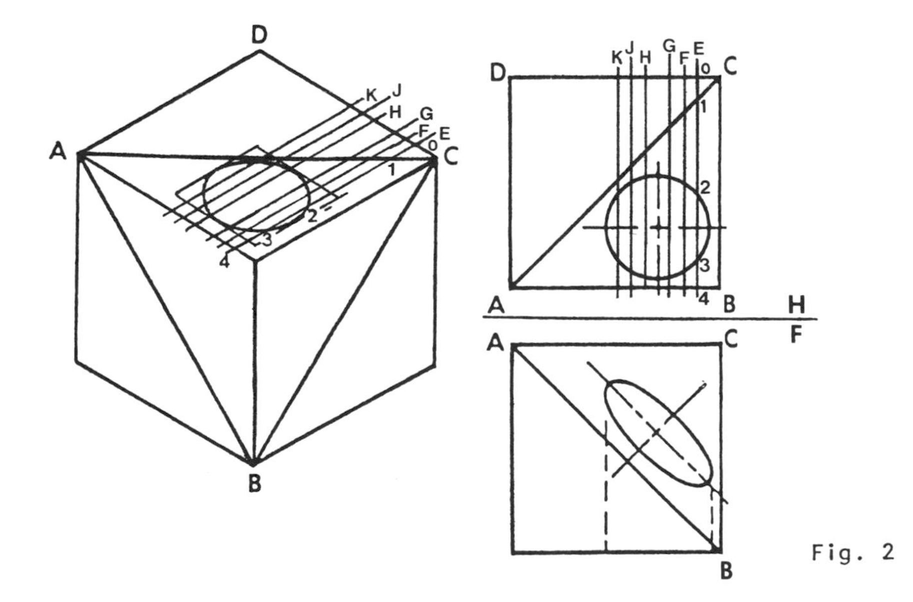

Fig. 2

Solution: Fig. 1 shows the given views of the block which has an inclined face with a circular hole cut through it. As in fig. 2, construct the block in isometric, omitting the actual view of the top of the hole, but drawing the box that encloses the top view of the hole on the top plane of the isometric box. Pass vertical cutting planes E, F, G, H, J, and K through the top view of the block. Each plane cuts the edge AC in one point, the circular view of the hole in two points, and the edge AB in one point. Plane E, for example, cuts points 1, 2, 3, and 4 from the block, as seen in fig. 2. Locate points 1, 2, 3, and 4 on the top plane of the isometric view. The trace of plane E on the face containing AB is line 4-4', constructed by projecting point 4 vertically downward to line AB, as in fig. 3. The trace of plane E on face ABC, the inclined face, is line 0-4'.

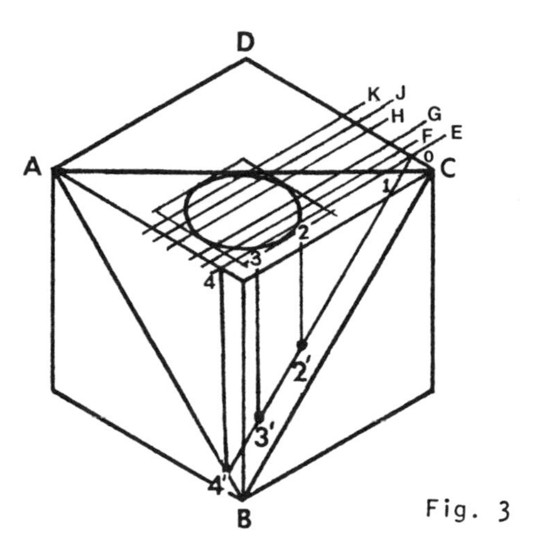

Fig. 3

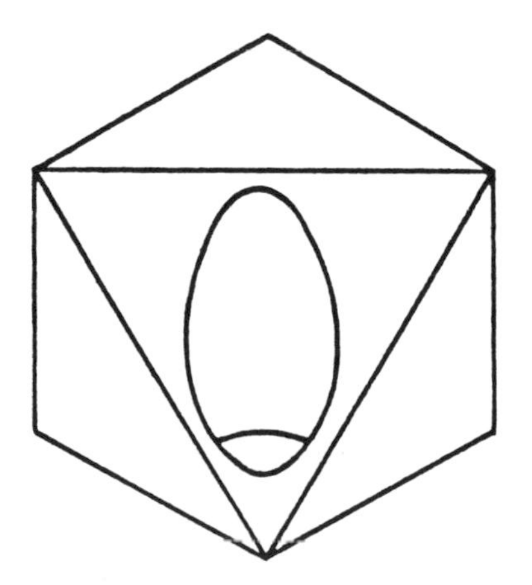

Fig. 4

Project points 2 and 3 to the trace 0-4' to points 2' and 3', locating these two points on the isometric view of the top of the hole. Locate additional points in the same manner and connect with a smooth curve to complete the isometric view of the block, as shown in fig. 4.

Find the isometric projection of a cube, using the auxiliary view method and the method of revolution.

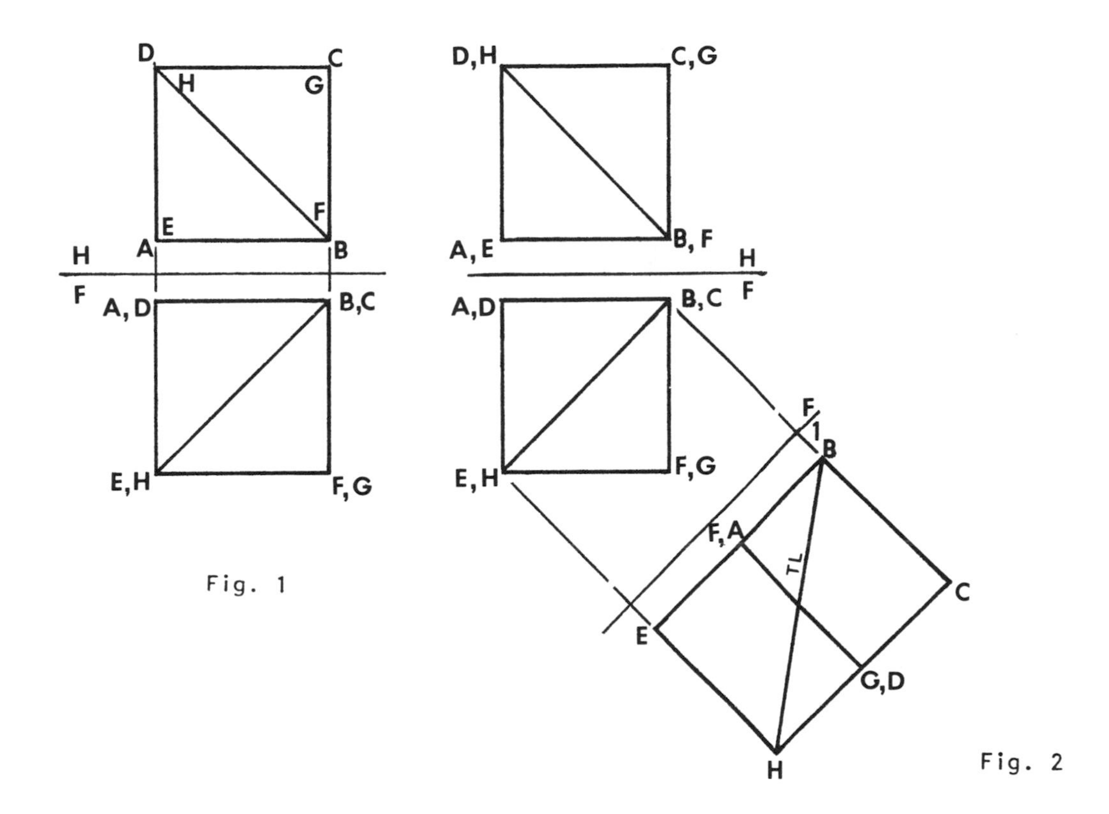

Fig. 1

Fig. 2

Solution: Fig. 1 shows the top and front views of the given cube with diagonal BH. The faces of the cube will appear equally inclined to the plane of projection when this diagonal appears as a point. The diagonal must first appear true length in order to appear as a point. As in fig. 2, construct auxiliary view 1 by placing the reference plane F-1 parallel to the diagonal BH in the front view. BH will appear true length in view 1 because the line of sight is perpendicular to BH in the front view. Construct auxiliary view 2 by making the line of sight parallel to the true length line BH in view 1 and the reference plane 1-2 perpendicular to BH, as shown in fig. 3. View 2 is the isometric projection of the cube since the body diagonal BH appears as a point in view 2. Each face, ABCD, BFGC, and AEFB, is equally inclined at 30° to the plane of projection, and the lengths of the edges are all about eight-tenths of their true lengths. Determine visibility.

A second method of obtaining the isometric projection of the cube is shown in figs. 4 through 6. Fig. 4 shows the top, front, and right side views of the given cube with diagonal 2-8. As in fig. 5, revolve the top view of the cube about an axis through point 2 until diagonal 2-8 is vertical, perpendicular to the reference plane H-F. Reconstruct the right side and front views of the cube.

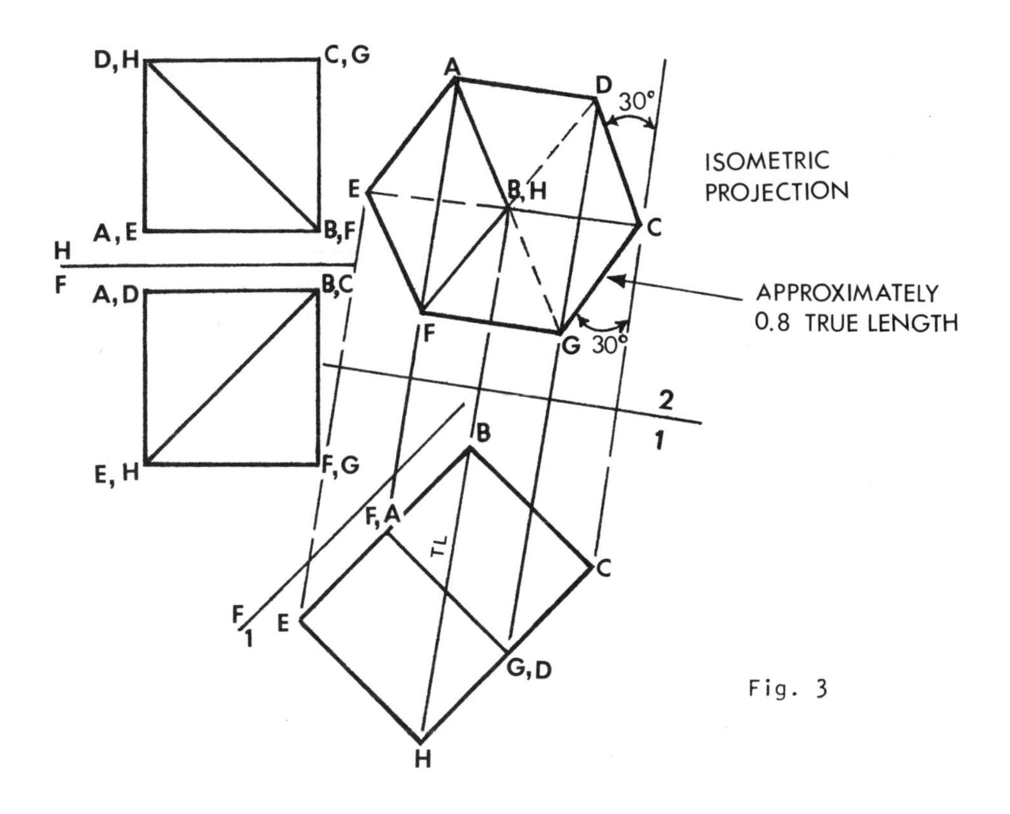

Fig. 3

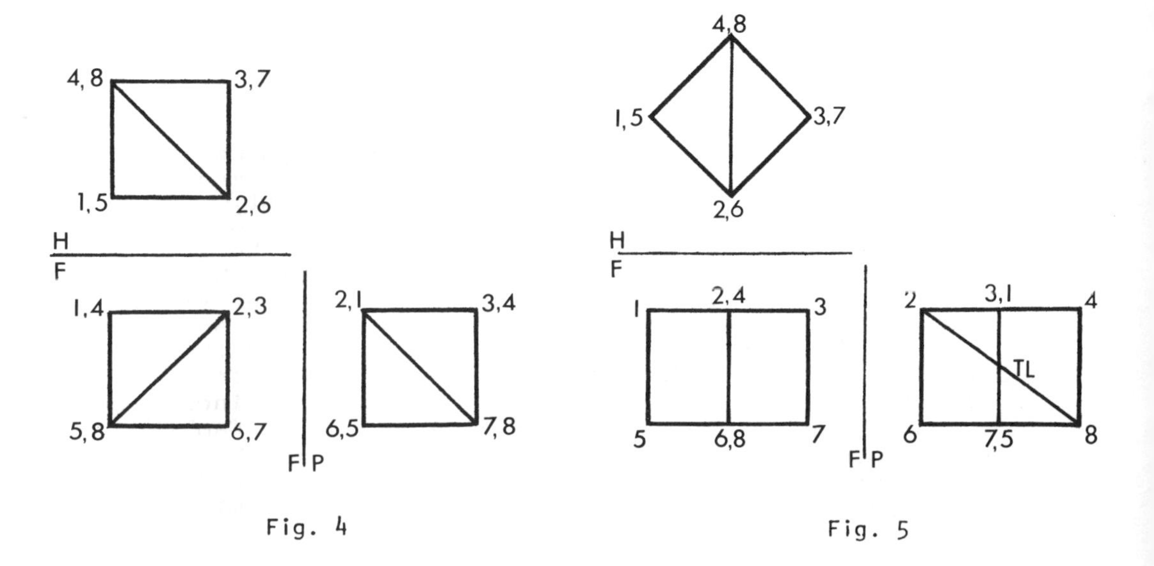

Fig. 4

Fig. 5

Since the diagonal is parallel to reference plane F-P in the front view, it is true length in the right side view. Revolve the right side view now about a vertical axis through point 2 until the diagonal 2-8 is perpendicular to plane F-P. Reconstruct the top and front views, as in fig. 6. Diagonal 2-8 appears as a point in the front view, which is the isometric projection of the cube. Determine visibility.

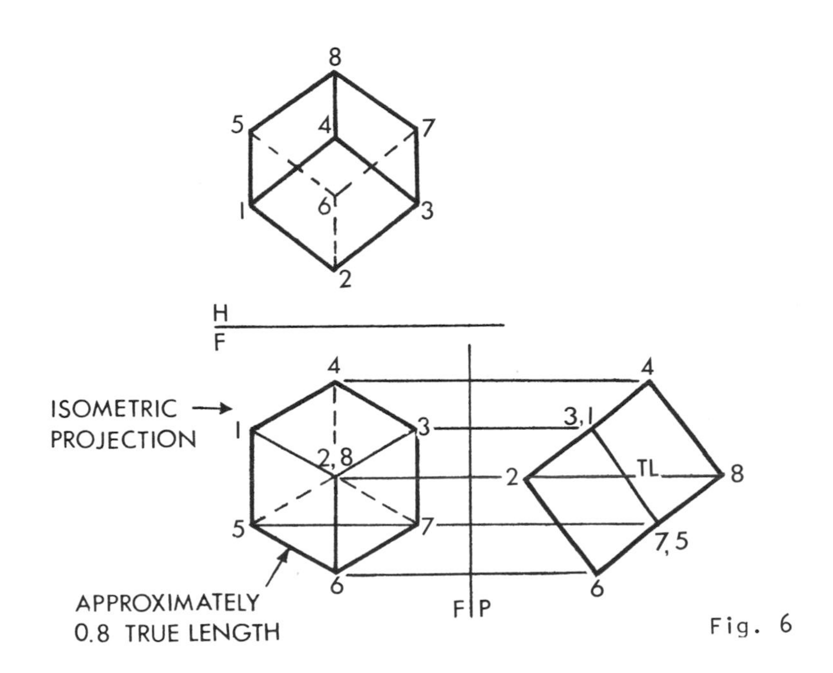

Fig. 6

• **PROBLEM 10-9**

Construct the isometric projection of the object whose top, front, and right side views are given in fig. 1.

Solution: Fig. 1 shows the given views of the object whose isometric projection is required. The procedure is exactly the same as though the isometric drawing were to be constructed. The key difference is, of course, the lengths of the lines composing the isometric view of the object. In an isometric drawing, the true lengths of the lines are used, but in an isometric projection only eight-tenths of the true lengths of the lines are used. As in fig. 2, construct the skeleton box to contain the object. The overall height, width, and depth of the isometric box are eight-tenths of the corresponding dimensions of the object.

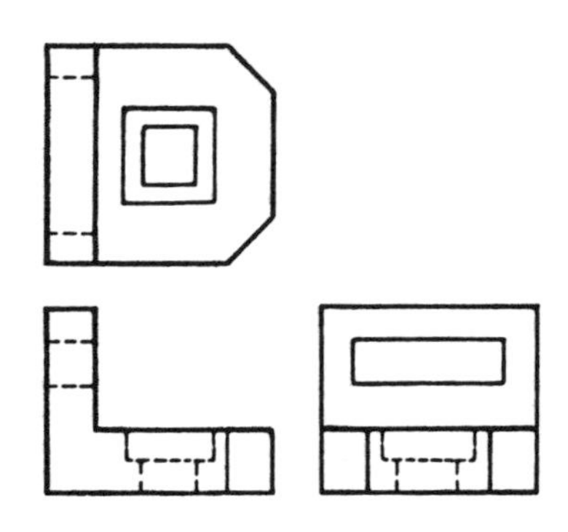

Fig. 1

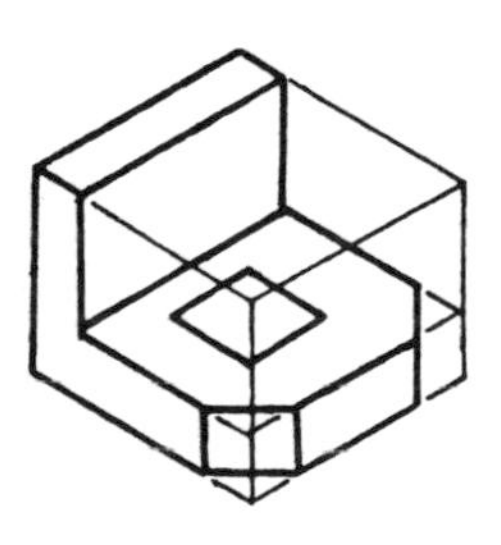

Fig. 2

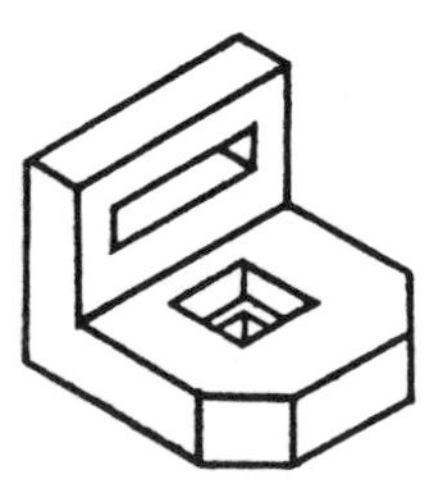

Fig. 3

Layout the vertical section of the object and the horizontal base with its angled corners. Outline the box in which the hole through the base is to be made. Complete the isometric projection, as shown in fig. 3, using 0.8 true length throughout.

DIMETRIC AND TRIMETRIC PROJECTION

• **PROBLEM** 10-10

Define dimetric projection and find the dimetric projection of a given cube.

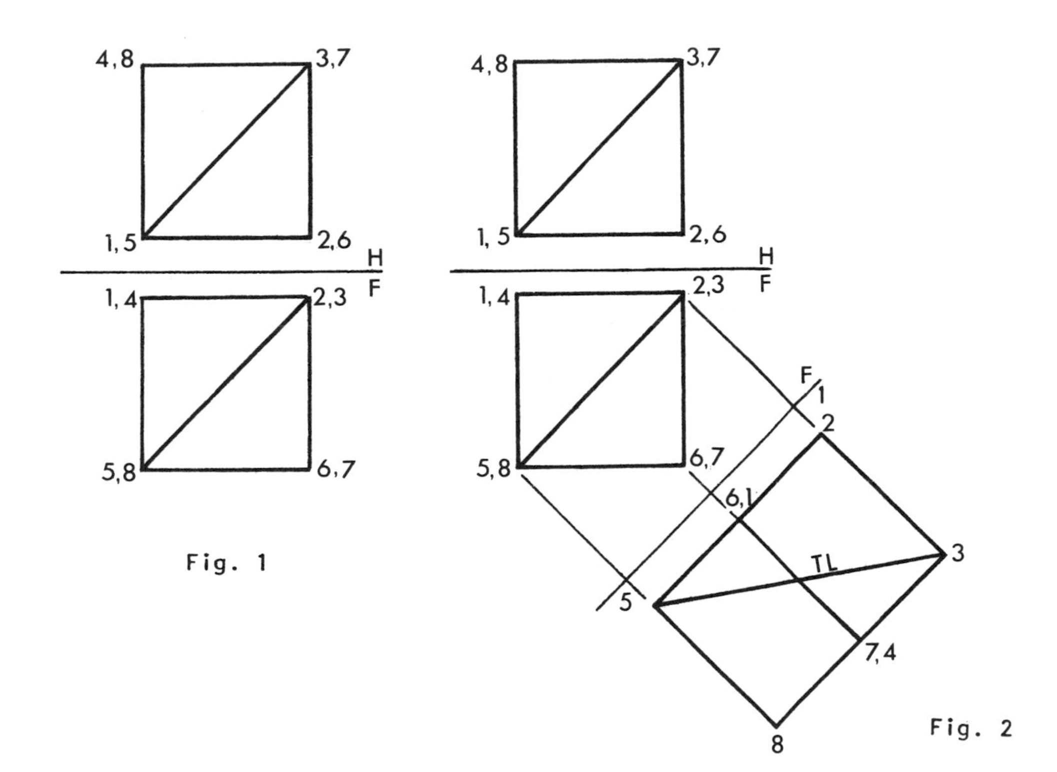

Fig. 1

Fig. 2

Solution: In the dimetric projection of an object, the object is placed in such a position that two faces of the object are equally inclined to the plane of projection and the third is not. Thus only two of the three axes of the object are equally foreshortened, and these two axes are separated by equal angles. An infinite number of angle combinations between the axes is possible, and a scale for measuring the foreshortened lengths must be made for each different angle.

A method for determining the dimetric projection of a cube is illustrated in figs. 1, 2, and 3. Fig. 1 shows the top and front views of the given cube with diagonal 3-5.

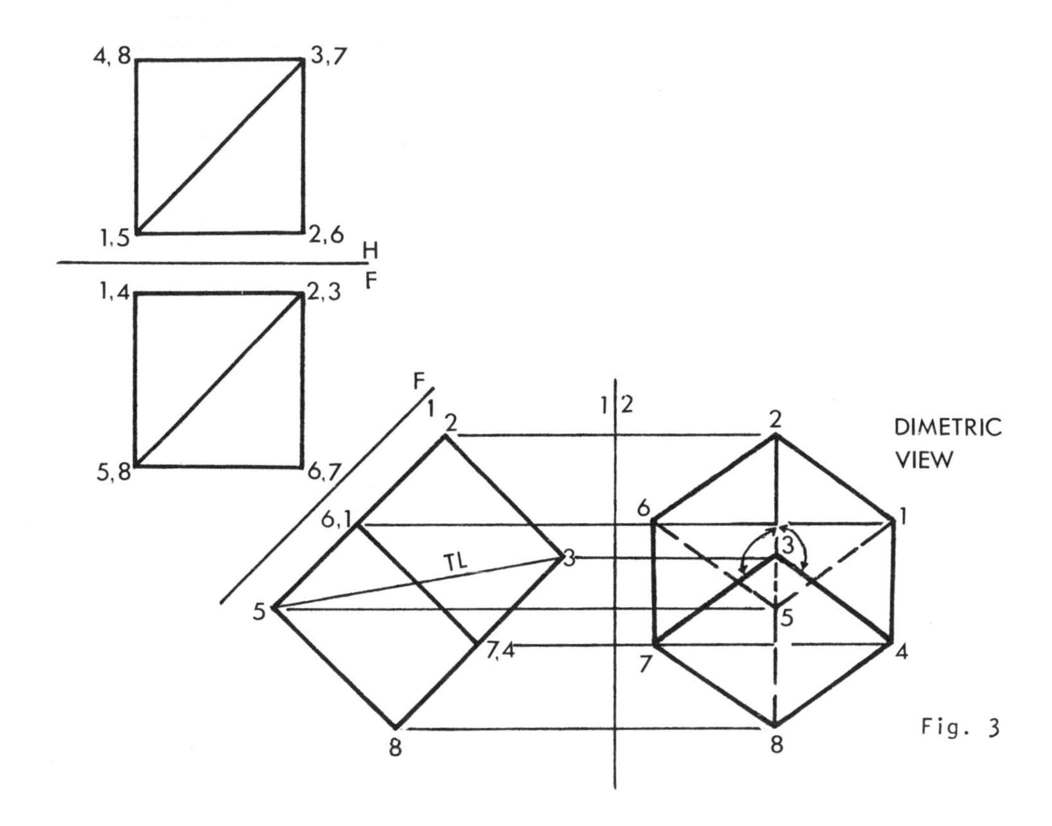

Fig. 3

Because any combination of angles can be used between the three axes as long as two of the angles are equal, the point view of the body diagonal is not necessary. The true length view of the diagonal is needed, however, to give an auxiliary view in which two faces of the cube are visible. Construct auxiliary view 1 by making the line of sight perpendicular to the diagonal 3-5 in the front view and the reference plane F-1 parallel to 3-5, as shown in fig. 2. The diagonal projects true length in view 1, and two faces of the cube are visible. As in fig. 3, construct a second auxiliary view in which two of the axes forming a corner of the cube are separated by equal angles. To do so, draw an arbitrary vertical reference plane 1-2 with which edges 2-3 and 3-8 make two different angles. Edge 2-3 is the line of intersection between perpendicular faces 6-7-3-2 and 1-4-3-2, so in view 2 the angle 7-3-2 will equal the angle 4-3-2. Complete view 2, as shown in fig. 3, and determine visibility. The angle 7-3-4 does not equal angles 7-3-2 and 4-3-2 since edge 3-8 makes a different angle with the reference plane 1-2 than edge 2-3. View 2 is the required dimetric projection of the given cube.

If a projection of an object were to be constructed without the use of auxiliary views, but with only the skeleton box, a scale to measure the foreshortened lengths on each axis would have to be made. Fig. 4, part "a" shows how such a scale is made. The standard, true length scale, AB, is placed at a 45° angle from the horizontal. The desired angles between the dimetric axes must be predetermined, recalling that two of the three angles must be equal, as shown in fig. 4, part "b". One axis, CB, is drawn at one of the two different angles, measured from a vertical downward from point B, and the other axis, DB, is drawn at the second angle, also measured from the vertical from B.

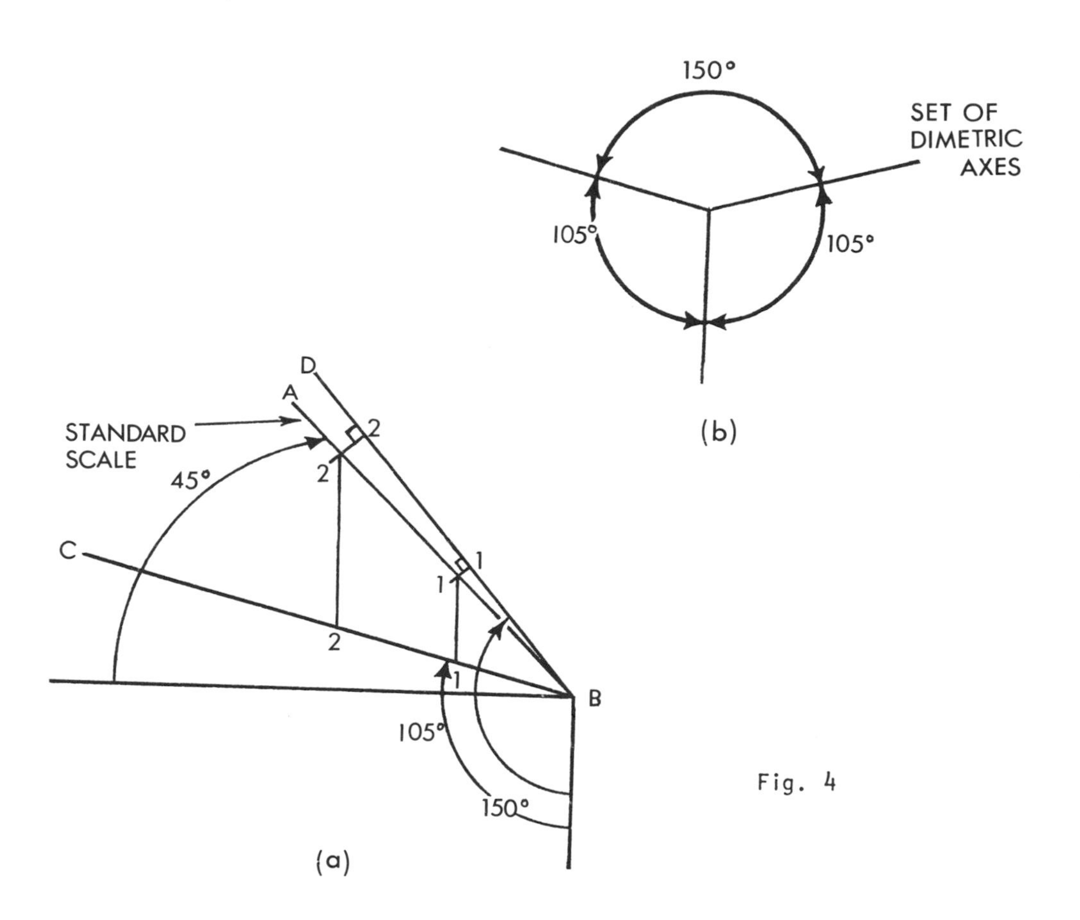

Fig. 4

Note that only two new scales are needed since the two axes that recede at equal angles use the same foreshortened scale. For the scale CB which is used for two axes, project vertically downward from each graduation of the standard scale to line CB. For the scale DB which is used for only one axis, draw projection lines from each graduation on the standard scale perpendicular to line DB. Scale CB is used for the width and depth axes, and scale DB is used for the height axis. Now a dimetric projection can be made without the need for auxiliary views.

• **PROBLEM 10-11**

Construct the dimetric projection of the given clip angle.

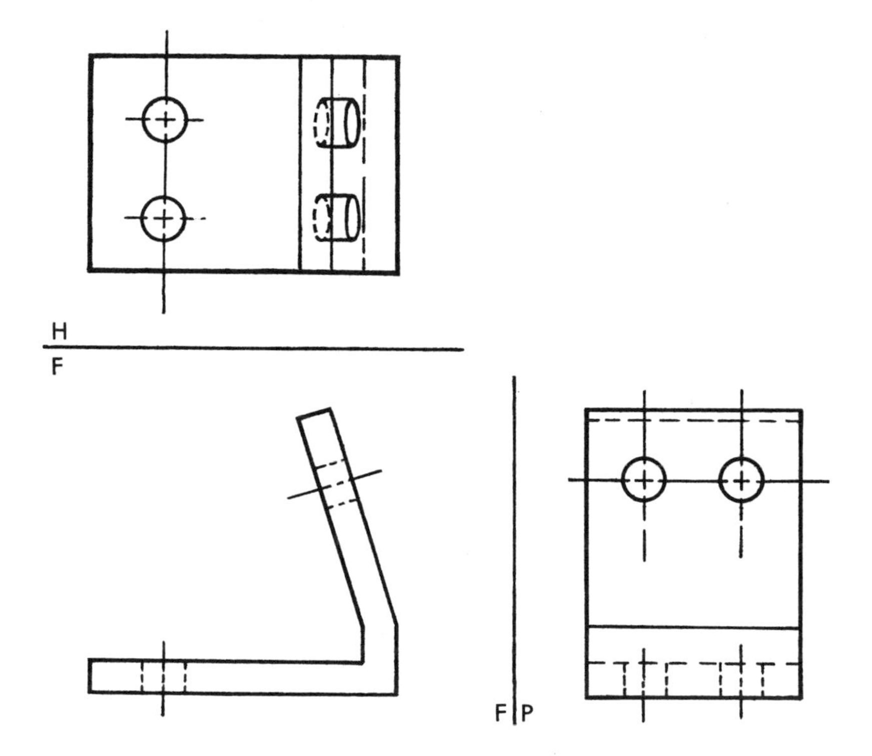

Fig. 1

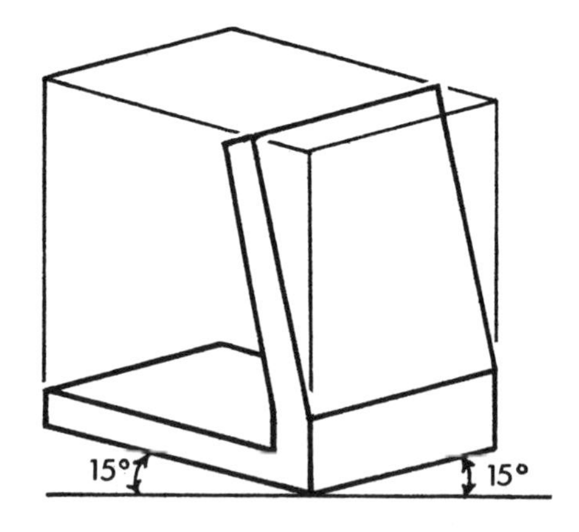

Fig. 2

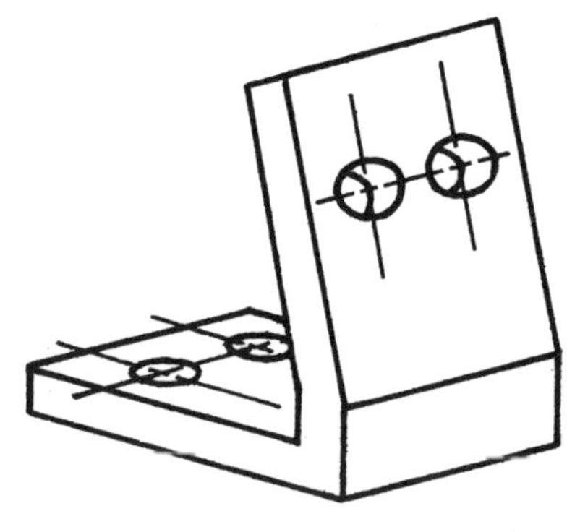

Fig. 3

Solution: Fig. 1 shows the top, front, and right side views of the given clip angle. Choose the angles to separate the three axes, recalling that two of the angles must be equal. The receding axes, as shown in fig. 2, for this solution are at a 15° angle from the horizontal. Construct the enclosing dimetric box, where the overall width and depth are both 0.73 true length and the overall height is .96 true length. Layout the angled portion of the clip, as well as the horizontal base, as shown in fig. 2. Locate the center lines for the circles which will appear as ellipses in the dimetric view. Note that an ellipse template may be used to draw the ellipses on the horizontal base since both receding axes have been equally foreshortened. An ellipse template may not be used, however, to draw the ellipses on the angled portion of the clip since the vertical axis was not foreshortened as much as the receding axes. Points must be plotted in the usual manner and connected with a smooth curve to construct these two ellipses. Darken the principal lines and erase the guide lines used. Fig. 3 shows the completed, required dimetric view of the clip angle. Note that hidden lines have been omitted, as is customary.

• PROBLEM 10-12

Construct the dimetric projection of the pyramidal object whose top and front views are given.

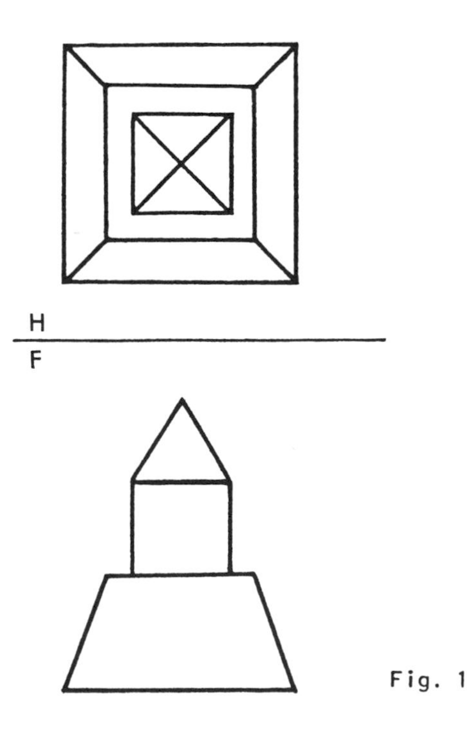

Fig. 1

Solution: Fig. 1 shows the given views of a pyramidal object which is symmetrical. In a dimetric projection of a solid, the solid is placed so that two of its faces are equally inclined to the plane of projection. It is not necessary, however, to make use of auxiliary views in the solution of this problem. A set of axes can be constructed, similar to those of an isometric view, with two of the axes separated by equal angles. As seen in fig. 2, an angle of 105° was chosen to separate two of the axes, and these receding axes would be at a 15° angle from the horizontal.

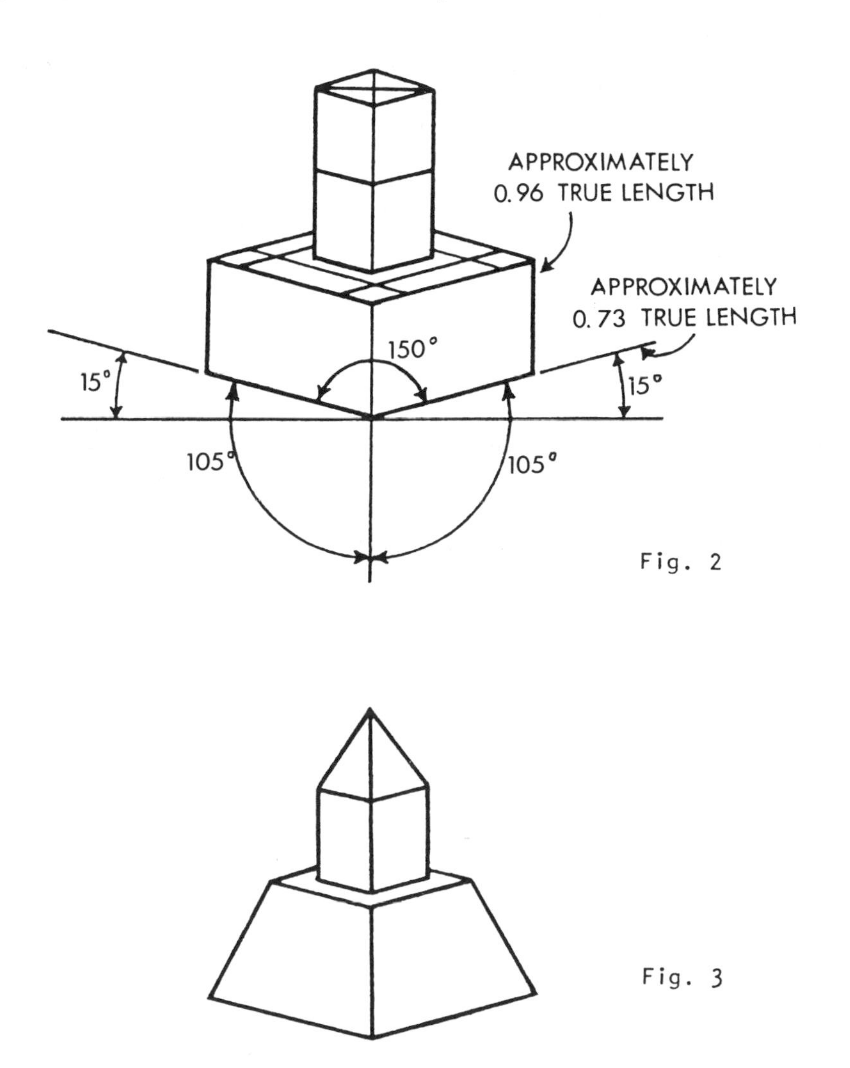

Fig. 2

Fig. 3

Note that any set of angles could have been chosen as long as two of the angles were equal. With the use of a foreshortened, dimetric scale, it can be seen that the lengths measured along the receding axes are approximately 0.73 of the true lengths and the lengths measured along the vertical axis are approximately .96 of the true lengths. As in fig. 2, construct the skeleton boxes to contain the base of the object and the peaked structure on top of the base. Complete the dimetric projection of the object as though it were an isometric view, bearing in mind the appropriate foreshortened lengths to be used. Fig. 3 shows this completed view.

Define trimetric projection, find the trimetric projection of a cube using the auxiliary view method and the method of revolution, and construct a trimetric scale.

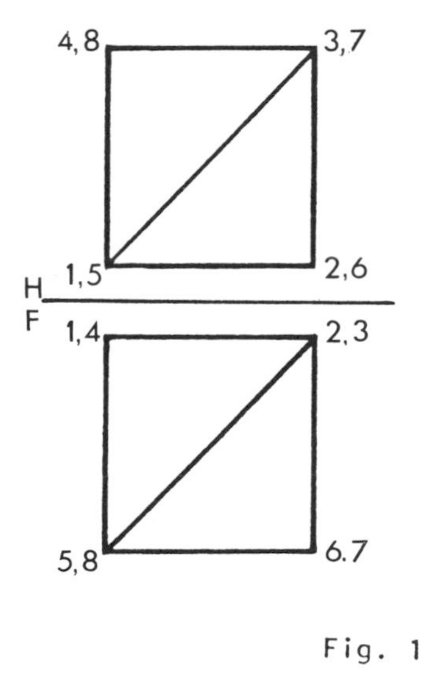

Fig. 1

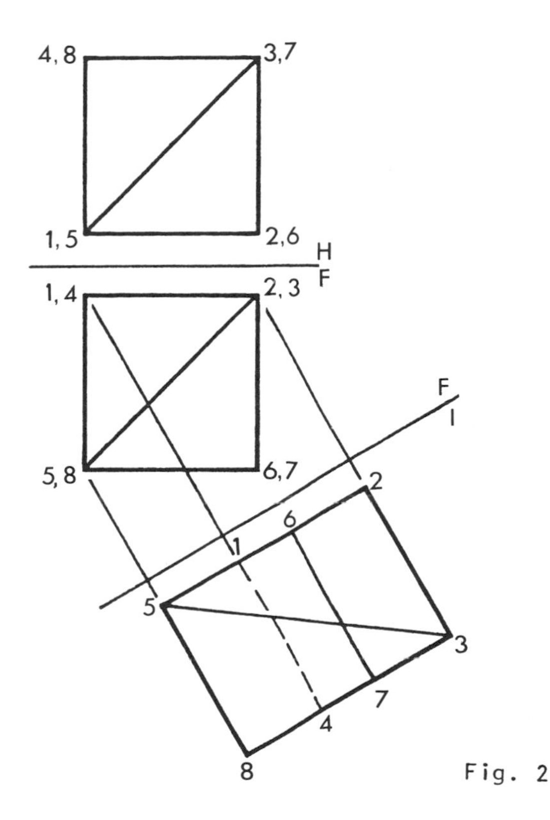

Fig. 2

Solution: In the trimetric projection of an object, the object is placed in a position where none of its faces are equally inclined to the plane of projection. Each of the three axes are separated by different angles, and lengths measured along these three axes are all foreshortened unequally. Many combinations of angles are possible, and a scale must be made for each axis.

The trimetric projection of an object can be obtained through the use of auxiliary views, as illustrated in figs. 1, 2, and 3. Fig. 1 shows the top and front views of the given cube with diagonal 3-5. Construct a first auxiliary view by placing an arbitrary reference plane, such as F-1 in fig. 2, not parallel to the diagonal. Two faces of the cube will be visible in view 1, as seen in fig. 2. Construct a second auxiliary view with lines of sight perpendicular to the diagonal of the cube and the reference plane 1-2 parallel to the diagonal, as in fig. 3. The diagonal will appear true length in view 2, and when the projected points 1 through 8 have been properly connected and visibility has been determined, the second auxiliary view is the trimetric projection of the cube. None of the faces are equally inclined to the plane of projection, and none of the axes have been equally foreshortened.

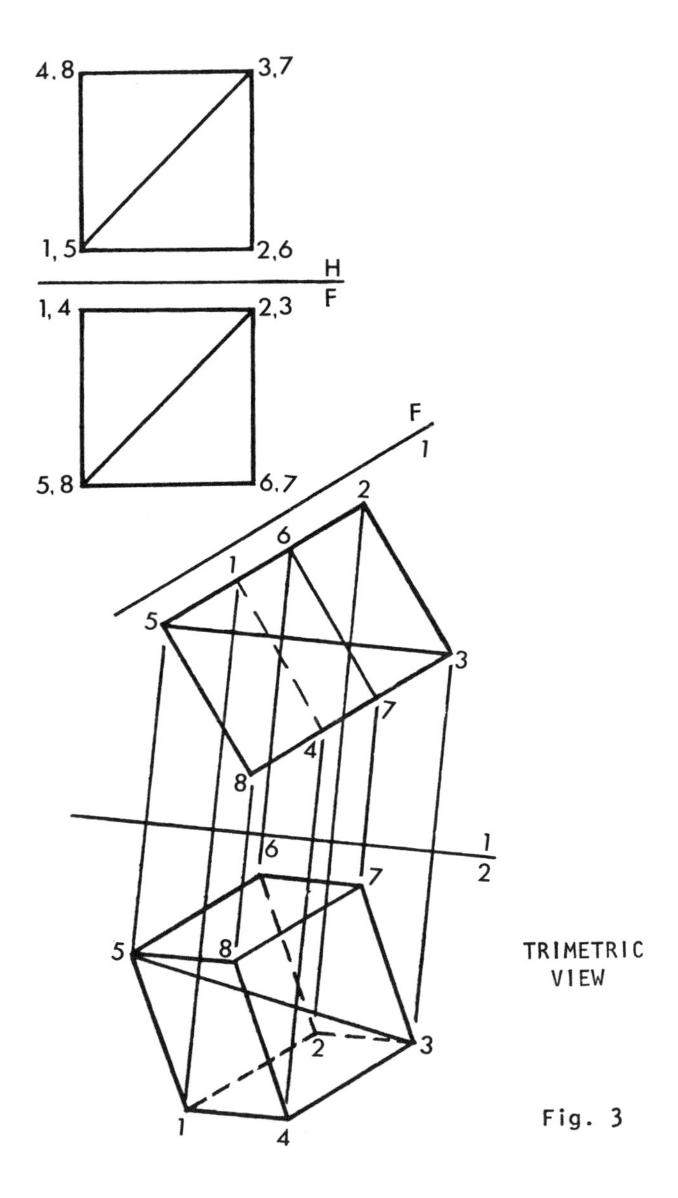

Fig. 3

A second method for finding the trimetric projection of an object employs revolutions, as illustrated in figs. 4, 5, and 6. Fig. 4 shows the top and front views of the given cube once again for clarity. Revolve the top view of the cube about an axis through point 1 through an arbitrary angle, 30° in this case. Project the revolved points to the front view to where they intersect horizontal projection lines from their corresponding original points, as shown in fig. 5. The original front view has been omitted for clarity. Two faces of the cube are visible in the front view now. Revolve the front view of the cube about an axis through point 3' through another arbitrary angle, 23° in this case. Project these twice-revolved points to the top view to where they intersect horizontal

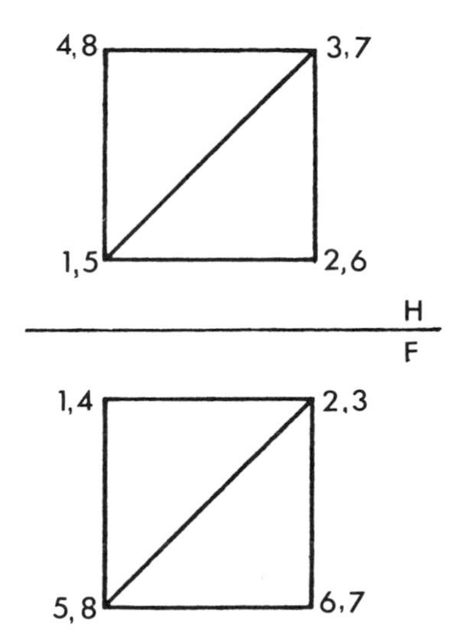

Fig. 4

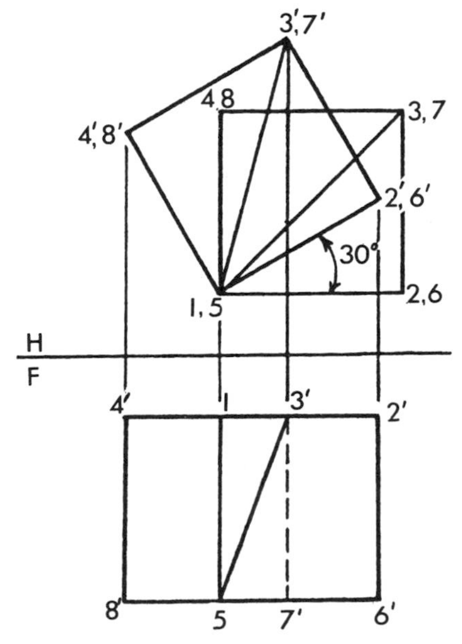

Fig. 5

projection lines from their corresponding once-revolved points, as shown in fig. 6. The first revolution has been omitted from the top view for clarity. This top view is now a trimetric projection of the cube.

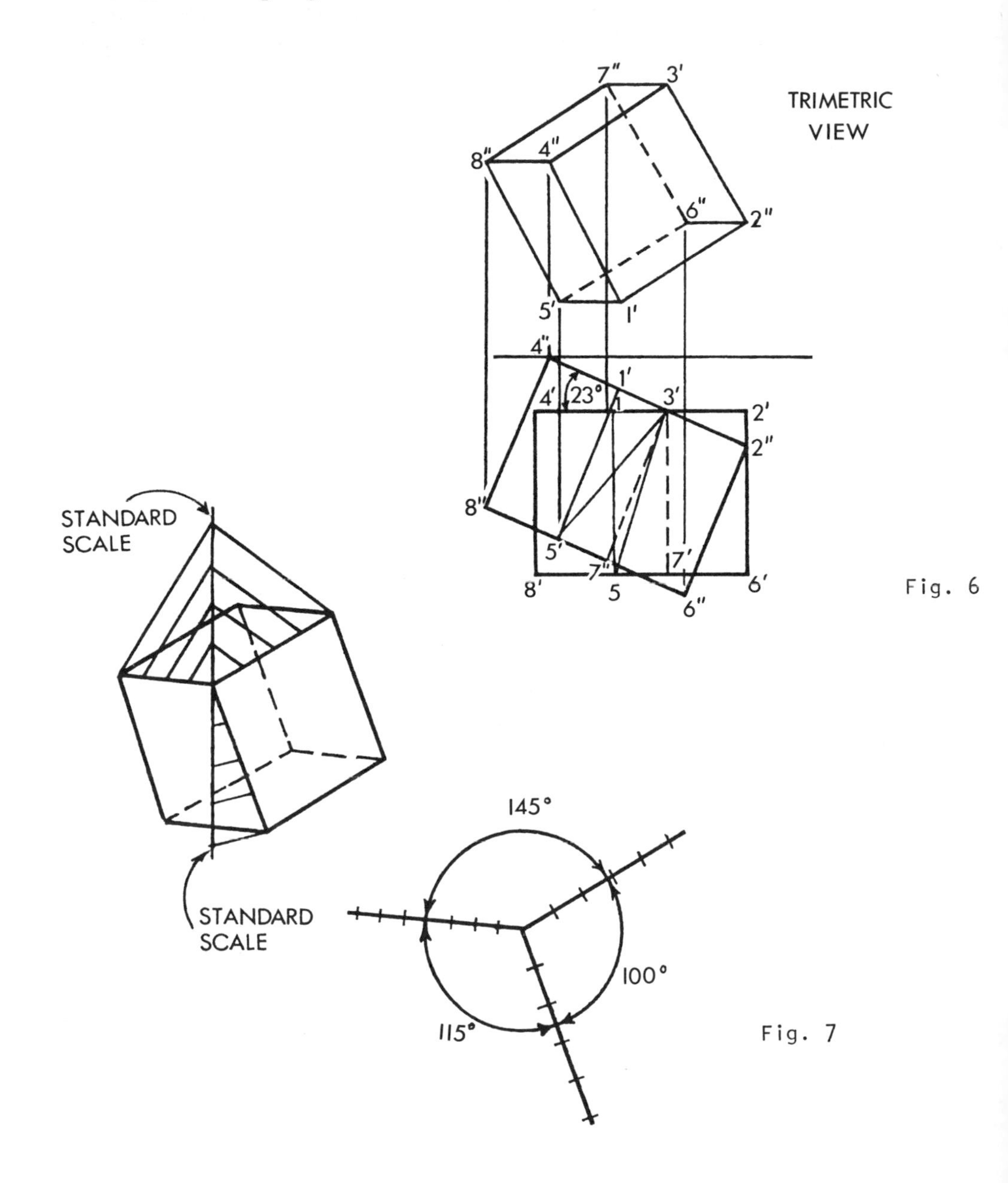

Fig. 6

Fig. 7

A scale can be easily constructed for use with a specific set of angles by the method shown in fig. 7, using the trimetric view found in fig. 3. Because all sides of a cube are equal, divide each side into an equal number of parts, four units in this case, by proportional divisions. The standard scale is used for this. These sides, or axes, can now be extended and scaled into more units for use in constructing trimetric pictorials with this particular set of angles. Fig. 7 shows the scaled axes separated by angles of 115°, 100°, and 145°.

Construct a trimetric pictorial of the block whose orthographic views are given.

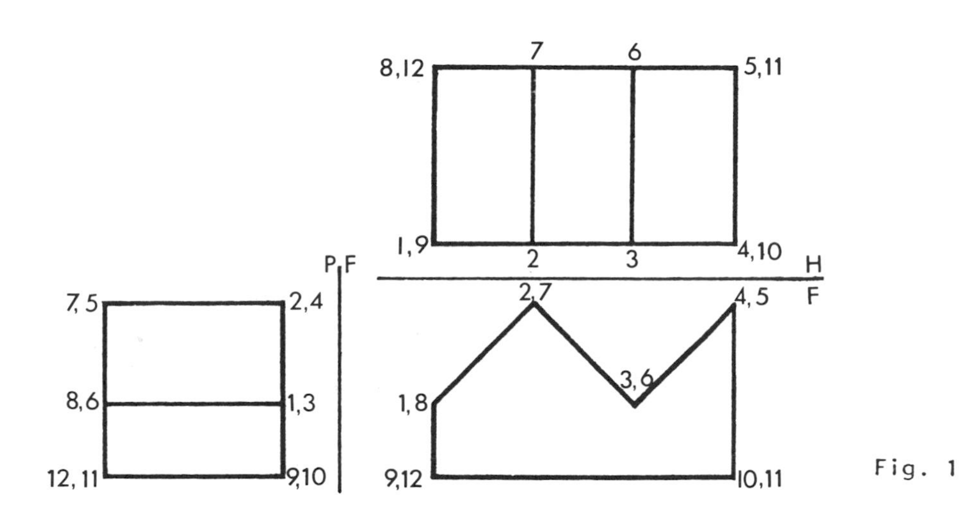

Fig. 1

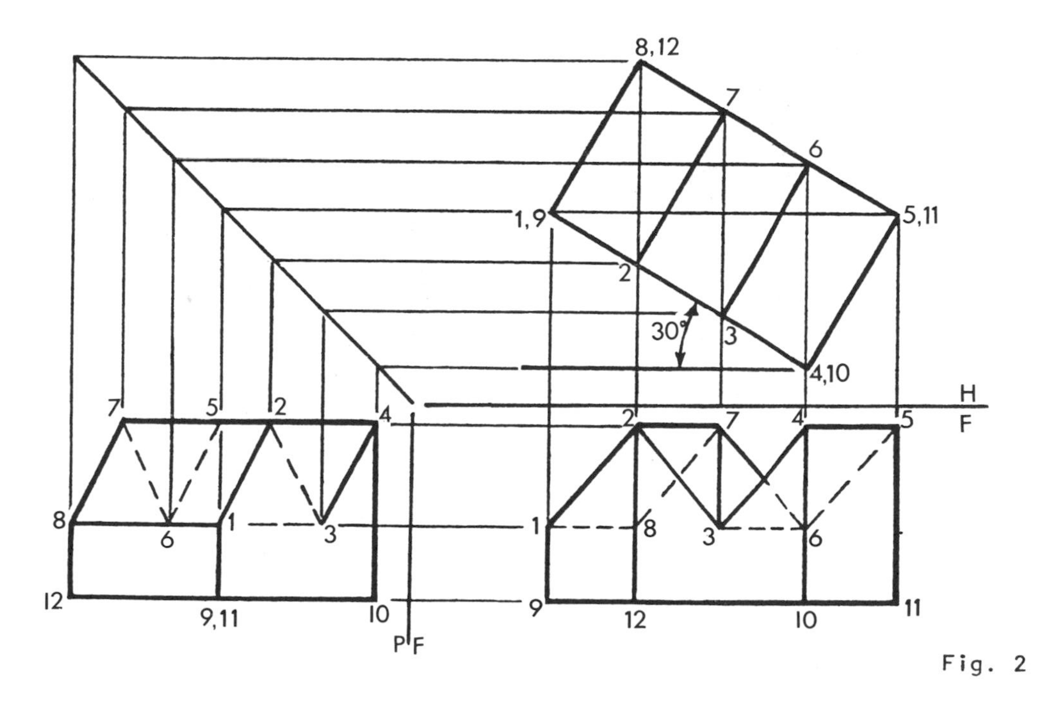

Fig. 2

Solution: Fig. 1 shows the given top, front, and left side views of the block, whose corner points are labeled 1 through 12. A trimetric projection is one in which the three axes are all separated by different angles. As in fig. 2, revolve the top view of the block about an axis through edge

4-10 an angle of 30°. Note that any arbitrary angle could have been chosen. The face 1-9-10-4-3-2 and the face 4-10-11-5 now are at different angles to the reference plane H-F. Project the revolved points to the front view to where they intersect horizontal projection lines from their corresponding original points, and complete the new front view. The previous front view has been omitted from fig. 2 for simplicity. Project from both the top and front views to construct the new left side view, seen also in fig. 2.

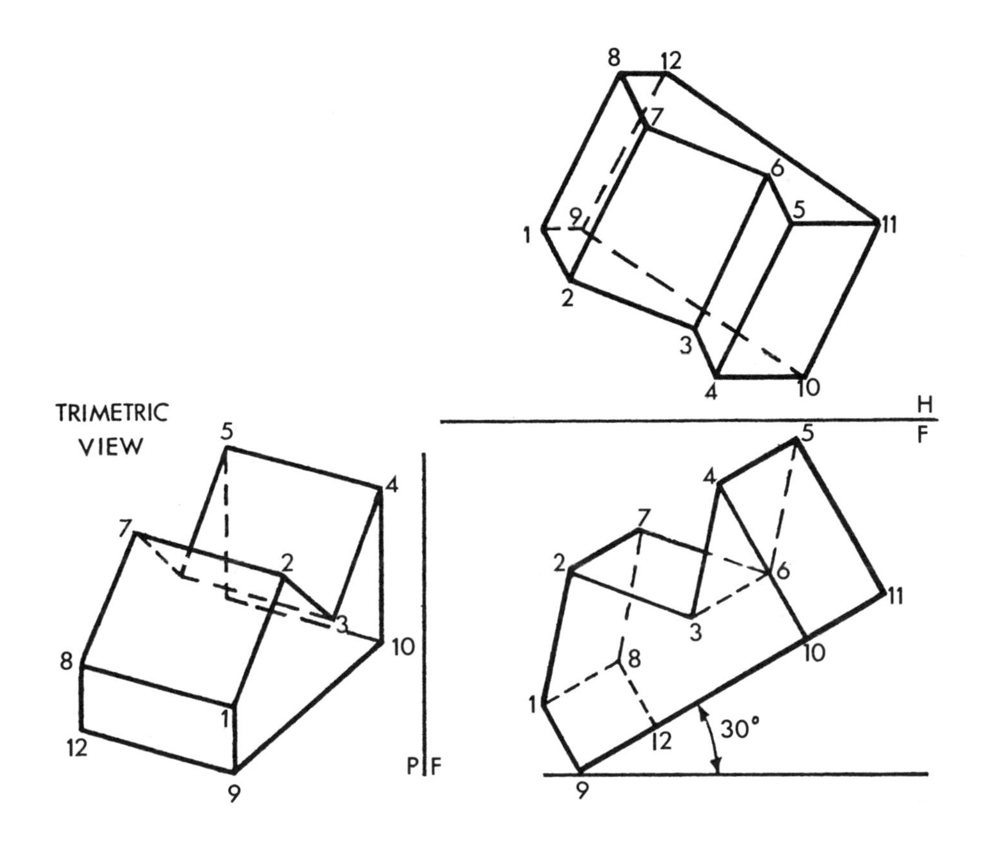

Fig. 3

Revolve the front view about an axis through point 9 through an angle of 30°, as shown in fig. 3. Project the twice-revolved points to the top view to where they intersect horizontal projection lines from their corresponding once-revolved points and complete the view. Project now from both the front and the top views to construct the new left side view. This view, as seen in fig. 3, is the required trimetric pictorial of the block.

Construct a trimetric pictorial of the block whose top and front views are given.

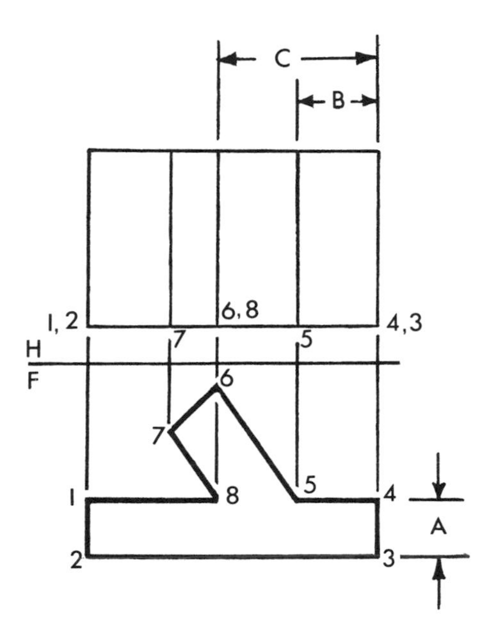

Fig. 1

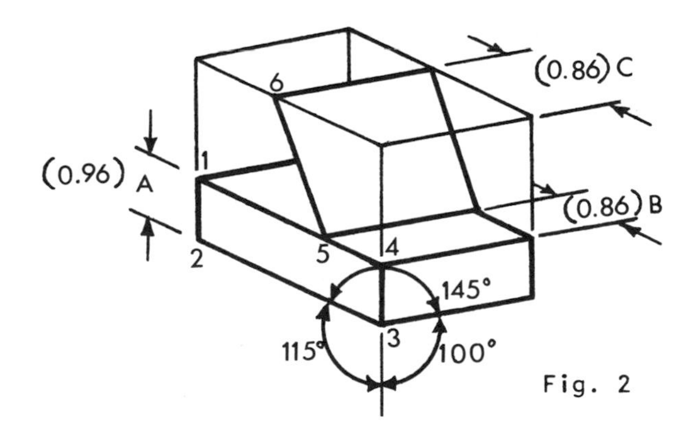

Fig. 2

Fig. 3

Solution: Fig. 1 shows the given top and front views of the block, which has an inclined portion on the base. A trimetric projection has no axis making the same angle with the plane of projection as any other axis. All three axes are foreshortened, but they are unequally foreshortened. As in fig. 2, construct the enclosing trimetric box, using appropriate scales for the set of angles chosen. In this case, angles of 100°, 115°, and 145° were used, and the scales were established by first finding the trimetric projection of a cube at the chosen angles, and then measuring the ratios of the lengths of the sides of the cube in that trimetric projection to the true lengths. The axis receding at a 25° angle is 0.59 true length, the axis receding at a 10° angle is 0.86 true length, and the vertical axis is 0.96 true length. Layout the base, points 1, 2, 3, 4, and 5, with the appropriate scales, and locate point 6 on the inclined face. Point 1 is 0.96 times the distance A above the 25° receding axis. Point 5 is at the same distance above the 25° axis as point 1, and also at a distance 0.86 B from the front corner. Point 6 is at a distance 0.86 C from the top front corner. Draw lines parallel to the 10° receding axis through points 5 and 6, and complete the inclined face as shown in fig. 2. Locate points 7 and 8 in the same manner as the previous points, and complete the trimetric projection by darkening principal lines and erasing the guide lines used. Fig. 3 shows this completed pictorial.

GENERAL AND CAVALIER OBLIQUE VIEWS

• PROBLEM 10-16

Construct a general oblique view of the object whose top, front, and right side views are given.

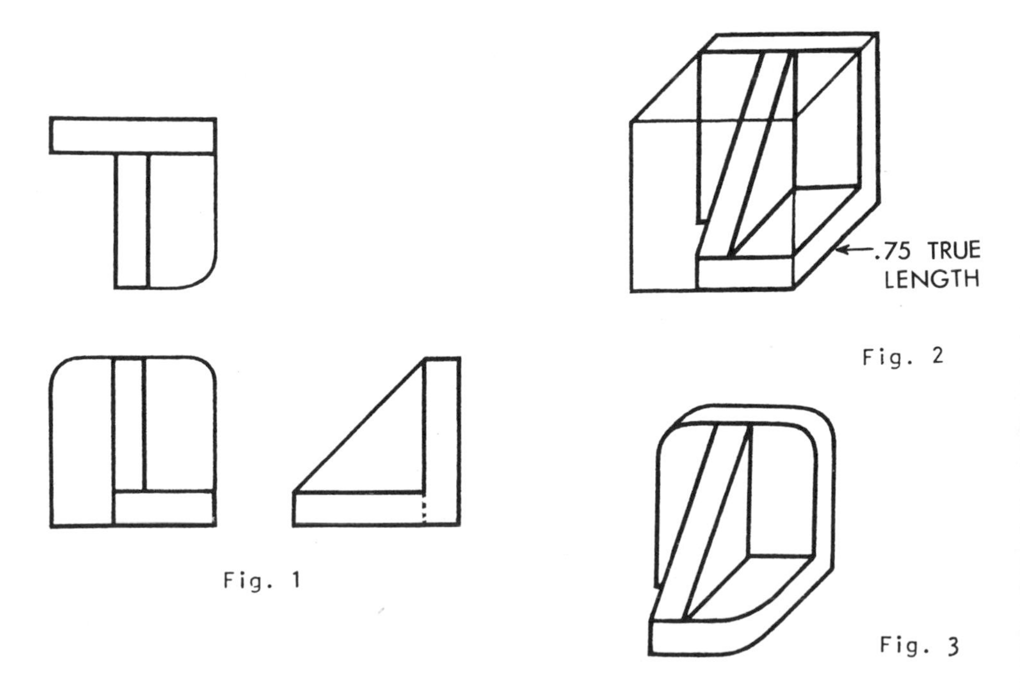

Fig. 1

Fig. 2

Fig. 3

Solution: Fig. 1 shows the given views of the solid figure. A general oblique drawing is one in which lengths measured along the receding axis are 0.75 of the true lengths. As in fig. 2, construct the enclosing oblique box of true height and width, as measured from the orthographic views, and depth of three-quarters true length. Layout the base and upright portions of the block, and construct the inclined faces by locating the endpoints of the limiting edge lines and connecting them. Construct the rounded edges with the use of circles for planes parallel to the frontal projection plane and ellipses for the front corners of the base. Erase guide lines and darken principal lines to complete the general oblique view, as shown in fig. 3.

• PROBLEM 10-17

Construct a general oblique view of the object whose top and front views are given.

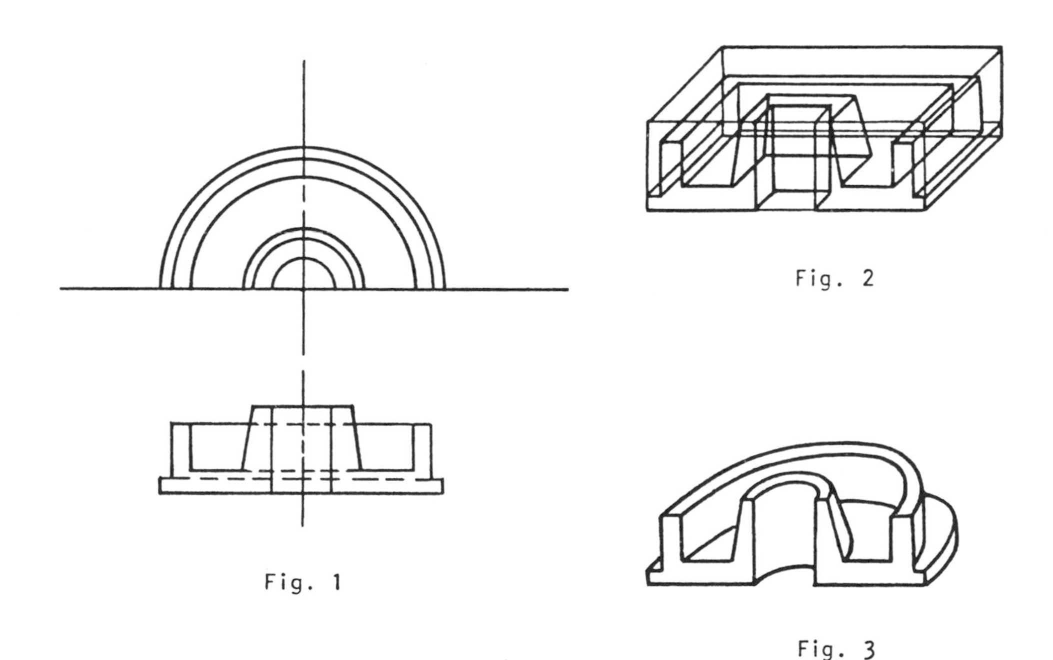

Fig. 1

Fig. 2

Fig. 3

Solution: Fig. 1 shows the given views of the semi-cylindrical solid. The object should be positioned so that the circular faces are parallel to the frontal plane of projection, however, as an exercise in plotting points for ellipses, place the long side of the front parallel to the frontal plane. The circular surfaces of this object will appear as ellipses in the oblique view but because the pictorial is to be a general oblique one, with the receding axis reduced, points for these surfaces must be plotted by the method previously described. As in fig. 2, construct the enclosing oblique box with true height and width and three-quarters true depth. Layout the base and upright portions of the object and construct boxes to enclose each circular surface. Plot points for these ellipses and connect the points with a smooth curve. Darken principal lines and erase guide lines for the complete general oblique view, as shown in fig. 3.

• PROBLEM 10-18

Construct a general oblique view of the H-block whose top and front views are given.

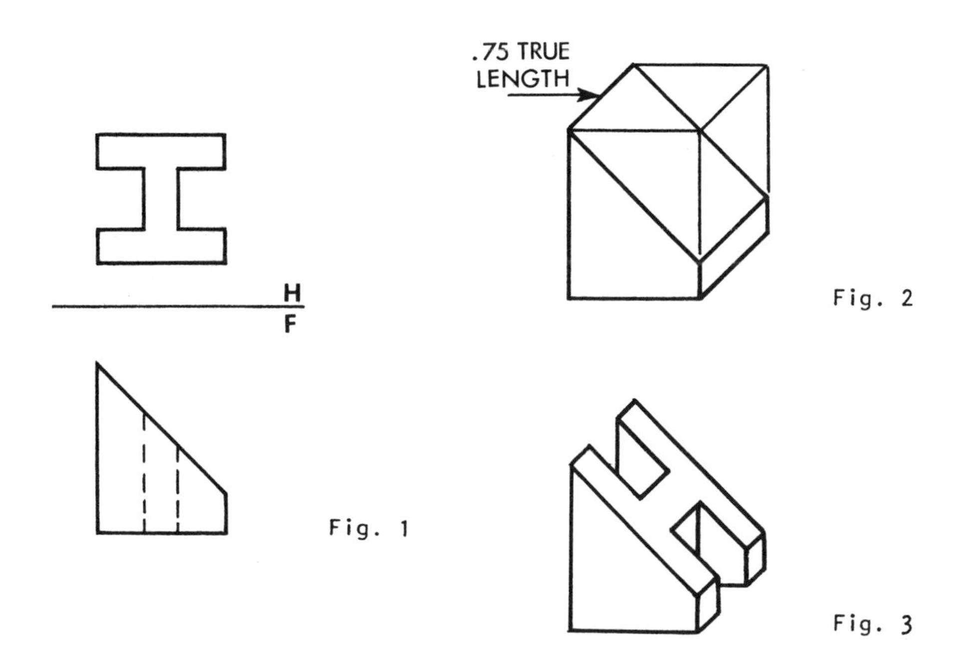

Fig. 1

Fig. 2

Fig. 3

Solution: Fig. 1 shows the given views of the H-block, which has an inclined surface. As in fig. 2, construct the enclosing oblique box with true height and width, and three-quarters true depth. Layout the face that is parallel to the frontal plane, and layout the inclined face. Complete the oblique view of the H-block by transferring distances from the orthographic views. Darken principal lines and erase guide lines, as shown in fig. 3.

• **PROBLEM 10-19**

Construct a Cavalier oblique view of the box jig, given the top, front, and right side views.

Solution: Fig. 1 shows the given views of the box jig. An oblique view of an object employs three axes, as does an isometric view, but only one axis is receding. One face of the object is drawn parallel to the plane of projection, along with any faces parallel to that one, and the rest are

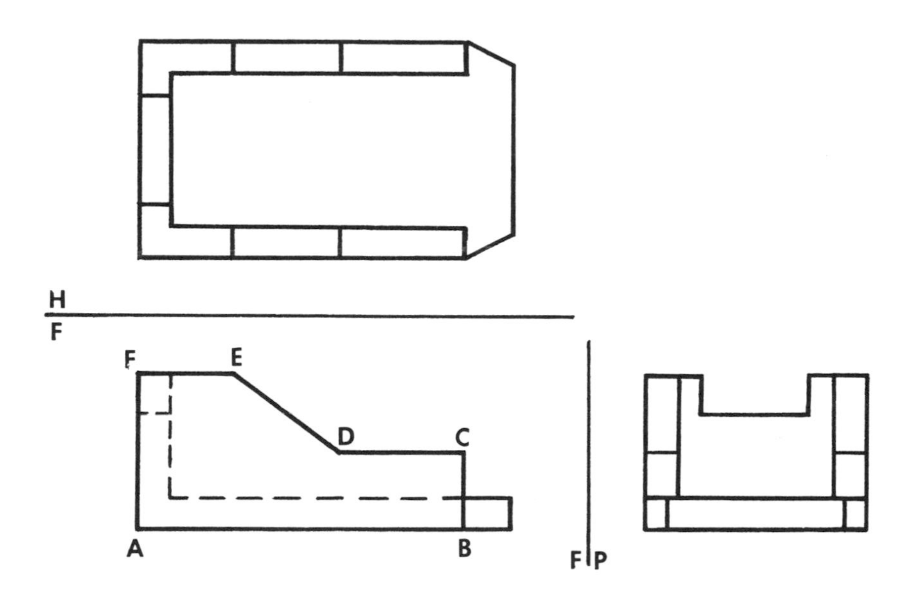

Fig. 1

constructed with the oblique axis. The axis that recedes usually does so at a 45° angle with the horizontal, although other angles will also form effective oblique pictorials. As shown in fig. 2, construct the enclosing oblique box with a vertical axis, a horizontal axis, and a 45° receding axis. The overall width, depth, and height of the box are the true length dimensions transferred from the orthographic views. When all true dimensions are used, the drawing is known as a Cavalier drawing. Locate face ABCDEF on the front plane of the oblique box, as shown in fig. 2. In constructing an oblique pictorial, the object is positioned in the most effective way to reduce the distortion. This means that, if the object is rectangular, as in this case, the face with the longer side is placed parallel to the plane of projection. If the object contains circular faces or planes, such as a cylinder, those faces are placed parallel to the plane of projection for ease of drawing. The dimensions are all true length, taken

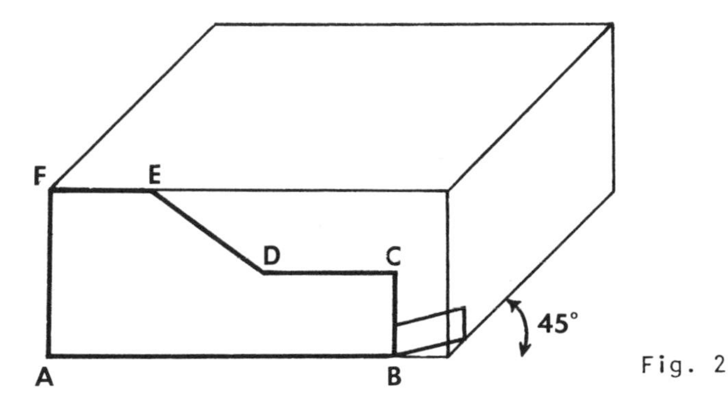

Fig. 2

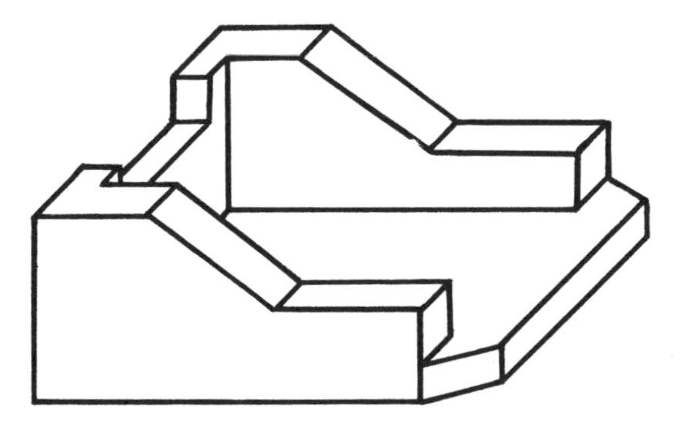

Fig. 3

from the orthographic views. The right corner is angled towards the back plane. All lines that are not on the oblique, vertical, or horizontal axes are constructed by locating their endpoints and connecting them. Layout the inclined face of the right corner in this manner. Complete the oblique view of the box jig by constructing the remaining faces, including all inclined ones, as shown in fig. 3. Darken principal lines and erase guide lines used.

Construct a Cavalier oblique view of the post footing, given the top, front, and right side views.

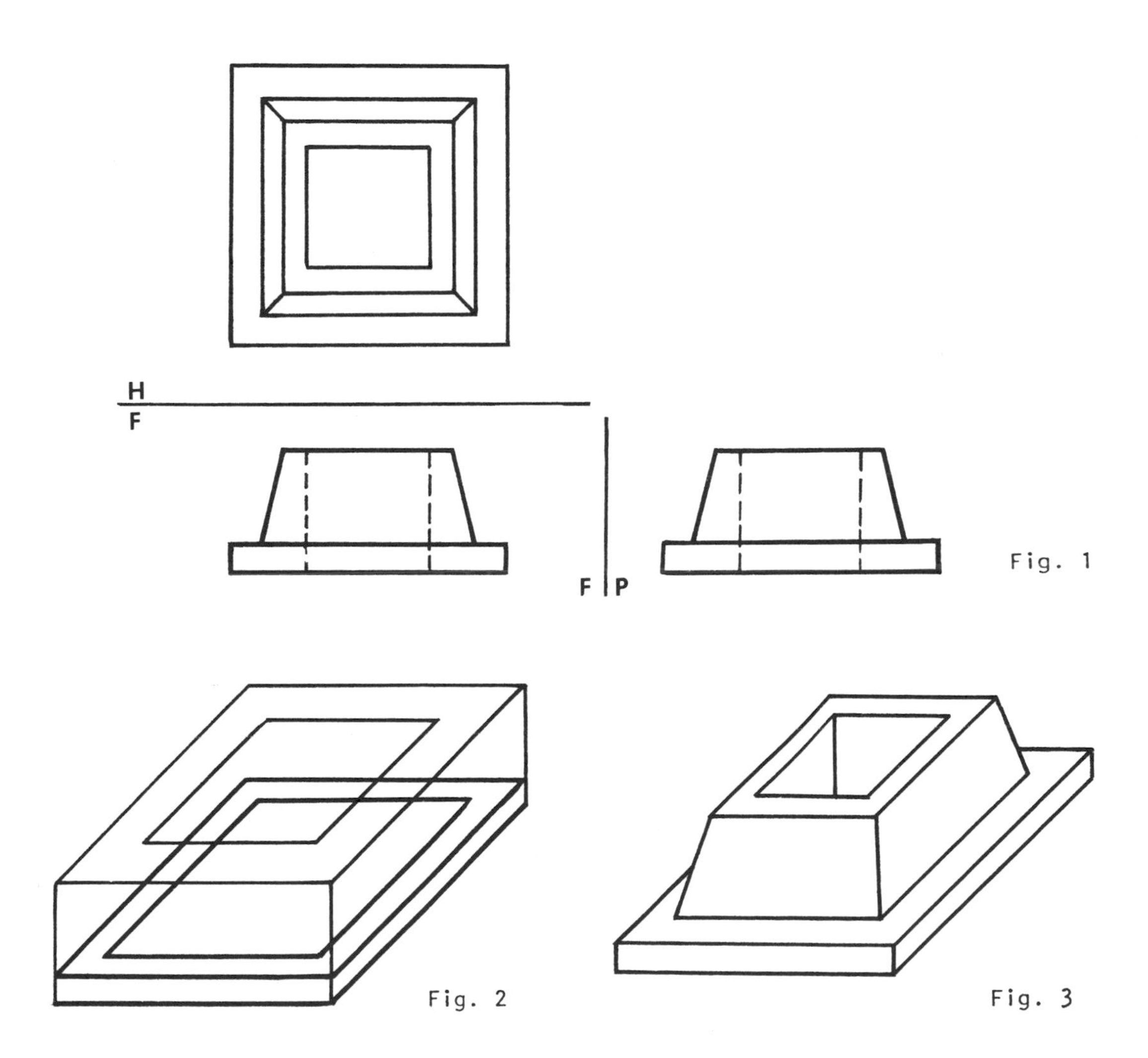

Fig. 1

Fig. 2

Fig. 3

Solution: Fig. 1 shows the given views of the post footing. The oblique box enclosing the object will have overall dimensions equal to those of the object, transferred from the orthographic views. As in fig. 2, construct the enclosing box, and layout the base. On the top plane of the box, layout the corner points for the top surface of the post footing by transferring true length distances from the top view. Locate the points where the inclined lines from the top corner points intersect the base by the same method. As in fig. 3, connect the endpoints of the inclined lines, and complete the oblique view by transferring the remaining faces to the oblique box. Darken the principal lines and erase the guide lines used.

• PROBLEM 10-21

Construct a Cavalier oblique view of the clutch, given the front and right side views.

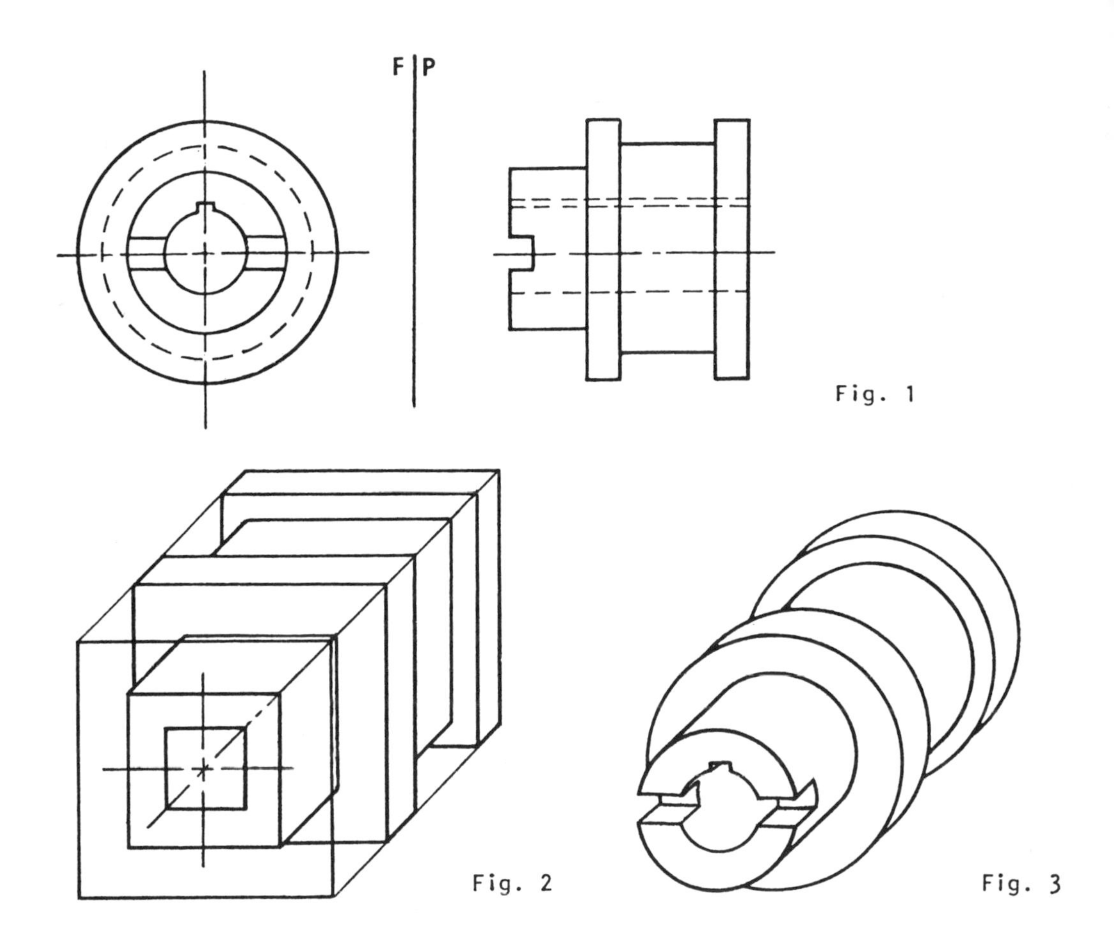

Fig. 1

Fig. 2

Fig. 3

Solution: Fig. 1 shows the given views of the clutch, which has many circular faces of different sizes. Because it is easier to construct circles than ellipses, position the clutch with these parallel circular faces parallel to the frontal plane of projection so that the faces appear true size and shape. As in fig. 2, construct the oblique box to enclose the clutch using the overall dimensions transferred true length from the orthographic views. Layout the faces of the clutch as though they were square instead of circular. Locate the centers of the squares, all of which are not shown in fig. 2 for clarity. Construct the circles and the keyway on the front face, and complete the required oblique view by darkening principal lines and erasing guide lines used. Fig. 3 shows the complete Cavalier pictorial.

• PROBLEM 10-22

> Construct a Cavalier oblique view of the yoke, given the top, front, and right side views.

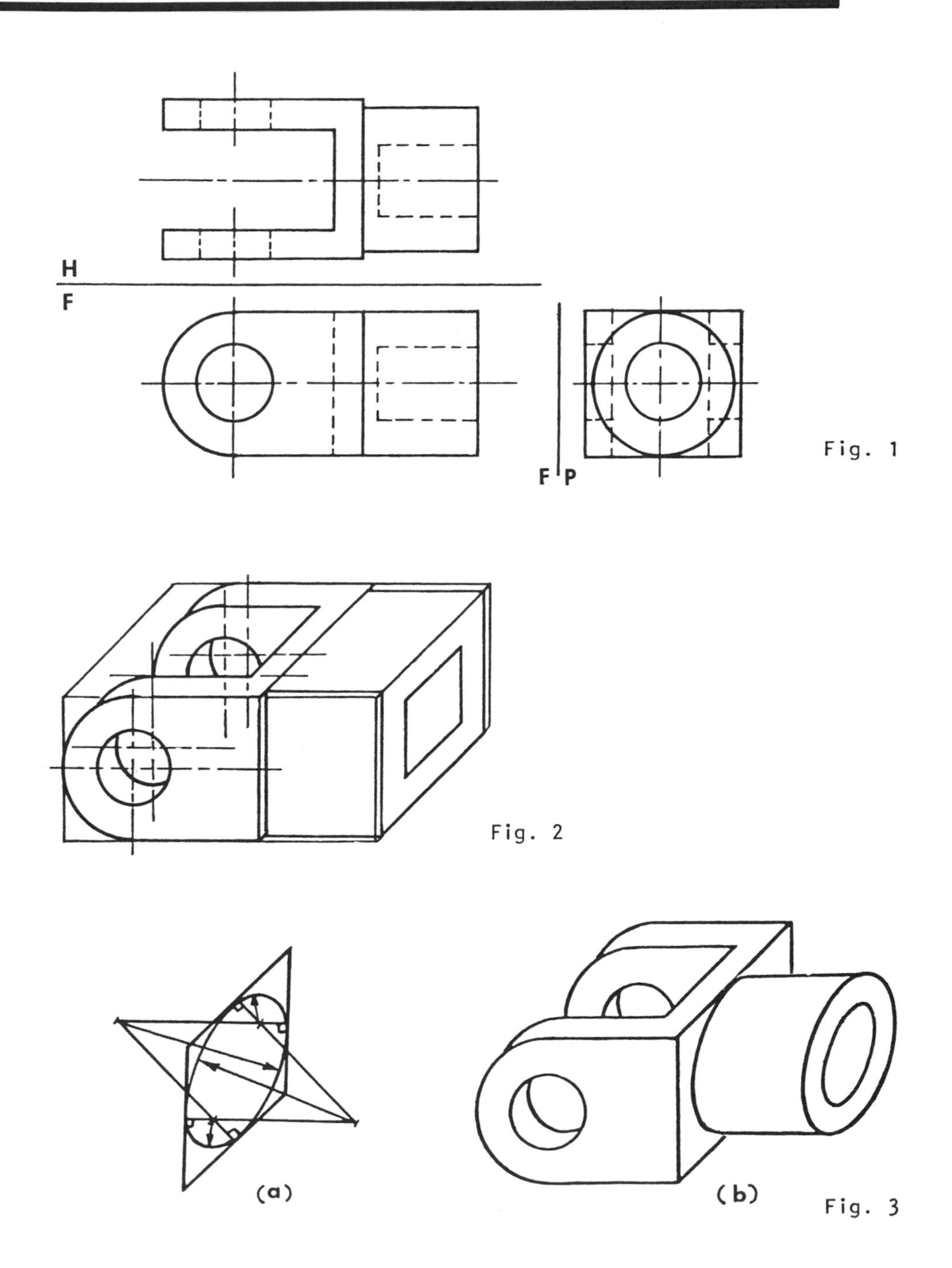

Fig. 1

Fig. 2

Fig. 3

Solution: Fig. 1 shows the given views of the yoke. This object has circular faces that lie parallel to two different axes. Since all these faces cannot be drawn parallel to the frontal plane of projection, a position must be chosen for the yoke to yield the most effective oblique view. Place the longest axis parallel to the frontal plane to minimize the distortion. Because this drawing is to be a Cavalier drawing, in which all lengths are true, the circles that lie parallel to the receding axis will appear as ellipses. As in fig. 2, construct the enclosing oblique box of true overall dimensions, and layout the circles and arcs that are parallel to the frontal plane. Layout the box to enclose the cylindrical portion of the yoke on the right end of the oblique box. As illustrated in fig. 3(a), the ellipses on the receding axis are constructed by the four-center method. The square enclosing the circle to be drawn as an ellipse is drawn appropriately in the oblique view. Perpendicular bisectors are constructed from each side of the oblique square to locate the centers for the construction of the arcs of the ellipse. As in fig. 3(b), complete the required oblique view of the yoke, and darken principal lines and erase guide lines used.

• PROBLEM 10-23

Construct a Cavalier oblique view of the object whose top and front views are given.

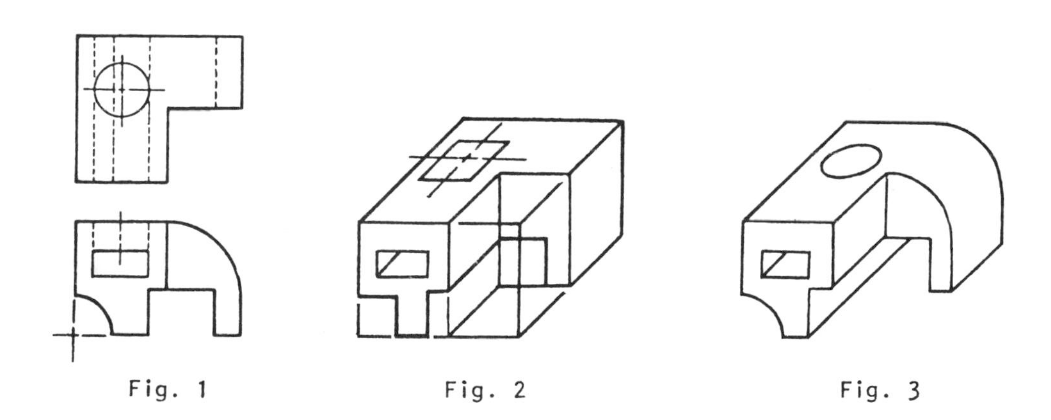

Fig. 1 Fig. 2 Fig. 3

Solution: Fig. 1 shows the given views of the object which has both circles and arcs on its various planes. As in fig. 2, construct the enclosing oblique box using dimensions equal to the overall height, width, and depth of the object, transferred from the orthographic views. Layout the visible planes of the object and outline the circles and arcs as though they were squares. Complete the oblique view by constructing the arcs, which are parallel to the frontal plane, as portions of circles, and the circle on the top plane as an ellipse. Darken principal lines and erase guide lines used, as shown in fig. 3. Note that the object was positioned in this manner because distortion is minimized in this case with the longest side parallel to the frontal plane of projection.

OBLIQUE PROJECTION

• PROBLEM 10-24

Construct an oblique projection of the block whose top and right side views are given.

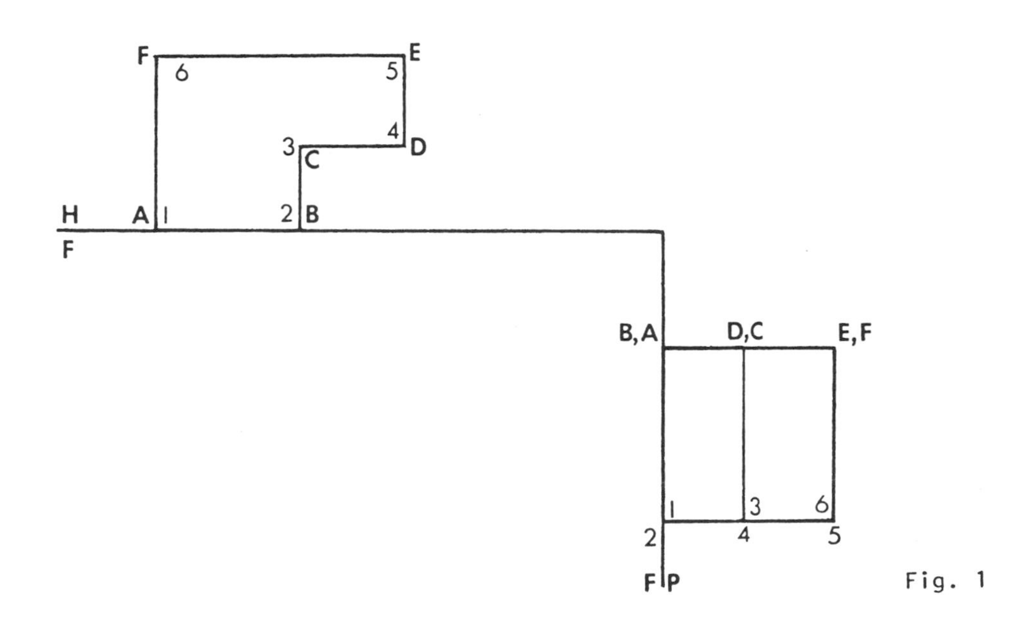

Fig. 1

Solution: Fig. 1 shows the given views of the L-shaped block. Oblique projection is a method of constructing a "three-dimensional" view or pictorial, using parallel projection lines that are oblique to the reference planes of the orthographic views. As illustrated in fig. 2, draw projection lines from points F, C, D, and E in the top view to reference plane H-F. These lines all make an angle of ϕ degrees with plane H-F, and therefore they are all parallel. Projectors from points A and B are not needed because these points already lie on the reference plane. Draw projection lines from points D(C), E(F), 4(3), and 5(6) in the right side view to reference plane F-P, all making an angle of θ degrees with plane F-P. Projectors from points B(A) and 2(1) are not needed because these points already lie on the reference, also. The face A-B-2-1 is true size and shape in the front view since it is parallel to both the H-F and the F-P reference planes. As in fig. 3, project the points where the projectors from points F, C, D, and E in the top view intersect plane H-F to the front view. Project the points where the projectors from points D(C), E(F), 4(3), and 5(6) in the right side view intersect plane F-P to the front view. The points where corresponding projectors intersect in the front view are the points of the edges of the block. Complete the front view, which is the required oblique projection, by connecting points that lie on the same edges, and determine visibility, as shown in fig. 3. Note that although face A-B-2-1 is true size and shape, the measurements along the receding lines are foreshortened, as is the case with axonometric projection. The angle θ, made between the F-P reference plane and the projectors from edge points of the right side view, should be less than 45 degrees to give a minimum amount of distortion in the oblique projection pictorial.

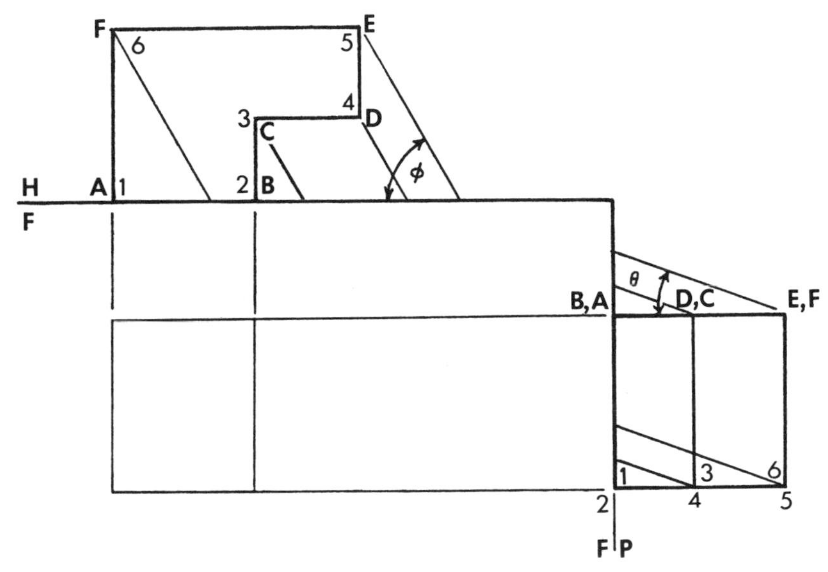

Fig. 2

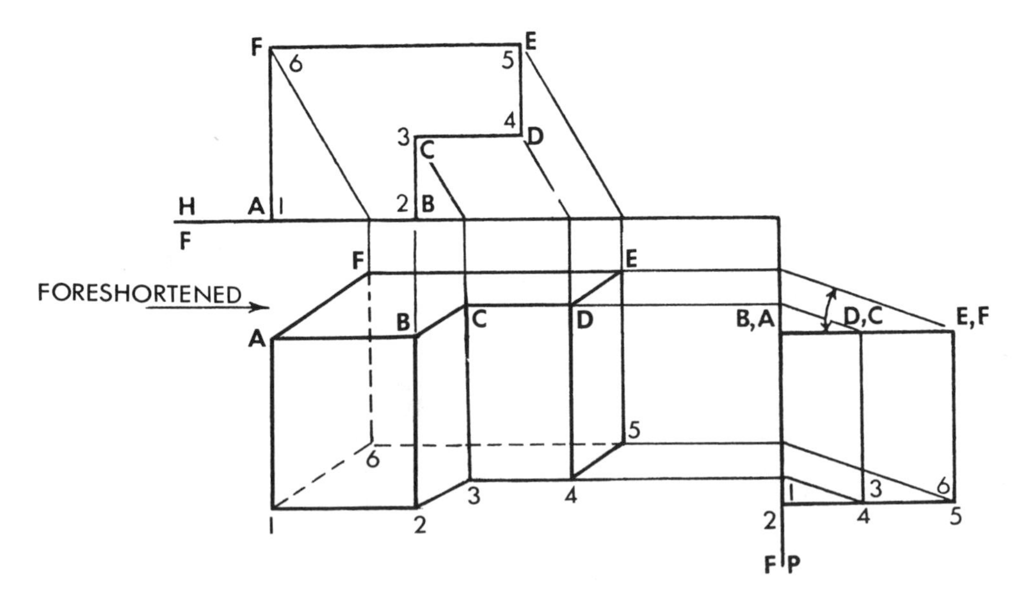

Fig. 3

• PROBLEM 10-25

Construct a cabinet oblique view of the pulley, whose front and right side views are given.

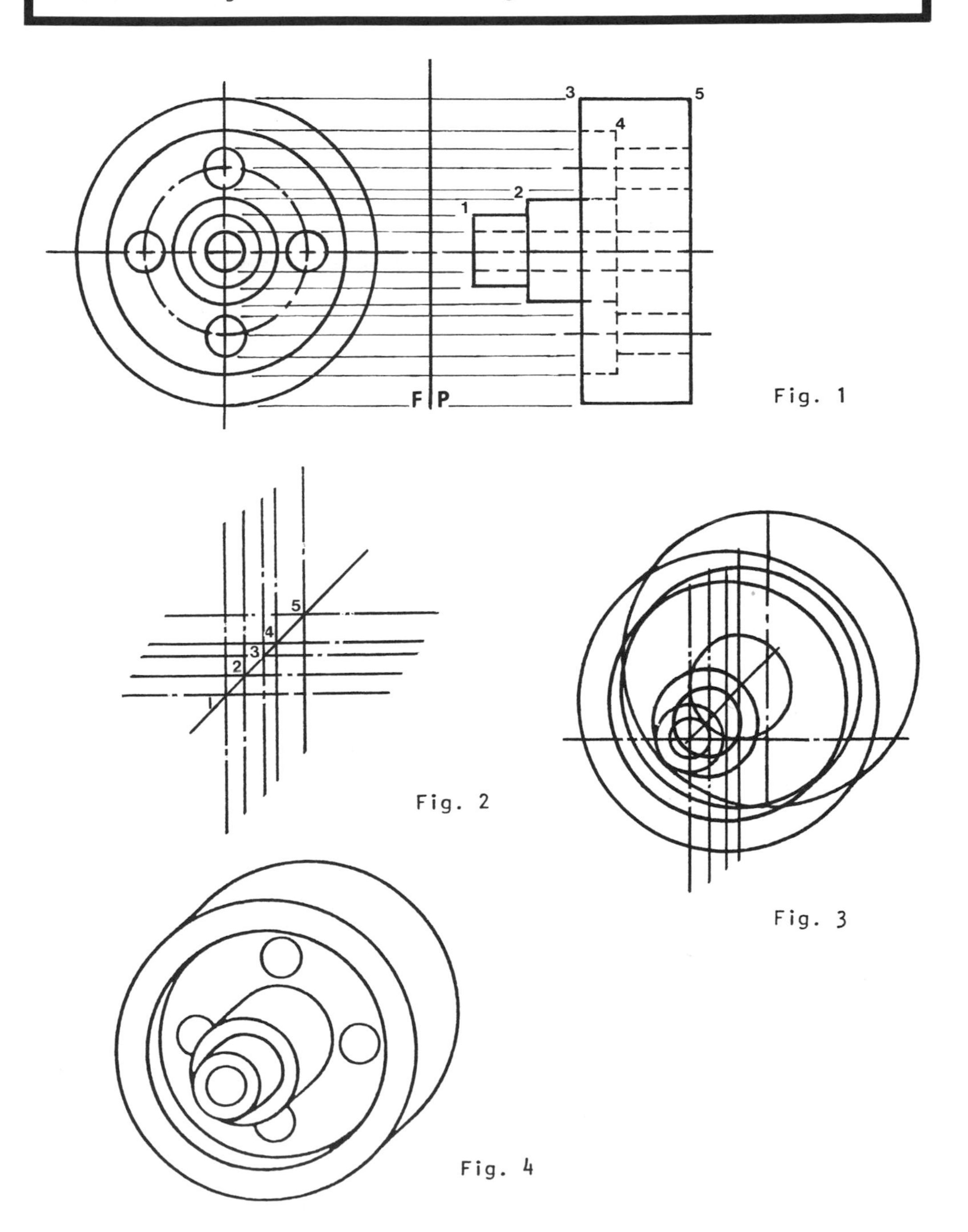

Fig. 1

Fig. 2

Fig. 3

Fig. 4

Solution: Fig. 1 shows the given views of the pulley, which has many circular surfaces. Position the pulley so that these surfaces, which are all parallel, are parallel to the frontal plane of projection. As shown in fig. 2, construct the center lines for the circular surfaces in oblique. As in fig. 3, construct the circles 1, 2, 3, 4, and 5 on the appropriate center lines with diameters as measured from the orthographic views. Complete the oblique view by connecting corresponding circles with tangent lines and adding the four small circles on surface 4. Erase the portions of surfaces that are not visible, and darken the principal lines, as shown in fig. 4.

• PROBLEM 10-26

Construct the cabinet oblique view of a given circle, using the method of plotting points.

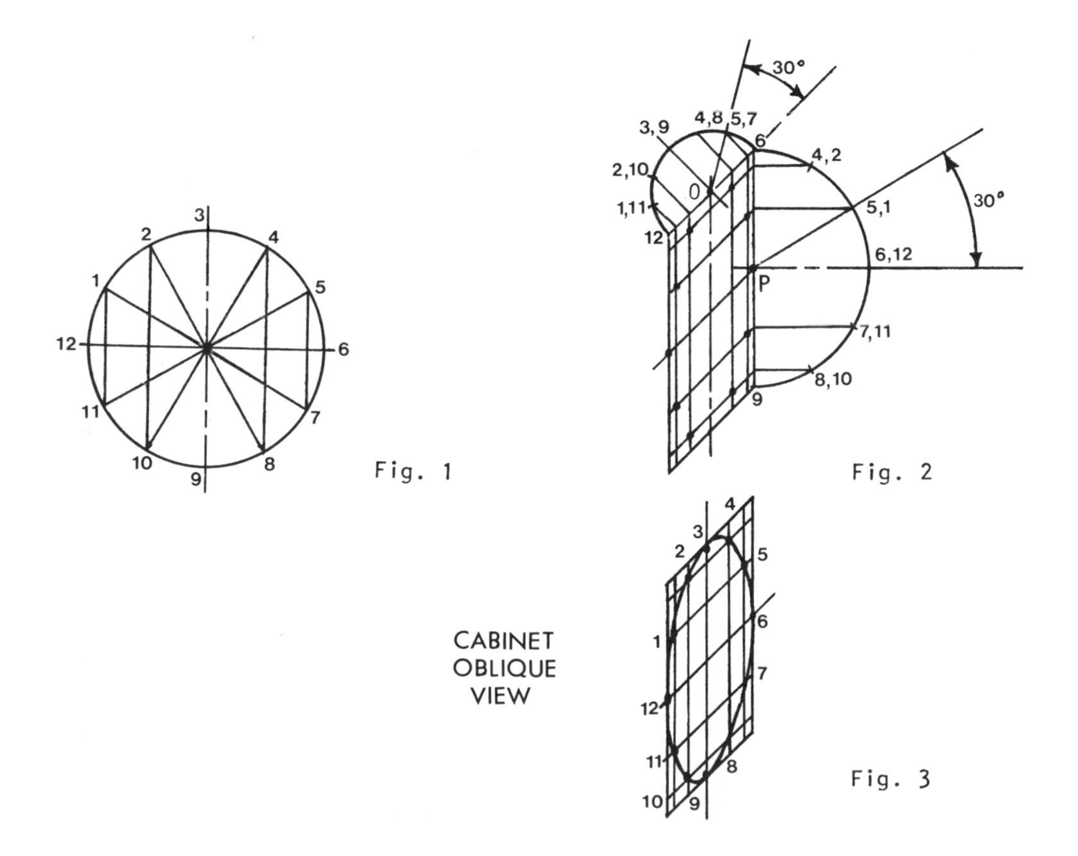

Solution: Fig. 1 shows the given circle with twelve points on the circle located at 30° intervals. A circle will appear as an ellipse in an oblique drawing if it lies in a plane parallel to the plane of the receding axis, or if it lies on the horizontal plane, which is formed with the receding axis. In a Cavalier oblique drawing, the four-center method or ellipse templates may be used to construct the ellipses because all lengths parallel to the axes are true lengths. These methods are completely invalid for use in a cabinet drawing, however, because the receding axis is reduced to half its true length while the vertical axis remains the same. Points must be plotted in order to yield a true view of the circle in oblique. As shown in fig. 2, construct the square which encloses the circle in oblique. The vertical sides are true length, but the receding sides are half-length. Locate the center lines of the circle. On one vertical side of the oblique rectangle, construct a semi-circle whose diameter is the length of the vertical side. On one receding side, construct another semi-circle whose diameter is the length of that side. Points 1 through 12 are labeled on each semi-circle at 30° intervals. Through the points on the "vertical" semi-circle draw horizontal lines to the vertical side of the rectangle, and through the points on the "receding" semi-circle draw lines perpendicular to and intersecting the receding side of the rectangle. Through the points where the horizontal lines of the "vertical" semi-circle intersect that side of the rectangle, construct lines parallel to the receding axis, and through the points where the inclined lines of the "receding" semi-circle intersect that side of the rectangle, construct vertical lines. These lines that form a grid across the oblique rectangle locate points on the ellipse. These points are the intersections of corresponding lines; for example, vertical line 4 and inclined line 4 intersect to locate a point of the ellipse. Connect these points with a smooth curve to complete the ellipse, as illustrated in fig. 3. The construction is neater if done on a piece of scrap paper and then the points transferred to the final drawing with dividers.

• PROBLEM 10-27

Construct a cabinet oblique view of the drill jig, given the top, front, and left side views.

Solution: Fig. 1 shows the given views of the drill jig. This object has circular portions on both the right side and the front. The planes containing the most circles, the left and right sides, should be parallel to the plane of projection, but if the receding axis containing the longest side of the drill jig were left full length, the distortion would be great. To avoid this, a cabinet drawing is employed. In a cabinet drawing, the receding axis is reduced to half-length, and all lines parallel to it are reduced to half their true length accordingly. As in fig. 2, construct the enclosing oblique box with true overall lengths along the horizontal and vertical axes, and half the true overall length along the receding axis. Draw the circles and semi-circles parallel to the front plane, and construct the box to enclose the circle and semi-circle on the receding side of the drill jig. Because the lengths measured along the receding axis are not true lengths, the four-center method or ellipse templates cannot be used to construct the ellipses. Instead points must be plotted.

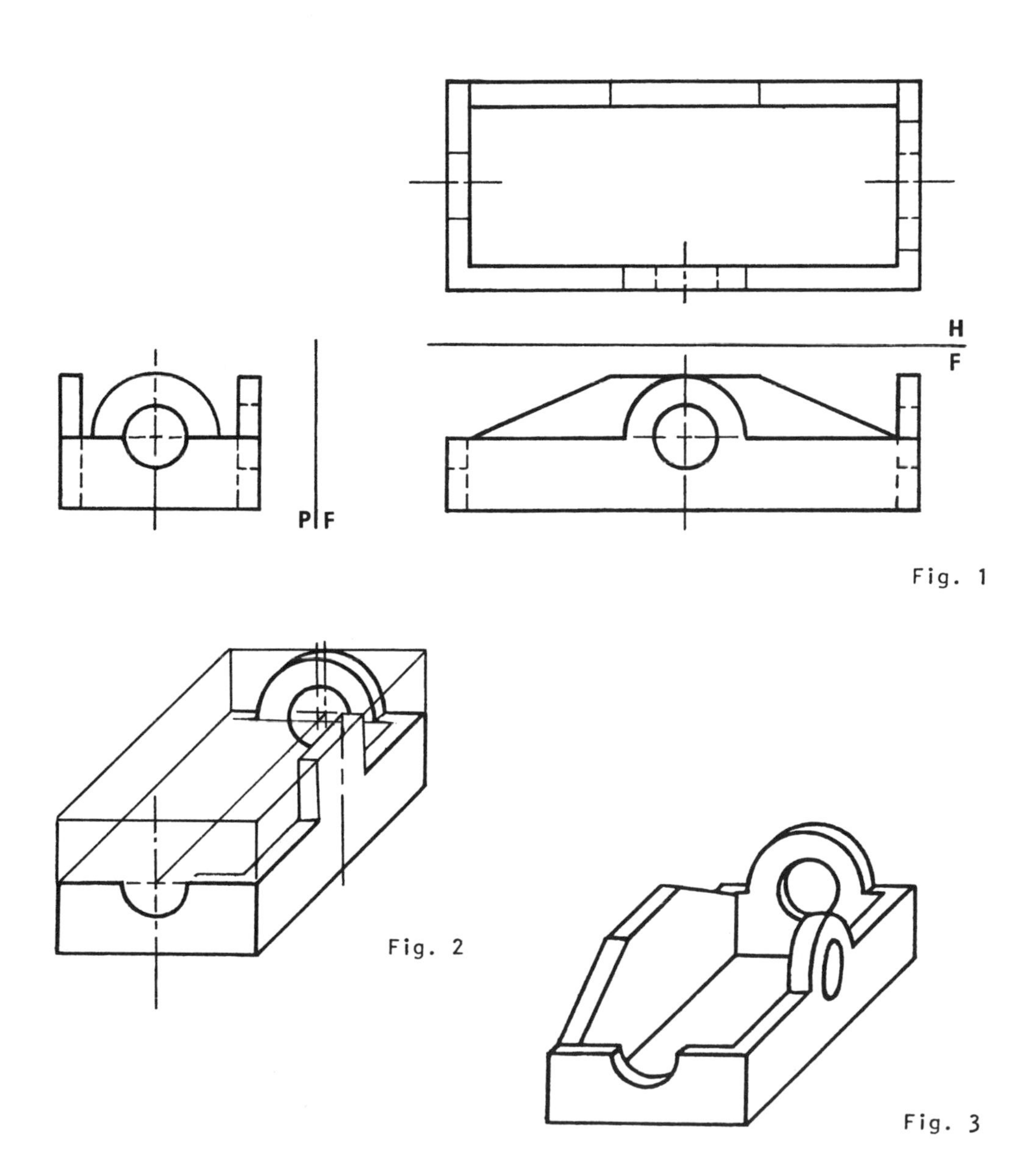

Fig. 1

Fig. 2

Fig. 3

Connect the points with a smooth curve for the oblique view of the circle and semi-circle, seen on the receding axis as elliptical. Darken principal lines and erase guide lines for the complete oblique pictorial of the drill jig, as illustrated in fig. 3.

CHAPTER 11

ONE AND TWO POINT PERSPECTIVE

PRINCIPLES OF PERSPECTIVE DRAWINGS

• PROBLEM 11-1

Describe one- and two-point perspective drawings and define or illustrate the following terms:

a) picture plane

b) horizon line

c) station point

d) vanishing point

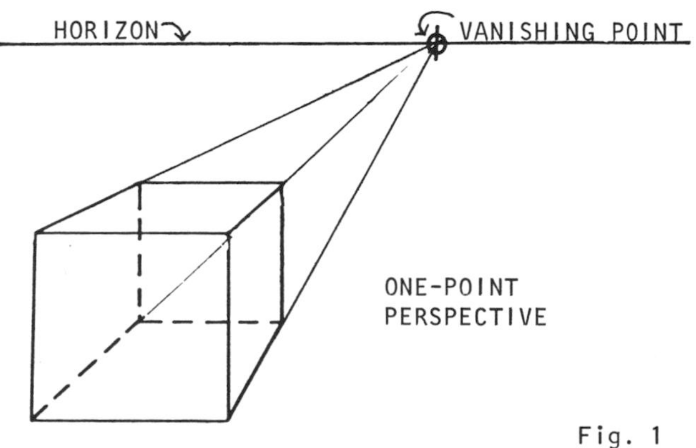

Fig. 1

Solution: A perspective view of an object is one in which parallel lines converge at a point as they recede from the observer. One- or two-point perspective views are so called depending on the number of points of convergence used in the view. As illustrated in fig. 1, one side of a block

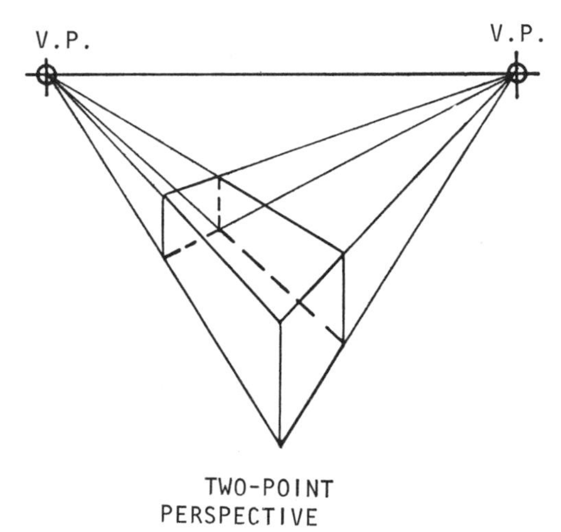

Fig. 2

lies on the plane of the observer, and this side appears true shape. The remaining sides of the block appear to recede, as their edges converge at a point (called the vanishing point) on the horizontal line drawn in the view (known as the horizon). This view of the block, in which

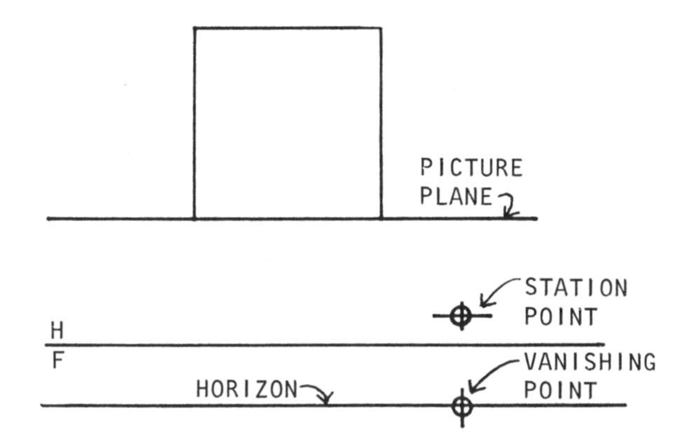

Fig. 3

parallel faces appear parallel to the observer, is called a one-point perspective view.

As illustrated in fig. 2, only one edge of the block lies on the plane of the observer. All faces of the block appear to recede, although vertical lines remain vertical since they are all parallel to the observer. The edges that are not vertical converge at two points on a horizontal line drawn in the view. This type of pictorial view is a two-point perspective view.

ONE POINT PERSPECTIVES

• PROBLEM 11-2

Construct a perspective of an object given the top and right side views, the station point, the ground line and the horizon shown in figure 1.

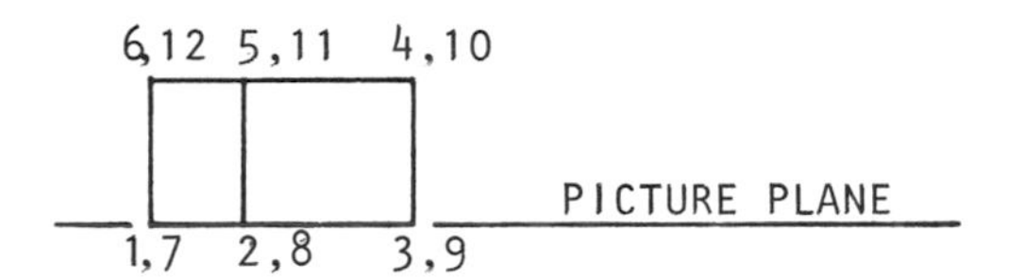

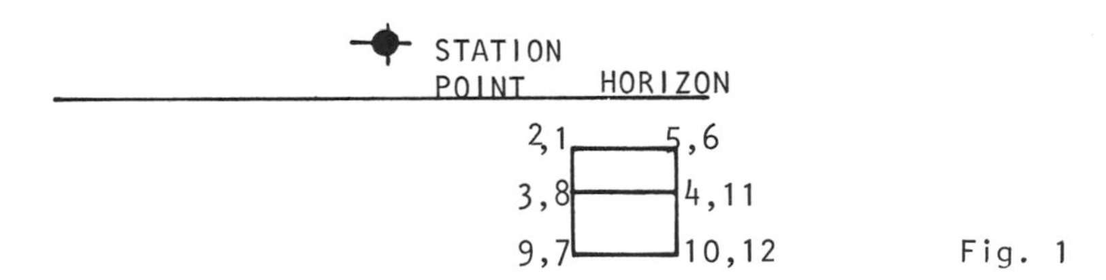

Fig. 1

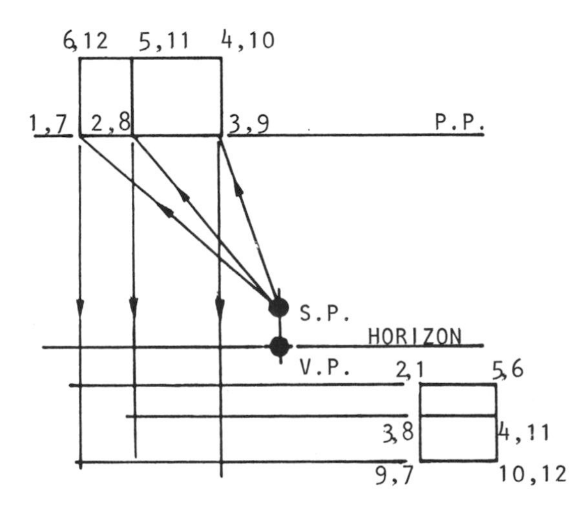

Fig. 2

Solution: First, draw projections from the top and side views to establish the front view of the object, as in figure 2. Next, locate the vanishing point. The vanishing point is on the horizon and it is the front view of the station point. It can therefore be found by dropping a vertical projector from the station point onto the horizon.

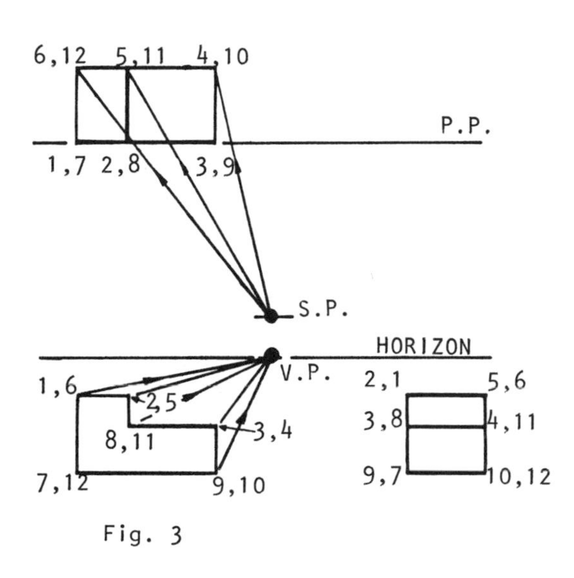

Fig. 3

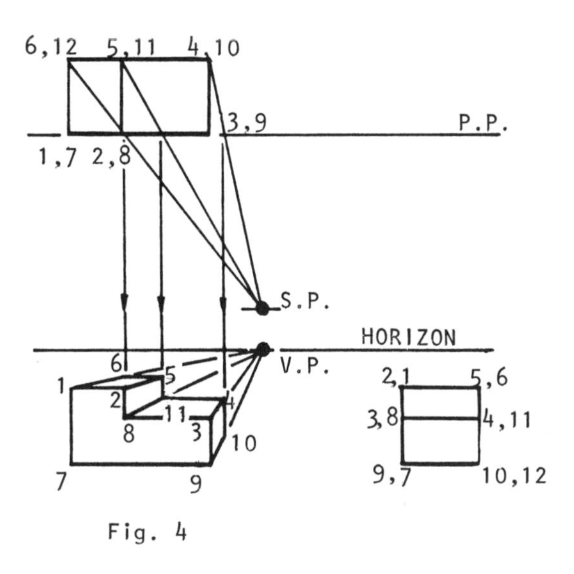

Fig. 4

From points other than those on the picture plane in the top view, draw projectors to the station point and from points in the front view, draw projectors to the vanishing point as shown in figure 3. Construct vertical projectors to the front view at the points where the projectors to the station point intersect the picture plane in the top view. The points at which these projectors cross their corresponding projectors to the vanishing point in the front view, will establish the complete perspective. See figure 4.

• PROBLEM 11-3

Construct a perspective of an object given the front and top view in figure 1.

Solution: In fig. 2, the edge view of the picture plane, which could have been placed anywhere in the top view, was conveniently constructed so that the object is resting upon it. The horizon, which could have been placed anywhere in the front view, was placed above the object. Often the top view is imposed on the front view in the area between the station point and the edge view of the picture plane and fig. 2 is spaced in this manner. The vanishing point was found by drawing a vertical projector from the station point to the horizon.

From points not resting on the picture plane in the

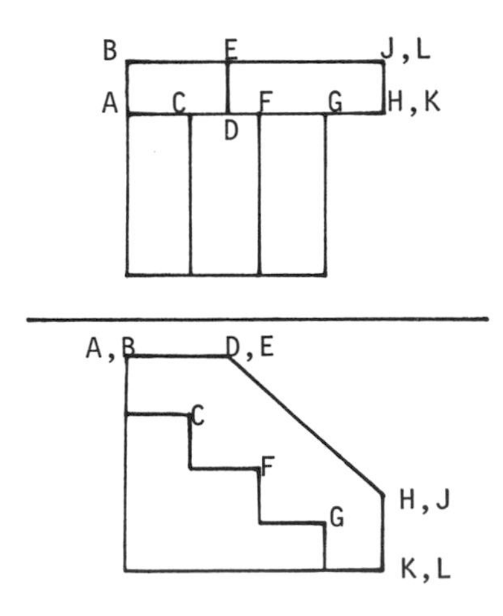

Fig. 1

top view, projectors are drawn to the station point (as shown in fig. 2) and from the points in the front view (as shown in fig. 3), projectors are drawn to the vanishing point. In fig. 2 notice that the entire figure was not included in the front view for clarity, although projectors from point A, B, point K, L, point H, J and point D, E were drawn to the vanishing point as were all other points (in fig. 3).

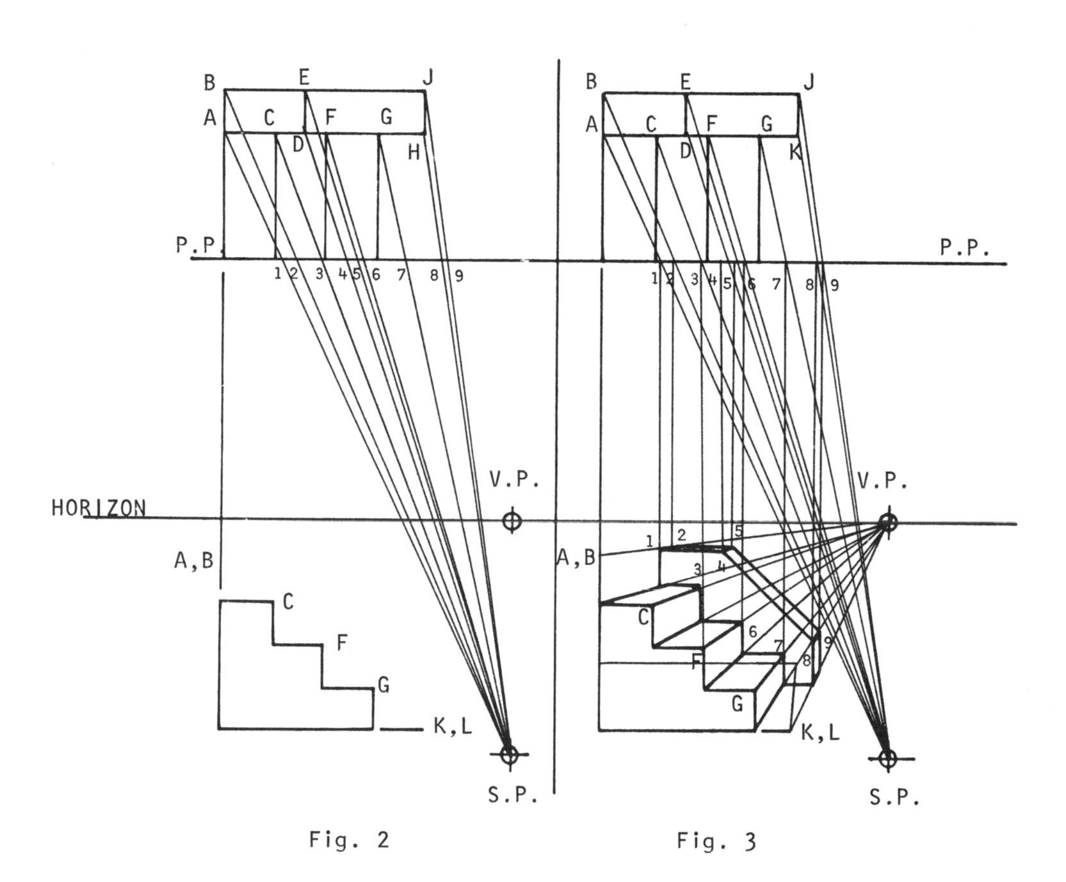

Fig. 2 Fig. 3

Vertical projections to the front view are constructed at the points where the projectors to the station point intersect the picture plane in the top view. The points at which these projectors cross their corresponding pro-

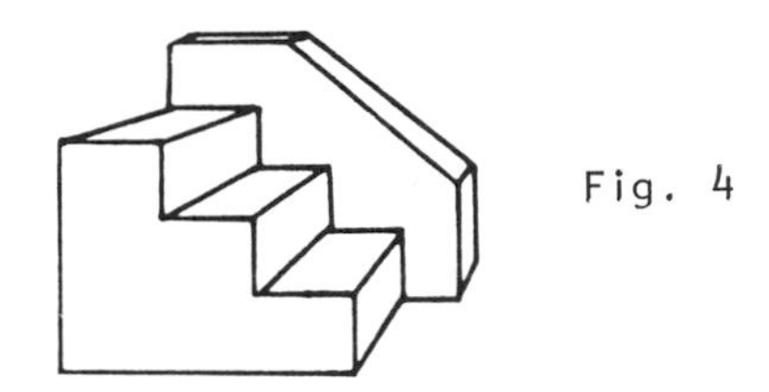

Fig. 4

jectors to the vanishing point in the front view, establish the complete perspective. See fig. 4.

• PROBLEM 11-4

A. Given the front and top views of an object, the horizon, the picture plane, the station point (A) and the vanishing point (A) in figure 1, construct a perspective of the object.

B. Construct another perspective of the same object in part (A) using station point (B) and vanishing point (B) in figure 1.

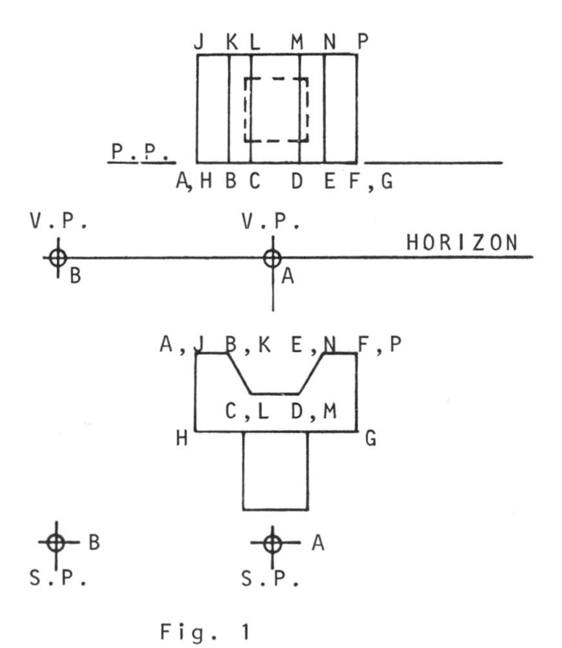

Fig. 1

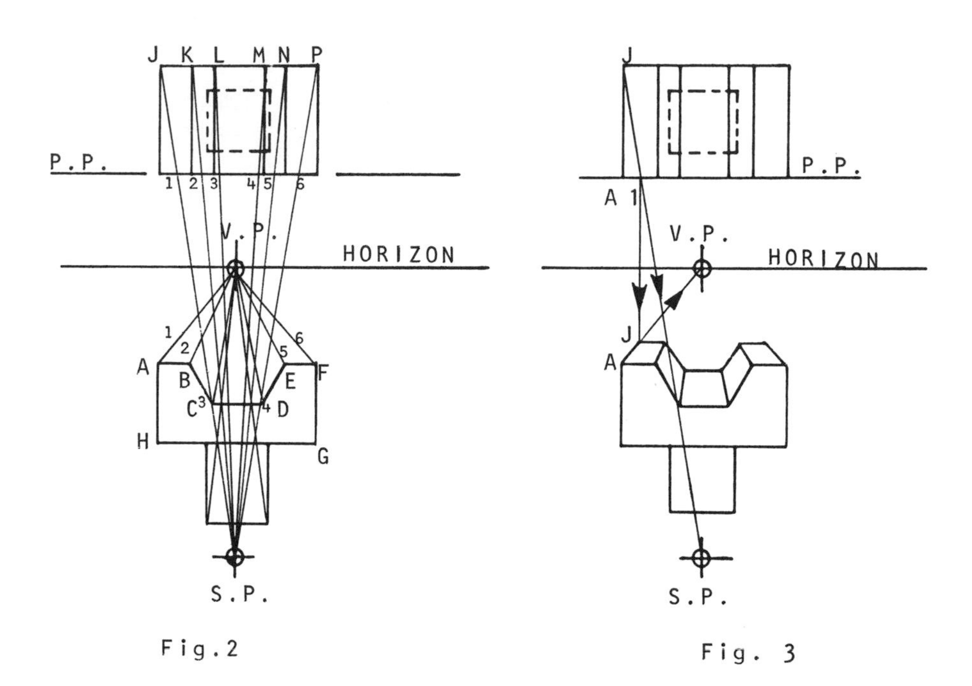

Fig.2

Fig. 3

Solution: A. From all points not on the picture plane in the top view, draw projectors to station point (A). From all points in the front view, draw projectors to vanishing point (A). Construct vertical projectors to the front view at the points where the projectors to station point (A) intersect the picture plane in the top view. The points at which these vertical projectors inter-

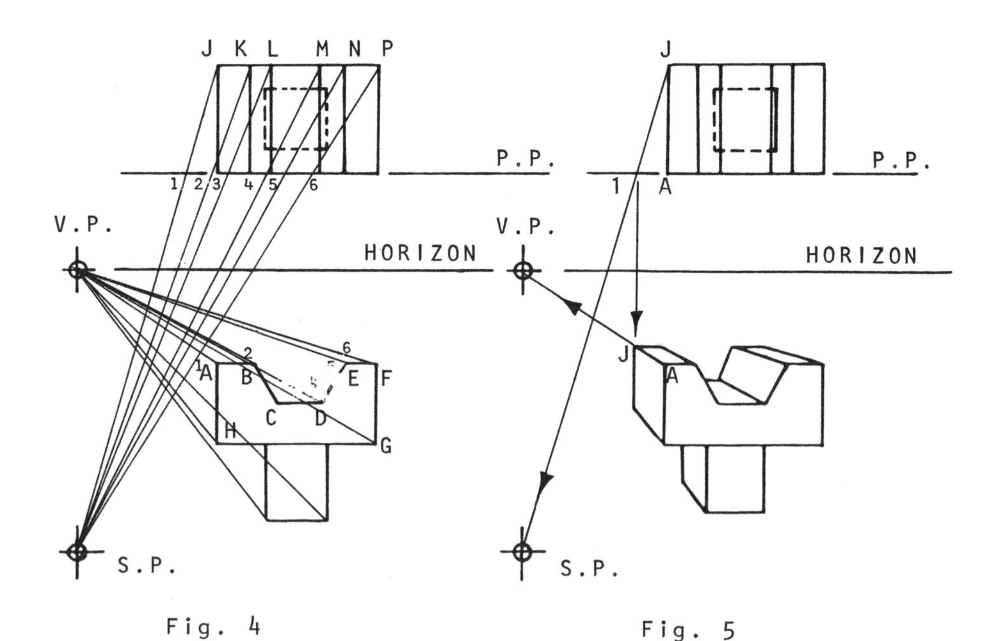

Fig. 4

Fig. 5

sect their corresponding projectors to vanishing point (A) in the front view, will establish the perspective.

B. The procedure for the construction of this perspective is the same as that explained above except vanishing point (A) and station point (A) are replaced by vanishing point (B) and station point (B).

• PROBLEM 11-5

Construct a perspective of an object given the object's top, front and right side views, the station point, the picture plane and the horizon shown in figure 1.

Solution: This perspective involves an object with a hole in it. Let us first complete the perspective as far as straight lines are concerned. This is done by constructing projectors from each point in the top view, not on the picture plane, to the station point and constructing projectors from points in the front view to the vanishing point. Then constructing vertical projectors to the front view from the points where the projectors to the station point intersect the picture plane in the top view. The points at which these vertical projectors cross their corresponding projectors to the vanishing point in the front view establish the perspective.

When constructing the perspective, draw projectors from points A, B, C and D to the station point and vanishing point in the corresponding views. Locate lines AB and CD on the perspective. The intersection of these two lines will indicate the center of the circle on the perspective, as in figure 2. As in figure 3, pass vertical lines through the object in the top view. Project these lines to the front view where they will appear as points. From these points draw projectors to the vanishing point. From the points in the top view, where the vertical lines intersect the circle, draw projectors to the station point. Draw vertical projectors to the front view at the points where the projectors to the station point intersect the picture plane in the top view. Connect the points where these vertical projectors intersect their corresponding projectors to the vanishing point in the front view with a smooth curve. Note that the accuracy of the curve will increase if more vertical lines are passed through the top view.

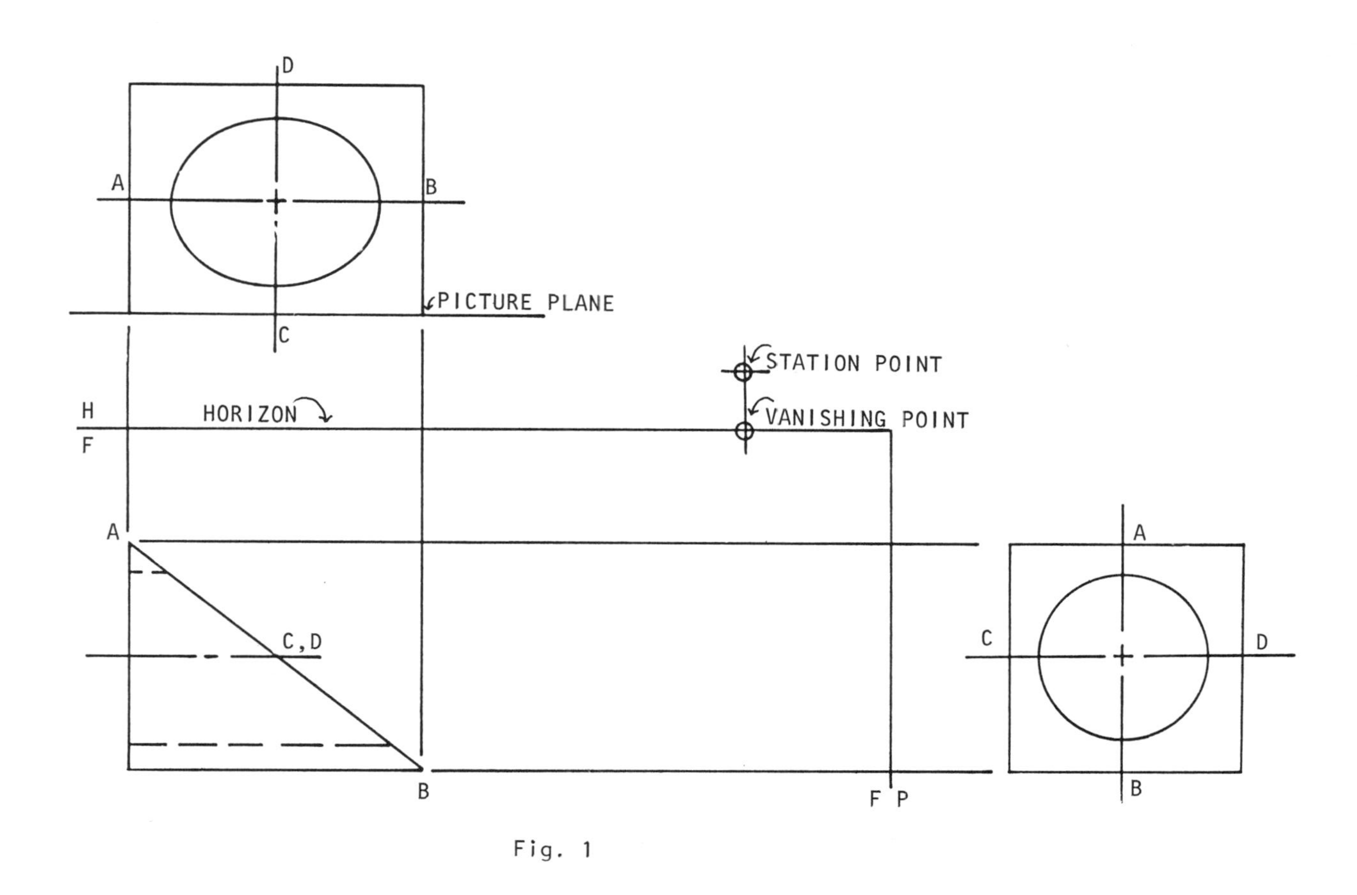
PICTURE PLANE
STATION POINT
VANISHING POINT
HORIZON
H
F
F P
A
B
C
D
C,D

Fig. 1

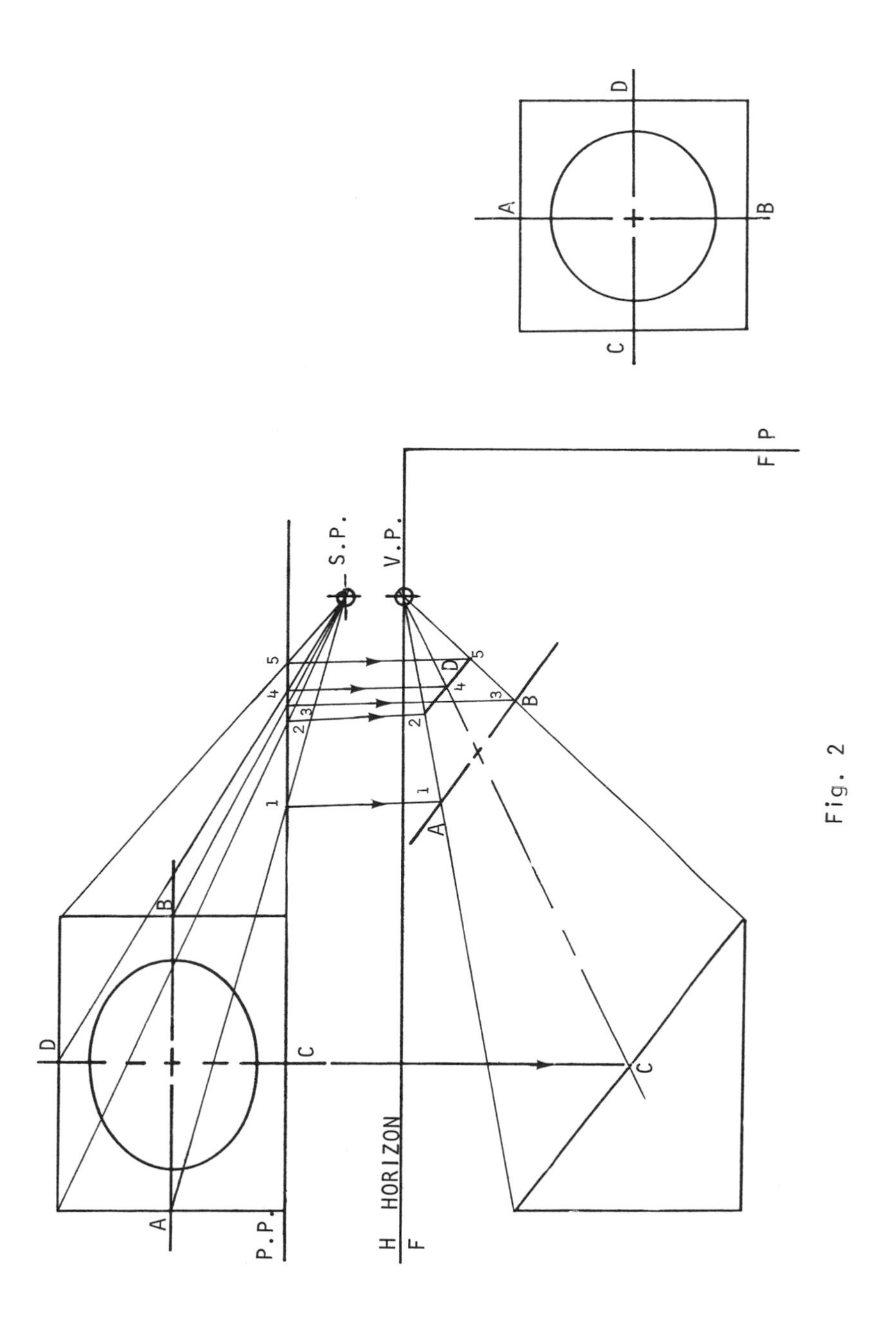

Fig. 2

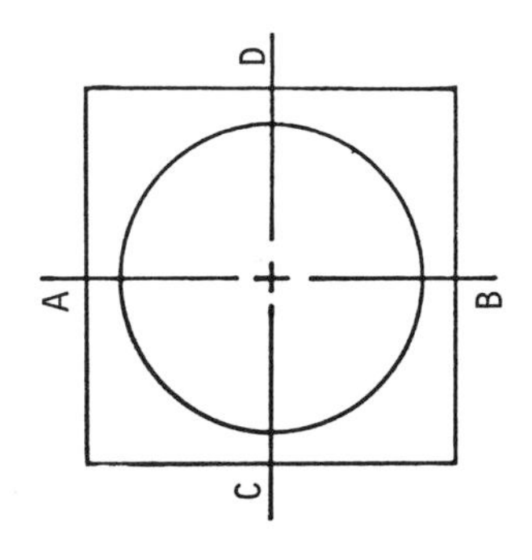

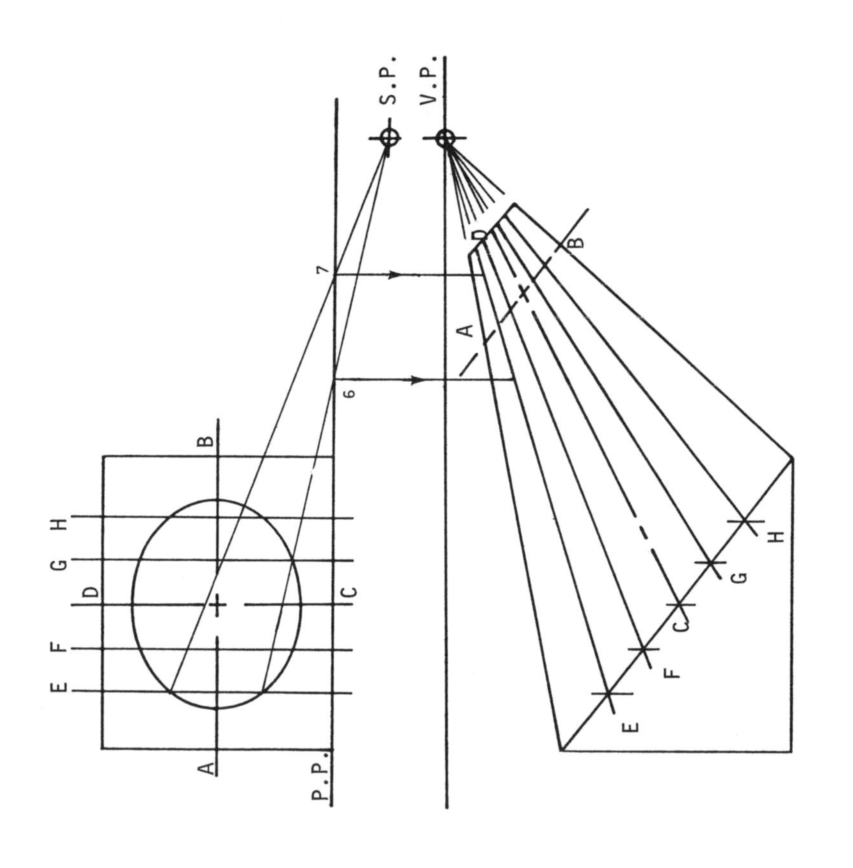

Fig. 3

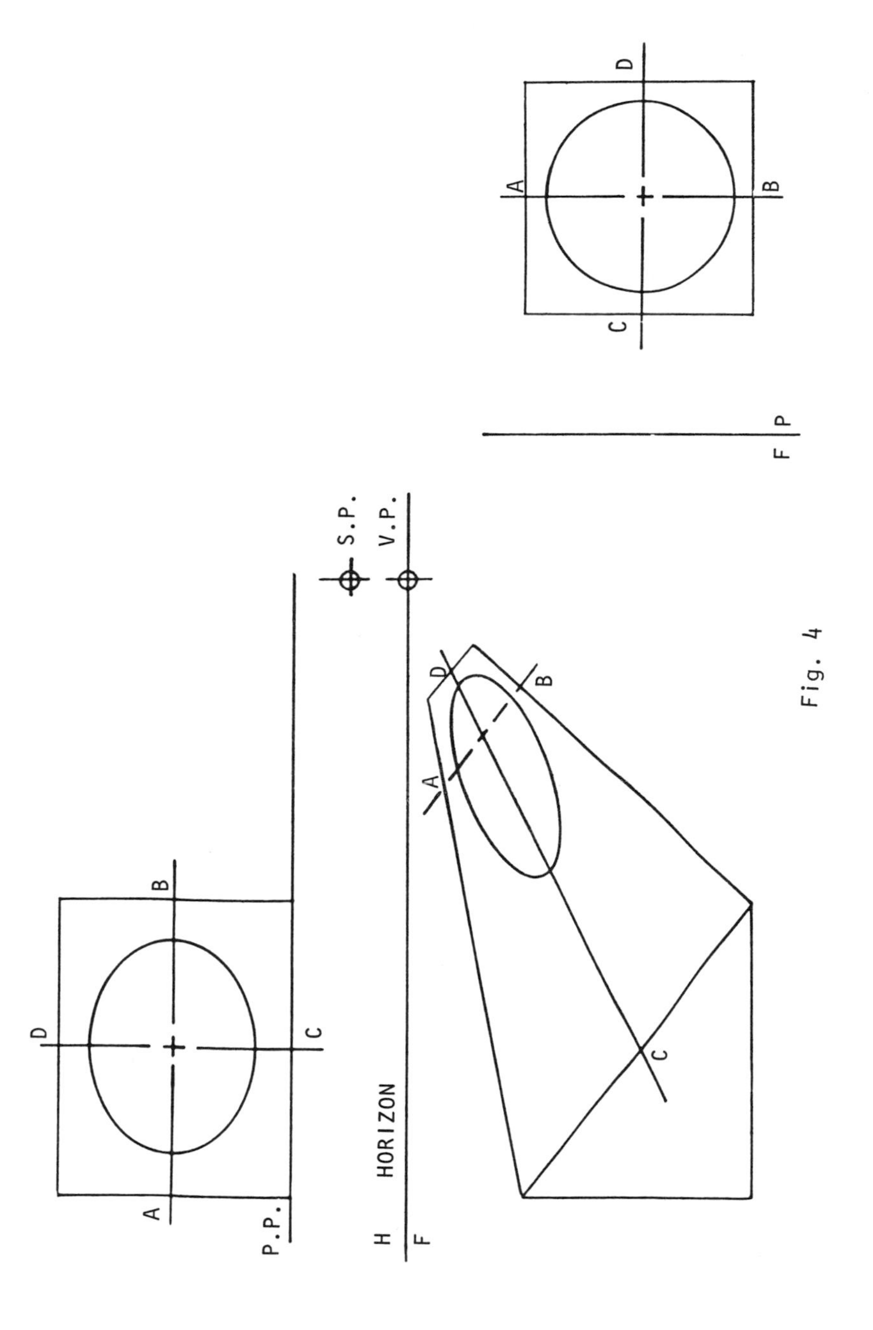

Fig. 4

Construct a perspective of an eccentric given the front, top and left side views, the horizon, the picture plane, the station point and vanishing point in figure 1.

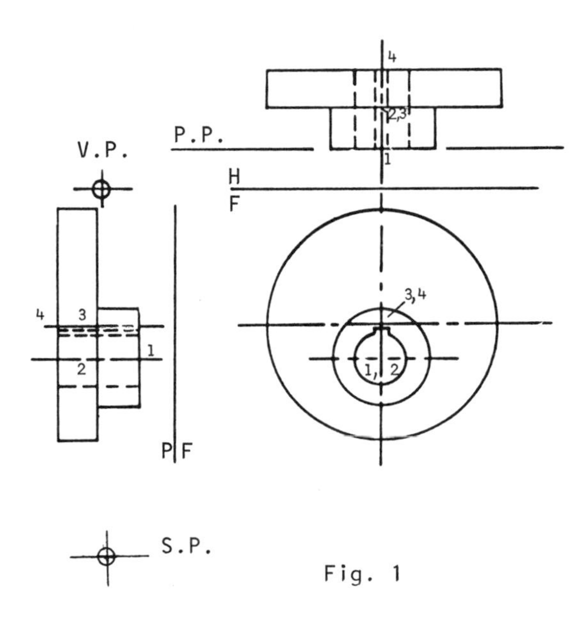

Fig. 1

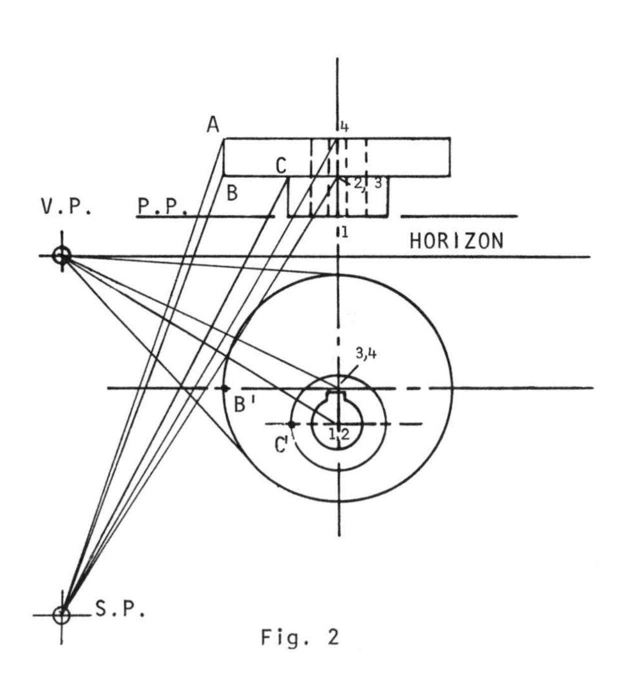

Fig. 2

Solution: Figure 1 shows an eccentric containing four circular edges parallel to the picture plane. Circle 1, at the front end of the hub and bore, has center 1, circle 2 has center 2 and radius 2C, circle 3 has center 3 and radius 3B, and circle 4 has center 4 and radius 4A. This can be seen by examining the front and top views in figure 2.

In a perspective drawing, only the surfaces which lie against the picture plane will be in true size. In this case, this will be circle 1. All of the other surfaces (circles 2, 3, and 4) will be smaller than true size.

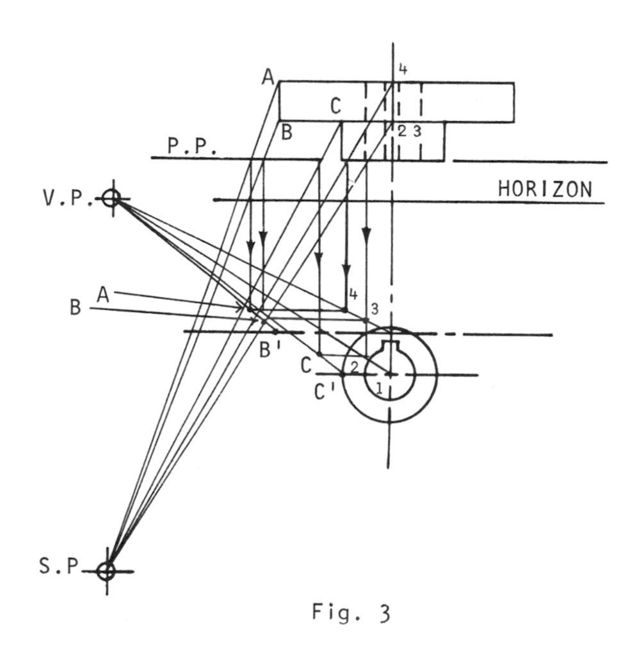

Fig. 3

Project points C, B, and A to the front view to C' and B' respectively (B' represents both points B and A in the front view). Line 2C' represents the radius of circle 2 and lines 3B' and 4B' represent the radii of circles 3 and 4 in the front view.

As in figure 3, leave out the front view of circle 3 (which coincides, in this view, with circle 4) for clarity. Draw projectors from the center point of each circle (point 1, 2 and point 3, 4) and from points C' and B' to the vanishing point, in the front view. In the top view, draw projectors from the center of each circle (point 1, point 4, and point 2, 3) and from points A, B, and C to the station point. Draw vertical projectors to the front view at the points where the projectors to the station point intersect the picture plane in the top view. Where these vertical projectors cross their corresponding projectors to the vanishing point in the front view, determine these points in the perspective. Connect points 2 and C, 3 and B, and 4 and A in the perspective, determining the size of the radii of circles 2, 3, and 4 in the perspective. Construct circle 3. This is done by drawing a circle of radius 3B centered at point 3 in the perspective. Now, circle 4 may be constructed. It lies behind circle 3 in the perspective and therefore only part of it may be viewed. To determine which part of it may be seen, construct projectors from the vanishing point tangent to circle 3 in

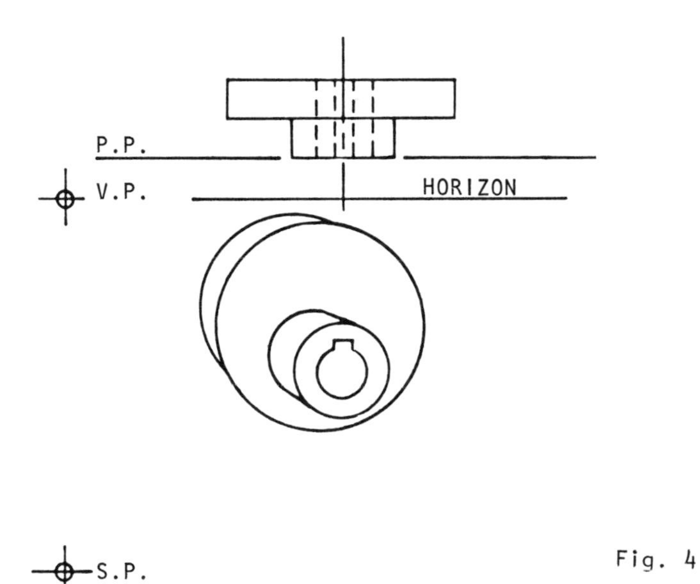

Fig. 4

the perspective. Draw circle 4 centered at point 4 with a radius of line 4A. The portion of circle 4 which lies within the area between the two projectors from the vanishing point and circle 3, may be seen. Lastly, construct circle 2 with radius of line 2C centered at point 2 in the perspective. The portion of circle 2 which can be viewed, lies between circle 1 and two projectors from the vanishing point tangent to circle 1. Erase all construction lines and darken areas which can be viewed. The complete perspective is shown in figure 4.

• PROBLEM 11-7

Construct a two point perspective of an object given the front and top views in figure 1.

Solution: Construct a horizontal picture plane in the

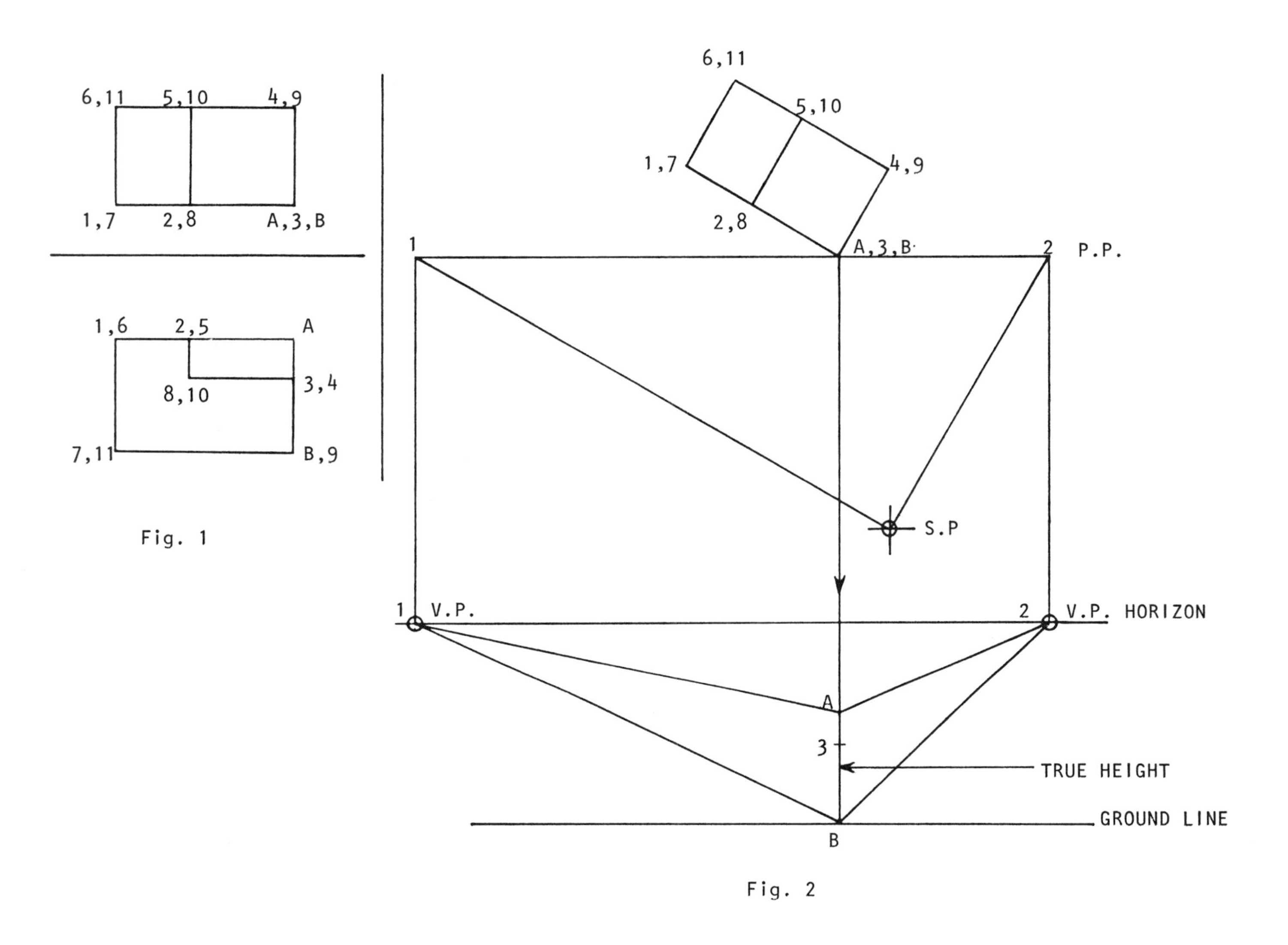
6,11
5,10
4,9
1,7
2,8
A,3,B
1,6
2,5
A
3,4
8,10
7,11
B,9
6,11
5,10
1,7
4,9
2,8
1
A,3,B
2
P.P.
S.P
1
V.P.
2
V.P.
HORIZON
A
3
TRUE HEIGHT
GROUND LINE
B

Fig. 1

Fig. 2

top view and horizon and ground line in the front view so the horizon is above the ground line. Rotate the top view of the object and place it on the picture plane so the forward edges of the object make angles of θ and $90-\theta$ with the picture plane respectively ($\theta = 30°$ is commonly used). Construct two lines from the station point to the picture plane parallel to the forward edges of the object (also making angles of θ and $90-\theta$). Construct vertical projectors to the front view from the points

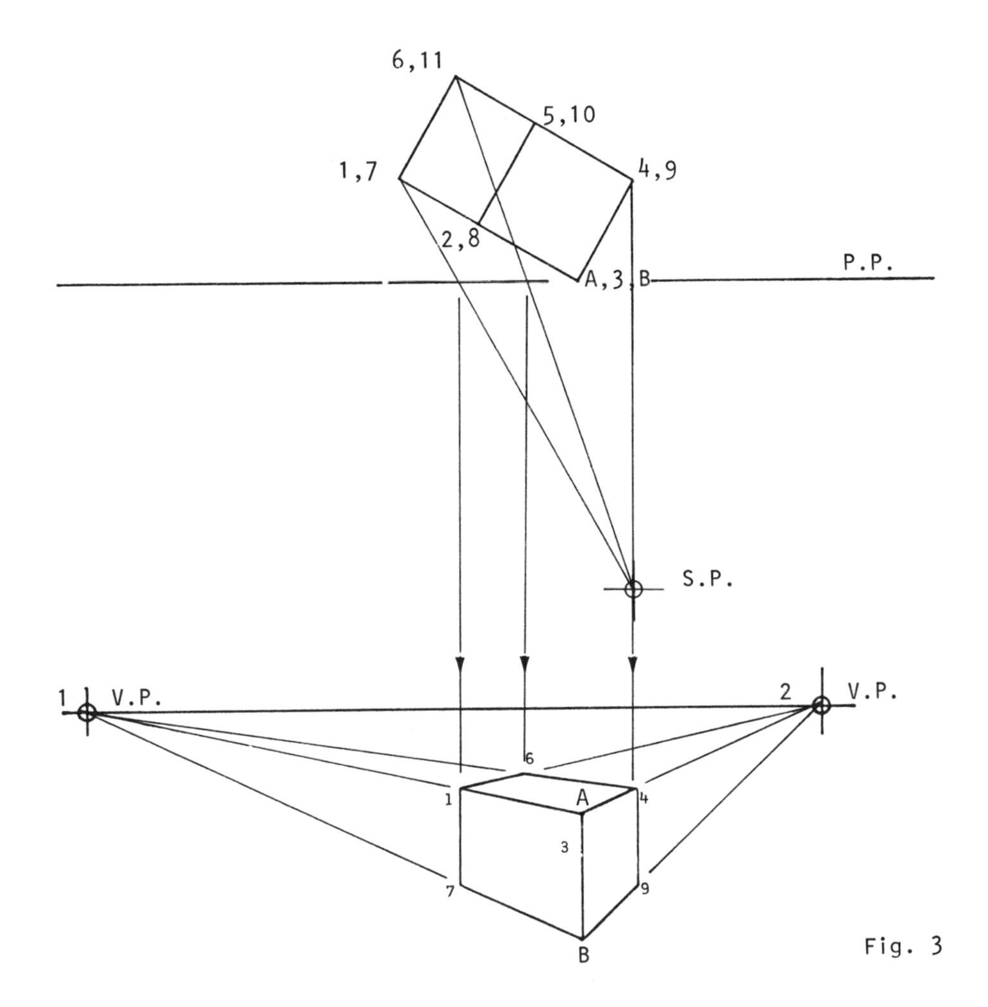

Fig. 3

of intersection of the picture plane and projectors from the station point in the top view. The points where these two vertical projectors intersect the horizon are the vanishing points. In a perspective, all lines on the picture plane will appear in true size. Construct vertical line A3B in true height (taken from the front view) on the ground line since AB is on the picture plane and therefore should appear in the true length. Draw projectors to the vanishing points from points A and B as in figure 2. From the top view, draw projectors from point 1,7 and point 4,9 to the station point. Construct vertical projectors to the front view from the points in the top view where the picture plane intersects the projectors to the station points. These will establish lines 1-7 and 4-9 in the front view. From points 4 and 1 in the front view draw projectors to the vanishing points. The intersection of these lines establishes point 6. This can be checked by drawing a projector from 6 to the station point and then

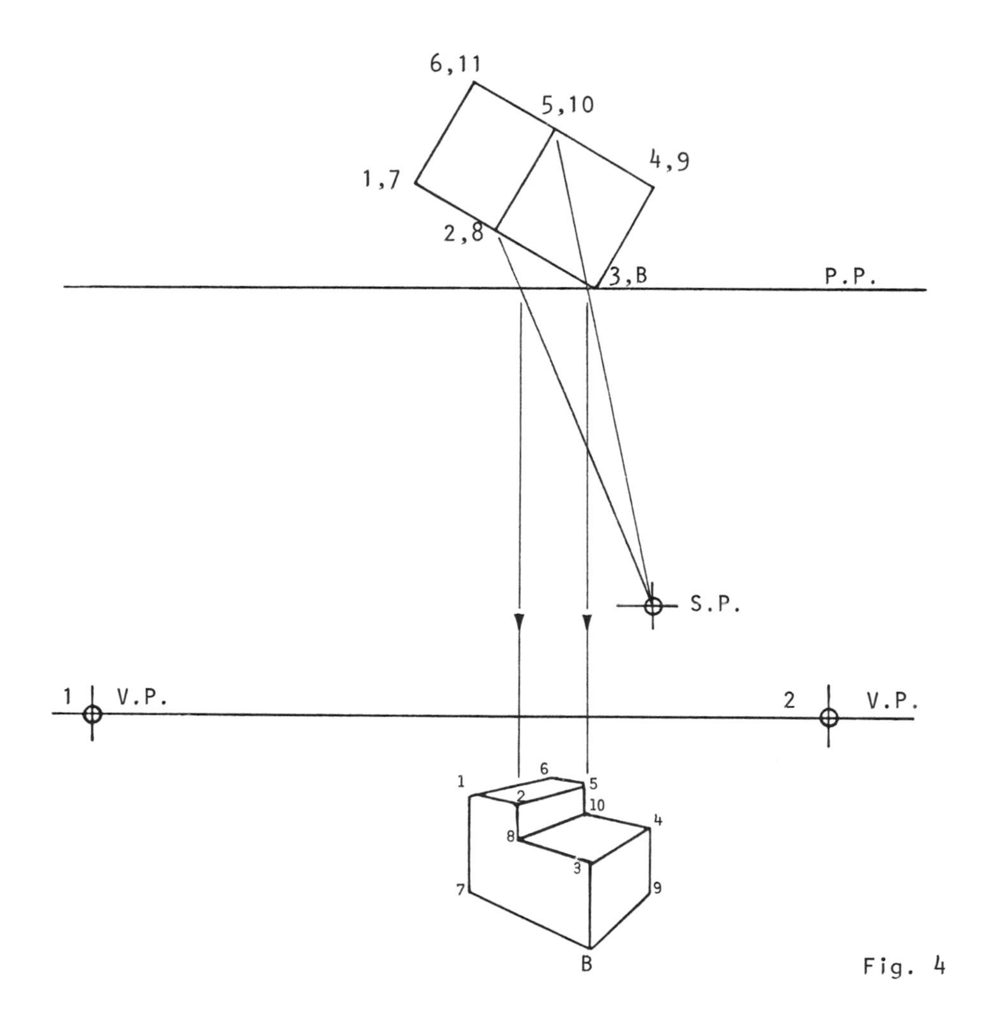

Fig. 4

dropping a vertical projector to the front view at the point where the picture plane and the projector from point 6 to the station point intersect. This establishes the box the figure is contained in (see figure 3). From point 3 on the line A3B in the front view draw projectors to the vanishing points. From point 2,8 and point 5,10 in the top view, draw projectors to the station point. At the point where these projectors in the top view cross the picture plane, construct vertical projectors to the front view. Where these projectors intersect lines A-1 and 6-4 (from figure 3) will establish points 2 and 5 in the perspective. Where these same vertical projectors intersect the projectors from 3 to the vanishing points will establish points 8 and 10 (see fig. 4). Connect points 2 and 5, 2 and 8, 8 and 10, and 10 and 5. Erase all construction lines. The complete perspective is shown in figure 4.

TWO POINT PERSPECTIVES

• PROBLEM 11-8

Draw a two point perspective of the house whose front and top views are given in the figure.

Solution: Draw a horizontal picture plane edge in the top view, and horizon and ground line in the front view so the horizon is above the ground line. Rotate the top view so the edges of the house make angles of θ and $90-\theta$ with the picture plane respectively (see figure). To locate the left and right vanishing points, draw two projectors from the station point to the picture plane at angles of θ and $90-\theta$ with a horizontal. These projectors will be parallel to the forward edges of the house. From the points where these two projectors intersect the picture plane, draw vertical projectors to the front view. The points where these vertical projectors to the front view cross the horizon are the two vanishing points (labeled V.P.L. and V.P.R. in the figure). Since point 1-2 is on the picture plane, in the top view, it will appear as a true length line in the perspective. Transfer, from the front view, the true height of line 1-2 to the perspective; also transfer the height of C and D onto the extended line 1-2 and label these C' and D'. Line 3-4 (which is not part of the house but useful for the construction of the perspective) is also on the picture plane and its true height can be transferred to the perspective from the front view (the height is the same as that of point A). Draw line 3-4, very lightly, as all construction lines are drawn. From points 1,2,3,4,C' and D' draw projectors to both the left and right vanishing points. These projectors represent the height of these points in the perspective. From each point in the top view draw projectors to the station point. From the points of intersection of these projectors to the station point and the picture plane, draw vertical projectors to the front view. Bring each vertical projector from the top view down until it intersects the corresponding

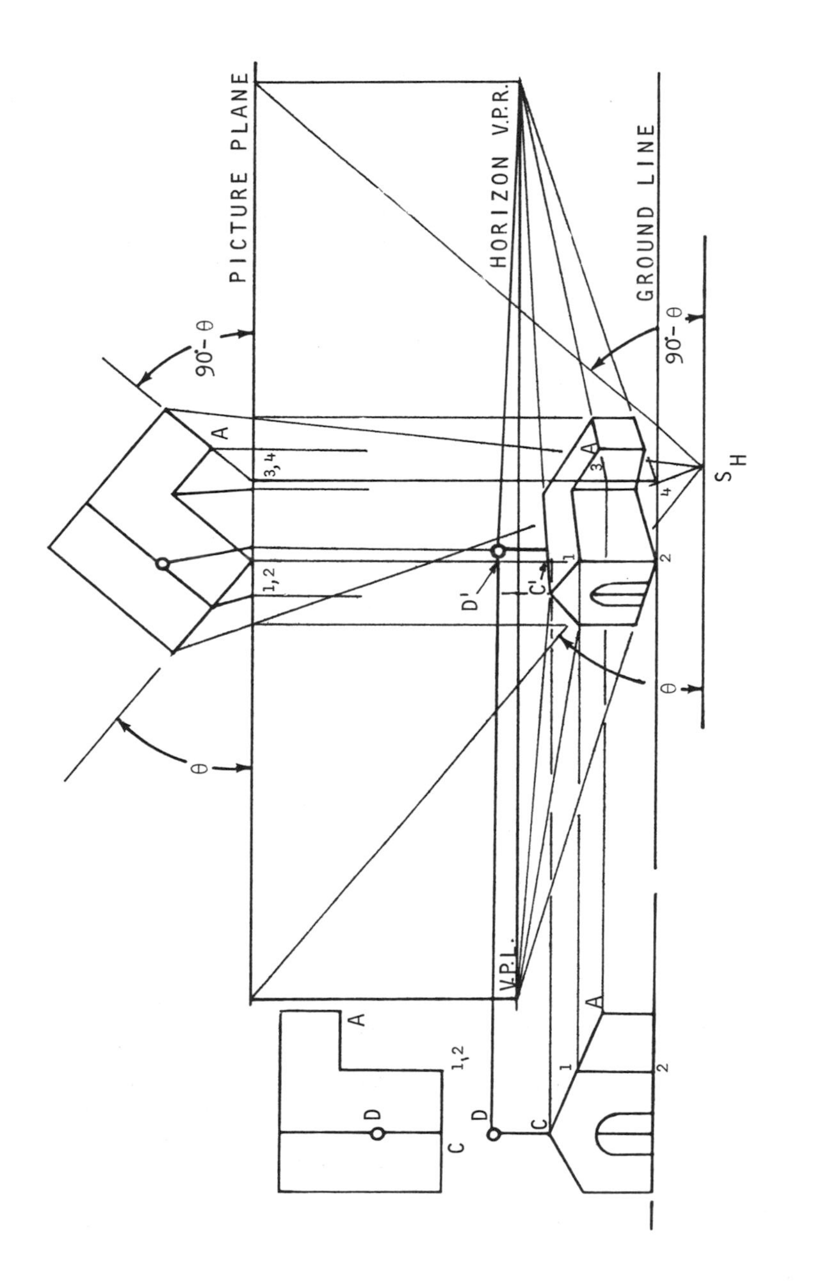

Fig. 1. Two-Point or Angular Perspective

projector to the vanishing points representing the height of that point in the front view. These points will determine the perspective.

Draw a two point perspective of an object given the front and top views in figure 1.

Solution: Construct a horizontal picture plane edge in the top view and horizon in the front view. Rotate the top view so the forward edges of the object make angles of θ and $90-\theta$ with the picture plane. In figure 2, $\theta = 30°$. Determine the vanishing points. This is done by constructing projections parallel to the edges of the object in the top view, from the station point, and constructing vertical projectors to the horizon from the points where the projectors from the station point cross the picture plane (see figure 2).

The object is a block with a hole in it. Construct the perspective of the block without the hole and then

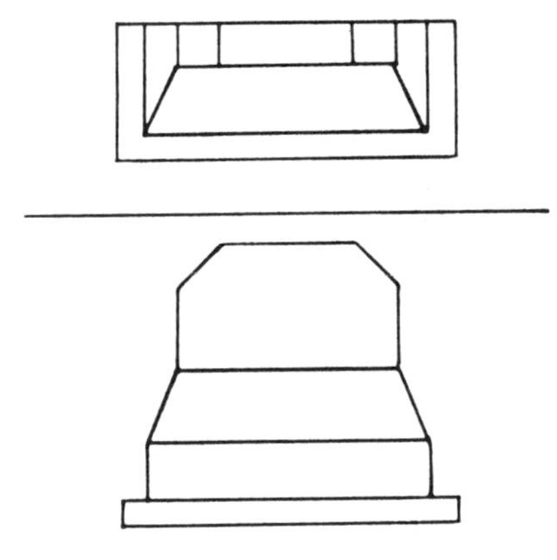

Fig. 1

construct the perspective of the hole within this block. From the front view transfer the true size of line A-B to the perspective. From points A and B draw projectors to the vanishing points. The projectors to $V.P._1$ represent the heights of points A and E and points F and B in the perspective. The projectors to $V.P._2$ represent the heights of points A and D and points B and C in the perspective. From the outer points of the box in the top view, draw projectors to the station point. At the intersection of these projectors and the picture plane, construct vertical projectors to the front view (see figure 2). The intersections of projectors 3 and 5 with the corresponding projectors to the vanishing point representing the heights of points E, F, D and C, in the front view, determine these points in the perspective. Draw projectors to the vanishing points from points D and E. The intersection of these projectors determines point H (projector 4 should also intersect point H in the front view). This establishes the block. To make the hole within the block, project the true heights of points H, I, J, and K from the front view, onto true length line AB. From these true height

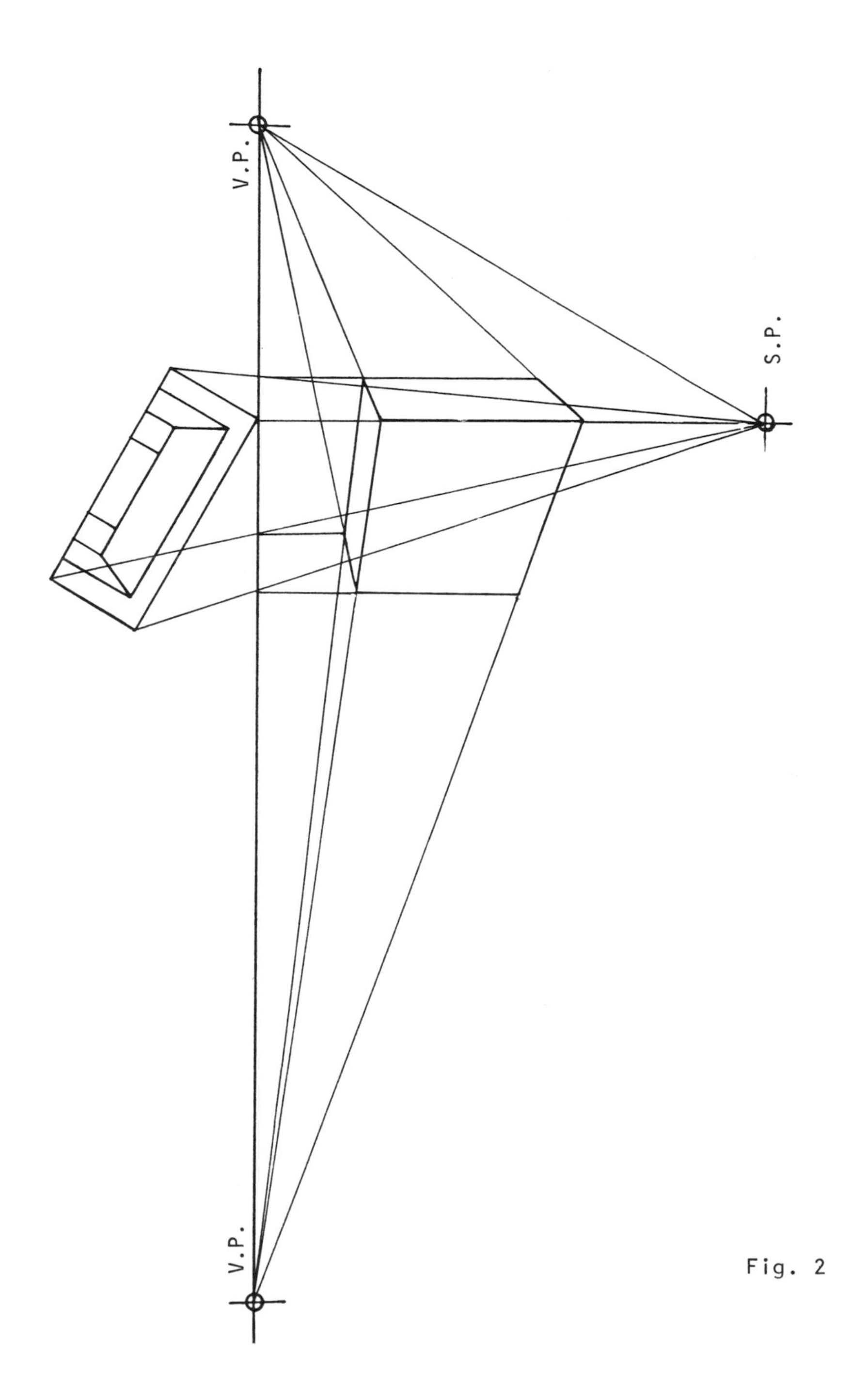

Fig. 2

points, construct projectors to point V.P.$_2$. From each point of the hole on the forward edge of the block, construct projectors to the station point. Construct vertical projectors to the front view at the points where these projectors cross the picture plane. The intersection of these vertical projectors and their

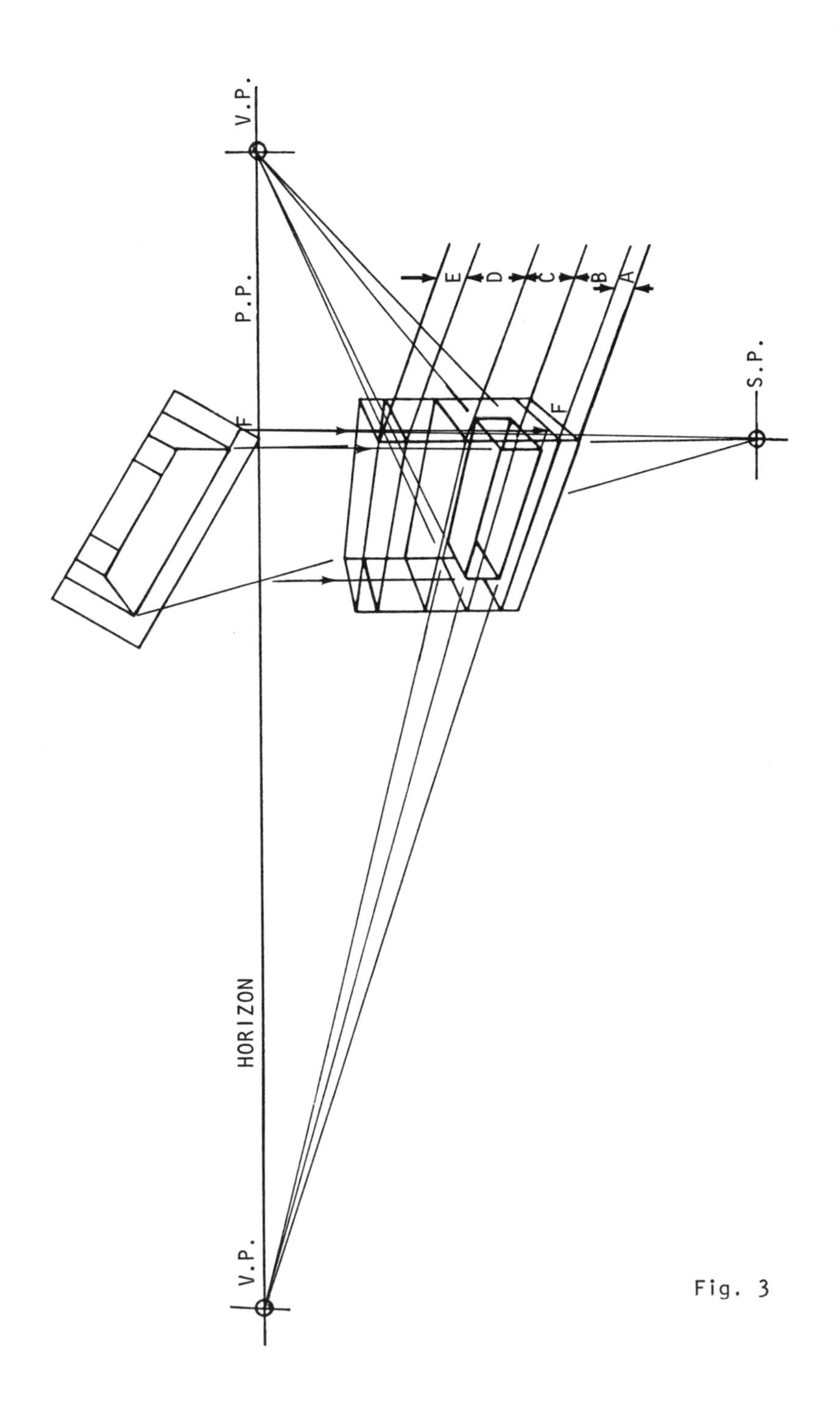

Fig. 3

corresponding projectors to V.P.$_2$ in the front view, determines the points in the perspective (see figure 3). From the true height points on line A-B draw projectors to V.P.$_1$. At the points where these projectors cross EF, construct projectors to V.P.$_2$. From the points of the hole on the back edge of the block, in the top view, construct lines to the station point (see fig. 4). At the intersection of these lines with the picture plane, construct projectors to the front view. The intersection of these projectors and their corresponding

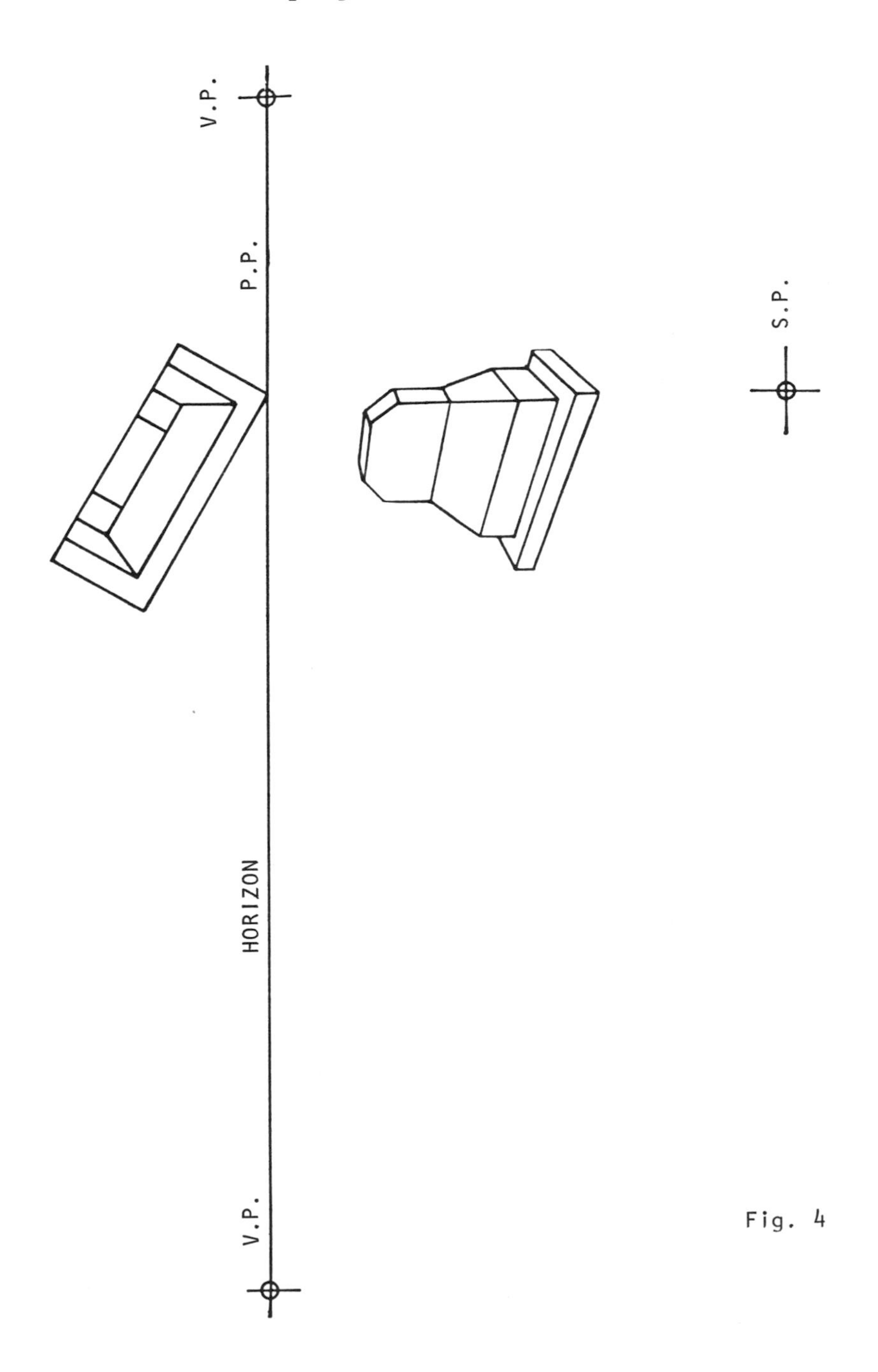

Fig. 4

projectors to $V.P._1$ in the front view determine these points in the perspective. Connect points and erase construction lines. The complete perspective is shown in figure 4.

• PROBLEM 11-10

> Draw a two point perspective of an object given the front and top views in figure 1.

Solution: In Figure 2 the horizon and picture plane coincide. The vanishing points are found by constructing projectors from the station point, to the horizon, parallel to the forward edges of the rotated top view of the object. To construct the complete perspective, the perspective of the box in which the object is contained is needed. The point of the object on the picture plane in the top view appears as a true length line

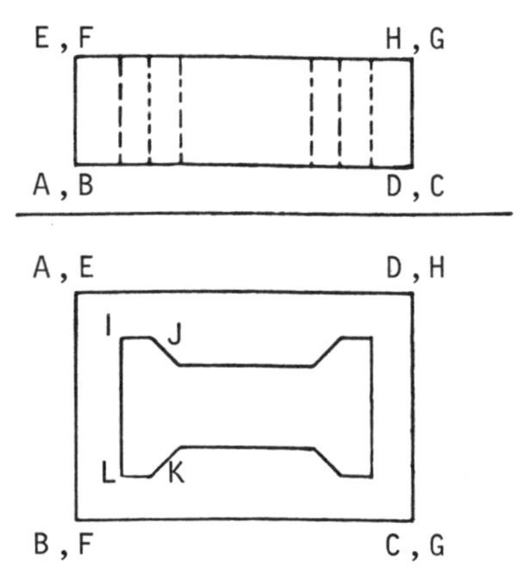

Fig. 1

in the perspective. From the front view, in Figure 1, the true height of the object is projected to the perspective. From the end points of this line, projectors to the vanishing points are drawn.

From the top view, projectors to the station point are constructed from outer points of the object. At the intersection of these projectors and the picture plane, vertical projectors are drawn to the front view. These projectors intersect their corresponding projectors in the front view and determine the perspective of the box in which the object is contained. On the true length line of this box, the true height (taken from the top view in figure 1) of each level of the object is laid out. At each level projectors are drawn to the vanishing points. These projectors establish the planes on which various levels of the object are contained; as shown in

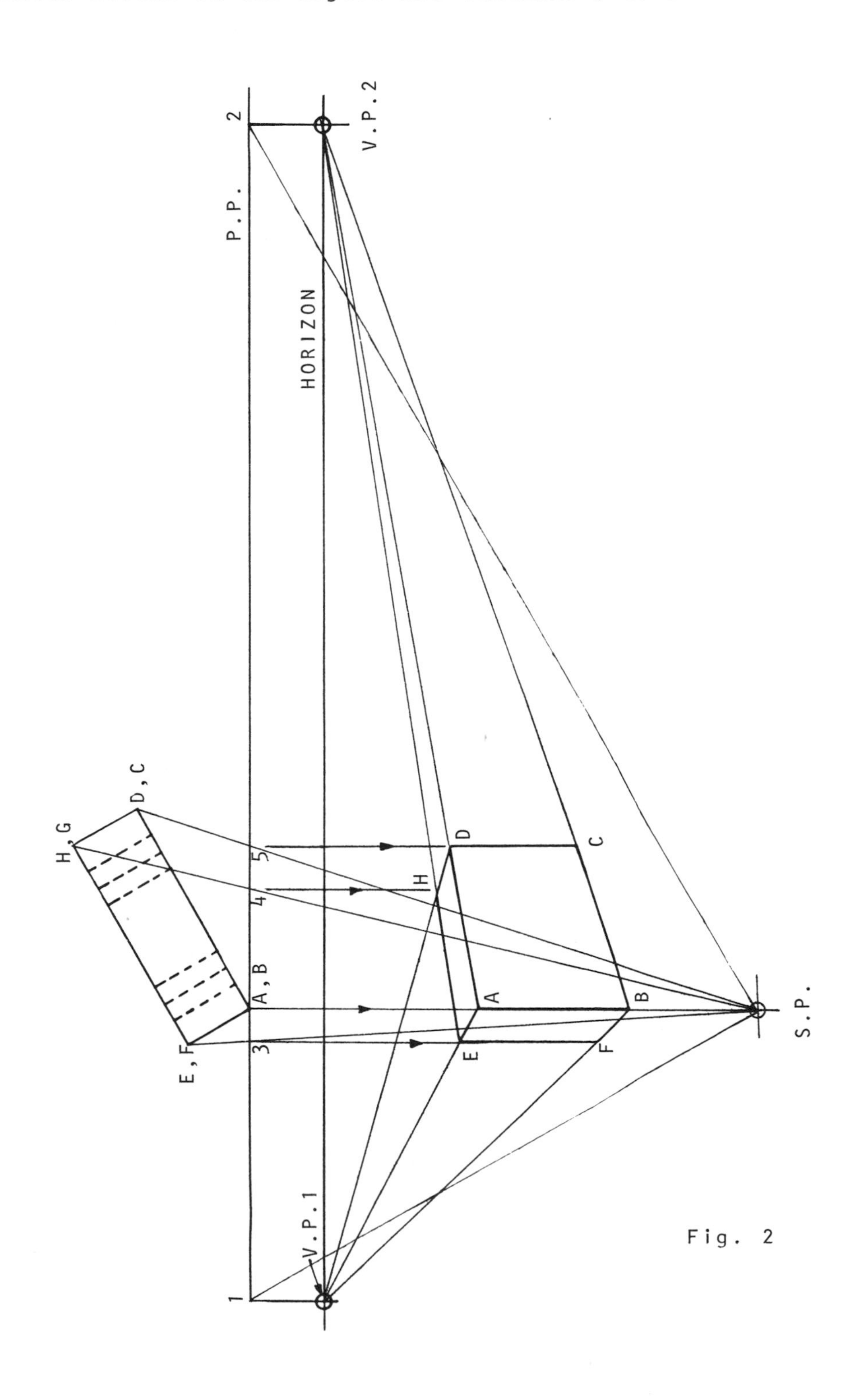

Fig. 2

figure 3. To construct the perspective, the edges in the top view are extended to the outer edges of the object. Projectors from the end points of these extended lines will locate their end points in the perspective view. The edges of the perspective will lie on the projectors to the vanishing points from these end points. This is illustrated in figure 3. In the top view an

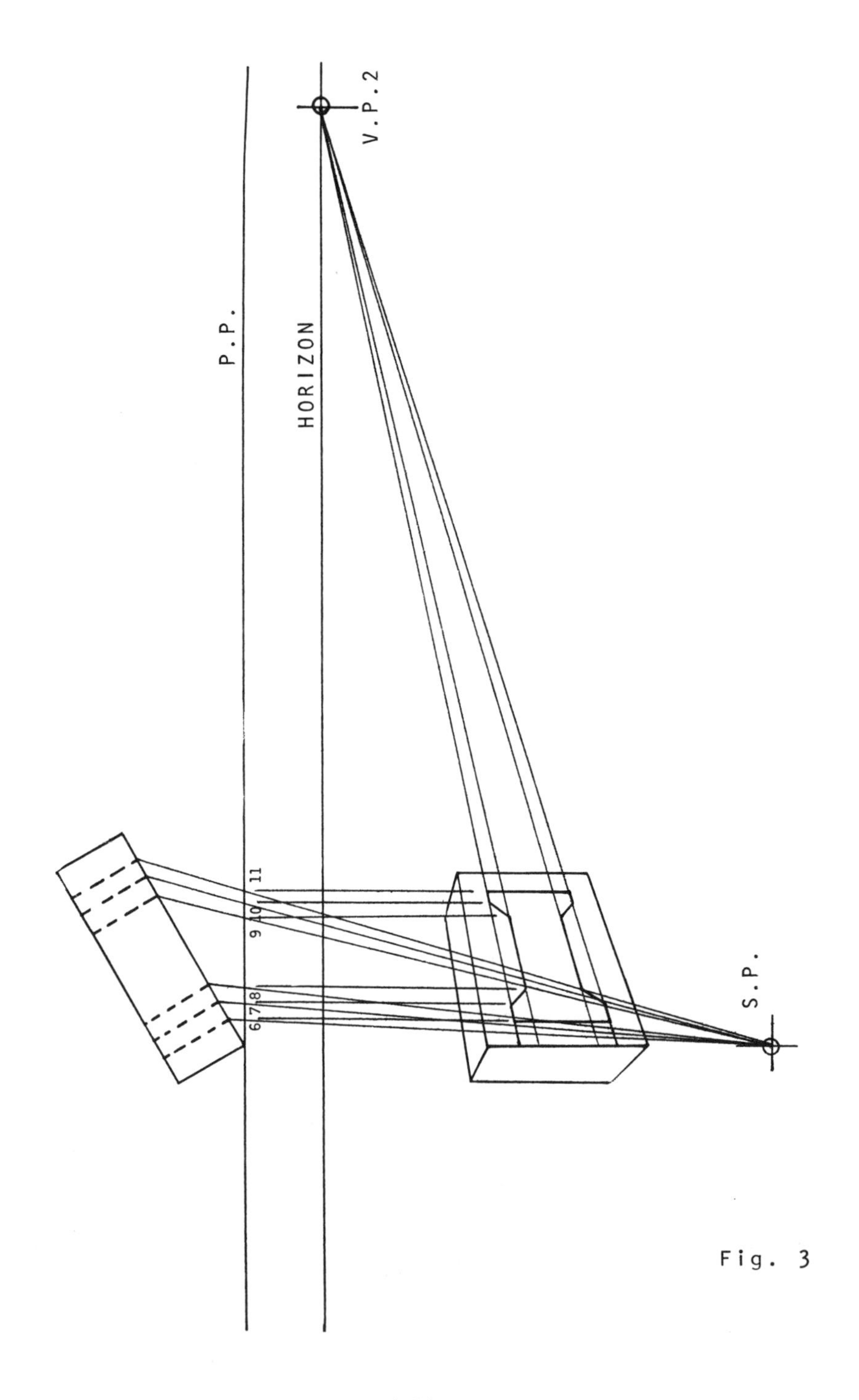

Fig. 3

edge was extended to point F. From point F a projector to the station point was drawn. At the intersection of this projector and the picture plane, a vertical projector was drawn to the front view. The point at which this vertical projector intersected the plane of height

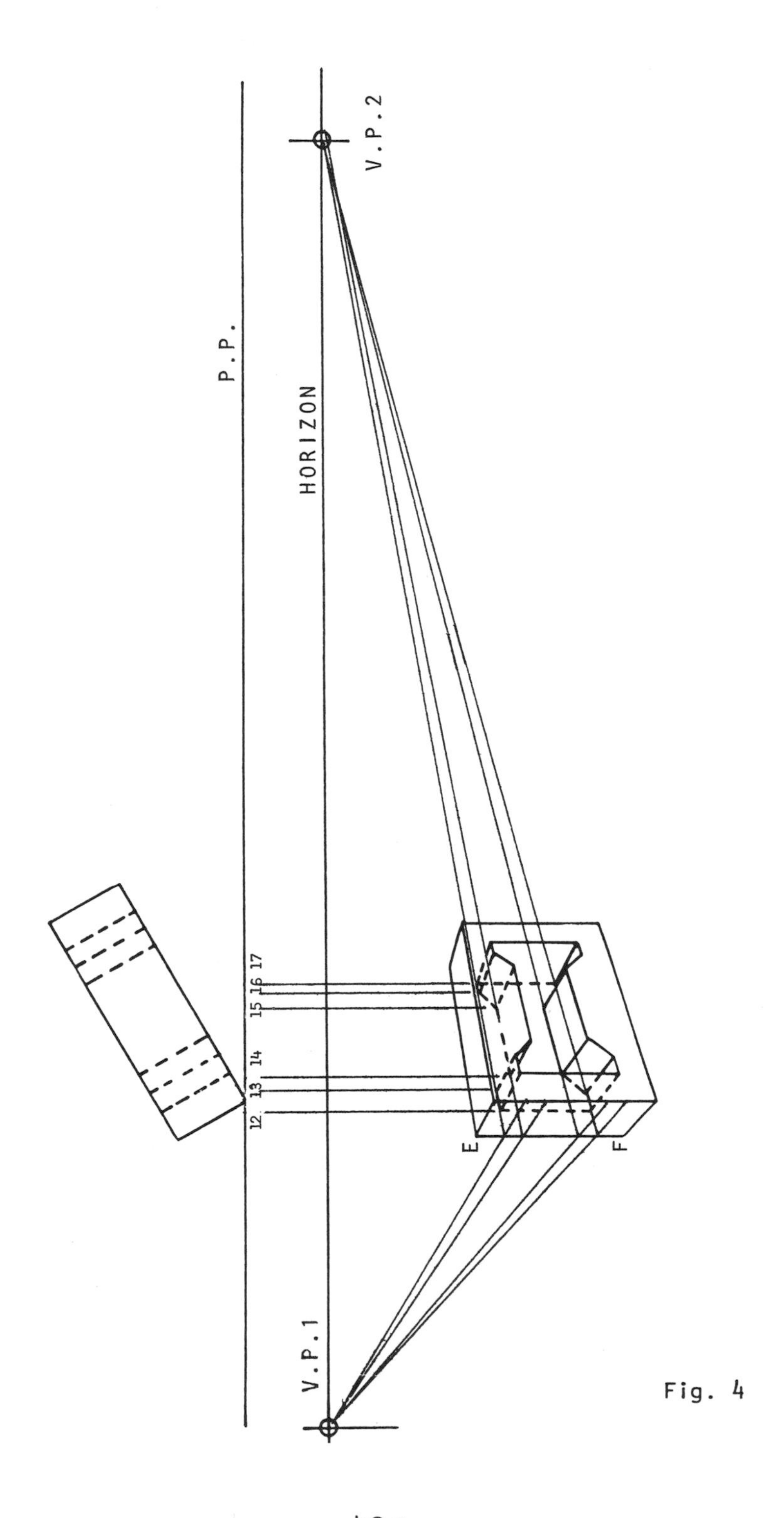

Fig. 4

A indicated point F in the perspective. From this point F, a projector was drawn to the vanishing point. The edge which was extended to point F lies on this projector from F in the perspective. Similar procedures are followed to construct the complete perspective in figure 4.

CHAPTER 12

DEVELOPMENTS

PRISMS AND PYRAMIDS

• PROBLEM 12-1

Develop the block described by the front and top views given in Fig. 1.

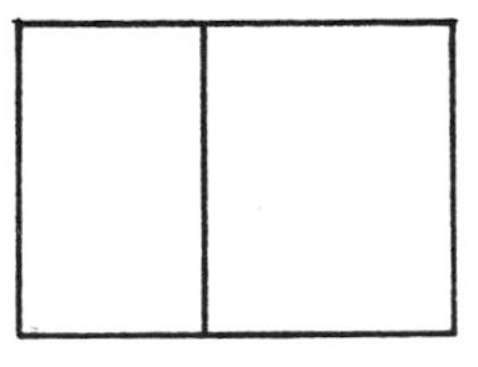

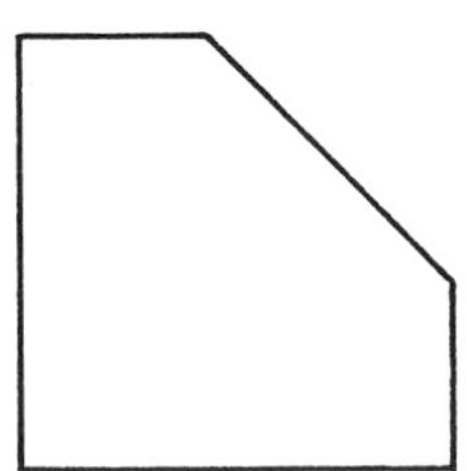

Fig. 1

Solution: Number all the points in the top view as an aid in developing the given block. Notice that all the lines in the given view appear true length and all the surfaces appear true shape, except for the inclined plane numbered 2-3-5-6 in Fig. 2. In order to find the true shape of this plane take an auxiliary view, a cutting plane parallel to the inclined plane. (See Fig. 2.)

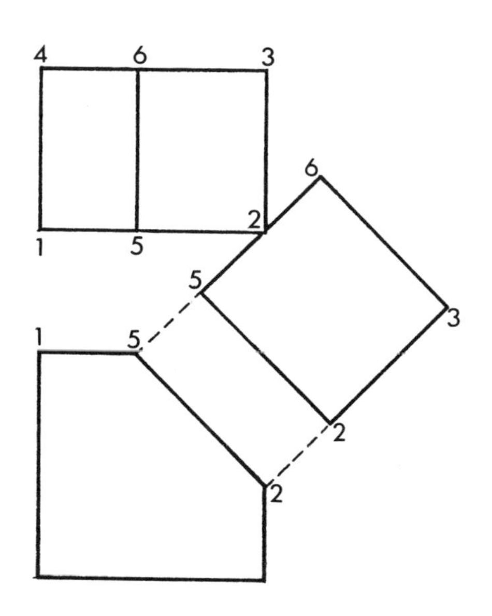

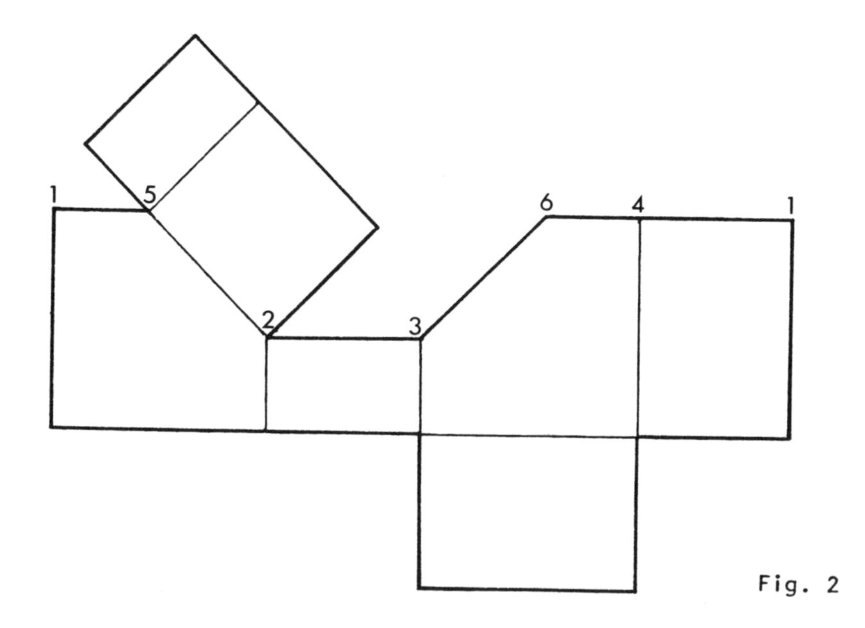

Fig. 2

In developing this block first project an extended line horizontally to the right, from the bottom edge of the front view. On this line mark off the distances of lines 1-2, 2-3, 3-4 and 4-1 in the same order as they appear in the top view. At each of these points draw vertical lines and then project all the horizontal lines in the front view out of the corresponding vertical line of the development. Lines 1-5 and 6-4 must be laid out on their corresponding locations. The bottom surface of the block can be drawn on any side of the extended line. The top surface is also matched to the corresponding points in the top of the development. Plane 2-3-5-6 is taken from the auxiliary view while plane 1-4-6-5 is taken directly from the top view. This is because it appears true shape in the given top view. (Refer to Fig. 2.)

• PROBLEM 12-2

Develop the right pyramid represented by the front and top views shown in Fig. 1.

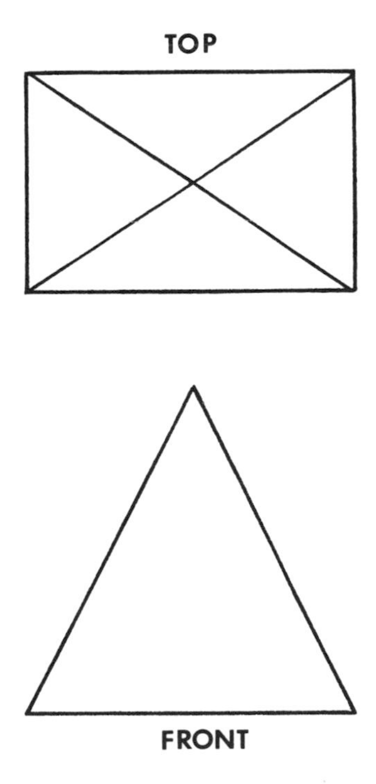

Fig. 1

Solution: In order to draw the development of the given pyramid, first determine the true shape of every plane in the pyramid. Notice that the given top view shows the base of the pyramid in true shape. This is true because the base appears in edge in the front view. Also notice that since the given object is a right pyramid, all the inclined edges have the same length. (Otherwise, the pyramid would be oblique.) Therefore it is necessary to find the true length of only one such line. In Fig. 2 all the corners of the pyramid have been labeled and line 5-2 was rotated in order to find its true length. To find the true length of line 5-2, rotate this line counterclockwise, in the top view, until the arc drawn intersects the horizontal line that goes through point "5". Project this intersection vertically down until this projection line intersects the horizontal line that goes through point "2" in the front view. (See Fig. 2.) The line marked "TL" is therefore the true length of line 5-2.

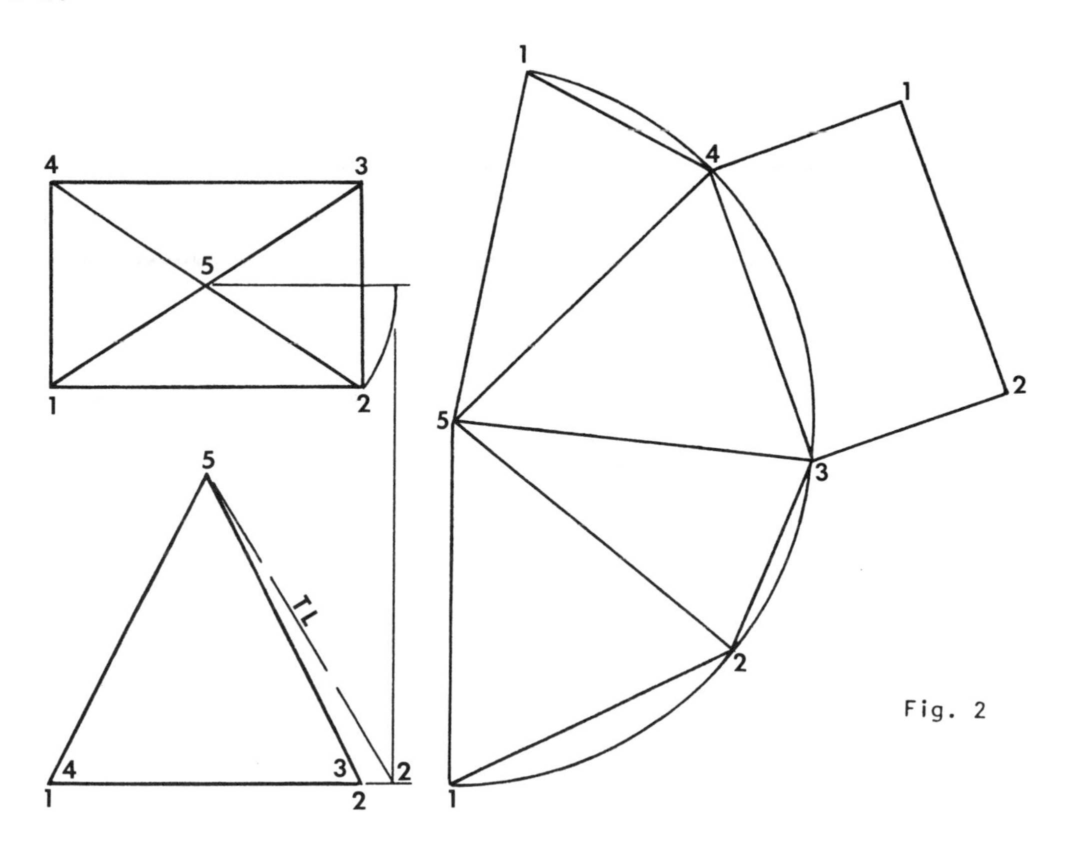

Fig. 2

When drawing the required development, notice that lines 5-1, 5-3 and 5-4 have the same length as line 5-2 whose true length was previously determined. Since all these lines have the same length we can conveniently choose the location of point "5". Use point "5" as the center to strike an arc with a radius equal to the true length of line 5-2. On this arc select a point and label it "1". Then take the lengths of lines 1-2, 2-3, 3-4 and 4-1 and mark them on the constructed arc. (Label each point as they are marked.) Draw lines between each of these points and point "5". Notice that point "1" must appear on both extremes of the development. To complete the development place the true shape of the base on any corresponding line. In the case of Fig. 2, line 4-3 in the development was matched with line 4-3 in the true shape (from the top view) of the rectangular base. This method of finding the true length of every side in each triangle is called the "triangulation method".

• PROBLEM 12-3

Develop the right prism whose front and top views are shown in the accompanying figure.

Solution: In order to develop the given right prism it is necessary to determine the true shape of each of the faces of the prism. Since the top plane A-B-C appears in edge in the front view, the given top view of this plane must be true shape. This means all the lines appear true length.

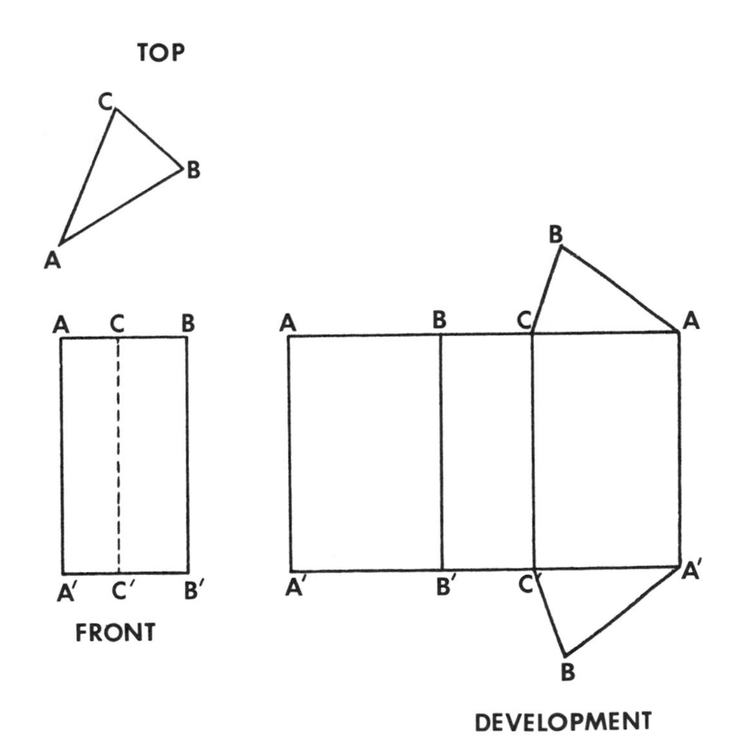

Because the given object is a right prism and the top view shows every vertical plane in edge, all the vertical lines in the front view also appear true length. Therefore, with the true shape of the top plane and the true length of the heights, the shape of every component of the prism has been determined.

To draw the required development proceed as follows: since the heights shown in the front view are all equal and true length, project two horizontal lines from the top and bottom edges. Draw a vertical line and label it A-A'. (Notice that this line was arbitrarily labeled.) Draw the parallel line B-B' at a distance equal to the true length of line A-B. Draw line C-C' and line A-A' at a distance equal to the true length of the lines B-C and C-A, respectively. Notice in the development the lines AB, BC and CA are taken directly from the given top view. To complete the development, place the top and bottom planes so that one of the lines in the plane coincides with the corresponding line in the development. In the accompanying figure the top and bottom planes were placed on lines C-A and C'-A', respectively. Notice that the bottom plane is congruent to the top plane because the object is a right prism.

• PROBLEM 12-4

Develop the truncated right prism represented by the front and top views shown in Fig. 1.

Solution: Label the corners of the top view as in Fig. 2. Notice that all the lines in both of the given views are shown true length.

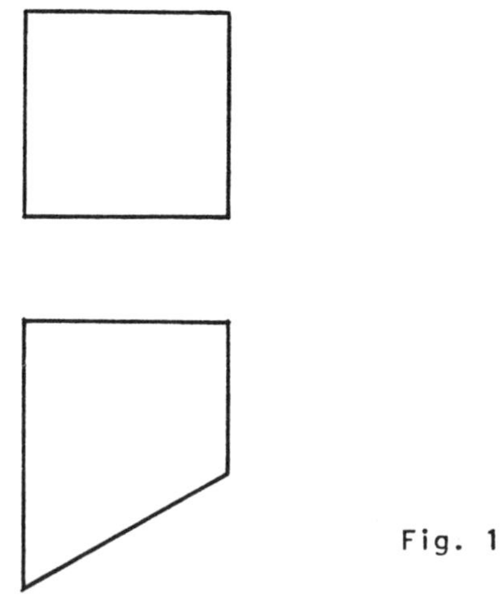

Fig. 1

Since the true length of all the lines of the object have been determined, draw the development of the prism as follows: first draw a horizontal line to represent the stretchout of the square. On this line lay out the distances corresponding to the four sides of the square. Start from the seam at point "1" in the top view and space off the distances 1-2, 2-3, 3-4, and 4-1 along the stretchout line.

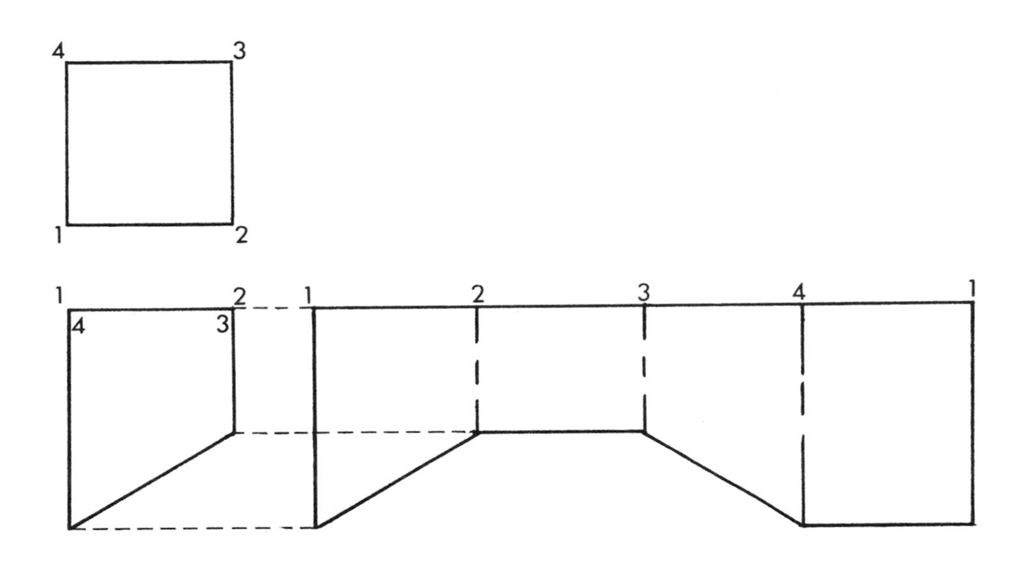

Fig. 2

Draw vertical lines at each of these points in the stretchout line and project lines from the points on the front view to intersect the corresponding vertical lines of the development drawing. Connect the points of intersection with straight lines in the same order as in the top view.

• PROBLEM 12-5

Develop the truncated hexagonal prism represented by the front and top views shown in Fig. 1.

Solution: In drawing any development always keep in mind that every component of the development must appear true shape. In the case of the given truncated prism notice that all the vertical lines in the front view are true length and that the top view represents the true shape of the base. By combining the true shape of the base and the true lengths of the heights it is possible to find the true shape of all the planes that form the truncated prism.

Before drawing the development, label all the points in the given views to aid in determining the correct development. Select a stretchout line projected perpendicular from the true length lines in the front view. In the case

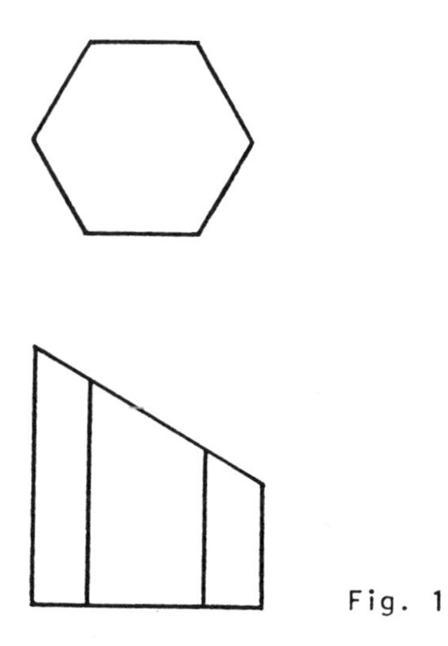

Fig. 1

of Fig. 2, the stretchout line was projected from the edge view of the base. On this stretchout line mark off the lengths of lines 1'-2', 2'-3', 3'-4', 4'-5', 5'-6' and 6'-1' and at each of the located points draw lines parallel

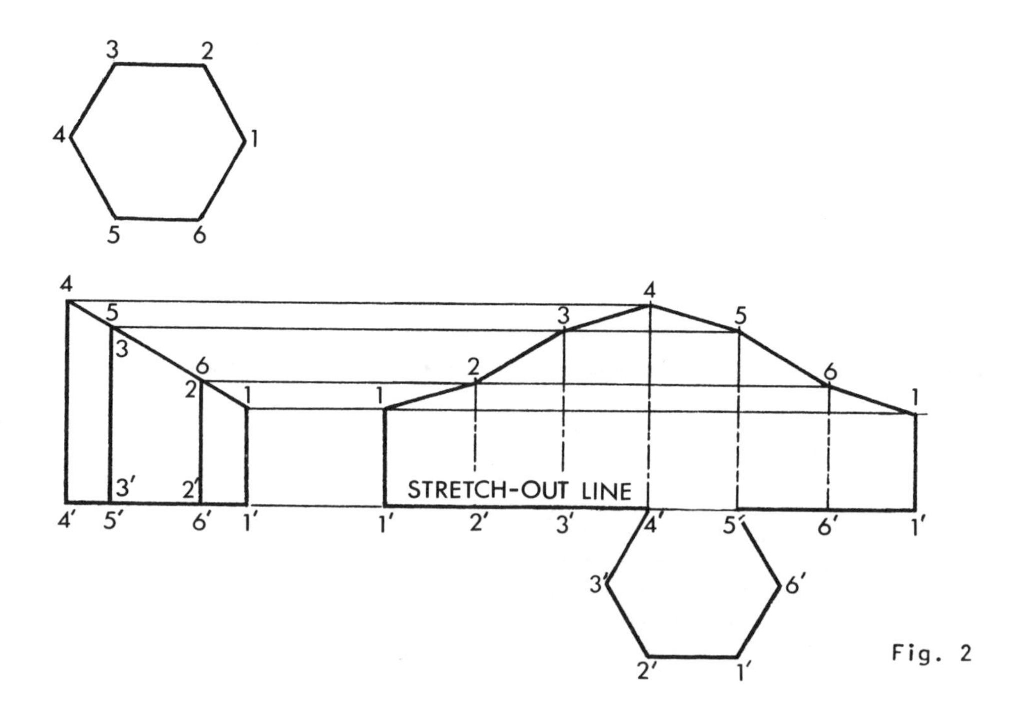

Fig. 2

to the true length lines of the front view. Project each point in the front view to the development, with projection lines parallel to the stretchout line. Number the intersections of the parallel lines in the development and the projection lines by their corresponding numbers. Join each of these numbers in order. Finally, place the true shape of the base at a convenient location in the development. In Fig. 2 the base was placed on line 4'-5' of the development coinciding with line 4'-5' of the base.

• PROBLEM 12-6

Develop the truncated hexagonal prism represented by the front and top views shown in Fig. 1.

Solution: The given top view of the prism represents any right view taken perpendicular to the sides of the prism. Therefore choose a cutting plane in the front view, which goes through all the sides of the prism, such as the one chosen in the front view of Fig. 2. Label all the corner

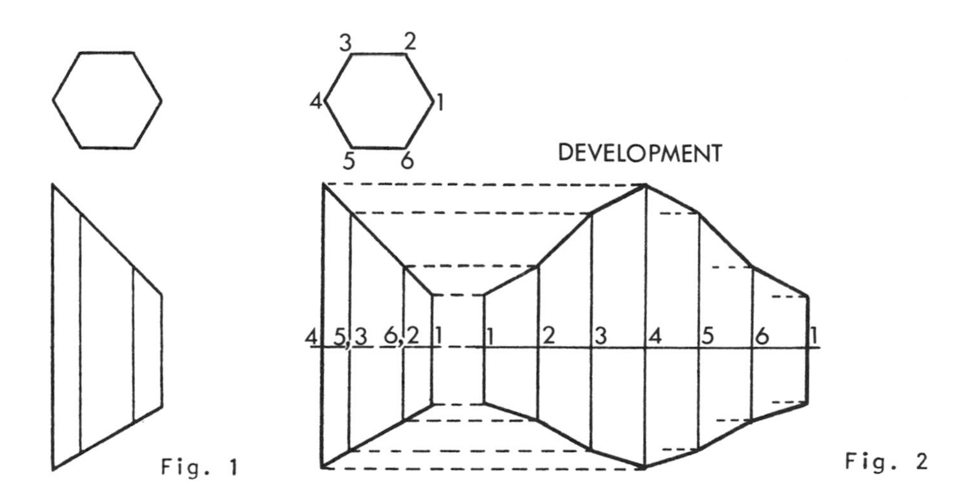

Fig. 1

Fig. 2

points of the right view and locate these points on the chosen cutting plane. Recall that a right view is used in order to obtain the true thickness of each side of the hexagonal prism as shown in the given top view. Notice that all the vertical lines in the front view appear true length. This is because the planes on which they lie appear in edge in the top view.

In developing the given prism, take advantage of the fact that all the vertical lines in the front view appear true length by projecting the extended line from the cutting plane constructed in the front view. On this extended line lay out the true lengths of the lines 1-2, 2-3, 3-4, 4-5, 5-6 and 6-1 as they appear in the top view. Draw vertical lines at each of the points located on the extended line. Project the points in the top and bottom edges from the front view onto the corresponding vertical lines of the development. When all these points have been located connect each two points as shown in the development of Fig. 2. Notice that the projections of the vertical lines in the front view can be made because all these lines appear true length in this specific view. No top or bottom faces are drawn because the prism was truncated by two inclined planes, one at the top and the other at the bottom.

• PROBLEM 12-7

Develop a layout development pattern for a hexagonal prism cut by a cranked plane.

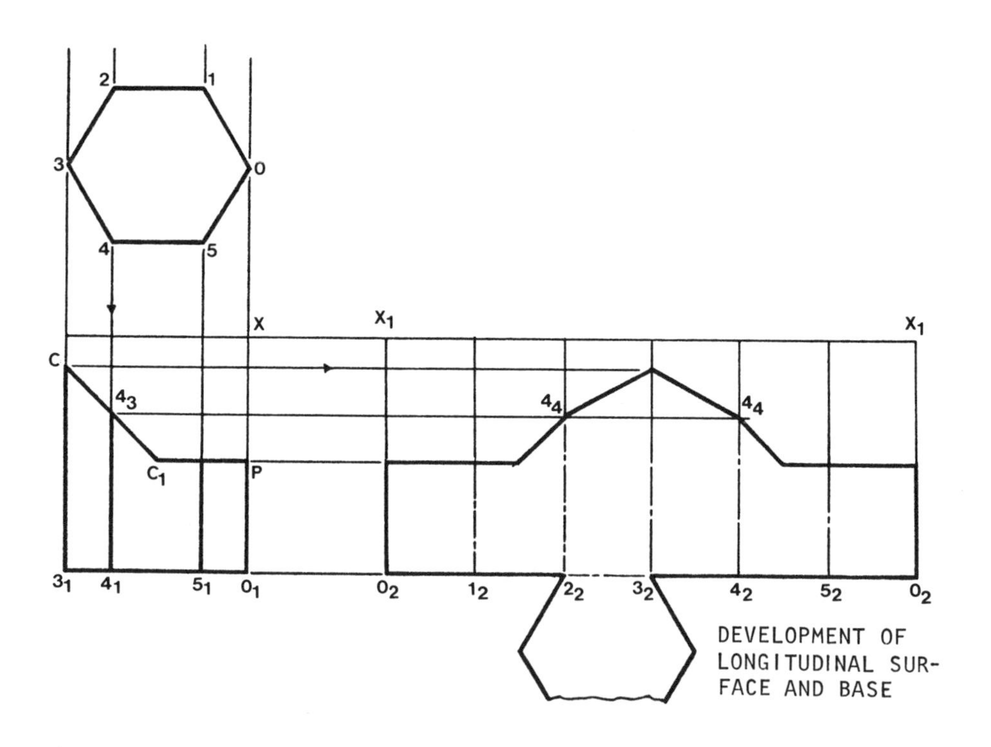

Solution: The figure shows a hexagonal prism cut by a cranked plane. Let this plane be called CC_1P. First draw the plan, numbering the six base corners as shown in the figure.

Imagine that the figure consists of a very thin film, marked with six equally spaced lines at each edge, it is to be cut along edge line PO_1 and opened to form a flat figure.

Extend the base of the prism, in the front view, and pick a convenient point on this extension line. On this point draw a line O_2,O_2 equal in length to the perimeter of the hexagon and project the height, X, from the elevation view. Connect corresponding points O_2 and X_1 with vertical lines to form the rectangle $X_1X_1O_2O_2$.

Draw vertical divisions to represent the corners of the hexagon and number them as shown. The divisions were easily found. Since the base of the prism is in edge in the front view, parallel to the H-F plane, it appears in true shape and size in the top view. Therefore lines 0-1, 1-2, 2-3 . . . all appear in true length in the top view.These lengths were laid out on line O_2 O_2 and labeled O_2-1_2, 1_2-2_2, 2_2-3_2 . . . as shown in the figure. through each point (1_2, 2_2, 3_2 . . .) a vertical line was passed and these are the vertical divisions. this gives the development drawing of the complete prism. the intersections of CC_1P and the hexagonal corners may now be projected onto the appropriate vertical lines on the development drawing, as shown in the figure, to give the required development surface of the prism below the cutting plane.

Note in the drawing that the base is included as part of the development. the hexagon can be positioned on any face of the development and its geometry will coincide with the plan view.

• PROBLEM 12-8

Develop the oblique rectangular prism represented by the front and top views shown in Fig. 1. The prism is closed at both the top and bottom faces.

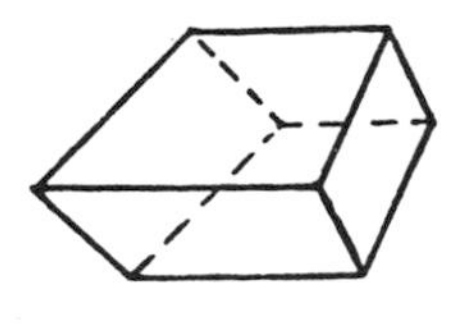

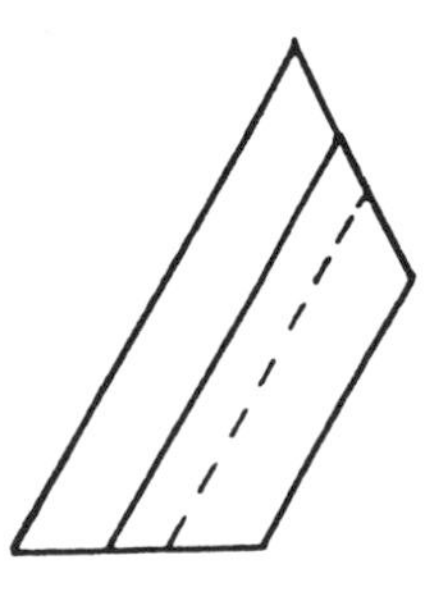

Fig. 1

Solution: Number all the corner points of the prism in both given views, as shown in Fig. 2. Find an auxiliary view whose cutting plane, F-1, is parallel to the top inclined face of the prism, in order to find the true shape of the top face. Then find another auxiliary view whose cutting plane, F-2, is perpendicular to the four side planes. This auxiliary view shows the true thickness of all the side planes. Notice that lines 8-4, 7-3, 5-1 and 6-2 in the top view appear parallel to the cutting plane F-H. Therefore, all these lines are true length in the front view. Also notice that the base plane 5-6-7-8 appears in edge in the front view and therefore it appears true shape in the top view.

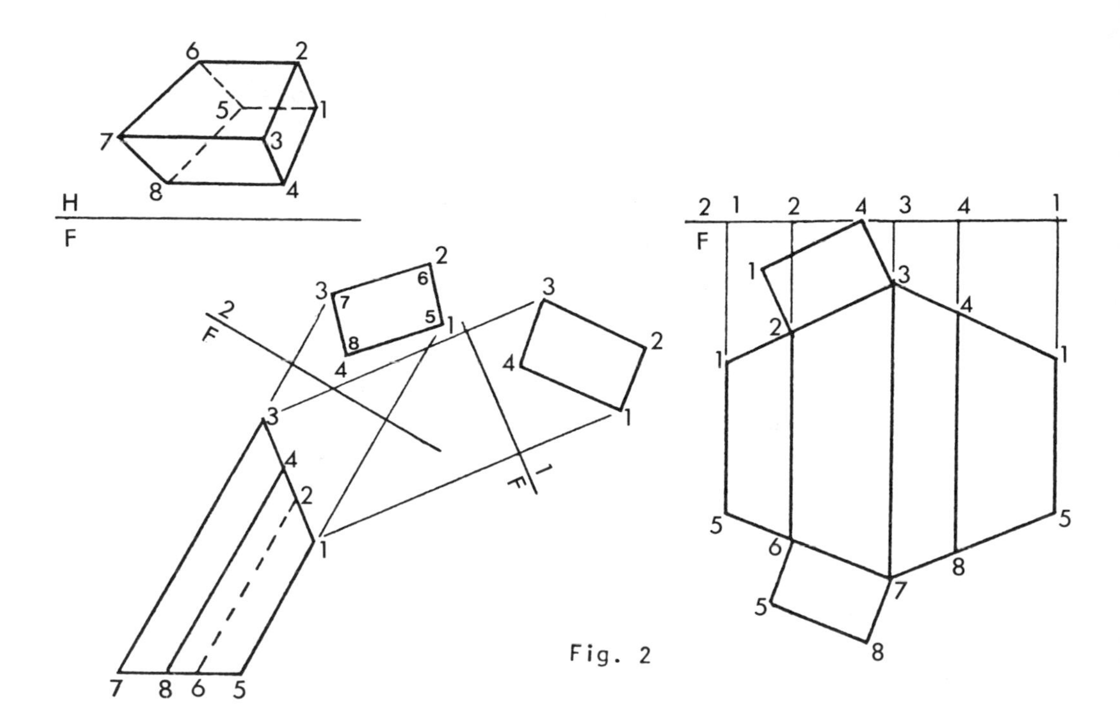

Fig. 2

In drawing the required development, first draw a stretch-out line such as the reference line F-2 used in the development shown in Fig. 2. On this stretch-out line mark off the true thickness of all the side planes of the prism as found in the auxiliary view projected onto the cutting plane F-2. At each of these points draw vertical lines and take measurements from reference line F-2 (in the front view) to each point in the front view. Lay out these measurements on the corresponding vertical line of the development. For example, to locate point "8" on the development measure its distance from reference line F-2 in the front view and then lay out this distance from the stretch-out line of the development along the corresponding vertical line, which goes through point "1" of the stretch-out line. Locate all the points in this manner and connect them. Place the true shape of the top plane, as found in the auxiliary view, projected onto the cutting plane F-1, so that one of the lines of the top plane coincides with one of the lines in the development. In the case of Fig. 2, the plane was placed on line 2-3. Finally, place the bottom plane 5-6-7-8, which appears true shape in the top view, in the same manner as the top plane (on line 6-7 for the development of Fig. 2.).

• PROBLEM 12-9

Develop the oblique prism represented by the front and top views shown in Fig. 1.

Solution: In developing an oblique prism first find an auxiliary view perpendicular to the sides of the prism. The perimeter of this view (represented by 1-2-3-4 in Fig. 2) represents the total length of the required development. By looking at the top view, it can be noticed

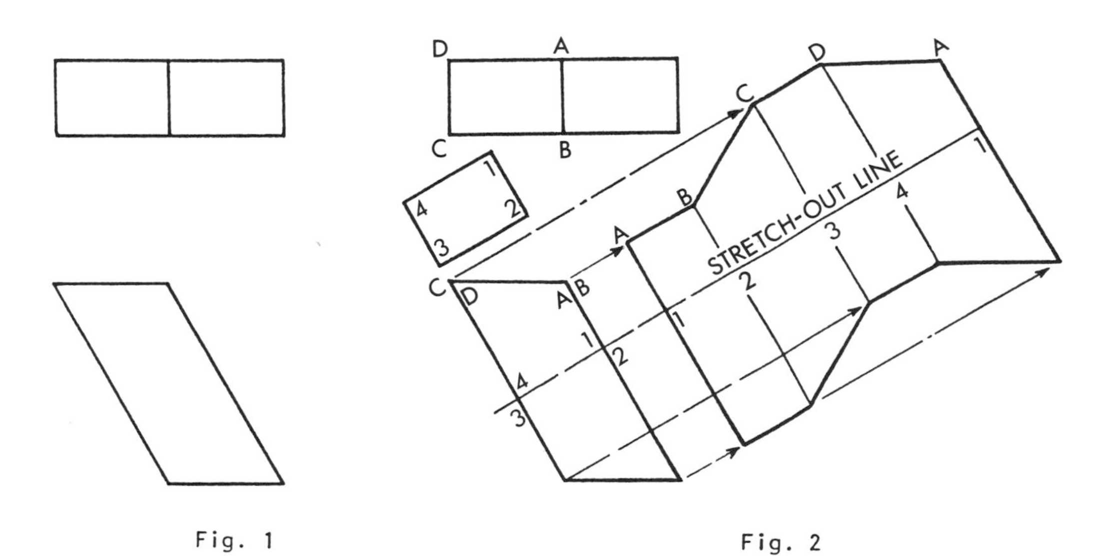

Fig. 1 Fig. 2

that the plane shown in the front view is vertical and parallel to the cutting plane between the top and front views. Therefore, the above mentioned plane appears true shape in the front view. For this reason we can take the stretch-out line perpendicular to the inclined lines in the front view, which appear true length. On the stretchout line select a point and label it "1." Starting from this point mark off the lengths of lines 1-2, 2-3, 3-4, and 4-1, as measured from the auxiliary view. Draw lines perpendicular to the stretch-out line at each of the above determined points. Then project all the points in the front view to the development. In Fig. 2 the four corners at the top of the prism were labeled A-B-C-D. Notice that points 1 and 2 are on the same plane as A and B, 2 and 3 on the same plane as B and C, etc. Therefore, lines A-B, B-C, C-D and D-A are projected on the same plane as lines 1-2, 2-3, 3-4 and 4-1, respectively in the development. (See Fig. 2.) Notice that unlike a right prism, an oblique prism does not unfold in a straight line. However, as in any development, all the sides of the prism appear true shape.

• PROBLEM 12-10

Develop the two intersecting prisms represented by the front, top and left side views shown in Fig. 1.

Solution: First label all the significant corner points on the three given views. In Fig. 2, the top corners of

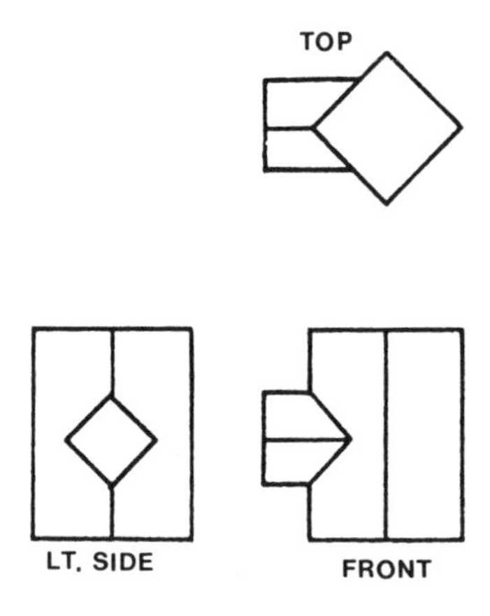

Fig. 1

the vertical prism were labeled 1, 2, 3, and 4, the front corners of the horizontal prism were labeled 5, 6, 7, and 8 and the points of intersection were labeled A, B, C, and D.

In developing the horizontal prism, first draw a horizontal line and on it lay out the lengths of the lines 5-6, 6-7, 7-8 and 8-5, which appear true length in the left side view. At each of these points draw vertical lines on which lay out the true lengths of lines 5-A, 6-B, 7-C and 8-D as they appear in the front or top view. Connect the lines A-B, B-C, C-D and D-A to complete the development. (See Fig. 2.) Notice that on this development the lines at the intersection now appear true length.

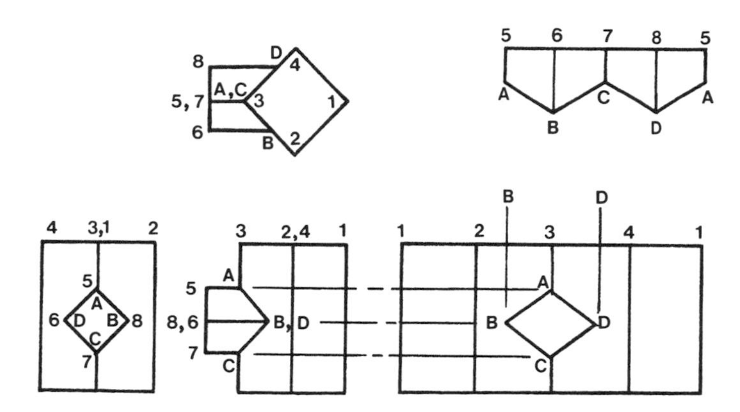

Fig. 2

To develop the vertical prism project a horizontal extended line from the top edge of the front view and on it lay out the true lengths of line 1-2, 2-3, 3-4 and 4-1 as they appear in the top view. At each of the points located draw vertical lines. Project every point in the front view horizontally to the development. Notice that points A and C lie on the edge going through point 3. Therefore, label them as such in the development. To locate points B and D, use the true lengths of lines A-B, B-C, C-D and D-A as determined in the development of the horizontal prism. The true length on these lines, however, can be found by rotation. Darken the outside edges, the hole A-B-C-D and the folding lines, in order to complete the required development.

• PROBLEM 12-11

Develop the object represented by the front and right side views shown in Fig. 1. The base and the top of the object are open.

Solution: From the two given views it can be noticed that the object consists of two vertical and two sloping planes.

The planes of the base and the top are both horizontal. Before drawing the required development, label the two given views as in Fig. 2 and find the true length of all the lines necessary to find the true shape of every plane in the object. The true length of these lines, however, can be found while developing the object. In finding the true shape of a specific plane it is a good policy to use the triangulation method. For example, to find the true shape of plane B-C-F-E draw the diagonal B-F and then find the true length of each of the three sides of both triangles, in order to find the true shape of each triangle. The two true shape triangles are placed so that line B-F coincides, thus forming the true shape of the plane in question.

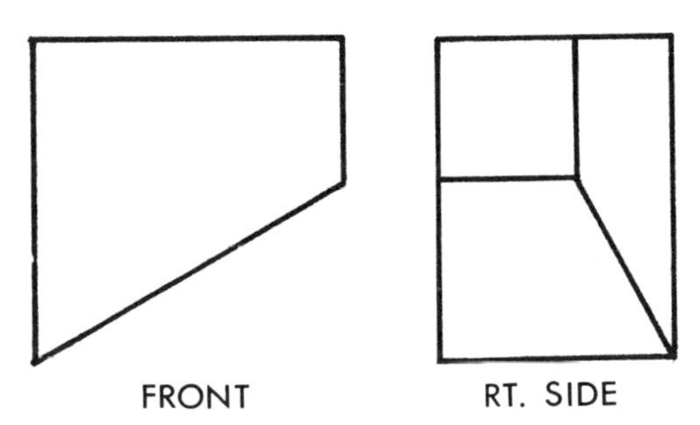

Fig. 1

In order to draw the required development, choose a starting line and a direction. In the case of Fig. 2, the development was started with the line B-E in a clockwise direction from the right side view. The order of drawing this development is as follows: first find the true length of line B-E (by rotation) and place it at a convenient location in the drawing. Use point "E" as the center and strike an arc with a compass of radius equal to the true length of line F-E, which appears true length in the side view. Find the true length of the diagonal B-F (by rotation) and using point "B" as center, strike an arc with radius equal to the true length of line B-F. Label the intersection of the two arcs "F." Then using the true length of line B-C (directly from the side view), strike an arc from point "B." Find the true length of line F-C and using this length as radius, strike another arc from point "F." This completes the construction of plane B-C-F-E.

Draw plane C-D-G-F in the same manner. Plane D-A-H-G can be drawn by reproducing the front view with line D-G of the front view coinciding with line D-G of the development. Notice that the front view is the true shape of plane D-A-H-G because all the lines that form this plane are parallel to the cutting plane between the front and side views. The last plane (A-B-E-H) can be drawn by noticing that both planes D-A-H-G and A-B-E-H are vertical.

Therefore, the partial development of these two planes is a straight line development. For this reason lines A-D and H-G of the development can be extended and on this extension lines A-B and H-E can be laid out, thus determining the last plane. An alternate way of constructing the last plane is by triangulation, finding the true length of the diagonal A-E. Recall that to find the true length of a line (by rotation) the line is rotated on one view

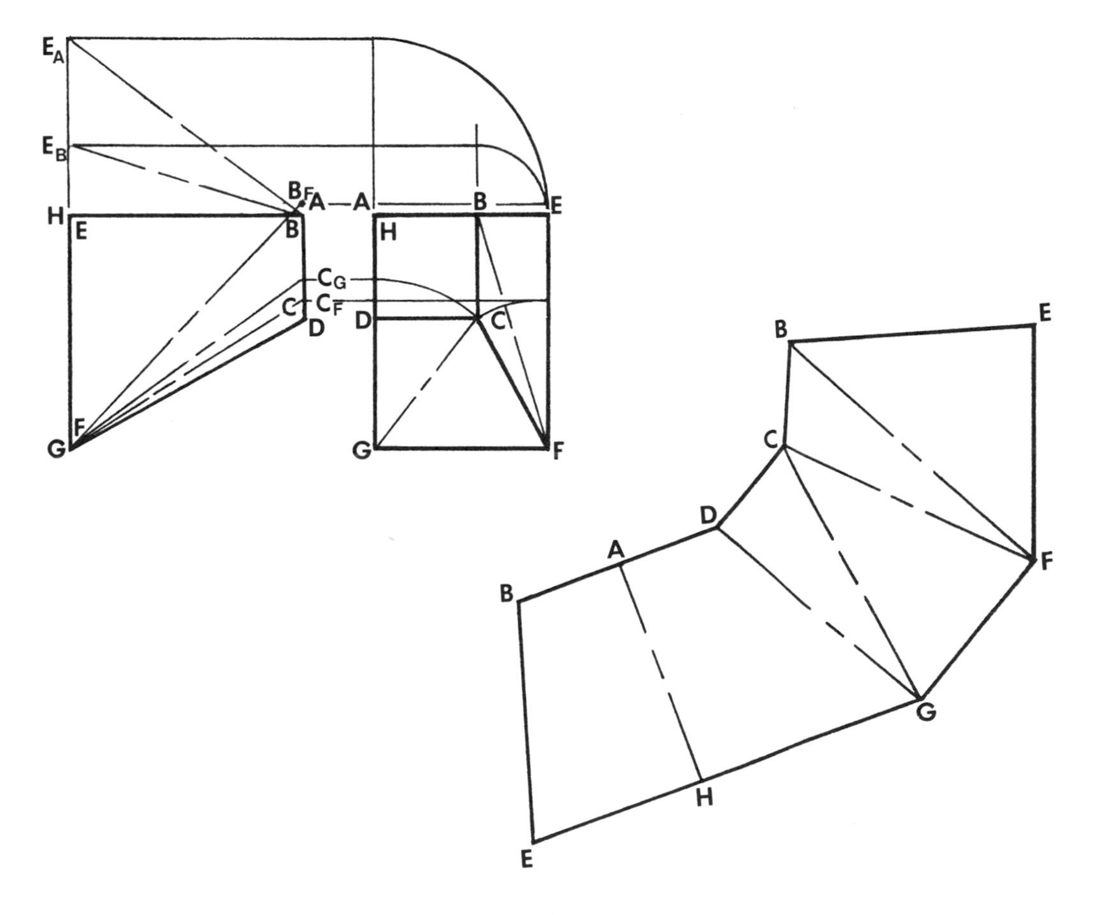

Fig. 2

until it is parallel to the cutting plane between this view and the adjacent view. The point which has been rotated is then projected to the adjacent view. The line joining the point of rotation and the new point is the true length of the line in question. For example, to find the true length of line B-E rotate point "E" about point "B" in the side view, to a position parallel to the cutting plane between the front and side view. Project point "E" to the front view (labeled E_a) and connect it to point "B" of the front view. The resulting line B-E_a is the true length of line B-E.

Produce a one-piece development of the sheet metal dish represented by the front and top views shown in Fig. 1.

Solution: Before attempting to draw the required development, we must make sure that we have the true shape of every component of the development. As an aid in better picturing the given object, the two given views have been labeled with the letters A through H in Fig. 2. It is important to notice that the given object is completely symmetrical. For this reason the lines A-E, B-F, G-C and

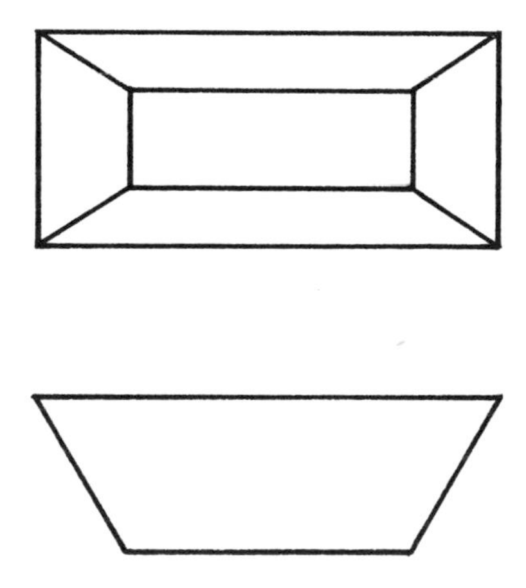

Fig. 1

H-D have the same length. Notice that none of these lines appear true length in either of the given views. In order to find their true length, rotate line F-B clockwise to a position parallel to the cutting plane F-H. The new position of point "B" has been labeled "B'" in the top view of Fig. 2. Project point "B'" from the top view, vertically down to the front view until it intersects the horizontal line going through point "B." Line F-B' in the front view is the true length of line F-B, whose length equals those of lines A-E, C-G and D-H. Realize that all the vertical and horizontal lines in the top view are true length and, therefore, the true length of every line in the given object has been obtained.

For convenience, the true length of the heights of each of the trapezoids on the sides of the dish were found as follows: in order to find the height of the trapezoids B-C-G-F and A-E-H-D, line G-I is drawn on both the front and top views. On the top view rotate line G-I to a position parallel to the cutting plane F-H. The position of point "I" was labeled "I'" in Fig. 2. Project the new point "I'" vertically down to the front view until it intersects the horizontal line going through point "I." The line H-I' in the front view is the true length of the height of the trapezoids in question.

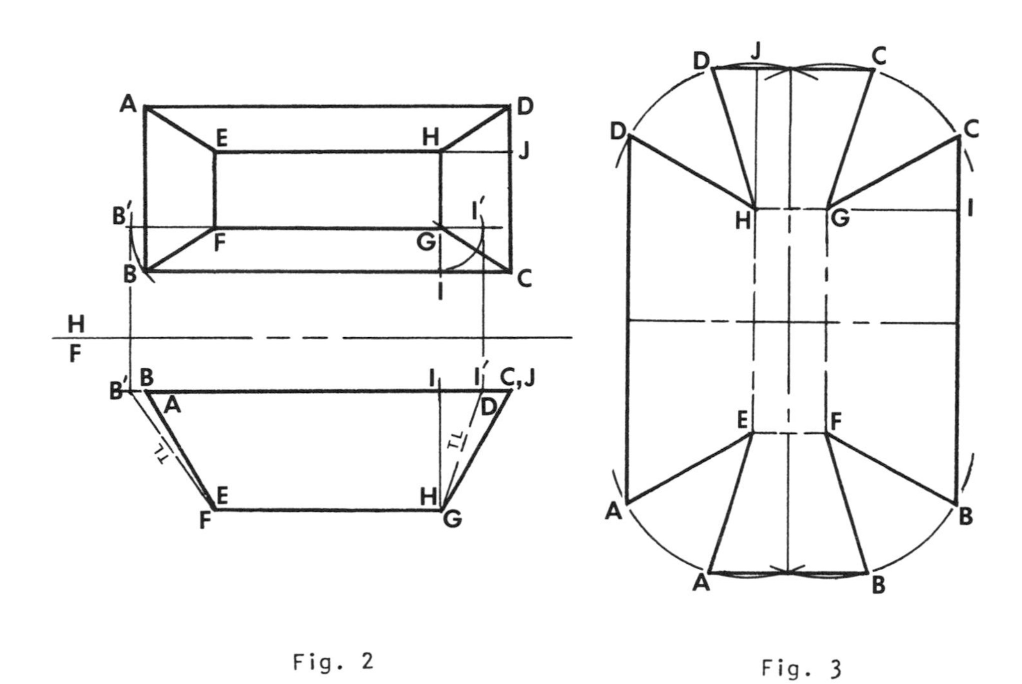

Fig. 2

Fig. 3

In order to find the true length of the trapezoids C-G-H-D and A-B-F-E, draw a horizontal line H-J in the top view. Notice that when point J is projected down to the front view, it coincides with points C and D. Since line H-J is already parallel to the cutting plane F-H, the line H-J in the front view is the true length of the height of the trapezoids in question.

In drawing the required development, first draw the rectangle E-F-G-H. Then from each of the corners of this rectangle draw an arc whose radius equals the true length of line F-B. Then draw lines D-C and A-B at a distance equal to the true length of line H-J from lines H-G and E-F, respectively. Include lines B-C and A-D parallel to lines G-F and H-E at a distance equal to the true length of line G-I. In drawing the lines in the outer edges of the development, keep in mind that the development is completely symmetrical. Finally, connect the lines E-A, F-B, G-C and H-D. Realize that the four trapezoids can be drawn without the use of their heights, by using the four sides whose true lengths were previously determined.

SPHERES

• PROBLEM 12-13

Draw the approximate development of a sphere using the orange-peel (or polycylindric) method.

Solution: The orange-peel method used to approximately develop a sphere consists of dividing the sphere into a number of sectors with planes that pass through the same diameter. This diameter appears as the center point in

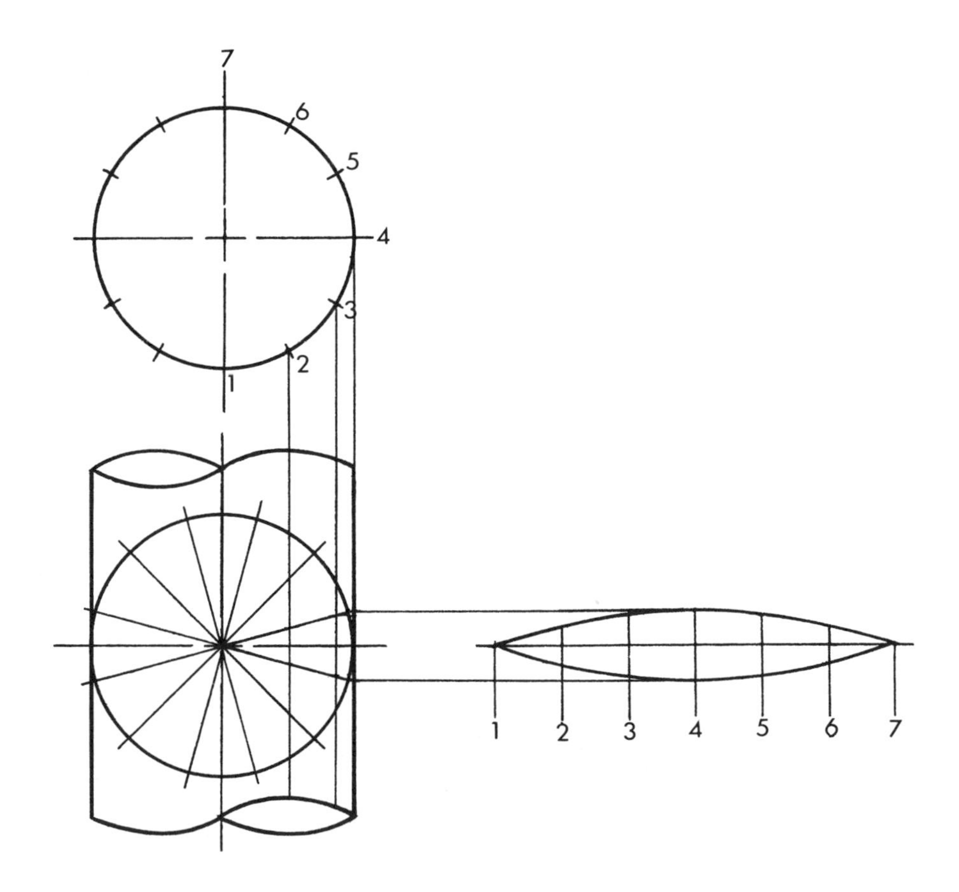

the front view of the accompanying figure. This method is also referred to as the polycylindric method because each small piece (also known as a gore) of the development is constructed as part of a cylinder.

To develop a sphere first construct a cylinder whose top view is the same as that of the sphere, which implies that it is tangent to the circle which represents the front view of the sphere. Divide the top view of this cylinder into a convenient number of equal divisions and project these divisions vertically down to the front view. Divide the front view of the sphere into a number of equal sectors, two of which are symmetrical with respect to the horizontal center line of this circle. The symmetry of these two sectors is desired because these sectors must be as close as possible to the constructed cylinder in order to get a better approximation. From the horizontal center line project an extended line horizontally to the right. On this line select a point to use as a reference to lay out the straight line distances between the divisions in the top view. Lay out the distances of only half of the circle, since the other half will be the identical development. In the accompanying figure notice that these points of division are numbered 1 through 7. On the extended line, at each point of division draw a vertical line. Project each intersection of the sector that is symmetrical with respect to the horizontal line and is to the right of the vertical center line, horizontally to the corresponding vertical line previously drawn on the development of this sector. To complete the development of the sector connect the points obtained by projection, with an irregular curve such as a French curve. To obtain the development of the whole sphere repeat the drawing of the above section eleven times for the eleven sectors of the front view of the divided sphere. These sections could be placed tangent to each other on a vertical line going through point "4" of the development as long as the extended line of each development is drawn horizontally.

• PROBLEM 12-14

Draw the approximate development of a sphere, using the polyconic (or zone) method.

Solution: As the method required in the problem implies, the sphere is determined by a series of intersecting cones. In the front and top views of the sphere described in the accompanying figure, the top hemisphere is determined by the frustums of three cones. Since the sphere is completely symmetrical, it is possible to use the same system of cones to determine the bottom hemisphere.

The first step in determining the development of the sphere by the polyconic method is to draw a convenient number of triangles tangent to the circle representing the front view with the upper vertex lying on the vertical center line of this circle. (See the accompanying figure.) Project the horizontal lines of these triangles vertically up to the top view and draw circles to represent the bases of the cones. Then, as in the development of a right cone, divide the top view into a convenient number of divisions.

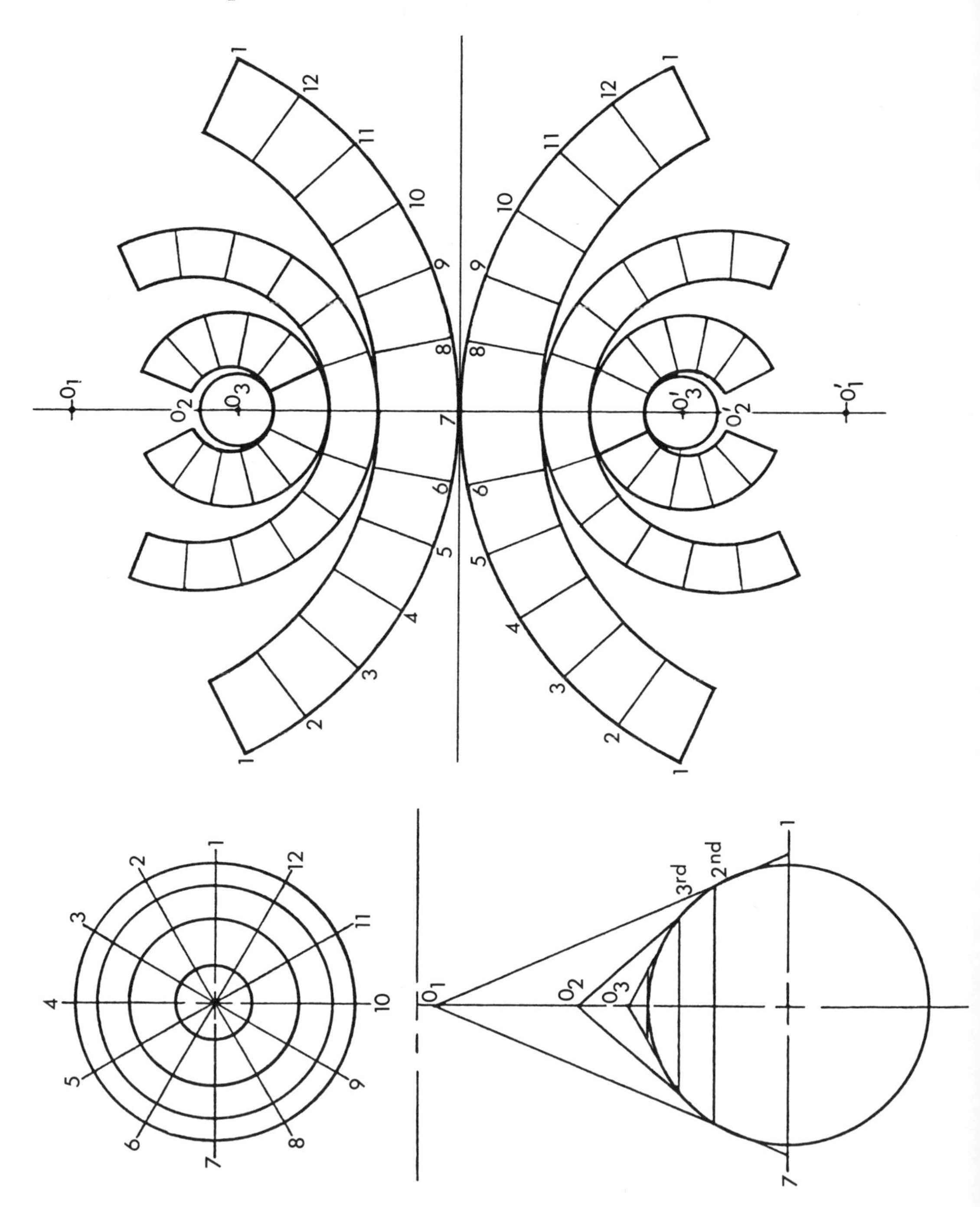

To begin the required development draw a vertical and a horizontal reference line as in the development of the accompanying figure and find the vertices of the cones 0_1, 0_2 and 0_3. To find the vertext "0_1" measure the distance of either line 0_1-1 or line 0_1-7 and lay it out on the vertical reference line from the intersection of the two reference lines. Recall that the side lines in the front view of a right cone are always true length. Using point "0_1" as the center, draw an arc going through the intersections of the two reference lines. Along either line 0_1-1 or 0_1-7 measure the distance that the next horizontal line is from point "0_1". Using this distance, draw another arc from point "0_1". Measure the straight line distance between any two divisions on either the outer or second outer circle in the top view and transfer these distances to the corresponding arc previously drawn on the development. Draw radial lines from point "0_1" to each of the division points laid out on the development. This completes the development of the frustum of the cone whose vertex is point "0_1". To continue the drawing locate point "0_2" by measuring the true length of each element line of the second cone as it appears in the front view. Lay the lines out along the vertical reference line from the top of the frustum already developed. Along the true length lines going through point "0_2," measure the distance that the third horizontal line is from point "0_2" and draw another arc with this distance, using point "0_2" as center. Measure the straight line distance between any two divisions of the circle which correspond to either the second or the third horizontal line and mark them off on the corresponding arc previously constructed on the development. Draw radial lines from point "0_2" to each of the points so located. Draw the development of the third frustum in the same way. Again, since the sphere is completely symmetrical, the bottom hemisphere can be drawn at the same time that the top hemisphere is drawn. For example, with the same measurement needed to locate point "0_1", it is possible to locate point $0_1'$ and with the same measurements used to draw the two arcs from "0_1", it is possible to draw the two arcs taken from point "$0_1'$" and so on.

To complete the development use the inner circle of the top view and place it tangent to the frustums already developed at both the top and the bottom hemispheres. This must be included so that the approximate sphere is completely closed.

CONES AND CYLINDERS

• PROBLEM 12-15

Draw a development pattern of a right cone, given the diameter of its base and the height.

Solution: Draw two views of the cone as shown in the

accompanying figure. The circumference can be measured in the top view and element AB in the front view represents the true length of all the elements of the right cone.

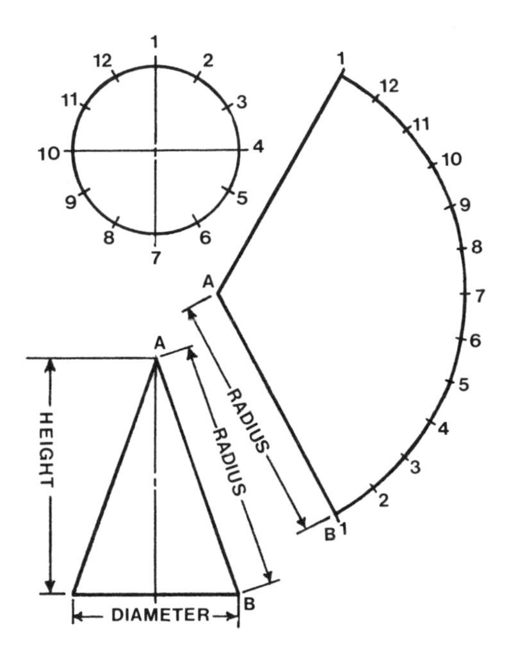

Mark off radius AB separately as shown and draw an arc of a circle with A as the center. Divide the top view in twelve equal parts and number the division points from 1 to 12. Starting from B in the pattern drawing, mark off lengths 1-2, 2-3, 3-4, etc. to obtain the surface development of the given cone. Notice that the development begins and ends with the same number (number "1").

• PROBLEM 12-16

Given the height and the diameters at the top and the base, draw a development pattern of a truncated right cone whose truncating plane is parallel to the base.

Solution: Draw the front view of the frustum from the given data and extend elements BD and CE to meet on the axis at point A. Draw a semicircle around the base line

BC and divide it into eight equal parts. Draw the development pattern of the imaginary cone with apex A as follows: pick a horizontal line with the length A-B. Using this same length strike an arc below and above the line A-B. This length can be used because line A-B appears true length in the front view. (This will be the case when the length of line A-D is used in the next step.) Measure the straight line distance between any two divisions in the semicircle shown in the front view and start laying out these distances along the constructed arcs. Number these marks as in the accompanying figure, eight divisions below the reference line and eight divisions above it. Notice that both points "8," above and below, correspond to point "C" from the front view.

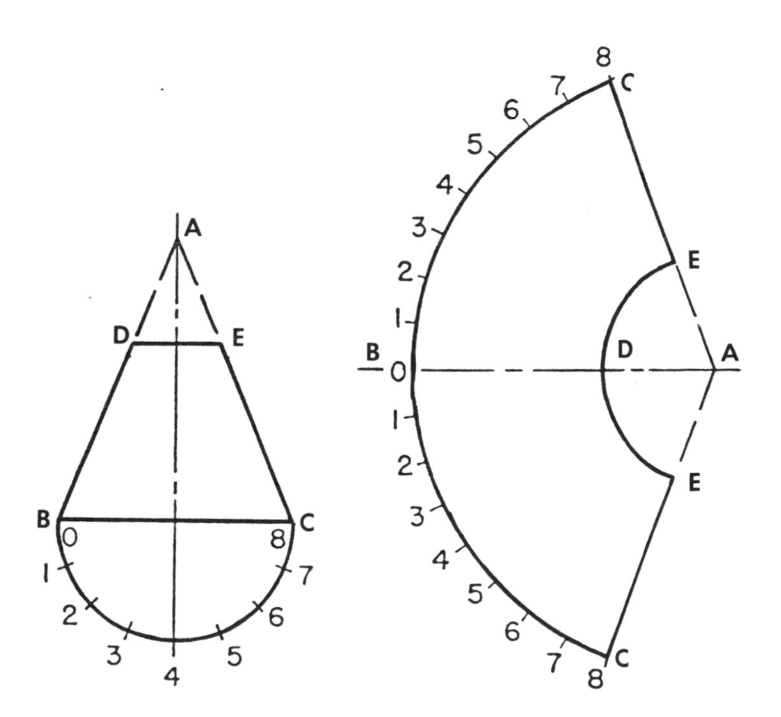

With point A as center in the pattern drawing and radius AD (from front view) strike an arc to cut sides A-C and A-B at points "E" and "D" respectively. The portion bounded by arcs CBC and EDE is the development pattern of the given frustum. Notice that eight divisions had to be located on each side of the reference line A-B because the semicircle has eight divisions. In order to account for the other half of the frustum, another eight divisions were needed.

• PROBLEM 12-17

> Draw the development of the truncated right cone represented by the front and top view shown in Fig. 1.

Solution: In order to develop the given truncated cone, first draw the development of the object as a cone in the following manner: divide the circular top view into small equal arcs. In Fig. 2 the circle was divided into twelve small arcs numbered 1 through 12. In the front view, extend the two side lines until they meet. Call this intersection "0". Then to start the development of the cone (without the truncated section) take the length of line 0-7 or 0-1, which appear true length in the front view, and draw an arc from a point "0" conveniently chosen. On this arc select a point and label it "1". Then take a compass and measure the straight line distance between any two divisions of the circular top view and using this measurement, mark off a series of points, starting from point "1", along the arc. Label these points (in order) one through twelve and then end with the number "1". (See Fig. 2.) From each of these points draw radial lines to point "0", thus determining the development of the cone.

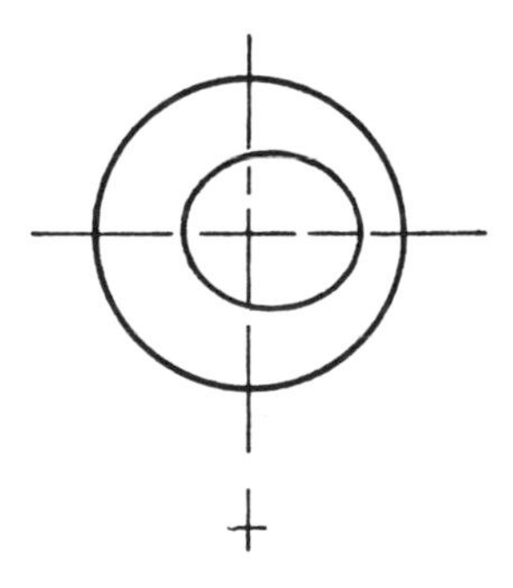

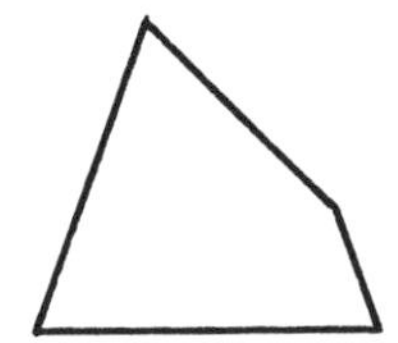

Fig. 1

To exclude the truncated part from the development of the cone, project the points 1 through 12 from the top view down to the edge view of the base. Number these points and draw radial lines from each of these points to point "0". Now label the intersection of the truncating plane and the radial lines by numbering them 1' through 12' in both the front and top views.

To obtain the true length of each of the lines 0-1', 0-2', 0-3', etc., project each of the intersections of the truncating plane and the radial lines horizontally to any of the two true length lines marked "TL" in the front view. In the case of Fig. 2 these points were projected to line 0-1. The true length of such lines is then measured from point "0" to the corresponding projection on line 0-1. For example, the true length of line 0-3' is measured from point "0" to the projection of point "3'" on line 0-1. Using the true length of the lines 0-1', 0-2', 0-3', etc., locate the points 1', 2', 3' . . . on the corresponding radial line of the development.

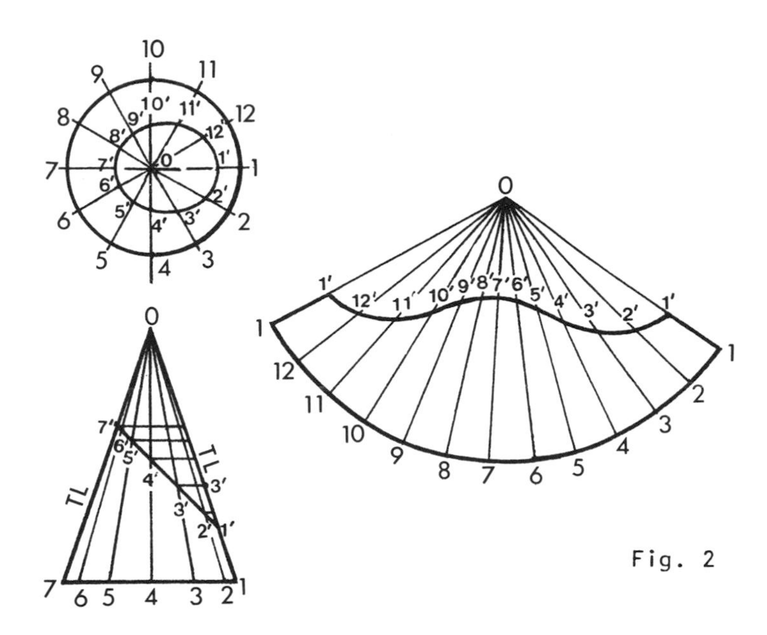

Fig. 2

When all these points have been located, connect them with an irregular curve such as a French curve.

• PROBLEM 12-18

Develop the surface of the right cone represented by the front and top views shown in Fig. 1.

Solution: Notice that the cut on the cone was made by a vertical plane and that the front view does not show the cut in true shape. In order to get a more accurate development it is better to work from a view which shows the cut in true shape. Realize that the development can be obtained from the given views; however, some inaccuracies will result.

Complete the circle in the top view and divide it into a convenient number of equal parts as shown in Fig. 2. Project the division points vertically down to the base of the front view and connect these points to the vertex of the cone. Draw a horizontal line, F-H, to represent the cutting plane between the front and top views. Construct an auxiliary view in which the cut on the cone will appear true shape. This can be accomplished by choosing the cutting plane H-l parallel to the edge view of the cut shown in the top view. Project all the

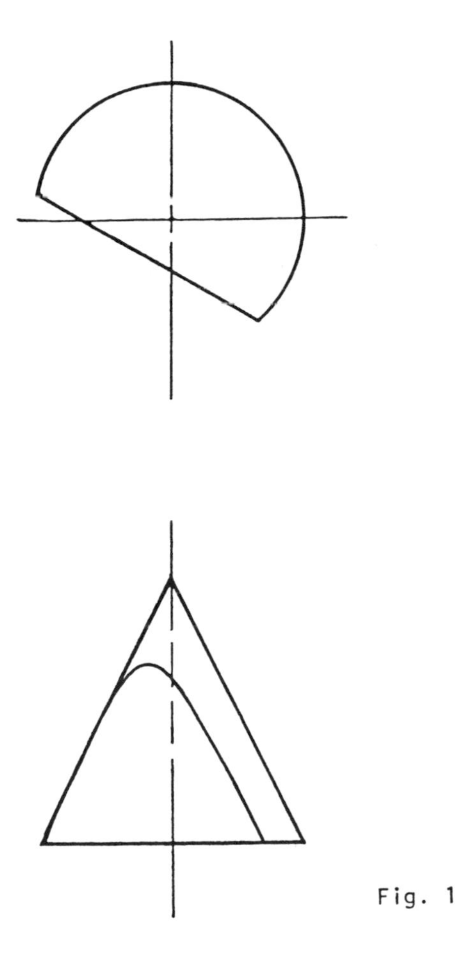

Fig. 1

points in the top view perpendicular to reference line H-l. Along these lines and from reference line H-l, lay out the distances corresponding to each point as measured from reference line F-H to the front view. A more accurate drawing of the cut can be drawn by projecting the division points from the top view onto the base of the auxiliary view and connecting these points to the vertex of the cone. Onto these lines the intersections of the edge view and the corresponding division lines are projected to establish points. To complete the drawing, these points are connected with an irregular curve. (See Fig. 2.)

In drawing the required development choose a point at a convenient location and label it "0". Using this point as the center strike an arc with radius equal to the true length of the line 0-1, as shown in the front view or lines 0-6 or 0-12, as shown in the auxiliary view.

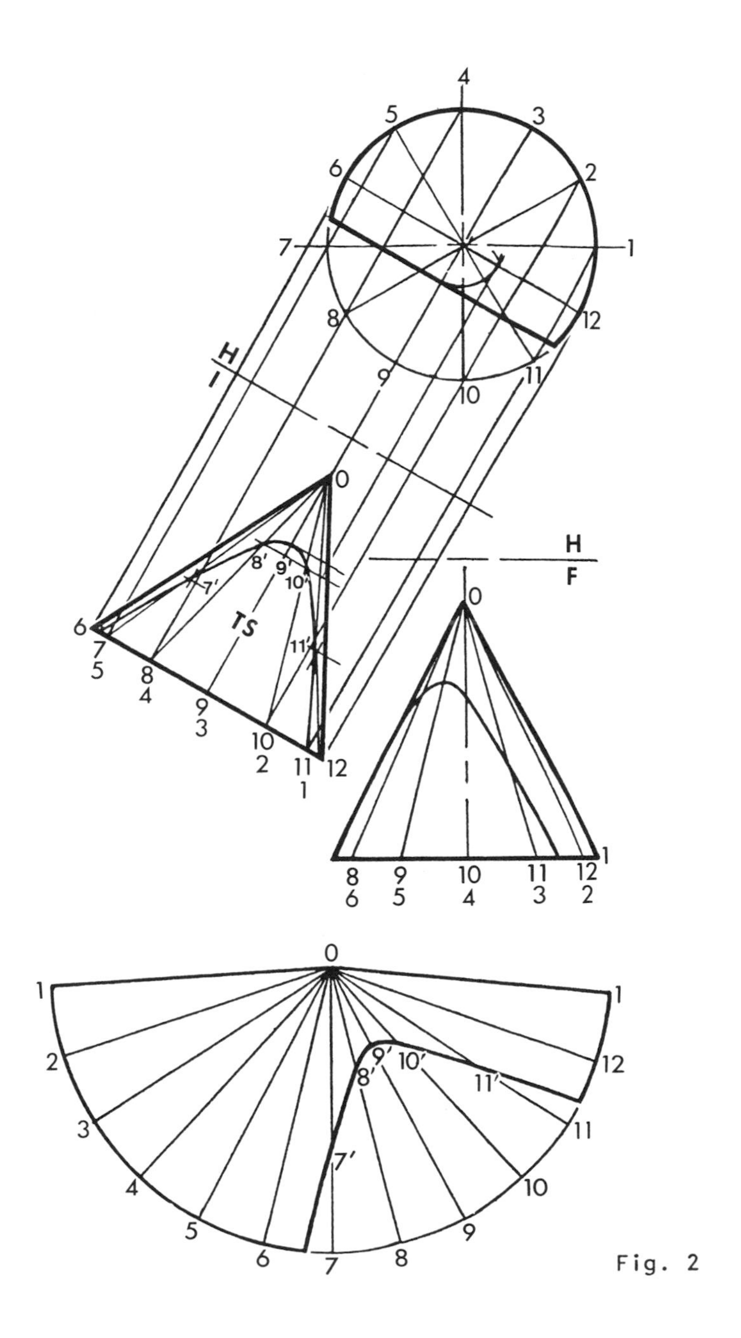

Fig. 2

Lay out all the divisions in the top view onto this arc and then draw radial lines from these points to point "0". In order to establish points on the development for the cut on the cone, find the true lengths of lines 0-7',

0-8', 0-9', 0-10' and 0-11' and lay them out on the corresponding radial line of the development. The true lengths of these points can be found by rotation. When the points 7', 8', 9', 10' and 11' have been established, measure the straight-line distance of the end points in the edge view of the cut from points 6 and 12 and lay out this distance from points 6 and 12 of the development. This establishes two more points for the cut on the cone. Finally, connect these points with an irregular curve to complete the required development.

• PROBLEM 12-19

A right cone is cut by two planes, one parallel to the base and the other at an angle to the base (see Fig. 1). Find the surface development of this truncated cone.

Solution: Construct the total development as though it were a complete cone that had not been modified. Divide the top view into a convenient number of equal arcs, as shown in Fig. 2. Notice that the two sides and the base of the cone were extended in order to complete the views

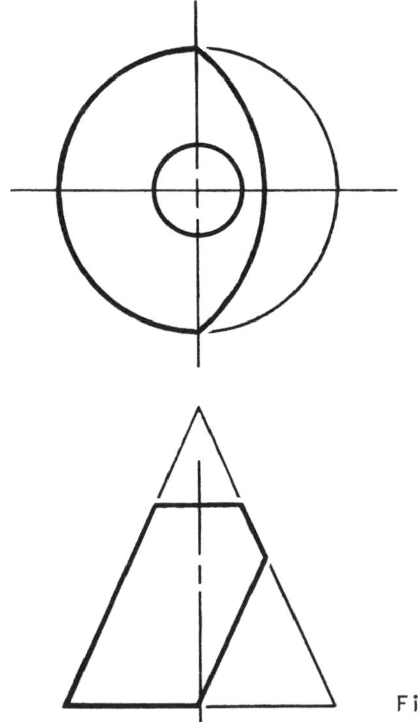

Fig. 1

for the right cone. Project the points of division from the top view onto the base of the front view and connect these points to the vertex of the cone. Then select a point "0" at a convenient location and strike an arc with radius equal to the length of line 0-1 or 0-7, which appear true length in the front view. On this arc select a point and label it "1". Then mark off all the divisions from the top view, as shown in the development of Fig. 2. This determines the development of the cone.

A conical section has been removed from the upper portion of the cone. This part of the pattern can be removed by constructing an arc in the development, using as the radius the true length line 0-7', which is found in the front view.

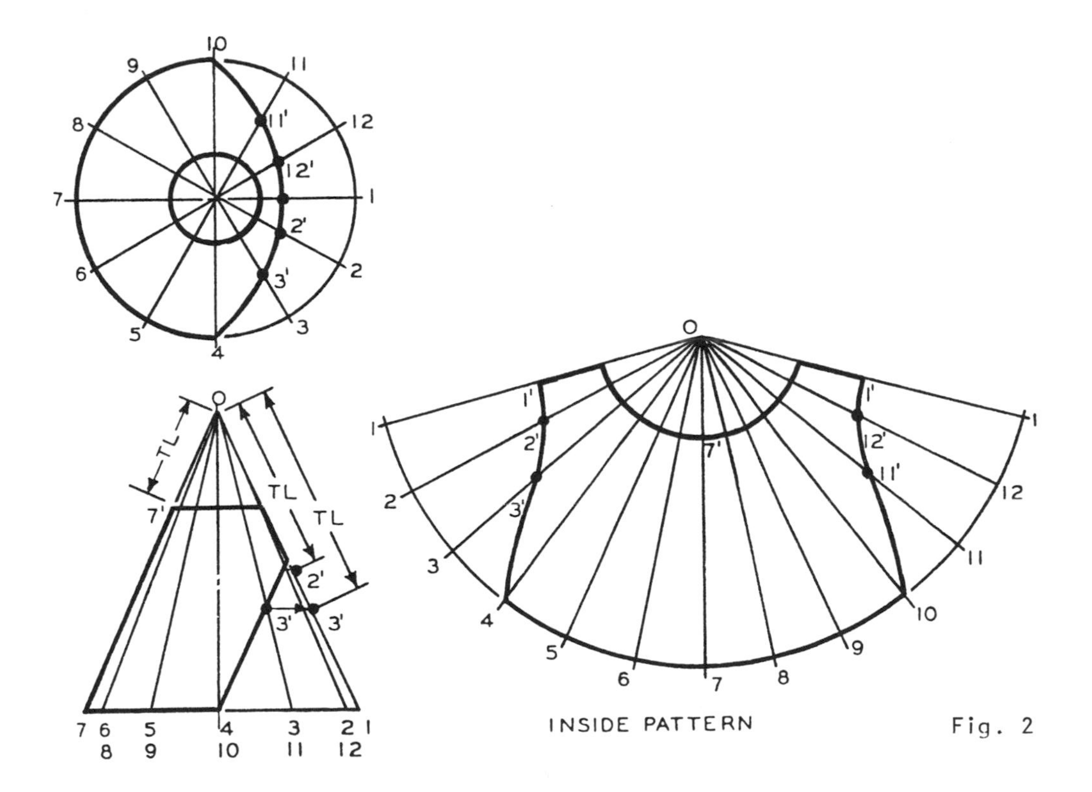

Fig. 2

The true length measurements from point 0 to the angled cutting plane are found by revolution. For example, point 3' in the top view is rotated counterclockwise to a position on the horizontal line going through the center of the top view. This point is then projected down to the front view, where it intersects a point on the true length line 0-1 which is on the same horizontal line as point "3'" before the rotation. In general, for any right cone the true length of a specific line measured from the vertex can be found by projecting the other endpoint of the line horizontally to any of the two true length side lines. For example, lines 0-2' and 0-3' are projected horizontally to the extreme element, 0-1 in the front view, where they will appear in true length. These distances are measured along their respective lines in the development to establish points through which the smooth curve will be drawn to outline the development, thus completing the required development.

• PROBLEM 12-20

Develop the oblique cone described by the front and top views shown in Fig. 1.

Solution: The method to be used in this type of problem is called triangulation. In this method the surface to be developed is broken up into a series of triangles and the development is then drawn by constructing the true shape of each triangle.

To triangulate the given oblique cone first divide the circle which represents the base of the cone into a convenient number of equal parts (twelve in the case of the top view of Fig. 2). Number each division point then draw a line connecting each of them and the vertex of the cone. Project the division points from the top view vertically down to the bottom edge of the front view. Number these points and connect them to point "0" in the front view.

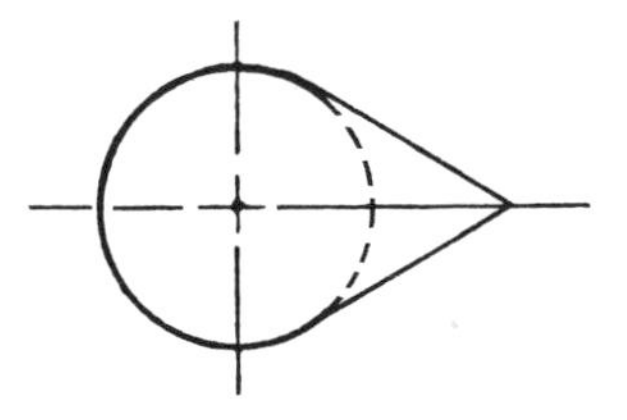

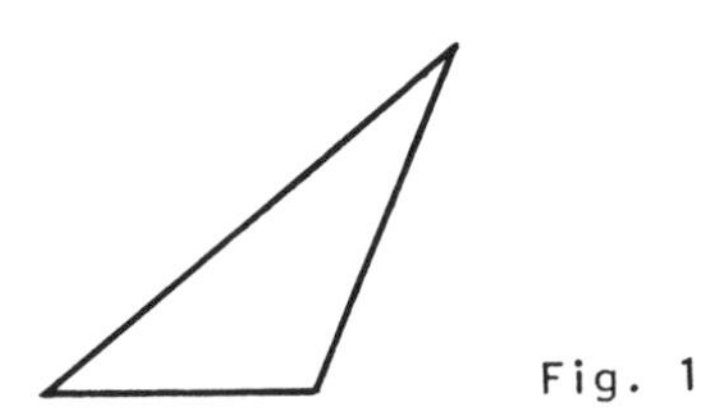

Fig. 1

In order to construct the true shape of a triangle the true lengths of all the three sides are needed, since the true angles are not known. Notice that lines 0-1 and

0-7 appear true length in the front view. This occurs because in the top view they appear parallel to the cutting plane between the front and the top views. To find the true length of the other lines rotate each line in the top view until they are parallel to the cutting plane between the front and top views. Another method is to rotate the lines onto the horizontal line going through point "0". In Fig. 2 points 2 through 6 were rotated counterclockwise to the horizontal line. It is necessary to rotate only these points because the top view is symmetrical with respect to the horizontal center line. When these points have been rotated project them down to a horizontal extension of the bottom edge of the front view. Label the new locations of the rotated points 2' through 6' and 8' through 12' and connect them to point "0". Lines 0-2', 0-3', etc. all appear true length as shown in the true length diagram of Fig. 2.

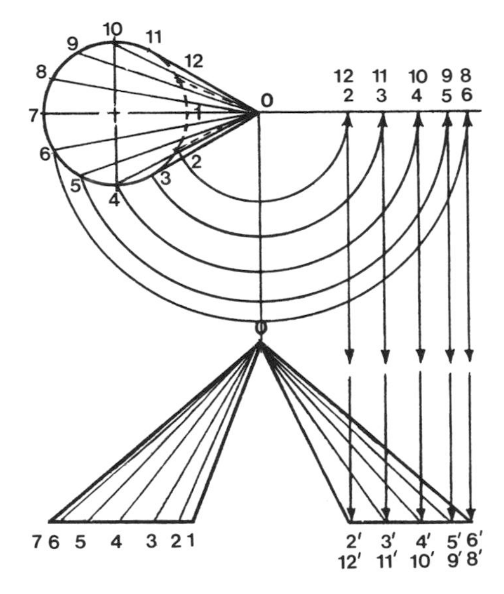

TRUE LENGTH DIAGRAM

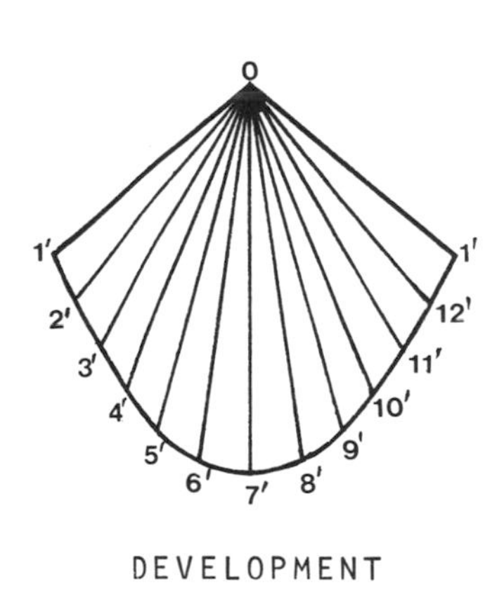

DEVELOPMENT

Fig. 2

After the true length of the line is determined, draw the development as follows: first draw a vertical line with the true length of line 0-7 at a convenient location. With a compass measure the straight line distance between any two divisions in the top view. From point "7" swing two arcs, one to the left and one to the right. Take another compass and measure the length of line 0-6', which is the same as 0-8', and swing two arcs from point "0" on the development to intersect the two arcs previously drawn. This will locate points 6' and 8' in the development. With the compass that measured the distance between each division swing arcs from both points 6' and 8'. Next, measure the length of line 0-5', which is the same as the length of line 0-9', and swing arcs from point "0" to locate points 5' and 9'. Continue this process of constructing the true shape of each triangle until all of them have been constructed. When all the points have been located in the development, connect them with an irregular curve to complete the development. (See Fig. 2.)

• PROBLEM 12-21

From the horizontal and frontal projections of the truncated oblique cone shown in Fig. 1, draw the lateral surface development of the cone.

Solution: The elements of an oblique cone are not all of

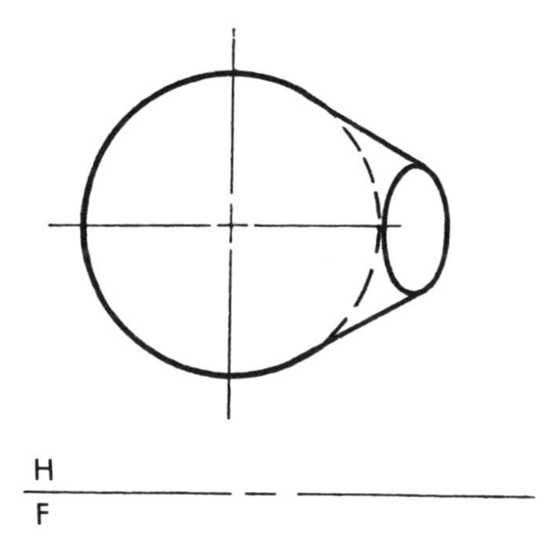

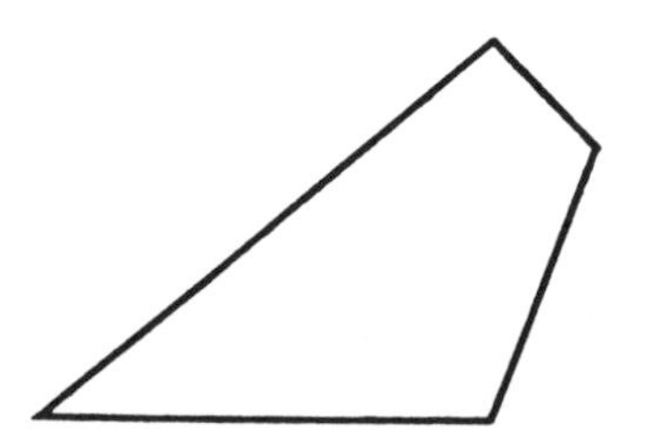

Fig. 1

equal length. Therefore the true length of each element must be determined before the surface can be developed.

In the two given views, extend the outer elements in order to locate vertex O. Divide the true shape of the bottom base, in the horizontal view, into a convenient number of equal divisions and number them as shown in Fig. 2. Notice that the circle in the top view is the

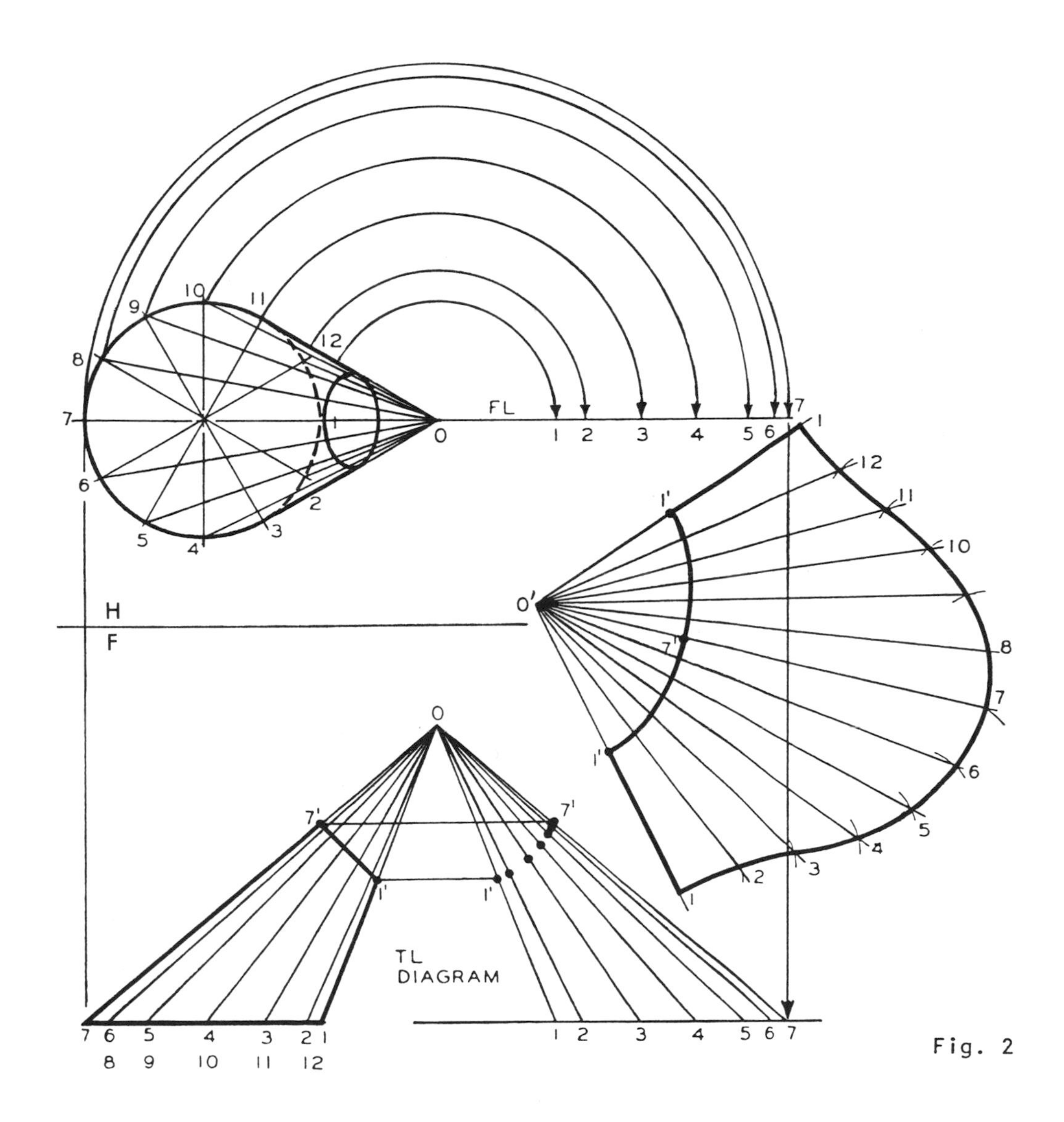

Fig. 2

true shape of the base because the top view is perpendicular to the front view and in the front view the base is shown as an edge. Construct elements to each numbered point of the circle by drawing straight lines from point "O" to each numbered point of the top view. Project the numbered points in the horizontal view vertically downward to the frontal view. In the frontal view, construct elements to each numbered point on the bottom base by connecting each point to the vertex "O".

Using the rotation procedure, determine the true lengths of the elements O-1 through O-12 with O serving as the center of rotation. Notice in the horizontal view, that the direction of rotation is clockwise and that all elements have been rotated into a position parallel to the cutting plane F-H. The true lengths of these elements are summarized in the true length diagram of Fig. 2. The true length diagram is drawn as follows: first each numbered point in the top view is rotated clockwise about point "O" until the arc intersects the horizontal line going through point "O". From this line project each point vertically down until it intersects the extension of the edge view of the base. The lines joining point "O" and these intersections are all true length.

To determine the location of points 1', 2', 3', etc. on the true length diagram, project each of these points in the front view horizontally to the right until they meet their corresponding true length line. For example, point "7'" is projected horizontally to the right up to the true length line O-7. Label this point "7'".

Starting at any convenient point O', begin the development by striking a small arc having a radius equal to the shortest element O-1 as determined in the true length diagram. Pick an arbitrary point on this arc and label it "1". From point 1, strike an arc having a radius equal to the chord distance 1-2 as it appears in the top view. From O', strike the true length of element O-2. Point 2 on the development can be found at the intersection of this arc with arc 1-2. Similarly, using distance 2-3 in the horizontal view and true length of element O-3, locate point 3 in Fig. 2. Repeat the procedure until all points have been located on the development and connect these points with a smooth curve. Notice that the distance from one point to another in the circle of the top view is the same throughout.

Referring to the true length diagram, measure the true length distances from O to points 1', 2', 3', etc. and lay them out on their respective elements in the development. On the development connect points 1', 2', 3', etc. with a smooth curve. This completes the required surface development of the given truncated oblique cone.

• PROBLEM 12-22

Develop the truncated oblique cone shown in Fig. 1, without finding its vertex.

<u>Solution</u>: The problem calls for a solution in which the vertex of the cone is not used. Notice that the cone is truncated by a plane parallel to the base and that both the base and the truncated face are circular.

In order to develop the given cone, first divide the two circles in the top view into small divisions. In the case of Fig. 2, twelve divisions were used. It is important that both circles are divided into the same number of sections. Label all the divisions. Notice that both circles were numbered in such a way that the same numbers were used for similar positions with the exception that the numbers in the base circle are primed. For example, divisions number "1" and "1'" are both in the extreme right of their respective circle.

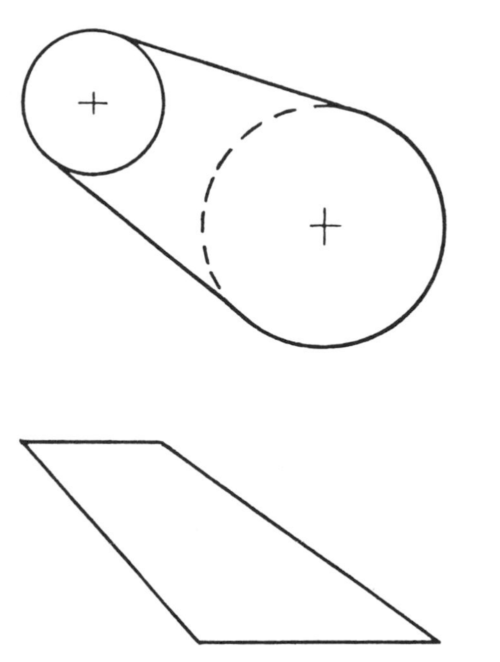

Fig. 1

After all the divisions have been properly labeled, connect each number with its corresponding primed number in the front view. For example, connect point "1" to "1'", point "2" to "2'", etc. (see Fig. 3). Notice that when these points are connected, they form a series of small trapezoids. Draw the diagonals of these trapezoids as shown for the trapezoids 9-9'-10-10' and 10-10'-11-11'. Recall that the development of an object shows all the parts in true shape. Therefore, the true length of the lines to be used in constructing the development must be found. The true length of these lines can be found by rotation. For example, to find the true length of line 1-1', take a compass and use point "1" as the center. Extend the compass to point "1'". Swing an arc up until it intersects a horizontal projection from point "1". Project this intersection vertically down to the base of the cone. Now join point 1 to the new point "1'". The line is the true length of line 1-1'. (See Fig. 3.) The true length of all the other lines can be found in the same manner.

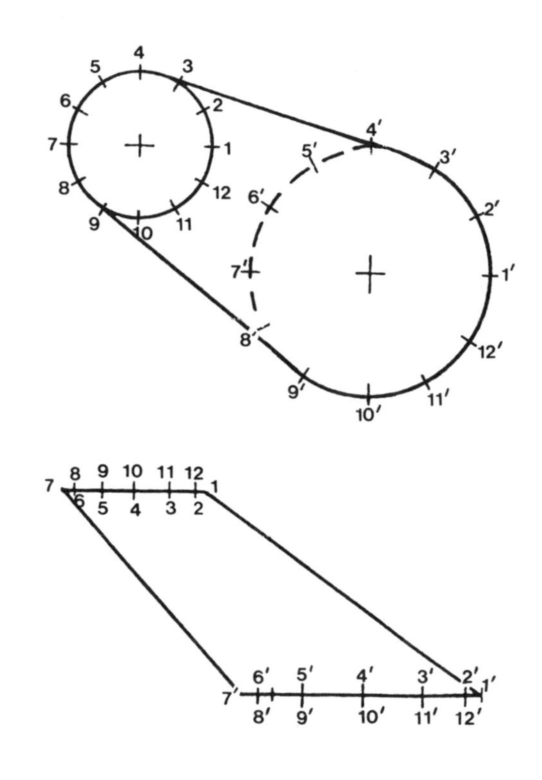

Fig. 2

To construct the required development, choose any line to begin. (In the case of Fig. 3 line 7-7' was chosen.) With the compass, measure the chord of any of the arcs used as the small divisions of the base circle. Then swing an arc from point "7'" with the above measurement. Find the true length of the diagonal 7-8' and with a compass swing an arc, using point "7" as the center, so as to intersect the small arc previously constructed. Label the intersection "8'". Measure the length of the chord of any of the arcs used as a small division of the circle that represents the truncated face. (For example, arc 1-2.) Using the above determined length as the radius, swing an arc from point "7". Find the true length of line 8-8' and determine point "8". Finish the development by using the same procedure that was used to determine points "8" and "8'". When all the points have been determined,

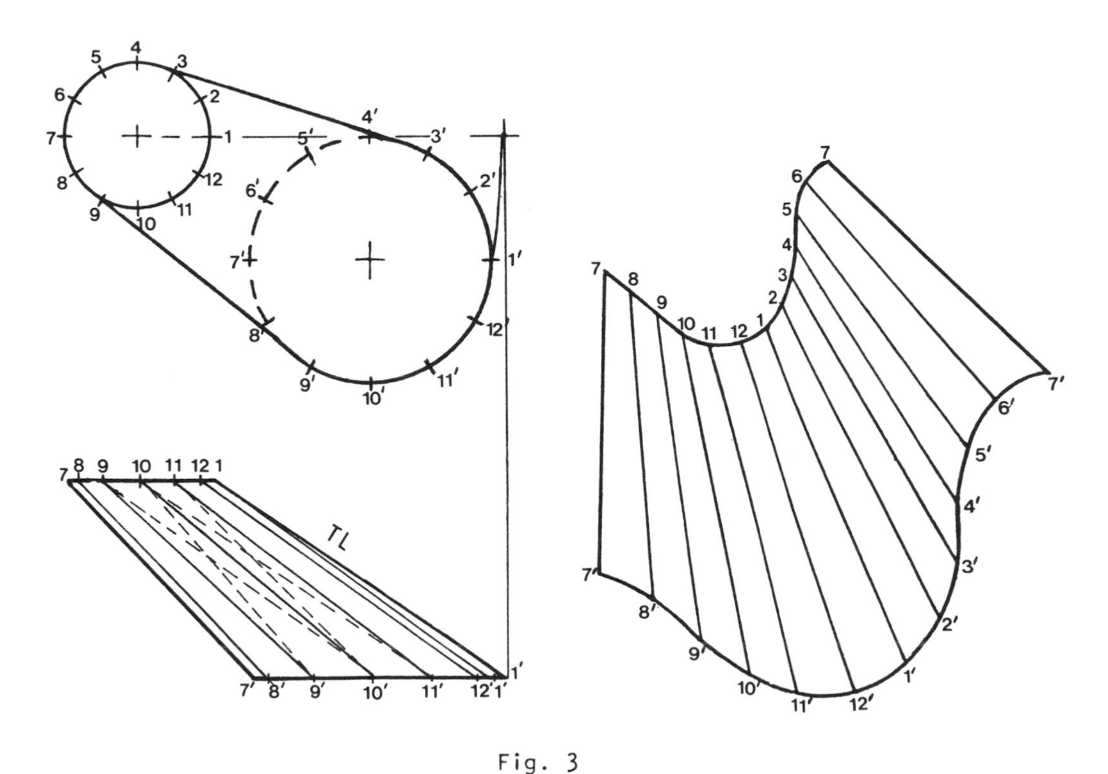

Fig. 3

use a French curve to connect the points. Notice that the development ends and begins with the same line (line 7-7').

• PROBLEM 12-23

Develop a right cylinder of given length and radius.

Solution: To develop a right cylinder, two opposite views are required. One of these views must show the diameter and the other must show the height. (See the accompanying figure.)

To make an approximate development, divide the circular view into a number of equal parts (the more divisions, the more accurate the development pattern) and number each of the divisions on the circular view (refer to the figure where 12 divisions have been made).

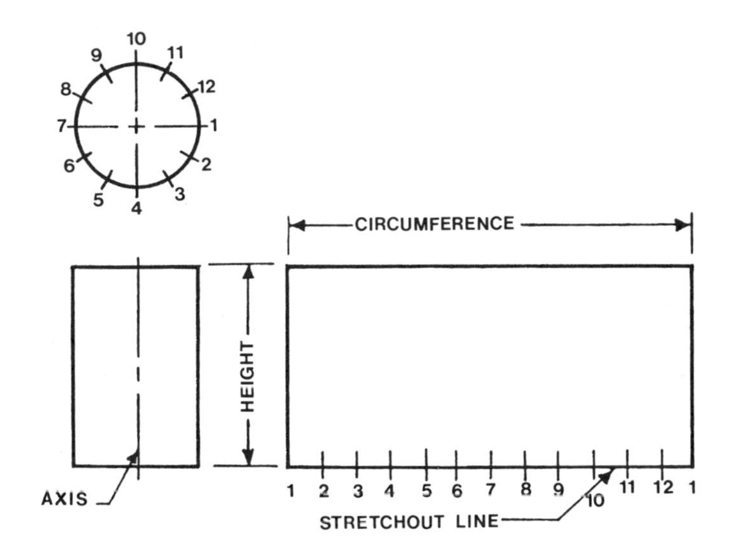

Draw the stretch out line which represents the bottom circumference of the cylinder. On the left end of the line, draw a perpendicular line of the same length as the height of the cylinder. Measure the straight line distance between two divisions on the circular view and lay out this distance on the stretchout line. Be certain to lay out as many divisions as there are on the circular view. This will give the approximate length of the circumference.

On the last division, draw a line perpendicular to the stretchout line. Draw the top edge of the cylinder parallel with the stretchout line at a distance equal to the height.

This method for obtaining the length of the circumference is approximate because the distances used are chordal distances and as you will recall a chord is shorter than its arc. The actual value of the circumference is $\pi \times$ Diameter.

• PROBLEM 12-24

Develop the truncated right cylinder represented by the front and top views shown in Fig. 1.

Solution: To make the development of this truncated cylinder, two adjacent views are necessary (commonly the top and front views). In order to construct the required development, divide the circular view into a number of equal parts. (In the case of Fig. 2, twelve divisions were used.) Project these points to the front view to obtain lines A-1, B-2, C-3, D-4, E-5, F-6 and G-7. (These lines are known as generators.)

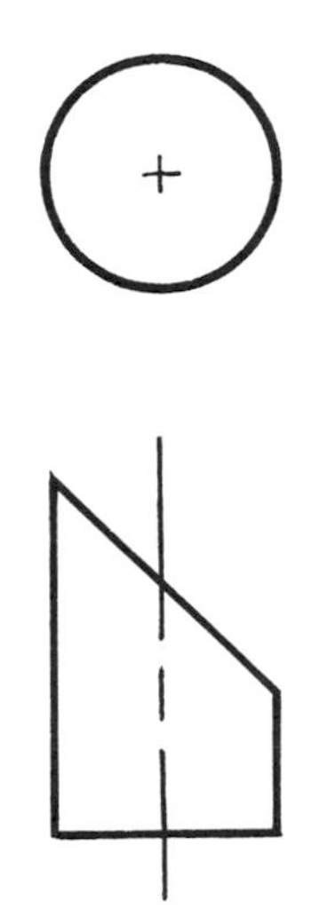

Fig. 1

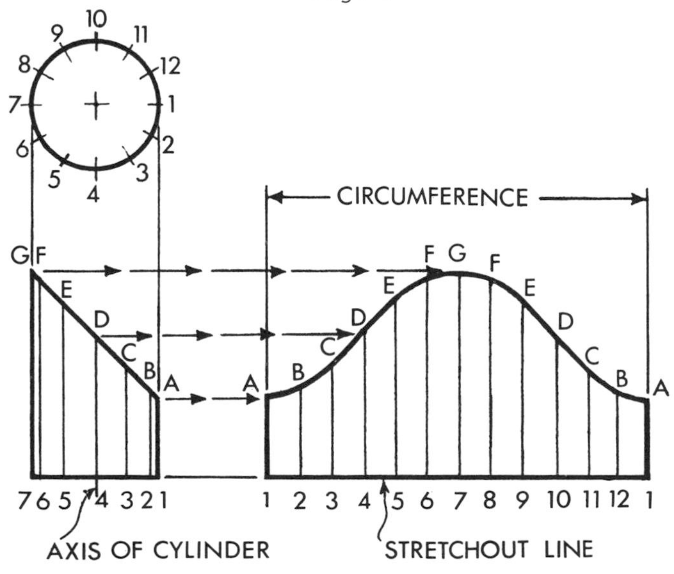

Fig. 2

Draw the stretchout line perpendicular to the axis of the cylinder. Notice that the axis of the cylinder coincides with line D-4. Measure the straight line distance between two divisions on the circular view and lay out this distance on the stretchout line for as many divisions as there are on the circular view. The total length of these divisions will add up to the approximate circumference of the cylinder. Draw perpendicular lines through each point on the stretchout line. (These lines are called ordinates.)

Project the points on the inclined edge in the front view horizontally to the right up to the corresponding ordinates. Connect the points of intersection thus obtained with a French curve. The space bounded by this curve and the stretchout line is the development pattern of the cylinder in question (line A-1 included).

• PROBLEM 12-25

Develop the cylinder which is cut by a cranked plane as described by the front and top views given in Fig. 1.

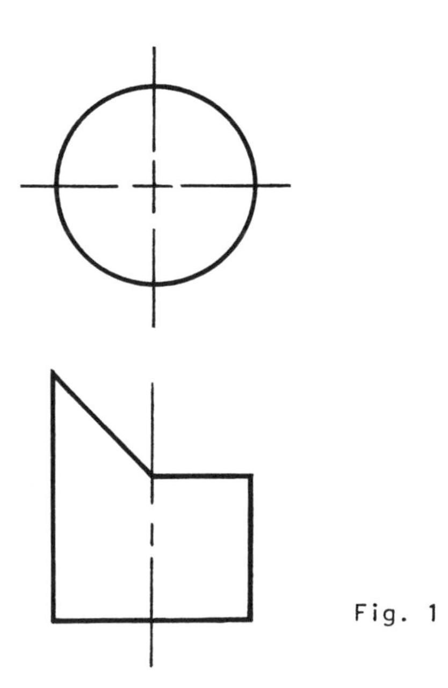

Fig. 1

Solution: To develop the given cylinder first divide the circular top view into a convenient number of divisions (refer to Fig. 2 where 12 divisions have been made). Number these divisions 1 through 12 and project them vertically down to the front view. Project an extended line from the bottom edge of the cylinder, as shown in the front view, horizontally to the right. Measure the distance between any two consecutive divisions in the top view and use it to lay out the location of each of the points 1 through 12 on the extended line. At each of these points draw a vertical line. Proceed to project all the points at the top and bottom surfaces of the cylinder onto the corresponding vertical (or element) line of the development. Connect the points 1 through 4 and 10, 11, 12 and 1 with straight lines and the rest with an irregular curve. At any point of the extended line, draw the circle that represents the bottom surface of the cylinder.

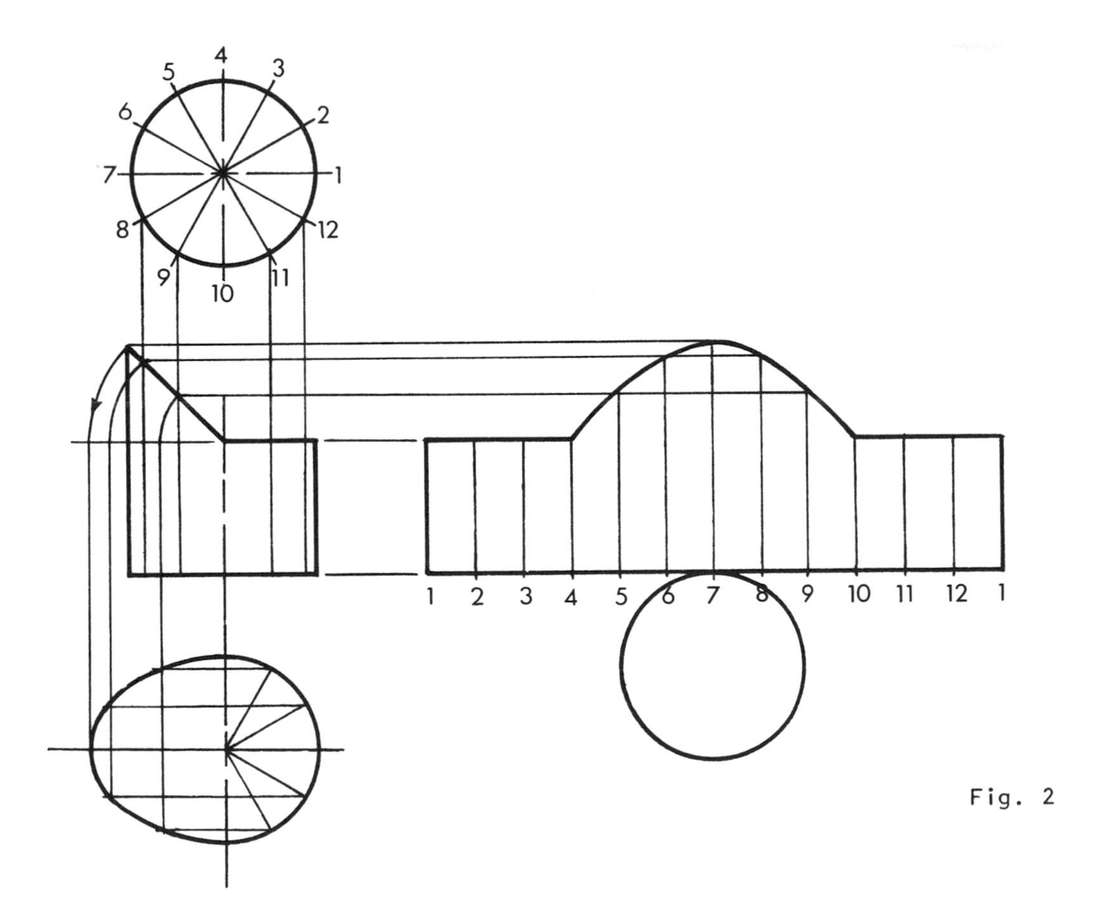

Fig. 2

To draw the development of the top surface of the cylinder as a whole, project the center line of the cylinder vertically down. At a convenient location draw the semi-circle which represents the right side of the top view, which appears true shape in the given top view. Divide this semi-circle in the same way that it was divided in the top view. Project each division point horizontally to the left. In order to find the true length of each segment determined by each point on the inclined cutting plane, rotate each segment in the front view counterclockwise to a position parallel to the cutting plane between the front and top views. Project these points vertically down to the corresponding horizontal line previously drawn on the development of the top surface. To complete the development, connect the points with an irregular curve. Notice that the true shape of the inclined plane could have been obtained by constructing an auxiliary view off the front view with a cutting plane parallel to the inclined plane.

• PROBLEM 12-26

From the front and top views given in Fig. 1, find the development of the oblique cylinder and its end pieces.

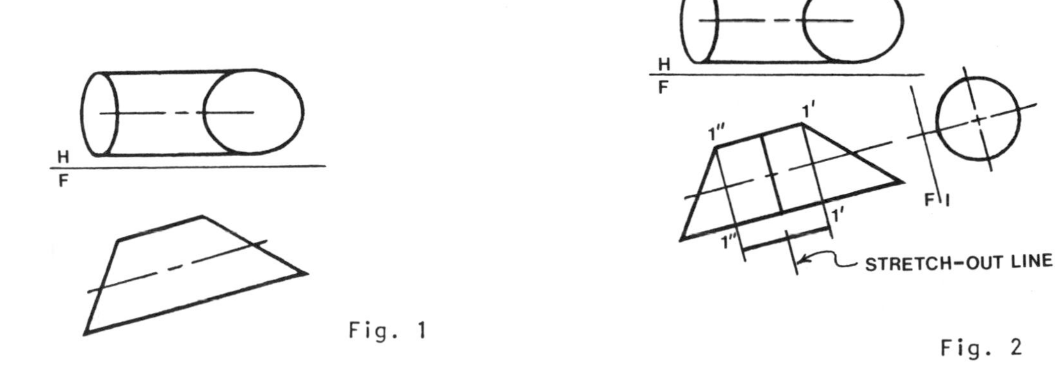

Fig. 1

Fig. 2

Solution: In order to properly divide the given cylinder, first draw an auxiliary view of the front view with reference line F-1 perpendicular to the true length axis of the given cylinder. (See Fig. 2.) The constructed view is a circle which represents the circumference of the cylinder. Divide this circle into small arcs. In the case of Fig. 3 the circle was divided into twelve small arcs.

Choose a stretchout line whose length is that of the circumference of the cylinder and perpendicular to the true length axis of the cylinder. Project the division points from the auxiliary view onto the front view. Number all the points as in Fig. 3. Measure the straight line distance between any two divisions of the circular view. Lay out this distance on the stretchout line, one for every division. At every point located on the stretchout line draw a line parallel to the axis of the cylinder. Number these lines, starting with point "1".

From the front view project the points previously located at each end of the cylinder onto the drawing of the development. Number the points determined by their corresponding projection lines, as in Fig. 3 and connect these points with a French curve. Notice that the right side end is numbered with primed numbers while the left side is numbered with double primed numbers (Fig. 4). Also notice that the lines on the development can be projected from the front view because they are all true length in the front view. This is because the cylinder appears to be parallel to the cutting plane in the top view.

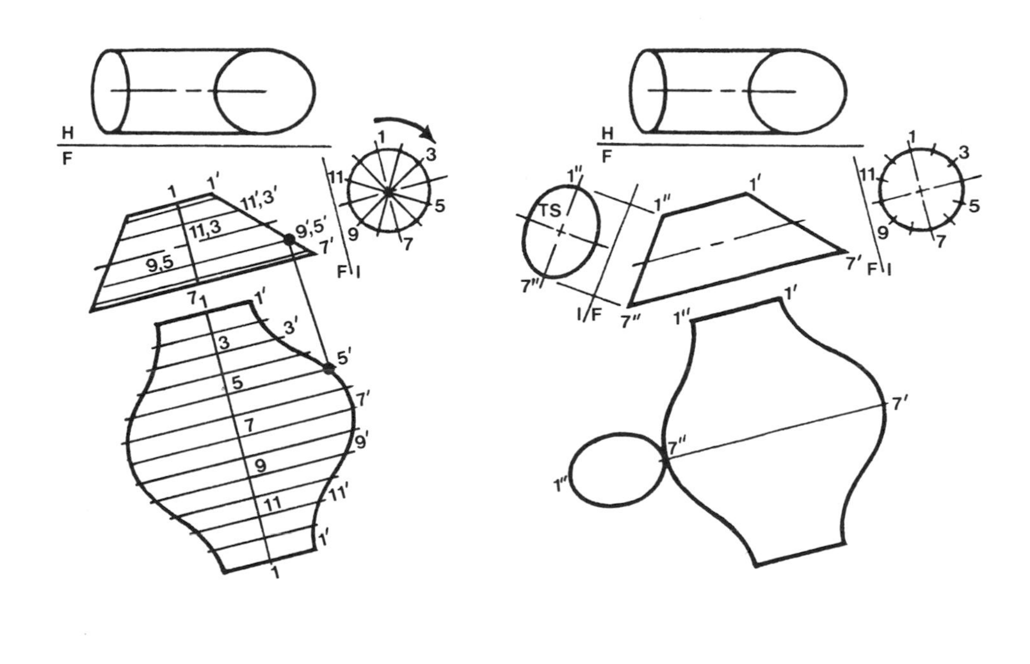

Fig. 3

Fig. 4

In order to draw the development of the two end pieces, two auxiliary views must be taken of the front view with reference lines (or cutting planes) parallel to each of the edge views of the end pieces. In Fig. 4 an auxiliary view was taken of the left end piece. The ellipse is true shape because the cutting plane F-1 was taken parallel to the edge view of the left end. The same procedure can be used in order to find the true shape of the right end piece. Notice that points "1"" and "7"" were marked on the true shape of the left end piece. The point "7"" on the ellipse was matched with point "7"" on the development of the cylinder. The same procedure would be followed for the right end.

Develop the cylinder and the hexagonal prism which intersect as described by the front and the top views given in Fig. 1.

Solution: In developing the cylinder first draw a right view to show the cylinder as a circle. (Refer to Fig. 2.) Divide this circle into a convenient number of equal divisions. Number these divisions and project them to the front view, parallel to the side edges of the cylinder. choose a cutting plane that intersects both of the side edges of the cylinder and is perpendicular to the center line. Extend this cutting plane (which is the edge view of the plane on which the right section lies) and take this extension to be the extended line for the cylinder.

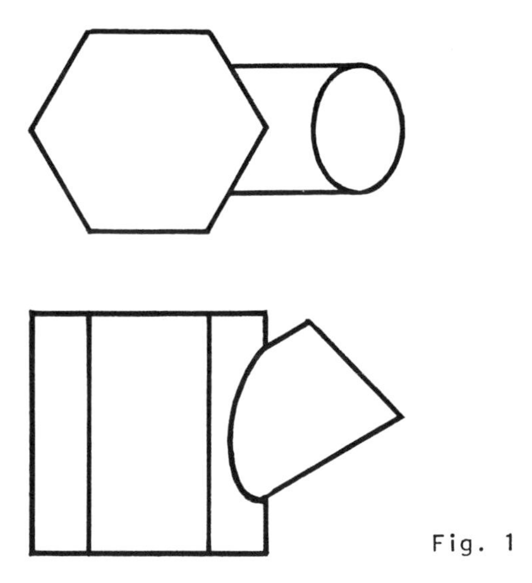

Fig. 1

On this extended line select a point and label it "1". Measure the straight line distance between any two divisions on the circular right section and for every division on this circle mark off this distance on the extended line. Number each of these points and at each one draw a line perpendicular to the extended line. Project each point in the top surface and in the intersection of the cylinder parallel to the extended line until the projection lines intersect the corresponding element line. Remember that the generators in the front view that were drawn from the right view of the cylinder are identified by the numbers of the points they were taken from. Therefore, when the point on the top surface of the cylinder is taken from the generator corresponding to points 3 and 11 and is projected to the development, the only points established by this projection are 3 and 11. All the other points are projected in the same manner. Finally, connect all the points established by the above method with an irregular curve, such as a French curve.

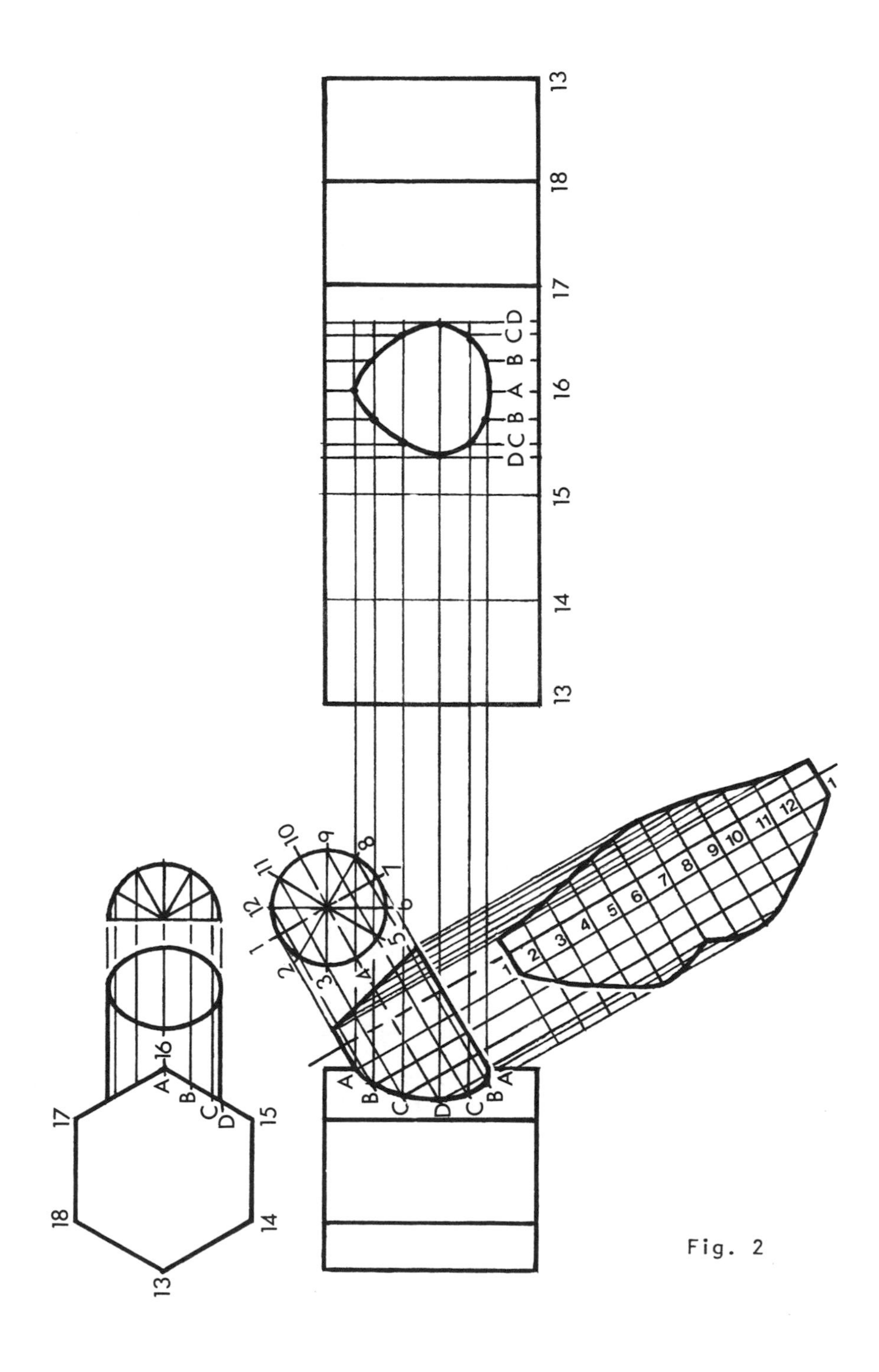

Fig. 2

To draw the development of the hexagonal prism, project an extended line horizontally from the bottom edge of the prism. On this extended line mark off the six distances of the hexagon in the top view and number them as in Fig. 2. Notice that the top view of the hexagonal prism is a right view and therefore the sides of the hexagon all appear true length in the top view. On each of the points, 13 through 18, draw vertical lines and project the top edge of the front view onto the development, in order to determine the outside edges of this development.

In order to include the hole at the intersection of the prism and the cylinder, draw a semicircle on the right side of the top view of the cylinder and divide it into a convenient number of divisions as shown in Fig. 2. Project these divisions onto lines 17-16 and 15-16 of the top view and letter the lower intersections A, B, C and D. Project these points vertically down to the front view. Notice that point "A" lies on the same edge as point "16"; therefore, mark off the distances of B, C and D from vertical line 16 of the developments, on both sides as taken from the top view. (See the development in Fig. 2.) Project points A, B, C and D from the front view horizontally to the corresponding element line of the development and finally, connect these points with an irregular curve. This completes the required developments.

• PROBLEM 12-28

Develop the two intersecting cylinders described by the front and top views shown in Fig. 1.

Solution: Since the lengths of the horizontal cylinder appear true length in the top view, the development of this cylinder should be projected from the top view while the vertical cylinder should be projected from the front view. This is done because the lengths of this cylinder appear true length in the front view.

To develop the horizontal cylinder first divide the circle in the front view into twelve equal parts and label them 1' through 12' as in Fig. 2. Project these divisions vertically up to the top view and number these projections at the front edge of the horizontal cylinder. Project an extended line horizontally from this front edge. On this extended line select a point and label it "1'". With a compass measure the distance between any two divisions in the front view and use it to lay out the circumference of the cylinder. On each of the points (points 1' through 12' in the development) draw vertical lines and project the curve of intersection between the vertical and horizontal cylinders horizontally to the development. This is done by projecting the points of intersection between the curve of intersection and the projection lines, to the corresponding element lines. For example, the point of the intersection between the intersecting curve and the generator for points 9' and 5' is projected horizontally onto the development until this projection line intersects the vertical lines marked 9' and 5' in the development. This establishes points 9' and 5' in the development. When points 1' through 12' have been established in the same manner, connect them with an irregular curve to complete the development of the horizontal cylinder.

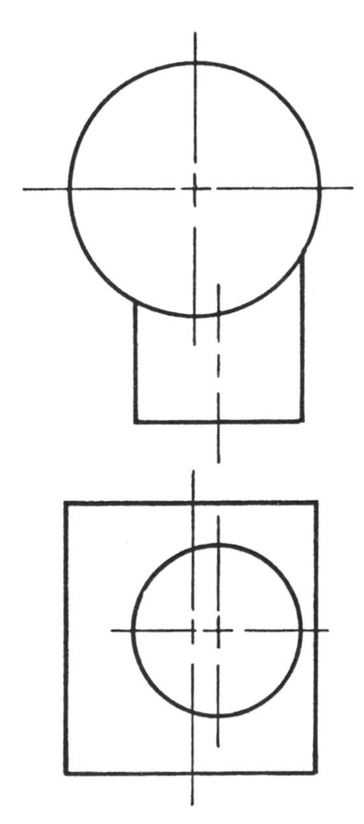

Fig. 1

In developing the vertical cylinder first divide the circle in the top view into twelve equal parts and label them 1 through 12 as in Fig. 2. Project these division points vertically down to the front view. Number these projection lines (or generators) in the bottom edges as in Fig. 2. Project an extended line from this bottom edge, then select a point on it at a convenient location and label it "1". Measure the distance between any two division points of the circle in the top view and use it to lay out the circumference of the vertical cylinder by locating points 1 through 12 on the extended line. On each of these points (1 through 12) of the development construct a vertical line. Project the top edge of the front view horizontally onto the development in order to complete its outside drawing.

To locate the points necessary to construct the hole in the development project the intersections between the circle in the front view and the generators, numbered 1 through 12, horizontally to the corresponding element line in the development. Notice that the only generator lines which actually intersect the circle are 3, 4, 5 and 6. When these intersections have been projected to element lines 3, 4, 5 and 6 of the development, connect all the points with an irregular curve to complete the development of the vertical cylinder.

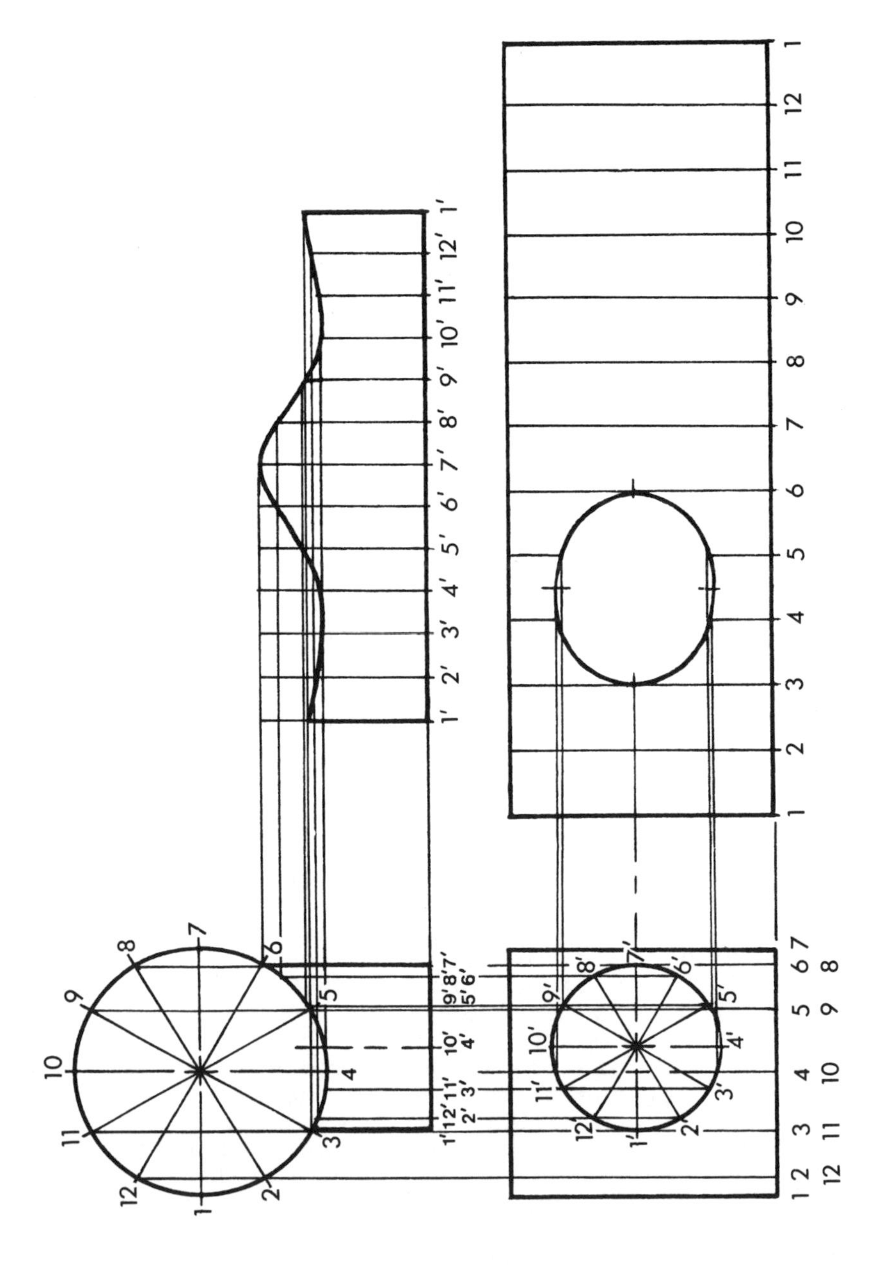

Fig. 2

Show the layout patterns (developments) for the 90° intersection of the two cylinders shown in the isometric drawing of Fig. 1.

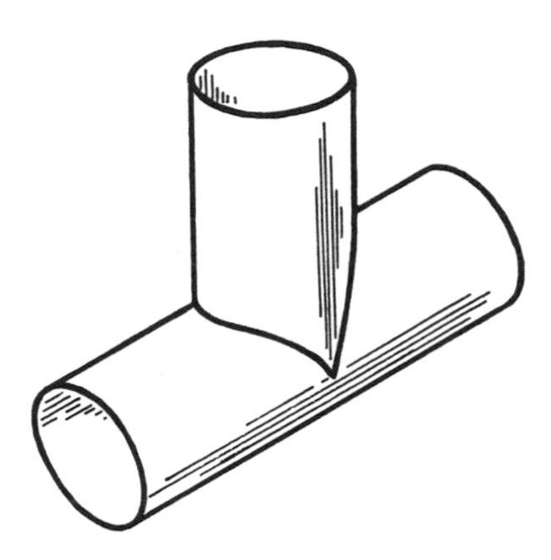

Fig. 1

Solution: Draw a side view of two cylinders as shown in fig. 2. Lines AB and CD represent the diameters of the vertical and horizontal cylinders, respectively.

On one end of each cylinder draw a semicircle and divide it into any number of equal parts, numbering the division points 1, 2, 3, 4, etc. In the case of Fig. 2, the semicircles were divided into eight equal parts. Draw generators from these points along the length of the cylinders. The points of intersection between these generators and the unfolding lines of the development will locate the curve of intersection of the two cylinders.

Draw a stretchout line A'B' for the development of the vertical cylinder and on it lay off the divisions spaced on the semicircle of the cylinder. Lay off twice as many divisions as there are on the semicircle, in order to compensate for the whole cylinder. Draw vertical lines (ordinates) through these points. From the points where the generators of the vertical cylinder intersect the generators of the horizontal cylinder, project horizontal lines to intersect the appropriate ordinates in the development of the vertical cylinder. A smooth curve drawn through points obtained will complete the development of the vertical cylinder.

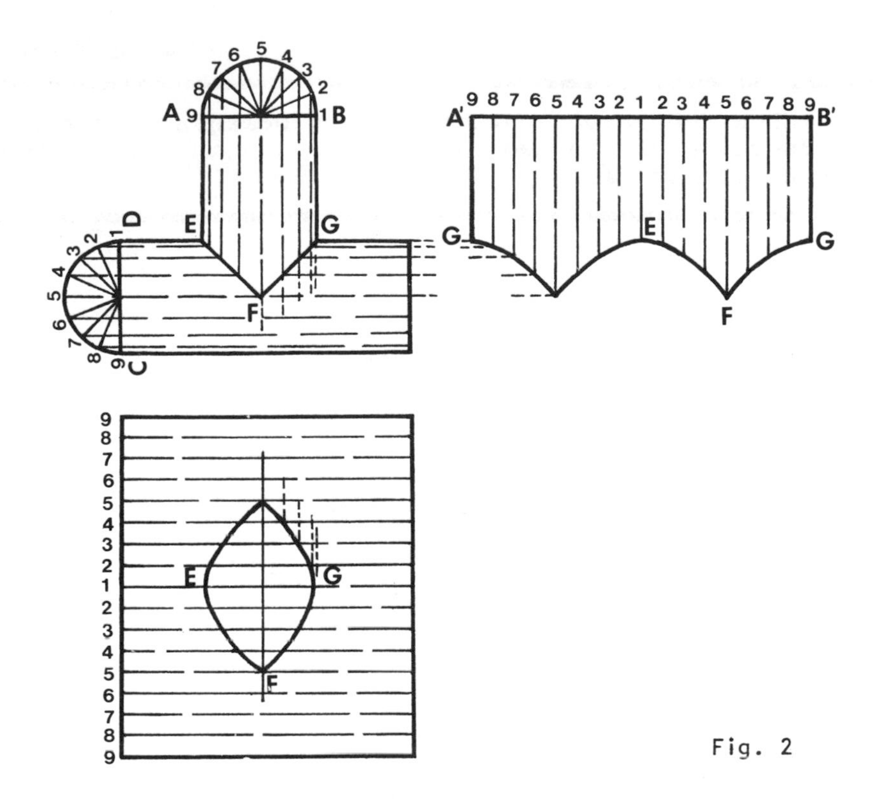

Fig. 2

In order to draw the development of the horizontal cylinder, project a stretchout from the diameter C-D. On this stretchout line mark off the divisions on the semicircle drawn on the horizontal cylinder. As before, make sure to double the amount of these divisions to compensate for the whole cylinder. On these points draw lines perpendicular to the stretchout line. From the points where the generators of the vertical and horizontal cylinders intersect, draw projection lines to intersect the appropriate element lines previously drawn. A smooth curve drawn through the points obtained will give the hole required in the developed sheet. To complete the development draw straight lines for the outside edges.

• PROBLEM 12-30

Draw the developments for the intersecting cylinders represented by the front and top views shown in Fig. 1.

Solution: Divide the circle in the given top view into a convenient number of divisions as shown in Fig. 2. In the front view draw a semicircle on the end edge of the sloping cylinder and divide it into a convenient number of divisions. Number all the division points constructed on the two given views. Notice that the side edges of the two cylinders appear true length in the front view.

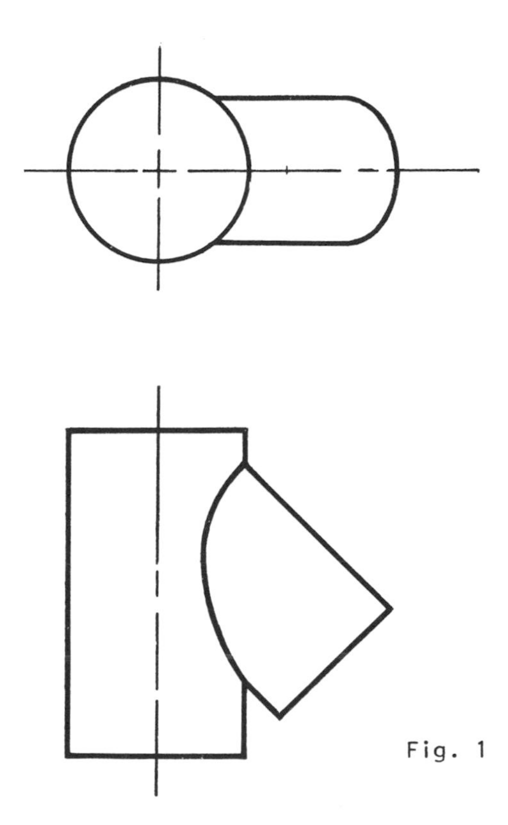

Fig. 1

In developing the vertical cylinder project an extended line from the bottom edge of the cylinder. On this line select a point and label it "1". Continue to locate numbers 1 through 12 by measuring their corresponding straight line distances in the top view and laying them out on the extended line. These distances must be measured consecutively, for example, from 1 to 2, from 2 to 3, etc. On the points located draw vertical lines and outline the outside edges of the development of the cylinder onto the development drawing. Project the division points from the top view vertically down to the front view to intersect the curve of intersection.

Project these points determined in the curve of intersection horizontally so as to cut the appropriate ordinate on the development drawing. A smooth curve drawn through the points obtained will give the hole required in the developed sheet. Notice that the projection lines retain the same numbers from where they were originally projected. For example, when points 2 and 12 in the top view are projected vertically down to the curve of intersection and this intersecting point is projected horizontally to the development drawing, the points which this projection line determines are still 2 and 12.

To draw a development of the sloping cylinder first produce an extended line which is perpendicular to the center line of the cylinder and is equal in length to the circumference of the cylinder. Divide this extended line into twice the number of equal parts as there are divisions in the semicircle constructed on the end edge of the cylinder. This is drawn twice in order to account for the other half of the cylinder. Label these new divisions and on each of them draw lines perpendicular to the extended line. Project every division point 1' through 7' parallel to the side edges of the cylinder so as to

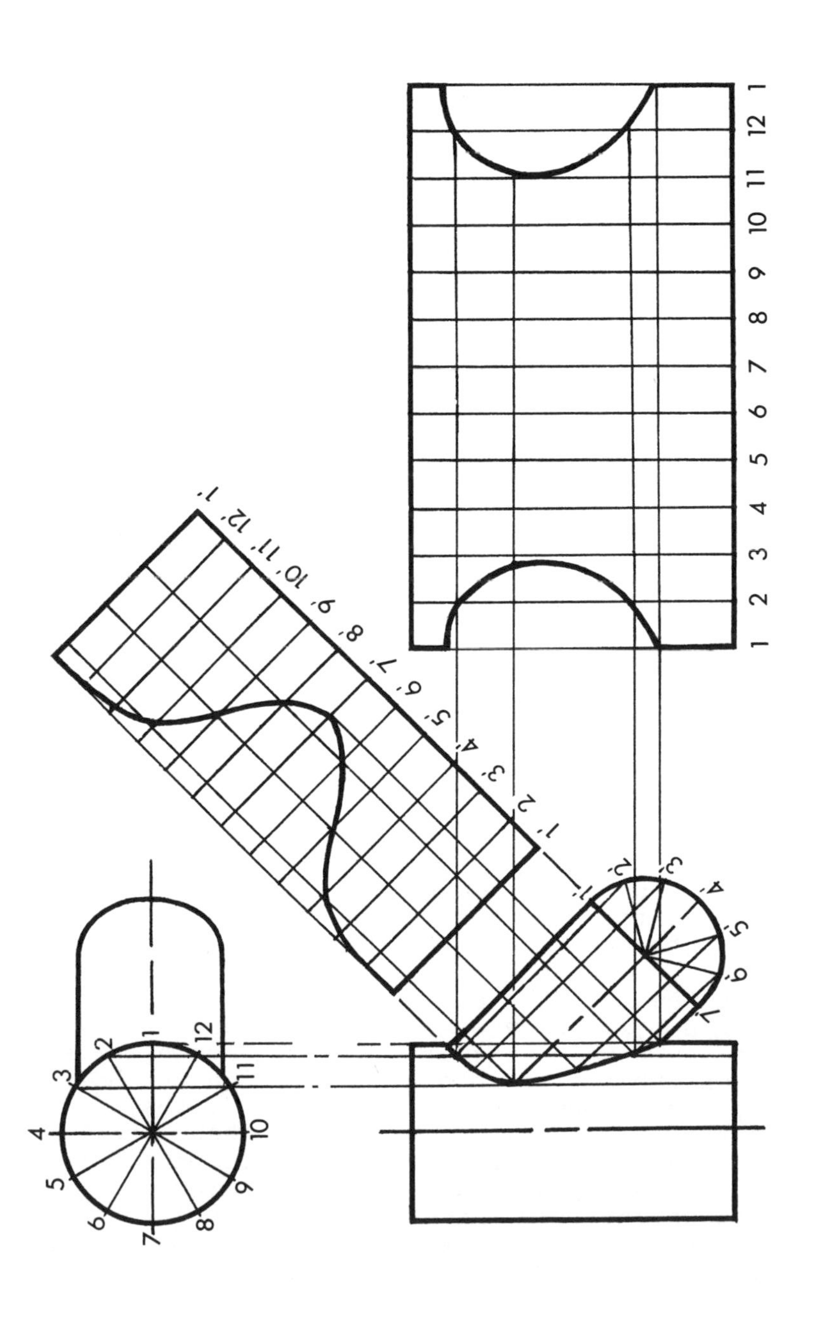

Fig. 2

intersect the intersection curve. Then project these new intersections parallel to the extended line to intersect the corresponding element line of the development. Connect the points so located with an irregular curve to complete the drawing. (See Fig. 2.)

• PROBLEM 12-31

Draw the layout patterns (developments) for the angle intersection of the two cylinders represented by the isometric drawing shown in Fig. 1.

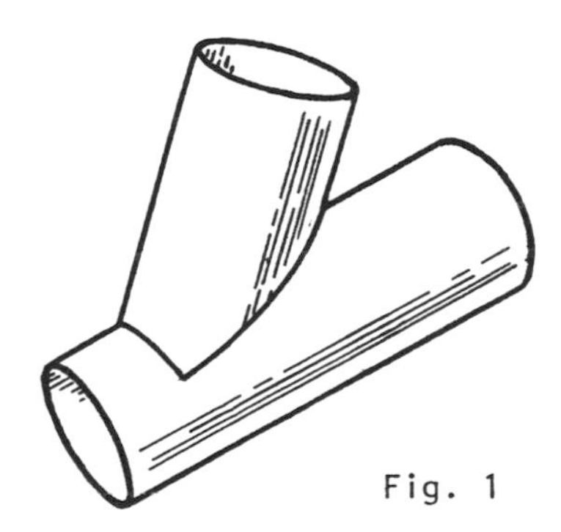

Fig. 1

Solution: Fig. 2a shows the front view of the two intersecting cylinders. On this front view draw a semicircle on each of the two cylinders and divide them into a convenient number of equal parts (Fig. 2a was divided into eight parts). Then draw generators at each division, parallel to the lengths of the cylinders.

To draw a development of cylinder A, as in Fig. 2b, first produce a stretchout line projected from diameter 1-8 and perpendicular to the center line of cylinder A. On this stretchout line mark off the divisions of the semicircle. Make sure that the number of divisions on the layout is twice as many as that in the semicircle, in order to compensate for the other half of the cylinder. At each of these points draw lines perpendicular to the stretchout line. Project the points lying in the curve of intersection of the two cylinders, parallel to the stretchout line until they intersect the corresponding element line on the development drawing. A smooth curve drawn through the points obtained will complete the development of cylinder A. (See Fig. 2b.)

To draw the development of cylinder B, as in Fig. 2c, first produce a stretchout line projected horizontally from the diameter 1-9 of cylinder B and use this line to mark off the divisions of the semicircle. Use the straight line distance between any two divisions as the approximate arc length. Make sure that the number of divisions on the stretchout line is twice that of the semicircle to account

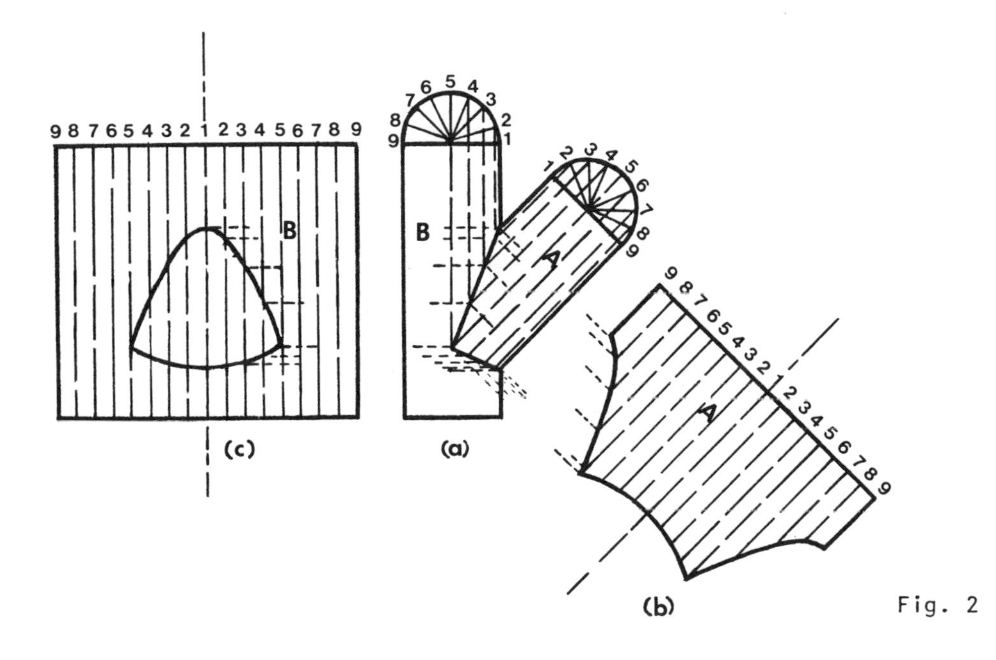

Fig. 2

for the rest of the cylinder. On each of these points draw lines perpendicular to the stretchout line. Project the points lying in the curve of intersection between the two cylinders, horizontally to the corresponding element line of the development drawing. A smooth curve drawn through the points obtained will give the hole required in the developed sheet. To complete the development draw the straight lines on the outside edges, which are projected directly from cylinder B.

• PROBLEM 12-32

Develop the three intersecting cylinders described by the top and front views shown in Fig. 1.

Solution: Each cylinder is developed separately, like any other truncated cylinder. For example, to develop the vertical cylinder in the top view first divide the circle that represents this cylinder in the front view into a convenient number of equal divisions. (Refer to Fig. 2 where 12 divisions have been made.) Number these divisions 1' through 12' and project them from the front view vertically up to the top view. Project an extended line horizontally to the right from the top edge of this cylinder in the top view. On this line select a point and label it "1'". Measure the straight line distance between any two divisions in the front view and lay out this distance from point "1'" to locate points 1' through 12'. On each of these points draw vertical lines below the extended line and project the points on the intersections of the cylinder horizontally onto the corresponding vertical line of the development. To complete the development of this cylinder connect these points with an irregular curve, for example, a French curve.

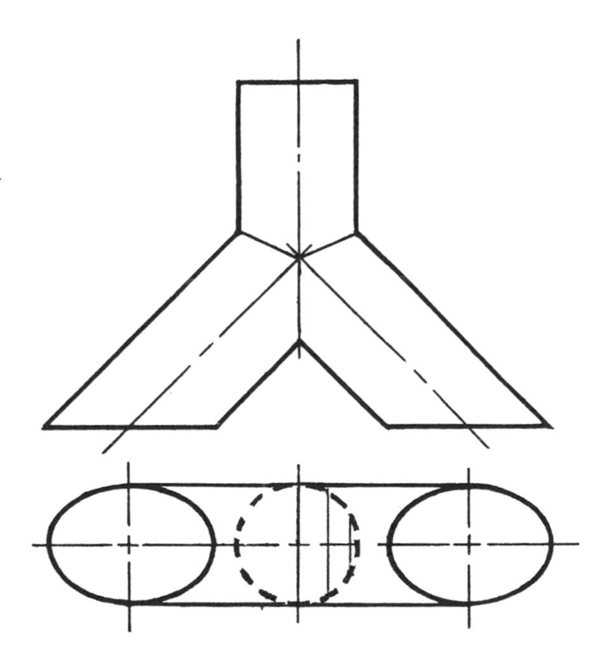

Fig. 1

In developing the other two cylinders notice that they are identical but opposite. Therefore only one of them needs to be developed. To develop the one on the right side, first draw a right view so the cylinder appears as a circle. Divide this circle into a convenient number of equal divisions. In Fig. 2, 12 divisions have been made. Number them 1 through 12 and project these divisions onto the front view, parallel to the side edges of this cylinder. Choose a cutting plane that intersects both side edges of the cylinder and is perpendicular to the center line. From this cutting plane project an extended line with projection lines parallel to the cutting plane. On this line pick a point and label it "1". Measure the straight line distance between any two divisions in the right view previously constructed and lay out this distance from point "1" on the extended line to locate points 1 through 12. At each of these points draw a line perpendicular to the extended line. Project all the points at the intersections of the cylinder and at the base, parallel to the extended line, onto the corresponding element line of the development. For example, when the point at the base of the cylinder determined by the projection line of points 9 and 5 is projected onto the development, it is projected to points 9 and 5 of the development. To complete the drawing of this development connect the points with an irregular curve.

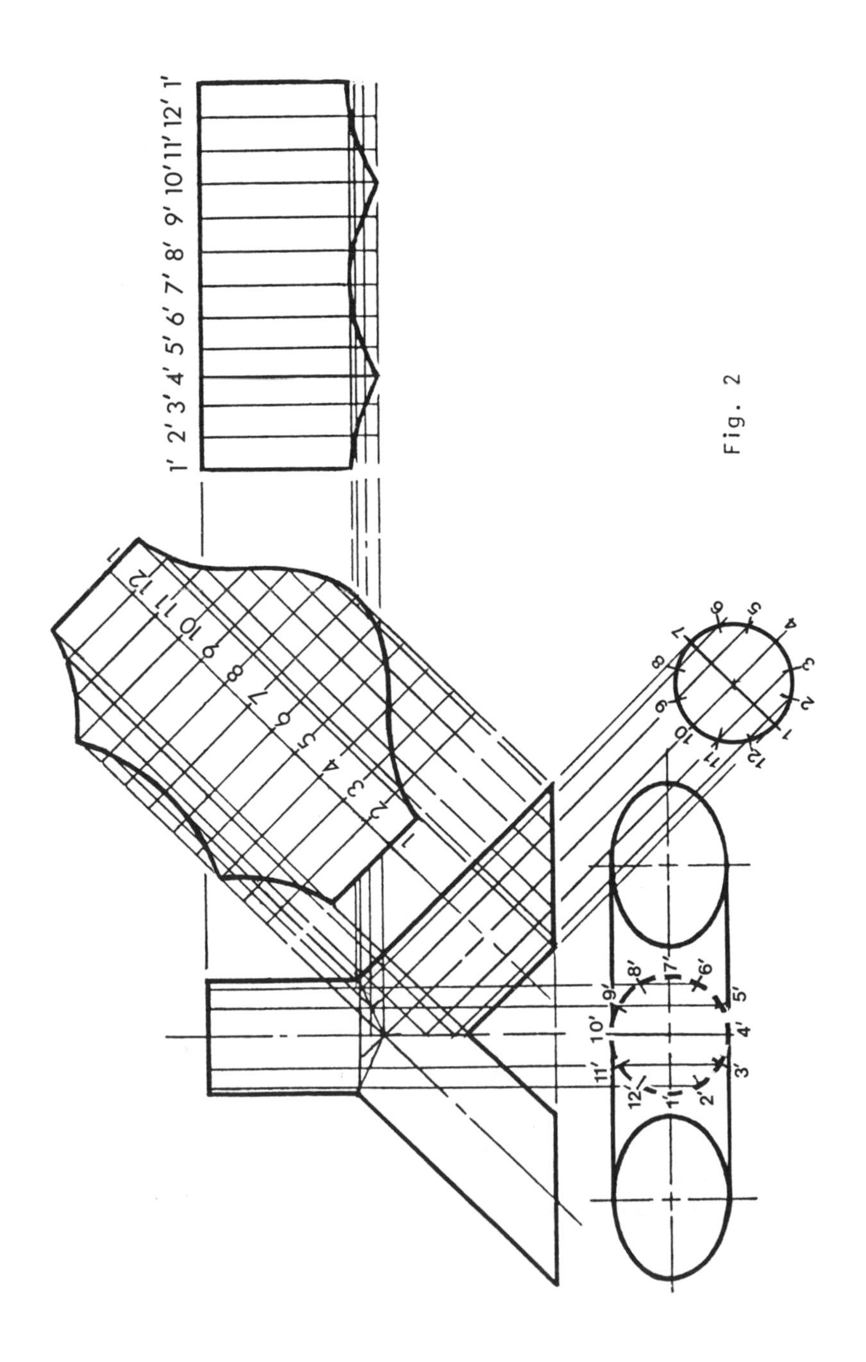

Fig. 2

THREE PIECE OFFSETS, TRANSITIONS, HOPPERS, ELBOWS, AND DOMES

• PROBLEM 12-33

Lay out the development of the two-piece angle, represented by the front view shown in Fig. 1, in the way that will save the greatest amount of material.

Solution: The problem calls for the most economical way of building the two-piece angle. To do this take advantage of the fact that the curve of intersection is the same for both pieces. This is true except when the one piece layout is cut off, then one of the pieces must be rotated 180 degrees in order to get the desired position of the given angle.

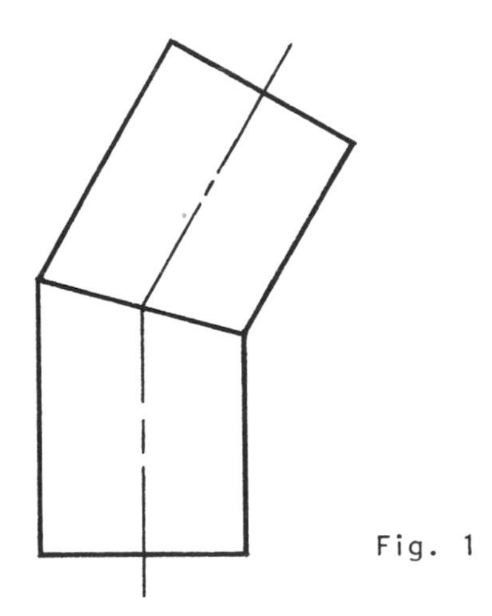

Fig. 1

In developing the given angle first construct a half view of the end of the pipe and divide it into a convenient number of equal parts as in the front view of Fig. 2. Number each division and project them to the front view, parallel to the side edges of the two pieces. Project an extended line horizontally from the bottom edge of the front view. On this extended line select a point at a convenient distance from the front view and label it "1". Measure the straight line distance between any two divisions on the semicircle constructed in the front view and lay out this distance on the extended line, one for every division on the pipe. Notice that the divisions 5 through 9 lie on the other half of the end view of the pipe. At each point located on the extended line draw a vertical line and then project every point at the intersection between the two pieces, horizontally onto the corresponding element line of the development. For example, when the intersection of the generator for divisions 12 and 8 and the line representing the intersection of the two cylinders is projected horizontally onto the development, the intersection of this projection line and the element lines for points 12 and 8 are marked to establish points. When points 1 through 12 have been established on the development connect them with an irregular curve, thus completing the development of the bottom piece.

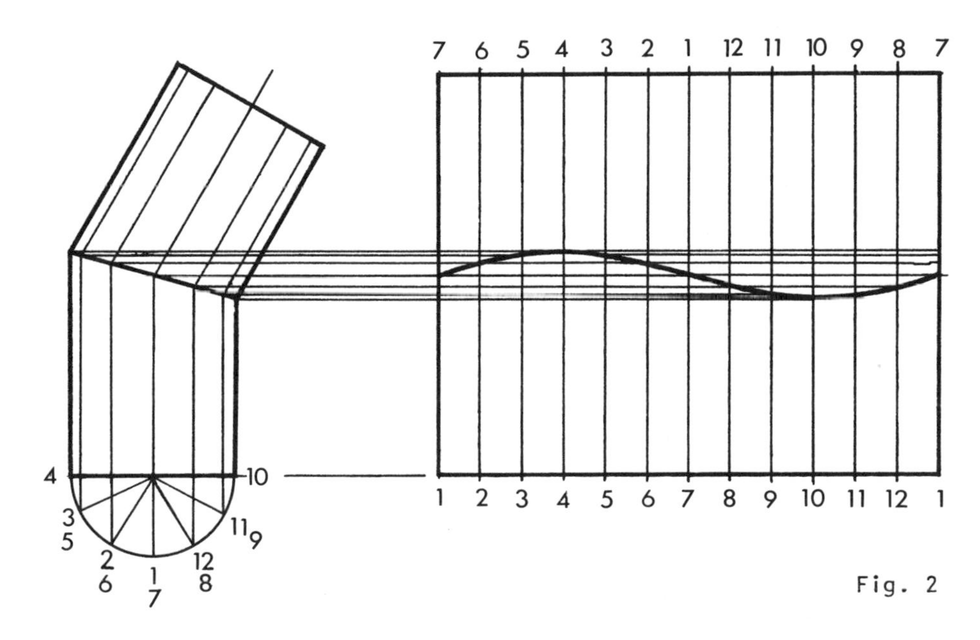

Fig. 2

To draw the development of the top piece, simply measure the length of one of its generators and lay out this distance vertically up from the proper point on the curve of the development of the bottom piece. On this point draw a horizontal line to complete the development of the top piece. To determine which is the proper point from where this selected distance is to be laid out, recall that one of the pieces is 180 degrees "out of phase." For example, point "11" is matched with the element line of point "9" because they are at 180 degrees from each other. (See the numbers on the top edge of the development.)

• PROBLEM 12-34

Draw a one-piece development of the three-piece offset represented by the front view shown in Fig. 1.

Solution: Construct a semicircle at the bottom edge of the

given front view and divide it into a convenient number of equal parts as shown in Fig. 2. Project the division point vertically up to the first piece, then parallel to the side of the second piece and then vertical to the last piece. Number each division including those corresponding to the other half of the circle.

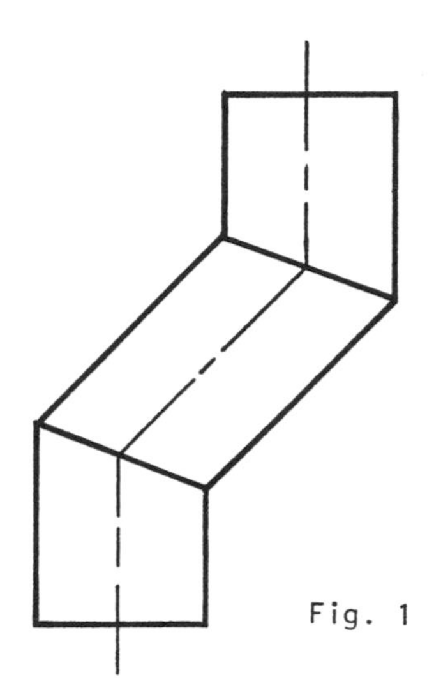

Fig. 1

In drawing the development first project an extended line from the bottom edge of the given front view perpendicular to the center lines of the top and bottom pieces. On this extended line select a point and label it "1".

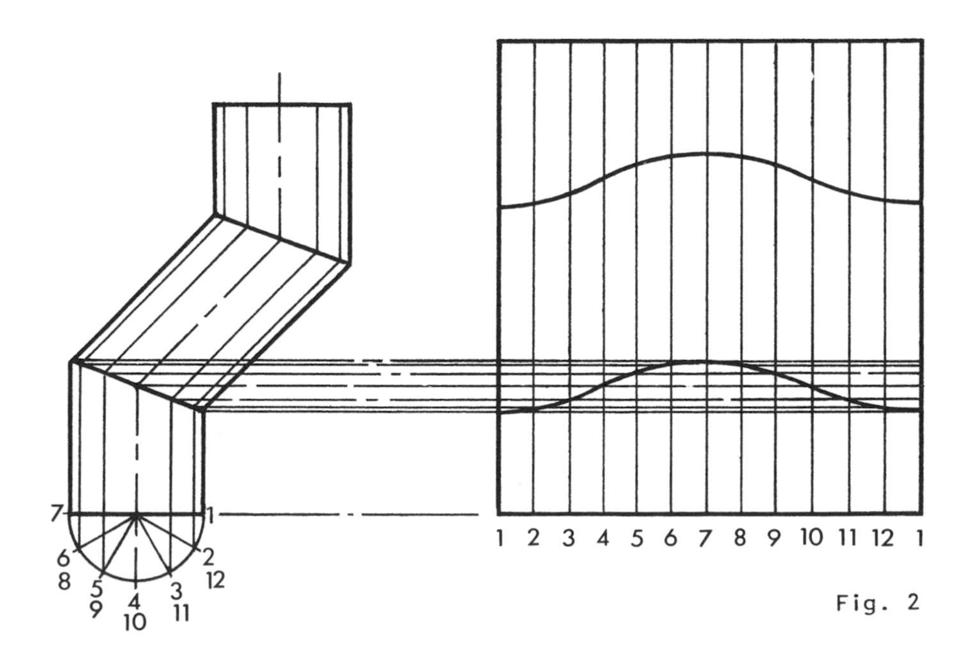

Fig. 2

Measure the straight line distance between any two divisions on the semicircle and lay these distances out on the extended line, in order to locate points 1 through 12. (See Fig. 2.) On these points draw vertical lines and project the points at the intersection between the bottom and middle pieces horizontally onto the corresponding element line of the development. Connect these points with an irregular curve to complete the development of the bottom piece.

To develop the middle piece notice that the length of all the generators is the same. Measure this true length as it appears in the front view and lay it out vertically from each point on the curve of the development of the bottom piece. Connect the points with an irregular curve to determine the development for the middle piece.

To draw the development of the top piece simply measure one of its heights and lay it out on the corresponding element line of the other two pieces. Through this point draw a horizontal line, thus completing the required one-piece development. Notice that when each piece is cut out from the development, the middle piece has to be rotated 180 degrees in order to get the correct offset. This means that point "1" of the top and bottom pieces will coincide with point "7" of the middle piece.

• PROBLEM 12-35

Develop the transition piece represented by the front and top views shown in Fig. 1.

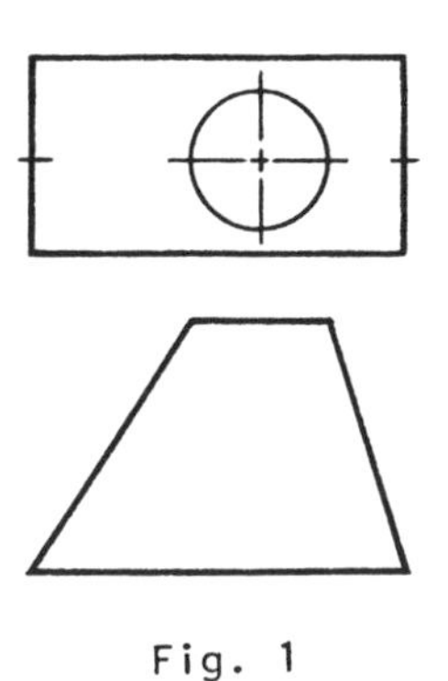

Fig. 1

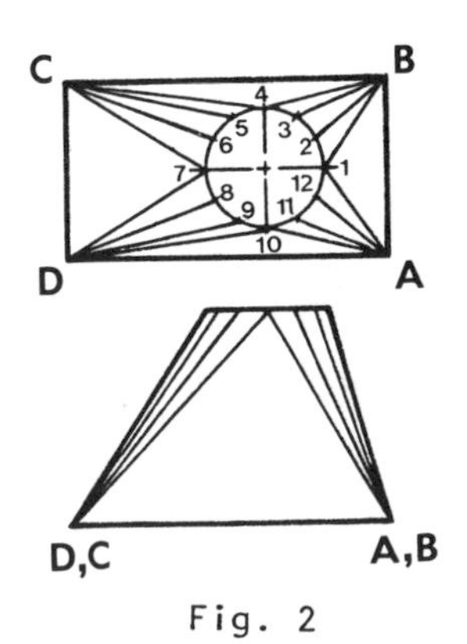

Fig. 2

Solution: The method used in developing this type of object is called triangulation. In this method the given views are broken up into triangles.

The circle in the given top view is divided into a convenient number of equal divisions as in Fig. 2. Number each division and letter the outside corners. Draw a line between each division and the closest outside corner so as to form a series of small triangles. Then find the true length of each triangle as follows: in the top view rotate each line drawn to the division points about the outside corner to which they are joined until the line is parallel to the cutting plane between the top and the front views. Project the position of the new endpoint vertically down until it intersects the horizontal line that goes through that point in the front view. For example, line A-1 in the top view Fig. 3 was rotated clockwise until the line became parallel to the cutting plane between the front and the top views. This cutting plane can be anywhere between the front and the top views as long as it is parallel to line A-D in the top view. When the line has been rotated, project point "1" down to the front view until it intersects the extension of the horizontal line going through the top edge of the front view (where point "1" is located). Join this new point "1" to point "A" of the front view. The new line A-1 in the front view is the true length of the original line A-1. In Fig. 3, points 7 through 10 were rotated about point "D" and points 10, 11, 12 and 1 were rotated about point "A". Lines D-7, D-8, D-9, D-10, A-10, A-11, A-12 and A-1 all appear true length in the front view after being rotated. The true lengths of the rest of the lines need not be determined because the top view is symmetrical about the horizontal center line. Therefore, the lines that are directly opposite have the same length. For example, line B-2 has the same length as A-12 and line C-5 has the same length as D-9.

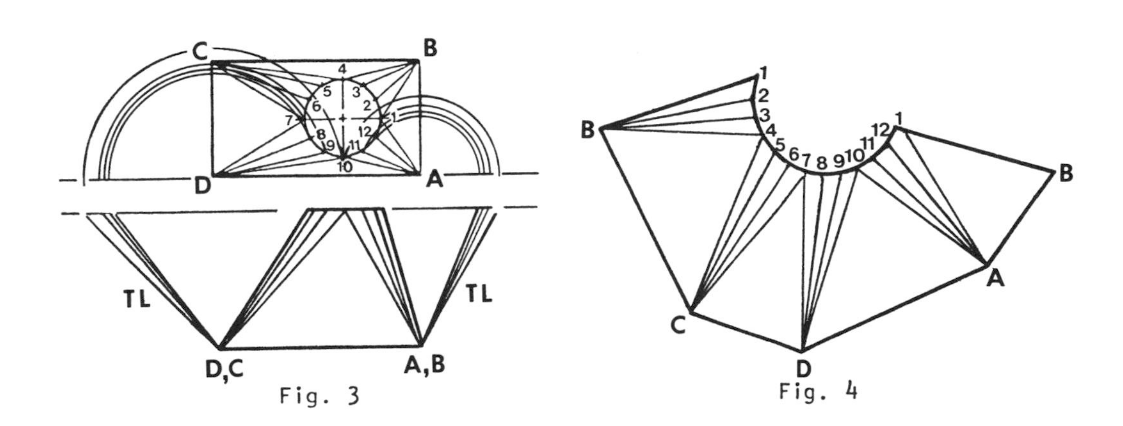

Fig. 3

Fig. 4

When all the true lengths of the lines of each triangle have been determined, draw the development as follows: first choose a starting line such as line B-1 in Fig. 4. With a compass, measure the straight line distance between the division points 1 and 2 and draw an arc, using this distance as the radius and point "1" as the center. From Fig. 3 take the true length of line B-2 and draw an arc using this distance as the radius and point "B" as the center. Draw this arc so that it intersects the arc previously drawn from point "1", thus determining point "2". From point "2" draw an arc of radius equal to the straight line distance between the division points 2 and 3. Take the true length of line B-3 from Fig. 3 and draw an arc from point "B" in order to determine point "3". Continue to construct the true shape of all the triangles in the top view in the same order they appear. (See Fig. 4.) Finally, connect the points 1 through 12 with an irregular curve, thus completing the required development.

• PROBLEM 12-36

Develop the hopper represented by the front and top views shown in Fig. 1.

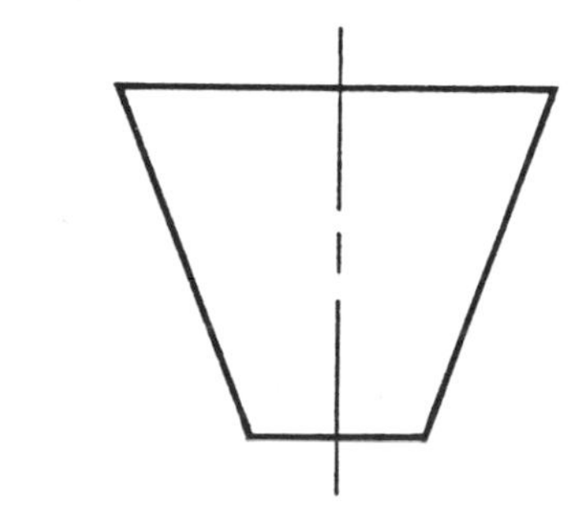

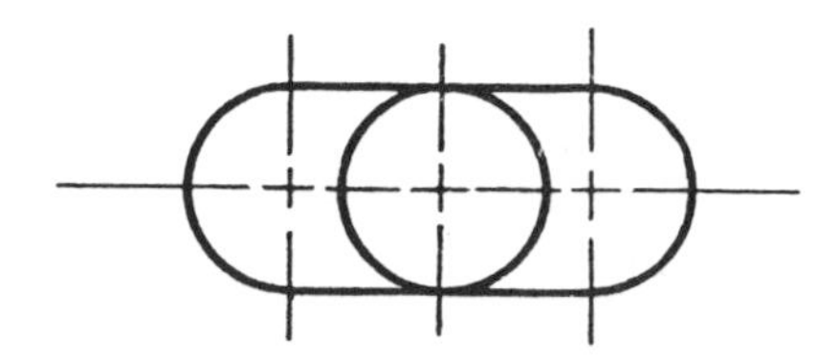

Fig. 1

Solution: Divide the circle and the semicircles in the front view into a convenient number of divisions as shown in Fig. 2. Label each division point and project these divisions vertically up to the top view. Connect the points located at the back edge of the hopper to the points in the front edge to form the set of parallel lines shown in the top view of Fig. 2. Notice that these parallel lines appear true length in the top view. This occurs because this connection of lines comes from lines which would appear parallel to the cutting plane between the front and top views if they were to be included in the front view. Also notice that the distance "W" marked in the front view of Fig. 2 is the same distance that all the other divisions are from one another.

To begin to draw the required development, first draw projection lines from all the points on the right side of the top view and those perpendicular to the parallel lines. Draw a line at a convenient distance from the top view and parallel to the parallel lines of the right side of the top view. Label this line 0_1-1 at the corresponding intersections. Measure the true length of line 0-1 from either the front or the top view. Use this distance as the radius to swing an arc from point "1" to intersect the projection line coming from point "0". This determines the true shape of triangle 0_1-0-1. Then use the distance "W" as a radius to strike arcs from one point to the other. For example, strike an arc from point "1" to the projection line coming from point "2". This determines point "2" in the development. From point "2" strike another arc that will intersect the projection line coming from point "3", thus determining point "3". Continue to locate points 4, 5, 6 and 7 in the same manner. At these points draw lines parallel to the previously drawn line 0_1-1 and end these lines at the corresponding projection lines taken from the front edge of the hopper.

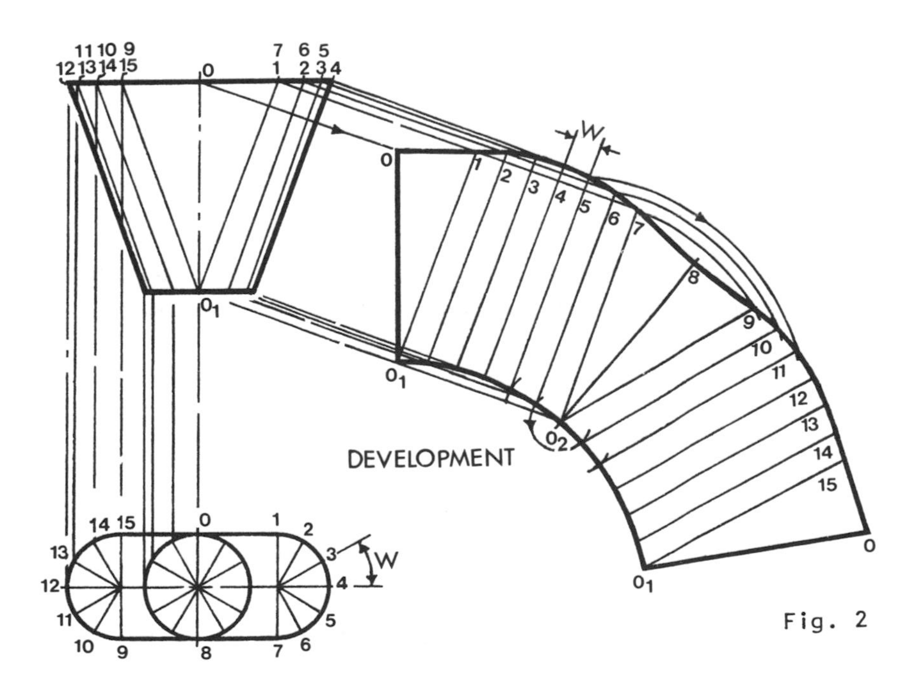

Fig. 2

To locate point "8", measure the true length of line 7-8 as it appears in the front view and use it as a radius to swing an arc from point "7" in the development. Then use the length of line 0_1-0 from the top view as a radius to swing an arc from point 0_2 in the development to intersect the arc previously drawn. This locates point "8" in the development. Notice that line 0_2-8 lies directly behind line 0_1-0 and therefore, the lengths of these two lines are equal.

To complete the development of the hopper, take advantage of the symmetrical shape of the hopper. The other half of the drawing can be done as the mirror image of the half already drawn. This is done by rotating points 0 through 7 about point 0_2, numbering the rotation of 7 as 9, the rotation of 6 as 10, etc., after determining each point. Take the true length of line 8-9 from the front view and use it as a radius to swing an arc from point 8, thus locating point 9. Then from point 9 draw an arc with radius "W" so as to intersect the rotation of point 6. This locates point 10. Follow the same procedure to locate points 11, 12, 13, 14 and 15. On these points draw lines parallel to line 0_2-9. To determine the length of these lines rotate all the points in the bottom part of the development about point 0_2 until they intersect the corresponding parallel line. Finally, locate point 0 by using the true lengths of lines 15-0 and 0_1-0 and form triangle 0_1-15-0. Connect points 1 through 7, 9 through 15 and the point at the bottom part of the development with an irregular curve. This completes the drawing.

• PROBLEM 12-37

Develop the four-piece round elbow shown in the figure. Pieces B and C are twice as long as pieces A and D (along the axis).

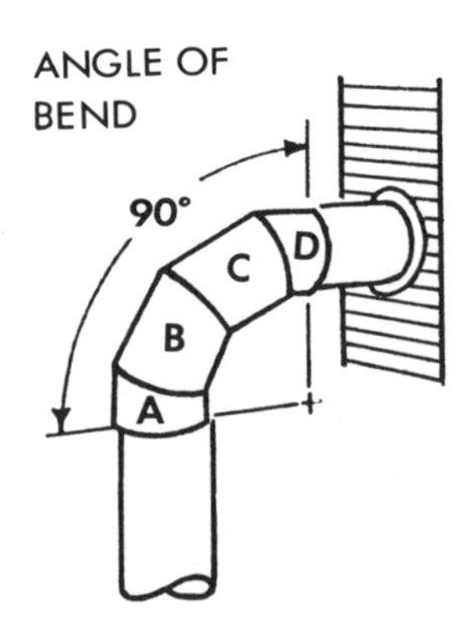

Solution: 1. Draw the heel and throat radii as in Fig. 1.

2. Divide the angle of bend into the proper number of pieces as in Fig. 2. The middle pieces of the elbow are twice as large as the end pieces. The angles are figured by using the following formula:

Number of spaces = (number of pieces x 2) - 2

= (4 x 2) - 2 = 6

The angle of bend, 90 degrees, is divided by 6. Each angle is 15 degrees.

3. Draw the 15 degree angles on the front view as in Fig. 2.

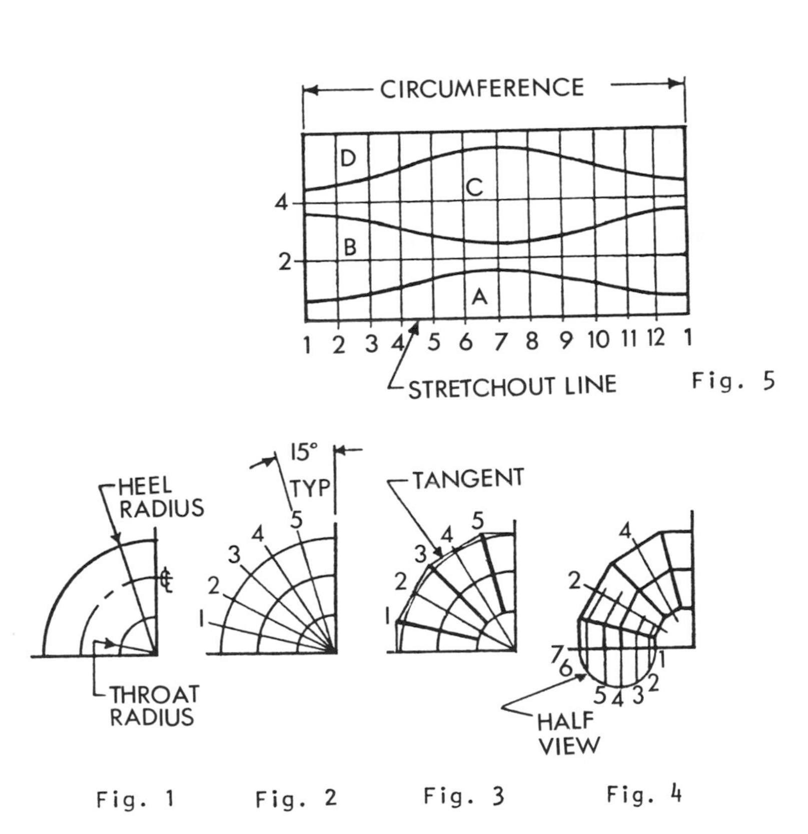

Fig. 5

Fig. 1 Fig. 2 Fig. 3 Fig. 4

4. Draw the miter lines. These are the places the pieces of pipe are to be joined. These are lines 1, 3 and 5, as shown in Fig. 3. Since the center pieces are twice as long as the ends, they are each 30 degrees.

5. Draw lines tangent to the arcs of each piece as in Fig. 3.

6. Draw a half view of the end of the pipe as in Fig. 4. Divide it into an equal number of parts.

7. Project these divisions to the front view. They are drawn parallel with the surface of each part of the elbow.

8. Lay out the stretchout line. The length is equal to the circumference of the pipe.

9. Lay out the distances between the points on the half circular view on the stretchout line. Finish the development using the same technique as explained for a truncated cylinder. Each piece of pipe is a truncated cylinder.

10. Lay out the pattern as shown in Fig. 5. When all four pieces are laid out together, each curve developed serves two pieces of the elbow. This also staggers the seams. The patterns A and C will have seams at the throat. Patterns D and B will have seams at the heel.

• PROBLEM 12-38

Draw the development of the octagonal dome described by the top and left side view given in Fig. 1.

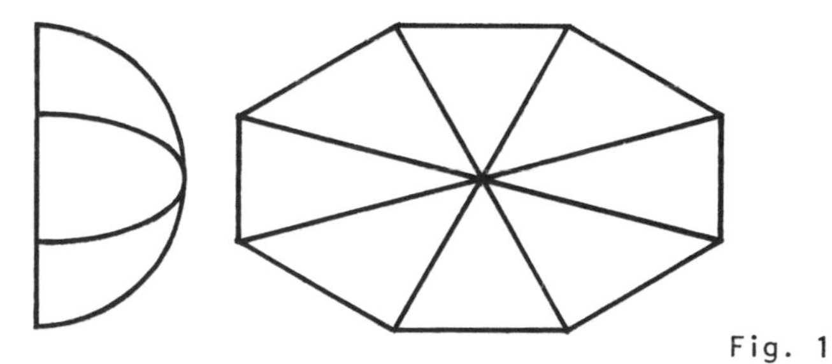

Fig. 1

Solution: It should be noticed that the given octagonal dome is completely symmetrical and therefore it need not be completely developed. The sides of the octagon which represent the outside edges of the top view all appear true length.

In preparing to develop this dome, first label the two given views as in Fig. 2. Then, for clarity, construct the front view of the dome as follows: first draw a number of vertical lines in the left side view in order to break up the given view. Project these lines vertically down to a 45 degree line constructed at a convenient distance below the left side view. Project the intersection of the 45 degree line and the vertical lines from the left view horizontally to the right. Then project every intersection in the left side view onto the corresponding line of the top view. Proceed to project the new intersections of the top view vertically down to meet the corresponding projection lines previously drawn. Draw the line for the base and use an irregular curve (such as a French curve) to connect the points that determine the rest of the curves in the front view. Label the front view by matching the projection lines.

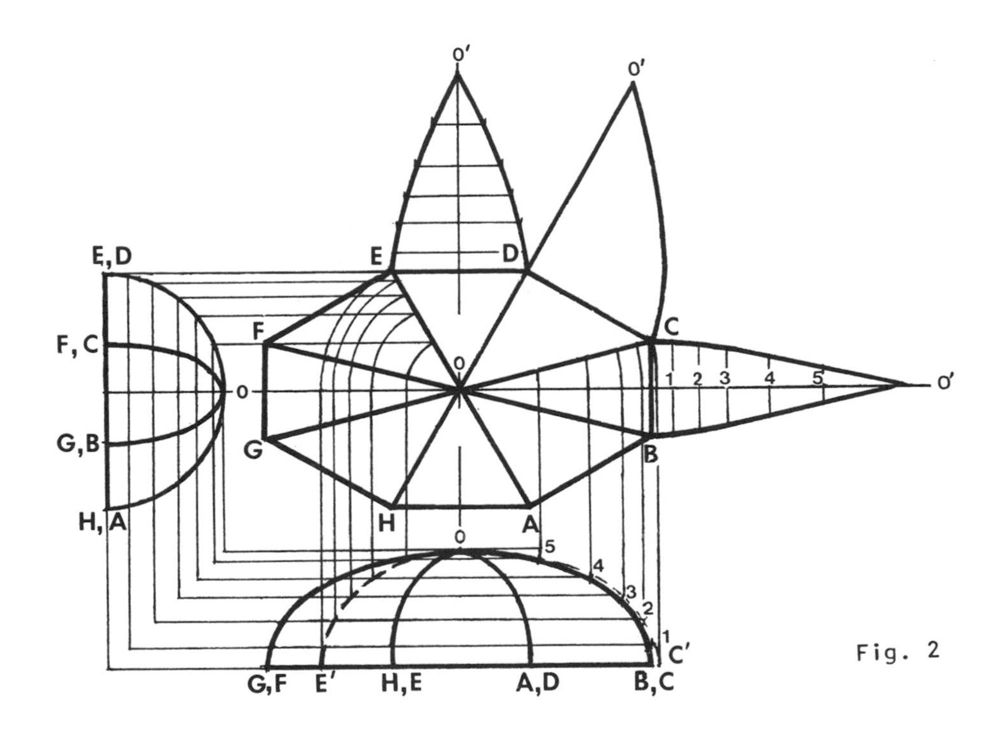

Fig. 2

In order to develop the section C-B-0, find the true length of either arc 0-B or 0-C. Notice that this section is symmetrical with respect to the horizontal center line as it appears in the top view. To find the true length of curve 0-C project all of the intersections of curve 0-C in the left side view with the vertical lines previously constructed horizontally onto line 0-C in the top view. Rotate these intersections to a position parallel to the frontal plane. Then project them down to the corresponding horizontal lines in the front view. Number these points 1 through 5 and connect the points with an irregular curve. In Fig. 2 curve 0-C' is the true length of curve 0-C. Since the given top view is the true shape, we can draw the rest of the development on it. Extend the horizontal center line of the top view to the right of side C-B. On the extension mark off the true distances C'-1, 1-2, 2-3, 3-4, 4-5 and 5-0 and draw vertical lines from each of these points. Next, project the intersections of the curves 0-C and 0-B with the vertical lines previously constructed in the left side view onto the corresponding vertical lines drawn on points 1 through 5. Connect these points with an irregular curve to complete the development of section C-B-0.

To develop section C-D-0, draw parallel lines perpendicular to side C-D going through each intersection of line 0-C and the projection lines coming from the side view. Next, along curve C-0' measure the distances C-1, 1-2, 2-3, 3-4, 4-5 and 5-0' and transfer them one by one from point C onto the next parallel line, and each following parallel line. Finally, connect these points to finish the development of this section. (See section C-D-0' in Fig. 2.)

To develop section D-E-0 first find the true length of the curve 0-E by rotating the intersection of 0-E and the horizontal projections from the left side view, to a position parallel to the frontal plane. Project them vertically down to the corresponding horizontal projection line and find the points of intersection. Connect the points in order to get the true length of curve 0-E' shown in Fig. 2. Extend the vertical center line in the top view above side E-D and lay out the true length of curve 0-E onto this extension by measuring the chordal distances between the points that were located. Draw horizontal lines at each of these points and proceed to project the intersection points of lines 0-E and 0-D, vertically up to the corresponding horizontal line. Connect these points with an irregular curve to complete the development of section D-E-0. Notice that the developments of sections E-F-0, G-H-0 and A-B-0 are the same as C-D-0; F-G-0 is the same as C-B-0, while A-H-0 is the same as E-D-0.

CHAPTER 13

FASTENERS AND SPRINGS

SCREW THREADS

• PROBLEM 13-1

Describe the common types of screw threads used in engineering design.

Solution: The drawings in Fig. 1 show the profiles of some of the more common types of threads, and their dimensions.

American Standard Thread (Fig. A). This thread is the standard thread in the United States. It was formerly known as the "United States Standard" (USS) of the "Sellers

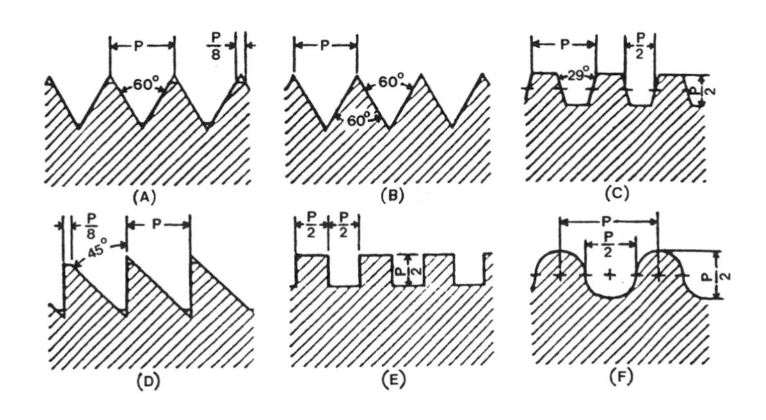

profile." The letter N, signifying "National," is used to designate this thread.

The American Standard includes five series of screw threads which differ from one another in pitch, or the number of threads per inch, for a given diameter. The series is comprised of (1) coarse thread (NC); (2) fine thread (NF); (3) 8-pitch thread; (4) 12-pitch thread; and (5) 16-pitch thread.

In addition to the five series of threads mentioned, there is another series of threads having the same profile as the American Standard. This thread is very fine and is called the Society of Automotive Engineers (SAE) thread, sometimes erroneously classified as the National Extra Fine (NEF) thread.

Sharp-V Thread (Fig. B). The sharp-V thread is similar to the American Standard thread except that the crests have not been cut off or the roots filled in. This thread is commonly used in boiler work or wherever it is necessary to have leak-proof joints.

Acme Thread (Fig. C). This thread is utilized for the transmission of power. Fig. C shows a profile of the thread and the necessary dimensions for laying it out. It is accepted practice to use a 30° angle in the drawing of the acme thread to represent the true thread angle of 29°. The point of failure in the square thread can be readily seen by comparing its profile with that of the acme thread. The comparison also shows how this defect has been eliminated in the acme thread by leaving more material at the critical point, thereby making the acme thread much stronger.

Buttress Thread (Fig. D). This is a special thread which permits the transmission of power in one direction only. It is used in the recoil mechanism of heavy guns and in the inertia starters of aircraft engines.

Square Thread (Fig. E). This thread is used for power transmission; it is found chiefly in jacks, vises, etc. It is an easy thread to cut in the lathe, and this accounts for its wide adoption. Its principal defect is that it shears at the root diameter. This shortcoming brought about a modification of the square thread which resulted in the development of the acme thread (Fig. C).

Knuckle Thread (Fig. F). This thread is the only one that can be both rolled and cast; though very crude, it has many uses. It is used, for example, on the bases of light bulbs and in the screw tops on glass jars.

Other Threads. There are many types of screw threads not illustrated in this chapter because they are not commonly used in the United States. One of these threads is the Whitworth or British Standard. This thread is similar to the American Standard thread, except that it has a thread angle of 55° and that the crests and roots are rounded off instead of being flat.

Another such thread is the International Standard thread, which has the same profile as the American Standard thread. The only difference is that it is measured in the metric system. It is standard throughout Europe, except in England, and is used in the United States on spark plugs.

• PROBLEM 13-2

Define all the terms used in discussing screw threads.

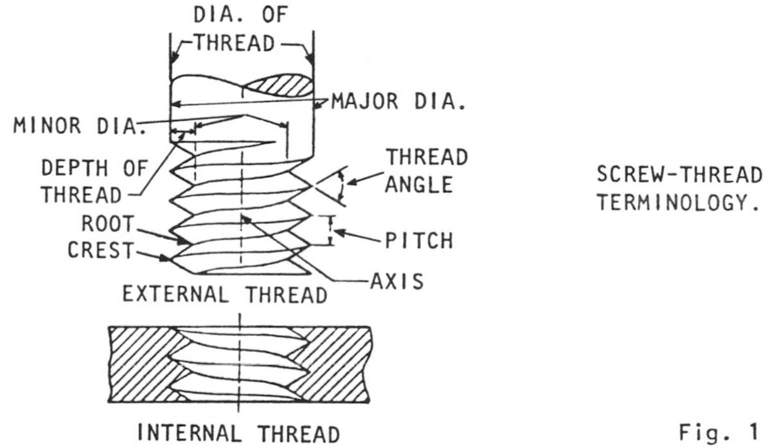

SCREW-THREAD TERMINOLOGY.

Fig. 1

Solution: Form. The profile shape (cross section) of the thread.

External thread. A thread on the outside of a member, Fig. 1.

Internal thread. A thread on the inside of a member, Fig. 1.

Axis. The longitudinal central line through the threaded part, Fig. 1.

Major diameter. The largest diameter of a screw thread, Fig. 1.

Minor diameter. The smallest diameter of a screw thread, Fig. 1.

Pitch diameter. The mean diameter between the major and minor diameters of the screw thread.

Pitch. The distance between corresponding points of consecutive threads measured parallel to the axis, Fig. 1.

Lead. The distance, parallel to the axis, that a screw advances in one complete revolution. See "multiple threads" and Fig. 2.

Crest. The top edge or surface joining the two sides of a thread, Fig. 1.

Root. The bottom edge or surface joining the sides of two adjacent threads, Fig. 1.

Depth of thread. The distance between crest and root measured normal to the axis, Fig. 1.

Thread angle. The dihedral angle between the sides of the thread, Fig. 1.

Right-hand thread. A thread that advances into engagement when turned clockwise. Threads are always right-hand unless otherwise specified.

Left-hand thread. A thread that advances into engagement when turned counterclockwise.

Single thread. A thread having any thread form cut on one helix of the cylinder, Fig. 2. Threads are always single unless otherwise specified. On a single thread the pitch and lead are equal.

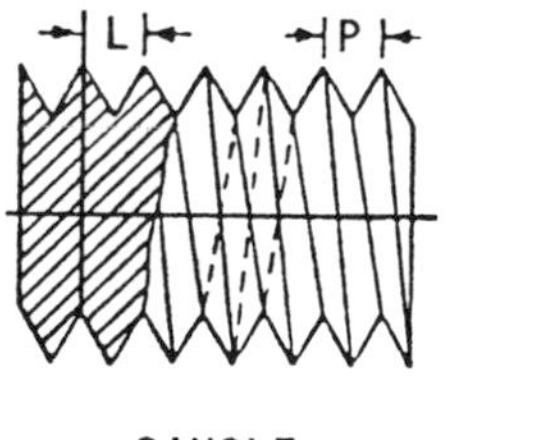

SINGLE

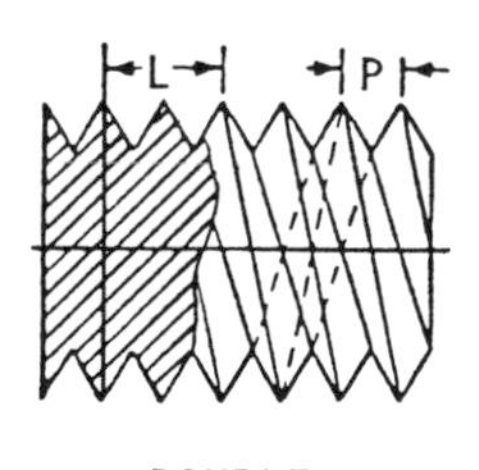

DOUBLE

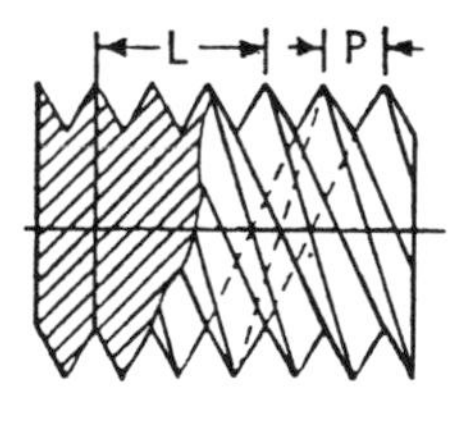

TRIPLE

SINGLE, DOUBLE, AND TRIPLE THREADS. Fig. 2

Multiple threads. A thread combination having the same form cut on two or more helices of the cylinder, Fig. 2. A more rapid advance is obtained without a coarser thread. On a double thread the lead is twice the pitch, and on a triple thread it is three times the pitch. Note that the helices of a double thread start 180° apart; those of a triple thread 120° apart. Fountain-pen caps often have quadruple threads, so that with a minimum of turning the cap is securely fastened to the barrel.

How are the screw-threads designated by notes? Discuss internal thread dimensioning with reference to a tap.

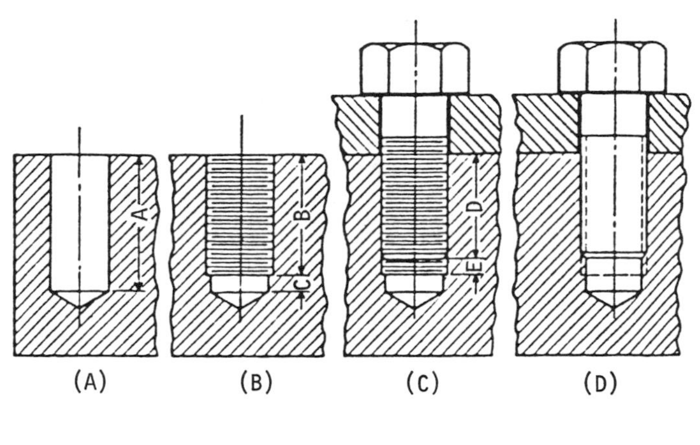

DIMENSIONING FOR INTERNAL THREADS.

Fig. 1

Solution: The type of thread is indicated by special notes.

External Threads. External screw threads are designated on a drawing, bill of material, etc. by listing the following items in order in a note.

1. Diameter of screw
2. Number of threads per inch
3. Type of thread (NC, NF, N, SAE, Square, etc.)
4. Class of fit
5. Right- or left-hand (RH, LH)
6. Multiple (single, double, triple, etc.)

A thread is assumed to be single and right-hand when it is not otherwise specified in a note.

Examples:

1/2-13 NC-2	1-10 N-2
3/4-16 NF-3	1¼-12N12-3
#10-24 NC-2	1½-6 NC-2, LH Double
1-5 Square	1-4 Acme, LH

Internal Threads. Internal threads, such as threaded holes, are designated by a note which lists the items in the following order.

1. Diameter of tap drill
2. Depth of tap drill
3. Diameter of thread
4. Thread specification
5. Class of fit
6. Depth of thread

Example: 7/8 Drill-2 Deep, 1-8 NC-2 x 1-3/4 Deep.

Fig. 1 shows the depth dimensions that are used when the threaded hole does not pass completely through a piece, or when the depth dimensions are not given in a note.

Fig. 1A shows the way in which the hole formed by the tap drill appears. The dimension A is usually taken as 1½D, where D is the major diameter of the thread. For soft metals, such as brass, aluminum, etc., this distance is increased to 2D; and for steel the distance is reduced to 1D.

Fig. 1B shows the same hole after the tap has been screwed into the hole made by the tap drill. Note that the thread does not extend to the bottom of the drilled hole. The reason for this is that the chips formed by

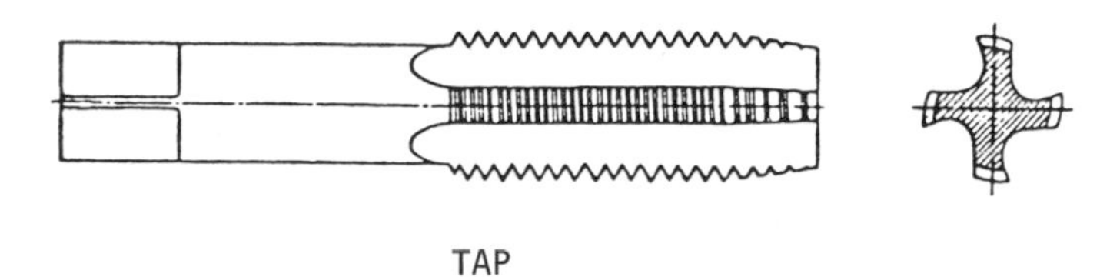

TAP

Fig. 2

the tap drill fall to the bottom of the hole and cause the tap drill to jam. Leaving clearance at the bottom eliminates this condition. The dimensions shown in Fig. 1B are taken so that A = B + C, and C is taken as ¼D.

Fig. 1C shows the tapped hole with a fastener in place. Note that the end of the fastener does not extend to the bottom of the threaded hole. The reason for this can be seen by inspecting the tap in Fig. 2, and observing how the first few threads cut by the tap are imperfect owing to the tapering of the tap at the bottom. The dimensions shown in Fig. 1C are taken so that B = D + E, and E is taken as ¼D. The American Standard regular thread symbol is used in Fig. 1, whereas Fig. 1D is the same type of fastener except that the American Standard simplified symbol is used instead.

Classify different types of screw-thread fits.

Solution: Loose Fit (Class 1) - Large Allowance. This fit provides for considerable freedom and embraces certain fits where accuracy is not essential. Examples: machined fits of agricultural and mining machinery; controlling apparatus for marine work; textile, rubber, candy, and bread machinery; general machinery of a similar grade; some ordnance material.

Free Fit (Class 2) - Liberal Allowance. For running fits with speeds of 600 r.p.m. or over, and journal pressures of 600 lbs. per sq. in. or over. Examples: dynamos, engines, many machine-tool parts, and some automotive parts.

Medium Fit (Class 3) - Medium Allowance. For running fits under 600 r.p.m. and with journal pressures less than 600 lbs. per sq. in.; also for sliding fits; and for the more accurate machine-tool and automotive parts.

Snug Fit (Class 4) - Zero Allowance. This is the closest fit which can be assembled by hand and necessitates work of considerable precision. It should be used where no perceptible shake is permissible and where moving parts are not intended to move freely under a load.

Wringing Fit (Class 5) - Zero to Negative Allowance. This is also known as a "tunking fit" and is practically metal-to-metal. Assembly is usually selective and not interchangeable.

Tight Fit (Class 6) - Slight Negative Allowance. Light pressure is required to assemble these fits and the parts are more or less permanently assembled, such as the fixed ends of studs for gears, pulleys, rocker arms, etc. These fits are used for drive fits in thin sections or extremely long fits in other sections and also for shrink fits on very light sections. Used in automotive, ordnance, and general machine manufacturing.

Medium Force Fit (Class 7) - Negative Allowance. Considerable pressure is required to assemble these fits and the parts are considered permanently assembled. These fits are used in fastening locomotive wheels, car wheels, armatures of dynamos and motors, and crank disks to their axles or shafts. They are also used for shrink fits on medium sections or long fits. These fits are the tightest which are recommended for cast-iron holes or external members as they stress cast iron to its elastic limit.

Heavy Force and Shrink Fit (Class 8) - Considerable Negative Allowance. These fits are used for steel holes where the metal can be highly stressed without exceeding its elastic limit. These fits cause excessive stress for cast-iron holes. Shrink fits are used where heavy force fits are impractical, as on locomotive wheel tires, heavy crank disks of large engines, etc.

• PROBLEM 13-5

What is the conventional way of representing screw threads?

Describe the procedure for drawing the following types of threads:

a) American Standard or Sharp-V Thread

b) Square Thread

c) Acme Thread

Solution: Before a screw thread can be drawn, certain information must be known, such as form of thread, major diameter, number of threads per inch, whether there is to be a single or multiple thread, and whether the thread is to be right- or left-handed.

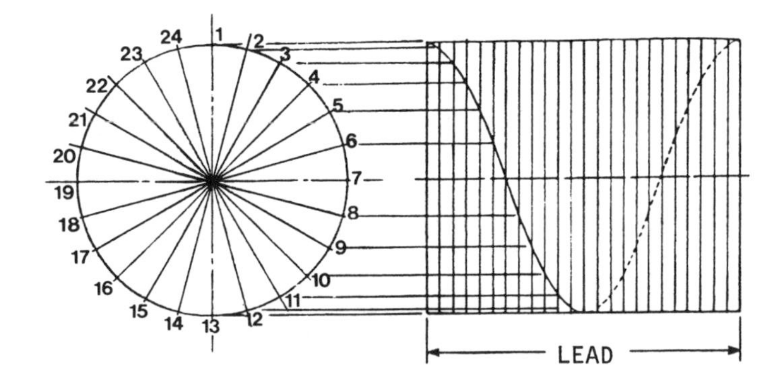

CONSTRUCTION OF HELIX.

Fig. 1

One method of making a screw thread is to move a cutting tool longitudinally along a shaft while the shaft is rotated. The thread thus cut takes the form of a helix, whose construction is shown in Fig. 1. It can be seen from Fig. 2 that one helix is formed for the major diameter of the thread, and another helix of different size for the minor diameter. It is obvious from this diagram that it would be very laborious to construct drawings of threads in this manner. Consequently, this type of representation is rarely used.

It has already been shown how laborious it would be to construct the helices for a true projection of a

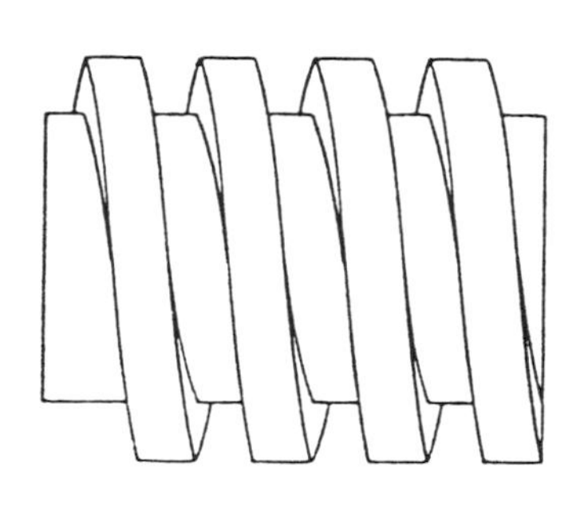
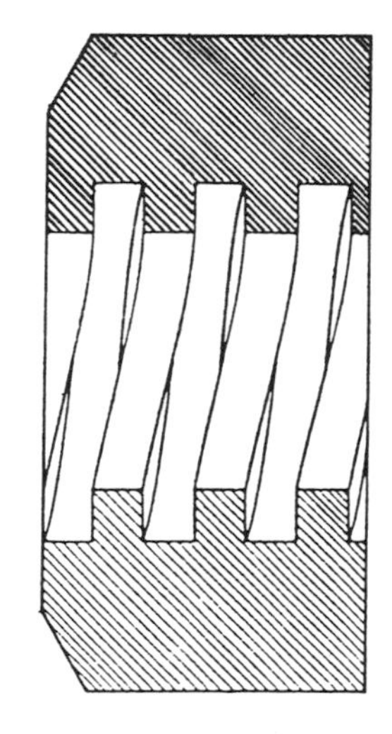

TRUE PROJECTION OF THREAD.

Fig. 2

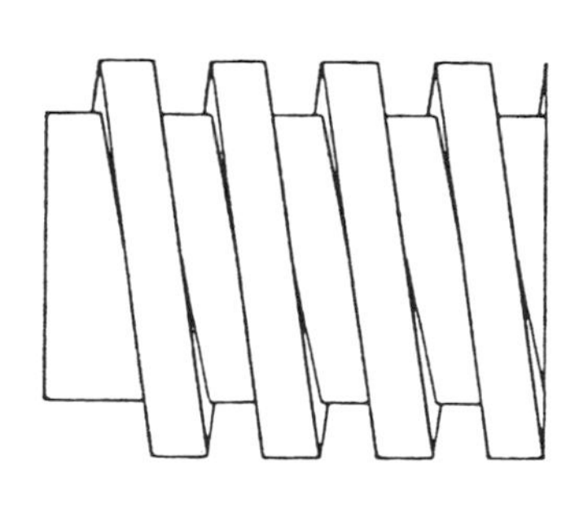
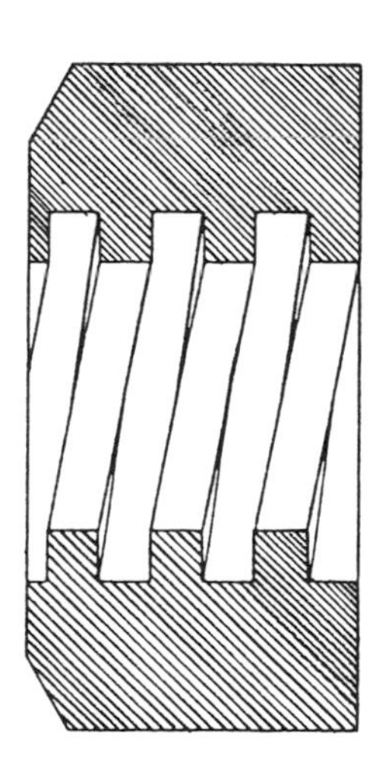

CONVENTIONAL REPRESENTATION OF THREAD.

Fig. 3

thread. In the conventional method of representing a thread, the helices are drawn as straight lines. Fig. 2 shows a screw thread as it actually appears, and Fig. 3 shows the same thread with the helices drawn as straight lines. By comparing these two diagrams, it can be seen that the difference between the two methods is so slight that construction of helices is not worth a great expenditure of time and energy.

a) American Standard or Sharp-V Thread. The American Standard thread and the sharp-V thread are both drawn in the same way. This is made possible by the elimination of the flat portions on the crests and roots of American Standard threads. These flat portions are really so small that they can be omitted in a drawing without being noticed.

Let it be required to draw an American Standard or sharp-V thread. The pitch distance is first laid out along the entire length of the thread, and lines are drawn through these points at angles of 60° as shown in Fig. 4(A). Since this is to be a single thread, a crest must appear directly opposite a root as indicated in Fig. 4(B). The next step is to connect the crests as illustrated in Fig. 4(C). The final step is to draw lines connecting the roots as shown in Fig. 4(D). If the drawing is to be inked, the lines joining the roots should be heavier than those connecting the crests.

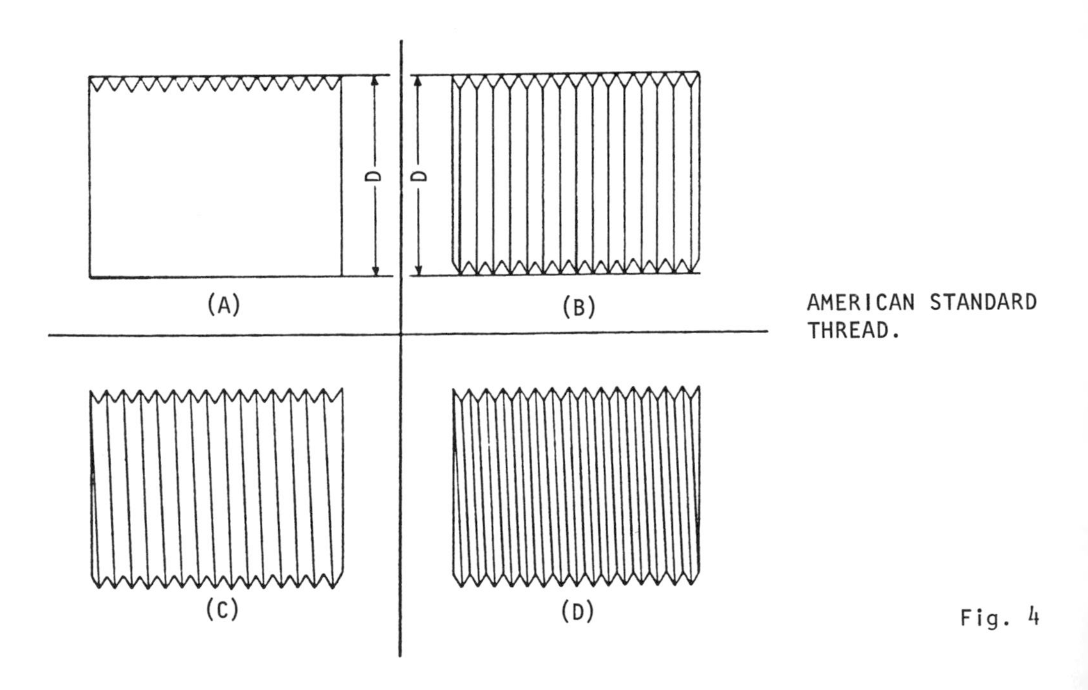

Fig. 4

b) Square Thread. Let it be required to draw a double, right-hand square thread. Draw lines a distance equal to P/2 in from the crest lines of the thread. Starting at any convenient point, lay off the distance P/2 along the length of the thread as shown in Fig. 5A. Since this is to be a double thread, a crest is to be shown opposite a crest. Fig. 5B shows how this profile appears with the construction lines removed. Fig. 5C shows the lines connecting the crests. The invisible lines were drawn in principally to show how certain visible lines were obtained. Fig. 5D shows the completed thread.

c) Acme Thread. Let it be required to draw a single, right-hand acme thread. Draw lines a distance of P/2

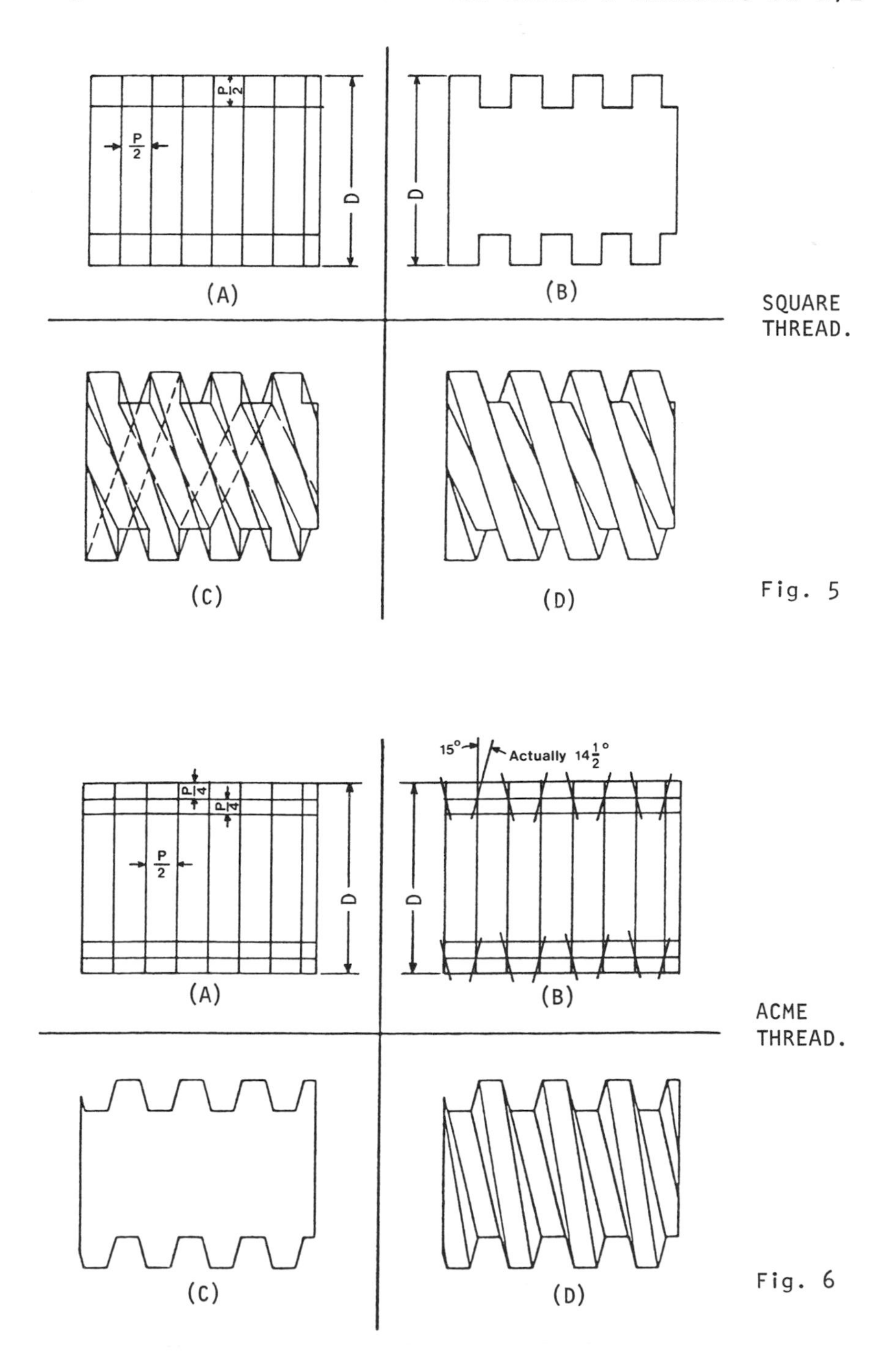

SQUARE THREAD.

Fig. 5

ACME THREAD.

Fig. 6

in from the crest lines. Also draw lines a distance of P/4 in from the crest lines. The length of the threaded portion is then divided by lines set a distance of P/2 apart, as shown in Fig. 6A. The next step is to draw lines, making angles of 75° with the horizontal, through the intersection of the lines that are a distance of P/4 in from the crest and the lines perpendicular to the axis of the screw, as shown in Fig. 6B. It should be noted that the 75° lines slope in the same direction at both ends of the vertical lines for a single thread. In the case of a double thread, the lines slope in opposite directions. The actual 14½° thread angle, shown in Fig. 6B, represents a single face of an acme thread. The 14½° angle adjacent to it makes a complete angle of 29° between the two thread faces. Fig. 6C shows the profile of the threads with construction lines removed. Fig. 6D shows the completed thread in which the lines joining the crests and roots have been added.

• PROBLEM 13-6

Give the graphic representation of thread symbols for

a) External Threads

b) Internal Threads

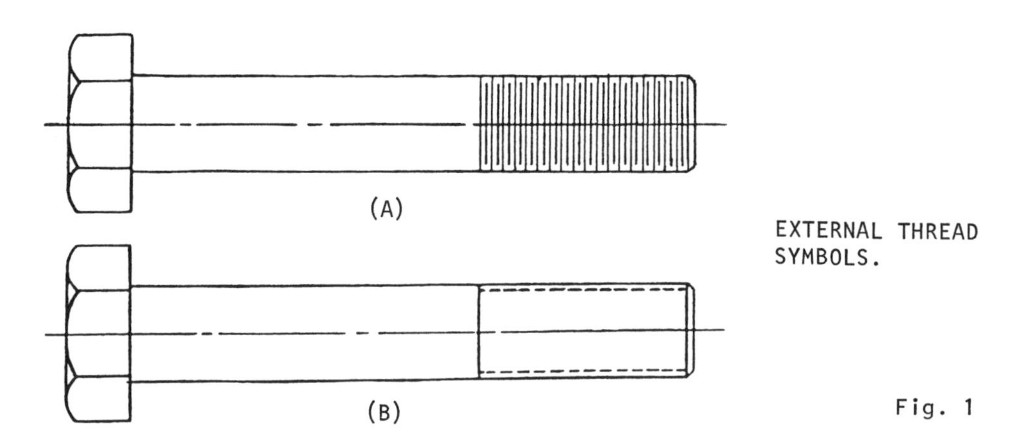

Fig. 1

<u>Solution</u>: When threads to be represented are shown less than 1" in diameter, it is recommended that the symbols

shown in Figs. 1A and 1B be used.

a) External Thread Symbols. Fig. 1A is known as the American Standards Association regular thread symbol, and it is recommended that this symbol be used on assembly drawings. The light lines representing the crests of the thread are not spaced according to scale. It is recom-

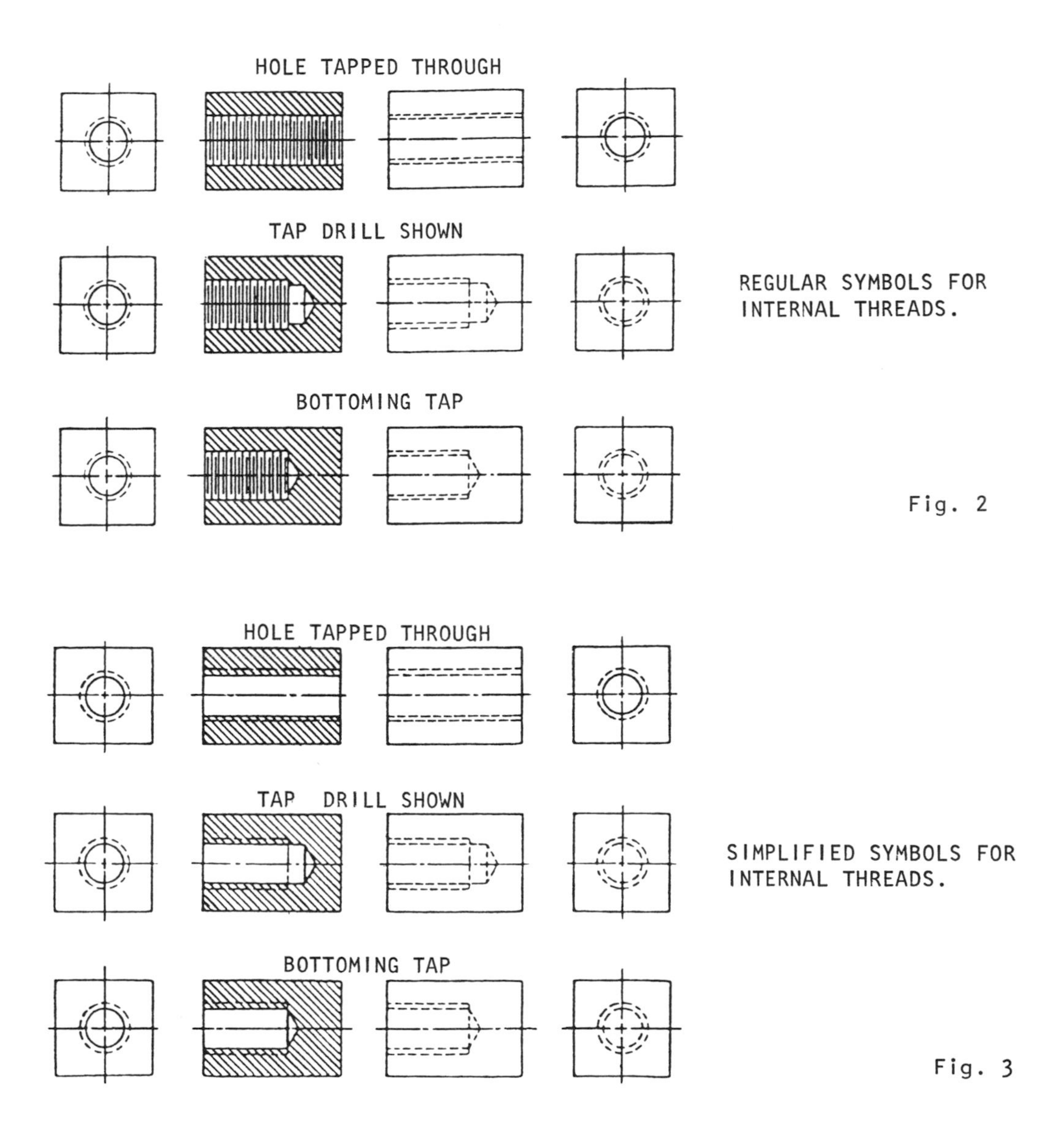

Fig. 2

Fig. 3

mended that these lines be spaced not closer than 3/32". The lines representing the roots of the thread are drawn heavier. These lines are spaced midway between the crest lines, and should terminate a minimum of 1/16" from the lines representing the major diameter of the thread.

Fig. 1B is the American Standards Association simplified thread symbol, and its use is recommended for detail drawings. The reason the use of these symbols is specified for a particular type of drawing is that anyone reading a detail drawing would be familiar with the various symbols used in that kind of drawing. This is not true of assembly drawings; therefore, the regular thread symbol (Fig. 1A) is used, since it more closely resembles a true thread than does the simplified thread symbol.

b) Internal Thread Symbols. Fig. 2 shows the American Standard regular thread symbols for use on internal threads. Fig. 3 shows the American Standard simplified thread symbols for use on internal threads.

BOLTS AND NUTS

• PROBLEM 13-7

Consider a 1-inch regular bolt, unfinished, with hexagonal head and hexagonal nut.

Draw the bolt and nut in terms of diameter, D.

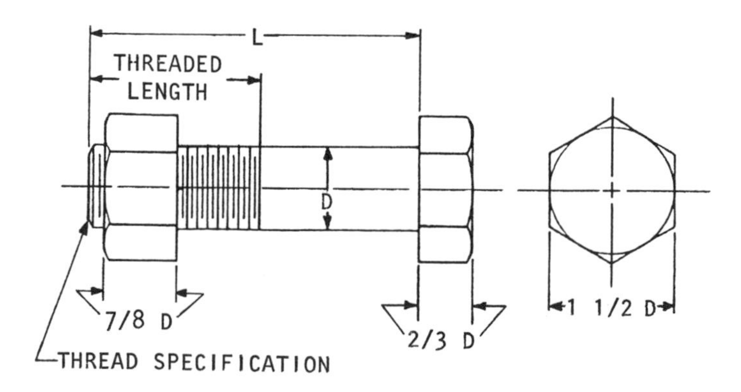

REGULAR BOLT, UNFINISHED, HEXAGONAL HEAD, AND HEXAGONAL NUT.

Solution: The following dimensions, taken from Standard tables, can be used for the head and the nut:

For the head:

a) width across flats = $1\frac{1}{2}$ inches (maximum)

b) nominal height = 43/64 inch

For the nut:

a) width across flats = 1½ inches

b) thickness = 7/8 inch (nominal size)

A very close approximation to the tabular dimensions (nominal) can be obtained from the following formulas, which are based on the diameter (D) of the bolt:

W, width across flats = 1½D.

H, height of head = 2/3D.

T, thickness of nut = 7/8D.

Knowledge of these relations will expedite the preparation of the drawing. The following figure shows the drawing of the bolt and nut in terms of D. It should be noted that bolt heads and nuts are drawn across corners in all views. This is in accordance with recognized practice in order to avoid confusion in distinguishing between hexagonal and square shapes.

• **PROBLEM 13-8**

Define the following standard nomenclature for the bolt heads or nuts and then show the detailed drawing which makes up the modern bolt.

(a) Unfinished bolt heads or nuts

(b) Semifinished bolt heads or nuts

(c) Finished bolt heads and nuts

(d) The washer face

(e) The height of head

(f) The thickness of nut

Solution: (a) Unfinished bolt heads or nuts are not machined or treated on any surface except for the threads.

(b) Semifinished bolt heads or nuts are machined or otherwise formed or treated on the bearing surface to provide: (1) a washer face [see (d) below] for bolt heads, and (2) either a washer face or a circular bearing surface, formed by chamfering the edges for the nut.

(c) Finished bolt heads and nuts are the same as the semifinished except for the nonbearing surface which is so treated as to provide a special appearance. The desired finish should be specified by the purchaser.

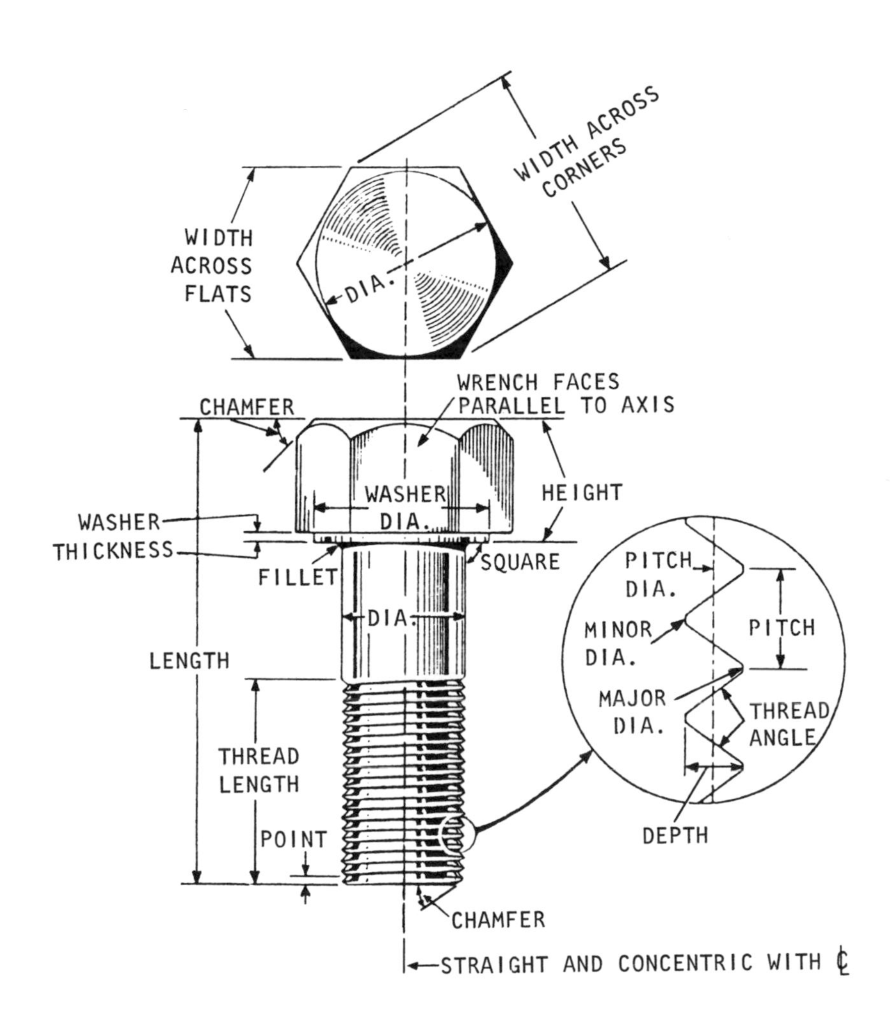

(d) The washer face is a circular boss, turned or otherwise produced on the bearing surface of a bolt head or nut to relieve the corners. Chamfering the corners of the nut will produce a circular bearing surface.

(e) The height of head is the overall distance from the top to the bearing surface and includes the thickness of the washer face where provided.

(f) The thickness of nut is the overall distance from the top to the bearing surface and also includes the thickness of the washer face where provided.

The detailed drawing which makes up the modern bolt is shown in the figure.

• PROBLEM 13-9

Consider a 3/4-inch heavy bolt, semifinished, hexagonal head and hexagonal nut. Use the appropriate formula, based on the diameter (D) of the bolt, to approximate the tabular dimensions (nominal) and then draw the bolt and nut in terms of (D).

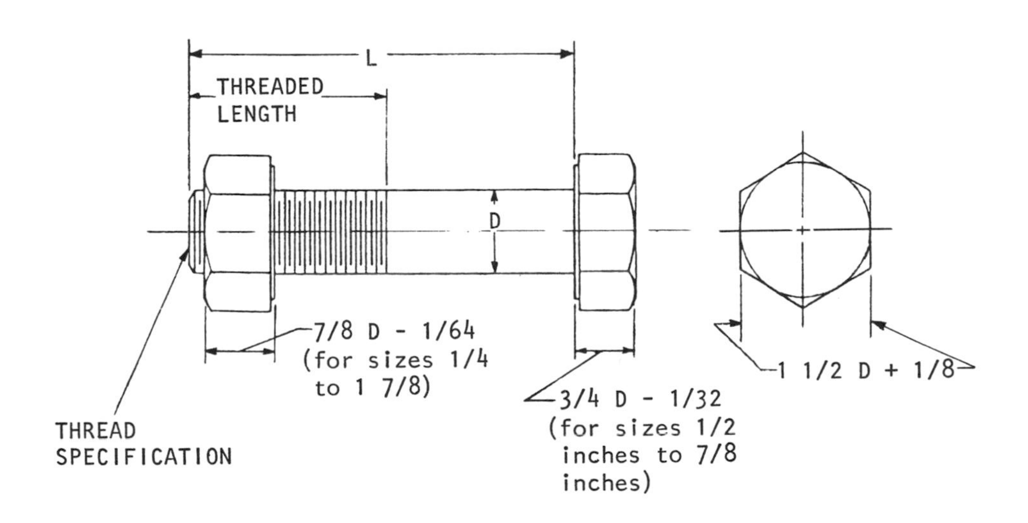

HEAVY BOLT, SEMIFINISHED, HEXAGONAL HEAD, AND HEXAGONAL NUT. Fig. 1

Solution: The following are the dimensions for the head and the nut specified in the problem taken from the standard tables.

For the head:

(a) width across flats = 1¼ inches (maximum)

(b) nominal height = 19/32 inch

For the nut:

(a) width across flats = 1¼ inches

(b) thickness = 47/64 inch (normal size)

For the preparation of the drawing the following formulas, based on the diameter, D, of the bolt, may be used:

W, width across flats = $1\frac{1}{2}$D + 1/8 inch.

H, height of head = 3/4D - 1/32 (for sizes 1/2 to 7/8 inch).

T, thickness of nut = 7/8D - 1/64 (for sizes $\frac{1}{4}$ to 1 7/8 inches).

The figure shows the drawing of the bolt and nut in terms of D.

• PROBLEM 13-10

The following information is given for three different kinds of bolts. Draw the bolts and show their dimensions:

(a) 3/4" x 2", 10-NC-2, SEMIFIN.HEX.HD.BOLT.

(b) 7/8" x $2\frac{1}{2}$", 14-UNF-2A, BRASS FIN.HEX.HD.BOLT.

(c) 3/4" x 2", 10-NC-2, HEAVY UNFIN.SQ.HD.BOLT.

Solution: It is recommended that bolt information be given by note, which includes diameter, length from bearing surface of head to end of shank, thread specification, kind of material (if other than steel), finish, shape of head, and nut (if different from head), and name (i.e., bolt), as in the above example.

The drawing should include the length of the bolt, threaded length, and thread identifications (see Figs. 1-a, 1-b, and 1-c). The minimum thread lengths for bolts which have been recommended are given in standard tables.

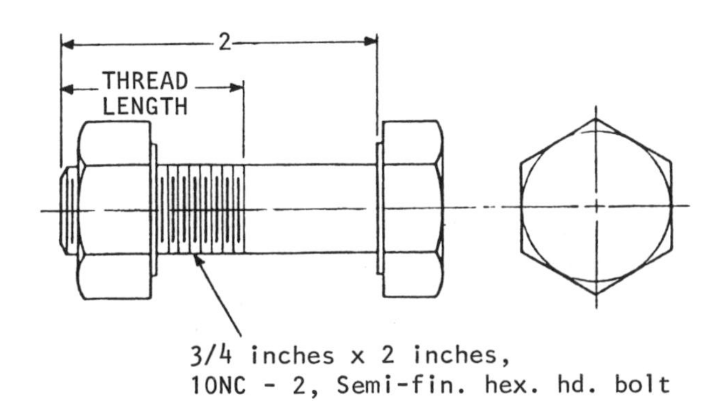

Fig. 1a

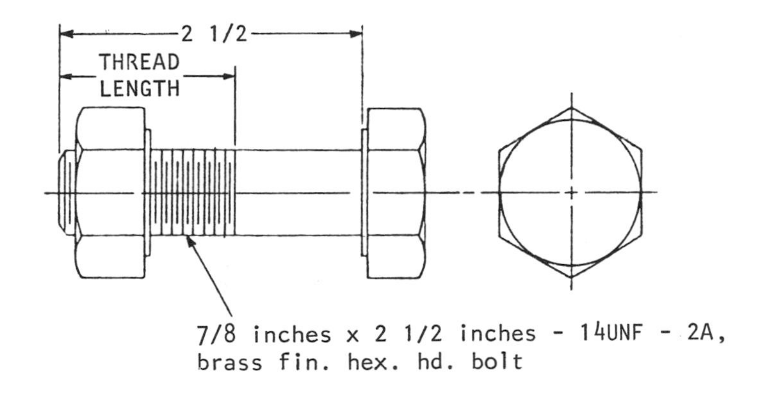

Fig. 1b

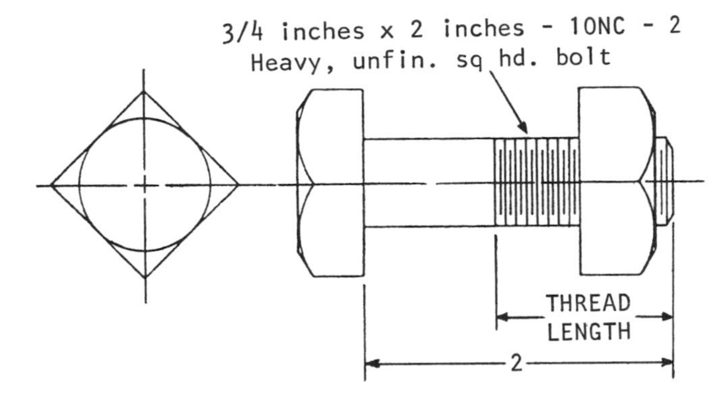

Fig. 1c

• PROBLEM 13-11

Explain and sketch two different types of stud bolts.

Solution: Stud bolts, for example, are made from round rods which are threaded at both ends. This type of bolt is used to fasten two pieces, one of which has a clear hole and the other a threaded hole. The bolt passes through the clear hole, is screwed into the threaded hole, and the pieces are then clamped by means of a nut screwed on the free end of the bolt. The figure shows types of alloy steel stud bolts commonly used with steel pipe flanges and pressure vessels. The bolt stud

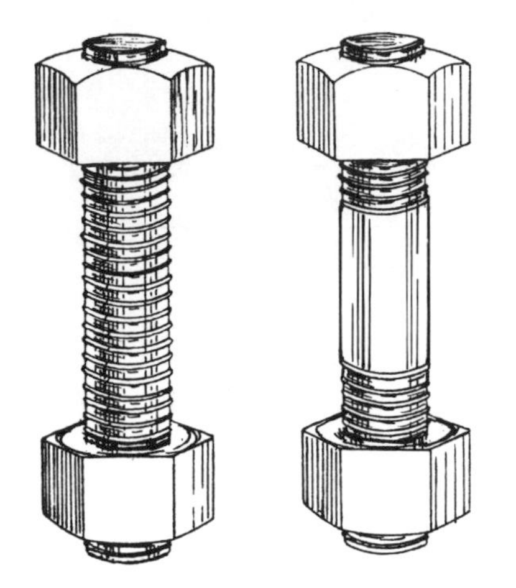

TWO TYPES OF ALLOY STEEL BOLT STUDS COMMONLY USED WITH STEEL PIPE FLANGES AND PRESSURE VESSELS. (FROM FASTENERS, VOL. 3, NO. 1. COURTESY INDUSTRIAL FASTENERS INSTITUTE.)

at left is threaded full length. The right bolt stud is threaded on ends.

• PROBLEM 13-12

Explain and draw (a) Plow bolts; (b) Track bolts.

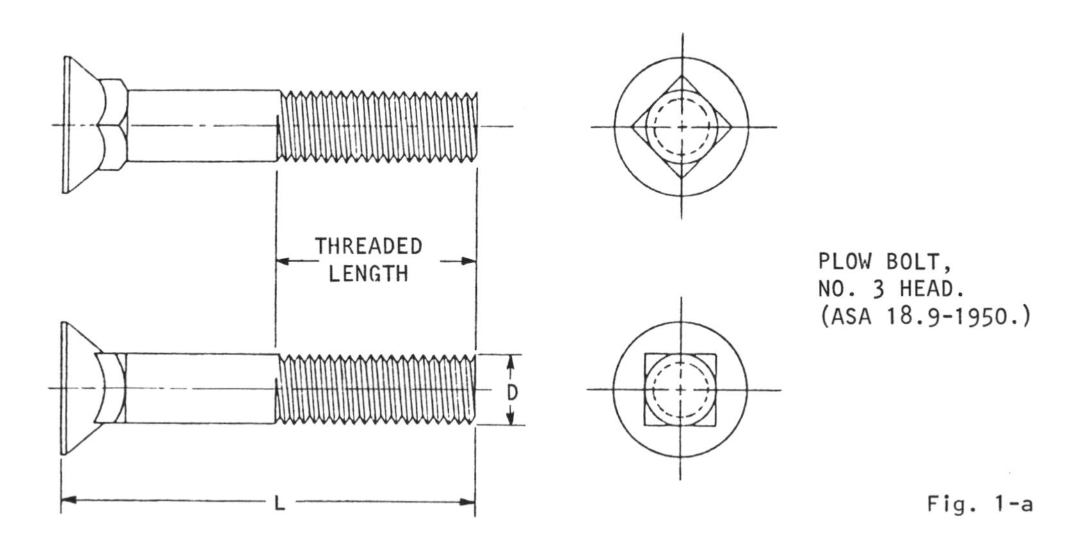

PLOW BOLT, NO. 3 HEAD. (ASA 18.9-1950.)

Fig. 1-a

Solution: Plow bolts are used in connection with the production of soil plows, cleaning machinery, tractors, corn cutters, etc. These have been standardized for both new designs and repair work. ASA B18.9-1950 recommends the use of numbers 3 and 7 regular heads for new designs only and the other forms for repair work. See Fig. 1-a.

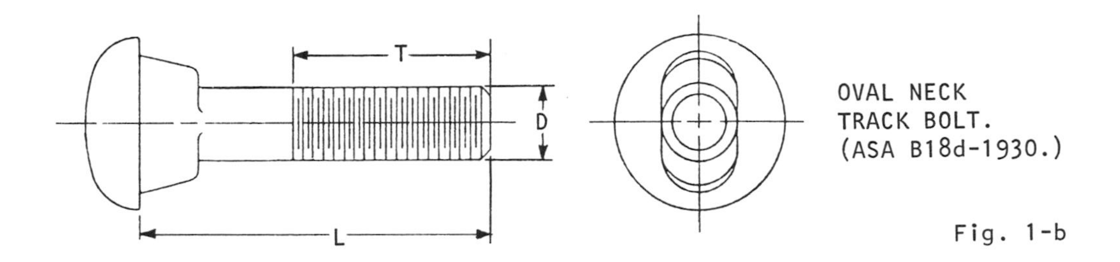

Fig. 1-b

Track bolts are used by railroads and electric railways in the United States and Canada. Recommended dimensions are presented in ASA B18d-1930. See Fig. 1-b.

• PROBLEM 13-13

Draw the following types of round unslotted-head bolts: (a) square-neck carriage bolt, (b) ribbed-neck carriage bolt, (c) button-head bolt, (d) step bolt, and (e) countersunk bolt.

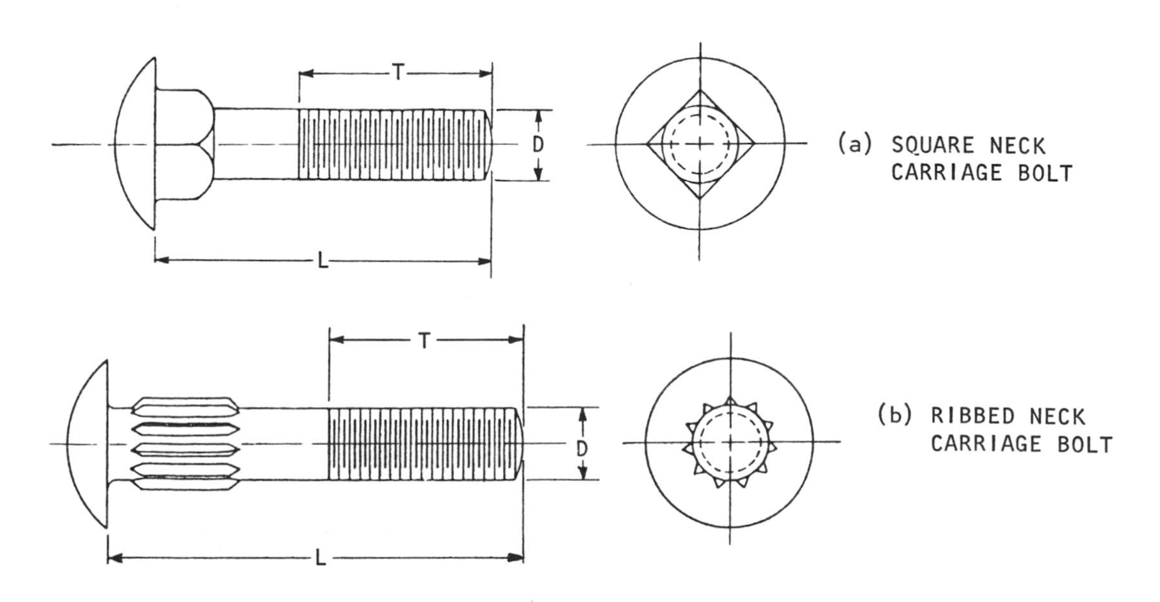

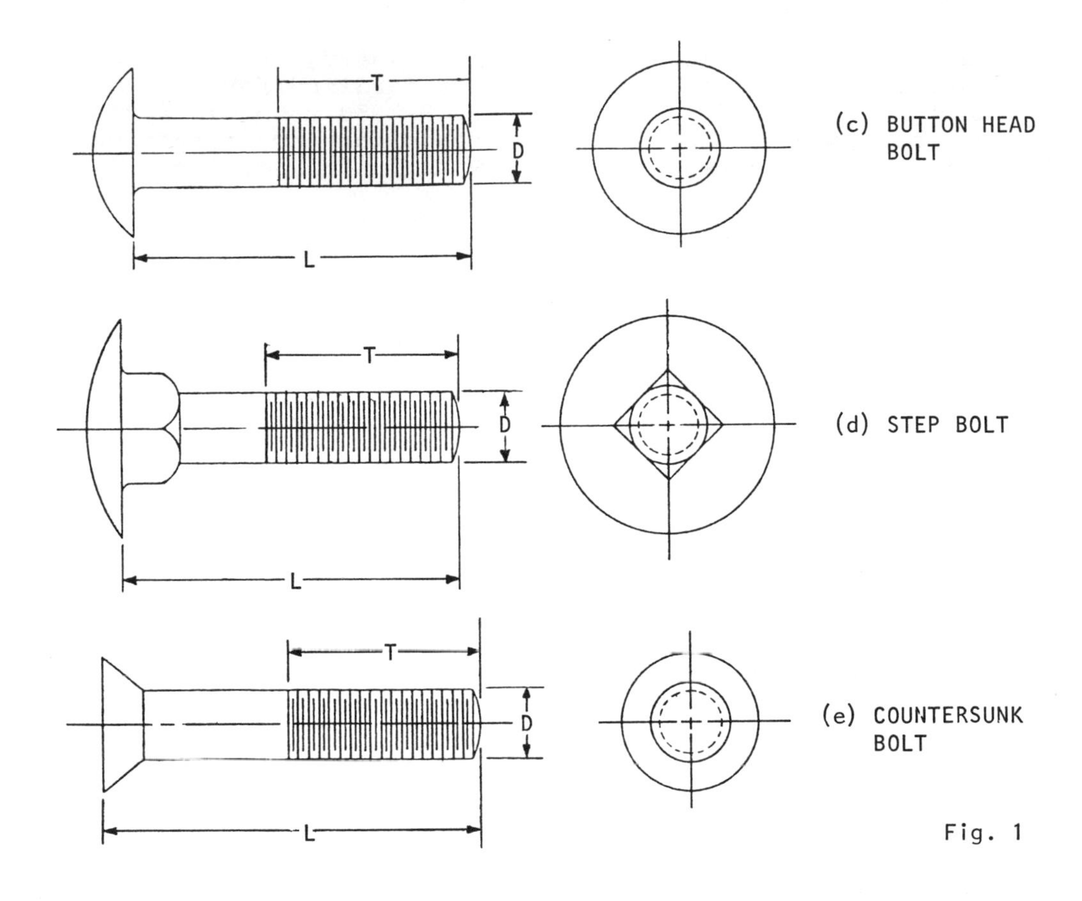

Fig. 1

Solution: Figures 1-a through 1-e show the five different types of round unslotted head bolts.

SCREWS, SPRINGS, AND RIVETS

• PROBLEM 13-14

Draw and explain the different kinds of following screws:

(a) cap screws

(b) machine screws

(c) set screws

Solution: (a) Cap screws are practically the same as through bolts, except for a greater length of thread, and are used without nuts. The shank passes through the hole of one piece and is screwed into the threaded hole of the second piece. Cap screws are used where removal of the pieces is infrequent.

The minimum length of thread is equal to 2D + ¼ inch, with a tolerance of + 2¼ pitch. If this is impossible, as

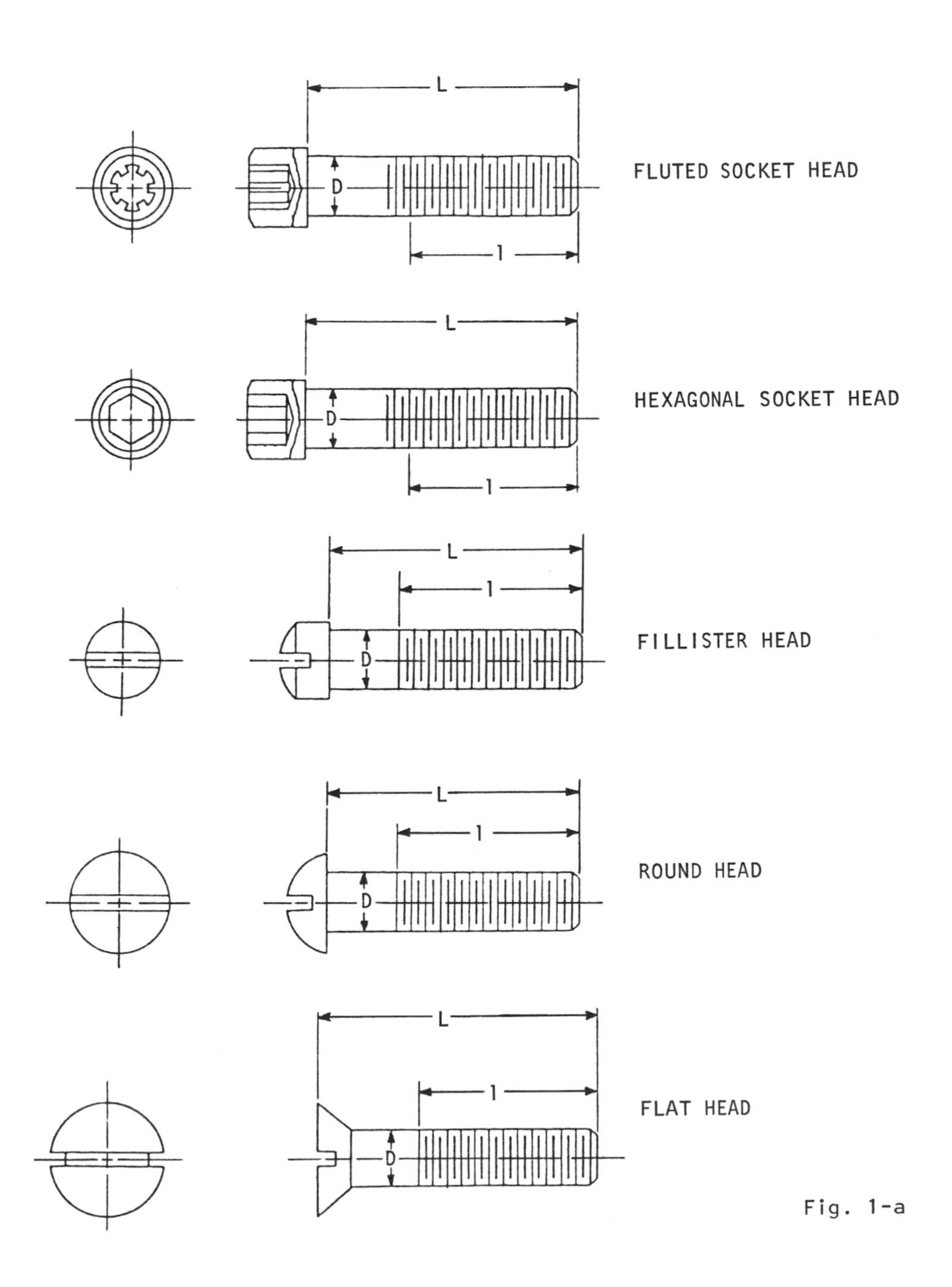

Fig. 1-a

in the case of short cap screws, the complete threads should extend to within 2½ threads of the head. The threads are either NC-3 or NF-3. Slotted head cap screws are machined but usually not heat treated.

Cap screws are available in a variety of heads, i.e., flat, fillister, round, hexagonal, and fluted-type socket heads. Dimensions for these are given in the tables of ASA B18.6-1947 and ASA B18.3-1947. Figure 1-a shows a few examples of cap screws.

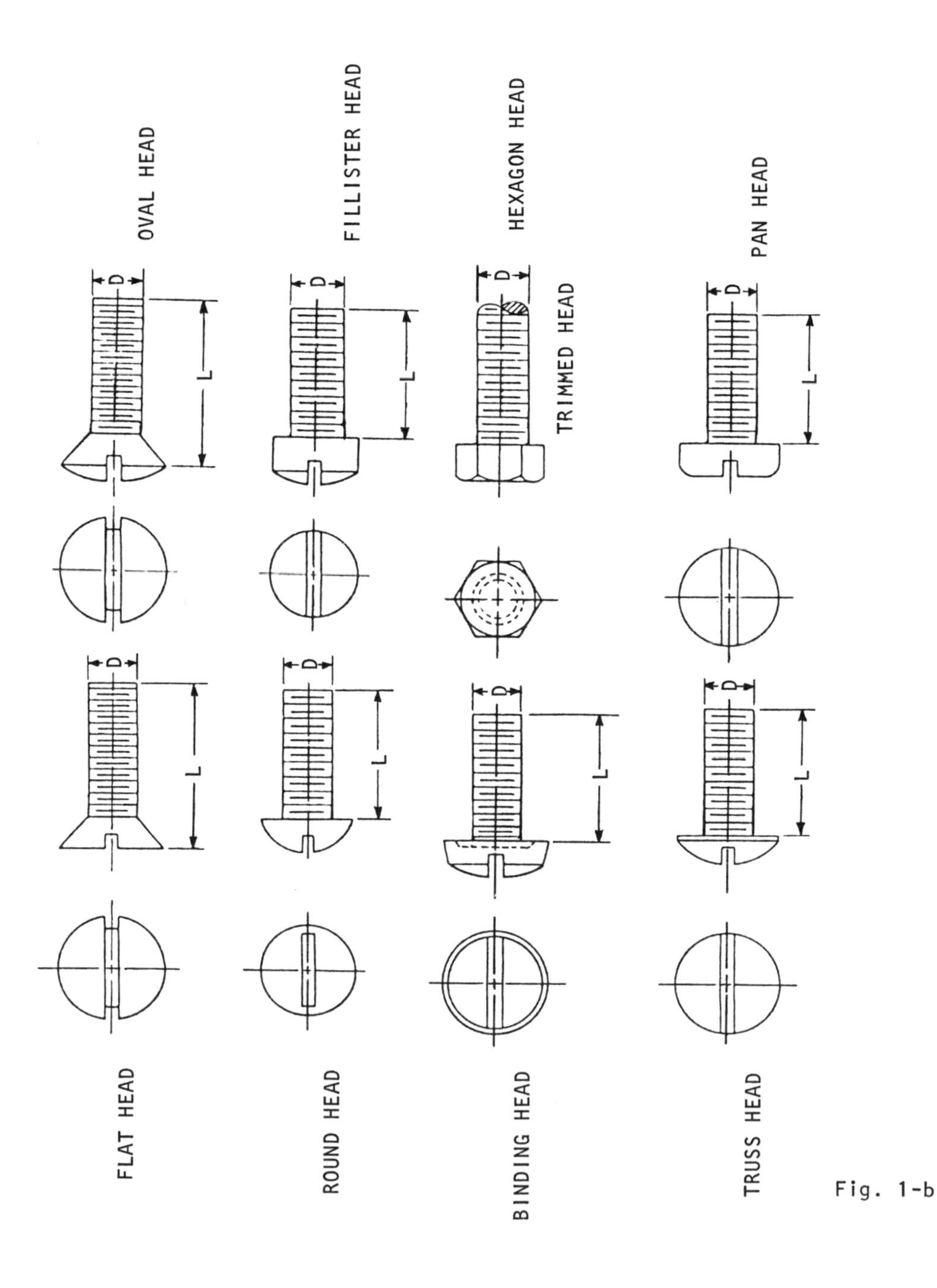

Fig. 1-b

(b) Machine screws are similar to cap screws except for size. In reality, they are small cap screws which are convenient to use in fastening relatively thin parts. On machine screws not over 2 inches long the complete threads extend to within two threads of the bearing surface of the head or closer, if practicable. Screws over 2 inches long have a minimum complete thread length of 1-3/4 inches. The screw threads on machine screws are either NC-2 or NF-2. These screws are supplied with a naturally bright finish and are not heat treated.

There are several standard machine screw heads according to ASA B18.6-1947. There are round, flat, fillister, oval, truss, binding, pan, and hexagon heads. Detailed dimensions are given in the tables of the above-mentioned

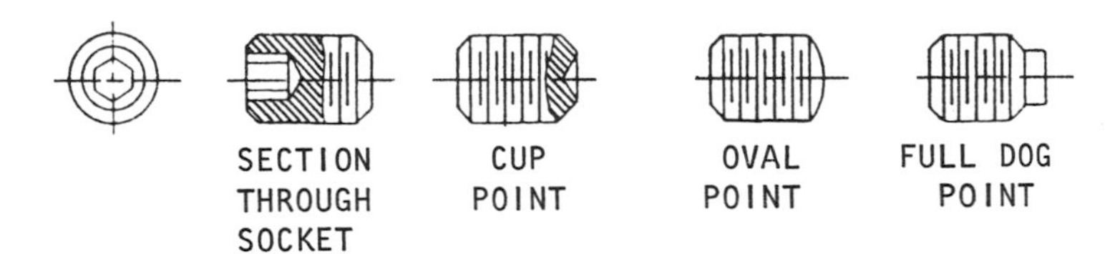

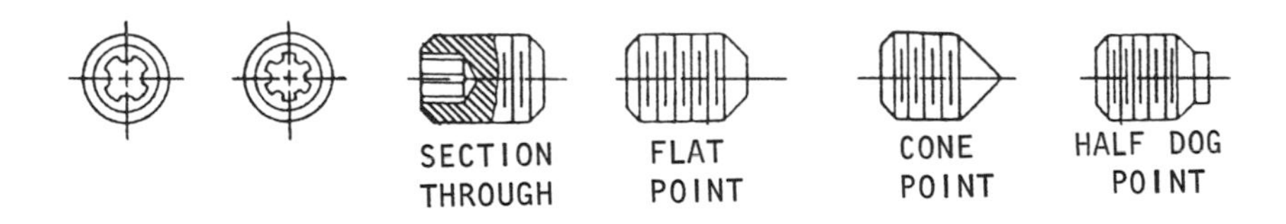

Fig. 1-c

standard. Figure 1-b shows examples of machine screws.

(c) Set screws are used to prevent relative motion between parts by entering the threaded hole of one part and setting the point against the other part, i.e., a pulley hub fastened to a shaft.

The American Standard, ASA B18.3-1947, includes hexagonal and fluted socket headless set screws with a variety of points: cup, flat, oval, cone, full-dog, and half-dog. Tables of this standard provide the necessary dimensions. Figure 1-c shows several socket headless set screws.

American Standard, ASA B18.6-1947, includes data for slotted headless set screws. The screw threads are either NC-2 or NF-2. Headless set screws are case hardened. The points are the same as those shown in Fig. 1-c.

• PROBLEM 13-15

Springs are generally classified under two divisions: wire springs and flat springs.

(a) What is a wire spring? Explain in detail the three different types of wire springs: (1) Compression, (2) Extension, (3) Torsion types.

(b) Define a flat spring.

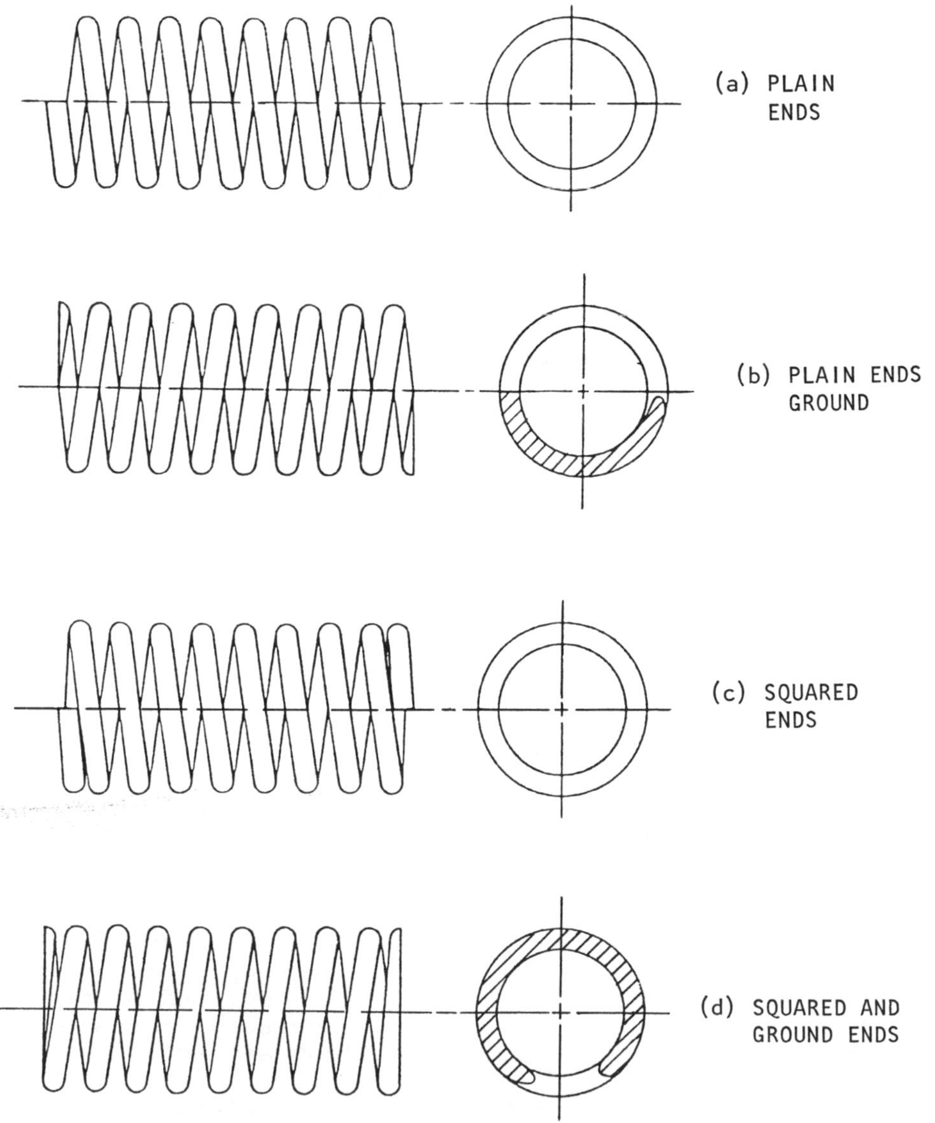

Fig. 1

Solution: (a) Wire springs may be helical or spiral and

may be made from round, square, or special-section wire. They are classified as (1) compression, (2) extension, or (3) torsion types.

1. Compression Springs. Compression springs are open-coiled helical springs that resist compression forces. The most common form has the same diameter throughout its entire length, and is known as a straight spring. Extensive use is also made of tapered and cone-shaped compression springs. The size of wire to be used is determined by the elastic limit and modulus of the material, loads, range of operation, and the type of application. The designer makes use of this information.

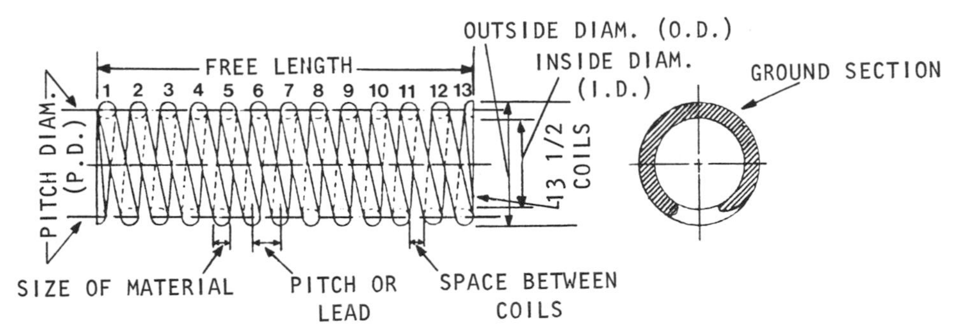

1. Material
2. O.D.
3. I.D.
4. Free length
5. Total number of coils
6. Type of ends
7. Solid length

Fig. 2

Specifications:

A. Material may be specified as "Spring Steel Wire." Full information should be given if a particular type or grade of material is desired.

B. Diameter Information. If the spring works in a hole, the outside diameter of the spring and the diameter of the hole should be given.

If the spring operates over a rod, the inside diameter of the spring and the diameter of the rod should be given.

If the spring works on a rod and in a hole, the diameters of the rod and hole should be given and also a suggested diameter of the spring.

C. Free length is the overall length of the spring in an unloaded condition.

D. The number of coils, whether active or total (preferably total) should be given.

E. The type of ends must be definitely specified as: (a) plain ends, (b) plain ends ground, (c) squared ends, or (d) squared and ground ends. This will eliminate the possibility of misinterpreting the type of ends shown on the drawing. See Fig. 1.

F. Solid length is the length of the spring with all coils closed. On ground springs this dimension is equal to the

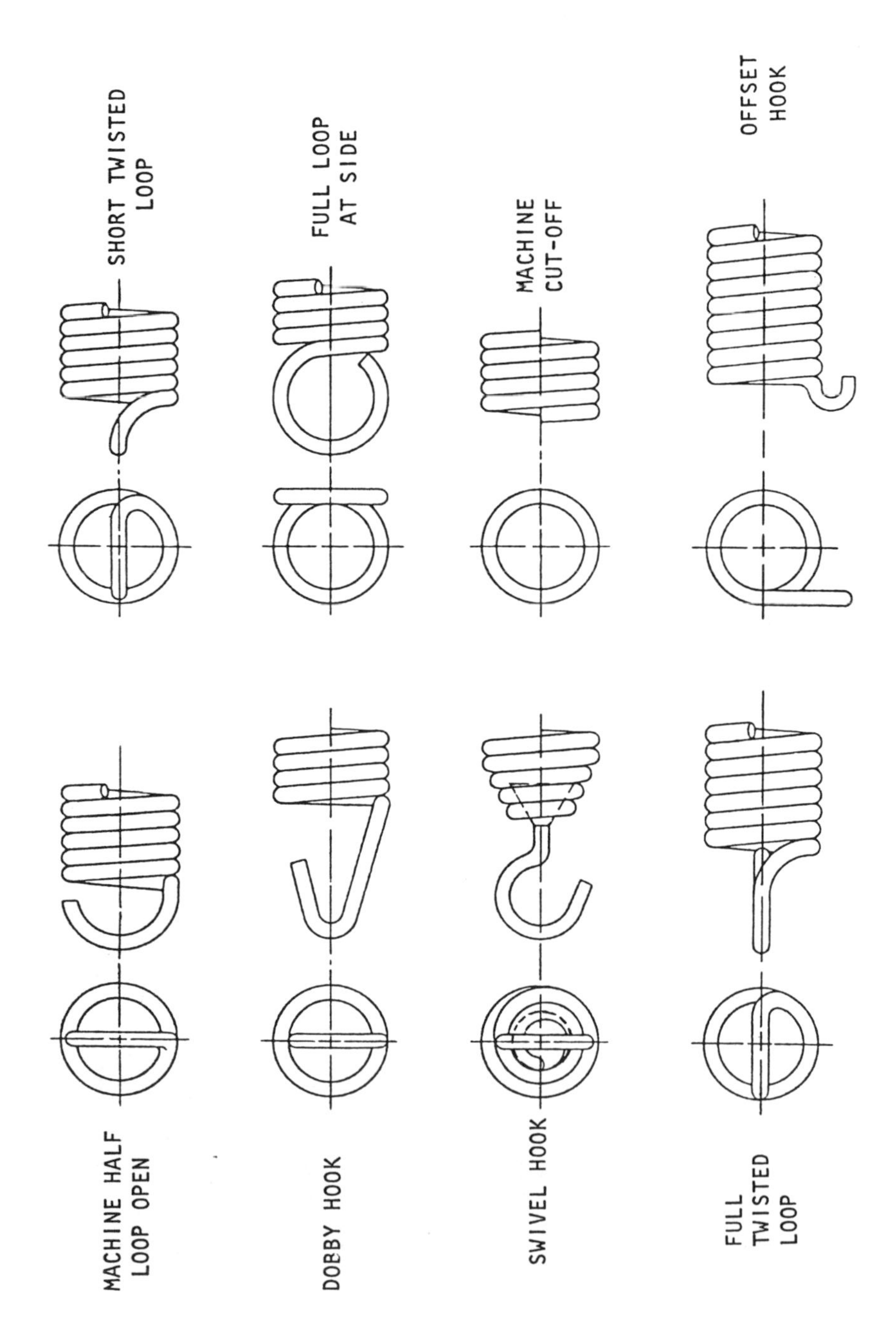

Fig. 3

total number of coils times the wire diameter. On springs with no end grinding the solid length is equal to the total number of coils times the wire diameter plus one extra wire diameter.

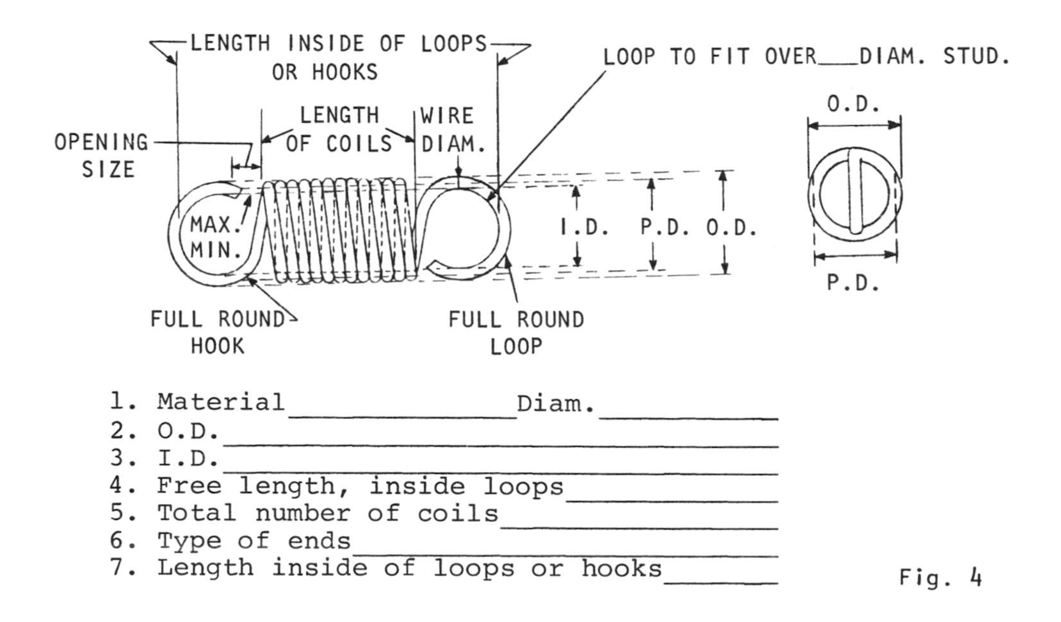

Fig. 4

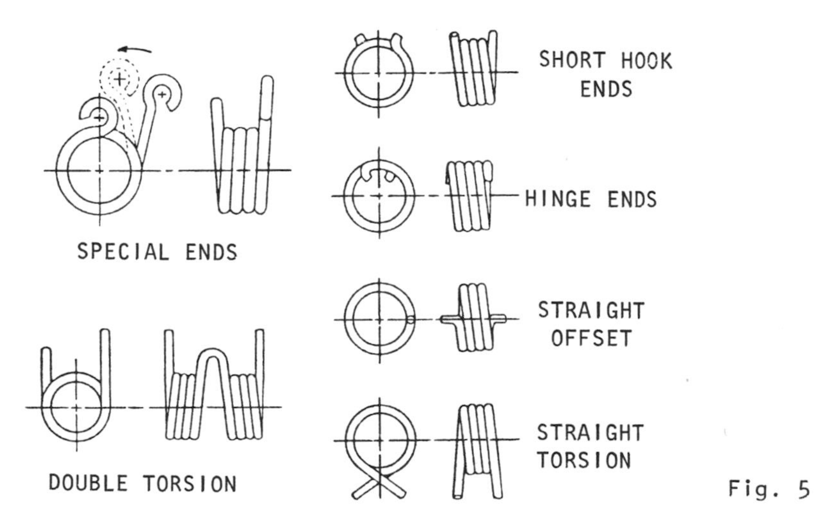

Fig. 5

The list of specifications with Fig. 2 shows the information that should be given to provide data for the compression springs.

2. Extension Springs. Extension springs are close-coiled helical springs that resist pulling forces. They are close

wound, in contact with each other, and made from round or square wire. The coils may be wound so tight as to require an effort to pull them apart. This coiling load is known as the initial tension which can be controlled to a certain extent.

Specifications include material, wire diameter, spring diameter, number of coils, free length, direction of coil, position of loops (if loops relative to each other must maintain a desired position, they may be specified as loops in line, at right angles, etc.), type of ends, initial tension, and loads. Figure 3 shows types of ends used on extension springs. Figure 4 shows the information that should be given to provide extension spring data for manufacturing purposes.

3. Torsion Springs. Torsion springs exert force along a circular path, thus providing a twist or a torque. Although compression and extension springs are subjected to torsional forces, a "torsion" spring is subjected to bending forces. The term "torsion spring" is applied to a helical spring loaded by torque. The ordinary spring hinge is typical of one of the common uses of torsion springs. Other applications, primarily to rotate parts or to cushion shock on rotating parts, include oven door springs on kitchen ranges, brush-holder springs on motors and generators, etc.

Torsion spring applications are so varied that the development of standard forms has not been found desirable. Designs should show the type of ends required (Fig. 5 shows a few of the commonly used forms), maximum outside and inside diameters, length, number of coils, size of wire, and the winding, whether left or right.

(b) Flat Springs can be defined as any spring made of flat or strip material. Two examples are:

1) Flat wire torsion-type springs made by winding flat stock in the form of a spiral.

2) Flat-coil springs known as clock or motor springs, made from strips of tempered steel wound up on an arbor and generally constrained in a drum or case. These springs are capable of storing up energy when wound and of delivering the stored energy in the form of torque through the arbor or shaft, or through the drum.

• PROBLEM 13-16

What is a rivet? (a) Explain and show the standard head shapes for large rivets. (b) Also define riveted joints and draw "butt joints" and "lap joints."

Solution: (a) Rivets are commonly used in permanently connecting members of structural frames and in the fabrication of tanks, boilers, containers, etc. Manufactured rivets consist of a cylindrical portion known as the body or "shank," and a "head" which is integral with the body. Rivets are designed by giving the diameter, length, and type of head, e.g., ½ inch x 3 inches button-head rivet.

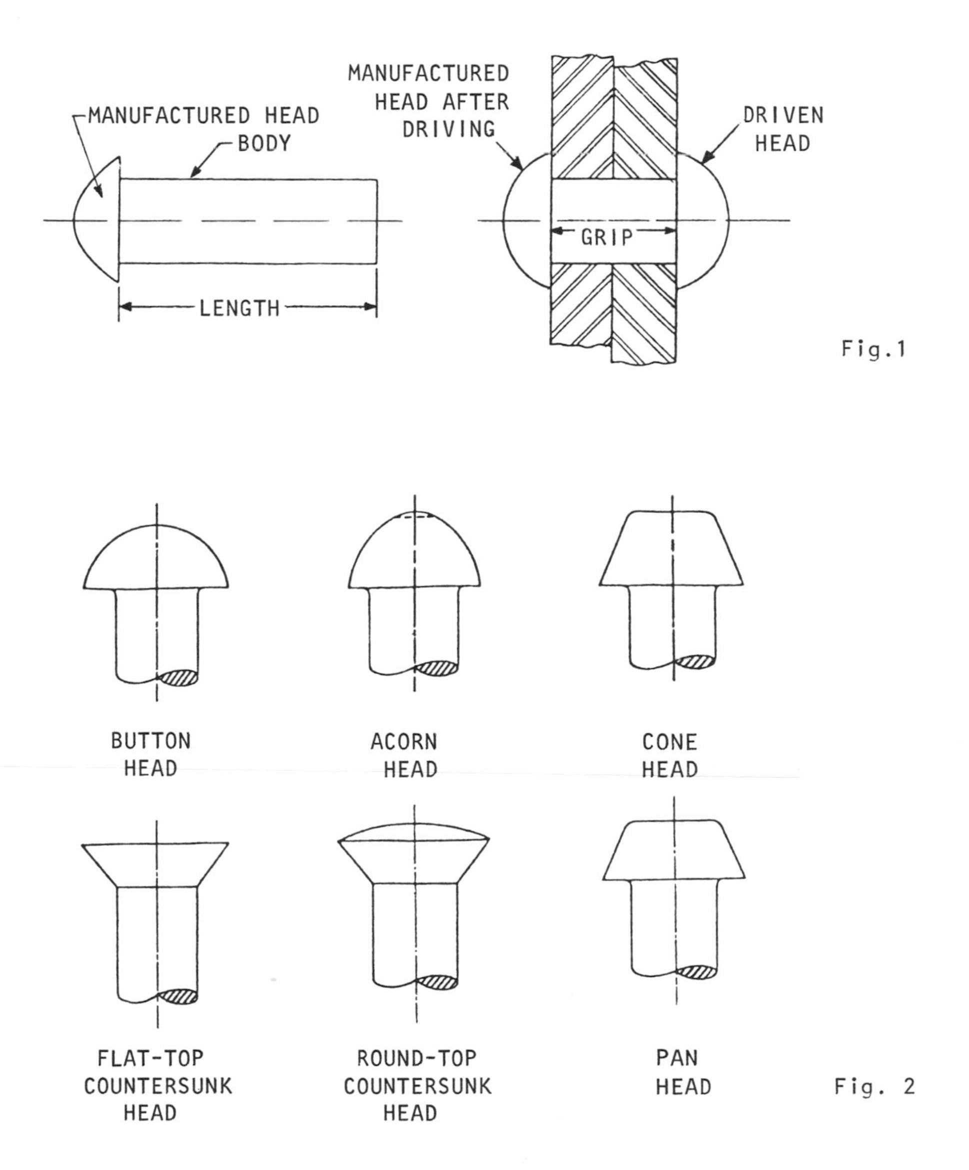

The length of a rivet is measured, in a line parallel with the axis of the rivet, from the largest diameter of the bearing surface of the head to the extreme end. The length of a rivet will depend upon its grip, which is the thickness of the metal plates held together by the rivet.

Figure 1 shows the nomenclature applied to a button-head rivet as manufactured, and after being driven.

Figure 2 shows standard head shapes for large rivets (½ inch to 1-3/4 inches inclusive) and also for small rivets (3/32 inch to 7/16 inch inclusive). Detailed information for the small rivets, however, is available in ASA B18a-1927.

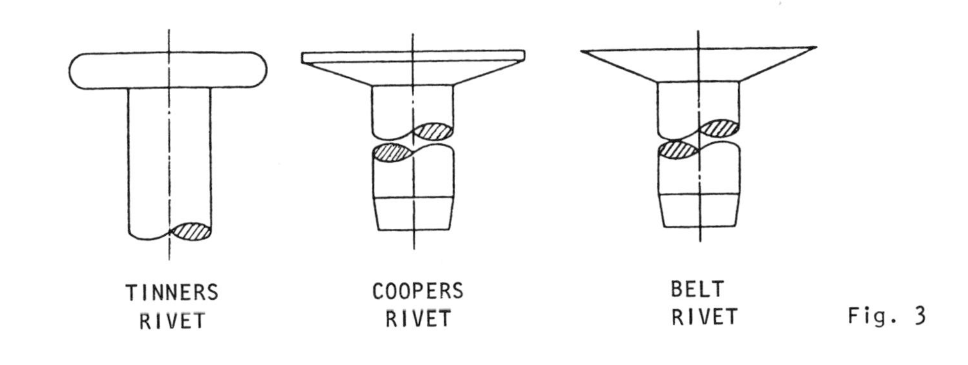

Fig. 3

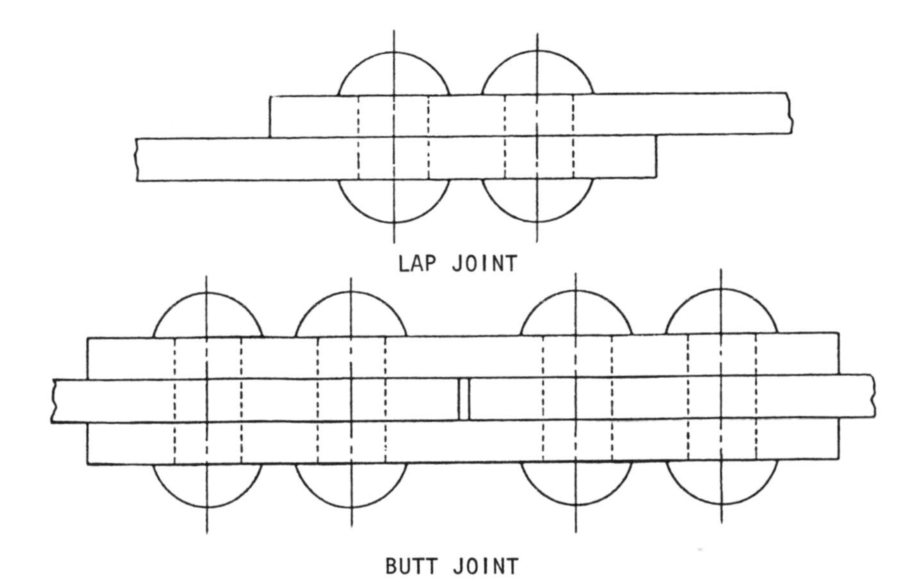

Fig. 4

In addition, Fig. 3 includes tinners' rivets (8 to 16 ounces), coopers' rivets (1 to 16 pounds) and belt rivets (No. 7 to No. 13).

(b) Riveted joints may be either butt joints or lap joints, depending upon the structure. Butt joints are commonly used in "girder splices," "column splices," etc. Lap joints are frequently used in tank fabrication and in hull construction. Figure 4 shows an example of a butt joint and a lap joint.

CHAPTER 14

CAMS AND GEARS

DISPLACEMENT DIAGRAMS

• PROBLEM 14-1

Define a cam and describe some of the most generally used types of cams.

Solution: Cams. A cam is a machine element with surface or groove formed to produce special or irregular motion in a second part, called a "follower." The shape of the cam is dependent upon the motion required and the type of follower that is used. The type of cam is dictated by the required relationship of the parts, and the motions of both.

Types of cams. The direction of motion of the follower with respect to the cam axis determines two general types, as follows: (1) radial or disk cams, in which the follower moves in a direction perpendicular to the cam axis, and (2) cylindrical or end cams, in which the follower moves parallel to the cam axis. Figure 1 shows at A a radial cam, with a roller follower held against the cam by gravity or by a spring. As the cam revolves the follower is raised and lowered. Followers are also made with pointed ends and with flat ends. B shows a face cam, with a roller follower at the end of an arm or link, the follower oscillating as the cam revolves. When the cam itself oscillates, the toe and wiper are used, as at C. The toe, or follower, may also be made in the form of a swinging arm.

A yoke or positive-motion cam is shown at D, the enclosed follower making possible the application of force in either direction. The sum of the two distances from the center of the cam to the points of contact must always be equal to the distance between the follower surfaces. The cylindrical groove cam at E and the end cam at F both

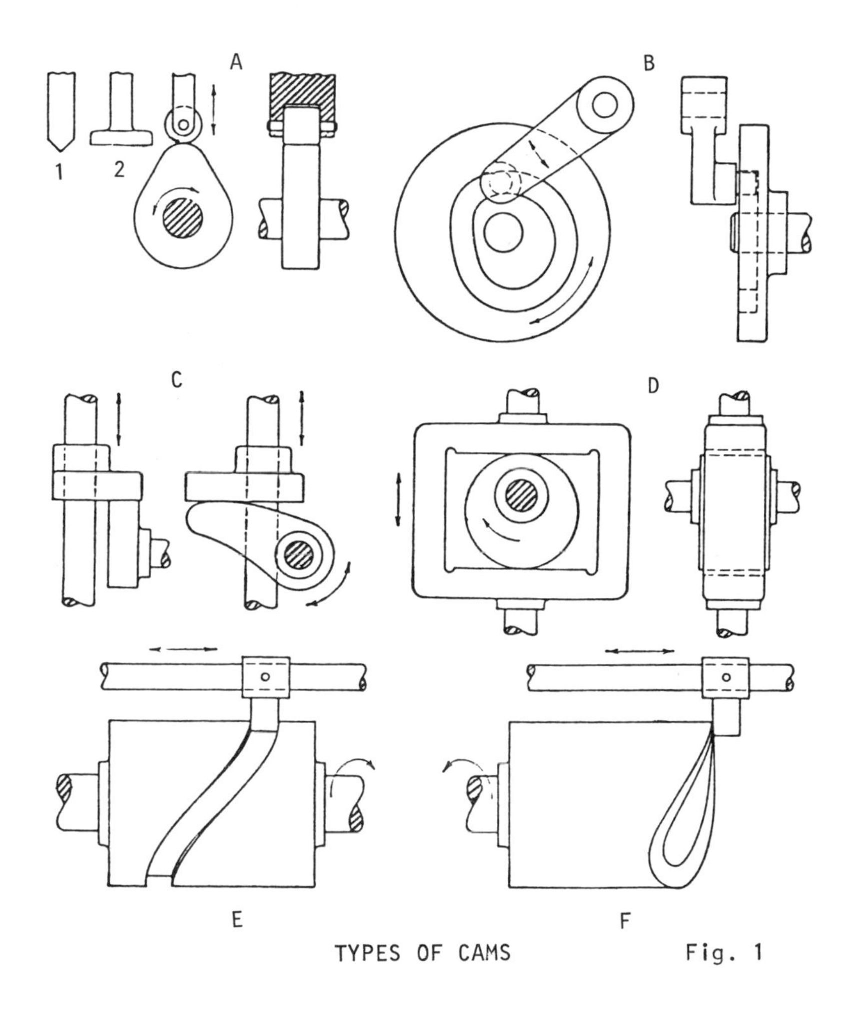

TYPES OF CAMS Fig. 1

move the follower parallel to the cam axis, force being applied to the follower in both directions with the groove cam, and in only one direction with the end cam.

• PROBLEM 14-2

What are cam (or displacement) diagrams?

Solution: Cams may be designed to move the follower with

constant velocity, acceleration, or harmonic motion. In many cases, combinations of these motions, together with surfaces arranged for sudden rise or fall, or to hold the follower stationary, go to make up the complete cam surface.

Cam diagrams. In studying the motion of the follower, a diagram showing the height of the follower for successive cam positions is useful and is frequently employed. The cam position is shown on the abscissa, the full 360° rotation of the cam being divided, generally, every 30° (inter-

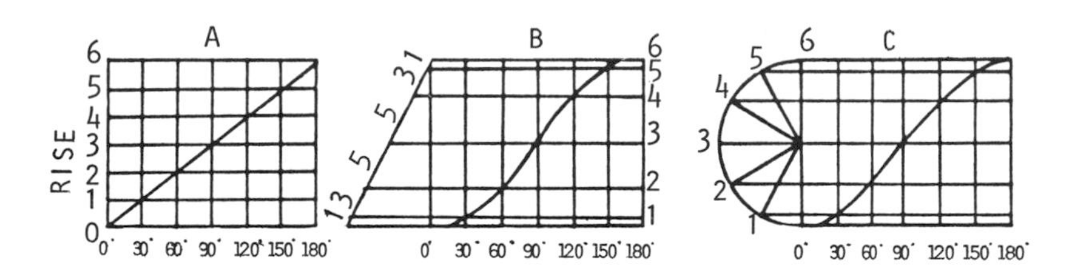

Position (Degrees)

METHODS OF PLOTTING CAM DIAGRAM; THREE KINDS OF MOTION. Fig. 1

mediate points may be used if necessary). The follower positions are shown on the ordinate, divided into the same number of parts as the abscissa. These diagrams are generally made to actual size.

Constant velocity gives a uniform rise and fall, and may be plotted as at A, Fig. 1, by laying off the cam positions on the abscissa, measuring the total follower movement on the ordinate and dividing it into the same number of parts as the abscissa. As the cam moves one unit of its rotation, the follower likewise moves one unit, producing the straight line of motion shown.

With constant acceleration, the distance traveled is proportional to the square of the time, or the total distance traveled is proportional to 1, 4, 9, 16, 25, etc., and if the increments of follower distance are made proportional to 1, 3, 5, 7, etc., the curve may be plotted as shown at B. Using a scale, divide the follower rise into the same number of parts as the abscissa, making the first part 1 unit, the second 3 units and so on. Plot points at the intersection of the coordinate lines, as shown. The curve at B accelerates and then decelerates to slow up the follower at the top of its rise.

Harmonic motion (sine curve) may be plotted as at C by measuring the rise and drawing a semicircle, dividing it into the same number of parts as the abscissa and projecting the points on the semicircle as ordinate lines.

Points are plotted at the intersection of the coordinate lines, as shown.

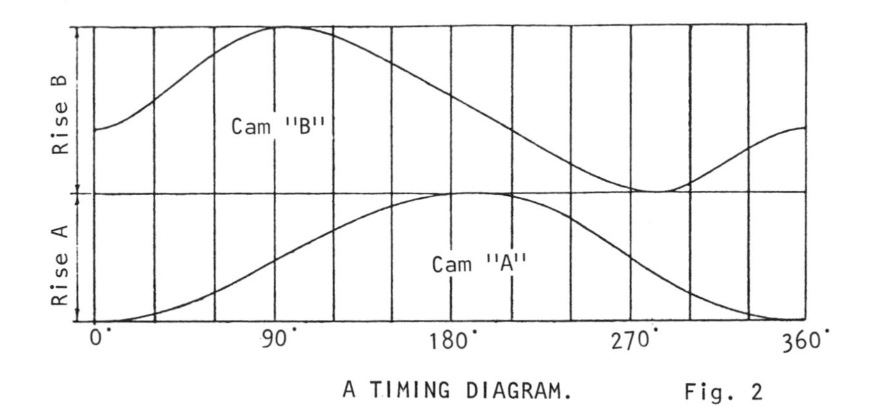

A TIMING DIAGRAM. Fig. 2

Timing diagrams. When two or more cams are used on the same machine and their functions are dependent on each other, the "timing" and relative motions of each may be studied by means of a diagram showing each follower curve. The curves may be superimposed, but a better method is to place on above the other as in Fig. 2.

• PROBLEM 14-3

Interpret the following cam displacement diagram drawn to exact scale (1" = 1"):

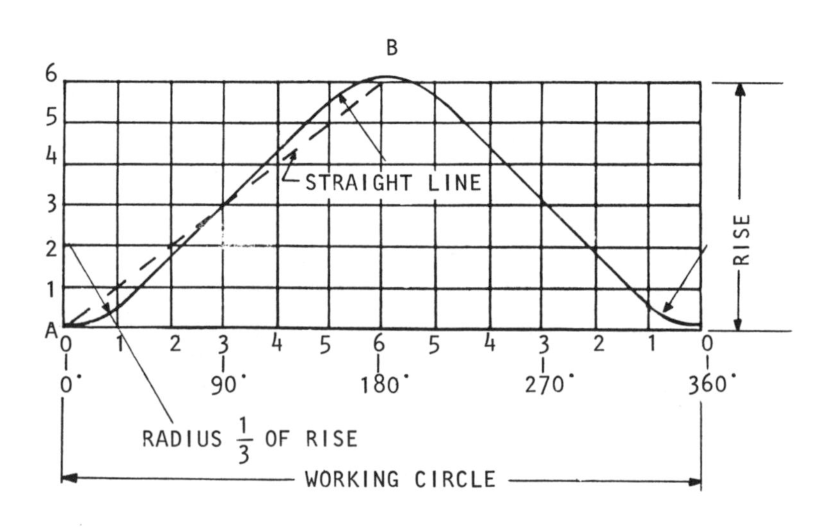

Solution: This is a diagram of uniform motion. Uniform motion means the follower rises and falls at a constant speed. In theory, the curve from A to B and B to O should have been a straight line, but it is common practice to draw an arc at each end of a uniform motion diagram. This eliminates an abrupt starting and stopping of the follower at the beginning and end of the interval. The curve has a radius of one-third the rise and is tangent to the straight line AB.

The follower rises uniformly one and one-half inches through 180° and falls to rest with the same uniform velocity through the remaining 180°. The follower does not dwell at any radial distance from the center of the cam.

• PROBLEM 14-4

Interpret the motion of the follower for the following displacement diagram:

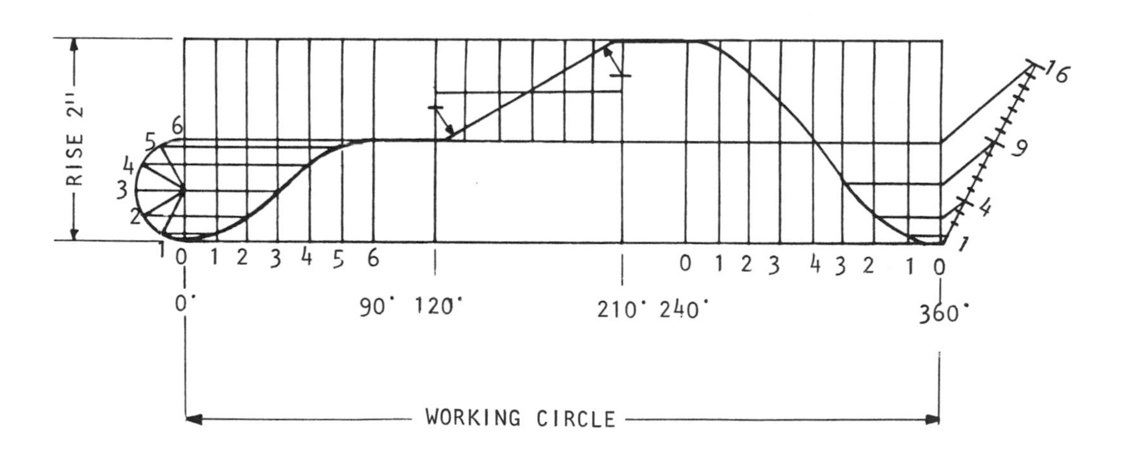

A DISPLACEMENT DIAGRAM FOR A CAM HAVING A COMBINATION OF MOTIONS.

Solution: 1. The follower rises one inch in 90 degrees using harmonic motion.

2. The follower dwells for 30°, i.e., during this period of time the follower does not move when the cam moves.

3. It then rises one inch in uniform motion as the cam turns 90 degrees.

4. At the top of the rise it dwells for 30 degrees.

5. It falls using uniform accelerated and decelerated motion through the remaining 120 degrees.

• PROBLEM 14-5

A cam rotating at a uniform speed is required to impart the motion described below:

(a) lift through 1-1/4" with uniform acceleration for 180° of cam rotation;

(b) return to initial position with uniform deceleration for the remaining 180°.

Draw the required displacement diagram.

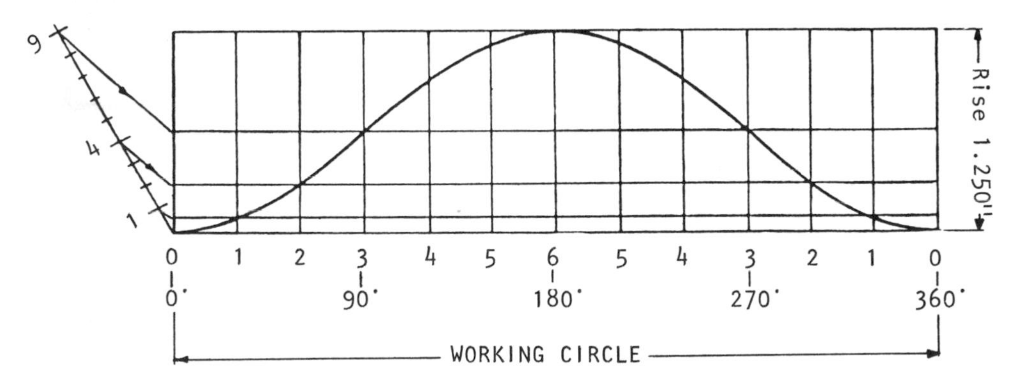

A UNIFORM ACCELERATED AND DECELERATED DISPLACEMENT DIAGRAM.

Solution: 1. Draw the length and rise of the diagram.

2. Divide the part of the length to be used for uniformly accelerated and decelerated motion into a number of equal parts.

3. Now divide half the rise into distances proportional to the square of the distances on half the length. Each point on the length is squared as 1^2, 2^2, and 3^2. This gives proportional parts as 1, 4, 9, and so forth, for marking the rise. Draw a line on any angle from Point 0. Mark off equal distances to equal the largest squared number on the rise. Draw a line from the largest number to the center of the rise. Draw lines parallel to this to the rise. Project these points on the rise across the diagram parallel to the length line. This gives one

half the curve.

The other half of the curve is the reverse of the half plotted. Points can be located using dividers.

• PROBLEM 14-6

A cam rotating at a uniform speed is required to impart the motion set out below:

(a) rise harmonically through 1 1/2" for 180° of cam rotation;

(b) return harmonically to initial position for the remainder 180°.

Draw the required displacement diagram.

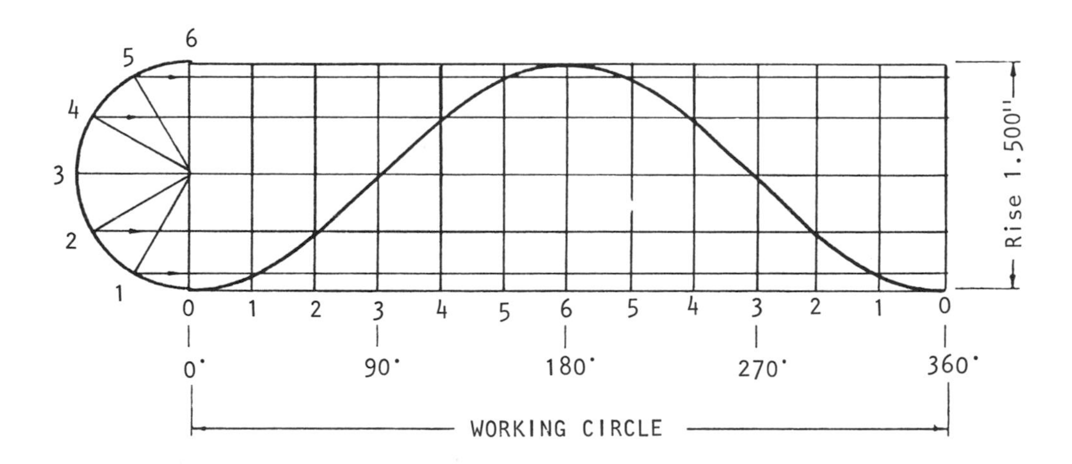

Solution: 1. Draw the length of the working circle.

2. Draw the rise line.

3. Construct a semicircle on the rise line. The radius equals half the rise.

4. Divide the semicircle into a number of equal parts. Number the points.

5. Divide the part of the diagram's length to be used for the harmonic rise into the same number of parts. Number the points. In the figure this is 180°.

6. Project the points on the semicircle across the

diagram. Where they intersect the line with the same number, is a point on the harmonic curve.

7. Connect the points in a smooth curve.

CAM PROFILES, LAYOUTS, AND FOLLOWERS

• PROBLEM 14-7

A uniform-velocity knife-edge follower is required to lift to its maximum (18 mm) in 90° of rotation, dwell for 90° and fall to rest in the remaining 180°. The minimum radius of the cam is 7 mm.

Construct the uniform-velocity displacement diagram and draw the required cam profile.

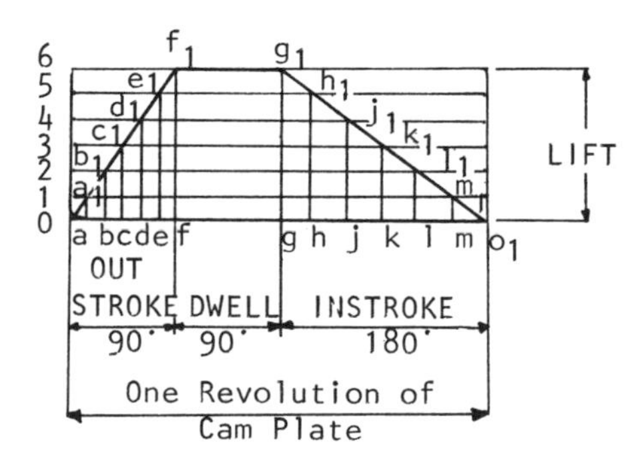

DISPLACEMENT DIAGRAM
INFORM VELOCITY. Fig. 1

Solution: To construct the displacement diagram, draw Oo_1 any convenient length and mark off the required angular movements of the cam. Subdivide the outstroke and instroke (18 mm) into a number of equal parts (six in this case) and label them as shown in Fig. 1. The intersections of the vertical and horizontal projections of these divisions will produce the required diagram. Note the method of showing the period of dwell.

To construct the cam profile, first draw a circle of radius 7 mm, the minimum radius of the cam.

Beginning at O, set off the outstroke such that angle OCf_1 is 90° as in Fig. 2. Divide this into six equal parts and label them as shown. Mark off distances aa_1, bb_1, etc., from the displacement diagram. A smooth curve from O to f_1 will give the cam outstroke. The

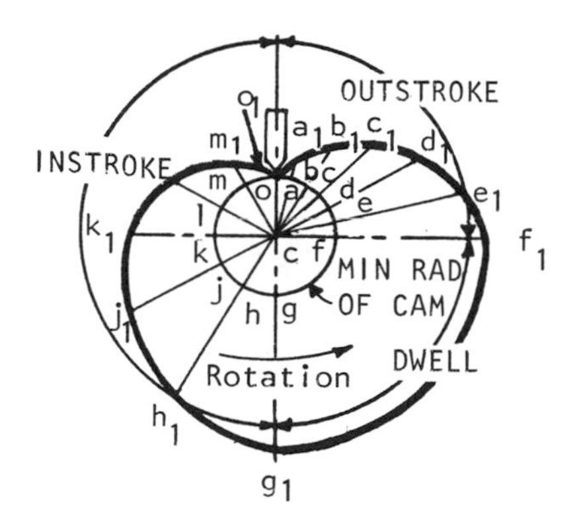

CAM PROFILE Fig. 2

period of dwell is the circular arc f_1g_1 concentric with the cam center. From g_1, the follower is required to fall to rest uniformly. Divide g_1 to O into six equal parts, mark off the instroke in similar manner to the outstroke, and join g_1 to O, by a smooth curve to complete the cam profile. It will be noted that the sharp changes of velocity at points O, f_1, g_1, and O_1 in the displacement diagram are smoothed out as curves.

• PROBLEM 14-8

It is required to lay out a cam profile for a cam with an oscillating roller follower pivoted 100 mm from the axis of rotation, the pivot lying on the cam center line, and satisfying the following conditions:

(a) the cam is to rotate clockwise;

(b) the follower is to lift through 12° for 120° of cam rotation;

(c) the follower is to dwell for 60° of cam rotation;

(d) the follower is to return to the initial position for 120° of cam rotation;

(e) the follower is to dwell for the remaining 60° of cam rotation.

The minimum radius of the cam is 50 mm and the roller diameter is 25 mm. The roller center is to lie on the vertical center line of the cam.

The motion is to be uniform velocity for both the outstroke and instroke.

Solution: Draw the center lines of the cam. With center

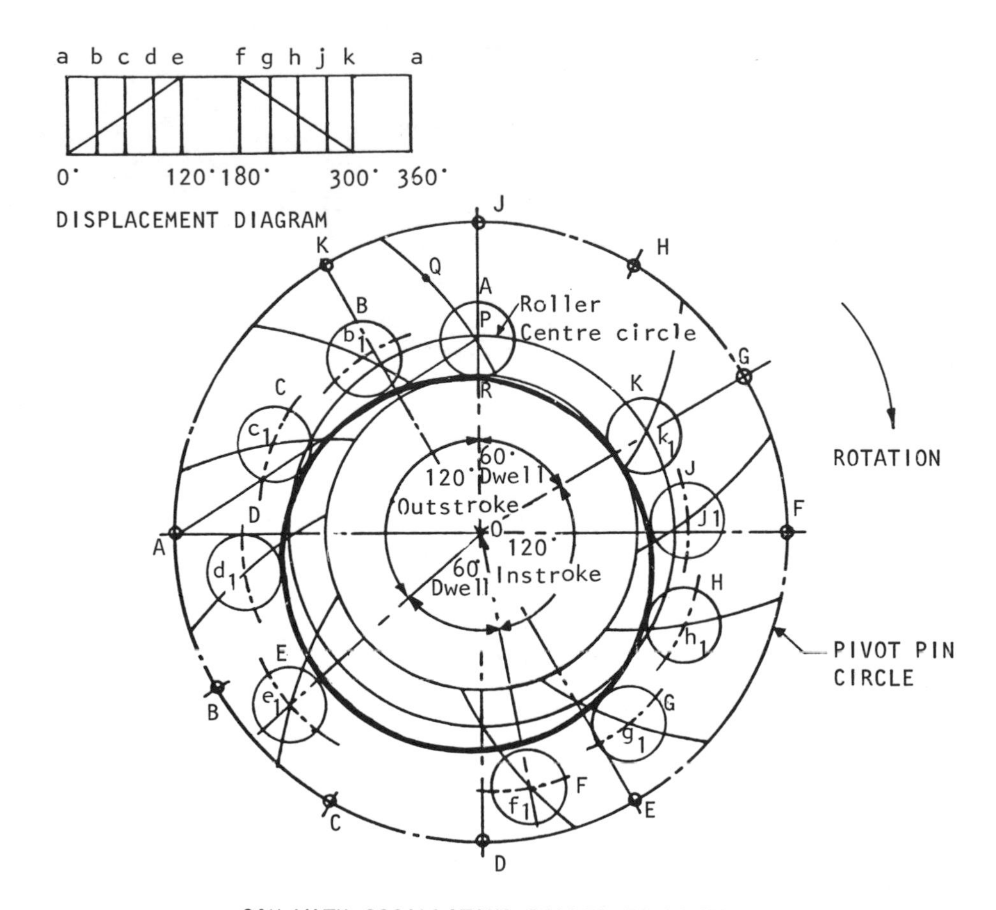

CAM WITH OSCILLATING ROLLER FOLLOWER.

O, and radius OA (100 mm), draw a chain-dotted circle. This circle represents the pivot-pin circle.

Let OR represent minimum radius of cam (50 mm) and draw a circle as shown: Let OP be the radius of the circle locating the center of the roller follower (OP = OR + half roller diameter = 62.5 mm). Draw a chain-dotted roller center circle with center O and radius OP.

Although the oscillating arm remains fixed about A, the method of construction is to fix the cam plate and rotate the follower, together with its arm, for each plotted point.

Mark off the outstroke on the pivot-pin circle such that angle AOE is 120°. Divide this into any convenient number of equal parts (four in this case) and label as shown. Mark off the dwell angle of 60°, and then mark off and subdivide the instroke angle of 120°, labelling as shown.

With center A, radius AP, draw an arc to pass through the roller center circle.

Draw similar arcs through other points obtained. The roller must lie along each of these arcs successively as the cam rotates.

Along the arc passing through point P and subtended from A, set off the total lift angle of 12°. This distance, PQ, should be used in setting out the displacement diagram, but the chord may be used, as the difference is so small that it will introduce very little error.

Construct the displacement diagram as shown.

Along each of the follower arcs, the rise and fall of the roller center may be traced measured outward from the roller center circle, as at bb_1, cc_1, dd_1, etc. Draw the roller follower for each point and from R connect with a smooth curve to give the required profile.

Note that the pivot-pin center A is connected to roller A, and center B is connected to roller B, and center C is connected to roller C, etc., and the roller centers themselves are denoted by b_1, c_1, d_1, etc.

• PROBLEM 14-9

A cam rotating anticlockwise at a uniform speed is required to impart the motion set out below:

(a) lift through 38 mm with uniform acceleration for 120° of cam rotation;

(b) return to initial position with uniform velocity for 150° of cam rotation;

(c) dwell for the remainder.

The minimum radius of the cam is 50 mm and the stroke of the follower is coincident with the center line of the cam.

Draw the cam profile for each of the following cases:

1) a knife-edge follower;

2) a roller follower of diameter 24 mm.

Solution: 1) Set out the displacement diagram as shown in Fig. 1, dividing the outstroke and instroke into six equal parts numbered from 0 to 6.

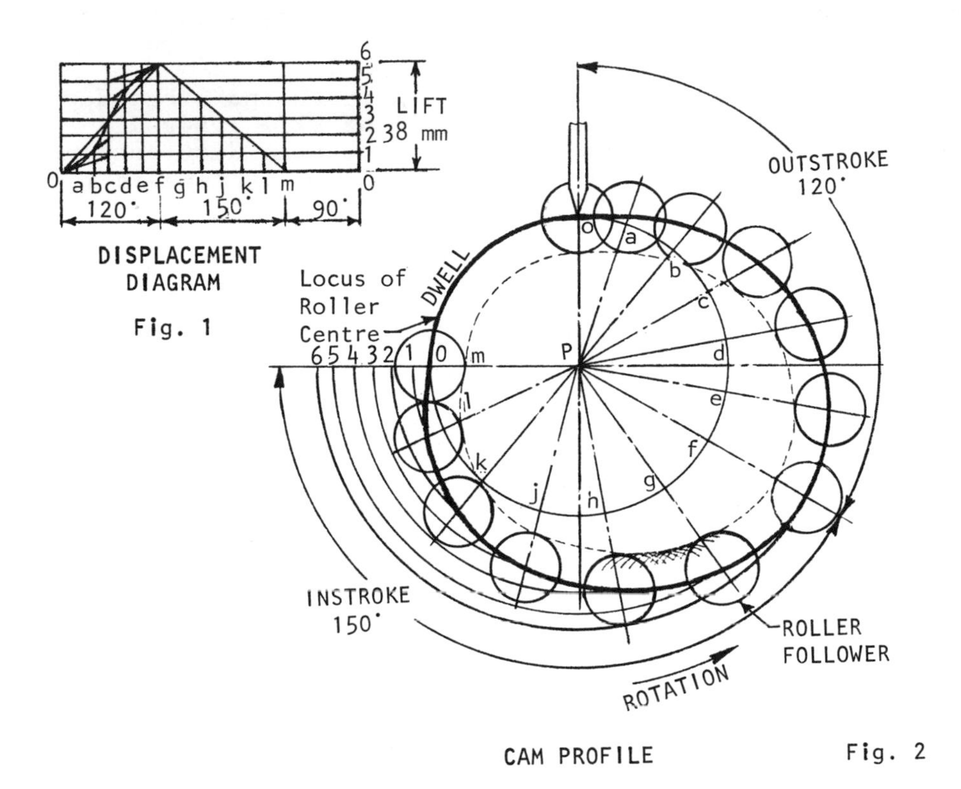

Draw the cam center lines and from their intersection P, set off the minimum radius of the cam PO.

Mark off the 120° of outstroke and 150° of instroke and divide each into six equal parts, labeling as shown.

Using dividers or compass, pick up increments of lift from the base of the displacement diagram to the curve, at points a, b, c, etc., and set off along the dividing arms from points a, b, c, etc., marked around the minimum radius circle.

Join the points thus obtained with a smooth curve, to give the cam profile for the knife-edge follower.

An alternative method of construction for the uniform velocity case is shown on the left-hand side of the cam. The instroke is divided into six equal parts and numbered 0, 1, 2, etc. These points are transferred as arcs struck from P, onto the appropriate dividing arms as shown.

The cam profile for the knife-edge follower is shown thick in Fig. 2.

2) When the friction and wear-reducing roller follower is fitted, the minimum radius of the cam now becomes the minimum distance of roller center to cam center and the follower displacement is determined by the displacement of its center. It is, therefore, necessary to plot the path, or locus, of the roller center, which in this example becomes the cam profile for the knife-edge follower.

With compass set at 12mm, the roller radius, draw a series of arcs or complete circles from the points of intersection of the dividing arms and the roller perimeter, thus clearly demonstrating the main reason for plotting the locus of the roller center. The cam profile for the roller follower is shown dotted in Fig. 2.

The greater the number of divisions used, the greater will be the accuracy of the cam profile. This is demonstrated between points h and g, where the number of closely spaced arcs are drawn to give the cam shape.

• PROBLEM 14-10

In the preceding problem, the stroke of the follower was coincident with the center line of the cam.

Draw the cam profiles for the knife-edge follower and roller-follower if, in this case, the follower is offset 19 mm from the cam center.

Solution: As the cam is required to rotate counterclockwise, the follower will be offset to the left to increase smoothness in operation.

Draw a circle of radius PR equal to 19 mm, the amount of offset required. Divide this circle into the required angles of outstroke and instroke as before, and set off the dividing arms as tangents, such that for each angular division turned through as the cam rotates the dividing arms will successively take up the tangential position RO.

Mark off the amounts of lift from the displacement diagram as before and draw a smooth curve through the points obtained to produce the required profile. (Fig. 1.)

Fig. 2 shows the treatment of the same cam for a roller follower where again the original profile becomes

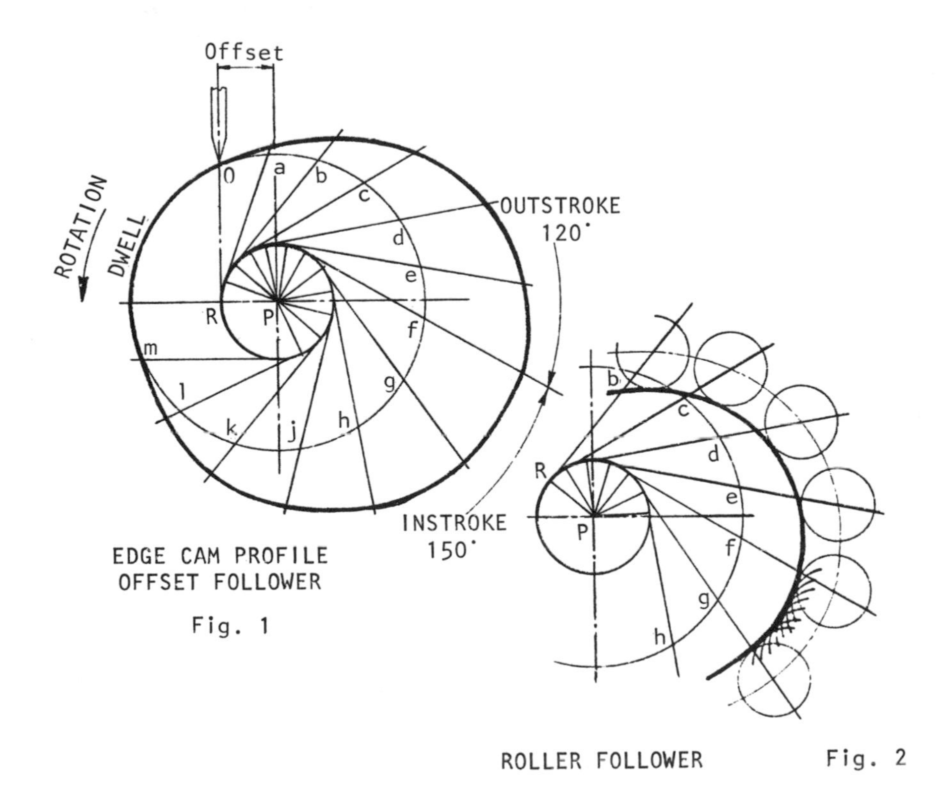

EDGE CAM PROFILE
OFFSET FOLLOWER

Fig. 1

ROLLER FOLLOWER Fig. 2

the locus of the roller center, from which the final cam shape may be obtained.

It will be seen that the use of an offset point follower imposes side loads on the follower guides and also bending of the follower itself, limiting the use of this mechanism.

• PROBLEM 14-11

For the following cam diagram, draw a plate cam profile assuming your own values for rise, base circle diameter, and shaft and roller follower dimensions.

Solution: The point C is the center of the shaft, and A is the lowest and B the highest position of the center of the roller follower. Divide the rise into six parts harmonically proportional. Divide the semicircle ADE into as many equal parts as there are spaces in the rise, and draw radial lines. With C as center and radius C1, draw an arc intersecting the first radial line at 1'.

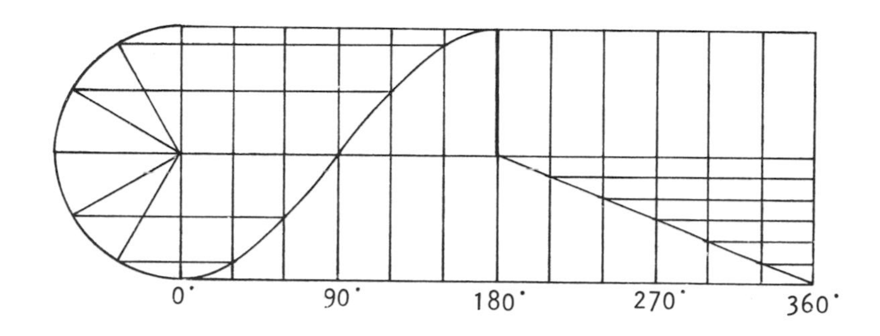

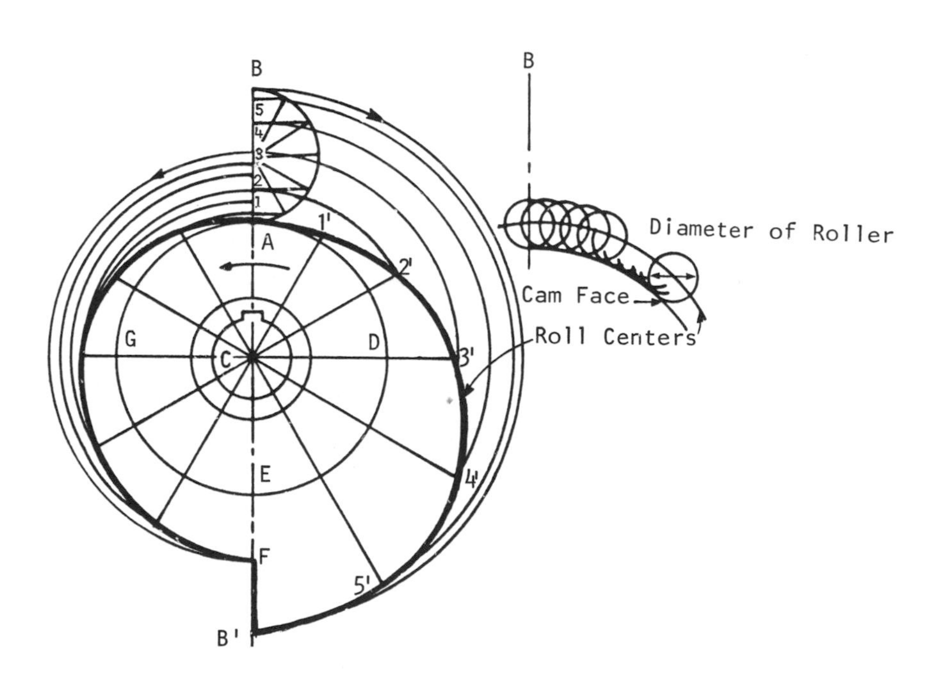

In the same way locate points 2', 3', etc., and draw a smooth curve through them. If the cam is revolved in the direction of the arrow, it will raise the follower with the desired harmonic motion. Draw B'F equal to one-half AB. Divide A3 into six equal parts and the arc EGA into six equal parts. Then for equal angles the follower must fall equal distances. Circle arcs drawn as indicated will locate the required points on the cam outline.

This outline is for the center of the roller; allowance for the roller size can be made by drawing the roller in its successive positions and then drawing a tangent curve as shown in the auxiliary figure.

• PROBLEM 14-12

Given the following cam data, draw the displacement diagram and cam profile.

Cam Data:

(a) Harmonic Motion Cam

(b) Flat Face Follower Oscillating Along Center Line of Cam

(c) Rise 1"

(d) Base Circle 1/2" Radius

(e) Hub 5/8" Diameter

(f) Shaft 3/8" Diameter

(g) Key 1/8" Square

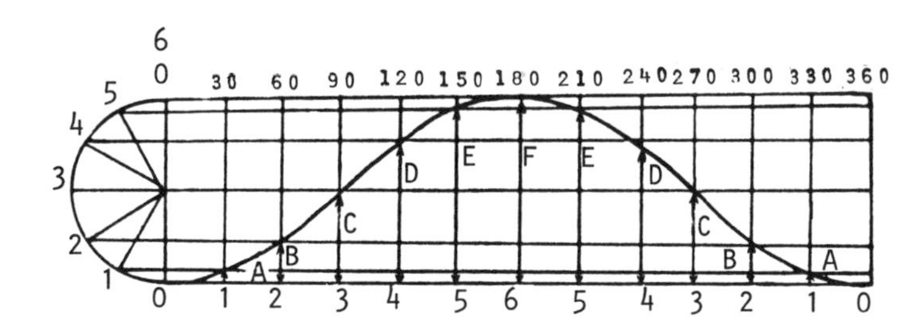

DISPLACEMENT DIAGRAM
STEP 1

Solution: 1. Draw the displacement diagram to the same scale as the cam drawing.

2. Locate the center of the cam shaft.

3. Draw the base circle. Its radius is equal to the distance from the center of the cam shaft to the face of the follower in its lowest position. This point is where the base circle crosses the center line of the follower.

4. Lay out the follower rise distances on the center line of the follower. These distances are found on the displacement diagram. Number in the same order as they are on the displacement diagram.

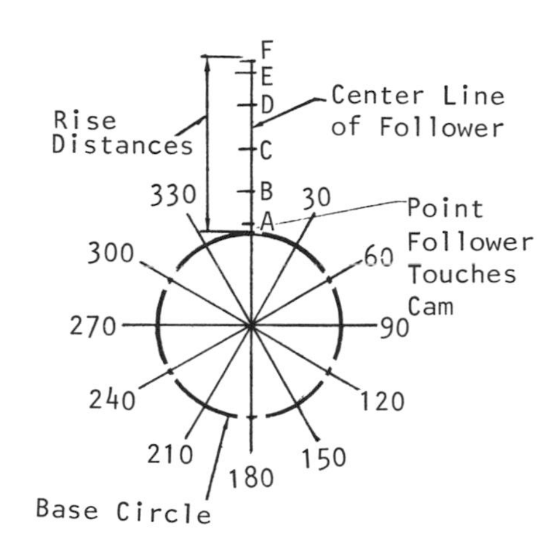

Steps 2-5

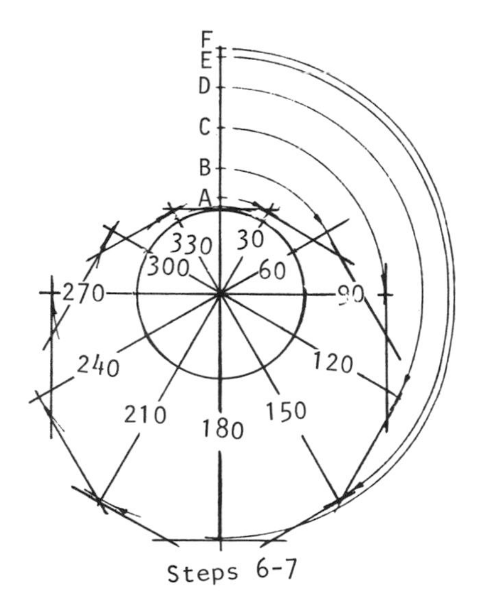

Steps 6-7

5. From the center of the cam shaft, draw lines at the same degrees as used on the displacement diagram. Label these. They are the center lines of the followers at each of these locations.

6. Using the center of the cam shaft, swing arcs from each rise distance until they cross the degree line on which they were measured on the displacement diagram.

7. At each of these points of intersection, draw a line perpendicular to the degree line. These represent the position of the follower at these points.

8. The cam profile is drawn by constructing a smooth curve tangent to these perpendicular lines. Notice that

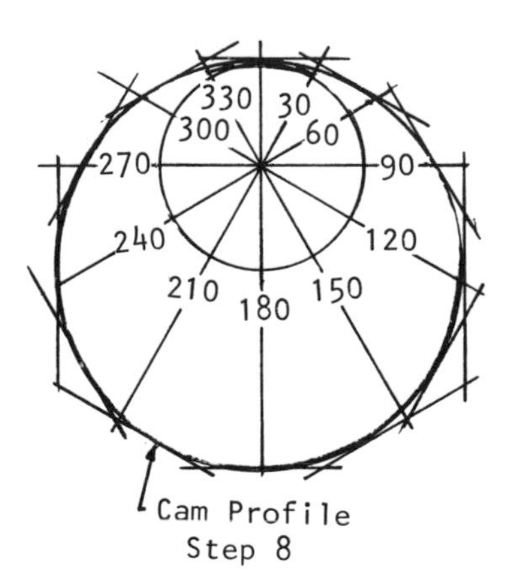

Step 8

the point of contact between the cam and the follower is not always along the center line. This is illustrated by the phantom followers drawn in a number of positions.

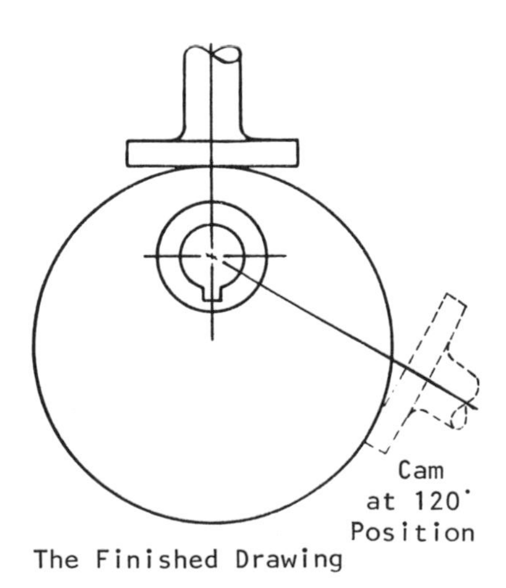

The Finished Drawing

9. Draw the shaft, hub, keyseat, and follower. The depth of the Cam keyslot will be half the height of the key, i.e., 1/16".

Given the following cam data, draw the displacement diagram and cam profile:

Cam Data:

(a) Harmonic Motion Cam

(b) Roller Follower Offset 7/16"

(c) Rise 1"

(d) Base Circle 1/2" Radius

(e) Hub 5/8" Diameter

(f) Shaft 3/8" Diameter

(g) Key 1/8" Square

(h) Follower Roller 3/8" Diameter

Solution: 1. Draw the displacement diagram to the same scale as the cam drawing.

2. Locate the center of the shaft on which the cam will operate.

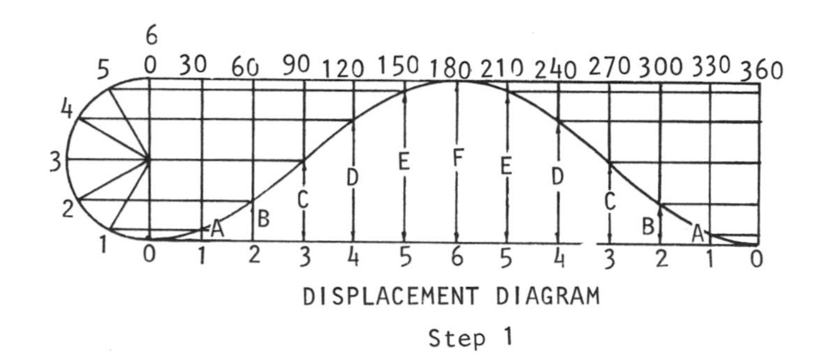

DISPLACEMENT DIAGRAM

Step 1

3. Draw the base circle from the center of the shaft. Its radius is equal to the distance from the center of the shaft to the center of the follower wheel at its lowest position. Where it crosses the center line of the follower is the lowest position. This is point O on the displacement diagram.

4. Draw the offset circle. It is drawn using the center of the shaft. The radius is equal to the offset.

5. Draw the center line of the follower.

6. Divide the offset circle into the same equal divisions used on the displacement diagram.

7. Number the divisions on the offset circle, beginning with zero and in a direction opposite the direction of the cam rotation. The 0 point is where the center line of the follower is tangent with the offset

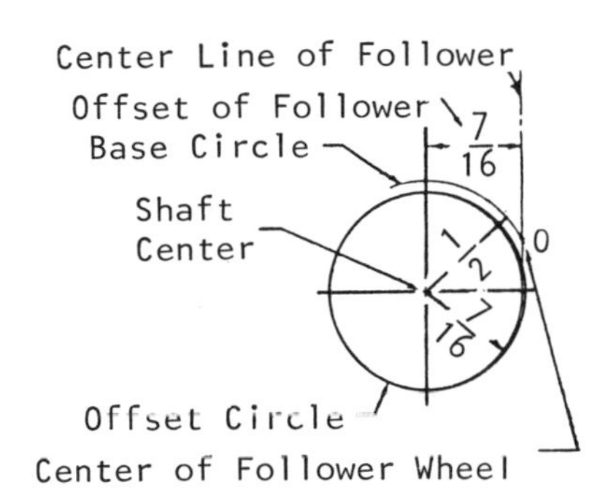

Steps 2-5

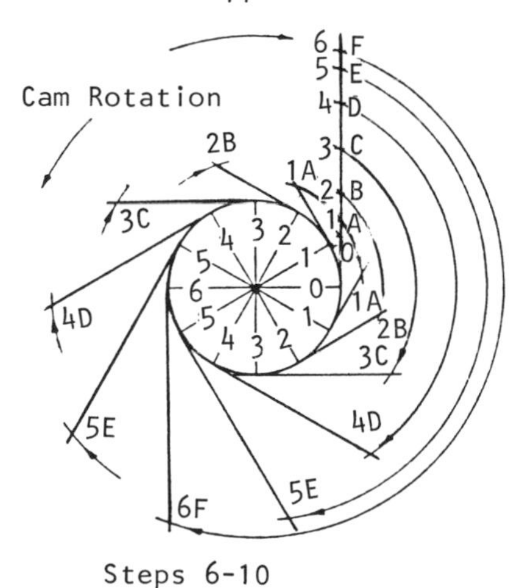

Steps 6-10

circle. Number each division the same as on the displacement diagram.

8. Draw lines through each point on the offset circle tangent to the circle. Number these the same as the points on the circle.

9. Lay out the rise distances for each point on the center line of the follower. These begin at the center

of the follower wheel. These distances are taken from the displacement diagram. Number each point as on the displacement diagram. These positions are the rise of the follower as the cam rotates.

10. Set a compass with a radius equal to the distance from the center of the cam shaft to point 1 on the center line of the follower. Swing an arc until this crosses tangent line 1. Repeat this for each point on the follower center line. This locates the center of the follower wheel in various positions around the cam.

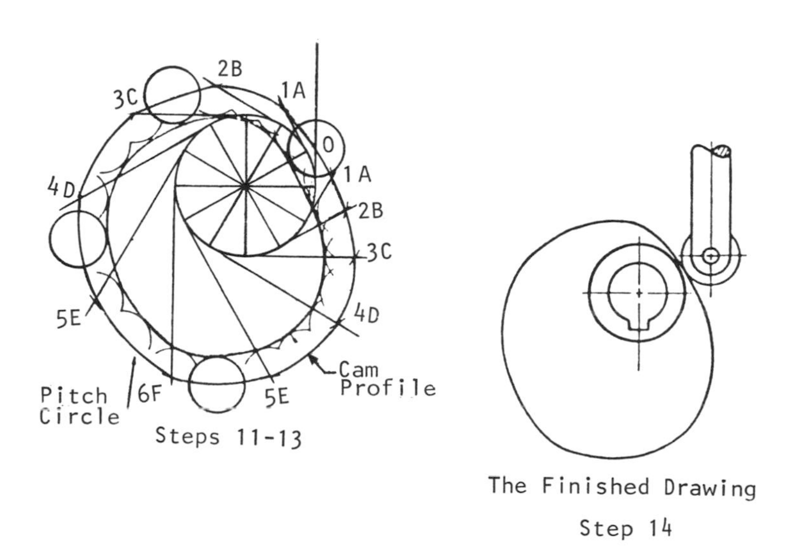

HOW TO DRAW A DISC CAM PROFILE WITH AN OFFSET FOLLOWER.

11. Connect these points with a smooth curve. This forms the pitch curve.

12. Set a compass to the radius of the follower wheel. Using the pitch curve as a center, swing a series of arcs. These arcs are tangent to the working face of the cam. The more arcs used, the more accurate the cam profile drawn.

13. Draw a smooth curve tangent to the cam wheel arcs. This forms the finished cam profile.

14. Draw hub, keyseat, and follower.

A follower for a cylindrical cam is required to travel with simple harmonic motion from datum zero to a maximum rise in 100° of cam rotation, dwell for 80° and return with simple harmonic motion to datum zero in the remaining 180° of cam rotation. Assuming your own values for the rise, base circle diameter and slot dimensions, draw a displacement diagram and suggest a layout for the cylindrical cam.

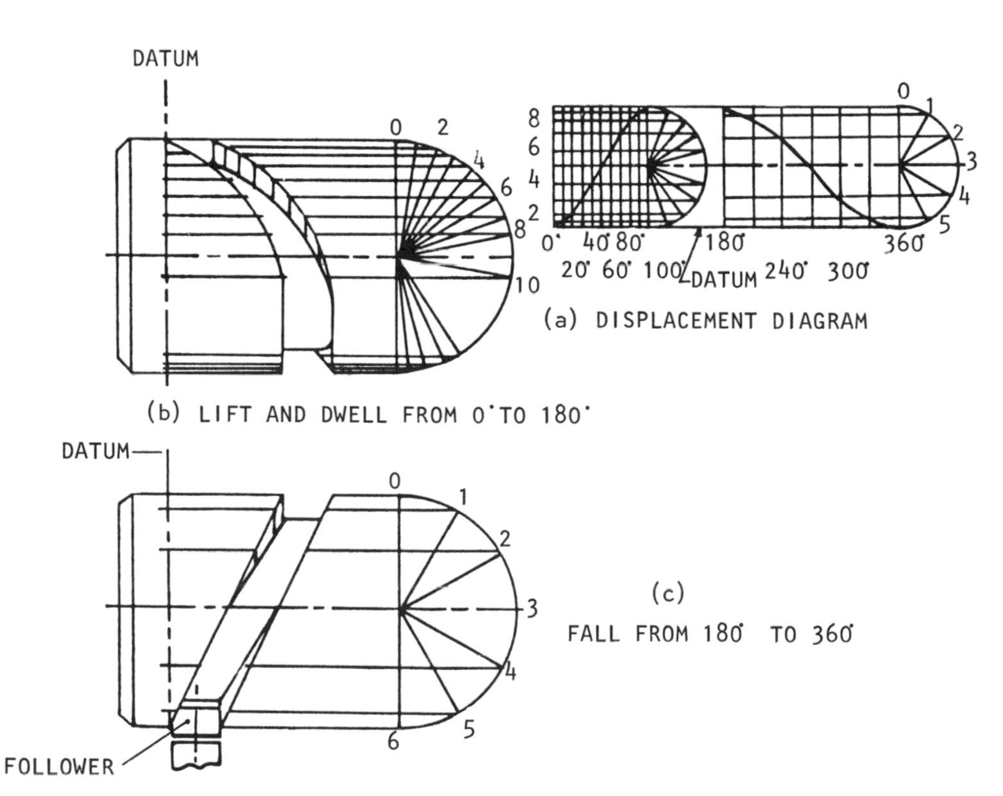

(a) DISPLACEMENT DIAGRAM

(b) LIFT AND DWELL FROM 0° TO 180°

(c) FALL FROM 180° TO 360°

Solution: Draw the displacement diagram, as in Fig. (a), to the same scale as the cam drawing.

Fig. (b) is drawn to cater for the first 180° of cam rotation. A datum line, representing the base line of the displacement diagram, is drawn at a convenient point on the cam surface, and a half end elevation is produced, divided to suit the conditions of the initial 180° of rotation. These divisions, numbered 0, 2, 4, 8, 8, and 10, are projected horizontally along the surface of the cam cylinder to intersect with the datum line. The 'lift' and dwell distances are transferred from the displacement diagram to the appropriate angular divisions on the cam surface, measured from the datum in both cases. The points produced on the cam will, when joined, give the locus of the follower center about which the follower diameter is drawn to produce the outer edges of the cam slot. When the depth and shape of the follower, or the required slot, are known, it is a straightforward task to draw the remainder of the cam slot. In the figure, the follower is assumed to be cylindrical and the depth of the slot is produced by dropping vertical lines from the intersections of the cam outer edges and angular displacement lines, equal in length to the slot depth. The dwell section is parallel to the datum line. A small part of the return section is also shown as the slot curves back towards the datum zero.

In Fig. (c) the 180° return is shown projected. This view is not a direct plan view of (b). The method of projection and the construction of the curves of the slot are as described above.

• PROBLEM 14-15

Lay out the shape of a disk cam to satisfy the following specifications:

Base circle diameter = 3 inches.
Roller follower diameter = 3/4 inch.
No offset.
Rise of follower = 1½ inches, with constantly accelerated and decelerated motion in first half of revolution of the cam.
Fall of follower = 1½ inches, with simple harmonic motion in second half of revolution of the cam.
Cam rotation--counterclockwise.

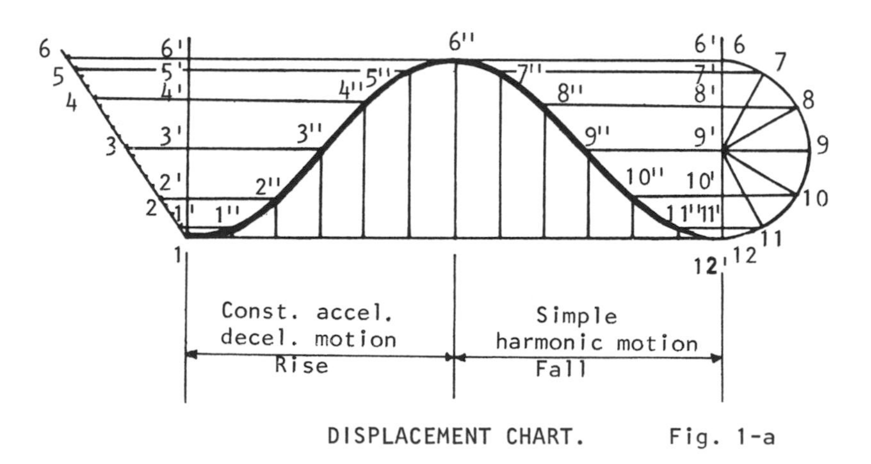

DISPLACEMENT CHART. Fig. 1-a

Solution: Figure 1-a shows the displacement chart. The displacement chart is now oriented in a convenient position in relation to the center of the roller at its lowest point.

See Fig. 1-b. On the vertical line drawn through the center of the roller, point C, the center of rotation of the cam is located at a distance of 1 7/8 inches (the sum of the radii of the base circle and the roller) below the center of the roller. Points 1", 2", etc., are projected horizontally to the axis of the follower. Now an arc is drawn with radius C-1 and center C, intersecting radial line n and locating

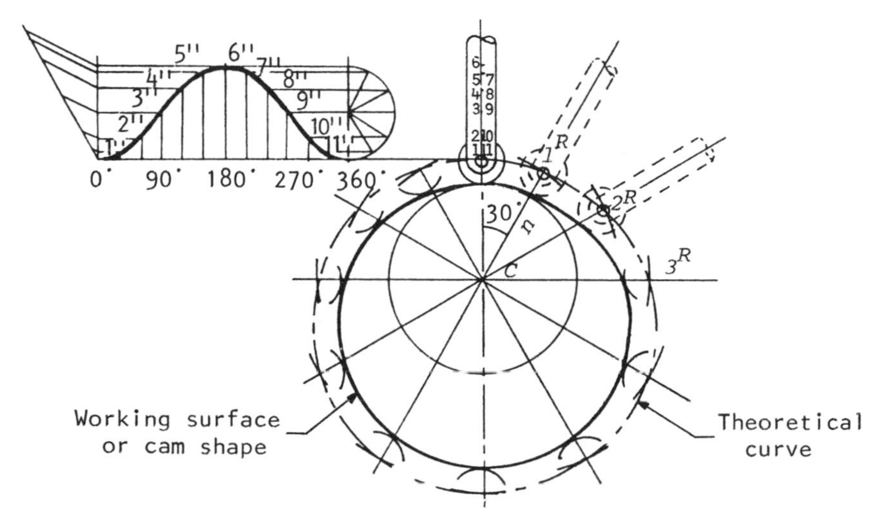

CAM SHAPE LAYOUT FOR ROLLER FOLLOWER. Fig. 1-b

point 1^R, which is the center of the roller if we suppose that the cam remains stationary and the follower moves clockwise (the cam actually moves counterclockwise and the follower remains in a vertical position). In a similar manner points 2^R, 3^R, etc., are located. The smooth curve passing through these points is the "theoretical curve," whereas the curve which is tangent to the roller positions is the "working surface" or the actual cam shape layout.

• PROBLEM 14-16

Lay out the shape of a disk cam to satisfy the following data:

Base circle diameter = 3 inches.
Flat follower.
No offset.
Rise = 1½ inches, with harmonic motion during the interval 0° to 90° of cam rotation.
Dwell during interval 90° to 180°.
Fall = 1½ inches, with constantly accelerated and decelerated motion--180° to 360°.
Cam rotation is clockwise.

Solution: The displacement chart is oriented as shown in the figure. Points 1", 2", etc., are again projected horizontally to the axis of the follower.

Since the base circle is 3 inches in diameter, point C, the center of rotation of the cam, is $1\frac{1}{2}$ inches below the lower surface of the follower. Point 1^R, for example, is located at the intersection of radial line m, with the arc having a radius equal to C-1 and having the center C. In a similar manner point 2^R is located at the intersection of radial line n and the arc of radius C-2; likewise for points 3^R, 4^R, etc.

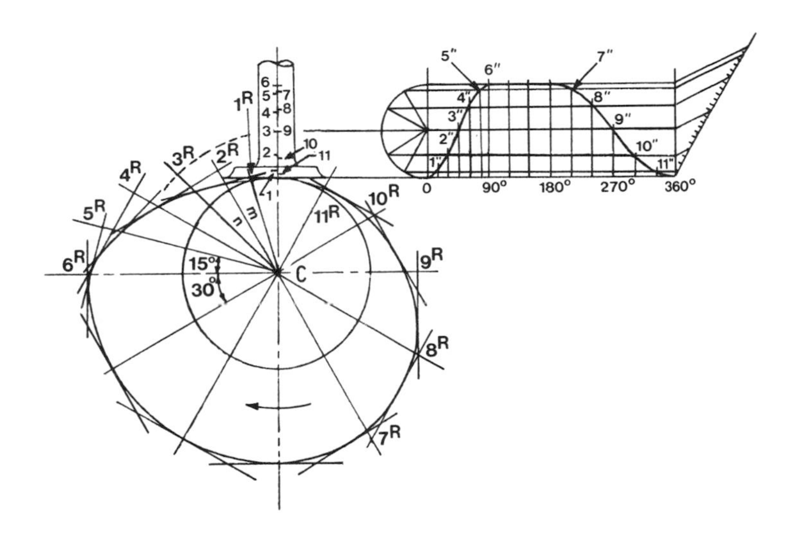

CAM SHAPE LAYOUT, FLAT FOLLOWER

Since the follower has a flat face, the cam surface will have to be tangent to the follower throughout the cam rotation. To accomplish this, we must first establish lines through points 1^R, 2^R, etc., that are respectively perpendicular to the corresponding radial lines and thus represent the flat face of the follower for different cam positions. These are shown in the figure. The smooth curve drawn tangent to these lines establishes the cam shape layout.

Lay out the shape of a disk cam to satisfy the following specifications:

Base circle diameter = 4 inches.
Knife-edge follower.
Offset of follower = 1 inch to the right of cam axis.
Rise = 1½ inches, with simple harmonic motion during the interval 0° to 180° of cam rotation.
Dwell during interval 180° to 270°.
Fall = 1½ inches, with constant acceleration and deceleration -270° to 360°.
Cam rotation is counterclockwise (when offset of follower is to the right, rotation of cam should be counterclockwise in order to reduce horizontal thrust).

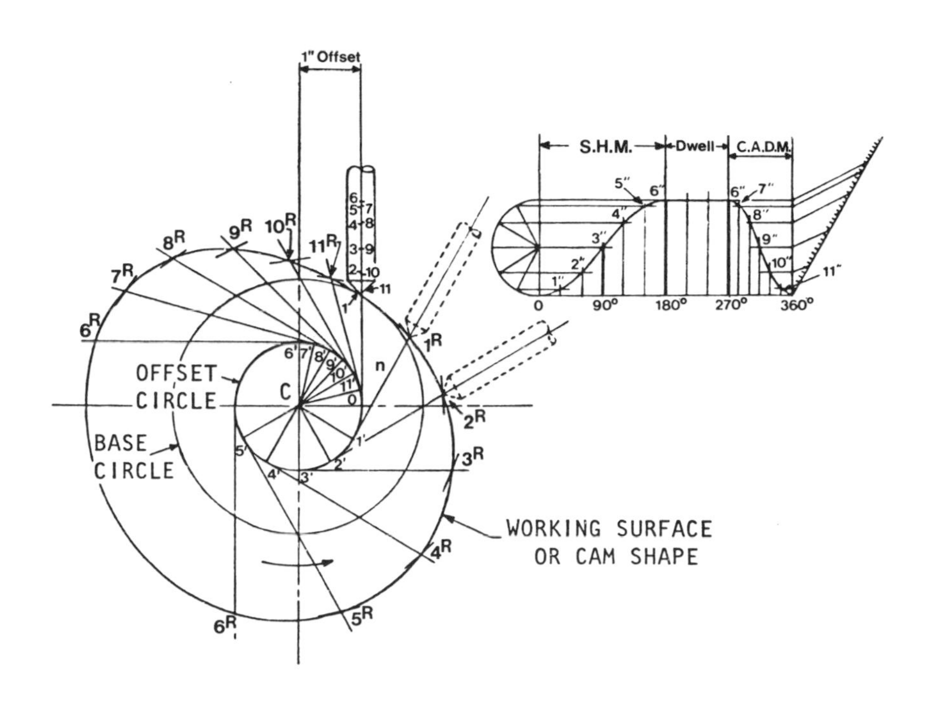

CAM SHAPE LAYOUT, OFFSET KNIFE-EDGE FOLLOWER

Solution: The displacement chart, which is oriented in a convenient position, is shown in the figure. With point C as the center of the two circles drawn, one the base circle, radius 2 inches, and the other the offset circle, radius 1 inch. The center line of the follower is 1 inch to the right of the cam axis. The intersection of the follower axis with the base circle is the lowest point on the cam. Points on the displacement chart are projected horizontally to the follower axis. Now let us suppose that the cam remains fixed and that the follower moves about the cam in a clockwise direction (since the cam rotation is counterclockwise).

To locate point 1^R on the cam surface, fist lay off a 30° angle on the offset circle and locate point 1' on the circumference. A tangent n to the offset circle drawn through point 1' is the center line of the follower. Now with center C and radius C-1, an arc is drawn to intersect line n at point 1^R. In a similar manner points 2^R, 3^R, etc., are located. The smooth curve drawn through these points is the working surface, or cam shape layout. It should be noted that the tangent to the offset circle serves to maintain the 1-inch offset.

• **PROBLEM 14-18**

The following cam-and-follower displacement data are given. It is required to determine (a) the velocity-time relationship and (b) the acceleration-time relationship, assuming that the constant rotational speed of the cam is 125 rpm.

S, follower displacement, in inches	0	0.1	0.5	1.0	1.5	1.9	2.0	2.0	2.0	2.0	1.5	0.5	0
θ, cam displacement, in degrees	0	30	60	90	120	150	180	210	240	270	300	330	360

Solution: Before solving the problem, the method of graphical differentiation is explained. Curves of a lower order are derived through the application of the principle that the ordinate at any point on the derived curve is equal to the slope of a tangent line at the corresponding point on the given curve. The slope of a curve at a point is the tangent of the angle with the X-axis formed by the tangent to the curve at the point. For all practical purposes, when constructing a derivative curve, the slope may be taken as the rise of the tangent line parallel to the Y-axis in one unit of distance along the X-axis, or the slope of the tangent equals y_1/k as shown in Fig. 1.

Figure 1 illustrates the application of this principle of graphical differentiation. The length of the ordinate y_1' at point A' on the derived curve is equal to the slope y_1/k at point A on the given curve as shown in (a). When the slope is zero as at point C, the length of the ordinate is zero and point C' lies on the X-axis for the derived curve. When the slope is negative, as shown at D, the ordinate is negative and lies below the X-axis.

A plot of the data (with θ converted to time in seconds) is shown in Fig. 2a. It should be observed that since the rotational speed is 125 rpm, one revolution of the cam takes

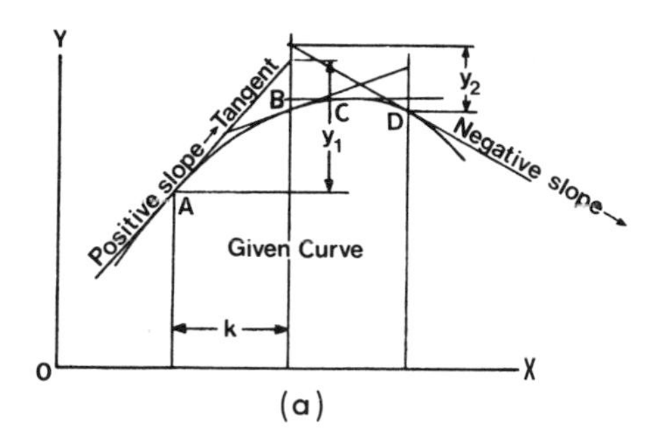

(a)

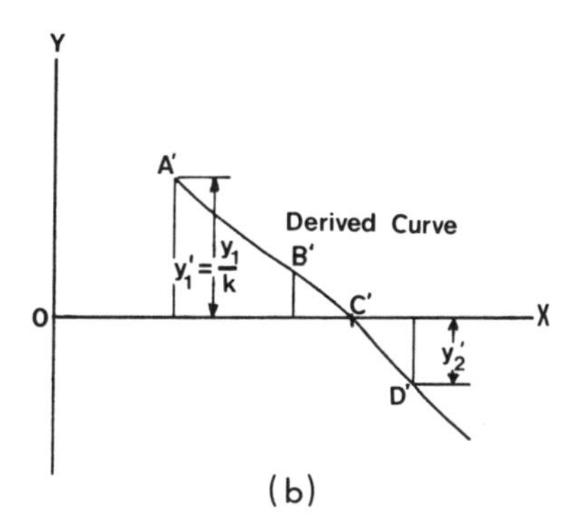

(b)

Fig. 1

place in 0.48 second. The time scale, t, is based upon this value.

The displacement-time curve is differentiated graphically as explained above to obtain the velocity-time curve shown in Fig. 2b. The latter curve is then differentiated to obtain the acceleration-time curve shown in Fig. 2C.

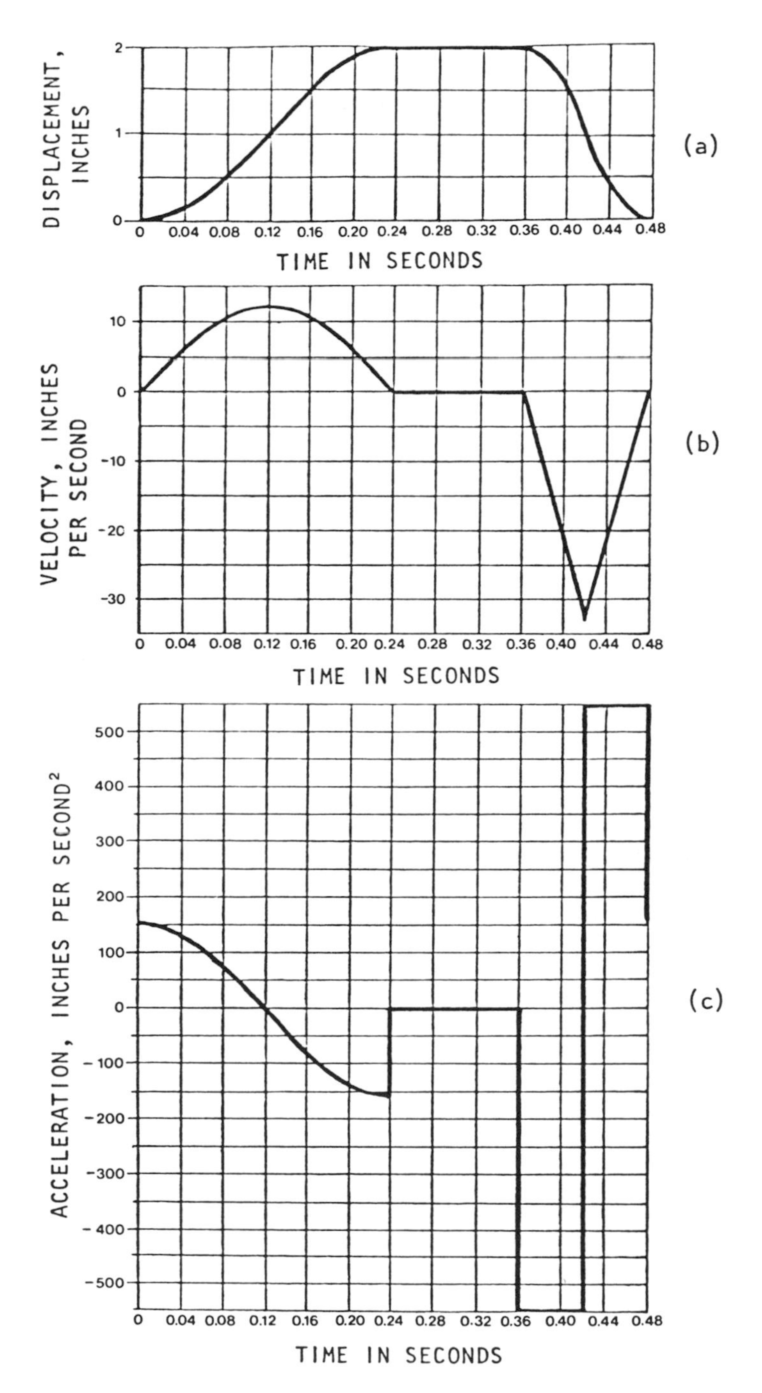

CAM DISPLACEMENT, VELOCITY, AND ACCELERATION Fig. 2

GEAR TERMINOLOGY

• PROBLEM 14-19

Explain the following gear terminology:

1. Pitch Circle
2. Addendum Circle
3. Root Circle
4. Base Circle
5. Pressure Angle
6. Chordal Addendum
7. Face Width

Solution: 1) Pitch Circle: An imaginary circle running through the point on the teeth at which the teeth from mating gears are tangent. The size of the gear is indicated by the pitch circle. For example, a six-inch gear has a pitch circle of six inches. Most of the gear dimensions are taken from the pitch circle.

2) Addendum Circle: A circle formed by the outside surface of the teeth. The diameter is found by adding twice the addendum to the pitch circle. This is sometimes called the outside diameter.

3) Root Circle: A circle passing through the bottom of the teeth.

4) Base Circle: A circle from which the tooth profile is generated. It is found by drawing a 14.1/2° or 20° line through the pitch point. The angle used depends upon the gear design.

5) Pressure Angle: The angle used to determine the tooth shape. It is the angle at which pressure from the tooth of one gear is passed to the tooth of another gear. Two angles are in general use. These are 14.1/2° and 20°. The 20° angle is more commonly used as it has smoother and quieter running characteristics, and a greater load-carrying ability.

6) Chordal Addendum: The distance from the top of the tooth to the chordal thickness.

7) Face Width. The width of the tooth measured from one face to the other.

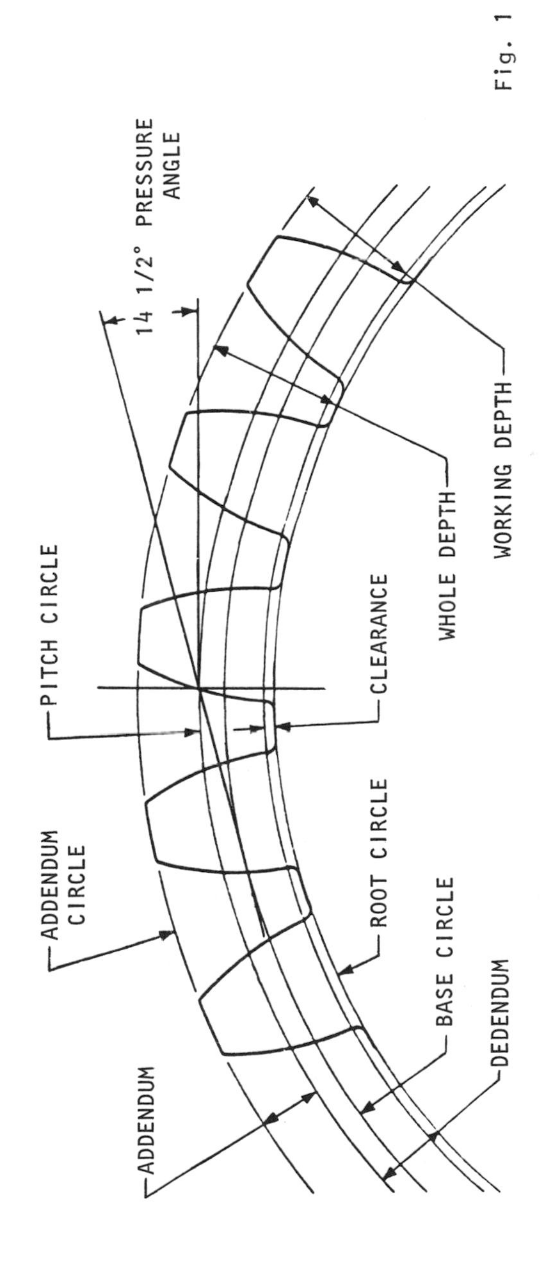
14 1/2° PRESSURE ANGLE
PITCH CIRCLE
ADDENDUM CIRCLE
ADDENDUM
CLEARANCE
ROOT CIRCLE
BASE CIRCLE
DEDENDUM
WHOLE DEPTH
WORKING DEPTH

Fig. 1

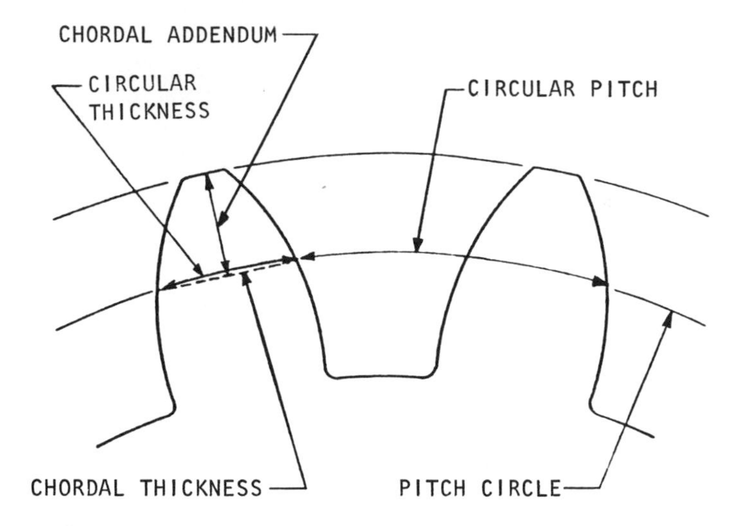

Fig. 2

GEAR APPLICATIONS

• PROBLEM 14-20

Define some of the more common terms and their symbols in the American Standard nomenclature for spur gearing.

Solution: The names of the various portions of a spur gear and the teeth are given below. The ANSI standard letter symbols and formulas for calculation are as follows:

N = number of teeth = $P_d \times D$

P_d = diametral pitch = number of teeth on the gear for each inch of pitch diameter = N/D

D = pitch diameter = N/P_d

p = circular pitch = length of the arc of the pitch-diameter circle subtended by a tooth and a tooth space = $\pi D/N = \pi/P_d$

t = circular (tooth) thickness = length of the arc of

the pitch-diameter circle subtended by a tooth = $p/2 = \pi D/2N = \pi/2P_d$

t_c = chordal thickness = length of the chord subtended by the circular thickness arc = D sin (90°/N)

a = addendum = radial distance between the pitch-diameter circle and the top of a tooth = a constant/P_d = for standard 14½° or 20° involute teeth, $1/P_d$

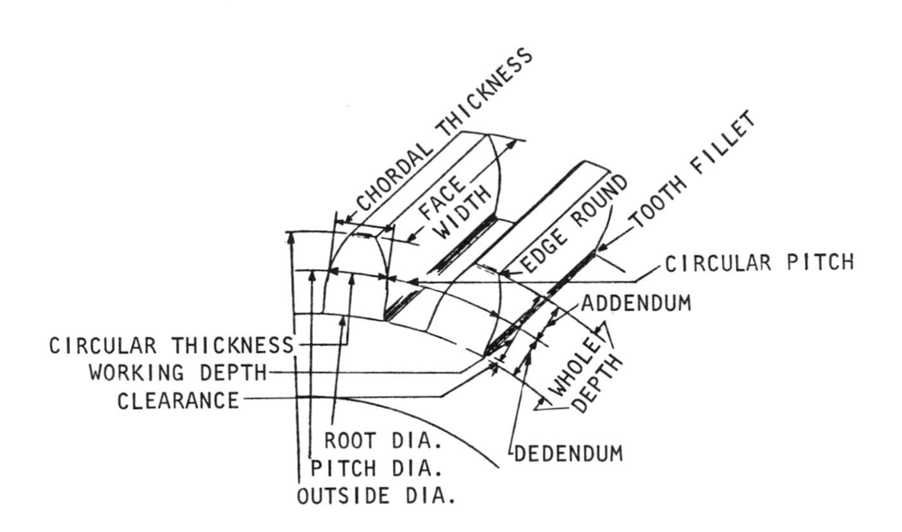

GEAR NOMENCLATURE

b = dedendum = radial distance between the pitch-diameter circle and the bottom of a tooth space = a constant/P_d = for standard 14½° or 20° involute teeth, $1.157/P_d$

c = clearance = radial distance between the top of a tooth and the bottom of a mating tooth space = a constant/P_d = for standard 14½° or 20° involute teeth, $0.157/P_d$

h_t = whole depth = radial distance between the top and bottom of a tooth = a + b = for standard 14½° or 20° involute teeth, $2.157/P_d$

h_k = working depth = greatest depth a tooth of one gear extends into a tooth space of a mating gear = 2a = for standard 14½° or 20° involute teeth, $2/P_d$

M = module = pitch diameter divided by number of teeth = D/N

D_o = outside diameter = diameter of the circle containing the top surfaces of the teeth = $D + 2a = (N + 2)/P_d$

D_R = root diameter = diameter of the circle containing the bottom surfaces of the tooth spaces = $D - 2b = (N - 2.314)/P_d$

F = face width = width of the tooth flank

f = tooth fillet = fillet joining the tooth flank and the bottom of the tooth space = $0.157/P_d$ max

r = edge round = radius of the circumferential edge of a gear tooth (to break the sharp corner)

n = revolutions per unit of time

In addition to the above basic letter symbols, subscripts G and P are used to denote gear and pinion, respectively. (When one of two mating gears is smaller than the other, it is known as the pinion.)

m = gear ratio = m_G for the gear = $N_G/N_p = n_p/n_G = D_G/D_p$

m = gear ratio = m_p for the pinion = $N_p/N_G = n_G/n_p = D_p/D_G$

(The pitch diameter and number of teeth are inversely proportional to speed.)

• PROBLEM 14-21

Describe the approximate circle-arc method for drawing the teeth of a standard involute-toothed spur gear.

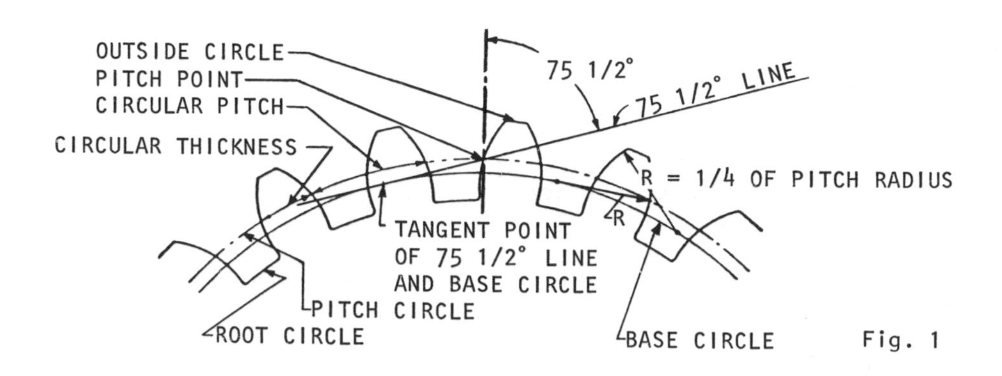

Fig. 1

Solution: To draw the teeth of a standard involute-toothed spur gear by an approximate circle-arc method, lay off the pitch circle, root circle, and outside circle. Start with the pitch point and divide the pitch circle into distances equal to the circular thickness. Through the pitch point draw a line of 75½° with the center line (for convenience the draftsman uses 75°). Draw the base circle tangent to the 75° line. With compasses set to a radius equal to one-fourth the radius of the pitch circle, describe arcs through

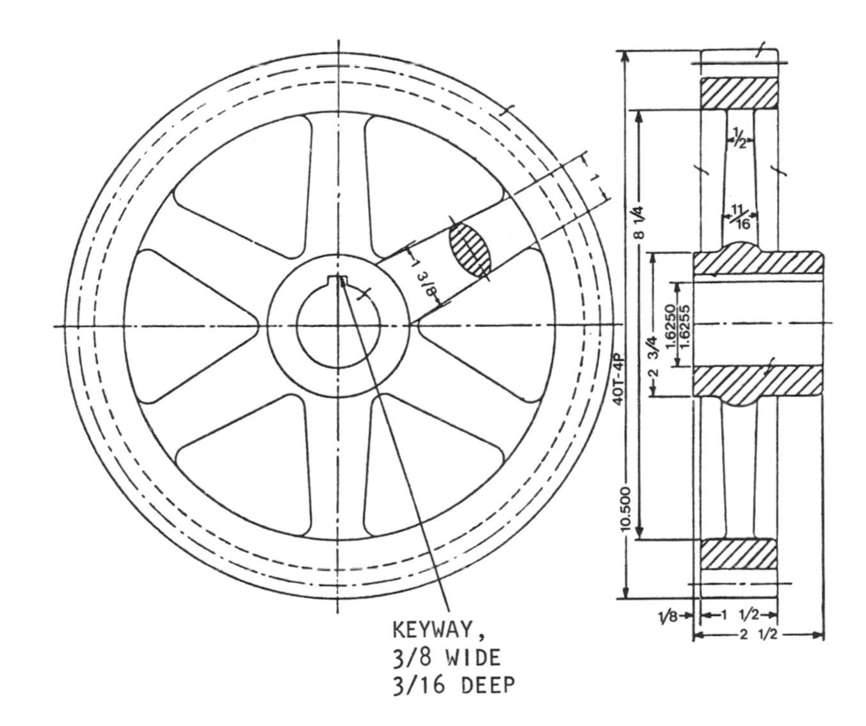

Fig. 2

the division points on the pitch circle, keeping the needle point on the base circle. Darken the arcs for the tops of the teeth and bottoms of the spaces and add the tooth fillets. For 16 or fewer teeth the radius value of one-fourth pitch radius must be increased to suit, in order to avoid the appearance of excessive undercut. For stub teeth the 75½° line is changed to 70°.

This method of drawing gear teeth is useful on display drawings, but on working drawings the tooth outlines are not drawn. Figure 2 illustrates the method of indicating the teeth and dimensioning a working drawing of a spur gear.

The dimensions necessary for the teeth of cut gears are: outside diameter, number of teeth, diametral pitch, and width of face.

Suppose that an order has been received as follows:

Required: Spur gear—steel.
$14\frac{1}{2}°$ pressure angle.
5 diametral pitch, 20 teeth.
Plain blank—2-inch face, ½-inch hub projection one side.
1¼-inch bore diameter—keyway and set screw.

Draw one sectional view of the spur gear to provide all the necessary information.

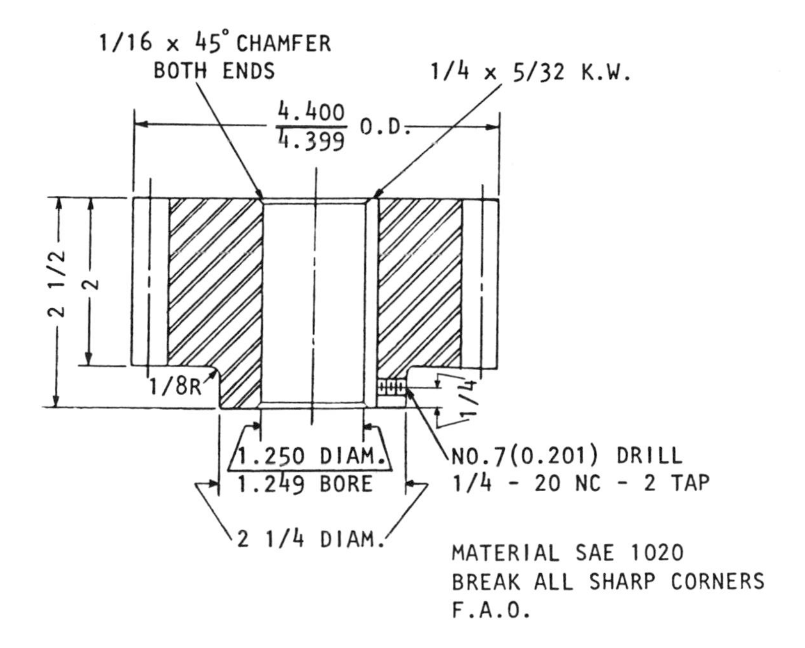

Solution: For the spur gear, it is necessary to compute gear dimensions from formulas using letter symbols which have been standardized (ASA B6.10-1950). Symbols, terms, and definitions, and useful formulas for spur gears are shown in the table below.

Spur Gear Formulas, Terms, and Definitions:

Term and Definition	Symbol	Formula
Diametral Pitch. The number of teeth per inch of pitch diameter .	P_d	$P_d = \frac{N}{D}$ (where N is number of teeth)

Circular Pitch. The distance along the pitch circle between corresponding profiles of adjacent teeth.	P	$p = \frac{\pi}{P_d} = \frac{\pi D}{N}$
Pitch Diameter. The diameter of the pitch circle.	D	$D = \frac{N}{P_d}$
Addendum. The height by which a tooth projects beyond the pitch circle.	a	$a = \frac{1}{P_d}$ (full-depth teeth) $a = \frac{0.8}{P_d}$ (for stub teeth)
Dedendum. Depth of tooth space below the pitch circle.	b	$b = \frac{1.157}{P_d}$ (full-depth teeth) $b = \frac{1}{P_d}$ (for stub teeth)
Whole Depth. Total depth of tooth space.	h_t	$h_t = a + b$
Outside Diameter. Diameter of the addendum circle.	D_o	$D_o = D + 2a$
Circular Thickness. The length of arc between the two sides of a gear tooth, on the pitch circle.	t	$t = \frac{P}{2}$
Chordal Thickness.* Length of chord subtending a circular-thickness arc.	t_c	$t_c = D \sin\left(\frac{90°}{N}\right)$
Chordal Addendum.* The height from the top of the tooth to the chord subtending the circular-thickness arc.	a_c	$a_c = a + \frac{D}{2}\left[1 - \cos\left(\frac{90°}{N}\right)\right]$
Center Distance. Distance between axes of mating gears.	C	$C = \frac{D_G + D_p}{2}$ (for external gears)

*From standard tables.

Before this gear can be made it will be necessary to determine the complete information for the machinist to cut the blank and for the operator of the gear-cutting machine to cut the teeth.

First, we shall make the following calculations, which are based on the above formulas.

(1) $D = \frac{N}{P_d} = \frac{20}{5} = 4.000$ inches

(2) $a = \frac{1}{P_d} = \frac{1}{5} = 0.200$ inch

(3) $D_o = d + 2a$

$= 4.000 + 2(0.200) = 4.400$ inches

(4) a_c = (from Table) = $1.0308 \div 5 = 0.2602$ inch

(5) t_c = (from Table) = $1.5692 \div 5 = 0.3138$ inch

(6) Keyway dimensions, $1/4 \times 5/32$ (taken from Standard tables) Use 1/4-inch set screw.

This information is used in preparing the working drawing, see the figure. It should be noted that one sectional view provides all the necessary information; therefore, the circular view is omitted. This represents good practice. The gear-cutting data are given in the table and are not repeated in the drawing dimensions.

• PROBLEM 14-23

Given N and D_o. Find P_d.

Solution: $D_o = D + 2a$

Substitute in terms of P_d,

$$D_o = \frac{N}{P_d} + \frac{2}{P_d}$$

$$= \frac{N + 2}{P_d}$$

Therefore,

$$P_d = \frac{N + 2}{OD}$$

• PROBLEM 14-24

Calculate D_p, D_G, N_p, N_G, D_{op} and D_{oG} for a standard 20° involute teeth spur gear, given:

a) Center distance = 7"

b) n_G = 500 rpm

n_p = 1,500 rpm

c) P_d = 6 .

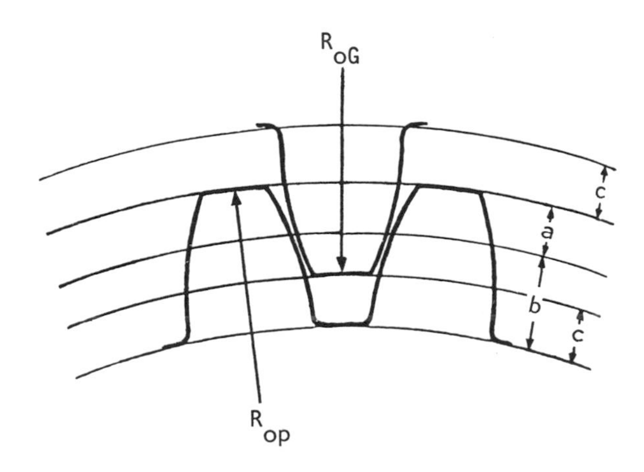

<u>Solution</u>:

$$m_G = \frac{n_p}{n_G}$$

$$= \frac{1500}{500} = \frac{3}{1}$$

$$= \frac{D_G}{D_p}$$

Therefore, $3D_p = D_G$

From the center distance,

$$D_p + D_G = 14"$$

Substituting for D_G in terms of D_p,

$$D_p + 3D_p = 14.$$

Solving $D_p = 3.50"$

Solving for $D_G = 3D_p$

$$= 3 \times 3.50 = 10.50"$$

Now, $N_p = P_d \times D_p$

$$= 6 \times 3.50 = 21$$

Similarly, $N_G = P_d \times D_G$

$$= 6 \times 10.50 = 63.$$

Now $D_{op} = (N_p + 2)/P_d$

$$= (21 + 2)/6$$

$$= 23/6 = 3.83"$$

$$D_{oG} = (N_G + 2)/P_d$$

$$= (63 + 2)/6$$

$$= 65/6 = 10.83"$$

These answers can be checked for accuracy:

Center distance =

$$R_{op} - (a_p + b_p) + c + R_{oG}$$

where R_o represents outside radius (half of D_o).

Therefore, Center Distance

$$= \frac{3.83}{2} - \left(\frac{1}{6} + \frac{1.157}{6}\right) + \frac{0.157}{6} + \frac{10.83}{2}$$

(values of a, b and c from definitions)

$$= 1.915 - (0.167 + 0.193)$$

$$+ 0.026 + 5.415$$

$$= 6.996 \approx 7.$$

• PROBLEM 14-25

> It is required to make a helical gear and pinion for a speed-reduction unit. The following sketch and information were submitted with the order:
>
> 20° stub teeth.
> 8 diametral pitch.
> 23° helix angle.
> Reduction ratio is 6:1.
> Pinion--SAE 4140--Integral with shaft.
> Gear--cast steel--20C. 6 spoke blank.

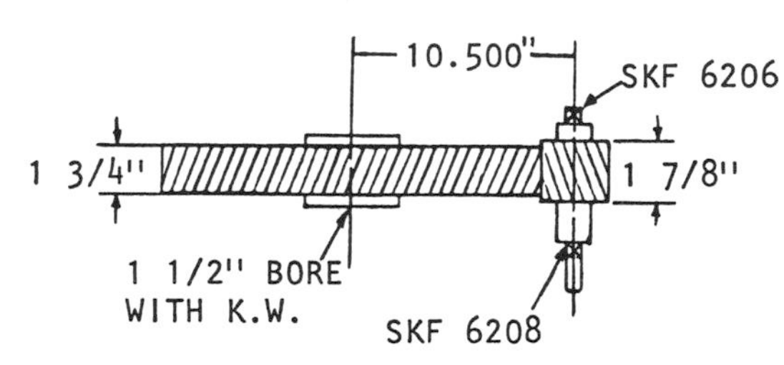

SKETCH FOR HELICAL GEAR AND PINION PROBLEM.

Solution: Before this gear can be made it will be necessary to use the formulas in the following table for helical gears.

Helical Gear Formulas, Terms, and Definitions:

Term and Definition	Symbol	Formula
Diametral Pitch. The number of teeth per inch of pitch diameter.	P_d	$P_d = \frac{N}{D}$ (where N is number of teeth)
Circular Pitch. The distance along the pitch circle between corresponding profiles of adjacent teeth.	p	$p = \frac{\pi}{P_d} = \frac{\pi D}{N}$
Pitch Diameter. The diameter of the pitch circle.	D	$D = \frac{N}{P_d}$
Addendum. The height by which a tooth projects beyond the pitch circle.	a	$a = \frac{1}{P_d}$ (full-depth teeth)
		$a = \frac{0.8}{P_d}$ (for stub teeth)

Term	Symbol	Formula
Dedendum. Depth of tooth space below the pitch circle.	b	$b = \frac{1.157}{P_d}$ (full-depth teeth) $b = \frac{1}{P_d}$ (for stub teeth)
Whole Depth. Total depth of tooth space.	h_t	$h_t = a + b$
Outside Diameter. Diameter of the addendum circle.	D_o	$D_o = D + 2a$
Circular Thickness. The length of arc between the two sides of a gear tooth, on the pitch circle.	t	$t = \frac{p}{2}$
Chordal Thickness.* Length of chord subtending a circular-thickness arc.	t_c	$t_c = D \sin\left(\frac{90°}{N}\right)$
Chordal Addendum.* The height from the top of the tooth to the chord subtending the circular-thickness arc.	a_c	$a_c = a + \frac{D}{2}\left[1 - \cos\left(\frac{90°}{N}\right)\right]$
Center Distance. Distance between axes of mating gears.	C	$C = \frac{D_a + D_p}{2}$ (for external gears)
Normal Diametral Pitch. The diametral pitch as calculated in the normal plane.	P_{nd}	$P_{nd} = \frac{P_d}{\cos\psi}$
Normal Circular Pitch. The circular pitch in the normal plane.	pn	$pn = p \cos\psi$
Lead. The axial advance of a helix for one complete turn, as in the teeth of helical gears and in the threads of cylindrical worms.	l	$l = \pi D \cot\psi$
Helix Angle. The angle between any helix and an element of its cylinder. (See sketch.) It is measured at the pitch diameter in helical gears and in worms.	ψ	

*From Standard table.

The necessary calculations are:

(1) $$C = \frac{D}{2} + \frac{d}{2}$$

(2) $$\text{Speed ratio} = \frac{N_G}{N_p} = \frac{D \times P_d}{d \times P_d} = \frac{D_G}{d}$$

From (1) and (2):

$$2C = d \times \text{speed ratio} + d$$

Therefore,

$$d = \frac{2 \times 10.500}{1 + 6} = 3.000 \text{ inches}$$

Then

$$D = 2\left(C - \frac{d}{2}\right)$$

$$= 2(10.500 - 1.5000) = 18.000 \text{ inches}$$

(3) $N_p = d \times P_d$ $\quad$ $N_G = D \times P_d$

$= 3 \times 8 = 24$ $\quad$ $= 18 \times 8 = 144$

(4) $$P_{nd} = \frac{P_d}{\cos\psi}$$

$$= \frac{8}{\cos 23°}$$

$$= 8.695$$

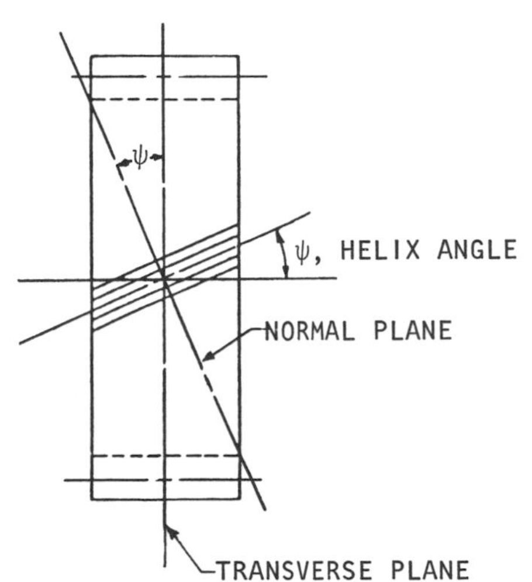

(5) $$a = \frac{0.8}{P_d}$$

$$= 0.100 \text{ inch}$$

(6) $$b = \frac{1}{P_d}$$

$$= 0.125 \text{ inch}$$

(7) $$D_o = D + 2a$$

$$D_o = 18.000 + 2 \times 0.100$$

$$= 18.200 \text{ inches}$$

$$d_o = 3.000 + 2 \times 0.100$$

$$= 3.200 \text{ inches}$$

(8) Chordal thickness* and chordal addendum from table.

(9) The SKF bearing catalogue is used to determine the shaft limits, shoulder dimensions and radius, and length of bearing seat.

*Chordal thickness should be measured on the normal of the tooth.

Summary

Term	Gear	Pinion
Number of teeth	144	24
Helix angle	23°	23°
Hand	Right	Left
Diametral pitch	8	8
Normal diametral pitch	8.695	8.695
Pitch diameter	18.000	3.000
Addendum	0.100	0.100
Dedendum	0.125	0.125
Outside diameter	18.200	3.200
Chordal thickness	0.1807	0.1805
Chordal addendum	0.1006	0.1032

The draftsman is now ready to make the drawings. The following figure shows the finished drawing of the pinion and shaft.

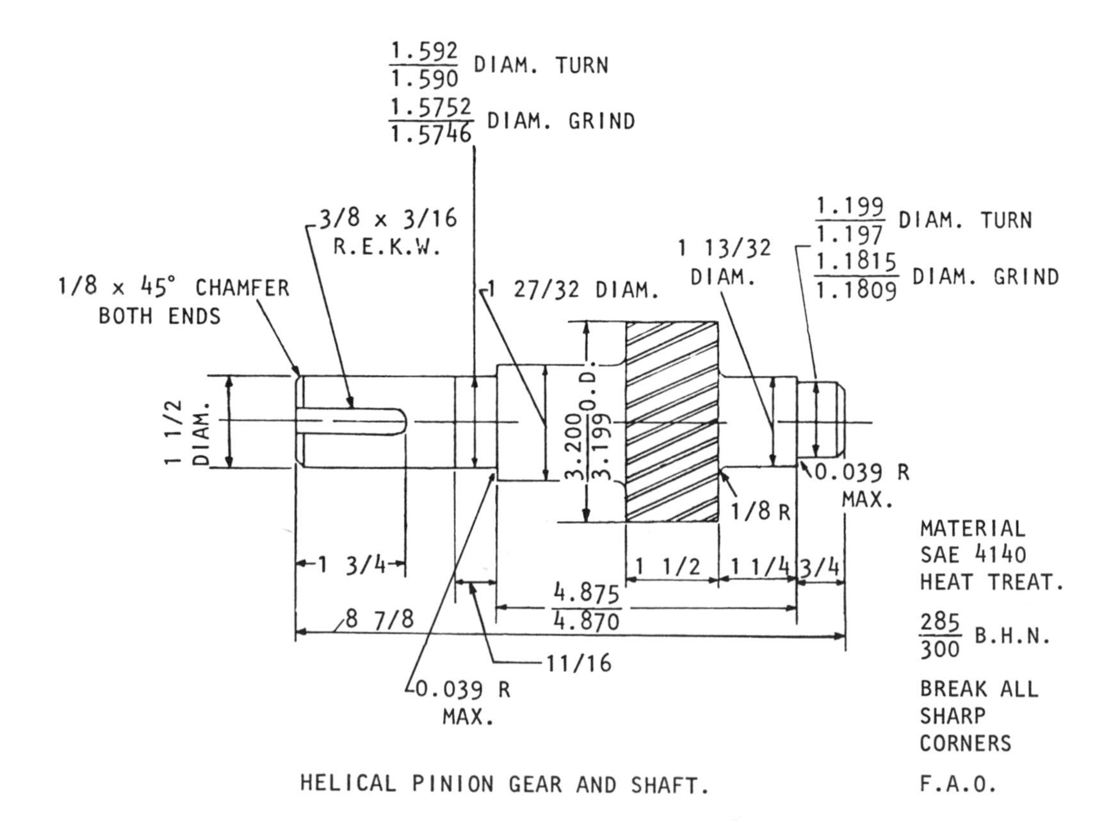

HELICAL PINION GEAR AND SHAFT.

The following data are given for a bevel gear layout:

Diametral pitch, 6;
14½° pressure angle;
1¼-inch face;
Speed ratio is 3:1
18T pinion (smaller gear).

Carry out the calculations to provide necessary information and draw the bevel gear and pinion bevel gear.

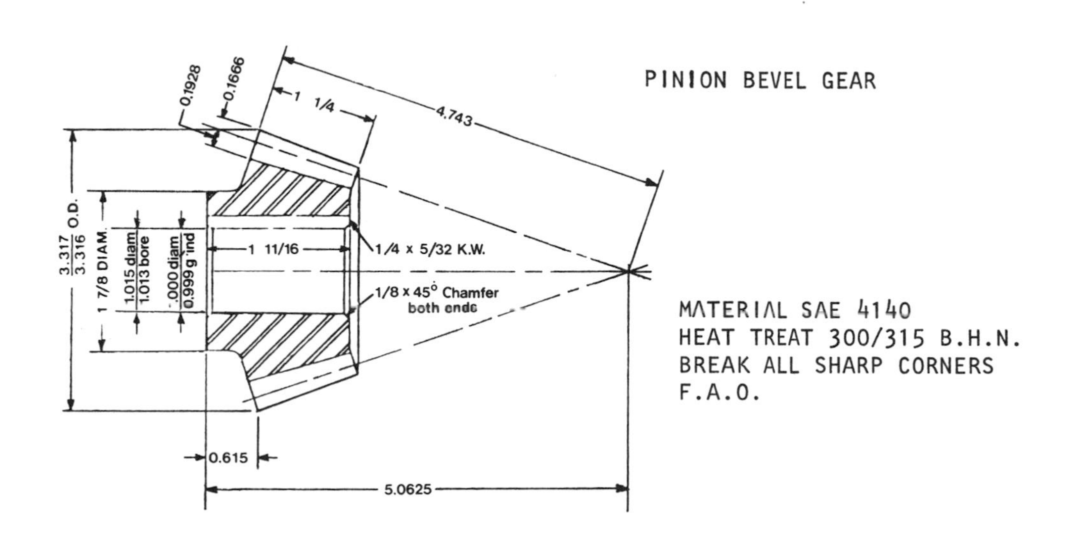

Solution: The table shows formulas, terms and definitions which should be used to calculate necessary information to sketch the bevel gear and pinion bevel gear in this problem.

Straight Bevel Gear Formulas, Terms, and Definitions

Terms and Definitions	Symbol	Formula
Pitch Diameter. The diameter of the pitch cone at the outer ends of the teeth.	(D,d)	$D = \frac{N}{P_d}$
Diametral Pitch. The number of teeth per inch of pitch diameter.	P_d	$P_d = \frac{N}{D}$
Circular Pitch. The distance on the pitch circle between corresponding profiles of adjacent teeth.	p	$p = \frac{\pi}{P_d} = \frac{\pi D}{N}$

ALL FILLETS AND ROUNDS 1/4 R.
BREAK ALL SHARP CORNERS.
MATERIAL CAST STEEL 0.40 CARBON.
HEAT TREAT 240/260 B.H.N.

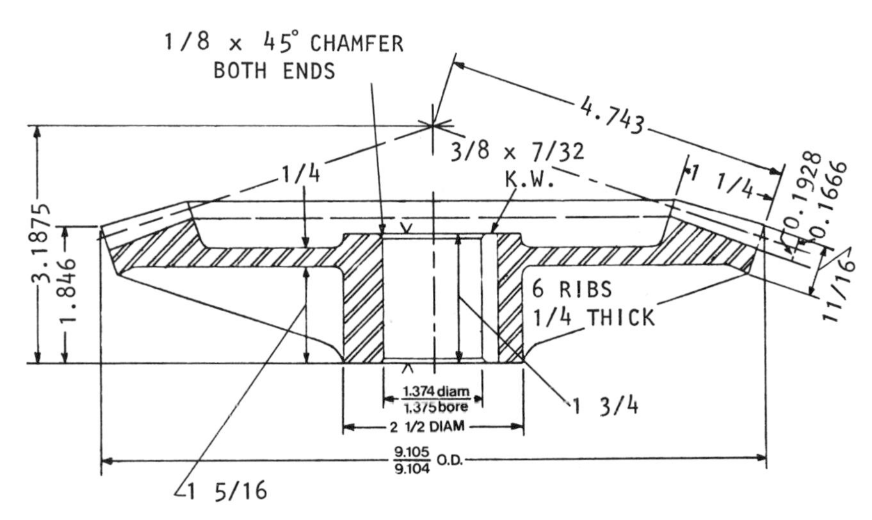

BEVEL GEAR

Cone Distance. Distance from the apex along an element of pitch cone to the outer ends of the teeth.	A	$A = \frac{1/2\ D}{\sin \Gamma}$
Addendum. The height by which the tooth extends beyond the pitch cone at the pitch diameter.	a	$a = \frac{1}{P_d}$
Dedendum. The depth of tooth space below the pitch cone at the pitch diameter.	b	$b = \frac{1.157}{P_d}$
Dedendum Angle. The angle between elements of the pitch cone and the root cone measured in an axial plane.	δ	$\delta = \tan^{-1} \frac{b}{A}$
Pitch Angle. The angle between an element of the pitch cone and its axis.	(Γ, γ)	$\tan \Gamma = \frac{N_G}{N_p}$; $\tan \gamma = \frac{N_p}{N_G}$ (for 90° shaft angle)
Outside Diameter. Diameter of largest face cone.	D_o	$D_o = D + 2a \cos(\Gamma \text{ or } \gamma)$

Term	Symbol	Formula
Face Angle. The angle between an element of face cone and its axis.	(Γ_0, γ_0)	$\Gamma_0 = (\Gamma + \delta)$; $\gamma_0 = (\gamma + \delta)$
Root Angle. The angle between an element of the root cone and its axis.	(Γ_R, γ_R)	$\Gamma_R = (\Gamma - \delta)$; $\gamma_R = (\gamma - \delta)$
Point Backing. Distance from base of the face cone to back surface of hub.	PB	PB = PLB + a sin(Γ or γ)
Mounting Distance. Distance from pitch cone apex to back surface of hub.	MD	MD = PLB + ($\frac{1}{2}$D tan Γ)
Pitch Line Backing. Distance from base of pitch cone to back surface of hub.	PLB	$PLB_p = \frac{D}{4} / \left(\frac{D}{d} + 1\right)$; $PLB_G = \frac{D}{4} - PLB_p$ (Typical)
Face. Width of gear blank measured along pitch cone surface.	F	F$\left(\text{should not exceed } \frac{A}{3}\right)$
Formative Number of Teeth. (Necessary information for selection of tooth cutter.)	N'	$N' = \frac{N}{\cos(\Gamma \text{ or } \gamma)}$
Chordal Thickness. Length of chord subtending a circular-thickness arc.	t_c	From Standard table. Use formative number of teeth, N'.
Chordal Addendum. Height from top of tooth to the chord subtending the circular-thickness arc.		From Standard table. Use formative number of teeth, N'.
Tooth Angle. The angle required for setting up the tooth-generating machine.	θ_t	$\theta_t = \tan^{-1}\left[\frac{\frac{P}{4} + b \tan\phi}{A}\right]$ where ϕ is the pressure angle

The following calculations will provide the necessary information.

Required	Formula	Gear	Pinion
Teeth in gear	Ratio $= N_G/N_p$	$N_G = 3 \times 18 = 54$	18 (Given)
Pitch diameter	$D = N/P_d$	$D = \frac{54}{6} = 9.00''$	$d = \frac{18}{6} = 3.000''$
Circular Pitch	$P = \pi/P_d$	$P = 3.1416/6 = 0.5236$	p = same as gear
Addendum	$a = 1/P_d$	$a = \frac{1}{6} = 0.1666$	a = same as gear
Dedendum	$b = 1.157/P_d$	$b = 1.157/6 = 0.1928$	b = same as gear
Pitch cone angle	$\Gamma = \tan^{-1} N_G/N_p$	$\Gamma = \tan^{-1} \frac{54}{18} = 71°34'$	$\gamma = \tan^{-1} \frac{18}{54} = 18°26'$
	$\gamma = \tan^{-1} N_p/N_G$		
Cone distance	$A = \frac{1}{2}D/\sin \Gamma$	$A = \frac{9}{2}/\sin 71°34' = 4.743$	A = same as gear
Dedendum angle	$\delta = \tan^{-1} b/A$	$\delta = \tan^{-1} \frac{0.1928}{4.743} = 2°20'$	δ = same as gear
Root angle	$\Gamma_R = \Gamma - \delta$	$\Gamma_R = 71°34' - 2°20' = 69°14'$	$\gamma_R = \gamma - \delta = 18°26' - 2°20' = 16°6'$
Face angle	$\Gamma_0 = \Gamma + \delta$	$\Gamma_0 = 71°34' + 2°20' = 73°54'$	$\gamma_0 = \gamma + \delta = 18°26' + 2°20' = 20°46'$
Outside diameter	$D_0 = D + 2a \cos(\Gamma)$	$D_0 = (9.000 + 2 \times 0.1666)$	$d_0 = 3.000 + 2 \times 0.1666 \cos 18°26'$
		$\cos 71°34' = 9.105''$	$= 3.317''$

Formative number of teeth	$N' = N/\cos \Gamma$	$N_G' = 54/\cos 71°34' = 170$	$N_p' = 18/\cos 18°26' = 19$
Chordal addendum	See Table	$1.0041/6 = 0.1673$	$1.0324/6 = 0.1721$
Chordal thickness	See Table	$1.5708/6 = 0.2618$	$1.5690/6 = 0.2615$
Pitch line backing	$PLB_p = \frac{D/4}{D/d + 1}$		$PLB_p = \frac{\frac{9}{4}}{\frac{9}{3} + 1} = 0.5625$
	$PLB_G = D/4 - PLB_p$	$PLB_G = \frac{9}{4} - 0.5625 = 1.6875$	
Point backing	$PB = PLB + (a \sin \Gamma)$	$PB = 1.6875 + 0.1666 \sin 71°34'$ $= 1.846$	$PB = 0.5625 + 0.1666 \sin 18°26'$ $= 0.615$
Keyway	See Table	$3/8 \times 7/32$	$1/4 \times 5/32$
Tooth angle	$\theta_t = \tan^{-1}\left[\frac{P/4 + (b \tan\phi)}{A}\right]$	$\theta_t = \tan^{-1}\left[\frac{\frac{0.5236}{4} + (0.1928 \quad \tan 14\frac{1}{2}°)}{4.743}\right]$ $= 2\ 11'$	Same as gear
Mounting distance	$MD = PLB + \left(\frac{D \tan \Gamma}{2}\right)$	$MD = 1.6875 + \left(\frac{9 \tan 71°34'}{2}\right)$ $= 3.1875$	$MD = 0.5625 + \left(\frac{3 \tan 18°26'}{2}\right)$ $= 5.0625$

A worm and worm-gear are required for a 50:1 speed reducer. The following specifications were included in the order:

Shafts are perpendicular center distance = 7.5625 inches.

Worm gear: 2½-inch face, 80° angle of contact.

Brass--solid web 1¼ inches thick, 9 5/8-inch rim diameter.

5 1/8-inch hub diameter, 4 inches through bore, equal projections.

2 3/4-inch $\begin{cases} + .000 \\ - .001 \end{cases}$ bore diameter, with keyway.

Worm: 3/4 circular pitch, 14½° pressure angle, single thread--R.H., .010 backlash.

Material--SAE 4140. Carburize and caseharden 1/16 inch deep.

See the following sketch.

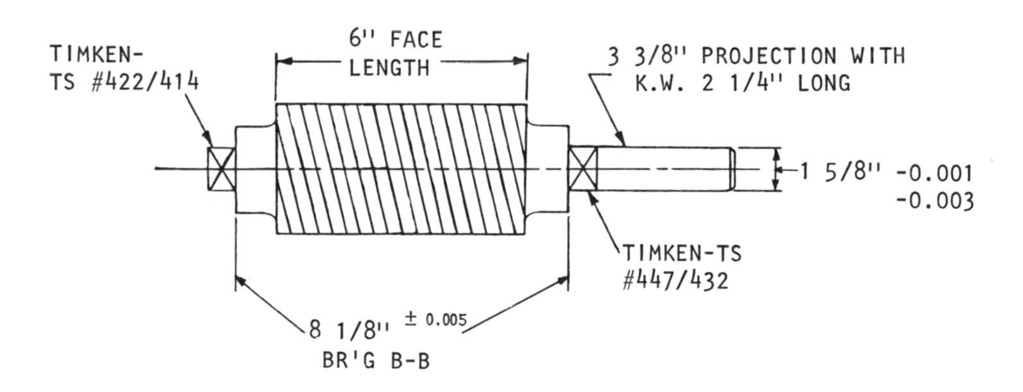

Solution: The American Gear Manufacturers Association recommends the following worm and worm-gear proportions to calculate the necessary information to make the required sketch.

Worm and Worm-Gear Formulas, Terms, and Symbols:

Terms for Worm	Symbol	Formula
Pitch diameter	d	$d = 2.4p + 1.1$ (recommended)
		$= 2\left[C - \frac{D}{2}\right]$

Addendum (chordal addendum)	a	$a = 0.318p = a_c$
Whole depth	h_t	$h_t = 0.686p$
Outside diameter	d_o	$d_o = d + 2a$
Helix angle	ψ	$\psi = \cot^{-1}\left(\frac{\pi d}{\ell}\right)$
Lead	ℓ	$\ell = p \times$ No. of threads
Chordal thickness	t_c	$t_c = \frac{p}{2}\cos\psi - B^*$
Face length	F_ℓ	$F_\ell = p\left(4.5 + \frac{N_G}{50}\right)$ (recommended)

Terms for Worm Gear	Symbol	Formula
Number of teeth	N	$N = \frac{\pi D}{p}$
Pitch diameter	D	$D = \frac{N_p}{\pi}$
Throat diameter	D_t	$D_t = (N + 2)\,\frac{p}{\pi}$
Helix angle	ψ	$\psi = \cot^{-1}\left(\frac{\pi d}{\ell}\right)$
Throat radius	TR	$TR = \frac{d}{2} - 0.318p$
Center distance	C	$C = \frac{1}{2}\,(D + d)$
Blank diameter	D_o	$D_o = D_t + 0.478p$
		$= D_t + 2TR\left(1 - \cos\frac{\alpha}{2}\right)$
Face width	F	$F = 2.38p + 0.25$
Chordal addendum	a_c	$a_c = a + \frac{D}{2}\left[1 - \cos\left(\frac{90°}{N}\right)\right]$
Chordal thickness	t_c	$t_c = D\sin\left(\frac{90°}{N}\right)\cos\psi$

*B, backlash, the amount by which the tooth space exceeds the thickness of the engaging tooth at the pitch circle.

To determine the additional required information the following calculations are done:

Part 1. For the worm gear, see Fig. 1

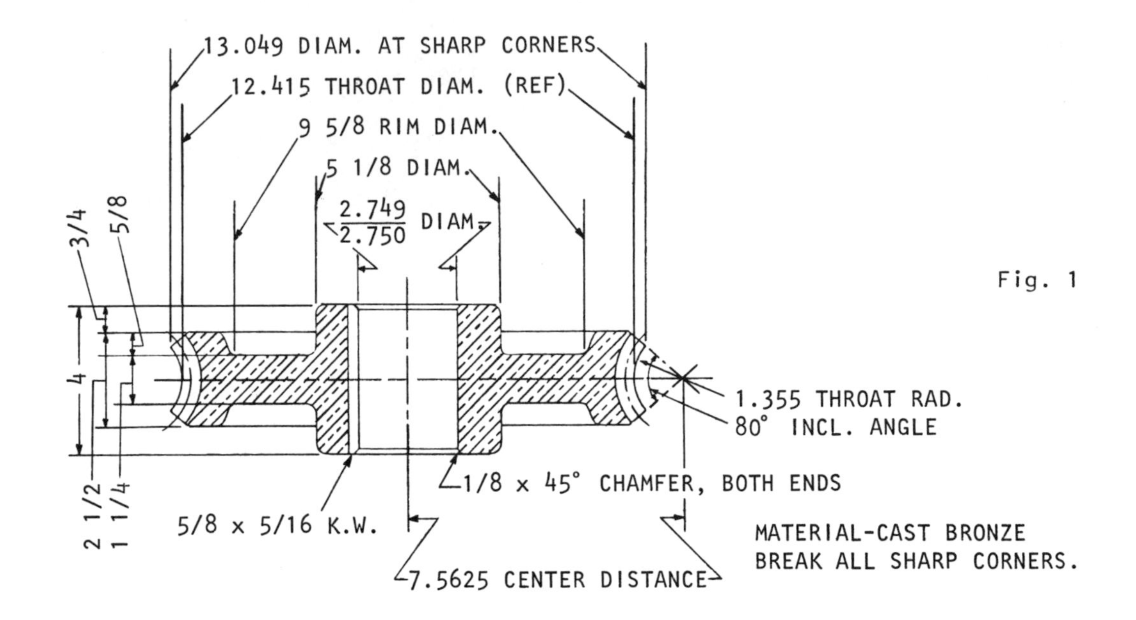

Fig. 1

Throat diameter, $D_t = (N + 2)\,\frac{p}{\pi}$,

$$\left(\text{where } N = \frac{\text{ratio}}{\text{No. threads for worm}} = \frac{50}{1}\right)$$

$$= (50 + 2)\,\frac{0.75}{\pi}$$

$$= 12.415 \text{ inches}$$

Pitch diameter, $D = \frac{N_p}{\pi}$

$$= \frac{50 \times 0.75}{\pi}$$

$$= 11.9375 \text{ inches}$$

Throat radius, $TR = \frac{d}{2} - 0.318p$ ($d = 3.1875$ inches)

$$= \frac{3.1875}{2} - 0.318 \times 0.75$$

$$= 1.355 \text{ inches}$$

Blank diameter, $D_o = 2TR\left(1 - \cos\frac{\alpha}{2}\right) + D_t$ (to sharp corners)

(α = angle of contact = 80°)

$$= 2 \times 1.355 \times 0.234 + 12.415$$

$$= 13.049 \text{ inches}$$

Helix angle, ψ = same as for worm (see below)

Chordal addendum = a_c = same as for spur (see Table)

$$a_c = 1.0123 \text{ (for 50T spur with } P_d = 1)$$

Therefore,

$$a_c = \frac{1.0123}{\pi/0.75}$$

$$= 0.2422 \text{ inch}$$

Chordal thickness, t_c = [t_c (for spur)] cos ψ

$$t_c = 1.5705 \text{ (for 50T spur with } P_d = 1)$$

Therefore,

$$t_c = \frac{1.5705}{\pi/0.75} \times 0.997$$

$$= 0.3750 \text{ inch}$$

Part 2. For the worm, see Fig. 2.

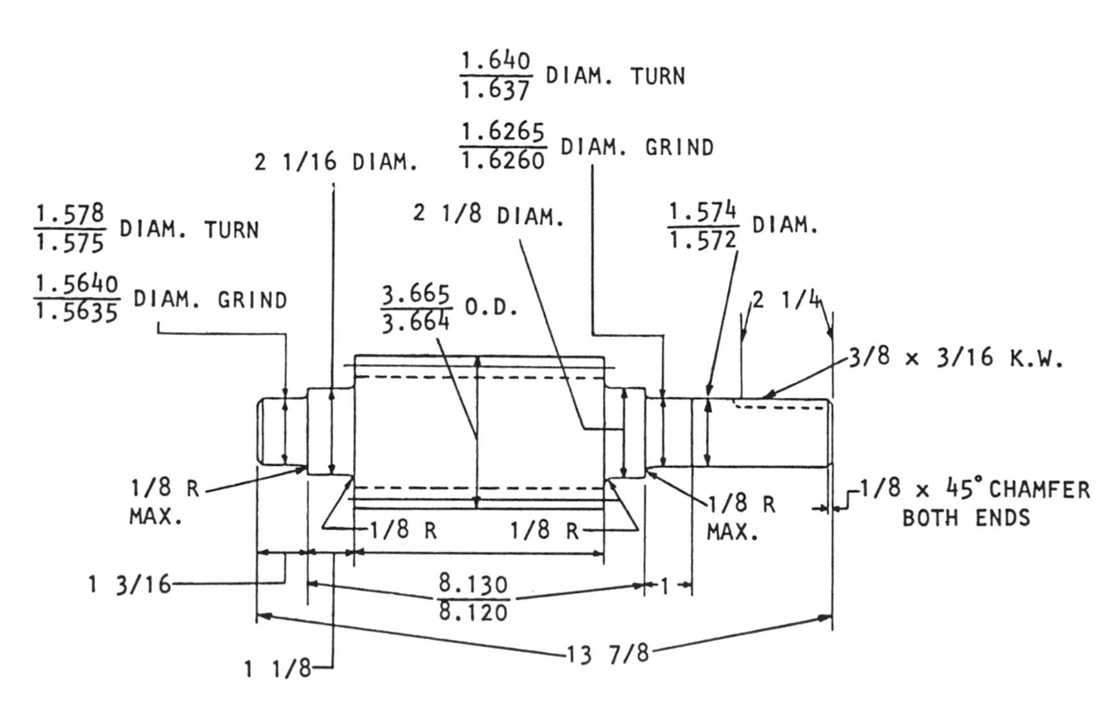

Fig. 2

Pitch diameter, $d = 2\left(C - \frac{D}{2}\right)$

$$= 2\left(7.5625 - \frac{11.9375}{2}\right)$$

$$= 3.1875 \text{ inches}$$

Helix angle, $\psi = \cot^{-1} \frac{\pi d}{\ell}$

$$\ell = p \times \text{No. threads} = \frac{3}{4}$$

Therefore,

$$\psi = \cot^{-1}\left(\frac{\pi \times 3.1875}{0.75}\right)$$

$$= 4°17'$$

Addendum, $a = 3.18p$

$$= 3.18 \times 0.75$$

$= 0.2385$ inch (Note: this is a chordal addendum.)

Whole depth, $h_t = 0.686p$

$$= 0.686 \times 0.75$$

$$= 0.5145 \text{ inch}$$

Chordal thickness, $t_c = \left(\frac{p}{2} \cos \psi - \text{backlash}\right)$

$$= \frac{0.75}{2} \times 0.997 - 0.010$$

$$= 0.364 \text{ inch}$$

Outside diameter, $d_o = d + 2a$

$$= 3.1875 + 2 \times 0.2385$$

$$= 3.6645 \text{ inches}$$

The following information was taken from The Timken Engineering Journal:

For bearing TS No. 422/414:

Bearing bore 1.5625 inches

Shaft tolerance = +0.0015 inch / +0.0010 inch

Shoulder diam. = 2 1/16 inches

Therefore,

Shaft limits $= \frac{1.5640}{1.5635}$

For bearing TS No. 447/432:

Bearing bore = 1.6250 inches

Shaft tolerance = $\begin{matrix} +0.0015 \text{ inch} \\ +0.0010 \text{ inch} \end{matrix}$

Shoulder diam. = 2 1/8 inches

Therefore,

Shaft limits $= \frac{1.6265}{1.6260}$

CHAPTER 15

VECTOR ANALYSIS

DEFINITIONS AND TERMINOLOGY OF VECTORS AND SCALARS

• **PROBLEM** 15-1

Distinguish between vector and scalar quantities.

Solution: A Scalar, or a scalar quantity, can be defined as a quantity that has a magnitude but no direction. Scalar quantities can be defined by a single number of units of a particular dimension, such as a temperature of 65° Celsius, a pressure of 50 Kg per square cm or a distance of 20 miles. Scalars follow the laws of ordinary addition, subtraction and multiplication.

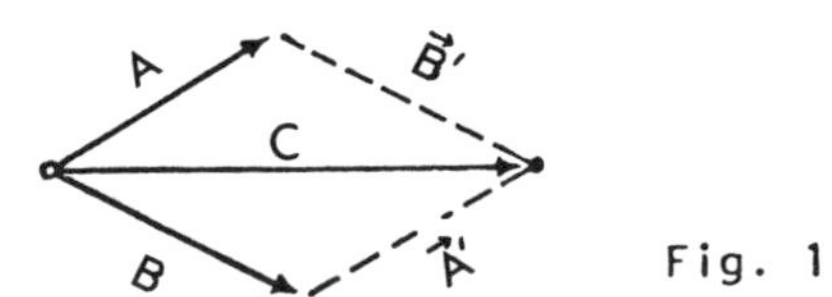

Fig. 1

Vectors, or vector quantities, can be defined as quantities that have magnitude and direction. Because vectors have magnitude and direction they follow different rules for multiplication, addition and subtraction which will be discussed.

Quantities, such as force and velocity, are vectors, thus they are completely specified only if both a direction and a magnitude are given. These quantities are represented graphically by an arrow to indicate direction, the magnitude being shown by the length of the arrow. Further, these quantities can be combined by the parallelogram law, where vector $\vec{C}$ the diagonal of the parallelogram formed by vectors $\vec{A}$ and $\vec{B}$ and dotted lines $\vec{A}'$ and $\vec{B}'$ which are lines parallel to $\vec{A}$ and $\vec{B}$ respectively (see figure 1) is equal to the sum of vectors $\vec{A}$ and $\vec{B}$, often called the vector sum of $\vec{A}$ and $\vec{B}$. This sum is represented as:

$$\vec{C} = \vec{A} + \vec{B}$$

Where vector $\vec{C}$ is the resultant vector with direction from the middle of the corner formed by vectors to the opposite side. An alternate method for combining vectors $\vec{A}$ and $\vec{B}$ is to place one vectors' tail on the head of the remaining vector without changing the magnitude or direction of either vector (see figs. 2a, 2b).

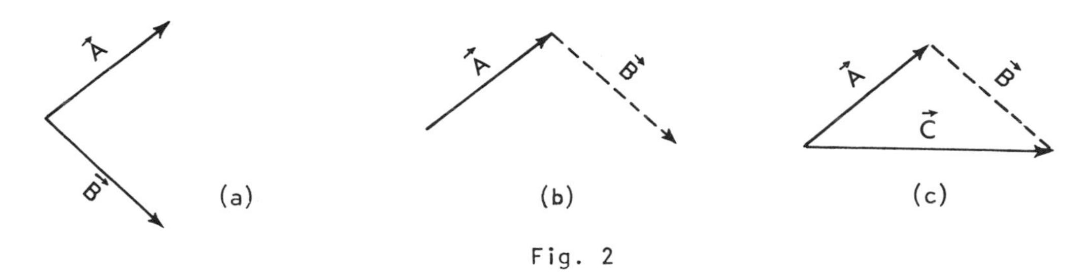

Fig. 2

(The dotted line represents vector $\vec{B}$'s tail on head of vector $\vec{A}$ without a change in $\vec{B}$'s magnitude or direction. The solid line is vector $\vec{A}$ which remained stationary.) Draw a line from the tail of the stationary vector (here vector $\vec{A}$) to the head of the vector which was moved. (See fig. 2c.) You now have the vector sum of $\vec{A}$ and $\vec{B}$ which is $\vec{C}$. Note that figure 2c is a triangle. This will always occur when combining 2 vectors with angles between them. This method can only be used when vectors $\vec{A}$ and $\vec{B}$ are drawn to scale. This method is sometimes called the triangle method.

• PROBLEM 15-2

Define various terms required for understanding the techniques of problem solving with vectors.

Solution: A force is a push or a pull that tends to produce motion. All forces have magnitude, direction, point of application, and sense. A force is represented by a rope being pulled in figure A.

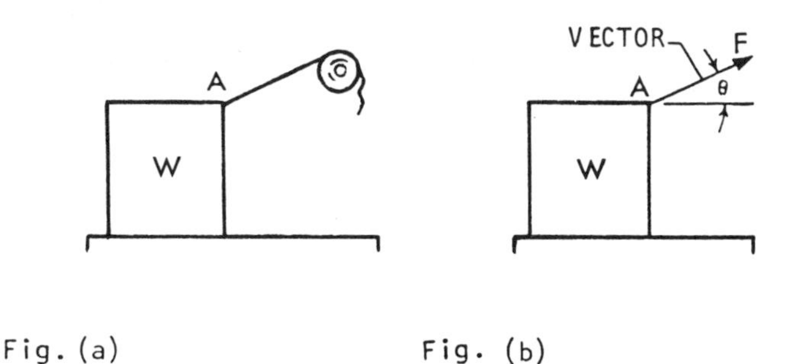

Fig. (a) Fig. (b)

THE REPRESENTATION OF A FORCE BY A VECTOR.

A vector is a graphical representation of a quantity of force which is drawn to scale to indicate magnitude, direction, sense, and point of application. The vector $\overline{F}$ shown in figure B represents the force of the rope pulling weight, W.

Magnitude is the amount of push or pull. Graphically, this is represented by the length of the vector line.

Direction is the inclination of the force with respect to a reference coordinate system. This is represented by angle θ in figure B.

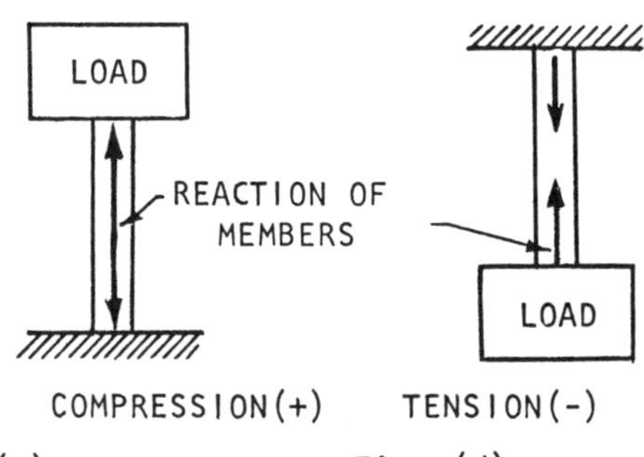

Fig. (c) Fig. (d)

A COMPARISION OF TENSION AND COMPRESSION IN A MEMBER.

The point of application is the point through which the force is applied on the object or member. This is point A in figure A.

Sense is either of the two opposite ways in which a force may be directed, i.e., towards or away from the point of application. The sense is shown by an arrowhead attached to one end of the vector line. In figure A, the sense of the force is away from point A. It is shown in figure B by the arrowhead at F.

Compression is the state created in a member by opposite forces acting upon a member. The member tends to be shortened by compression (figure C). Compression is represented by a plus sing (+).

Tension is the state created in a member by subjecting it to opposite pulling forces. A member tends to be stretched by tension, as shown in figure D. Tension is represented by a minus sign (-).

A force system is a combination of all forces acting on a given object.

The resultant is a single force that can replace all forces.

The equilibrant is the opposite of a resultant. It is the single force that can be used to counterbalance all forces of a force system.

Components are individual forces which, if combined, would result in a given single force. For example in figure E, A and B are components of resultant R.

A space diagram depicts the physical relationship between structural members. The force system in figure F is given as a space diagram.

A vector diagram is composed of vectors which are scaled to their appropriate lengths to represent the forces

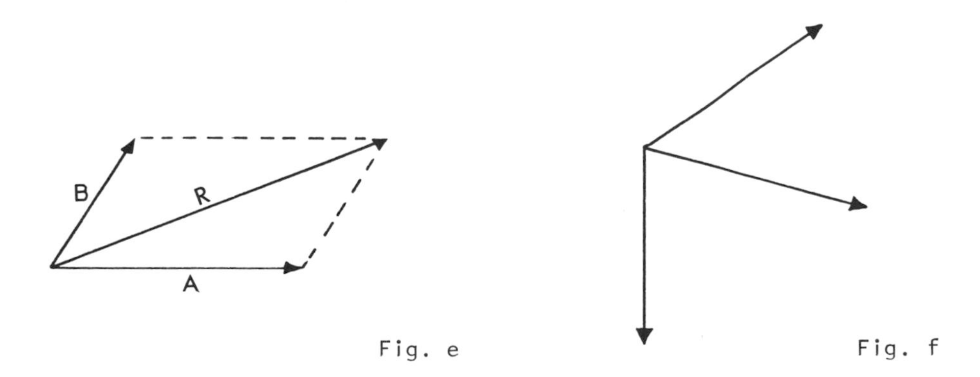

Fig. e

Fig. f

within a given system. A vector diagram may be a polygon or a parallelogram.

A coplanar vector system consists of vectors which lie in the same plane.

A noncoplanar vector system consists of vectors which lie in different planes.

Concurrent vector system is one in which all vectors intersect at a common point.

Nonconcurrent vector system is one in which the vectors do not intersect at a common point.

Parallel vector system (a special case of nonconcurrent vector system) is one in which all vectors are parallel.

TWO AND THREE VECTOR SYSTEMS

• PROBLEM 15-3

Two men find it necessary to move a filing cabinet. One man pushes with a force of 120 lbs. The other man pushes with a force of 90 lbs. Find the resultant force graphically.

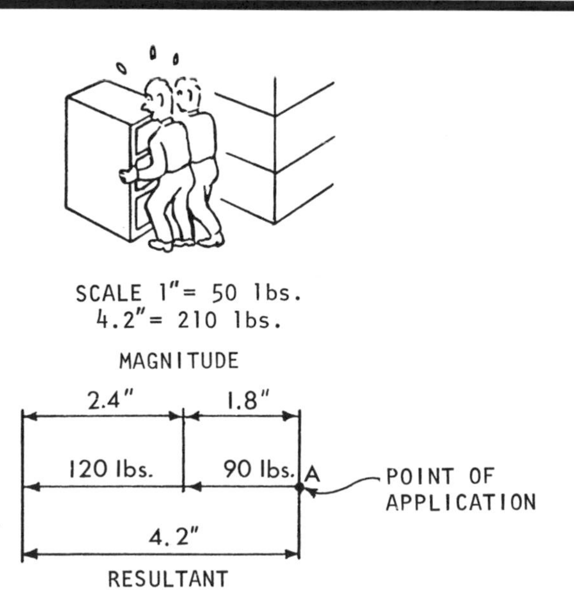

Solution: This is a simple example of addition of coplanar forces. Both forces are applied in the same direction and therefore can be added.

Select the scale of 50 lbs = 1". Therefore, 90 lbs = 1.8" and 120 lbs = 2.4". The point of application A represents the cabinet. At A, draw the vector of 90 lbs (= 1.8" length) magnitude in the direction that the cabinet is being moved. At the tip of this vector draw another vector of 120 lbs (= 2.4") magnitude in the same direction as the 90 lbs vector. On adding, it is found that the resultant vector has a length of 4.2" in the same direction. On the scale selected 4.2" is equal to 210 lbs which is the combined force with which the cabinet is being moved.

• PROBLEM 15-4

During a picnic two groups of men decided to have a tug of war. The total force exerted by team B was 1000 lbs. Team A put forth a total force of 1500 lbs. Add the two forces graphically.

Solution: Select a scale of 500 lbs = 1". At A draw a force vector of 1500 lbs (= 3") magnitude in the direction of team A. At the tip of this vector draw a

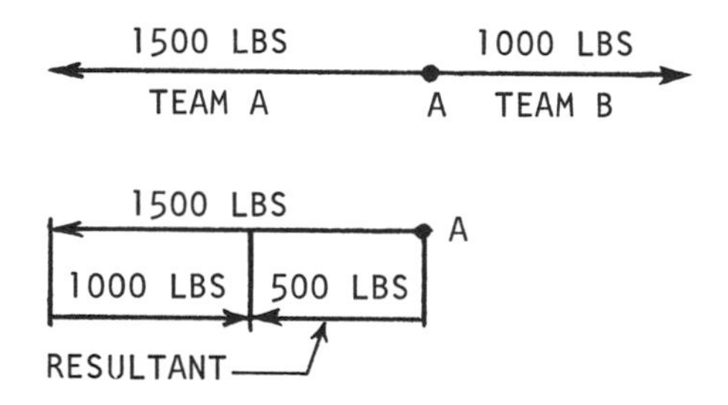

FORCES ACTING IN OPPOSITE DIRECTIONS CAN BE ADDED GRAPHICALLY.

force vector of 1000 lbs (= 2") magnitude in the direction of team B. The resultant as can be seen from the figure is equal to 500 lbs (= 1") in magnitude having the same direction and sense as the force exerted by team A.

The figure shows a system of concurrent coplanar forces D_a and D_b. Find the resultant of the two forces by the parallelogram method.

Solution: To find the resultant force and direction of pull on PA, add the vectors graphically. To do this, construct a parallelogram using sides D_a and D_b. Then draw a diagonal from the point of application. The diagonal is the resultant force. The magnitude of the resultant is found by measuring the length of the diagonal.

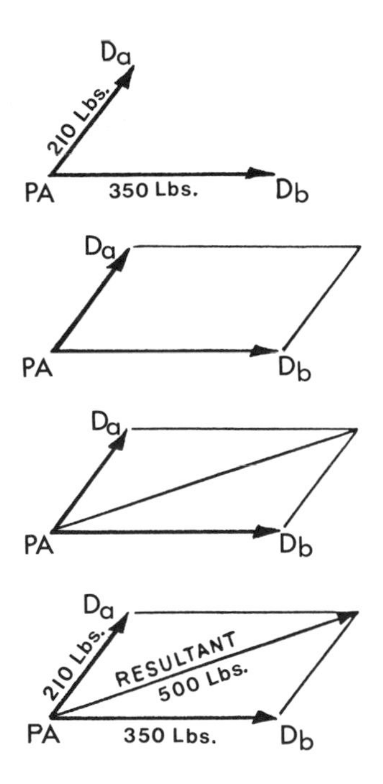

Suppose we use a scale of 100 lbs = 1", then the length of D_a will be 2.1" and the length of D_b will be 3.5". The length of the diagonal of the parallelogram constructed using D_a and D_b as sides will be measured as 5". This 5" represents 500 lbs on the scale selected. Hence, a force of 500 lbs applied in the direction of the diagonal would replace the forces of 210 lbs and 350 lbs applied in the direction shown in the problem.

• PROBLEM 15-6

An airplane is flying 125 mph on a north to south course according to its air speed indicator and compass readings. A cross wind of 30 mph is blowing S 60°W. What are the ground speed and course of the airplane?

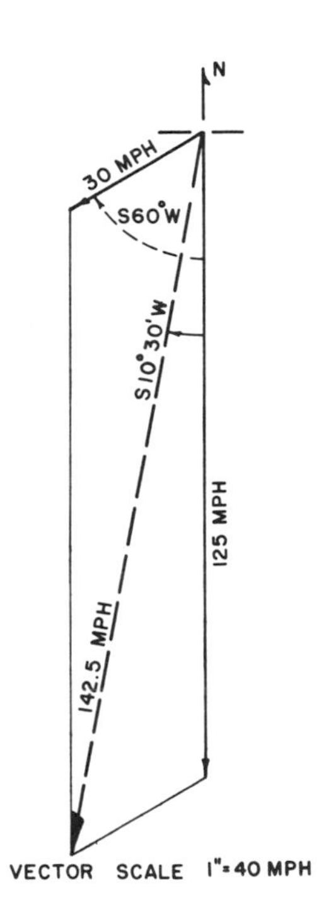

Solution: Select a vector scale of 1" = 40 MPH. Draw a vector in the north to south direction to represent the velocity and direction of the aircraft. Draw the wind velocity vector as given.

By the parallelogram method, we obtain the result that the airplane is flying in the direction S10°30'W with a speed of 142.5 mph.

• PROBLEM 15-7

A plane is flying due west at an air speed of 200 mph. The wind is blowing 75 mph from a southeasterly direction (315 degrees from north). Considering these two factors, what will be the air speed and direction of the plane?

Solution: By plotting the direction and magnitude

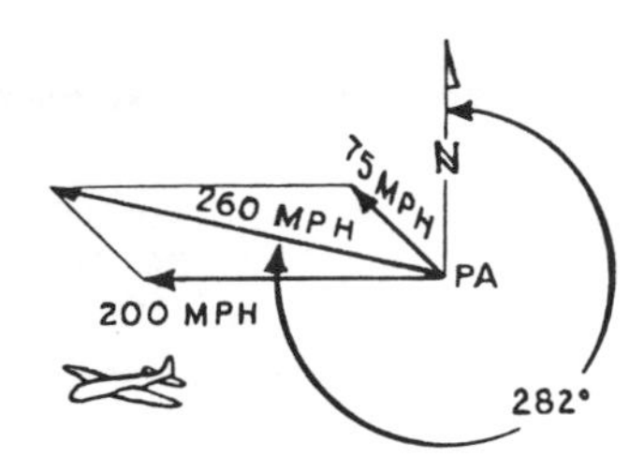

(speed) of the plane and the wind, a parallelogram can be drawn. A diagonal from the PA becomes the resultant. In the illustration, the resultant and speed would be 260 mph at 282 degrees or N48°W.

• PROBLEM 15-8

Two airplanes in level flight at an elevation of 10,000 feet are at the positions shown by points A and B in the figure. Airplane A is flying N45°E at a ground speed of 400 mph and airplane B is flying due west at a ground speed of 300 mph. Find the distance between the two airplanes when they are closest to each other, and also the distance each airplane travels to reach that position.

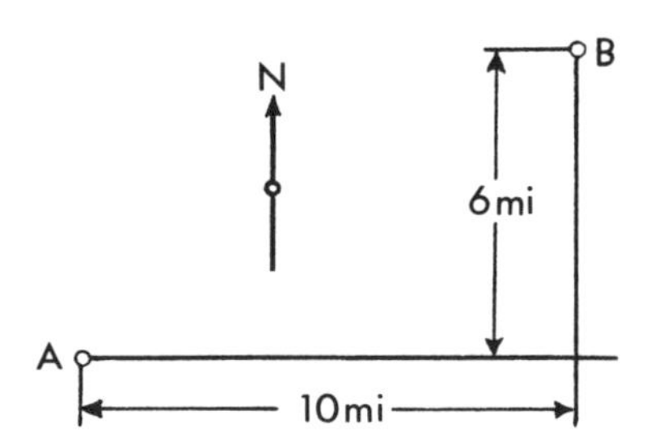

Solution: As shown in the second figure, the positions of airplanes A and B are plotted, using the distance scale. Next, using the velocity scale, the velocity vector of airplane B is reversed and added to the velocity vector of airplane A, giving a resultant relative velocity of 650 mph. Airplane B is then temporarily considered to remain stationary while A travels in the direction shown by this resultant. The perpendicular distance from B to point C on the line of the resultant shows that the clearance between the two airplanes is 1.1 miles. A line is then drawn from C parallel to the direction of motion of B until it intersects the actual direction of flight of A at point D. A line from D drawn parallel to BC intersects

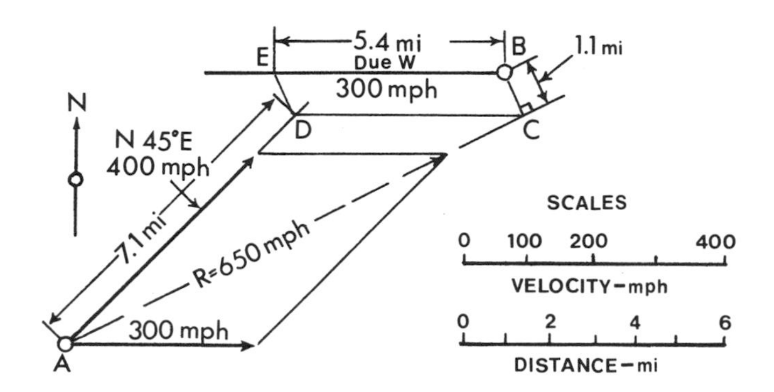

the direction of flight of airplane B at point E. Points D and E represent the positions of the two airplanes when they are closest to each other. The distances from A to D (7.1 miles) and from B to E (5.4 miles) are then scaled to find the distance that each airplane traveled to reach this position.

• PROBLEM 15-9

If a rope, AMB in fig. 1, suspended from a ceiling supports a 100 pound weight, determine the pull in MA and MB.

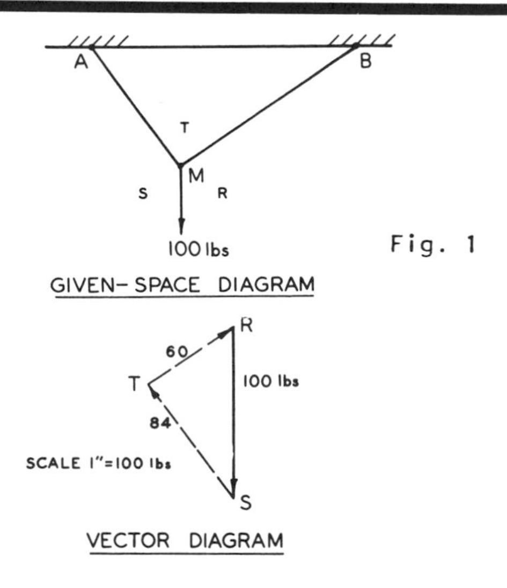

Fig. 1

Solution: If the supporting point M is to remain at rest, the pull of the 100 lb weight must be balanced by two forces whose lines of action lie along ropes MA and MB and whose resultant is equal to 100 lbs pointing upward because the system is in equilibrium. Using the triangle method since the resultant is known and the line of action of the forces, it is possible to find the direction and magnitude of each force. Letting 1" = 100 lbs construct the vectors. Call the resultant vector AB (see fig. 2a). Draw vectors which cause resultant AB with correct lines of action (a) shown in fig. 1) such that all three vectors form the triangle in figure 2b, pt. C is where vectors $\overline{BC}$ and $\overline{AC}$ intersect

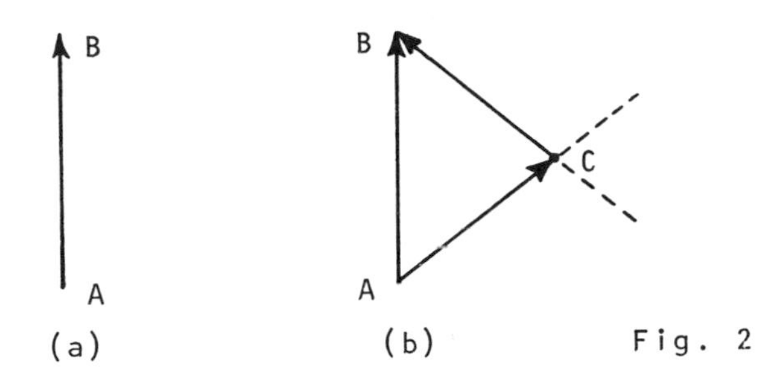

Fig. 2

closing the triangle. Now measure each using scale 1" = 100 lbs to determine each length. Directions are determined according to triangle rule, keeping in mind that vector $\overline{AB}$ is resultant and vectors $\overline{BC}$ and $\overline{AC}$ are components.

• PROBLEM 15-10

> The following figure shows a 300 pound weight supported by two cables. Find the forces in the two cables that will support the weight.

Solution: It can be noticed that the system is at rest (equilibrium) therefore, the cables must have a resultant in the opposite direction of the weight with the same magnitude. Construct the picture with a vector pointing upward and its components in the lines of action of the cables such that all three form a triangle (see fig. 2). Let 300 lbs = 1".

Point p is where the vector lines intersect the closing triangle. The directions of the vectors are determined

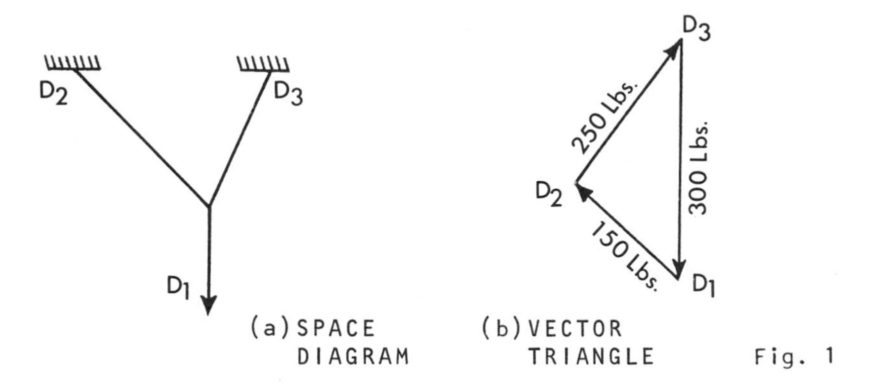

(a) SPACE DIAGRAM (b) VECTOR TRIANGLE Fig. 1

A VECTOR TRIANGLE IS USED TO FIND THE FORCES NECESSARY TO HOLD THE 300-POUND SIGN IN EQUILIBRIUM.

by the triangle method, but keep in mind which are components and which are resultants. The magnitudes are measured by measuring the length of the components to

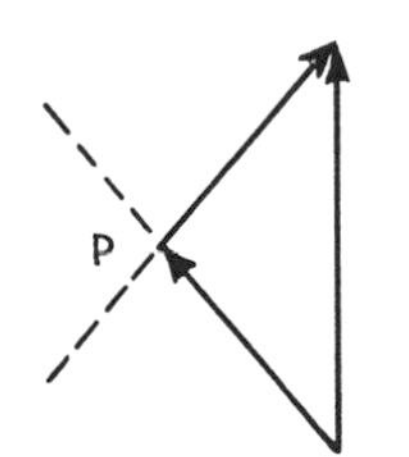

Fig. 2

the scale 300 lbs = 1". This results in the magnitudes being 150 lbs and 250 lbs respectively.

• PROBLEM 15-11

> In the figure a simple structure consisting of two members connected at C is supporting a single load, F, of 50 lbs. Determine the magnitude of the axial force imposed on each of the members. Does this force apply a compressive or tensile stress to the member?

Solution: In figure (a) force F is laid out to scale parallel to its direction in the space diagram. An equilibrant E equal and opposite to F is then laid out as shown. This force must be the resultant of the axial forces in members CA and CB. A parallelogram is therefore constructed with E as a diagonal by drawing lines through each end of E parallel to members CA and CB. The values of the axial forces in members CA (CA) and CB (CB) are then scaled and found to be 39 pounds and 43 pounds, respectively. By comparing the vector diagram with the space diagram and imagining the force CA to be acting at A, it can be seen that CA is acting away from the joint at C. This tends to stretch the member CA, indicating that CA is in tension. Similarly, if CB is imagined to be acting at B, member BC is seen to be in compression.

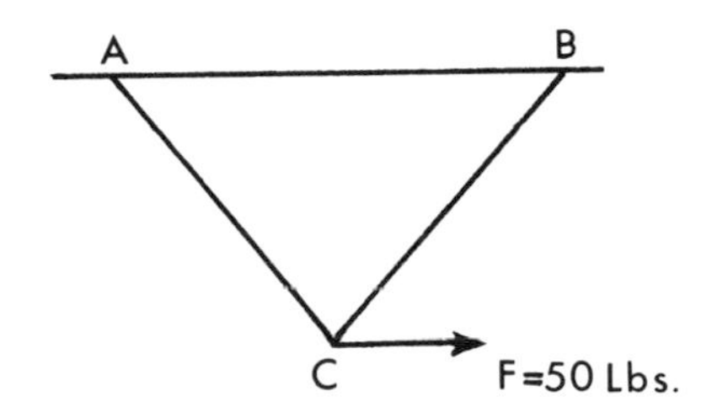

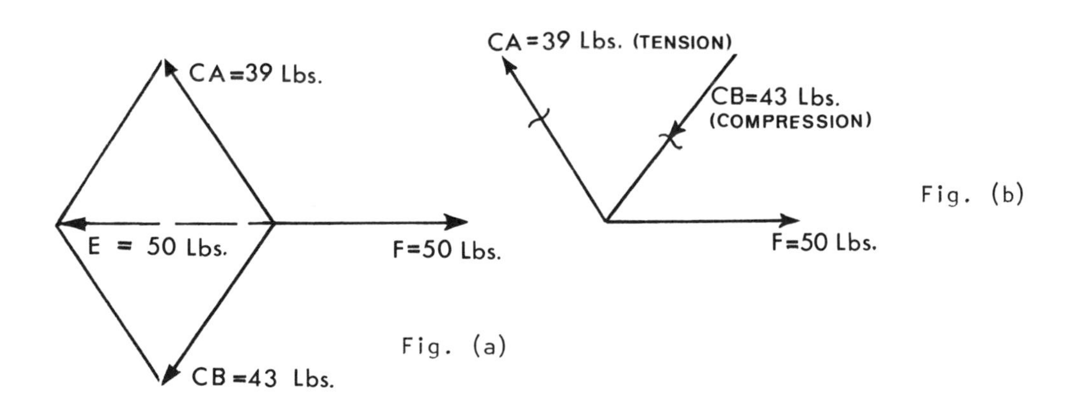

Fig. (a)

Fig. (b)

The student may find that he can determine tension and compression more easily by drawing a free-body diagram, as shown in figure (b). In the free-body diagram, the members AC and BC are considered to have been cut off, the portions not drawn being replaced by the forces CA and CB. These forces have exactly the same effect as did the original members; that is, they keep the joint in equilibrium. On the free-body diagram, the member AC is seen to be in tension, since the arrow representing CA points away from the joint and therefore indicates that the force is tending to lengthen member AC. Similarly, the member BC is in compression because the arrow representing CB points toward the joint, indicating that the force tends to shorten member BC. It should be noted that the directions on the free-body diagram can be approximate, since they are used only to determine tension and compression.

• PROBLEM 15-12

> In the situation in the given figure, a sign hangs from a support above a building entrance. Find the forces acting on the cable and support.

Solution: To resolve the forces, first draw the vertical component in its correct magnitude. Now draw the support and the cable in their proper direction. These are the

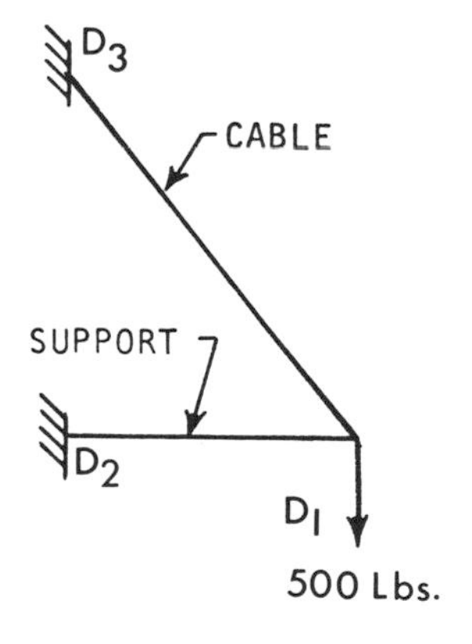

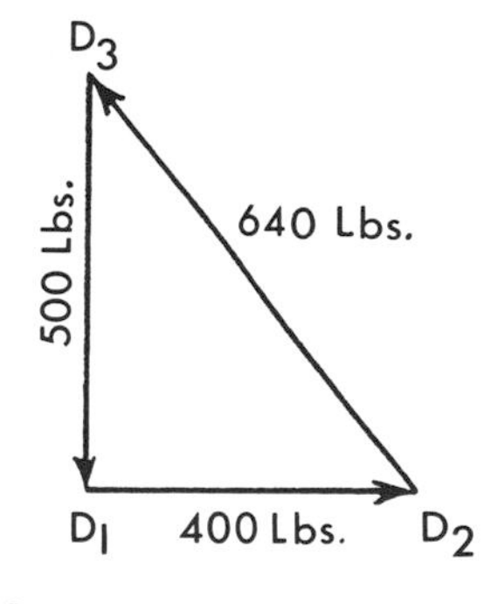

Ⓐ SPACE DIAGRAM Ⓑ VECTOR TRIANGLE

THE FORCES IN THE CABLE AND SUPPORT ARE FOUND BY DRAWING A VECTOR TRIANGLE.

unknown forces in the diagram. The force polygon will now close since the forces are in equilibrium. Scaling the lines will indicate the forces acting on the cable and support.

• PROBLEM 15-13

Split a component with a magnitude of 400 pounds into two vectors with the directions shown in this figure.

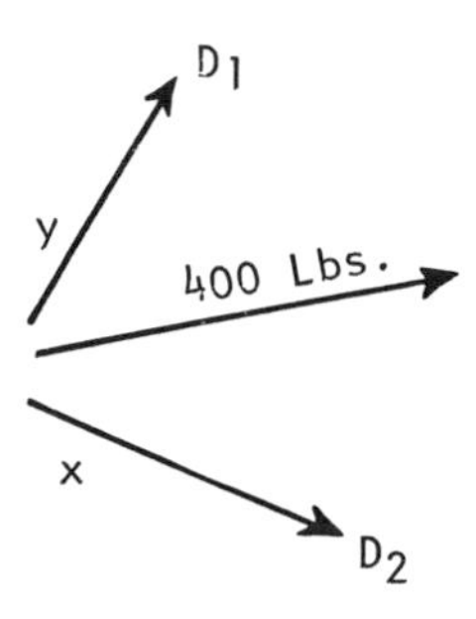

Solution: The parallelogram method can be used. The component is the diagonal. Connect lines representing the vectors wanted with the ends of the component. These must be in the desired direction. Draw a parallelogram using these as sides. This marks the

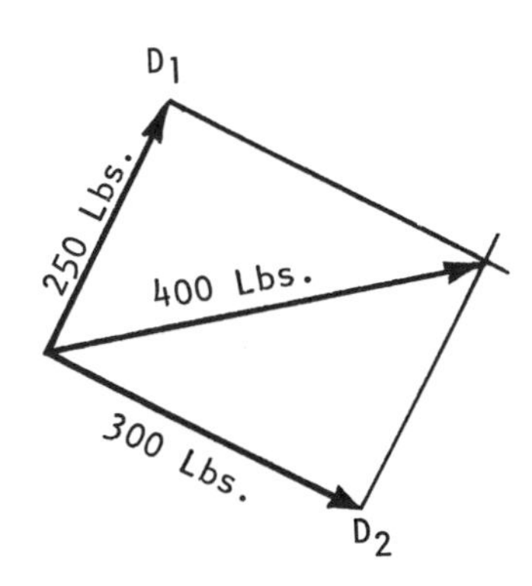

vectors to length. Measure them using the same scale used to draw the component. This gives the magnitude of each vector.

• PROBLEM 15-14

A stationary wheel of 40" diameter and weighing 150 lbs is to be pushed and moved over a 4" high rigidly fixed block on the ground. How much force will be required to make the wheel rise over the block?

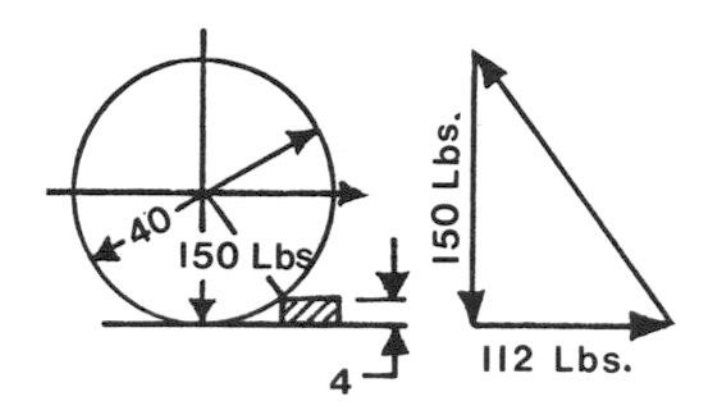

Solution: First draw the known force to a known scale. This is the downward force of 150 lbs. Next draw the other two forces parallel to the forces they represent. A horizontal line representing the pushing force and an inclined line representing the resisting force at the point where the wheel intersects the block. The horizontal member of the force triangle is the amount needed to move the wheel over the block. Measure this member to the same scale used to draw the known force and the result will be 112 lbs.

• PROBLEM 15-15

By using the parallelogram law, find the resultant and equilibrant of three coplanar concurrent-vectors $\bar{A}$, $\bar{B}$, and $\bar{C}$, shown in fig. 1.

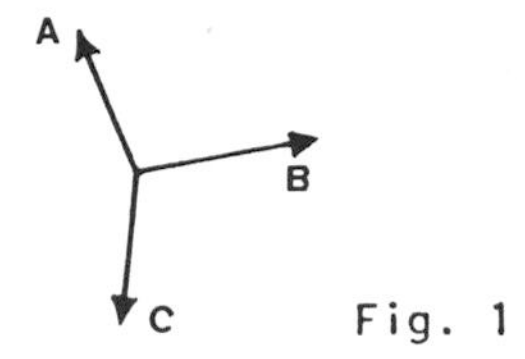

Fig. 1

Solution: In this problem the original system consists of more than two vectors, the resultant can be found by using the parallelogram law to perform successive additions. As shown in figure 2(a) vectors $\bar{A}$ and $\bar{B}$ are added, by the parallelogram law, to obtain their resultant R_1. Then R_1 is added to vector $\bar{C}$ to obtain R_2 which is the final resultant of vectors $\bar{A}$, $\bar{B}$ and $\bar{C}$. The force necessary to balance the system or equilibrant $\bar{E}$ of vectors $\bar{A}$, $\bar{B}$ and $\bar{C}$ is equal in magnitude and opposite direction to final Resultant R_2, see fig. 2(a). Note that the resultant and equilibrant remain unchanged when the order of addition is changed. For example, as shown in figure 2(b), vectors $\bar{B}$ and $\bar{C}$ are added to give R_1, which is then added to $\bar{A}$ to obtain the final resultant R_2. The equilibrant is also shown in the figure.

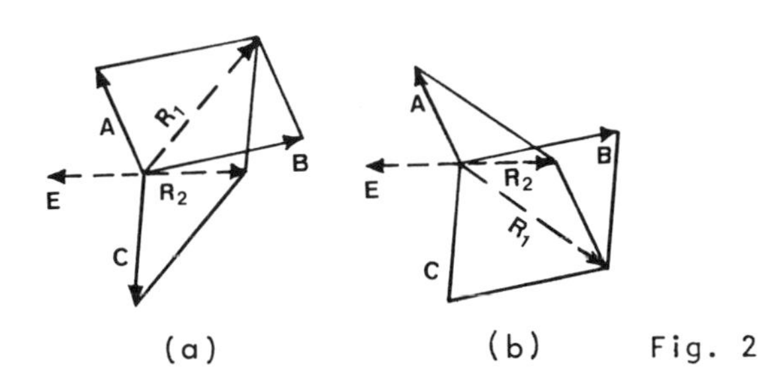

Fig. 2

• PROBLEM 15-16

Figure 1 shows a system of concurrent-coplanar vectors $\bar{D}_1$, $\bar{D}_2$ and $\bar{D}_3$. Use the parallelogram law and successive additions to find the final resultant of this vector system.

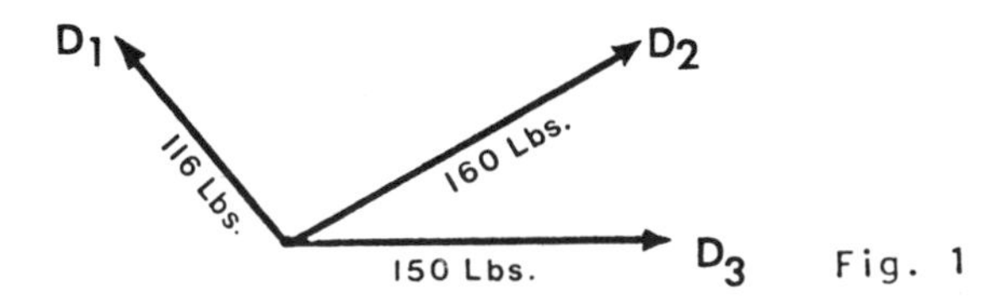

Fig. 1

Solution: First find the resultant of two of the vectors. In figure 2(a), vectors $\bar{D}_1$ and $\bar{D}_2$ are used. Their

resultant is 180 pounds. Then this resultant is made into a parallelogram with vector D_3. See figure 2(b). The resultant is 270 pounds. This is the resultant for the three vectors $\overline{D}_1$, $\overline{D}_2$ and $\overline{D}_3$.

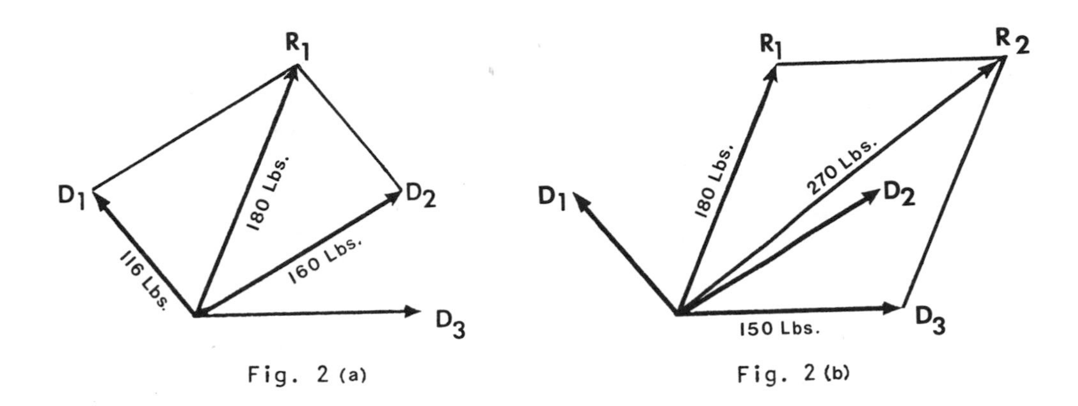

Fig. 2 (a) Fig. 2 (b)

• PROBLEM 15-17

> Find the resultant and equilibrant of three coplanar concurrent vectors $\overline{A}$, $\overline{B}$ and $\overline{C}$ in the figure by using a vector polygon.

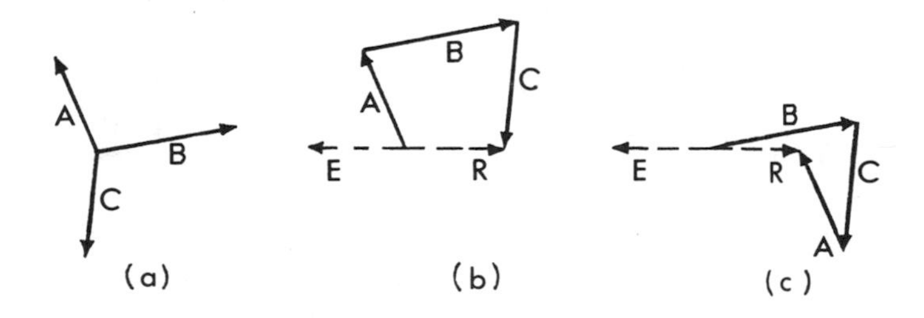

RESULTANT AND EQUILIBRANT OF MORE THAN TWO COPLANAR CONCURRENT VECTORS USING A VECTOR POLYGON

Solution: (1) Plot vector $\overline{A}$.

(2) Add vector $\bar{B}$ with its tail at the tip of vector $\bar{A}$.

(3) Draw vector $\bar{C}$ with its tail at the tip of vector $\bar{B}$.

(4) Connect the tail of vector $\bar{A}$ to the tip of vector $\bar{C}$. This represents the resultant $\bar{R}$ (fig. (c)).

(5) The equilibrant $\bar{E}$ is equal and opposite to $\bar{R}$.

(6) In figure (b), the vector polygon is drawn by connecting vectors $\bar{B}$, $\bar{C}$ and $\bar{A}$ in tail-to-tip order. Here again the resultant and equilibrant remain the same when the order of addition is changed.

• PROBLEM 15-18

The figure shows a space diagram with three vectors drawn to scale. Is this diagram in equilibrium?

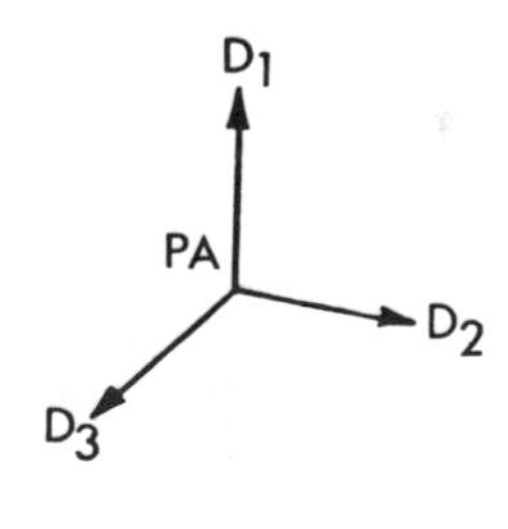

Ⓐ SPACE DIAGRAM

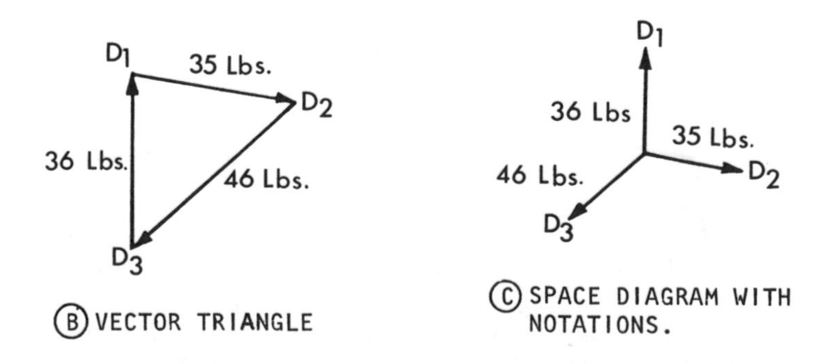

Ⓑ VECTOR TRIANGLE

Ⓒ SPACE DIAGRAM WITH NOTATIONS.

Solution: The solution can be found by constructing a vector triangle, shown in the figure. Each side is drawn with the original magnitude and direction. An attempt is made to form a triangle. If a triangle can be formed, with all vectors head to tail, the diagram is in equilibrium. This is slightly different from the triangle rule because you are no longer looking for a resultant but instead seek an equilibrant which is in the opposite direction of resultant.

The graphically constructed triangle proves that the diagram is in equilibrium.

• PROBLEM 15-19

Force vectors $\bar{A}$, $\bar{B}$ and $\bar{C}$ are shown drawn to scale in the figure. Find the resultant and equilibrant of this noncoplanar, concurrent vector system using:

1. The Parallelogram Method

2. The Vector Polygon Method.

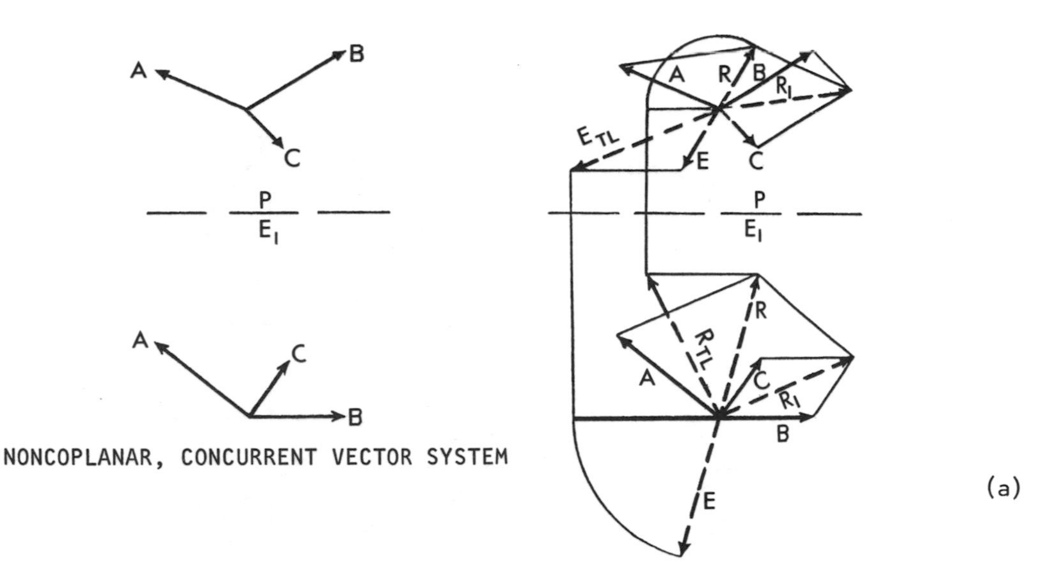

Solution: The principles used in finding the resultants and equilibrants of noncoplanar, concurrent vector systems are exactly the same as in coplanar systems, except that a third dimension is involved. This means that two orthographic views are required to show the positions of the vectors in space, and, similarly, to show the shape of the corresponding vector diagram.

Parallelogram Method

The P and E_1 views of force vectors A, B, and C are shown drawn to scale in figure (a) and the resultant and equilibrant are to be found. The parallelogram law is applied separately in each view, first to add B and C to give R_1 and then to add R_1 and A to obtain the resultant R. The accuracy of the construction can be checked by projection between views. The equilibrant E is equal and opposite to R. Since neither R nor E appears true length in either view, they are revolved as shown to find their true lengths (R_{TL}) and (E_{TL}).

Vector Polygon Method

The space diagram of force vectors, A, B, and C is shown drawn to scale in figure (b) and the resultant and equilibrant are to be found. In the plan view of the vector diagram, each vector is drawn parallel to the corresponding vector in the plan view of the space diagram. Similarly, in the E_1 view of the vector diagram, each vector is drawn parallel to the corresponding vector in the E_1 view of the space diagram, with the vectors connected in the same tail-to-tip order in each view. The resultant R is then drawn from the tail of the first given vector to the tip of the last in each view, and its true length found by revolution. The equilibrant, equal and opposite to R, is not shown.

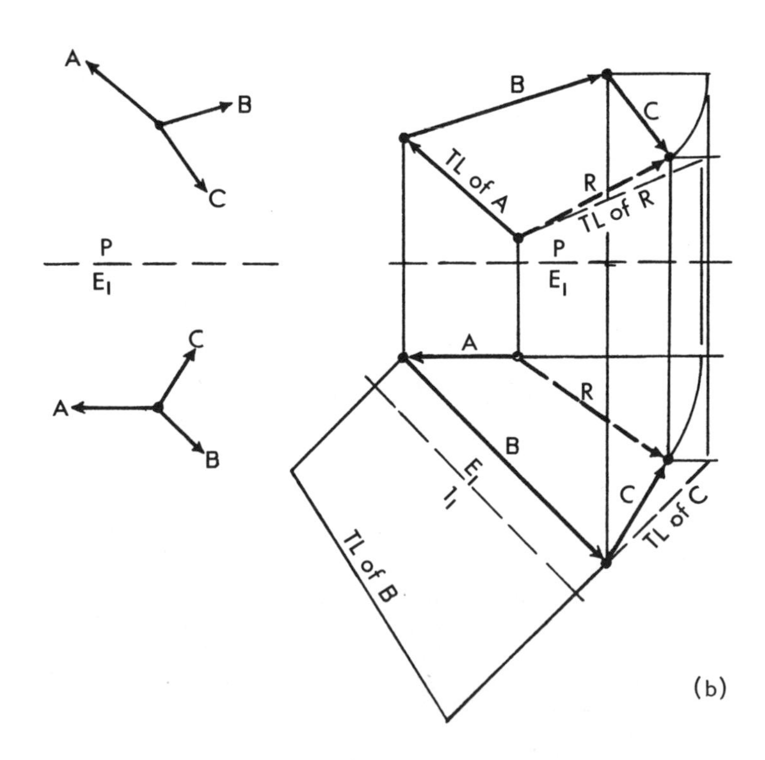

• PROBLEM 15-20

A 3-D space frame supporting a 300 lb. load is shown in the figure. Find the axial forces AD, BD, and CD by the parallelogram law.

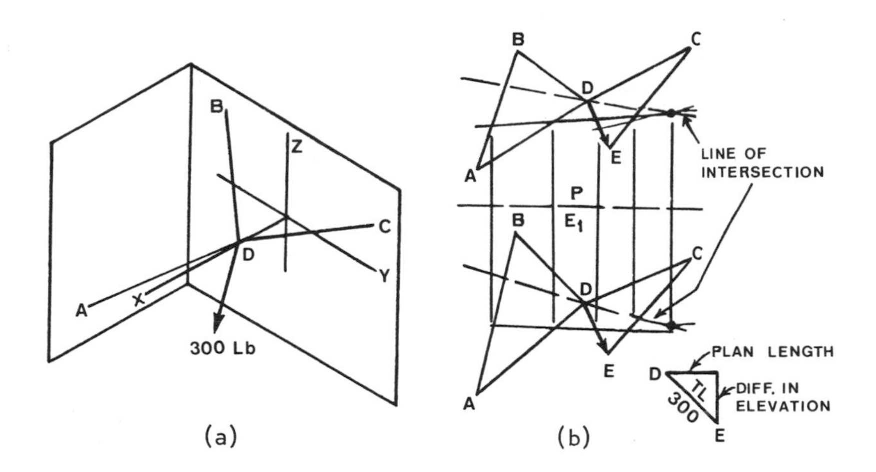

(a) (b)

Solution: The space frame and applied load are laid out to scale [Figure 1(b)], using triangulation to obtain the projections of the 300-pound load (DE).

When four forces are in equilibrium, the resultant of any two of the forces must be equal, opposite to, and collinear (along the same line) with the resultant of the other two forces. Thus, for example, the resultant of the given load DE and the force in the member CD must be equal, opposite to, and collinear with the force in members AD and BD. The resultant of forces DE and CD lies in plane CDE and the resultant of AD and BD lies in plane ADB. In order for these resultants to be collinear, they must both lie on the line of intersection of planes CDE and ADB.

Point D is one point on this line of intersection, since all the forces are concurrent at D. In Figure 1(b), a second point is found by the cutting-plane method. (A third point obtained as a check has been omitted for clarity.)

The resultant R_1 of the load DE and force CD lies along this line of intersection. In Figure 1(c), R_1 is found separately in each view by drawing a line through E parallel to line CD. Force CD is then found by drawing a line parallel to DE to complete the parallelogram. Note that the accuracy of construction of the parallelograms can be checked by projection between views.

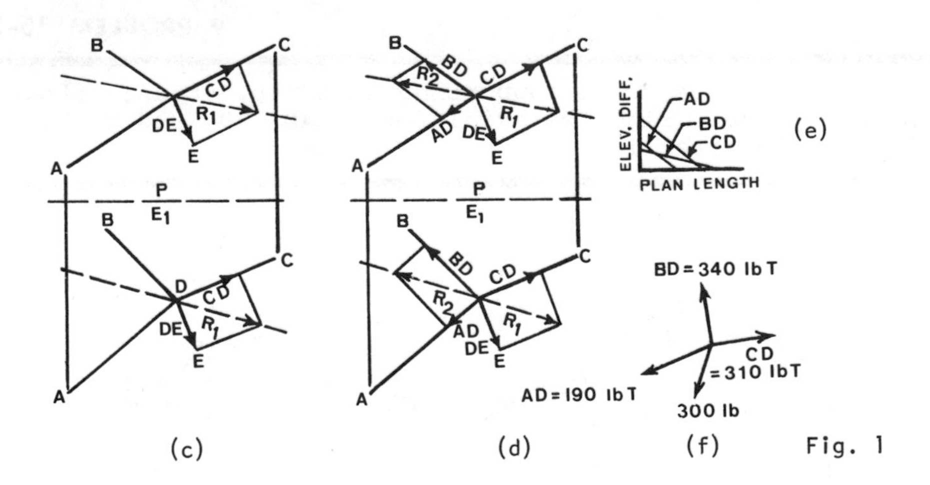

Fig. 1

The resultant R_2 of forces AD and BD is then drawn equal and opposite to R_1 [Figure 1(d)], and a parallelogram is constructed separately in each view to obtain forces AD and BD. The true lengths of vectors AD, BD, and CD can be

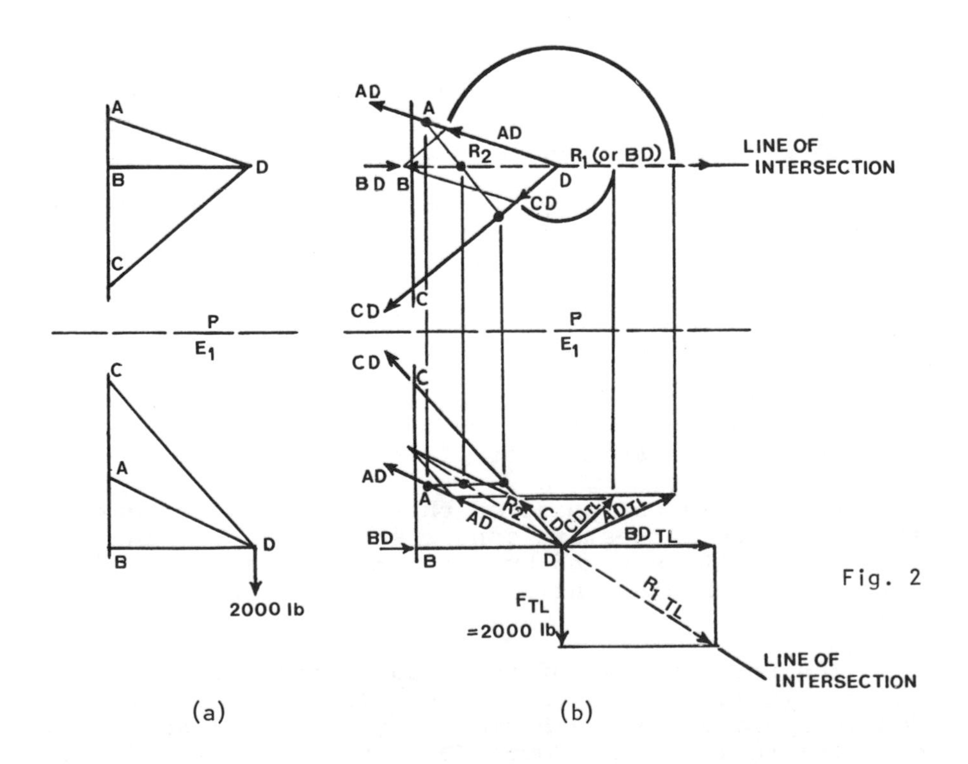

Fig. 2

found by triangulation, as shown in Figure 1(e), by revolution, or by drawing additional views. In the three-dimensional free-body diagram [Figure 1(f)], all three members are seen to be in tension.

The procedure shown step by step in Figure 1 normally would be accomplished completely in one set of related views. The normal procedure is illustrated in finding the forces in members AD, BD, and CD of the frame given in Figure 2(a). As shown in Figure 2(b), the frame and the 2000 lb load are drawn to scale. The line of intersection of plane ACD and the plane formed by BD and the load is found. A parallelogram is constructed in view E_1 with the load and BD as sides and R_1 as a diagonal. R_2 is then drawn equal and opposite to R_1, and parallelograms are constructed in both views to obtain AD and CD. Finally, the true lengths are found by revolution and the forces are transmitted to the ends of the members to determine tension or compression.

• PROBLEM 15-21

Horizontal and frontal projections of a tripod, the legs OA, OB, and OC of which are subjected to an upward vertical force of 100 lbs., are shown in the figure. Find the magnitude and type (tensile or compressive) of the loads acting in each leg of the tripod.

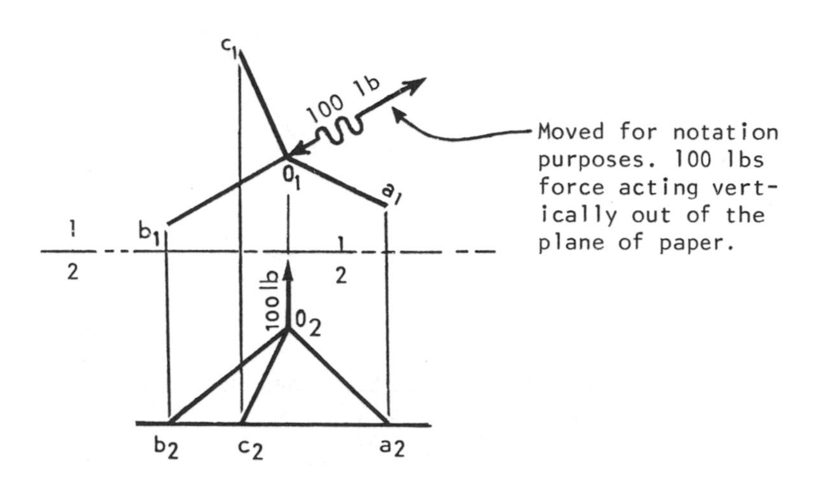

Solution: a. Through leg o_1a_1, construct a vertical plane X-X (which appears as an edge in view #1 of the space diagram) that contains the line of action of the applied 100-lb load. (The load appears as a point in view #1.) Plane X-X intersects the plane formed by legs o_1c_1 and o_1b_1 (the line of intersection is indicated as o_1r_1 in view #1, and as o_2r_2 in view #2).

b. The line of intersection OR is the line of action of the combined forces in legs OB and OC. (They are represented as F_c.) Now determine the magnitudes of the forces F_1 (leg OA) and F_c (legs OB and OC).

Construct vector diagram #1 so that the vectors are parallel to the line of action of forces F_1 (leg OA) and F_c (leg OR). The result is a coplanar vector diagram, which appears as an edge in view #1.

c. Revolve the edge view of vector diagram #1 parallel to RL 1-2. The true shape of the diagram--a triangle--appears in view #2. The magnitude of the vectors representing F_1 = 70 lb and F_c = 75 lb can now be measured.

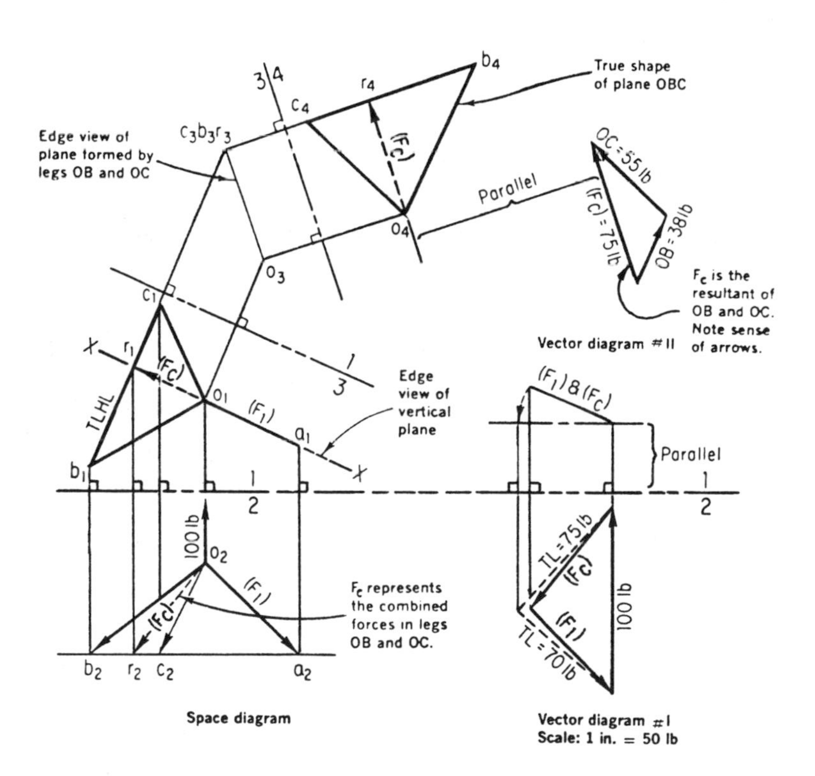

d. In the space diagram, determine the true shape view of the plane formed by legs OB and OC. (You should view the true length horizontal line b_1c_1 as a point in order to determine the edge view $o_3b_3c_3$ of the plane in view #3.) The true shape of this plane appears in view #4. Included in the true shape is the line of intersection OR between the vertical plane X-X and the plane formed by legs OB and OC. (This line of intersection is indicated as o_4r_4 and is the direction of the combined force F_c.) From view #4 of the space diagram, construct vector diagram #II, utilizing the magnitude of the combined force F_c determined in the first vector diagram. The vectors in vector diagram

#II are parallel to the direction of F_C (o_4r_4) in view #4 and to the legs o_4b_4 and o_4c_4 in view #4. Since the true shape of the plane formed by legs OB and OC appears, the vector diagram #II is also true shape and therefore all the vectors appear true length. You find by measuring that the vector representing the force in leg OB is equal to 38 lb and the vector representing the force in leg OC is equal to 55 lb.

c. Determine the types of load in the tripod legs by transferring the arrow senses from the vector diagrams to the space diagram (this was done in view #2). From this you can see that all the tripod legs are under tensile loads.

(Note: In this example, remember that both vector diagrams #I and #II deal with coplanar force systems. This approach is possible because two unknown forces were combined into one unknown force.)

• PROBLEM 15-22

Tripod frame abc is acted upon by a 1000 pound force with direction w. Find the forces in members a, b and c.

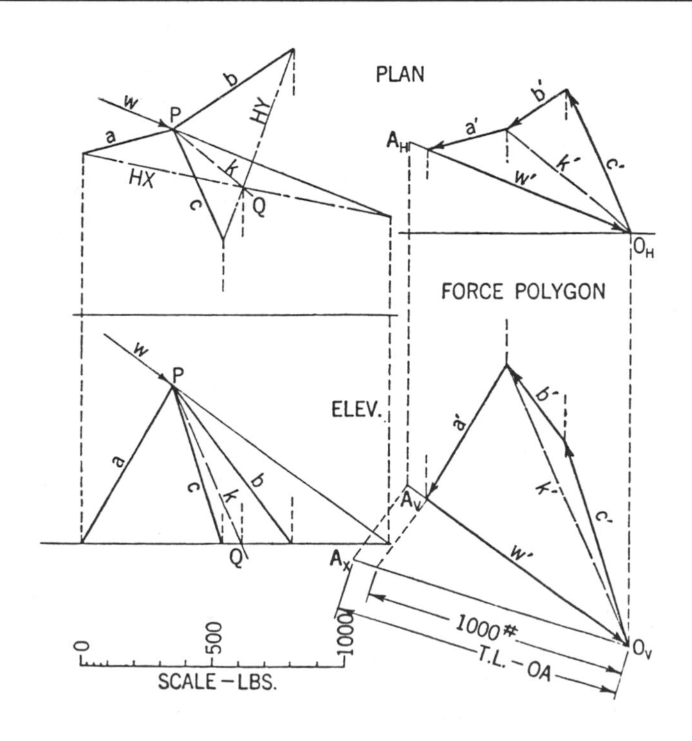

Solution: The resultant of forces having directions w and a, is equal and opposite to the resultant of the forces in the directions b and c. Since this resultant must lie in the plane of w and a, and also the plane of b and c, it lies along the line k which is the intersection of these planes.

One point on k is given by the intersection P of a, b, and c. A second point Q is found by passing a horizontal plane to cut the given members. The plane aw cuts this H-plane in HX; the plane bc cuts it in HY. Point Q is determined by the intersection of HX and HY.

Construction of force polygon. Assume the top and front views of a point of space O. Through O draw line OA of any convenient length and parallel to w, making $O_V A_V$ parallel to w_V and $O_H A_H$ parallel to w_H. Lay off on OA a distance representing 1000 pounds, using a suitable force scale. This distance is laid off on the normal view of OA and then projected to $O_V A_V$ and $O_H A_H$, given the vector w'. The arrowhead is at O.

Complete the force triangle w'k'a' by drawing vectors k' and a' parallel to frame members a and k respectively.

Complete the force triangle k'b'c' by drawing vectors b' and c' parallel to frame members b and c respectively.

The true lengths of vectors a', b', and c' (not shown here) represent the forces in frame members a, b, and c. Since the system is assumed to be in equilibrium, the arrows have the same sense in each view of the force polygon.

• PROBLEM 15-23

Given is the top and front views of a three-member frame which is attached to a vertical wall in such a way that it can support a maximum weight of 600 lbs. Find the stresses in the structural members.

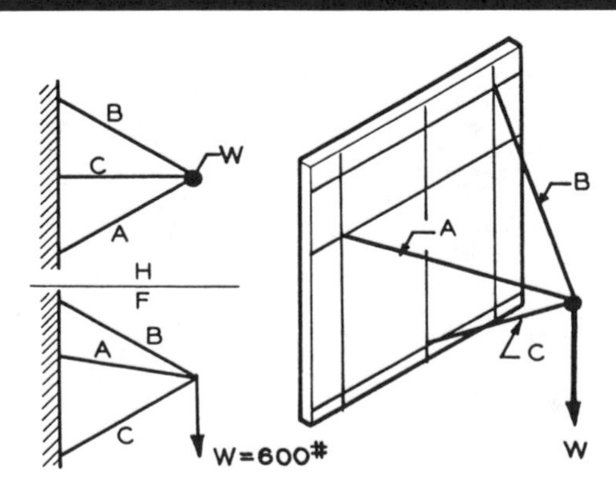

Solution: 1. To limit the unknowns to two, construct an

auxiliary view to find two vectors lying in the edge view of a plane. (See figure 1.) Use the auxiliary view and top view in the remainder of the problem. Draw a vector polygon parallel to the members in the auxiliary view (see figure 1) in which w = 600 lbs is the only known vector. Sketch a free-body diagram for preliminary analysis.

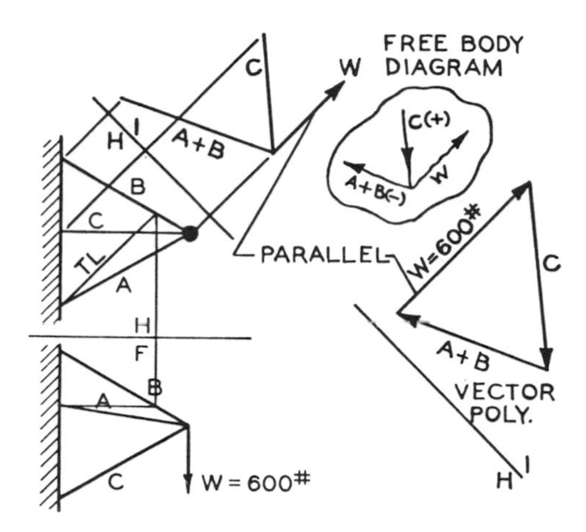

Fig. 1

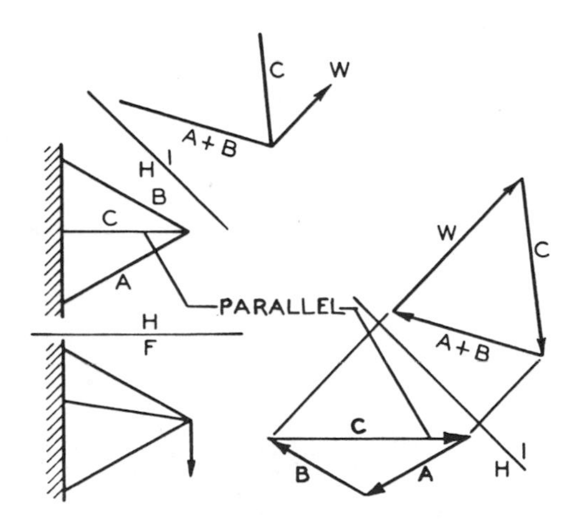

Fig. 2

2. Construct an orthographic projection of the view of the vector polygon found in step 1 so that its vectors are parallel to the members in the top view. (See figure 2). The reference plane between the two views is parallel to the H-1 plane.

3. Project the intersection of vectors A and B in the horizontal view of the vector polygon to the auxiliary view polygon (see figure 3) to establish the lengths of

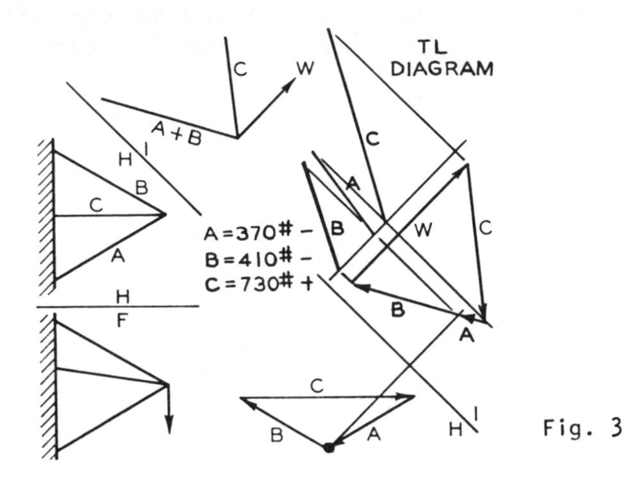

Fig. 3

vectors A and B. Determine the true lengths of all vectors in a true-length diagram and measure them to determine their stresses.

• PROBLEM 15-24

Given an externally loaded tripod frame as shown in the figure, determine the type and magnitude of the stress present in each member.

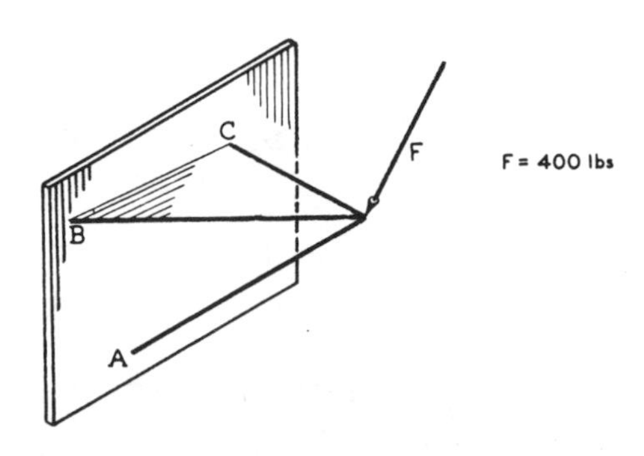

Solution: 1. Determine the projections of magnitude of the given load, F.

2. Let members OB and OC form a plane BOC.

3. Through the free end of the load vector, R, construct a line parallel to member OA.

4. Determine the exact location of the piercing point P of line OA and the plane BOC.

5. Through piercing point P draw an action line parallel to member OC until it intersects the line of action of member OB.

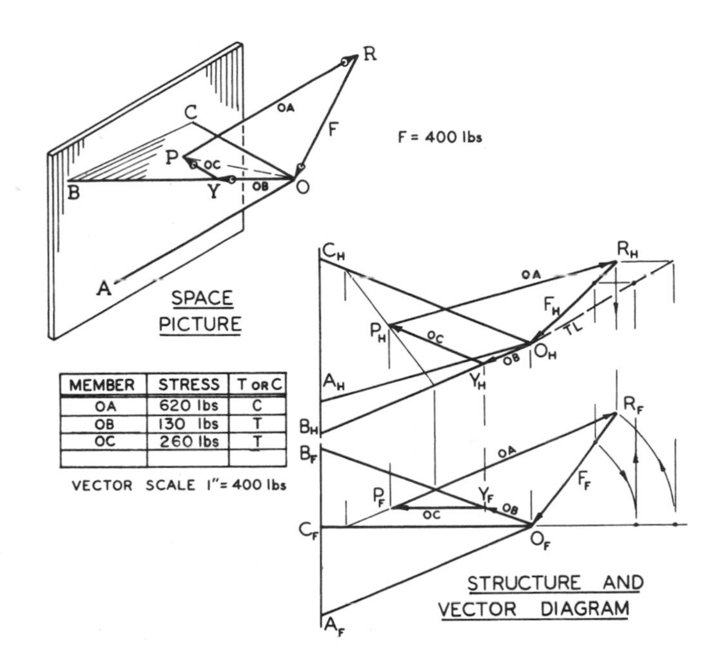

MEMBER	STRESS	T OR C
OA	620 lbs	C
OB	130 lbs	T
OC	260 lbs	T

6. Proceeding around the vector diagram in the direction indicated by the arrowhead of the given load vector, place arrows on the vectors, OB, OC, OA.

7. Determine the true magnitudes represented by vectors OB, OC, OA. Construction left for the student.

8. Determine kind of stress in each member.

9. Tabulate the results.

• PROBLEM 15-25

Find the forces acting on the members a, b, and c of the frame which supports the weight W.

Solution: Draw the front and side views of vector OR parallel to W and equal in length to 3000 pounds on the force scale. Through point O draw lines a, b, and c parallel to frame members a, b, and c respectively. Construct the views of the parallelepiped. The magnitudes of the forces

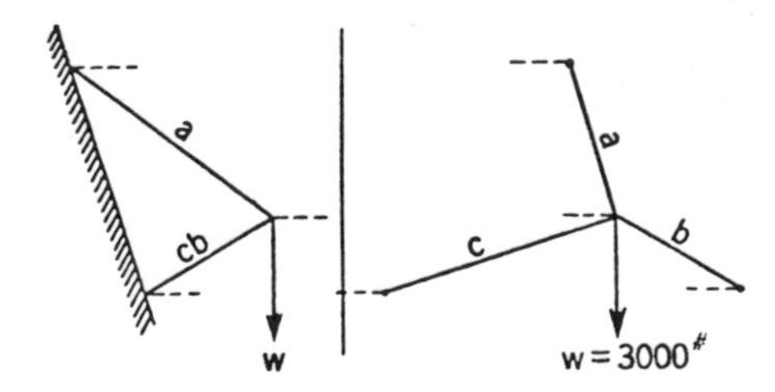

in members a, b, and c are represented by OQ, QS, and RS respectively. The true lengths of these vectors give the magnitudes of the components.

On examining the completed drawing, it will be seen that the force problem can be solved by drawing a skew quadrilateral having three sides parallel to, and in the same direction as, the three given vectors. The fourth and closing side of the quadrilateral is the balancing

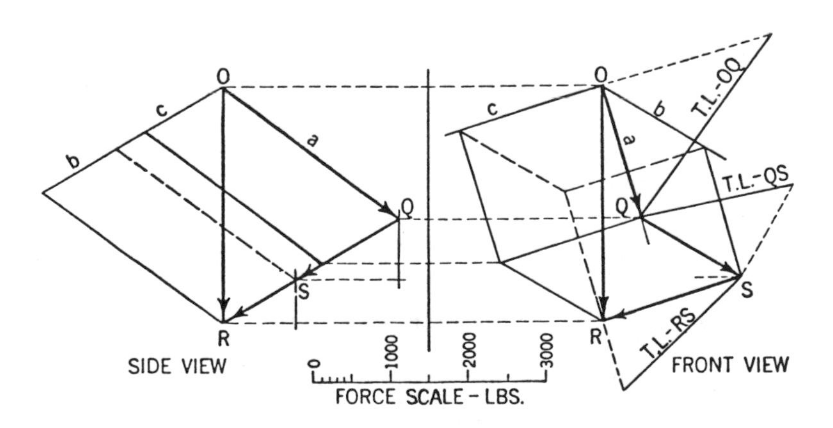

force. Such a figure is called a force polygon. When the vectors forming a force polygon point in the same direction around the polygon, the forces are in equilibrium. If one vector is drawn in a contrary direction, it represents the resultant of the remaining forces. The arrows shown in the figure represent the components of OR.

• PROBLEM 15-26

Given are the top and front views of a structural crane upon which a known force of 10,000 lbs is applied in the direction shown at point 0. Find the stresses in each of the structural members, 0-3, 0-2, and 0-1.

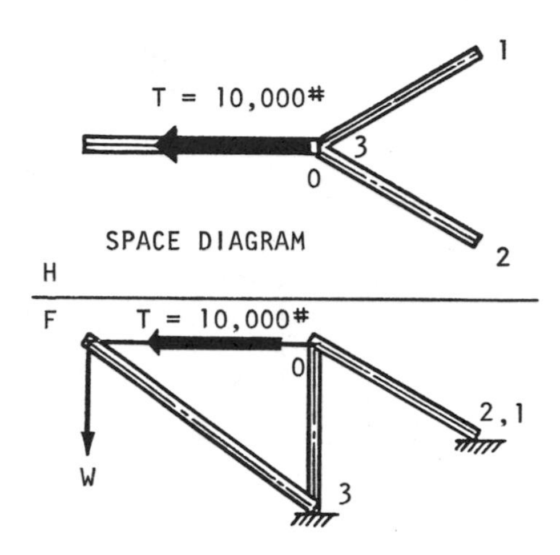

Solution: 1. Construct a vector polygon in the front view by drawing force T and the other forces as vectors (see fig. 1). Note that one side of the polygon represents the summation of two vectors, 0-2 and 0-1. Construct the top view of the polygon orthographically by drawing vectors 0-2 and 0-1 parallel to their top view. Vector 0-3 appears as a point in the top view.

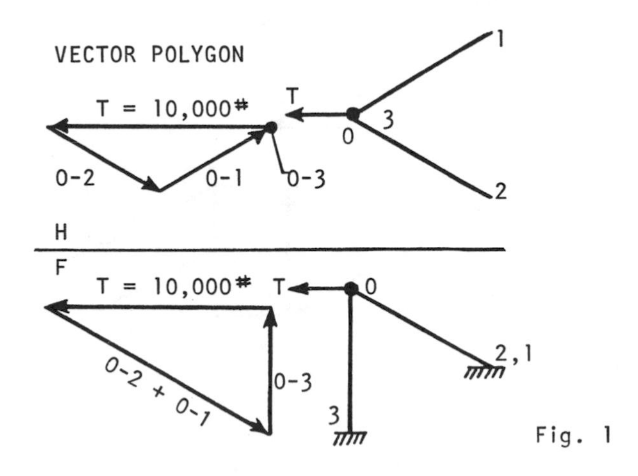

Fig. 1

2. Project the point of intersection of 0-2 and 0-1 in the top view to the front view (as done in fig. 2) to sep-

arate these vectors. Draw all vectors head-to-tail. Referring to fig. 2 you will note the sense of vectors 0-1 and 0-2 is away from pt. O causing tension while 0-3 is towards pt. O causing compression.

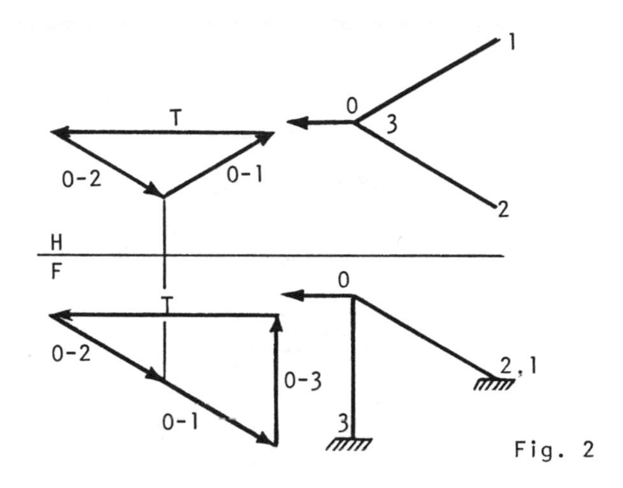

Fig. 2

3. The completed top and front views found in step 2 do not give the true lengths of vectors 0-2 and 0-1 since they are oblique. Determine the true lengths of these

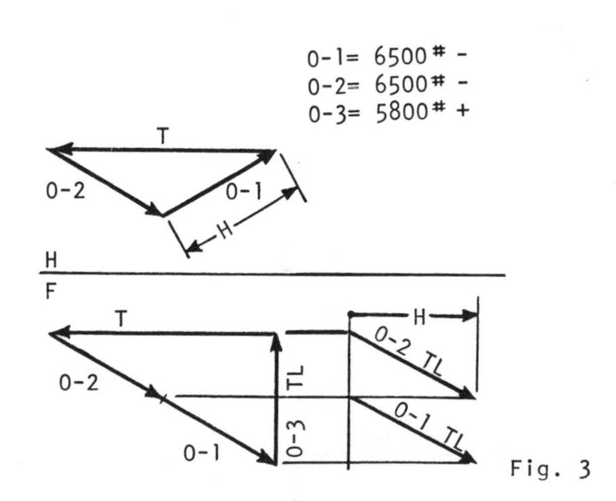

Fig. 3

lines by revolving the lines and drawing a true-length diagram. Then scale these lines to find the stresses in each member.

• PROBLEM 15-27

Fig.(a) shows three unknown concurrent noncoplanar forces and their known resultant. Determine the magnitude of OA, OB and OC.

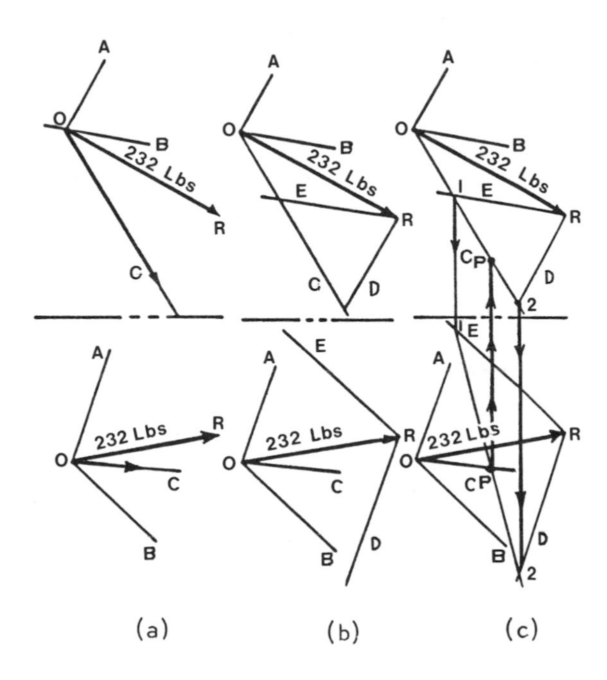

Solution: In order to find the components, begin by drawing the opposite end of the parallelepiped. See Part B of the figure. This is begun at the end of the resultant. In both views, draw two sides of the base, RE and RD, parallel to OB and OA. The third step is to determine where the edge, OC, pierces the base of the parallelepiped. Part C of the figure shows the application of determining where a line pierces a plane. Where line OC pierces the base is the length of one of the edges of the parallelepiped. Next the base can be completed from point C_p, shown in Fig. D. The magnitudes of forces OA, OB, and OC can be scaled at this time; however, the parallelepiped is usually completed, Part E, and visibility determined,

Part F. To obtain the magnitudes of each force, OA, OB, and OC must be revolved parallel to a plane of projection.

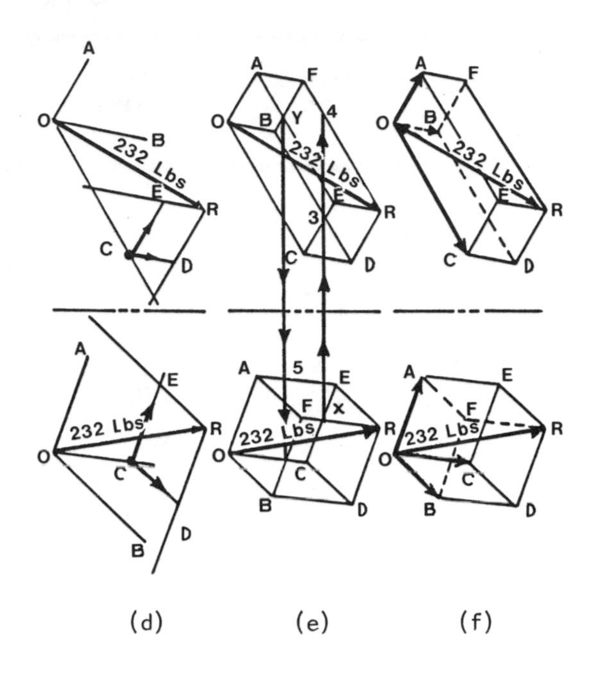

(d) (e) (f)

• PROBLEM 15-28

Determine the magnitude and direction of the resultant of the system of concurrent, noncoplanar forces shown in the figure.

Solution: The projection of the forces in space upon any plane must form a system of forces in equilibrium. Before the vector polygon can be constructed, it is necessary to determine the projected magnitudes of the given forces. In other words, lay off the given magnitude of each force on the true-length view of its line of action, and then show the projections of these magnitudes on the direction lines of the given force diagram, Fig. (b).

The members of the vector chain are then constructed, Fig. (c), parallel to the lines of action and equal to the projected magnitudes shown in the corresponding views of the force diagram. In order that the system of forces shall be in equilibrium a closing link is drawn in the vector diagrams to show the line of action and the projected magnitude of either the resultant or the equilibrant of the system. The direction of the resultant, Fig. (c),

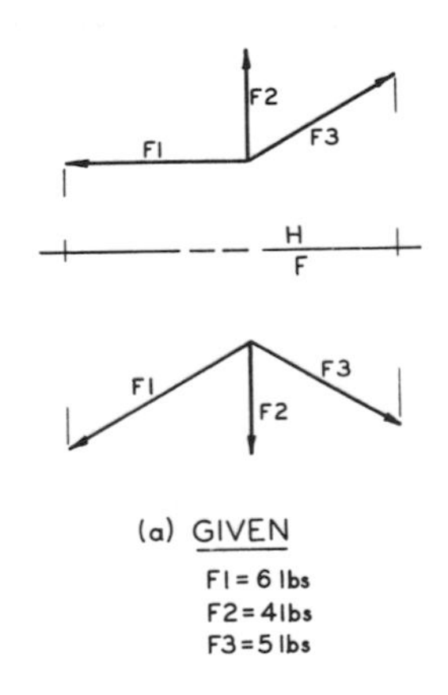

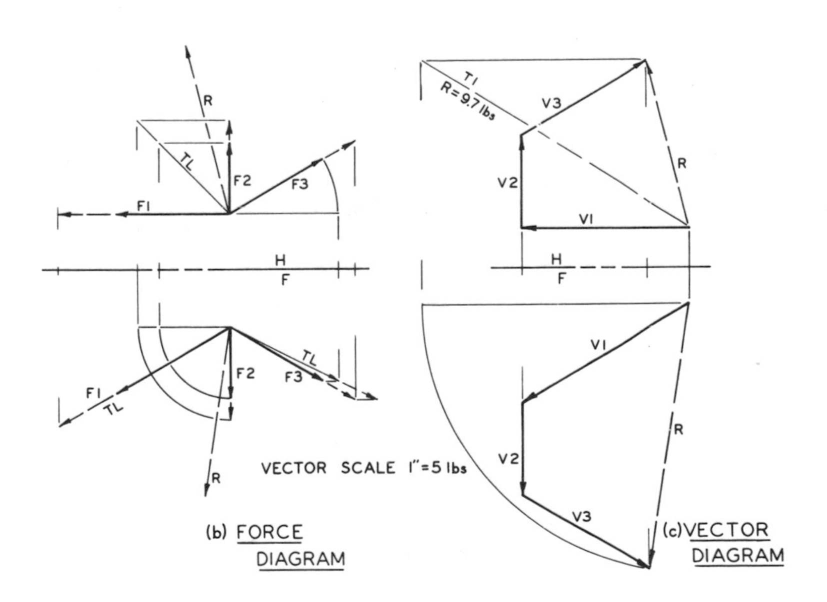

opposes that of the chain, while the equilibrant continues the direction of the rest of the vectors in the chain. The true magnitude of the resultant can be found by rotation or by the auxiliary view method. The line of action and the direction of the resultant are then applied to the original force diagram to complete the solution of the given problem.

BOW'S NOTATION

• PROBLEM 15-29

Explain Bow's notation with the aid of an example.

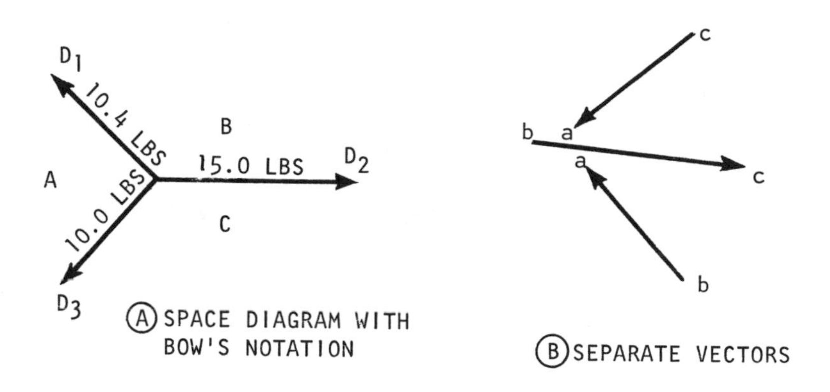

(A) SPACE DIAGRAM WITH BOW'S NOTATION

(B) SEPARATE VECTORS

Solution: Bow's notation is a system of designating coplanar forces. This method of noting forces acting in the same plane is widely used in engineering practice. Bow's notation simply uses capital letters between lines of action. Any line of action or force can be designated by

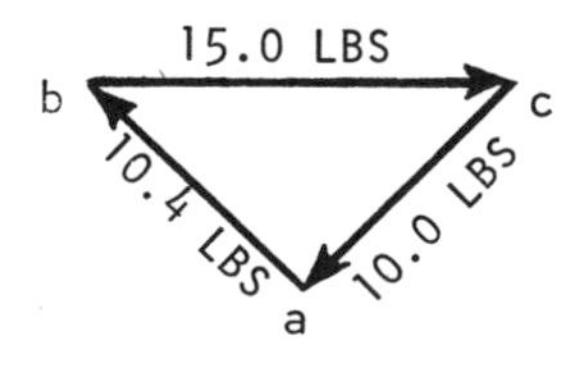

(C) VECTOR POLYGON

THESE THREE FORCES ARE IN EQUILIBRIUM WITH BOW'S NOTATION

the letters in the adjacent spaces. In the illustration, force D_1 is referred to as AB. Force D_2 is called BC. Force D_3 is identified as CA. The letters are usually placed between the forces in a clockwise manner. The first letter notes the tail end (A). The arrow end is noted by the second letter (B). The tip end of force D_2

is C and the tail end is B. Thus the force is called BC. It is standard practice to read the forces in a clockwise manner. Instead of A, B, C, etc., numerals such as 1, 2, 3, etc. may also be used.

For complicated structures, letters may be used along with numbers.

• PROBLEM 15-30

A three dimensional frame supporting a 2000 lb load, and the corresponding resultant vector system are shown in the figure. Use

(a) Edge view method, and

(b) Point view method

to show that the values given for the vectors are correct.

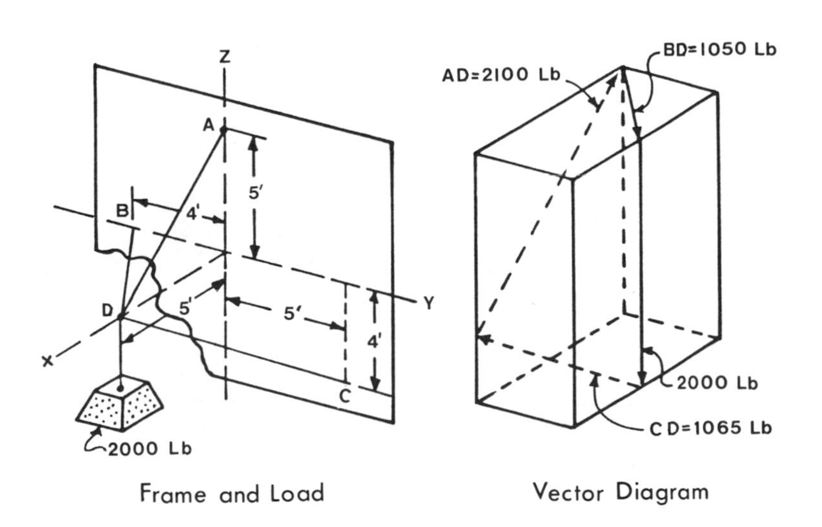

Frame and Load

Vector Diagram

Solution: Edge-view method:

The P and E_1 views of the frame are first drawn to scale as shown in Figure 1(a), the load being represented by an arrow.

The plane containing members AD and CD is seen as an edge in view I_1, where the unknowns therefore appear in only two directions.

Note: In drawing the vector diagram, it is absolutely essential that, in both views, (1) the vectors be kept in the same sequence and (2) the two vectors in the plane that appears as an edge be kept consecutive (that is, adjacent to each other).

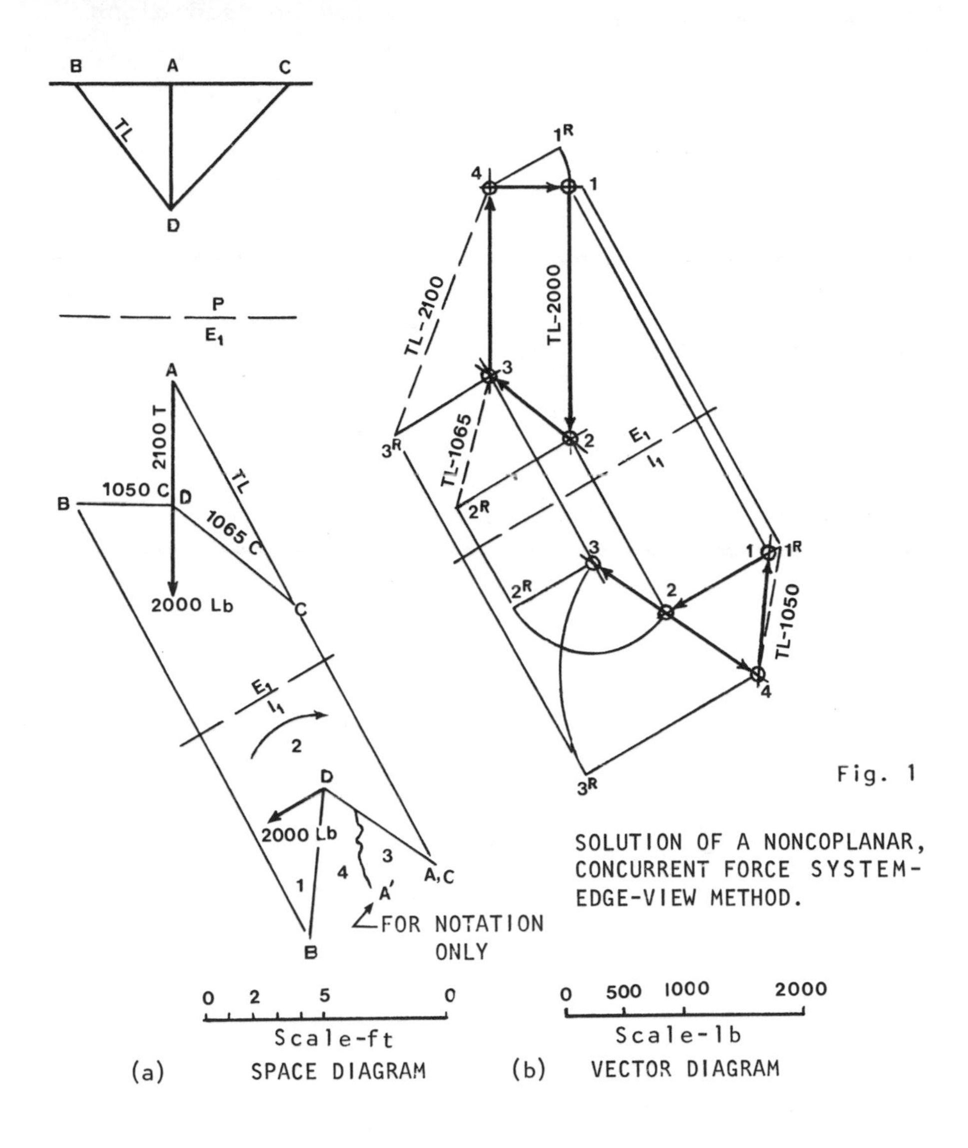

Fig. 1

SOLUTION OF A NONCOPLANAR, CONCURRENT FORCE SYSTEM-EDGE-VIEW METHOD.

(a) SPACE DIAGRAM (b) VECTOR DIAGRAM

To insure this, Bow's notation is applied in the view showing the plane as an edge (view I_1). A number is placed in each space between members and the load. To allow a number to be placed between members AD and CD, member AD is considered to be moved to the position (A'D) shown by the wavy line. This position is for purposes of notation only, and does not change the actual direction of member AD. The curved arrow is drawn to indicate the order in which the vectors are to be connected in the vector diagram. The direction can be either clockwise (as shown) or counterclockwise, but once the choice is made, the vectors must be laid out in that order in both views.

The values of the unknown forces are determined by drawing the E_1 and I_1 views of the vector diagram as shown in Figure 1(b). A rotation line is established parallel to the $\frac{E_1}{I_1}$ rotation line of the space diagram so that directions for the vectors can be established parallel to the corresponding members of the space diagram.

The step-by-step development of the vector diagram is illustrated in Figure 2.

First, as shown in Figure 2(a), a line is drawn, in each view of the vector diagram, parallel to the known load in the corresponding view of the space diagram; Figure 1(a). The known load is laid out to scale in view E_1, where it appears true length, and is then projected into view I_1. In view I_1 of Figure 1(a), where Bow's notation was applied, the load is between numbers 1 and 2 as we read around the joint in the chosen clockwise direction. Thus, the direction of the vector is established (by Bow's notation) as extending from number 1 to number 2 in the direction indicated by the arrowhead. Therefore,

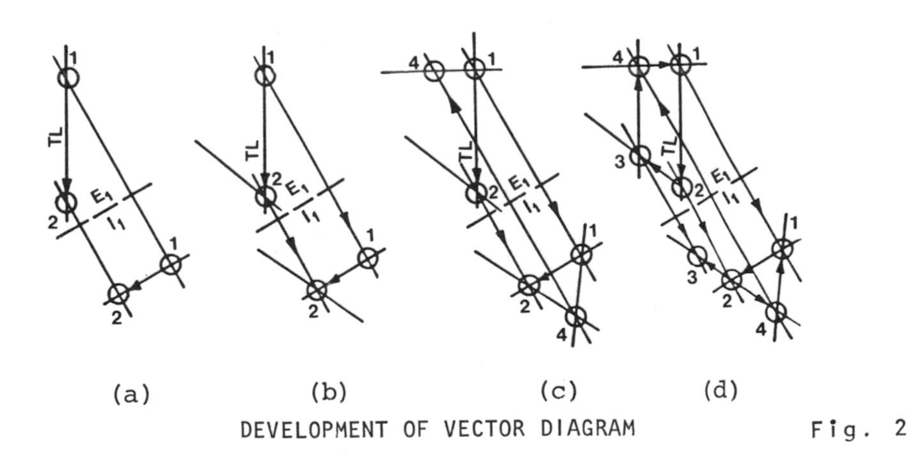

DEVELOPMENT OF VECTOR DIAGRAM Fig. 2

the number 1 is placed at the tail of the vector and the number 2 at its tip in both views of the vector diagram.

Next, as shown in Figure 2(b), a line parallel to member CD is drawn through point 2 in each view. Point 3 must lie somewhere along this line because member CD is between numbers 2 and 3 of Bow's notation. Since the length of vector 2-3 is unknown, the exact position of point 3 must be established by locating its intersection with vector 3-4. Clearly, since point 3 is not known, vector 3-4 cannot be drawn until point 4 has been located.

In view I_1, point 4 must lie along the line drawn through point 2, because member AD, which appears between numbers 3 and 4, coincides with member CD in view I_1 of the space diagram. Also, point 4 is on vector 4-1, which is shown by Bow's notation to be parallel to member BD. Therefore, as shown in Figure 2(c), a line is drawn through point 1 parallel to member BD in both views. Point 4 is found in view I_1 at the intersection of this line with the line previously drawn through point 2, and is then located in the plan view by projection.

Next, as shown in Figure 2(d), a line is drawn parallel to member AD, through point 4 in view E_1, to intersect the line previously drawn through point 2, thus locating point 3. Point 3 is then found in view I_1 by projection. The vector diagram is completed by adding arrowheads as shown, since the vectors must be connected tip to tail in the vector polygon.

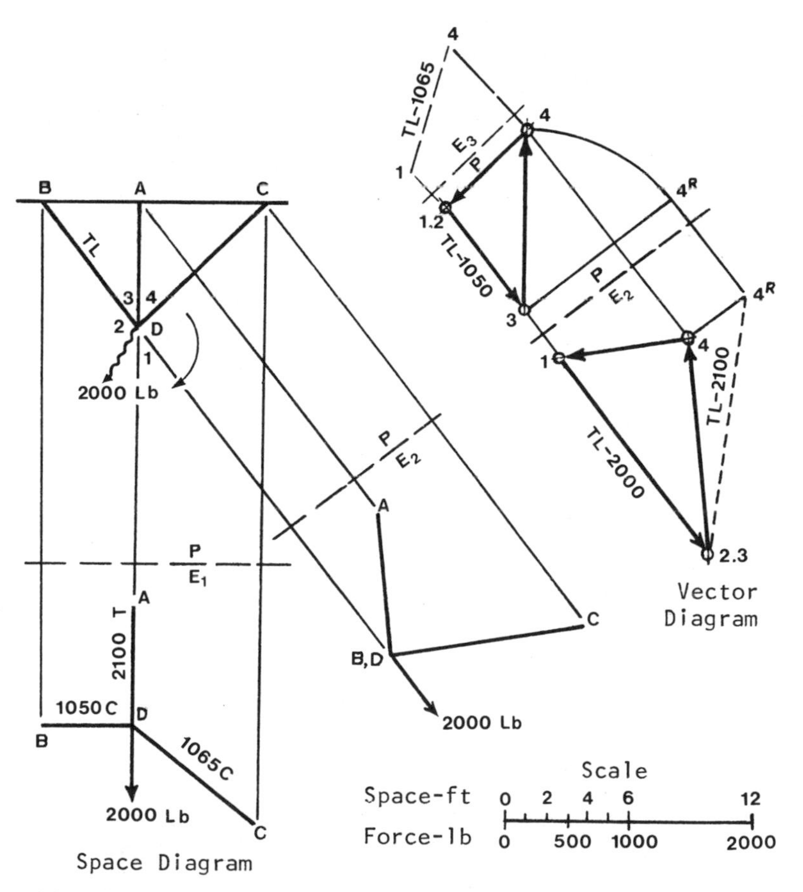

SOLUTION OF A NONCOPLANAR, CONCURRENT FORCE SYSTEM-POINT-VIEW METHOD. Fig. 3

The true length of each vector is found by revolution as shown in Figure 1, and the value is scaled. For easy reference to the answers, the values are placed along the members in one view of the space diagram.

The final step is to determine whether the stress in each member is tension or compression by comparing directions in the vector and space diagrams (Figure 1). For example, the stress in member CD is shown by the vector from 2 to 3. In either view, the arrow on vector 2-3 is seen to point along CD toward the joint at D, thus indicating that CD is in compression. Similarly, the stress in AD is tension, since vector 3-4 points away from the joint, and the stress in BD is compression because vector 4-1 points toward the joint. The letters T for tension and C for compression are placed after the values of the stress on the space diagram, thus completing the solution.

Point-view method:

The P and E_1 views of the space diagram are drawn (Figure 3) using the same data as in the preceding example. Since BD is in its true length in the plan view, view E_2 is drawn to show BD as a point. The unknowns therefore appear in only two directions in view E_2.

Bow's notation is applied in the plan view, where the 2000-pound load is moved to one side for purposes of notation. The curved arrow is drawn to show that the vectors will be connected in clockwise sequence in the vector diagram.

The rotation line $\frac{P}{E_2}$ for the vector diagram is drawn parallel to the corresponding rotation line for the space diagram. Vector 1-2, representing the known load, is laid out to scale in view E_2, where it appears in true length. Vector 2-3 appears as a point in view E_2, and perpendicular to the rotation line in the plan. In view E_2, lines are drawn through point 3 parallel to AD and through point 1 parallel to CD, thus locating point 4 by intersection. A line parallel to CD is then drawn through point 1 in the plan view, and point 4 is located by projection from view E_2. Next, point 3 is located by drawing a line parallel to AD through point 4 in the plan view, and the vector diagram is completed by adding arrowheads as shown. The value of the stress in BD is found by scaling vector 2-3 in the plan, where it appears true length. The true lengths of the other vectors are found by revolution (as shown for vector 3-4) or by additional views (as shown for vector 4-1). Finally, tension or compression is determined as in the preceding example, and the stresses are recorded on one view of the space diagram.

• PROBLEM 15-31

Horizontal and frontal projections of an externally loaded tripod, the legs of which are OA, OB, and OC, acted upon by an external downward vertical load with a 100-lb magnitude are shown in the given figure. Find the magnitude and type of the load (tensile or compressive) carried by each leg of the tripod.

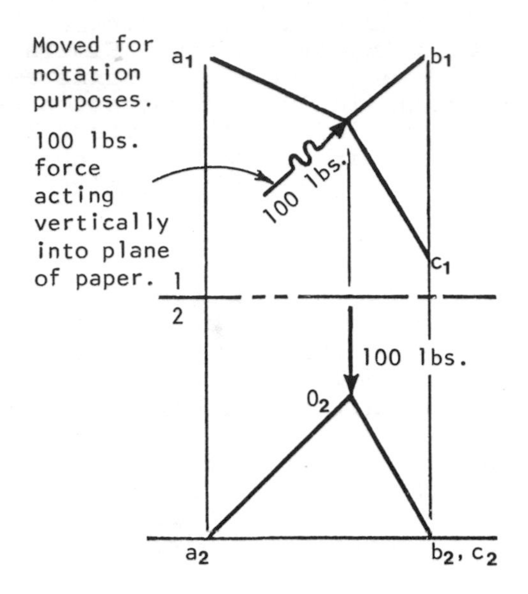

Solution: a. In the space diagram in Fig. 1, the plane formed by legs OB and OC appears as an edge $o_2b_2c_2$ (view #2). Apply Bow's notation in view #2. (Number consecutively the spaces on either side of the known load, and then number the spaces on either side of each tripod leg.) For purposes of notation, leg o_2c_2 was moved to a general position indicated by $o_2c_{2_m}$.

b. Summarize Bow's notation in view #1, indicating o_1a_1 between #2 and #3, o_1b_1 between #4 and #1, and o_1c_1 between #3 and #4.

c. Using a convenient scale (1 in. = 50 lb has been indicated in Fig. 1) construct a vector diagram in which the vectors are parallel to the given external load and to the respective legs of the tripod.

[Note: The following steps sequentially develop the procedure for drawing the complete vector diagram.]

d. Referring to Fig. 2, construct a vector representing the applied 100-lb load. This vector will start at some convenient point #1 (as indicated in view #2) and end at point #2. Its horizontal projection is a point indicated as #1.2.

e. Referring to Fig. 3, in view #2, construct a line from point #2 parallel to leg o_2a_2. (The length of this line is at this stage indefinite.)

f. Referring to Fig. 4, in view #2, construct a line from point #1 parallel to legs o_2b_2 and o_2c_2 (in this view they are behind each other and therefore have the same direction). Extend this line until it intersects the line

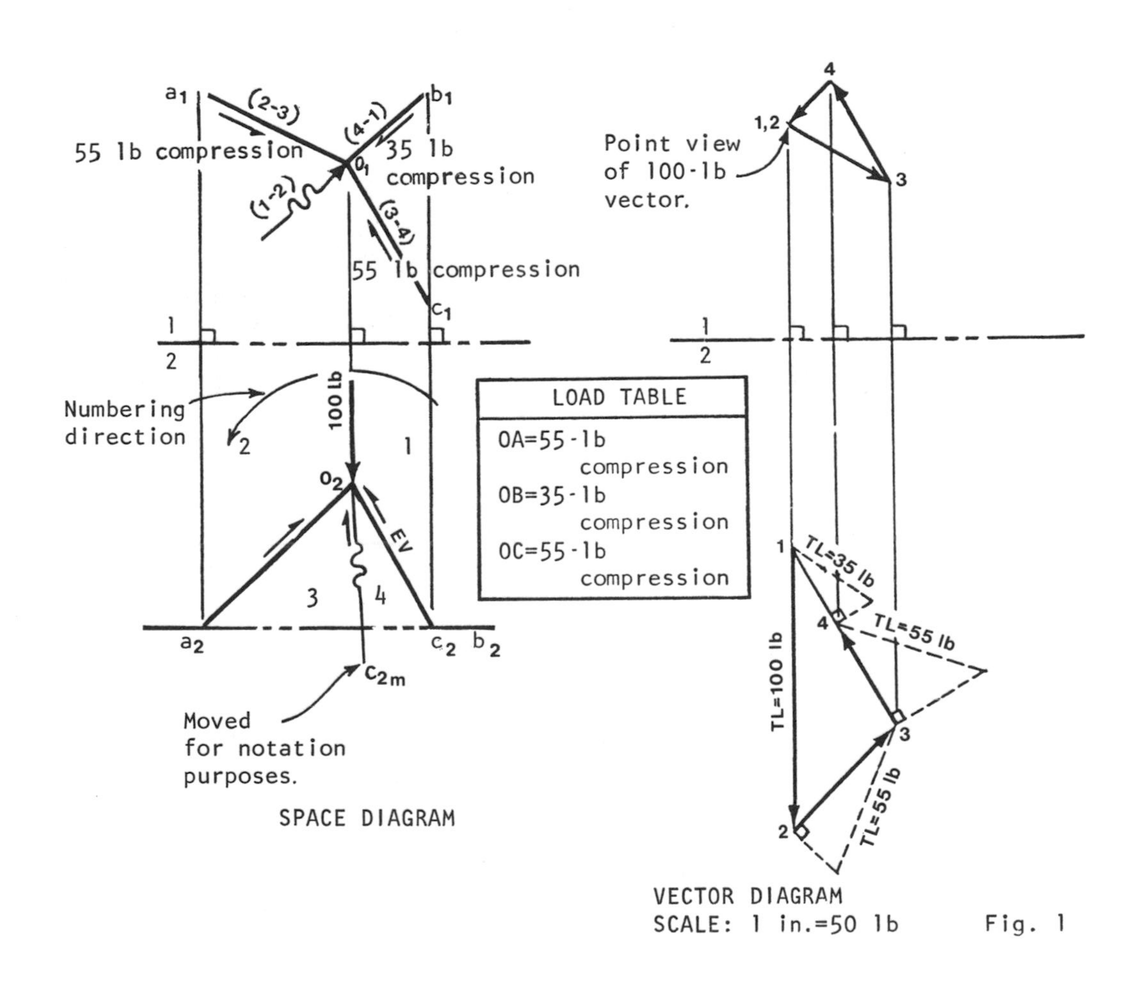

Fig. 1

2-3 (which is parallel to leg o_2a_2 in view #2). The point of intersection of these two lines locates #3 in the vector diagram. In view #2, note that the vector diagram has closed, in spite of the fact that #4 has not yet been located as a specific point. You must obtain additional information from view #1 in order to locate #4.

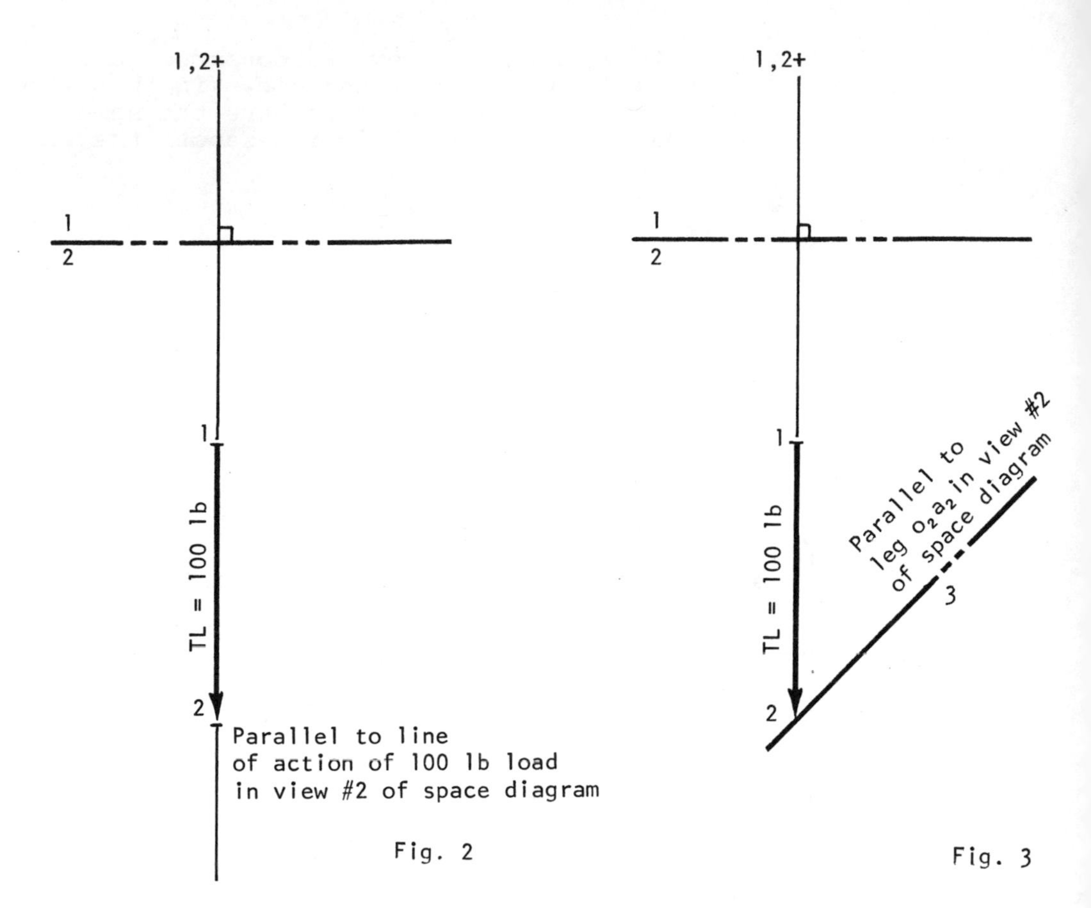

Fig. 2

Fig. 3

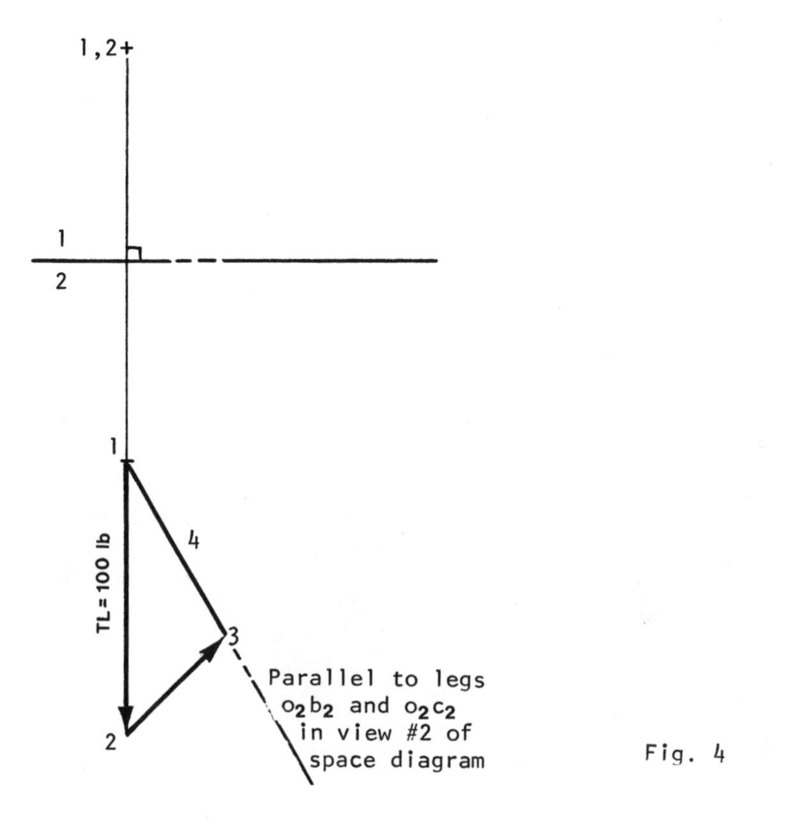

Fig. 4

g. Referring to Fig. 5, in view #1, construct a line from the point view of the 100-lb force (indicated as #1.2) parallel to leg o_1a_1. (The length of this line is indefinite.)

h. Referring to Fig. 5, from view #2, project #3 upward to view #1 to determine point #3.

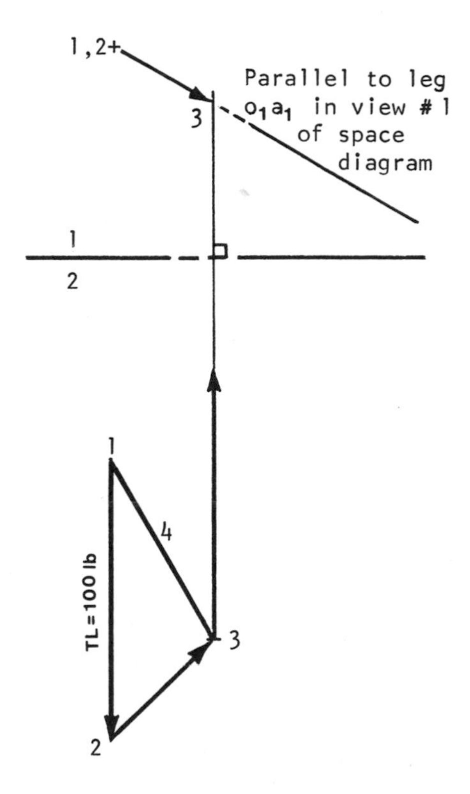

Fig. 5

i. Referring to Fig. 6, in view #1, construct a line from point #3 parallel to leg o_1c_1.

j. Referring to Fig. 7, from the point view of the applied 100-lb load (indicated as #1,2 in view #1 of the vector diagram) construct a line parallel to leg o_1b_1 in the space diagram. The point of intersection of the lines parallel to legs o_1c_1 and o_1b_1 locates point #4 in view #1 of the vector diagram. Project point #4 from view #1 downward to view #2 in the vector diagram.

k. Referring to Fig. 7, the sense of the applied 100-lb load determines the sense of the remaining vectors. To indicate a condition of equilibrium, the sense of the vectors must follow each other. This means that the arrows will be pointing toward #2, #3, #4, and #1. You can see that the vector diagram closes,indicating a resultant of zero magnitude, and therefore, a condition of equilibrium for the given tripod. (Show the sense of the vectors in both views.)

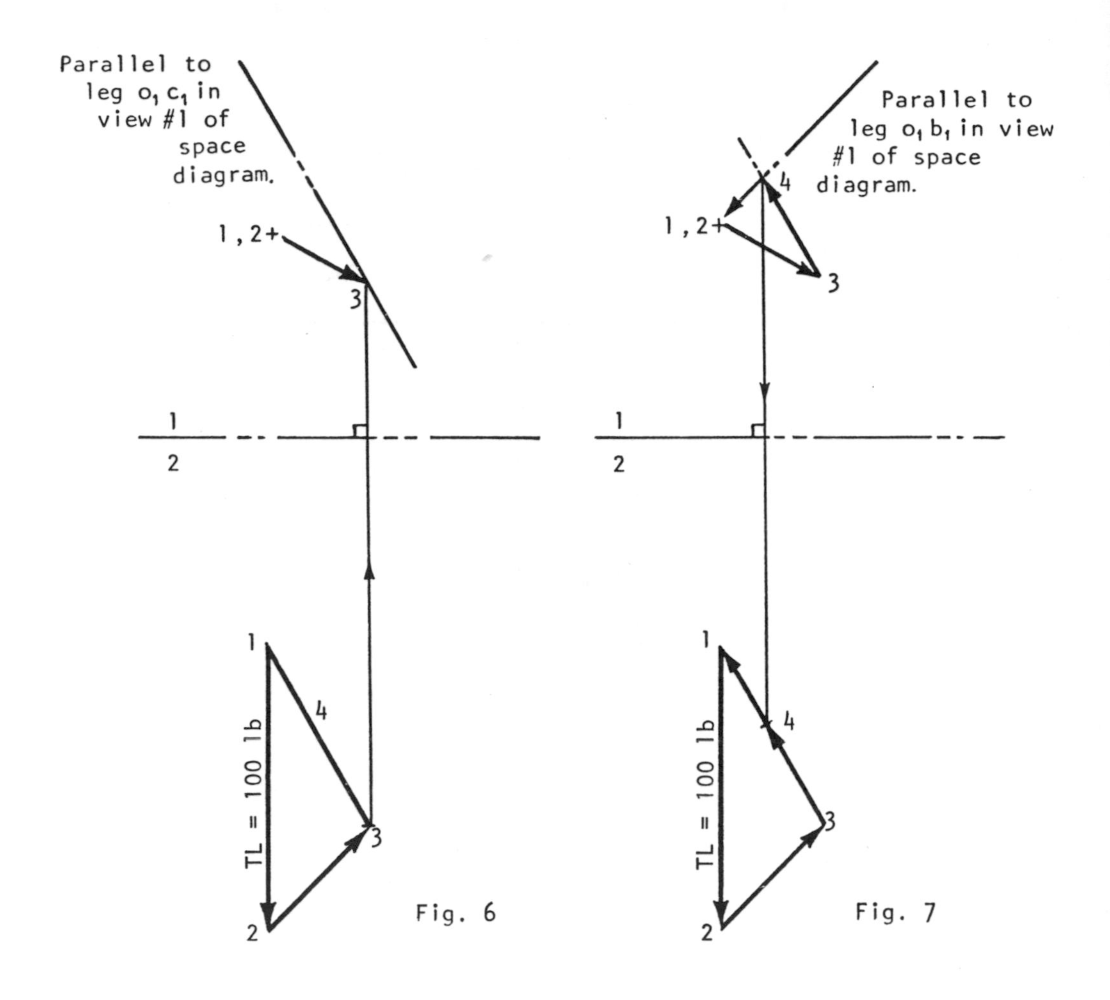

Fig. 6

Fig. 7

l. Referring again to Fig. 1, in the vector diagram, determine the true length of each vector using the short-cut method. The loads carried by each leg of the tripod are summarized as: leg OA = 55 lb, leg OB = 35 lb, and leg OC = 55 lb.

m. To determine whether the loads are tensile or compressive, transfer the sense of each vector to its respective leg in either view #1 or #2. The sense of a vector acting toward the point of application of the external 100-lb vertical load indicates that the leg is in compression. From the space diagram in Fig. 1, you can see that the sense of all the vectors act toward the point of application; therefore, all the tripod legs are in compression.

• PROBLEM 15-32

Determine the type and magnitude of stress in the members of a tripod structure supporting a 500 lb. vertical load.

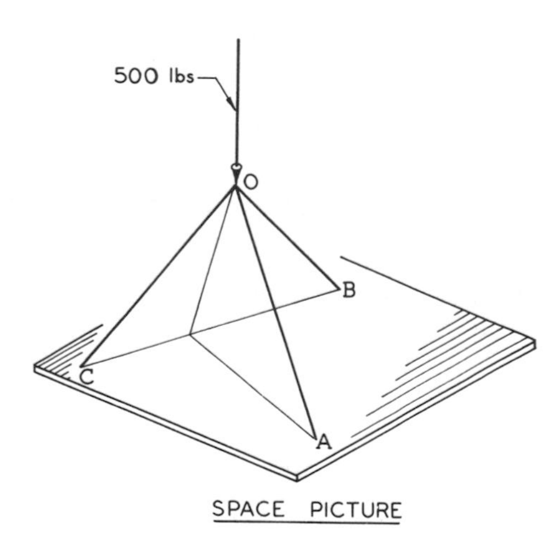

SPACE PICTURE

Solution: If the structure is to be in equilibrium, the vectors, representing the load and the reactions, must form a closed vector chain.

The magnitude and direction of the applied load, 500 lbs., and the directions only of the three supporting members are known. The known load must be distributed into the supporting members if the structure is to be in equilibrium.

In the given orthographic views of the structure diagram, two of the supporting members lie in a plane which appears as an edge in the front view. Because of this, two views of a closed vector diagram can be constructed immediately.

1. Apply Bow's Notation only to the view showing the edge view of the plane containing two of the members, in this case the front view. Starting with the known load and proceeding through the single member OA and through the edge view of the plane OBC, apply the letters RSTU as shown in figure. For this problem the direction of travel around the point of concurrency O is clockwise.

2. Construct two views, H' and F', of a vector diagram such that the directions of the lines of action and projectors are parallel to the given H and F views of the structure in the following manner:

   a. Construct two views H' and F' of the vector RS representing the given load. Note that, since the load is vertical, the vector appears in true magnitude in the F' view and as a point in the H' view.

   b. Construct, through S_F, a line of action parallel to the single member O_FA_F. This line of action contains the point T.

   c. Close the F' view by constructing, through point R, a line of action parallel to the edge view of the plane OBC. This line of action contains the points T and U. Point T is now fixed in the H' and F' views.

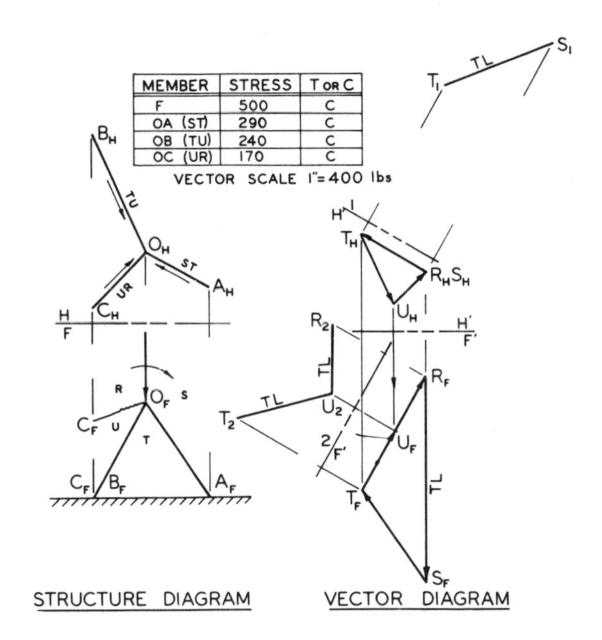

   d. Close the H' view by drawing through T parallel to the H view of member OB. This line of action contains point U. Draw through R a line parallel to member OC. Since this line also contains point U, its location is now fixed.

   e. Complete the F' view by projecting point U onto the line parallel to the edge view of the plane.

   f. Commencing with the known force RS, place arrows at the terminal points T and U. This establishes the direction of vectors R to S, S to T, T to U and U to R.

g. Find the true magnitudes of the vectors ST, TU, and UR.

h. Transfer the direction of the vector arrows, ST, TU, and UR, to corresponding members of the structure diagram. If the arrow points toward the point of concurrency, the member is in compression; away from the point, tension.

i. Tabulate.

• PROBLEM 15-33

Determine the type and magnitude of the stress in the members of a tripod with a load of 100 pounds applied as shown in the figure.

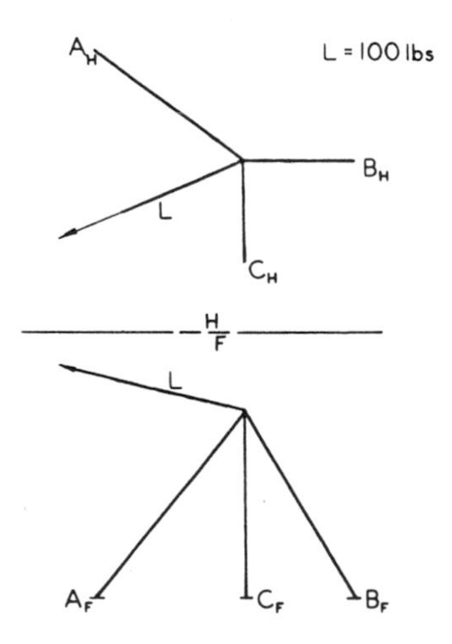

Solution: 1. Determine the projections of magnitude of the load.

2. Construct an auxiliary view showing the edge of a plane containing two of the unknowns.

3. Apply Bow's notation to the constructed auxiliary view.

4. Construct the two projections of the vector polygon respectively parallel to the auxiliary and its adjacent view. The vector representing the load is parallel to and equal in length to the projected magnitudes found in the auxiliary and its adjacent view.

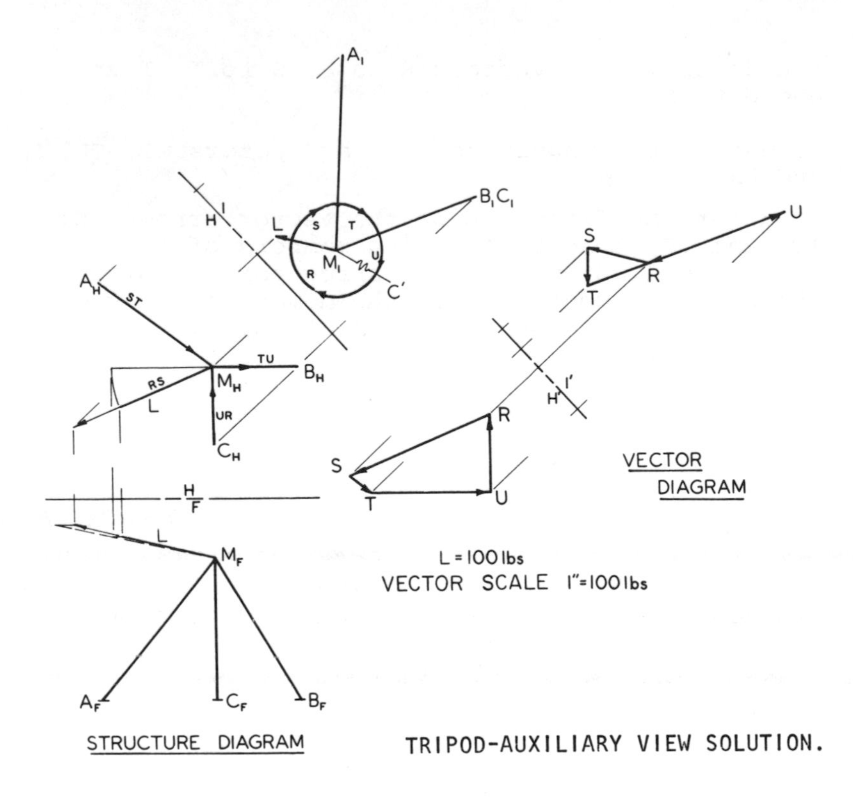

TRIPOD-AUXILIARY VIEW SOLUTION.

5. Determine the true magnitudes of the stress in the members of the tripod.

6. Determine the kind of stress in the members (compression or tension).

7. Tabulate all results.

FORCES

• PROBLEM 15-34

Given two parallel nonconcurrent forces F1 (60 lbs.) and F2 (40 lbs.) applied to a rigid body as shown in the figure. Determine:

a. the magnitude and direction of the equilibrant.

b. the location of the equilibrant on the rigid body.

Solution: 1. Apply Bow's notation.

2. Construct the vector polygon, to some appropriate scale. Note that the polygon is a straight line since the forces are parallel. The magnitude of the equilibrant E will be the vector sum, CA, and its direction will be from C to A.

3. From an arbitrarily selected pole point P, draw straight lines (rays) to the points A, B and C of the vector polygon. The resulting figure is sometimes called a ray diagram.

The vector AB has thus been resolved into two components AP and PB, and the vector BC into BP and PC. Notice that components PB and BP are equal in magnitude but opposite in direction, thus cancelling out and leaving AP and PC as the components of the equilibrant CA.

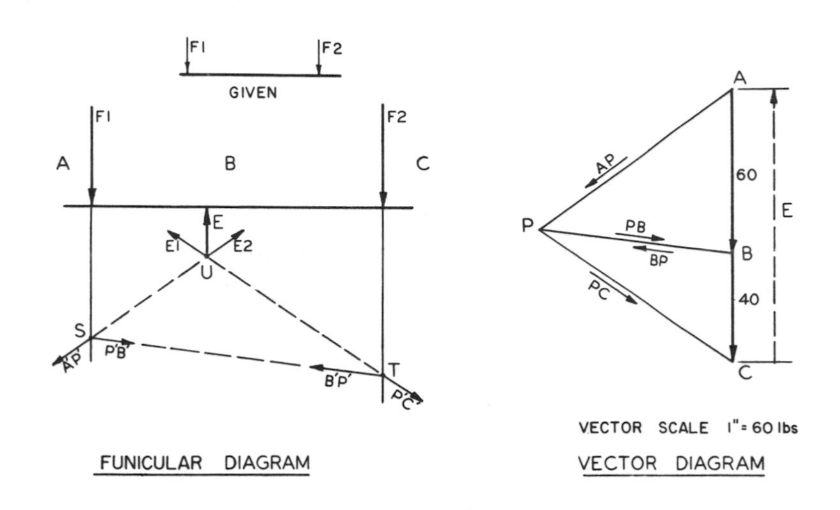

4. Construct the funicular diagram or string diagram in order to locate the exact position of the equilibrant on the given body.

a. Through any point S on F1 extended, construct two lines of action A'P' and P'B' respectively parallel to the vectors AP and PB.

b. Through a point T on F2 extended, construct two lines of action B'P' and P'C' respectively parallel to the vectors BP and PC.

Note: The point T must be located so that the action line B'P' is collinear with the action line P'B' through point S in order that the system shall be in equilibrium.

5. Determine the position of the line of action of the equilibrant.

The position must be such that two components, E1 and E2, of the equilibrant are respectively collinear with P'C' and A'P' if the system is to be in equilibrium. Therefore the line of action of the equilibrant, E, must be parallel to CA and pass through the point U formed at the intersection of the lines A'P' and P'C' extended.

• PROBLEM 15-35

A beam is loaded with three parallel, unequal loads as shown in the given figure. Find the reactions at the supports and the total resultant that will replace the parallel loads.

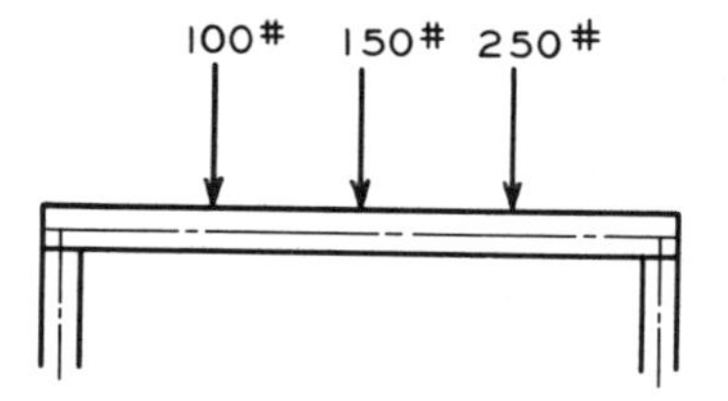

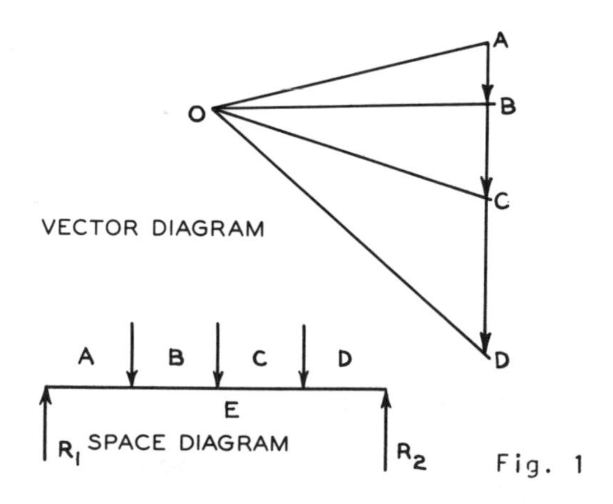

Fig. 1

<u>Solution</u>: 1. Apply Bow's notation. Find the graphical summation of the vectors by drawing them head-to-tail in a vector diagram at a convenient scale. Locate pole O at a convenient location and draw rays from point O to each end of the vectors. See fig. 1.

2. Extend the lines of force in the space diagram, and draw a funicular diagram with ray oa in the A-space parallel to OA, ob in the B-space parallel to OB, oc in the C-space parallel to OC and od in the D-space parallel to OD. (See fig. 2.) The last ray, which is drawn to close the diagram, is oe. Transfer this ray to the vector polygon and use it to locate point E, thus establishing the lengths of R_1 and R_2 which are EA and DE, respectively.

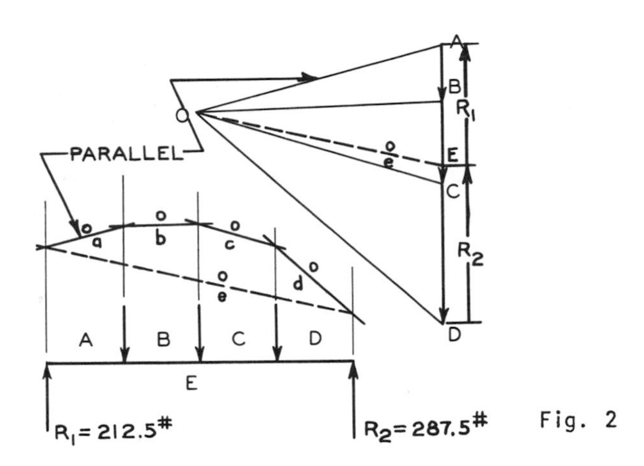

Fig. 2

3. The resultant of the three downward forces will be equal to their graphical summation, line AD. (See fig. 3.)

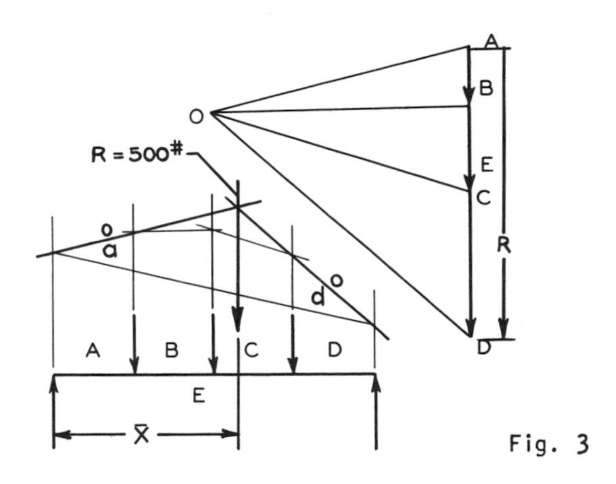

Fig. 3

Locate the resultant by extending rays oa and od in the funicular diagram to a point of intersection. The resultant R = 500 lb, will act through this point in a downward direction. $\overline{X}$ is a locating dimension.

• PROBLEM 15-36

Given two nonparallel forces F1 (50 lbs.) and F2 (30 lbs.) acting on a rigid member, pinned at one end and supported by a roller at the other. (Fig. 1.) Determine the reactions at the pinned joint and at the roller.

Solution: The line of action of the reaction at the roller joint must be perpendicular to the rigid member since this joint cannot withstand any force parallel to the member. The line of action of the reaction at the pinned joint is unknown since this joint withstands force in any direction. However, if the member is to be in equilibrium, the line of action of any reaction at the pinned joint must pass through the pin.

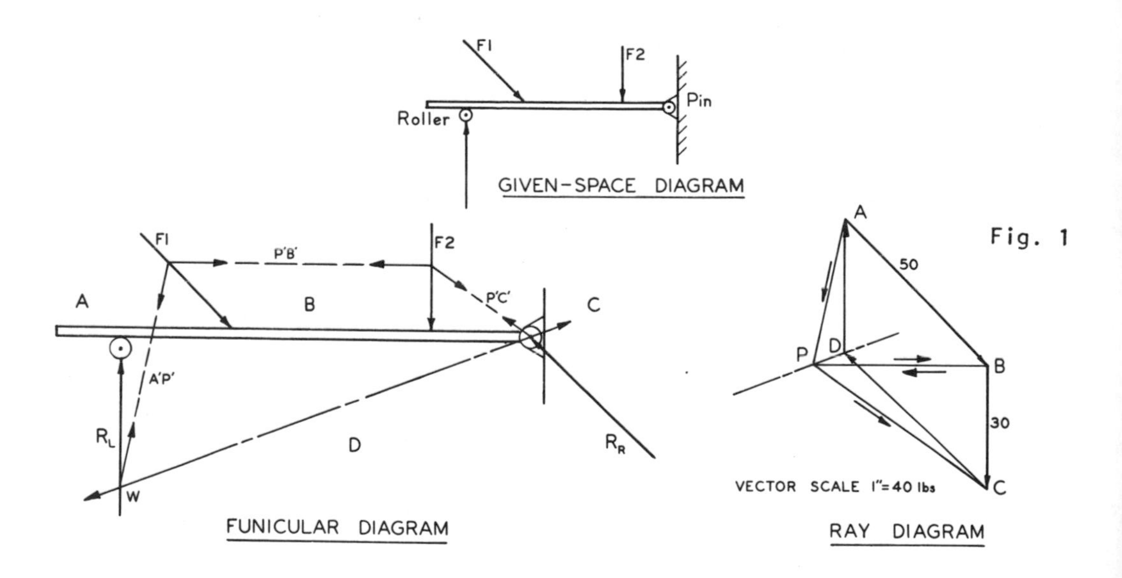

Fig. 1

1. Apply Bow's notation.

2. Construct the partial vector polygon AB, BC.

3. From pole point P construct rays to points A, B, and C.

4. Starting at the pin, construct the funicular diagram by drawing lines of action P'C', P'B', A'P' parallel to components PC, PB, PA of the ray diagram. The vector polygon and the funicular diagram must form closed circuits if the given member is to be in

equilibrium. The vector polygon cannot be correctly closed until the line of action of the reaction at the pinned joint is found.

5. Close the funicular diagram by a straight line from the pinned joint through the point W.

   This closing line represents the collinear lines of action of two opposing, equal components of the desired reactions.

6. Construct, through P of the vector diagram, a line parallel to the closing line of the funicular diagram.

   The vector polygon can now be closed since the line of action and direction of one reaction is known.

7. Construct the vector DA, parallel to R_L.

   This vector represents the reaction at the roller joint.

8. Construct the straight line CD.

   This line represents the line of action and magnitude of the reaction at the pinned joint. Its direction is from C to D.

9. Determine the magnitudes of the reactions.

   Since this is a coplanar system, the vectors CD and DA can be scaled directly.

• PROBLEM 15-37

Given are three nonparallel, nonconcurrent coplanar forces F1 (10 lbs.), F2 (15 lbs.) and F3 (8 lbs.) acting on the free body shown in the figure. Find the line of action, point of application and magnitude of the resultant, R.

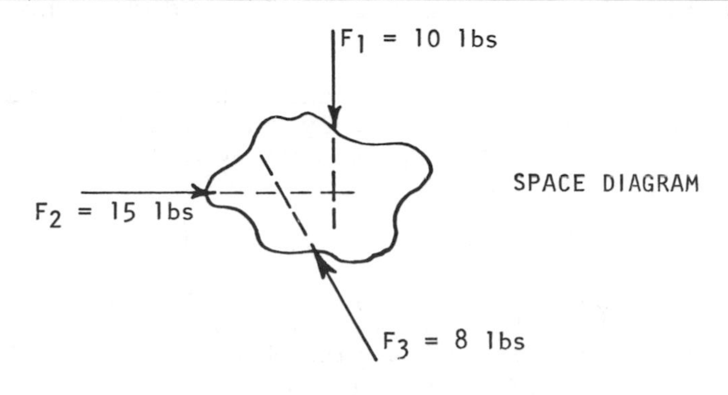

Solution: 1. Apply Bow's notation.

2. Construct a vector polygon to some appropriate scale. The magnitude of the resultant R will be the vector AD, and its direction will be from A to D.

3. From an arbitrary selected pole, point P, draw straight lines (rays) to the points A, B, C and D of the vector polygon.

The vector AB has thus been resolved into two components AP and PB, the vector BC into BP and PC, and the vector CD into CP and PD. Notice that components PB and BP are equal in magnitude but opposite in direction, thus canceling out. Similarly, PC and CP cancel out, leaving AP and PD as the components of resultant AD.

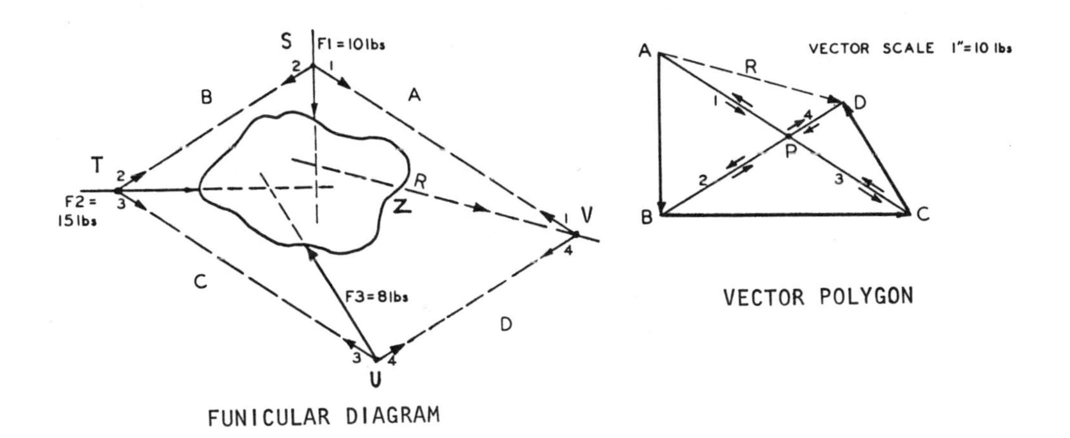

4. Construct the funicular or string diagram in order to locate the exact position of the resultant on the given free body.

(a) Through any point S on F1 (in given space diagram), construct two lines of action 1 and 2 respectively parallel to vectors AP and PB. Extend line of action 2 to cut vector F2 (extended if necessary) at T.

(b) Through point T on F2, construct two lines of action 2 and 3 respectively parallel to vectors BP and PC. Extend the line of action 3 to cut vector F3 (extended if necessary) at U.

Note: Lines of action 2 in (a) and (b) will be collinear.

(c) Through point U on F3, construct lines of action 3 and 4 respectively parallel to vectors CP and PD. Extend the line of action 4 to cut line of action 1 extended [Ref (a)] at V.

Note: Lines of action 3 in (b) and (c) will be collinear.

5. The line of action of the resultant, R, will be parallel to AD and pass through V. The resultant, R, will act at point Z indicated on the free body.

• PROBLEM 15-38

Given the necessary views of a Derrick lifting a known load, determine the type and magnitude of stress in the structure members.

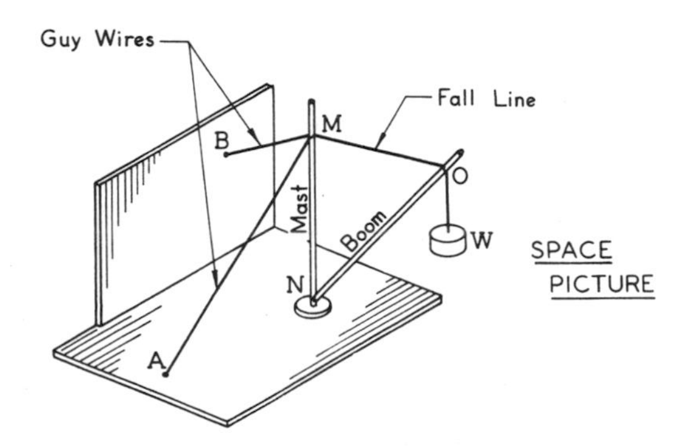

Solution: This problem is solved by separating it into two parts:

Part 1: A coplanar, concurrent system of forces consisting of the boom, fall line and weight acting at the common point O.

Part 2: A noncoplanar, concurrent system of forces consisting of the guy wires, mast and fall line acting at the common point M.

Part 1: Resolve the given load W into two components whose lines of action are parallel to the boom and the fall line. See vector diagram (a). The magnitude and line of action of the stress in the fall line becomes the known force of Part 2. Notice that the fall line is under tension.

Part 2: A point view of one unknown is needed. Since the mast is vertical, its horizontal projection is a point which lies in any one of several edge views. Plane MNB was chosen for this illustrated solution.

1. Apply Bow's notation to the view showing the mast as a point.

2. Construct the two view vector diagram, Fig. (b). Remember that the known load is in tension, therefore RS is equal and opposite to JG.

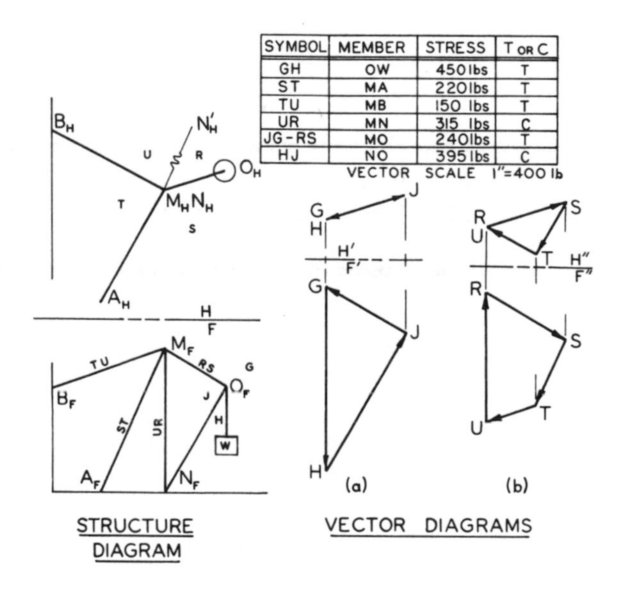

SYMBOL	MEMBER	STRESS	T OR C
GH	OW	450 lbs	T
ST	MA	220 lbs	T
TU	MB	150 lbs	T
UR	MN	315 lbs	C
JG-RS	MO	240 lbs	T
HJ	NO	395 lbs	C

3. Determine true magnitudes of stress in the members.

4. Determine kind of stress in the members.

5. Tabulate results.

• PROBLEM 15-39

> Find the stresses in each member of the Fink truss loaded as shown and indicate whether each member is in compression or tension.

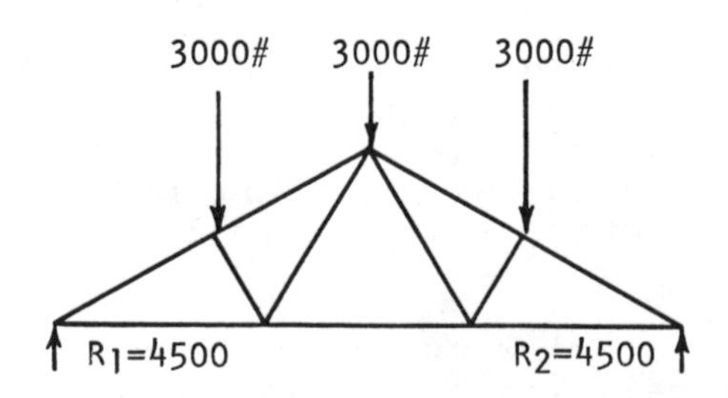

Solution: 1. Label the portions of supports between the

outer forces of the truss with letters and the internal portions with numbers, using Bow's notation. Add the given load vectors graphically in a stress diagram, and sketch a free-body diagram of the first joint to be analyzed.

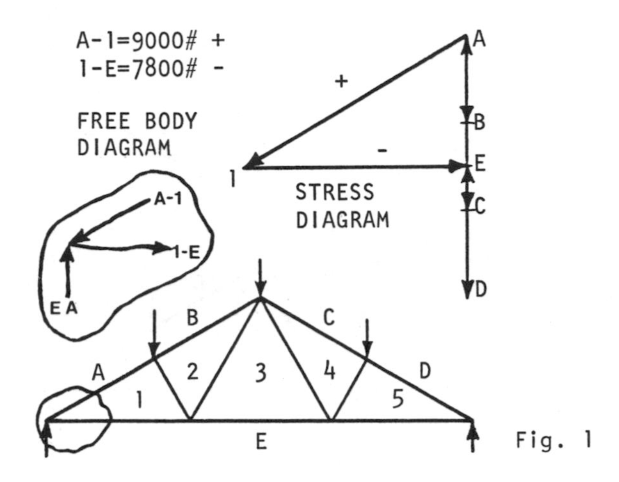

Fig. 1

lyzed. Using vectors EA, A-1, and 1-E drawn head to tail, draw a vector diagram to find their magnitudes. Vector A-1 is in compression (+) because its sense is toward the joint, and 1-E is in tension (-) because its sense is away from the joint.

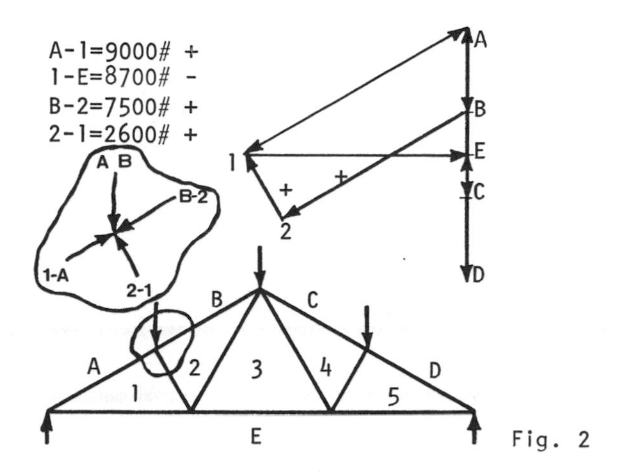

Fig. 2

2. Draw a sketch of the next joint to be analyzed. Since AB and A-1 are known, we have to determine only two unknowns, 2-1 and B-2. Draw these parallel to their direction, head to tail, in the stress diagram using the exist-

ing vectors found in step 1. Vectors B-2 and 2-1 are in compression since each has a sense toward the joint. Note that vector A-1 becomes 1-A when read in a clockwise direction.

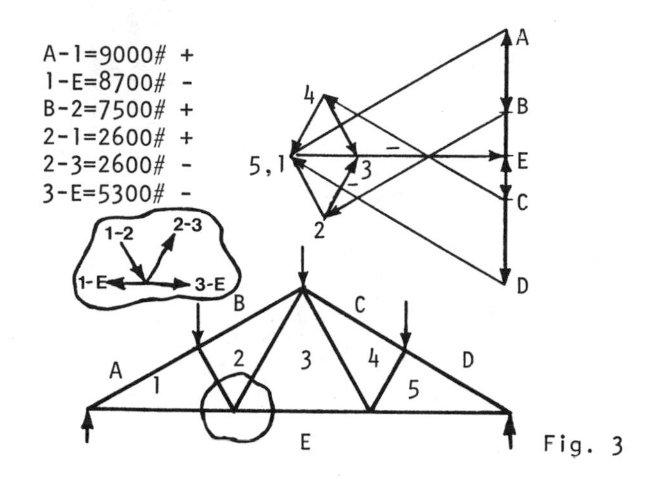

Fig. 3

3. Sketch a free-body diagram of the next joint to be analyzed. The unknowns in this case are 2-3 and 3-E. Determine the true length of these members in the stress diagram by drawing vectors parallel to given members to find point 3. Vectors 2-3 and 3-E are in tension because they act away from the joint. This same process is repeated to find the stress of the members on the opposite side.

COUPLES

• PROBLEM 15-40

What is a couple?

Solution: Two noncollinear forces equal in magnitude, parallel, but acting in opposite directions are called a couple. It is evident that the vector sum of these two forces is zero ($\Sigma F = 0$). However, the moment (ΣM) does not equal zero. It is equal to the product of one of the forces (F) times the length (d) of the common perpendicular (M = F•d). If a system of nonconcurrent forces or vectors is in equilibrium both ΣF and ΣM must equal zero. In the sketch above, the couple will spin the disc unless a couple of equal but opposite moment is inserted to restrain it.

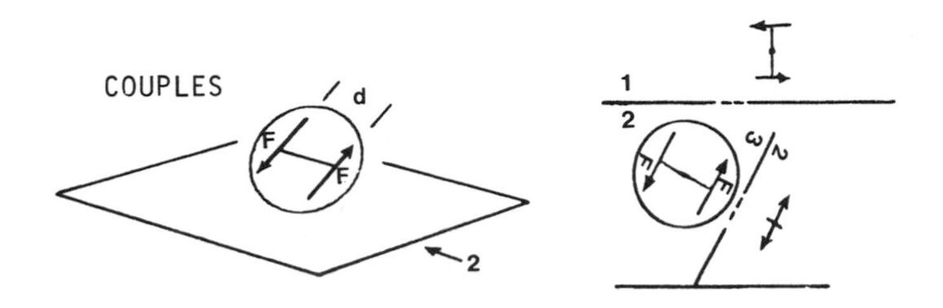

The moment of the couple is the same (F•d) for any position on the plane of the couple. Consider a moment taken at any point O in the plane of the couple, then F(Y + d) - F•Y = F•d.

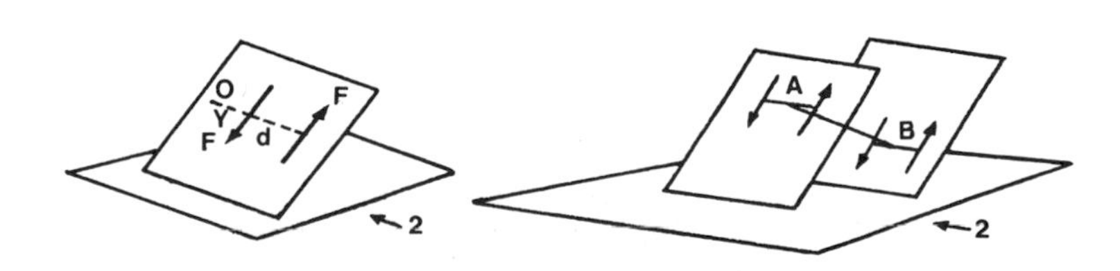

The couple may be moved to any parallel plane without changing its effect. The sketch shows that the line AB, perpendicular to the plane of the couple, will be subjected to the same turning moment by placing the same couple at either the end A or B or at any other point on the line.

• PROBLEM 15-41

A series of parallel forces is applied to a beam in the figure. The spaces between the vectors are labeled with letters which follow Bow's notation. Determine the resultant.

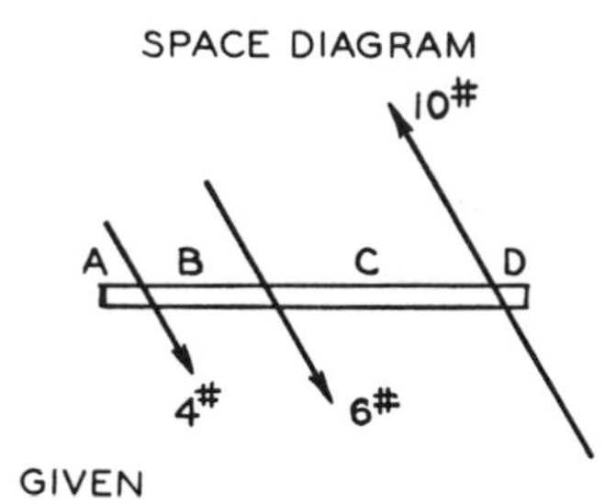

Solution: 1. After constructing a vector diagram, we have a straight line which is parallel to the direction of the forces and which closes at point A in such a way that the forces are equal in both directions. We then locate pole point O and draw the strings of a funicular diagram. (See figure 1.)

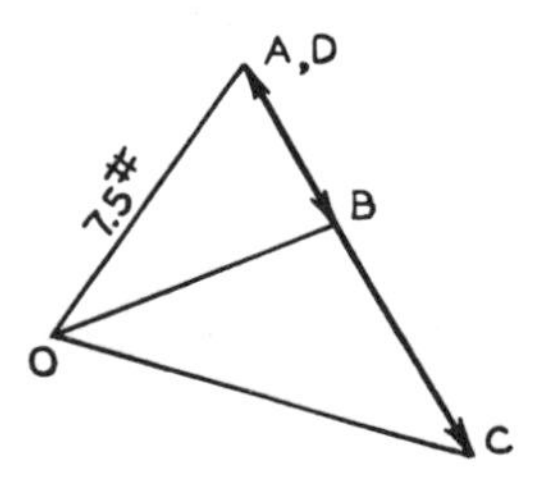

Fig. 1

2. The rays are transferred to the space diagram, where they are drawn in their respective spaces. For example, o-c is drawn in the C-space between vectors BC and CD. The last two rays, o-d and o-a do not close at a common point, but are found to be parallel; the result is therefore a couple. The distance between the forces of the

couple is the perpendicular distance, E, between rays o-a and o-d in the space diagram, using the scale of the space diagram. The magnitude of the force is the scaled distance from point O to A and D in the vector diagram, using the scale of the vector diagram. The moment of the couple is equal to 7.5 lb × E in a counterclockwise direction. (See figure 2.)

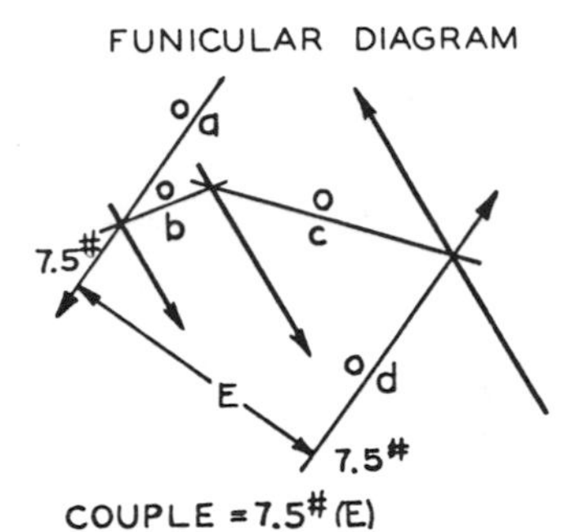

Fig. 2

CHAPTER 16

GRAPHS AND CHARTS

RATE AND TRAVEL

• PROBLEM 16-1

Suppose A does a unit of work in three days. B does a similar unit of work in five days. If they work together on a like unit, how long will it take? (By Graphical Method)

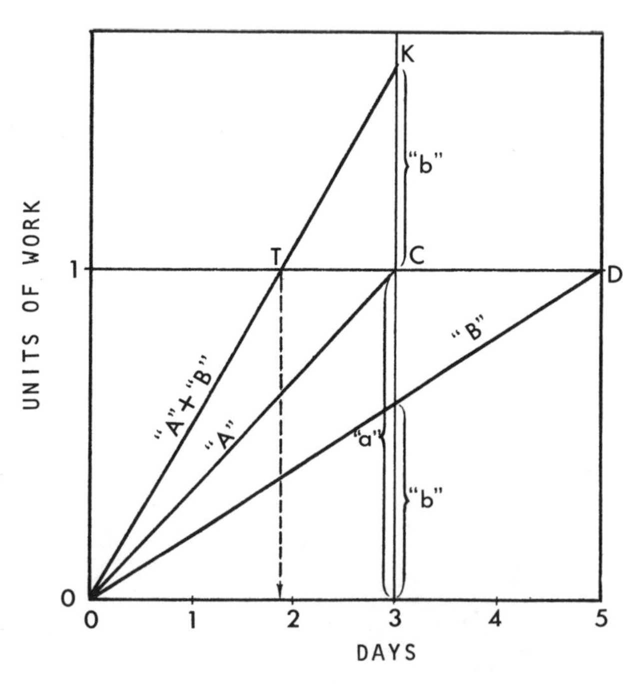

GRAPHICAL SOLUTION OF WORK PROBLEM.

Solution: The following figure shows the graphical solution of this problem:

The graphic representation of the performances by A and B is shown by the lines marked "A" and "B". Line "A" is located by joining the origin, O, with point C, and line "B" by joining points O and D. Now at any time, say three days, A and B together complete ("a" + "b") units of work. If we connect points O and K (K was established by adding graphically "b" to "a") we will have established line OK, which shows the rate at which A and B work together. The intersection of line OK with the horizontal line drawn through point 1 on the Units of Work scale locates point T. The x-value of this point is 1 7/8 days which is the required time to complete a unit of work if A and B work together.

• PROBLEM 16-2

Suppose A works for two days at a rate shown graphically in the following figure, fails to report for work the third day, and then completes the unit of work in the next two days, as shown in the Fig. B's performance, including an absence on the afternoon of the second day, is also shown. Had they worked together on a unit of work (assuming the same work history), how long would it have taken?

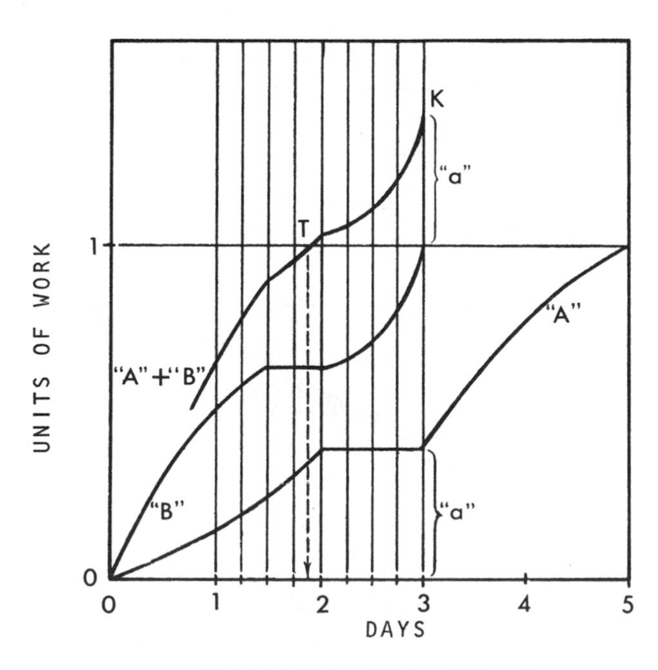

GRAPHICAL SOLUTION OF MODIFIED WORK PROBLEM.

Solution: To obtain graphical solution it is necessary only to introduce a number of vertical lines between 1 and 3 of the time scale, and on each of them to add graphically segments such as "a" to the corresponding segments representing the performance of B. The curve drawn through the newly established points, such as point K, intersects the horizontal line drawn through point 1 of the Units of Work scale in point T. A vertical line drawn through point T intersects the time scale at 1.8 days, the required time.

• PROBLEM 16-3

An automobile travels at a speed of s miles per hour at any time, t seconds, in accordance with the following data:

s	0	2.9	6.0	9.3	11.2	11.6
t	0	5	10	15	20	25

Find the number of miles traveled in 25 seconds.

Solution: First the data are plotted and a smooth curve is drawn through the points, as shown in Fig. 1-a. It should be noted that the horizontal scale is also graduated in hours. This is very convenient since we are seeking the result in terms of the miles traveled. The area under the curve represents this distance. The area will be found by a combination of graphical and numerical methods.

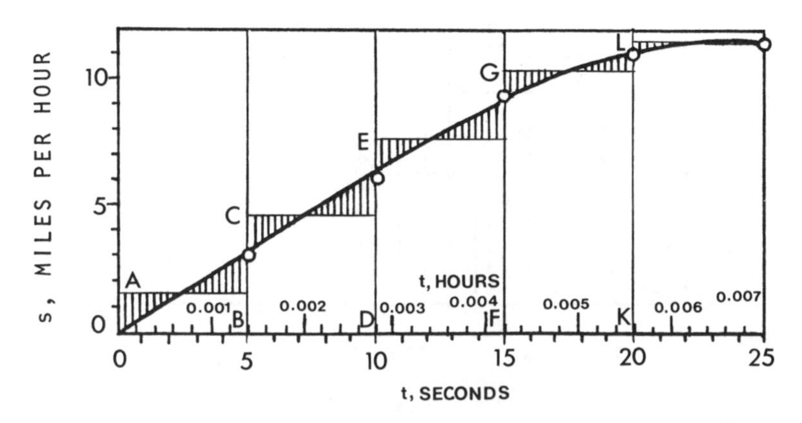

Fig. 1a

GRAPHICAL ADDITION APPLIED TO TRAVEL PROBLEM.

First divide the area into strips of equal width, since the change in curvature is not great (the widths should depend upon the sharpness or flatness of the curve, i.e., narrow for sharp portions of the curve). Horizontal lines, such as the one passing through point A, are introduced across each strip so that the area above the curve is very nearly balanced by the area below the curve. Placement of the horizontals is easily judged by eye. Now, the sum of the rectangles is a very close approximation to the area under the curve.

O A,B C,D E,F G,K L Fig. 1b

GRAPHICAL ADDITION OF SEGMENTS SHOWN IN FIG. 1a

If the sum of the segments OA, BC, DE, FG, and KL is measured in terms of the miles per hour scale (Fig. 1-b), then this result multiplied by 0.00138

$$\left(\frac{\text{mile}}{\text{hour}} = 0.000277 \frac{\text{mile}}{\text{second}},\ 5 \times .000277 = .00138 \text{ mile/5 sec.}\right)$$

the width of the strip, in hours, is the required distance traveled. In this case line 0 . . . L measured 36 units of the miles per hour scale; hence the distance traveled is 36 x 0.00138 = 0.05 mile.

GRAPHICAL ARITHMETIC AND EQUATION SOLVING

• PROBLEM 16-4

Part a: Determine graphically the product of a,b,c, and d.

Part b: Obtain the product of 2, 1/2, 3, and 3/4.

Solution: Part A:

Suppose it is determined graphically the product of a,b,c, and d.

Let $a \times b = R$ (1)

and $R \times c = S$ (2)

and $S \times d = T$ (3)

where T is the result.

The first equation may be rewritten as

$$\frac{1}{a} = \frac{b}{R} \quad (4)$$

The second equation may be rewritten in a similar manner as

$$\frac{1}{c} = \frac{R}{S} \quad (5)$$

and, finally, the third equation as

$$\frac{1}{d} = \frac{S}{T} \quad (6)$$

Graphically, equation 4 can be solved as follows: suppose we construct right triangle ABC so that side AB is one unit long and side BC is b units long (see Fig. 1).

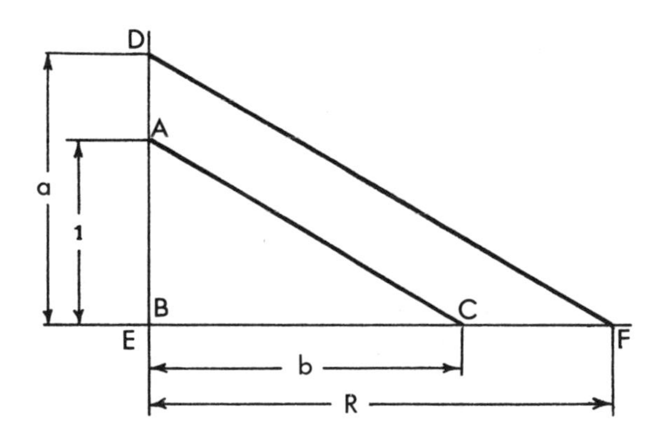

GRAPHICAL MULTIPLICATION, R=ab. Fig. 1

Now suppose DE, which is a units long, is laid off as shown in the figure. A line drawn through point D parallel to AC intersects the horizontal line at point F. The length of side EF is the product of a and b, since

$$\frac{AB}{BC} = \frac{DE}{EF} \quad \text{or} \quad \frac{1}{b} = \frac{a}{EF} \quad \text{or} \quad EF = R = a \times b$$

Therefore, R is the product of a and b.

In a similar manner equations (5) and (6) may be solved. Figure 2 shows the solution of equation (5) and Fig. 3 of equation (6).

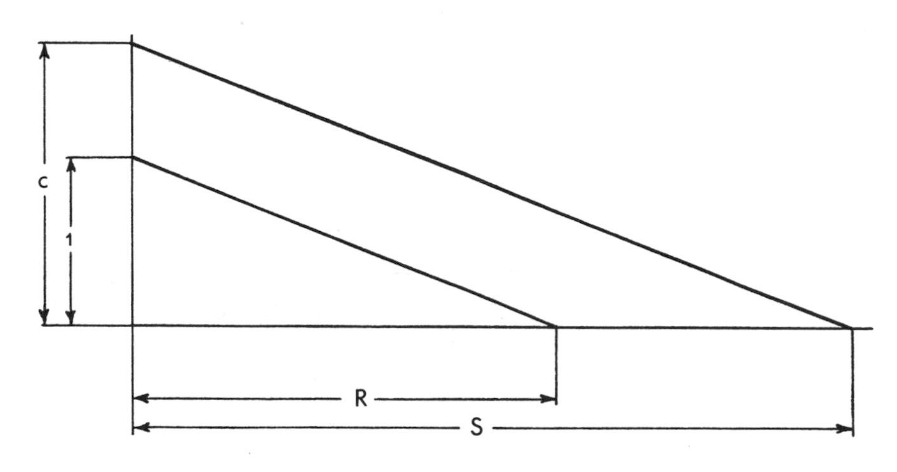

GRAPHICAL MULTIPLICATION, S = Rc. Fig. 2

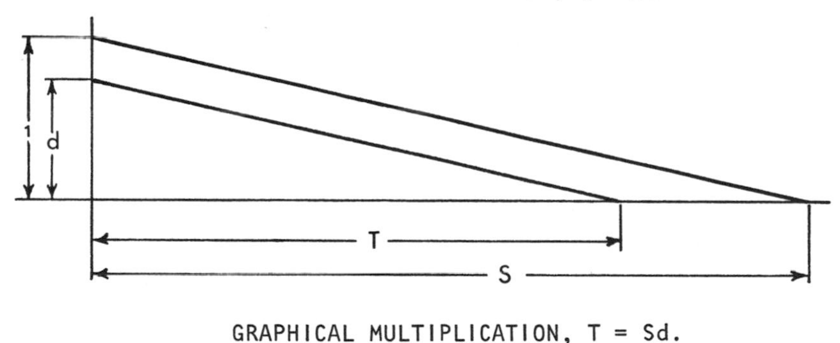

GRAPHICAL MULTIPLICATION, T = Sd. Fig. 3

Now, instead of drawing three separate figures, we may combine them as shown in Fig. 4.

Part B:

Consider the following figure (Fig. 5).

(a) Lay off one unit on the vertical axis.

(b) Now lay off three units on the same axis (the largest given value is used in order to confine the solution to a desired maximum distance on the vertical axis).

(c) Lay off ½ unit on the horizontal axis.

(d) Connect the points marked ½ and 1 and then draw a parallel line through the point marked 3. The intersection of this line with the horizontal axis is point R.

(e) Join points R and 1.

(f) Now lay off two units on the vertical axis and draw a line through point 2 parallel to line R-1, thus locating point S on the horizontal axis.

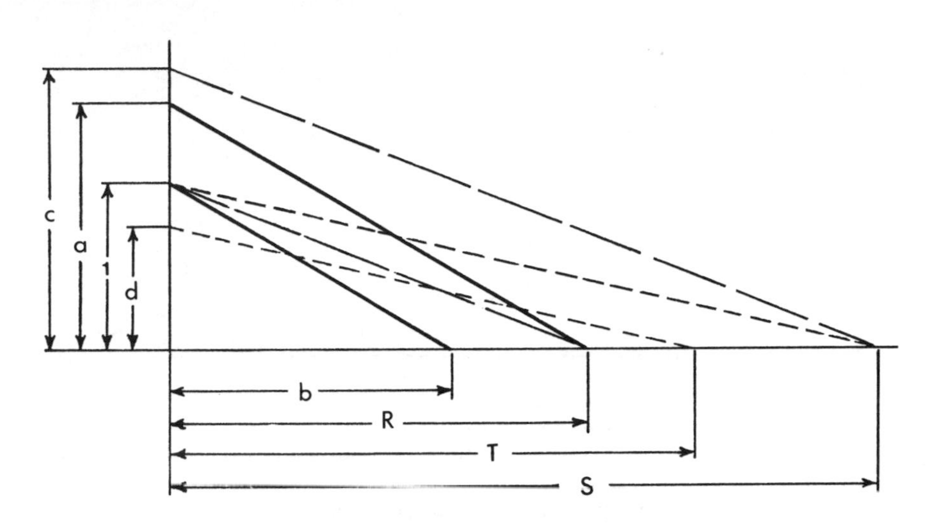

GRAPHICAL MULTIPLICATION, $a \times b \times c \times d = T$.
(COMBINATION OF FIGURES 1, 2, AND 3) Fig. 4

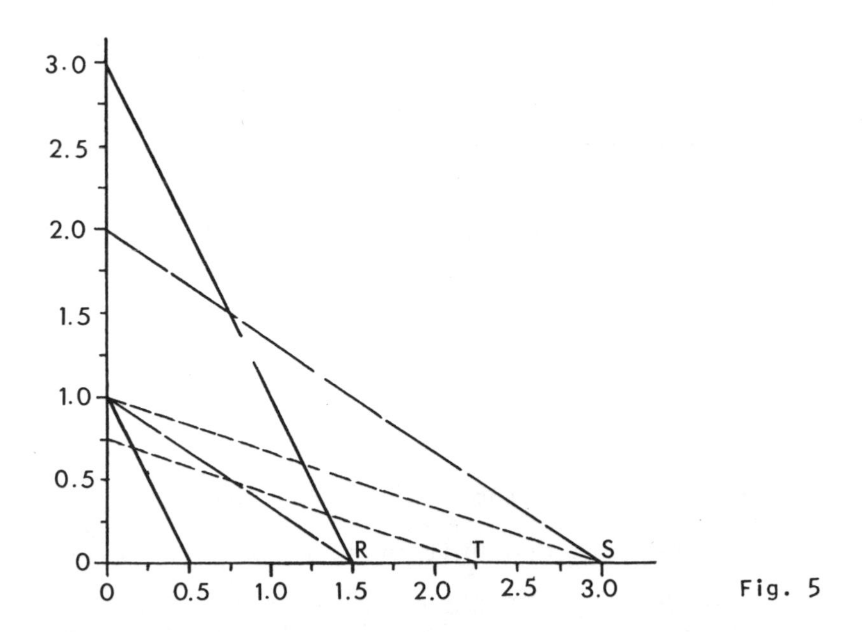

Fig. 5

GRAPHICAL MULTIPLICATION, $2 \times 1/2 \times 3 \times 3/4 = 2\ 1/4$.

(g) Finally join points S and 1 and then draw a parallel line through the point 3/4, thus locating T on the horizontal axis. Distance O-T or $2\frac{1}{4}$ units is the required product.

• PROBLEM 16-5

By Graphical Method, determine the quotient

$$AB = \frac{BC}{EF}$$

where BC = 3 units and EF = 2 units.

Solution: Consider the following figure:

a. Lay off one unit on the vertical axis ED = 1.

b. Lay off two units on the horizontal axis EF = 2.

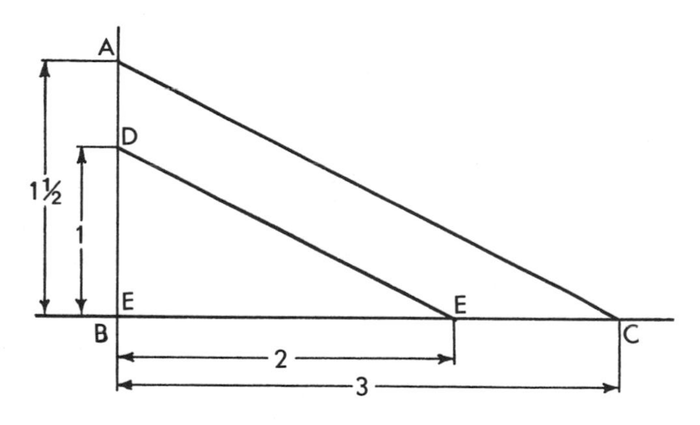

GRAPHICAL DIVISION.

EXAMPLE: $\frac{3}{2} = 1\frac{1}{2}$. FROM $\frac{BC}{EF} = \frac{AB}{BC}$; $\frac{3}{2} = \frac{AB}{1} = 1\frac{1}{2}$.

c. Now lay off three units on the same axis BC = 3.

d. Connect the points D and F and then draw a parallel line through the point c. The intersection of this line with the vertical axis is point A. The distance AB = 1½ unit is the required quotient, since in both triangles, EDF and ABC, we know that

$$\frac{AB}{DE} = \frac{BC}{EF}$$

or

$$\frac{AB}{1} = \frac{3}{2}$$

or AB = 1½ units.

• PROBLEM 16-6

Obtain the value of T by Graphical Method:

a.

$$T = \frac{a \times b}{c}$$

b.

$$T = \frac{abc}{de} = \frac{a}{d} \times \frac{b}{e} \times c$$

Solution: Part A.

Consider the following relation:

$$\frac{a \times b}{c} = T$$

This equation can be rewritten as

$$\frac{a}{c} = \frac{T}{b}$$

If we lay off distance oa on the y-axis (Fig. a-1) and distance oc on the x-axis, respectively equal to the values of a and c, then a line drawn through b (distance ob = value of b) parallel to the line ac, will intersect the y-axis in the value of T.

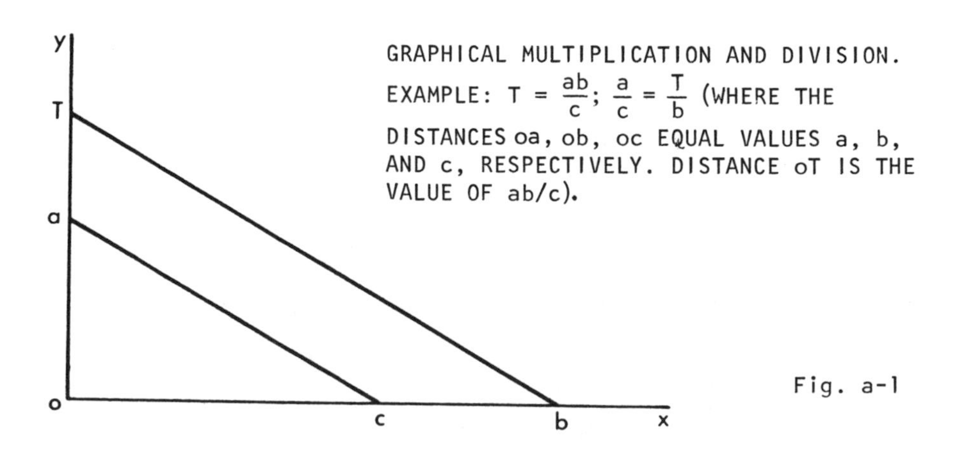

GRAPHICAL MULTIPLICATION AND DIVISION. EXAMPLE: $T = \frac{ab}{c}$; $\frac{a}{c} = \frac{T}{b}$ (WHERE THE DISTANCES oa, ob, oc EQUAL VALUES a, b, AND c, RESPECTIVELY. DISTANCE oT IS THE VALUE OF ab/c).

Fig. a-1

This is easily seen, since we know that

$$\frac{oa}{oc} = \frac{oT}{ob}$$

and

$$oT = \frac{oa \times ob}{oc}$$

or

$$T = \frac{a \times b}{c} .$$

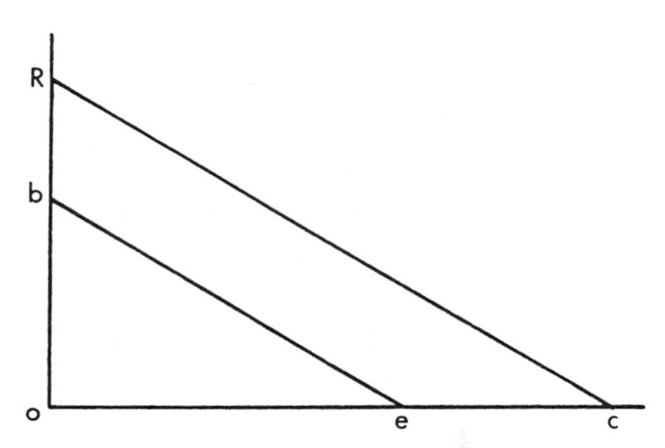

GRAPHICAL MULTIPLICATION AND DIVISION. Fig. b-1
EQUATION: $R = \frac{bc}{e}$.

Part B.

Consider the relation:

$$T = \frac{abc}{de} = \frac{a}{d} \times \frac{b}{e} \times c$$

Let $$\frac{bc}{e} = R \qquad (1)$$

and

$$T = \frac{a}{d} R \qquad (2)$$

As in Part A, equation (1) can be rewritten

$$\frac{b}{e} = \frac{R}{c} .$$

The graphical solution of this equation is shown in Fig. b-1. Now equation (2) can be rewritten as

$$\frac{a}{d} = \frac{T}{R} .$$

The graphical solution of this equation is shown in Fig. b-2. Of course, both figures can be combined to form a

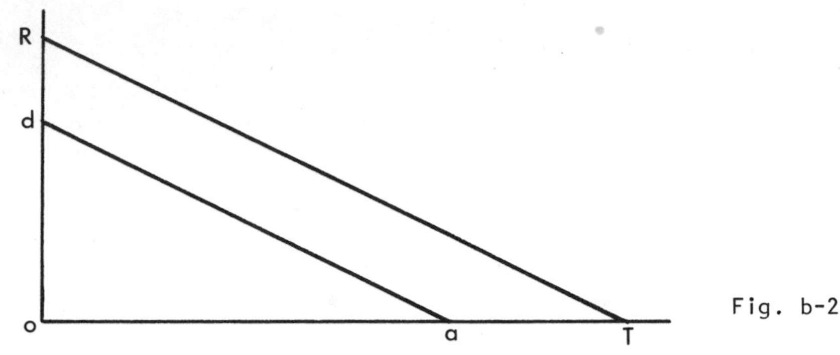

GRAPHICAL MULTIPLICATION AND DIVISION. EQUATION: $T = \frac{aR}{d}$.

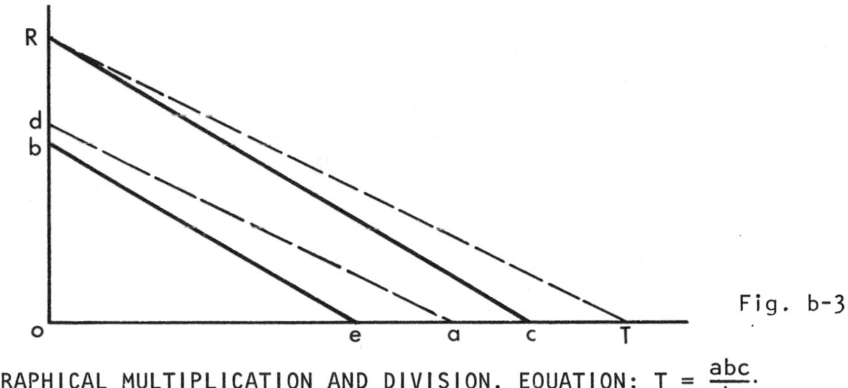

GRAPHICAL MULTIPLICATION AND DIVISION. EQUATION: $T = \frac{abc}{de}$.
(COMBINATION OF FIGURES b-1 AND b-2)

single drawing for the graphical solution of the given equation,

$$T = \frac{abc}{de} .$$

This is shown in Fig. b-3.

• PROBLEM 16-7

By Graphical Method determine:

a. The resistance R equivalent to two parallel resistances R_1 and R_2 . See fig. a-1.

b. The resistance R equivalent to three parallel resistances R_1, R_2, and R_3. See fig. b-1.

Solution: Part a.

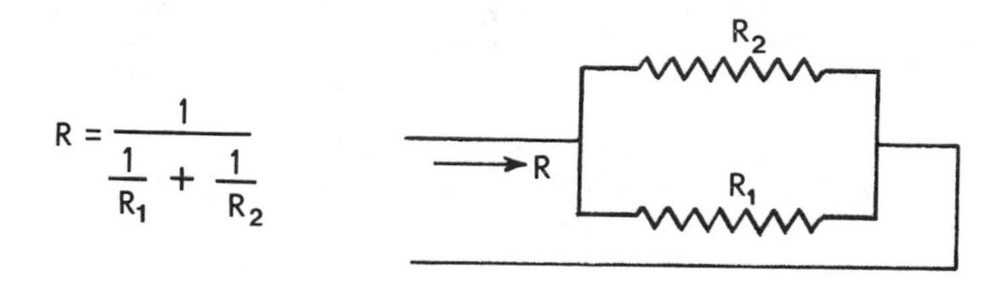

Fig. a1

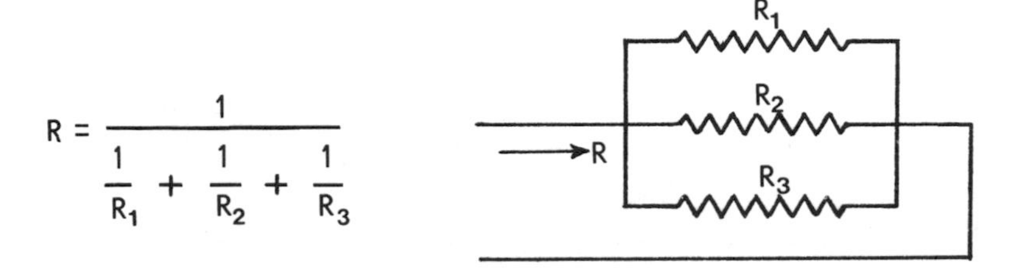

Fig. b-1

To find the graphical solution of equation

$$R = \frac{R_1 R_2}{R_1 + R_2}$$

We should follow these steps:

1. Lay off R_1 unit on the vertical axis OA = R_1. See fig. a2.

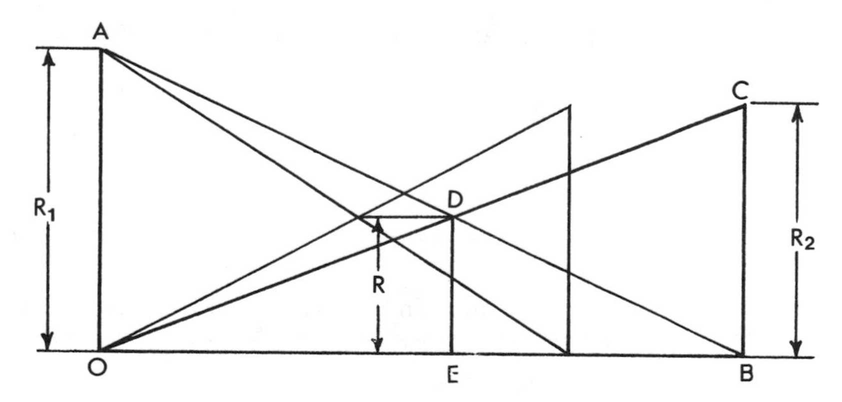

GRAPHICAL SOLUTION OF EQUATION $R = \frac{R_1 \ R_2}{R_1 + R_2}$.

Fig. a-2

2. Lay off R_2 unit on the vertical axis, right side, BC = R_2

3. Connect points C to O and A to B, and then locate point D, the intersection of lines OC and AB.

The vertical distance, DE, from point D to base line OB (any convenient length) measured to the same scale used in laying off R_1 and R_2 is the required value of R.

That the above construction is correct can be easily shown. Consider Fig. a-3 which is essentially the same as Fig. a-2. Let us draw a horizontal line through point D to intersect lines AO and BC in points F and G, respectively.

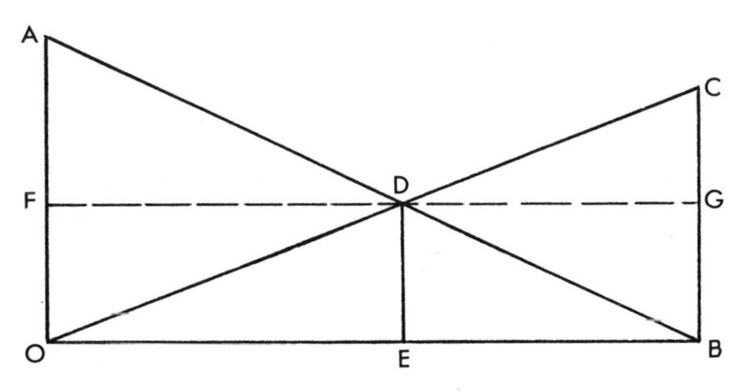

CONSTRUCTION USED TO SHOW THAT $DE = \frac{0A \times BC}{0A + BC}$. Fig. a-3

Now, since triangles AFD and DEB are similar,

$$\frac{AF}{DE} = \frac{FD}{EB}$$

Likewise, triangles DEO and CGD are similar. Hence,

$$\frac{DE}{CG} = \frac{EO}{DG} = \frac{FD}{EB}$$

Therefore,

$$\frac{AF}{DE} = \frac{DE}{CG} \quad \text{or} \quad \overline{DE}^2 = AF \times CG$$

Now if we compare Figs. a-2 and a-3, and keep in mind the above relation, $\overline{DE}^2 = AF \times CG$, we see that

$$R^2 = (R_1 - R)(R_2 - R)$$

or

$$R^2 = R_1R_2 - R_1R - RR_2 + R^2$$

or

$$R = \frac{R_1R_2}{R_1 + R_2}$$

Part b.

Graphically this part is solved similar to part a. Figure b-2 shows the graphical solution of the equation

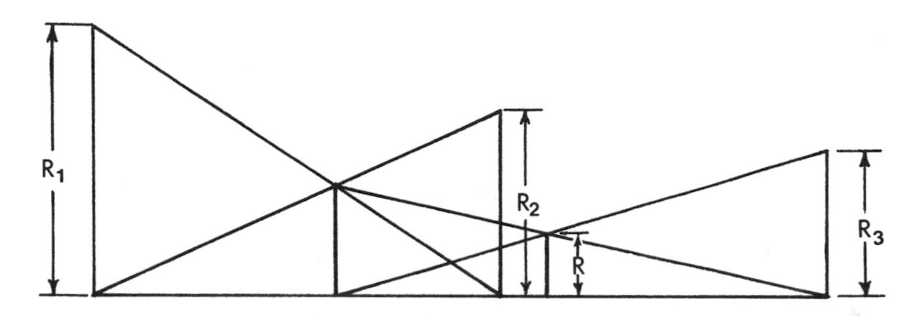

GRAPHICAL SOLUTION OF THE EQUATION $R = \frac{1}{\frac{1}{R_1} + \frac{1}{R_2} + \frac{1}{R_3}}$. Fig. b-2

$$R = \frac{1}{\frac{1}{R_1} + \frac{1}{R_2} + \frac{1}{R_3}} .$$

• PROBLEM 16-8

By graphical method, solve the following equations for values of x and y which satisfy both equations

$$x + 2y = 8$$

$$-2x + 5y = 11$$

Solution: Each of these equations may be represented graphically in the following manner. See figure:

1. First set up x- and y-axes at right angles to each other and then graduate the axes as shown.

2. Now study the first equation. When $x = 0$, $y = 4$. The point 4 on the y-axis is designated by the letter A. This point is said to have coordinates (0,4) which simply means the x-distance is 0, and the y-distance is 4. When $y = 0$, $x = 8$ (point B). The line which joins points A and B is the graphical representation of the expression

$x + 2y = 8$. In a similar manner, we may locate points C and D from the second equation. When $x = 0$, $y = 11/5$ (point C), and when $y = 0$, $x = -11/2$ (point D). The line which joins points C and D is the graphical representation of the expression $-2x + 5y + 11 = 0$.

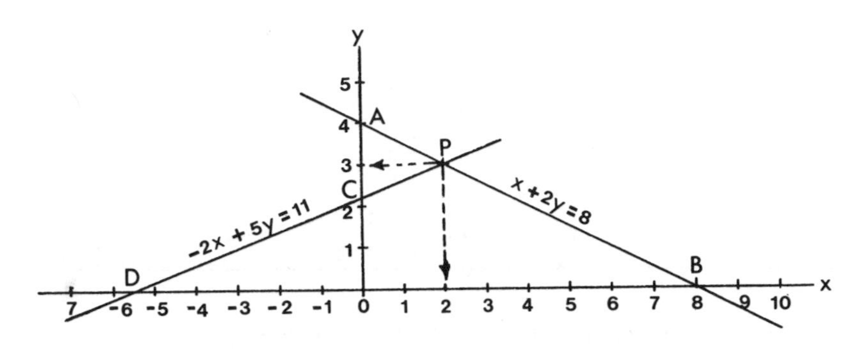

GRAPHICAL SOLUTION OF EQUATIONS $x + 2y = 8$, AND $-2x + 5y = 11$ FOR VALUES OF x AND y THAT SATISFY BOTH EQUATIONS. ANSWER: $x = 2$; $y=3$.

The intersection of lines AB and CD, namely point P, is common to both lines so that it must have x- and y-coordinates which satisfy both equations. Graphically these coordinates are $x = 2$ and $y = 3$.

• PROBLEM 16-9

Obtain graphically the values of x which satisfy the equation:

$$x^2 - 5x + 6 = 0$$

Solution: One approach to the solution would be to let $y = x^2 - 5x + 6$ and then plot the curve. This means, simply, the determination of the coordinates of points which lie on the curve.

For example, when $x = 0$, $y = 6$

when $x = 1$, $y = 2$

when $x = 2$, $y = 0$

etc.

We could arrange a convenient table of the values of y for the different values of x, thus

Point	A	B	C	...	F	G	...
x	0	1	2	...	-1	-2	...
y	6	2	0	...	12	20	...

The smooth curve (Fig. 1) drawn through the points is the graphical representation of the given equation.

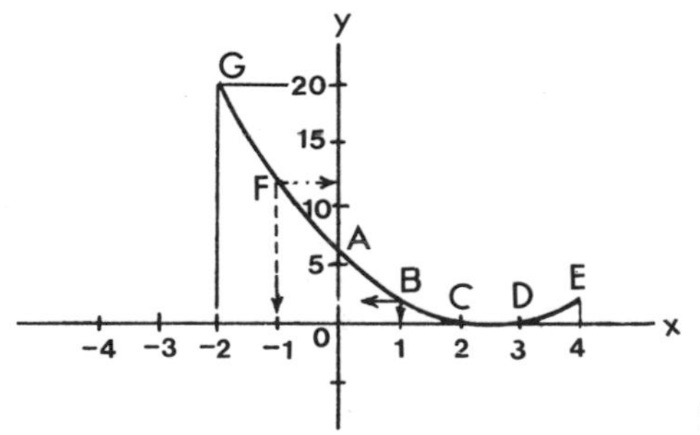

Fig. 1

It is clearly seen that when y = o, x = 2 and x = 3 (points C and D). These values of x satisfy the given equation.

A second approach makes the solution possible in a simple manner. The equation

$$x^2 - 5x + 6 = 0$$

may be rewritten as

$$y = x^2 \quad (1)$$

and $$y = 5x - 6 \quad (2)$$

The graphical simultaneous solution of these two equations is easily obtained. Plot first the equation $y = x^2$ and then the equation $y = 5x - 6$. These are shown in Fig. 2. The intersections of the curve and line, namely points P and Q, have x-coordinates 2 and 3, respectively. These are the

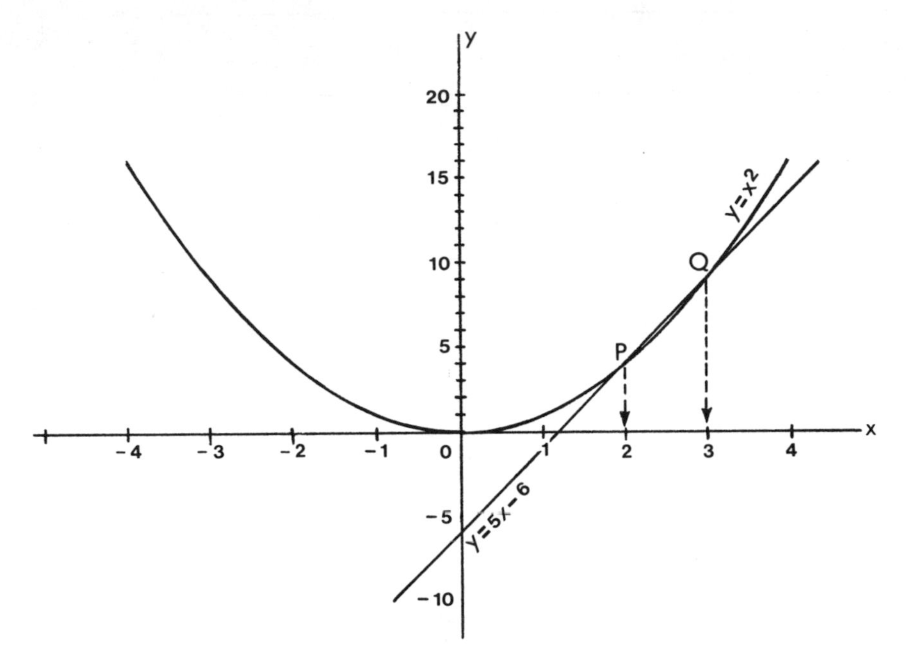

GRAPHICAL SOLUTION OF EQUATION OF $x^2 - 5x+6 = 0$.
ROOTS ARE $x = 2$ AND $x = 3$. Fig. 2

values of x which satisfy the given equation $x^2 - 5x + 6 = 0$.

• PROBLEM 16-10

A stone is dropped into a well. Seven seconds later the sound of impact with the water is heard at the ground surface. How deep is the well?

Solution: Suppose we make a graph having two curves, which represent (1) the stone's drop to the water surface and (2) the travel of the sound of the impact. (See Fig.) Now the intersection of the two curves, point P, has a distance coordinate of 660 feet, which is the depth of the well. Actually this problem involves nothing more than the simultaneous solution of two equations.

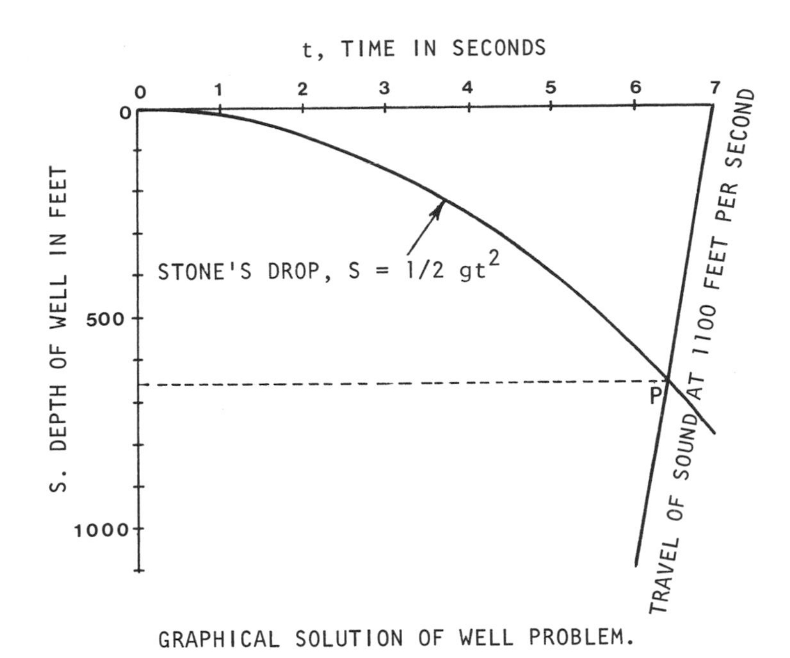

GRAPHICAL SOLUTION OF WELL PROBLEM.

1. Equation of motion: $S = \frac{1}{2} gt^2$

2. Equation of sound: $S = \alpha t$

where α : 1100 feet/second.

CLOCKS, MIXTURES, ACCELERATION, DIFFERENTIATION, AND MAP CONTOUR

• PROBLEM 16-11

(a) At what time between 5 and 6 o'clock are the hands of the clock together?

(b) At what times are they 15 minutes apart?

Solution: Let us first prepare a graph which shows the hour-hand scale along the x-axis and the minute-hand scale along the y-axis. We know that the minute hand travels twelve times as fast as the hour hand. Graphically, the travel of the minute hand is shown by line AB. The hour hand meanwhile travels between clock readings 5 and 6; hence the hour-hand travel is shown by line CD. The intersection of lines AB and CD locates point P, whose x- or y-coordinate shows that the hands of the clock are together at 5:27.

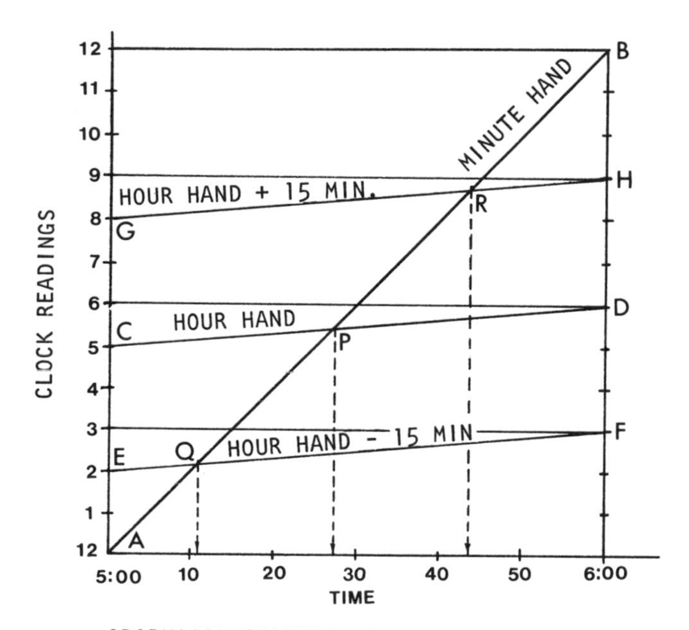

GRAPHICAL SOLUTION OF CLOCK PROBLEMS.

Part (b) of the given problem has two solutions. One solution is given by the coordinates of point Q, and the other by the coordinates of point R where point Q is the intersection of EF and AB and point R is the intersection GH and AB. EF and GH are both parallel to CD being minus and plus 15 minutes apart from CD respectively.

• PROBLEM 16-12

How many gallons of cream (30% fat) should be mixed with what number of gallons of milk (4% fat) to yield 25 gallons of cream containing 20% fat?

Solution: First we prepare a chart with the vertical

axis graduated to represent the number of gallons of milk or cream, and the horizontal axis to represent the number of gallons of butterfat. Now we can establish the 4%, 20%, and 30% butterfat lines. The total yield is to be 25 gallons of cream containing 20% fat. This means that the butterfat content is 5 gallons, hence the horizontal line through the 25 on the vertical scale intersects the 20% line at point A, the butterfat coordinate of which is 5 gallons.

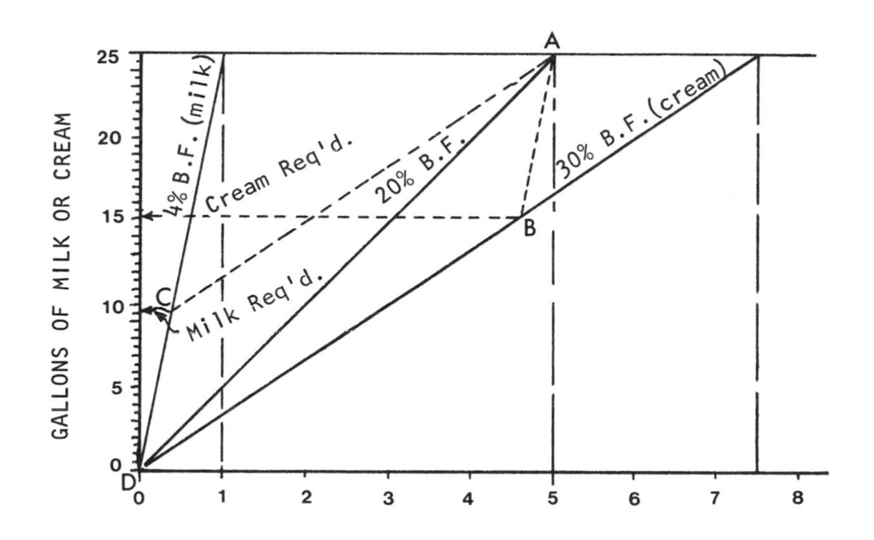

Gallons of butterfat

GRAPHICAL SOLUTION OF MIXTURE PROBLEM.

A line drawn through point A parallel to the 4% line intersects the 30% line at point B, whose cream coordinate is 15.4 gallons.

Similarly a line drawn through point A parallel to the 30% line intersects the 4% line at point C, whose milk coordinate is 9.6 gallons.

Why is the construction shown correct? It can be seen that our problem is merely to locate point C on the 4% line so that a line through C at a 30% slope will pass through point A, which represents 25 gallons of the required mixture. This was accomplished, simply, by drawing through point A a line parallel to the 30% line. Study of the parallelogram ACDB verifies the fact that the butterfat coordinate of point C added to the butterfat coordinate of point B is equal to 5, which represents the number of gallons of butterfat in the 25-gallon mixture.

• PROBLEM 16-13

Suppose the following data are given, where t represents time in seconds and S displacement in feet. Determine the acceleration curve.

t, in seconds	0	1	2	3	4	5	6	7	8
S, in feet	0	1	8	27	64	125	216	343	512

Solution: A plot of the data is shown. The increment along the t-axis is made equal to 0.5 second (h = 0.5) in the figure shown. We measure the ordinates, to the smooth curve passing through the plotted points, at the end of each interval. For convenience we prepare a table for the corresponding values of S and t, and for the necessary computations of accelerations. This is shown in table 1.

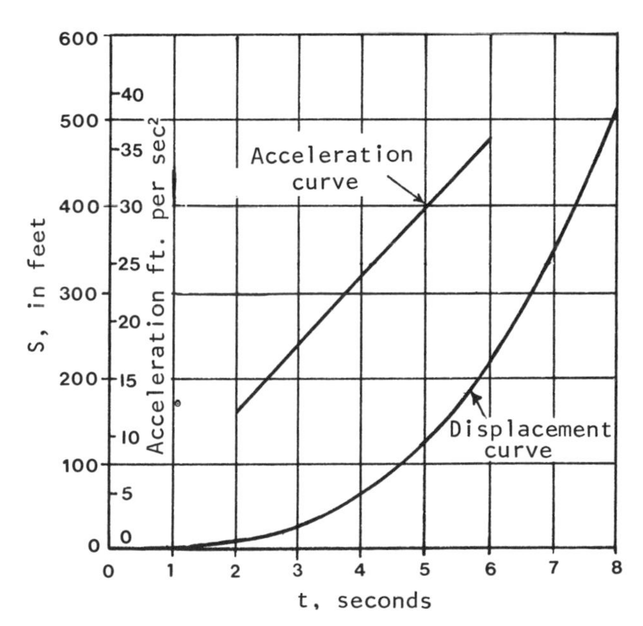

GRAPHO-NUMERICAL METHOD TO DETERMINE ACCELERATION CURVE FROM GIVEN DISPLACEMENT CURVE.

If we plot the acceleration values as in the figure above, we observe that the curve through these points is a straight line, showing that the acceleration varies uniformly. In general the length of the acceleration curve can be increased by taking a smaller value of h. For an experiment that involves a large number of points, the calculations may be performed by an automatic computing machine, thus reducing calculation time.

t	S	$-2S_n$	S_{n+4}	S_{n-4}	$a = \frac{S_{n+4} + S_{n-4} - 2S_n}{4}$
0	0	0	8		
0.5	0.125	-0.25	15.6		
1.0	1	-2	27		
1.5	3.38	-6.76	42.8		
2.0	8	-16	64	0	a = (64 + 0 - 16)/4 = 12
2.5	15.6	-31.2	91.1	0.125	a = (91.1 + 0.125 - 31.2)/4 = 15
3.0	27	-54	125	1	a = (125 + 1 - 54)/4 = 18
3.5	42.8	-85.6	166	3.38	a = (166 + 3.38 - 85.6)/4 = 21
4.0	64	-128	216	8	a = (216 + 8 - 128)/4 = 24
4.5	91.1	-182.2	275	15.6	a = (275 + 15.6 - 182.2)/4 = 27
5.0	125	-250	343	27	a = (343 + 27 - 256)/4 = 30
5.5	166	-332	422	42.8	a = (422 + 42.8 - 332)/4 = 33
6.0	216	-432	512	64	a = (512 + 64 - 432)/4 = 36
6.5	275	-550		91.1	
7.0	343	-686		125	
7.5	422	-844		166	
8.0	512	-1024		216	

TABLE 1

The data selected for the above demonstration were based upon the formula $S = t^3$. Students acquainted with elementary calculus would, of course, know that, since $v = ds/dt$ and $a = dv/dt$, the corresponding relations would be $v = 3t^2$ and $a = 6t$.

• PROBLEM 16-14

The following experimental data relates displacement, s, in feet, to time, t, in seconds. Find (a) the velocity-time relationship and (b) the acceleration-time relationship.

t	1	2	3	4	5	6	7	8	9	10
s	1.1	3.1	3.9	4.2	3.9	3.3	2.7	2.1	1.8	2.1

Solution: A plot of the given data is shown in the following figure. This curve which shows displacement with

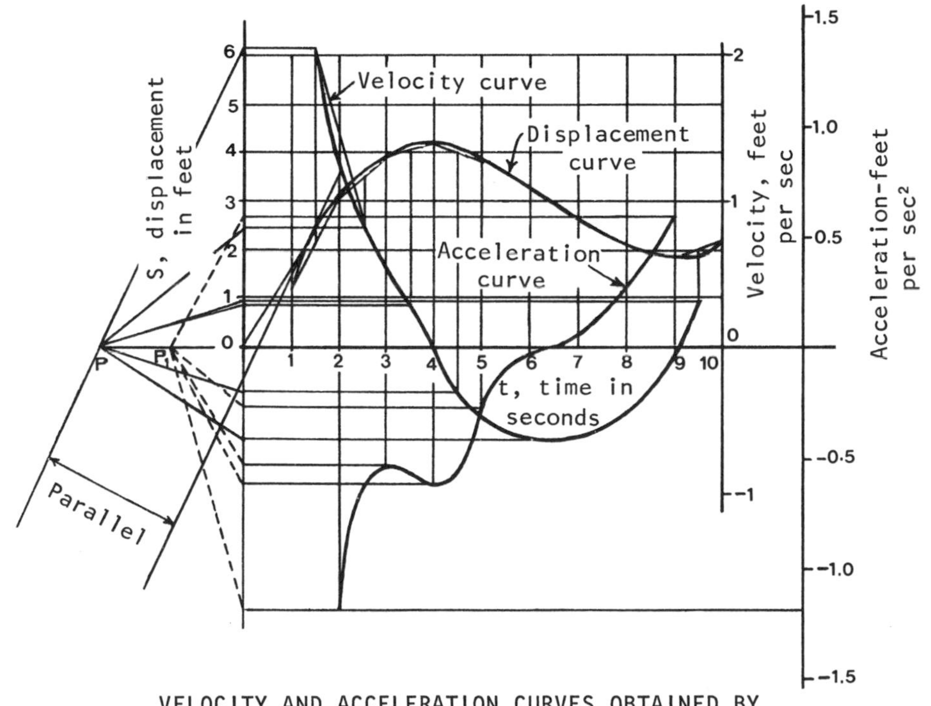

VELOCITY AND ACCELERATION CURVES OBTAINED BY GRAPHICAL DIFFERENTIATION OF DISPLACEMENT CURVE.

respect to time is called a displacement curve. To find the velocity time relation by the plot method, the following steps are used:

(1) A pole distance must be selected. In this problem a convenient pole distance of 3 is selected, point P. Remember that the distance in abscissa unit determines the ordinate scale ratio for both curves.

(2) With the location of the pole point P established, rays are drawn, parallel to the straight lines of the given curve until they intersect the y-axis for the derived curve.

(3) These points of intersection are projected horizontally to determine the derived curve which is shown in the following figure.

This curve shows the velocity relationship, since the rate of change of s with respect to t is velocity. Expressed algebraically, this is $ds/dt = v$.

Similarly, if we use the velocity-time curve as the given curve and a pole distance OP_1 equal to 1-1/2 units, we obtain the second derived curve, which shows the acceleration-time relationship. Expressed algebraically, this is $dv/dt = a$, where a represents acceleration.

The velocity scale is obtained by dividing the original s scale by the pole distance 3, whereas the acceleration scale is obtained by dividing the velocity scale by the pole distance, 1-1/2.

• PROBLEM 16-15

Consider the contour map shown in Fig. below. A lake is to be formed in the depression by filling it with water to elevation 35. (a) How many acre-feet of water will be required (neglecting water absorbed by the ground); (b) if the water level drops to elevation 25, how many acre-feet must be added to restore the lake level to elevation 35? (Note: An acre-foot of water is that volume required to cover an acre with one foot of water or is 43,560 cubic feet of water.)

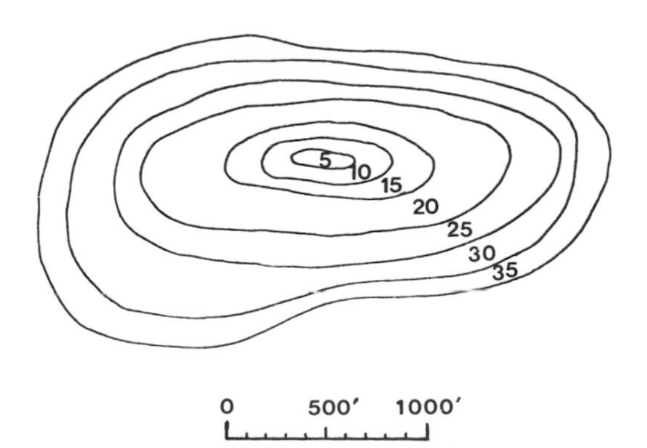

CONTOUR MAP FOR LAKE PROBLEM.

Solution: The area bounded by each contour is first

determined by the planimeter method. The values thus obtained are shown in the table below:

Contour	5	10	15	20	25	30	35
Area, thousands of square feet	37.5	112.5	313	1025	1710	2650	3450
Area, acres	0.86	2.58	7.17	23.4	39.2	60.7	79.0

A plot of these values is shown in the following figure. The total volume, 840 acre-feet, is obtained by graphically integrating the area under the curve which passes through the plotted points.

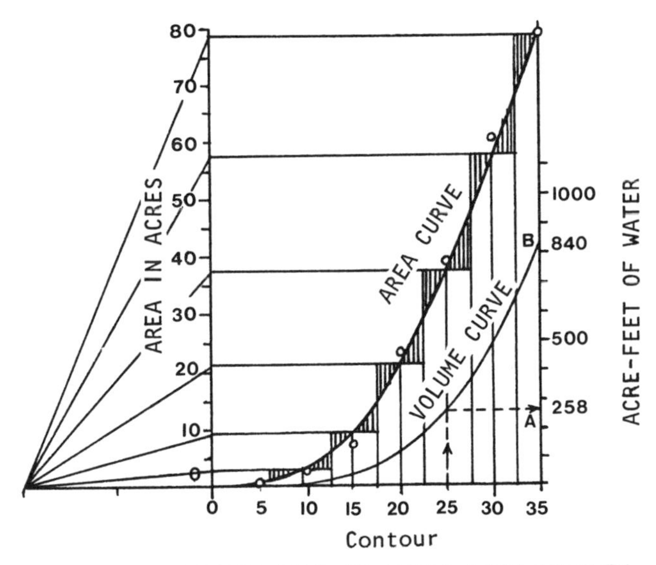

GRAPHICAL INTEGRATION USED TO SOLVE LAKE PROBLEM.

The volume of water necessary to restore the level from elevation 25 to elevation 35 is represented by segment AB, the length of which can now be laid off from zero on the acre-feet scale and read directly from the scale.

• PROBLEM 16-16

Divide the irregularly shaped lot shown in the following figure into three equal areas by lines that are perpendicular to the front property line OC.

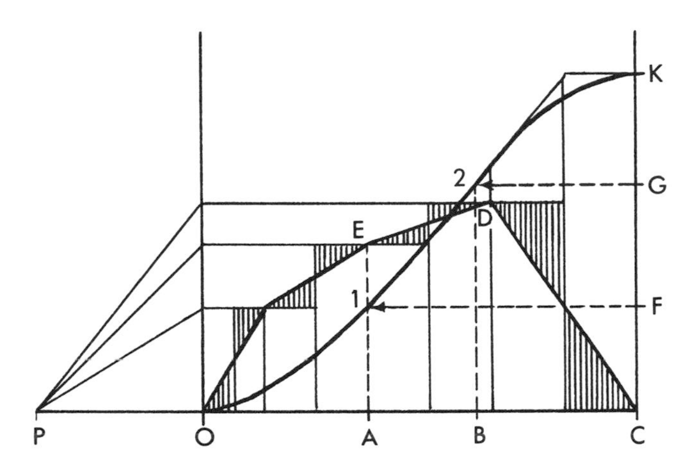

GRAPHICAL INTEGRATION USED TO DIVIDE AN IRREGULAR-SHAPED LOT INTO THREE EQUAL AREAS BY LINES PERPENDICULAR TO PROPERTY LINE OC.

Solution: The integral curve is found by using the alternative construction of the pole-and-ray method. Distance CK, which represents the total area, is divided into three equal segments CF, FG, and GK. Horizontal lines drawn through points F and G to the integral curve establish points 1 and 2, respectively. Vertical lines AE and BD drawn through these points divide the original area into three lots having equal areas.

• PROBLEM 16-17

A hoist hook, shown in figure 1a has a rated capacity of two tons. It is desired to determine the work done in opening the hook when loads varying from 8000 pounds to 22,000 pounds are applied. The table below shows the test results of elongation of the hook by the various loads:

Elongation, in inches	0	0.32	0.39	0.59	0.74	0.88	1.3	1.7	2.2
Load thousands of pounds	8	12	15	17.5	20	21	22	22	21

Figure 1b shows the hook with the maximum load applied.

FIG. 1-a: HOIST HOOK UNLOADED.

FIG. 1-b: HOIST HOOK WITH MAXIMUM LOAD.

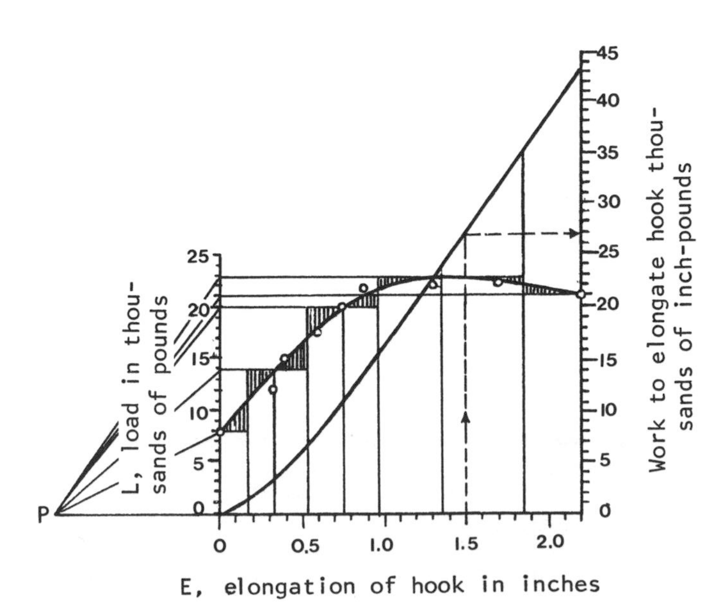

HOIST HOOK DATA AND INTEGRAL CURVE. Fig. 2

Solution: A plot of the test data and a curve which is representative of the plotted points are shown in Fig. 2. The work required to elongate the hook 2.2 inches is represented by the area under the curve. This can be determined by graphical integration. Once the curve of integration is established, we can easily determine the work required to elongate the hook any distance between the limits shown. For example, the work done for an elongation of 1.5 inches is 27,200 inch-pounds.

• PROBLEM 16-18

An automobile starting from rest attains a speed of 25 mph in 20 seconds. (a) What is the acceleration? (b) What is the distance traveled? Use: (1) the calculation method. (2) the graphical method.

Solution: Part 1, calculation method:

Let V_0 = initial velocity ft/sec.
V_1 = final velocity ft/sec.
$\bar{a}$ = acceleration ft/sec.^{2}
S = distance ft

We know that

$$V_0 = 0$$

$$V_1 = \frac{25 \times 5280}{3600} = 36.7 \text{ ft/sec.}$$

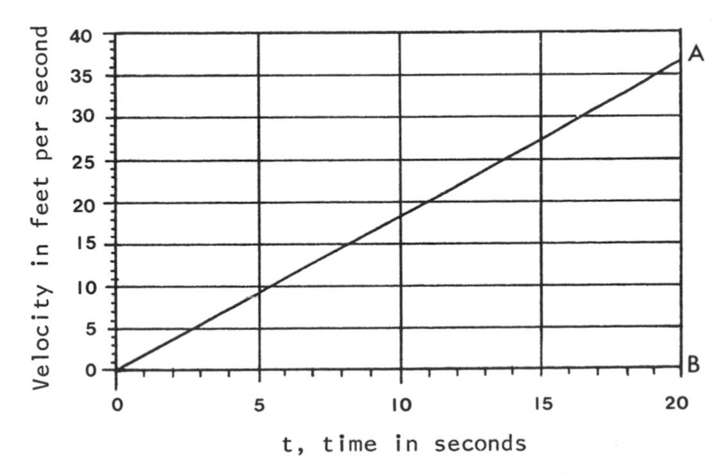

GRAPHICAL SOLUTION OF TRAVEL PROBLEM

Therefore, $\bar{a} = \frac{V_1 - V_0}{t} = \frac{36.7 - 0}{20} = 1.833$ ft/sec.^{2}

Now $S = V_0 t + \frac{1}{2} at^2$

or $S = 0 + \frac{1}{2} \times 1.833 \times 400 = 367$ ft.

Part 2: The relationship between speed and time is shown the figure. We know that acceleration is the rate of change of velocity with respect to time, or a = dV/dt; therefore, the slope of line OA is the acceleration, which is simply 367/20 = 1.83 feet per second^2 (average acceleration).

We also know that the distance traveled is the summation of the small displacements V dt, or the integral of velocity with respect to time; i.e.,

$S = \int V\,dt$. Graphically, this is simply the area under the curve OA, or in this example the area of the triangle OAB, which is equal to $\frac{36.7 \times 20}{2} = 367$ feet (assuming constant acceleration).

• PROBLEM 16-19

An automobile starting from rest attains a speed of 12 mph during the first 5 seconds (low gear), remains at that speed for 1 second (shifting from low to second), attains a speed of 20 mph during the next 5 seconds (driving in second), remains at that speed for 1 second (shifting from second to high), attains a speed of 35 mph in the next 8 seconds, and continues to operate at that speed for 5 seconds more. What is the distance traveled?

Solution: A graphical representation of the above information is shown in the following figure, where the given data are satisfied and the curve appears reasonable for the acceleration periods.

If we integrate the curve graphically, we obtain the integral curve, which enables us to read not only the distance, 805 feet, traveled at the end of 25 seconds, but also the distance traveled at the end of any time between zero and 25 seconds.

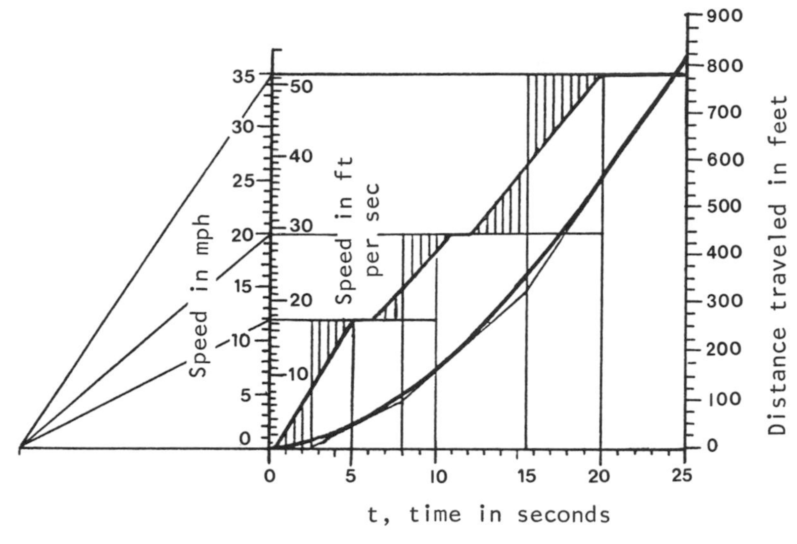

GRAPHICAL SOLUTION OF TRAVEL PROBLEM

CENTROIDS

• PROBLEM 16-20

Consider the area under the curve shown in Fig. 1. Part (1) - Define the centroid of the area and determine the coordinates $\bar{x}$ and $\bar{y}$ of the centroid. Part (2) - From the result of Part (1), determine graphically the centroid of the area under the curve which is obtained based upon the following data:

x	0	1.5	3	4.5	6	7.5
y	6	7	9	9.5	9	8

Solution: Part 1. The centroid of the area is that point about which the algebraic sum of all the moments of infinitesimal areas of the surface is zero. Now determine the coordinates $\bar{x}$ and $\bar{y}$ of the centroid.

The coordinate x is determined by taking moments about the y-axis and then dividing the sum of the moments of all strips (width dx) by the total area.

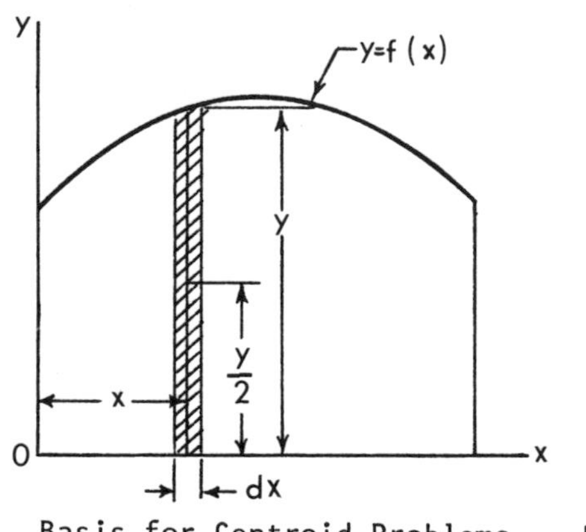

Basis for Centroid Problems Fig. 1

The moment about the y-axis of a typical strip such as y dx is equal to xy dx. The summation of the moments of all strips is $\int xy\,dx$. Therefore the x-coordinate of the centroid is

$$\bar{x} = \frac{\int xy\,dx}{\int y\,dx}, \text{ where } \int y\,dx \text{ is the total area.}$$

By taking moments about the x-axis, we find that the moment of strip y dx is equal to $(y\,dx)(y/2) = \frac{1}{2}y^2dx$, where $y/2$ is the moment arm. Therefore, the y-coordinate of the centroid is

$$\bar{y} = \frac{\frac{1}{2}\int y^2\,dx}{\int y\,dx}.$$

Part 2. A plot of the data given is shown in Fig. 2.

Since $\bar{x} = \dfrac{\int xy\,dx}{\int y\,dx}$, we can plot a curve whose ordinates are (xy) and whose abscissas are values of x, as shown in the table below:

x	0	1.5	3	4.5	6	7.5
xy	0	10.5	27	43	54	60

Now the area under each curve can be determined by the use of a planimeter or by graphical integration. Once this is done

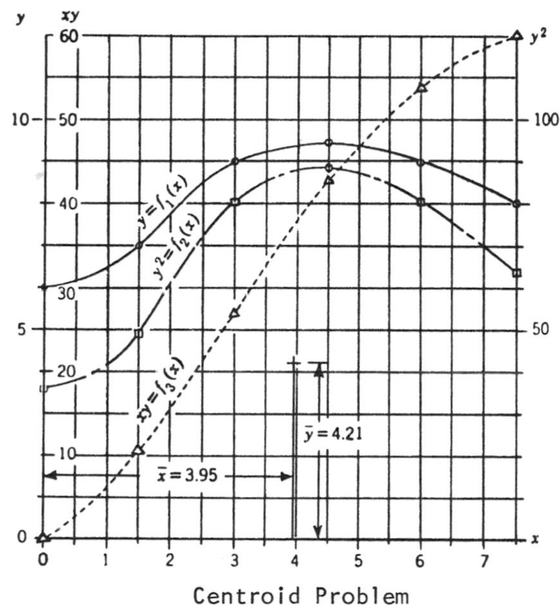

Centroid Problem Fig. 2

$$\bar{x} = \frac{\text{area under the curve } xy = f_3(x)}{\text{area under the curve } y = f_1(x)} = 3.95$$

In a similar manner, since

$$\bar{y} = \frac{\frac{1}{2}\int y^2 dx}{\int y dx},$$ if we plot the curve $y^2 = f_2(x)$ we can determine the area under this curve and then evaluate

$$\bar{y} = \frac{\frac{1}{2} \text{ area under the curve } y^2 = f_2(x)}{\text{area under the curve } y = f_1(x)} = 4.21$$

Figure 2 also includes the curve $y^2 = f_2(x)$, which is based upon the coordinates

x	0	1.5	3	4.5	6	7.5
y^2	36	49	81	89	81	64

• PROBLEM 16-21

Part (a): Consider the area bounded by the curve shown in Fig. 1. Determine the location of its centroid.

Part (b): And then determine the moment of inertia of the area shown in this figure about axis m.

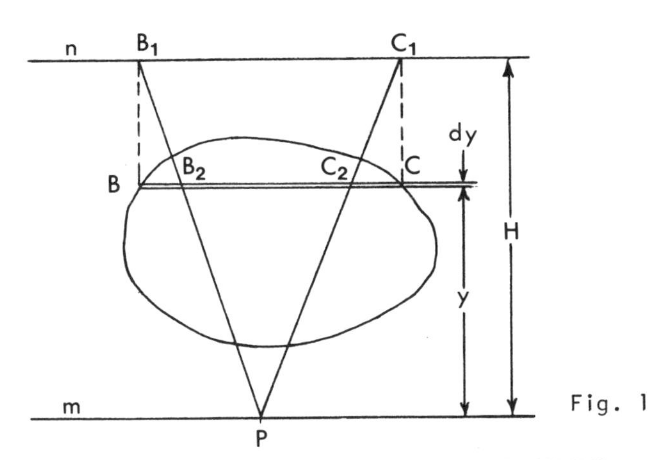

GRAPHO-MECHANICAL SOLUTION FOR CENTROID PROBLEMS

Solution: Part (a). First draw parallel lines m and n a convenient distance (H) apart. Now introduce line BC parallel to m and project points B and C to line n by perpendiculars to line n. These points are shown as B_1 and C_1, respectively. Lines PB_1 and PC_1 (where P is any point on line m) intersect line BC at points B_2 and C_2. From similar triangles PB_1C_1 and PB_2C_2 it follows that

$$\frac{B_2C_2}{B_1C_1} = \frac{y}{H},$$

or

$$\frac{dA_1}{dA} = \frac{y}{H}$$

if we let $dA_1 = B_2C_2 \times dy$, and $dA = BC \times dy$. Therefore,

$$A_1 = \int dA_1 = \frac{1}{H} \int ydA$$

and

$$\bar{y} = \frac{A_1 H}{A}$$

since

$$\bar{y} = \int \frac{ydA}{A}.$$

If the construction shown in Fig. 1 is repeated for additional parallels to BC, the area bounding the curve which passes through such points as B_2C_2 will be area A_1, which can be determined by the planimeter method or one of the methods of graphical integration. This is also true for area A.

It should be clear that parallel lines m and n could be drawn in any direction. If the construction is repeated for a new orientation of parallels m and n, another centroidal distance ($\bar{y}_1$) can be found with respect to the new position of line m. Then the centroid is the point in which the $\bar{y}$-line intersects the $\bar{y}_1$-line.

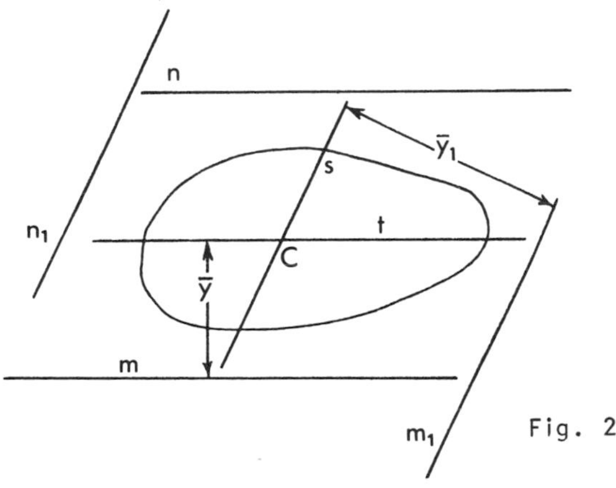

GRAPHO-MECHANICAL METHOD FOR LOCATING CENTROID.

For example, let us assume that $\bar{y}$ for the area shown in Fig. 2 has been found with respect to the horizontal position of lines m and n. Now suppose that $\bar{y}_1$ has been determined for the new positions of lines m and n as indicated by m_1 and n_1. The intersection of lines s and t is the centroid, C.

Part (b). The moment of inertia of this area about axis m is the summation of all the products formed by multiplying every elementary area by the square of its distance from the axis.

First let us repeat the construction shown in Fig. 1 to locate points B_2 and C_2. Now project these points to line n to locate points B_3 and C_3 respectively. Lines PB_3 and PC_3 will intersect line BC in points B_4 and C_4, which are on the boundary of area A_2.

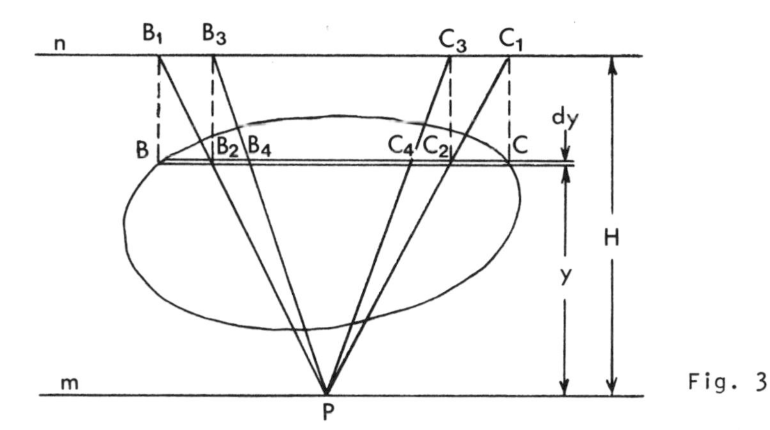

GRAPHO-MECHANICAL SOLUTION FOR MOMENT OF INERTIA PROBLEMS

From similar triangles PB_3C_3 and PB_4C_4 it follows that

$$\frac{B_4C_4}{B_3C_3} = \frac{y}{H} \quad ,$$

or

$$\frac{dA_2}{dA_1} = \frac{y}{H}$$

$$A_2 = \frac{1}{H} \int ydA_1 = \frac{1}{H^2} \int y^2 dA, \text{ since } dA_1 = \frac{ydA}{H}$$

(See Part (a).)

Therefore, the moment of inertia about axis m is A_2H^2. Area A can be determined by the planimeter method or by one of the methods of graphical integration.

Now let us consider the section shown in Fig. 4. It is required to determine its centroid. The construction

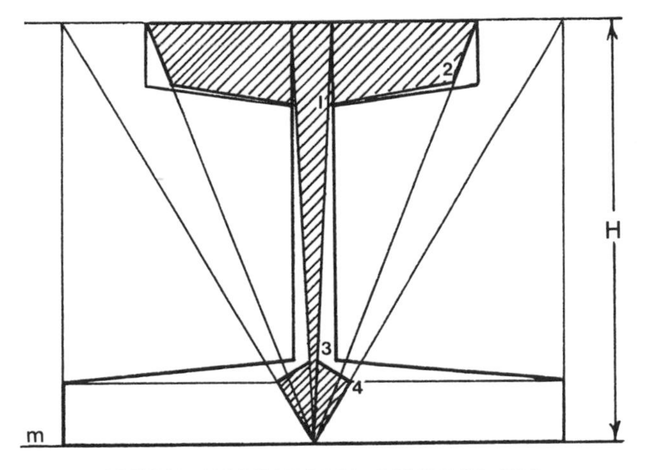

Fig. 4

GRAPHO-MECHANICAL SOLUTION FOR LOCATION OF CENTROID OF SECTION.

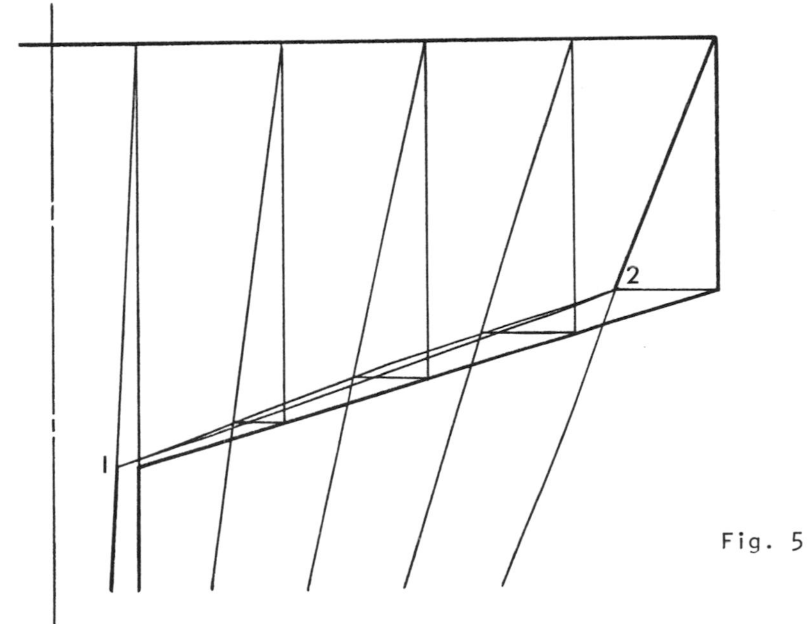

Fig. 5

ENLARGEMENT OF PORTION OF FIG. 4

used in Part B yields the shaded area A_1, which can be evaluated by the planimeter method. The $\bar{y}$ value for the centroid is

$$\frac{A_1 H}{A} ,$$

and $\bar{x}$ is zero if the y-axis is coincident with the axis of symmetry.

If area A_1 is regarded as the given section and the process repeated to determine area A_2, the moment of inertia about axis m is A_2H^2.

It should be noted that actually line 1-2 and line 3-4 are not straight lines, but in the above example are drawn as straight lines since they do not deviate appreciably from the correct curve. Figure 5 shows an enlarged drawing of the portion of the area that affects line 1-2, to illustrate the difference between the line 1-2 and the correct curve 1-2.

CHAPTER 17

DIMENSIONING

DIMENSIONING PROCEDURES

• PROBLEM 17-1

List the procedures for dimensioning and illustrate these procedures with an example.

Solution: If a simple procedure of planning the dimensioning is followed, observing the standards outlined in this solution, the resulting finished drawing will contain accurate, clear, complete, and readable specifications.

The following steps of procedure should be followed:

1. Use the minimum number of orthographic views necessary for complete shape delineation of the object. Choose a scale for the drawing and a reasonable spacing of views on the sheet to provide ample room for dimensions.

2. For each view, list the features whose contour shape is best delineated by that view.

3. Show the size dimensions (usually two) of each contour shape and the coordinate locations with respect to the most important surface or reference center line.

4. Add overall sizes, notes, and title.

Referring to the figure, the above steps are illustrated.

Step 1. Three views of the Stop Block are drawn, each being necessary to show critical contour.

Step 2. a. "H" projection shows:

Contour detail of two countersunk holes.
Slot in right end.

b. "F" projection shows:

Contour of the L-shaped cut.
Relative heights.

c. "P" projection shows:

Shape of beveled corner.

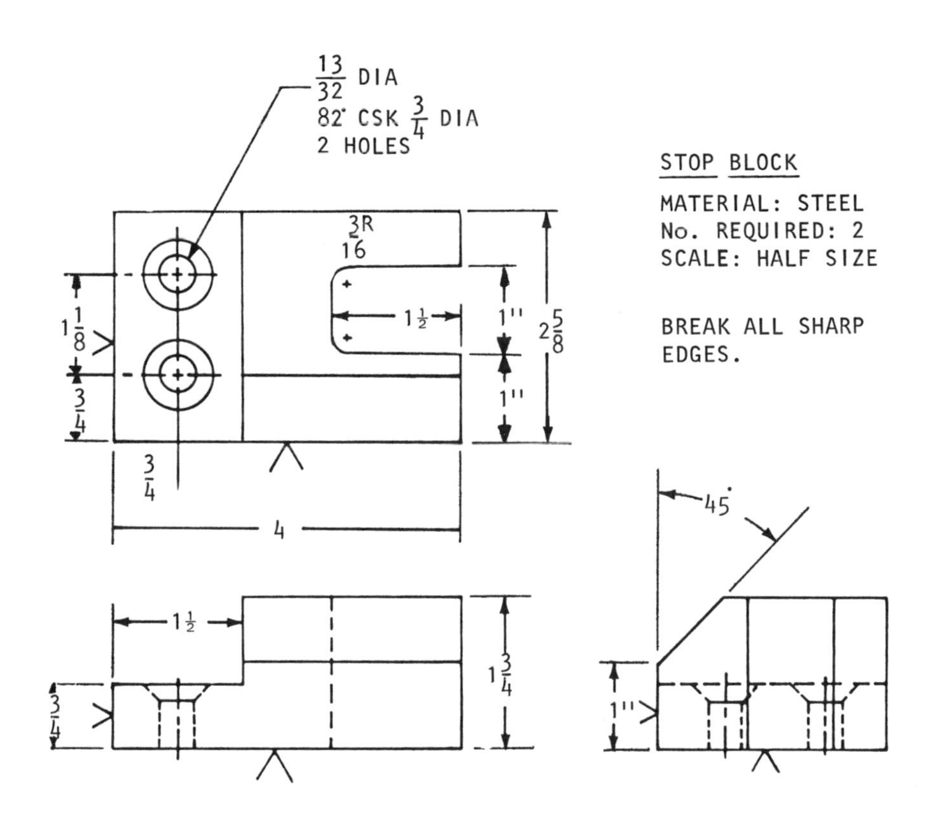

Step 3. a. Dimensions are placed on "H":
Holes--three dimensions of location (3/4, 3/4, 1 1/8") and the size specification.

Slot--three size dimensions (1" x 1 1/2), 3/16 R. for fillet, and one location (1").

b. Dimensions are placed on "F":

Size of the left "L" shaped portion--(3/4 height and 1 1/2 length).

c. Dimensions are placed on "P:"

Location height to the bevel (1").
Angle of the bevel (45°).

Step 4. Overall sizes added (2 5/8 x 4 x 1 3/4 hight).

Title added, including name of object, material, number required, scale, and general notes.

• PROBLEM 17-2

Stress the importance of notes (to supplement the dimensions on a drawing) by illustrating with an example.

Solution: Acquaintance with machine-shop methods and practices is of invaluable aid in helping the draftsman to place the proper notes on a drawing. An example of the value of knowing machine-shop practice is illustrated by Fig. 1, a diagram showing several holes. If the draftsman dimensions these holes by giving only their diameters, the mechanic making the object will not know which of the various methods of production to use--as, for example, drilling, reaming, coring, or boring. The draftsman, in cases of this type, should state specifically by a note or notes the method to be used in the making of these holes.

Notes should always be expressed in a definite form. Fig. 1 shows the accepted form of notes for various types of holes. The size of the hole is stated first, then the operation to be used in making it. Where several operations are performed on the same hole, the steps are listed in the order in which they should be made. Fig. 1 also shows methods of indicating counterbored and countersunk holes for screws. Either method of notation is correct.

In the case of drilled holes, in numbered or lettered sizes, the number or letter of the drill should be followed by its decimal equivalent in parentheses. (See Fig. 1.)

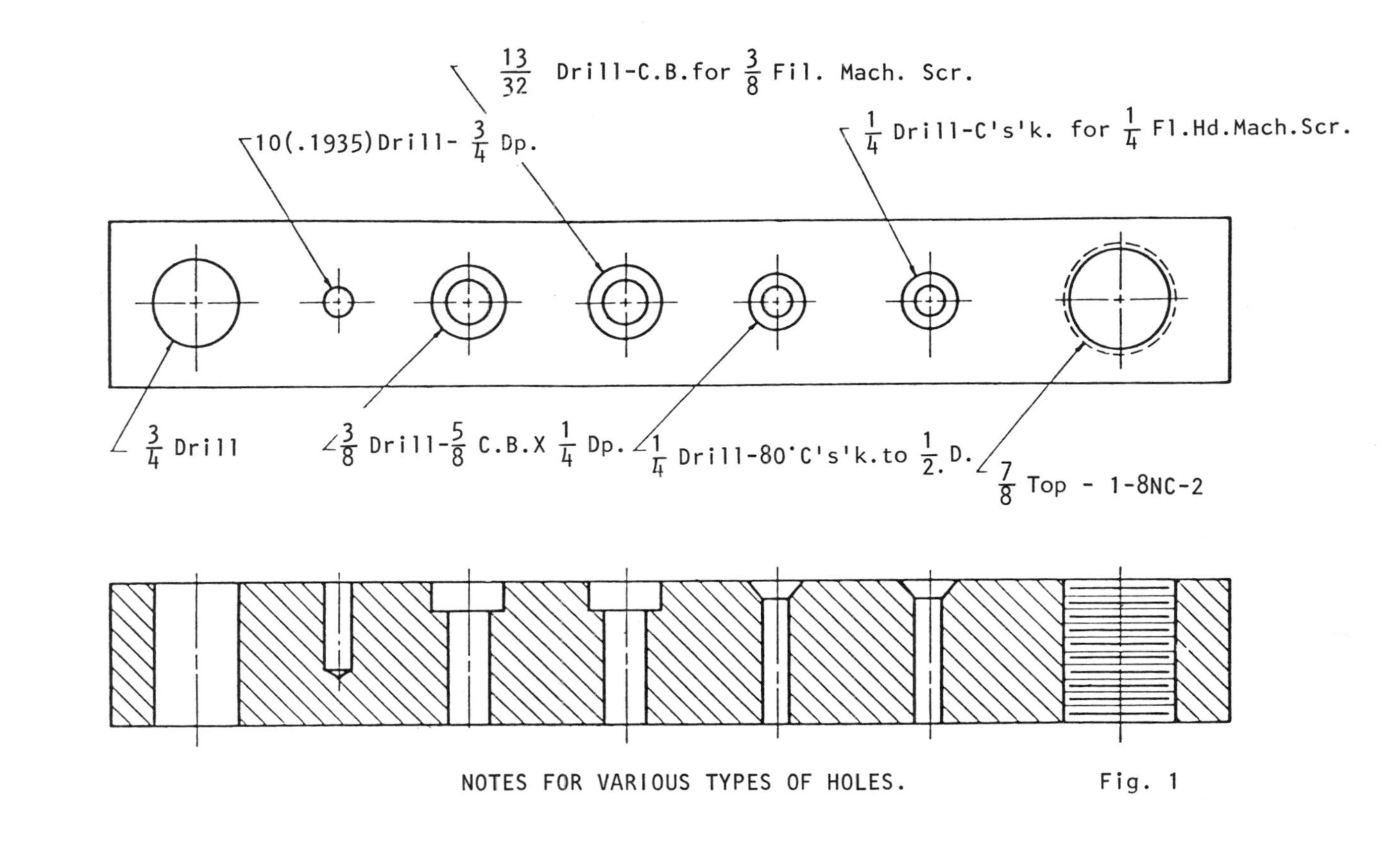

NOTES FOR VARIOUS TYPES OF HOLES. Fig. 1

• **PROBLEM 17-3**

Illustrate the method of dimensioning angles and describe the methods of dimensioning in limited spaces.

Solution: Fig. 1 shows several recommended methods of dimensioning angles.

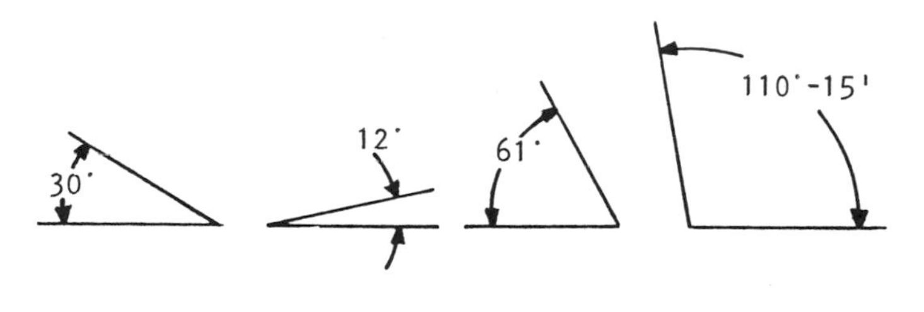

Fig. 1

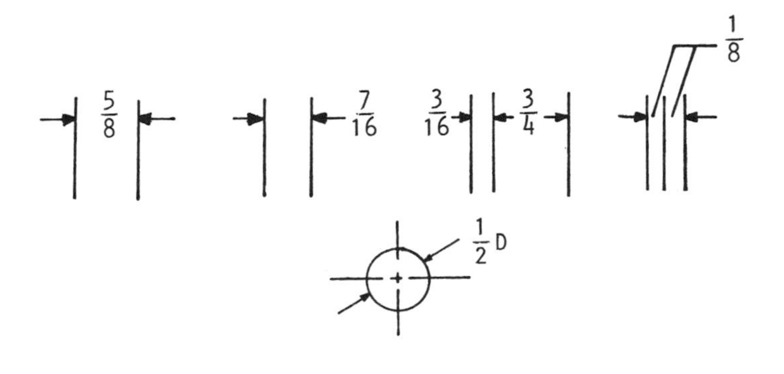

Fig. 2

Dimensions should never be crowded into small spaces. Fig. 2 shows several suggested methods of dimensioning where the space is limited.

STAGGERED AND BASE LINE DIMENSIONING

• PROBLEM 17-4

Dimension the following figure using staggered dimensions.

Solution: In many cases where there are a series of parallel dimensions, it is desirable to have the breaks in the dimension lines staggered to give sufficient space for the numerals and for clarity in reading. Fig. 1(b) illustrates this rule.

DIMENSIONING

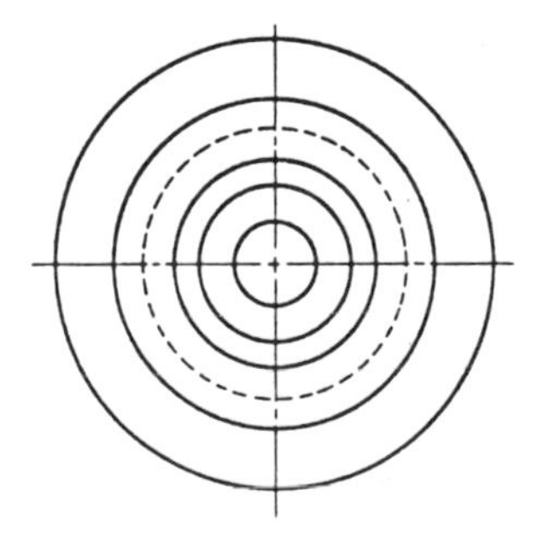

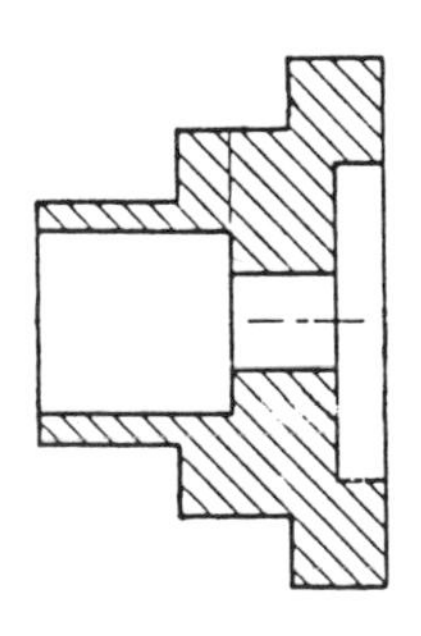

STAGGERED DIMENSIONS

Fig. 1a.

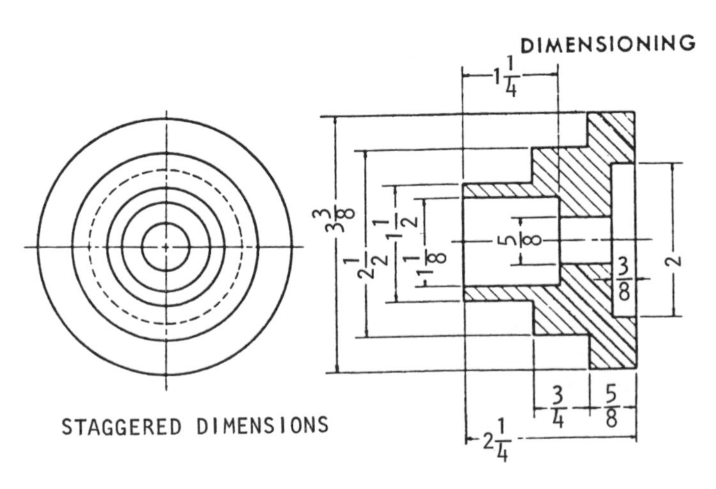

Fig. 1b

• PROBLEM 17-5

Illustrate base-line dimensioning in the given figure:

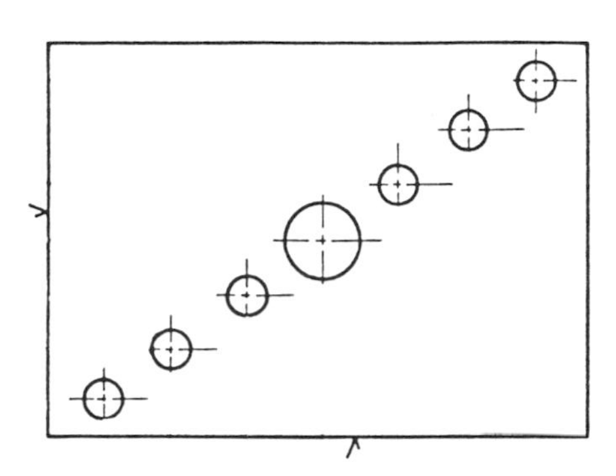

Fig. 1

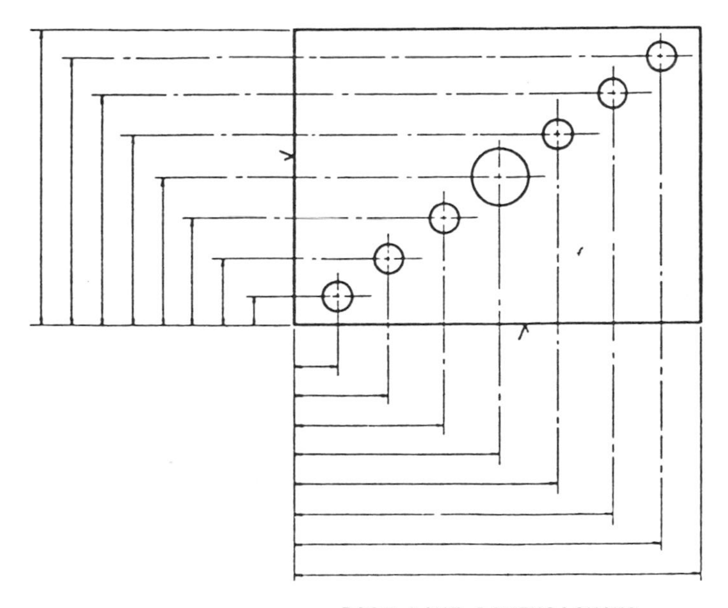

BASE-LINE DIMENSIONING

Fig. 2

Solution: In certain types of precise works, dimensions are sometimes referred to as "center line" or "base line." The purpose of base-line dimensioning is to prevent cumulative errors, for each dimension is independent of all others. Fig. 2 shows an example of base-line dimensioning. The dimensions used in this kind of work are always expressed in decimals, since fractional dimensions are not sufficiently accurate.

DIMENSIONING OF ARCS AND CURVES

• PROBLEM 17-6

Dimension the following arcs at 1, 1/2, 1/4 and 1/8 inches in radius respectively.

Solution: Fig. 2 shows the approved method of dimensioning arcs. The numeral giving the length of the radius

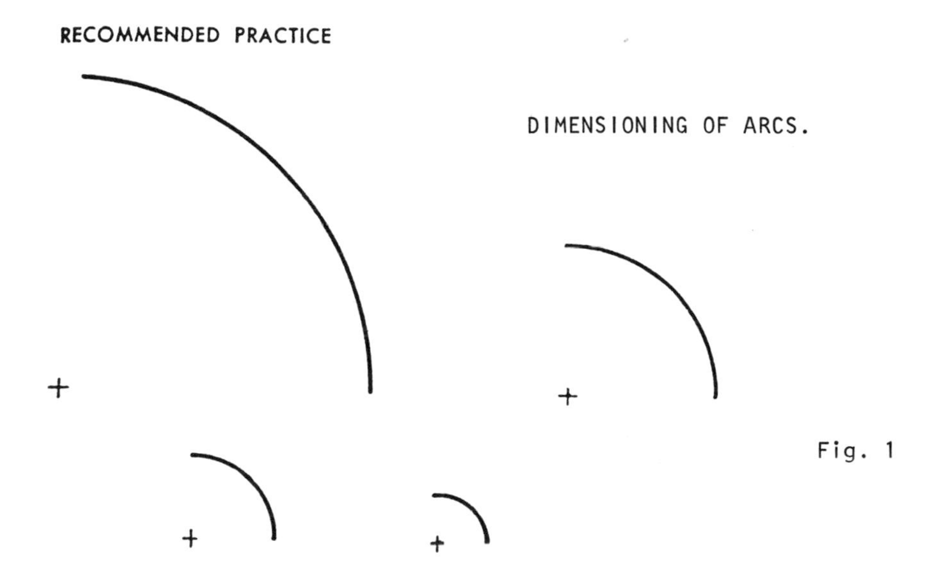

Fig. 1

should always be followed by R, which is the standard symbol for "radius." The dimension line should have only one arrowhead, which should touch the arc. The center of the arc should be clearly marked.

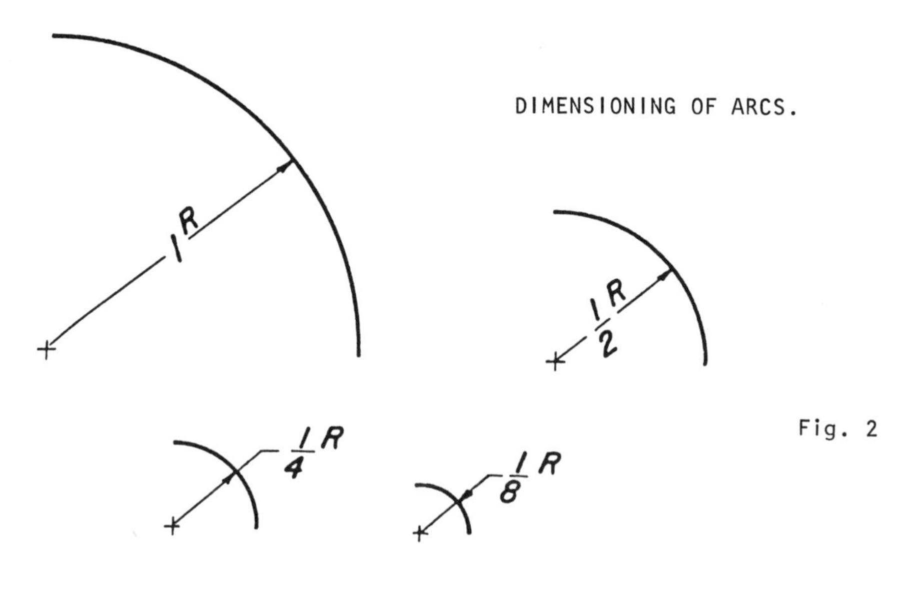

Fig. 2

• PROBLEM 17-7

Illustrate the method of dimensioning of the following figures.

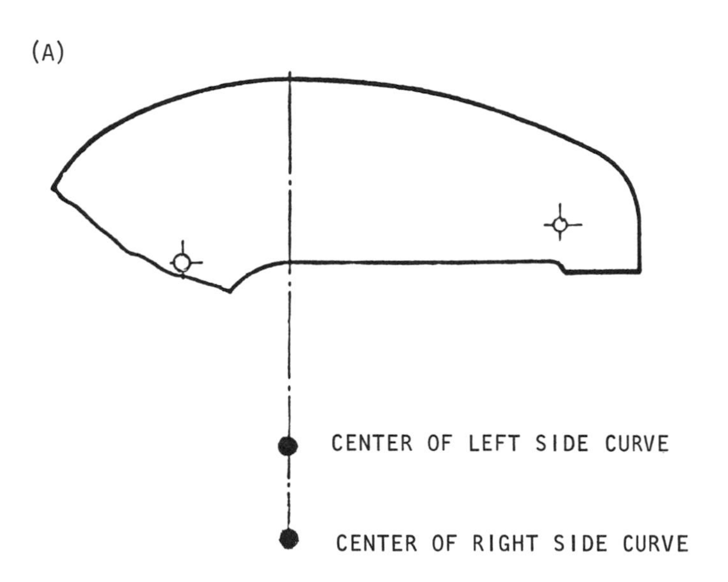

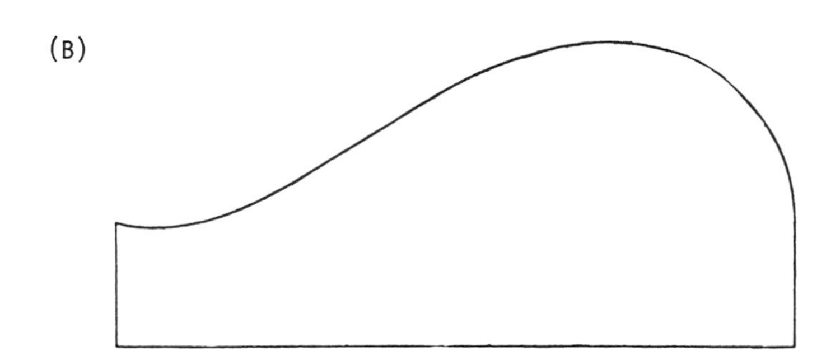

Solution: The method of dimensioning curved lines recommended by the American Standards Association is shown in Figs. 1 and 2. Figs. 3 and 4 show other methods of dimensioning curved lines which are widely used in industry.

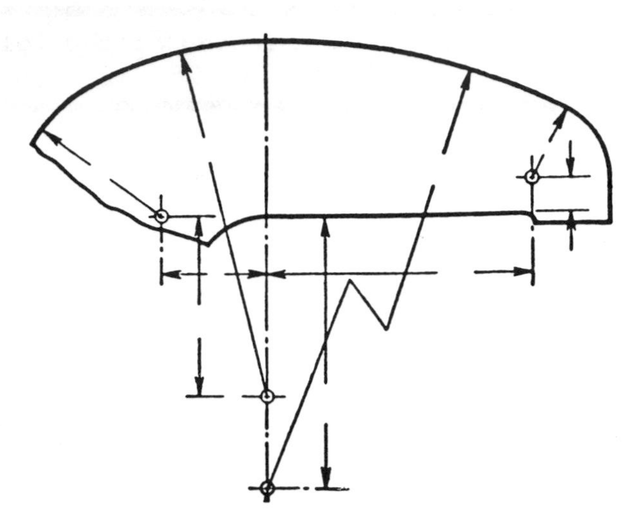

DIMENSIONING OF CURVED LINES WITH RADII. Fig. 1

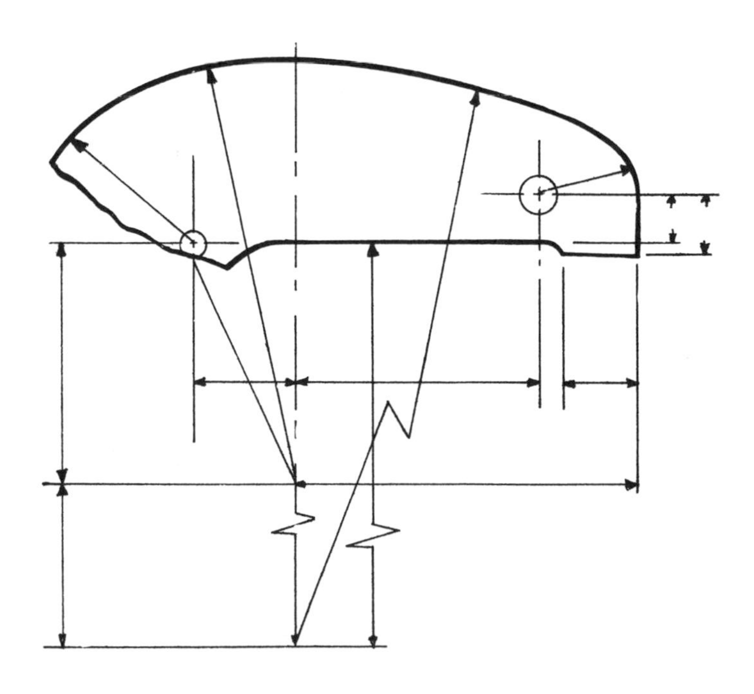

COMMON METHOD OF DIMENSIONING CURVED LINES. Fig. 2

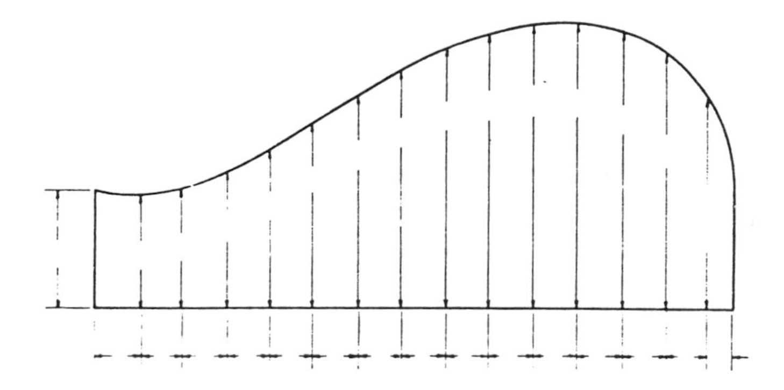

DIMENSIONING OF CURVED LINES WITH OFFSETS. Fig. 3

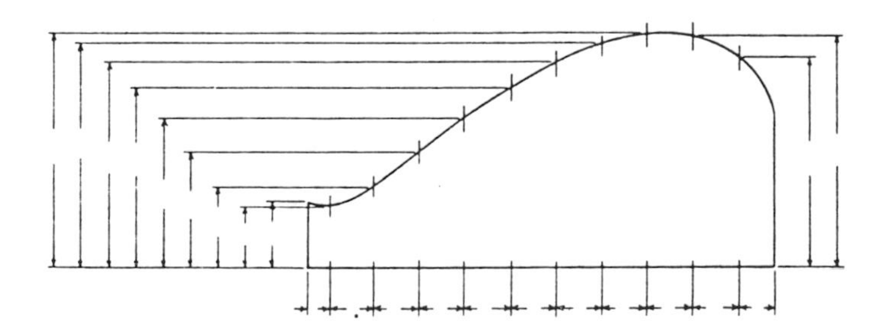

COMMON METHOD OF DIMENSIONING CURVED LINES. Fig. 4

APPLICATIONS

• PROBLEM 17-8

A. Illustrate location dimensions in the following figures:

B. Completely dimension the following:

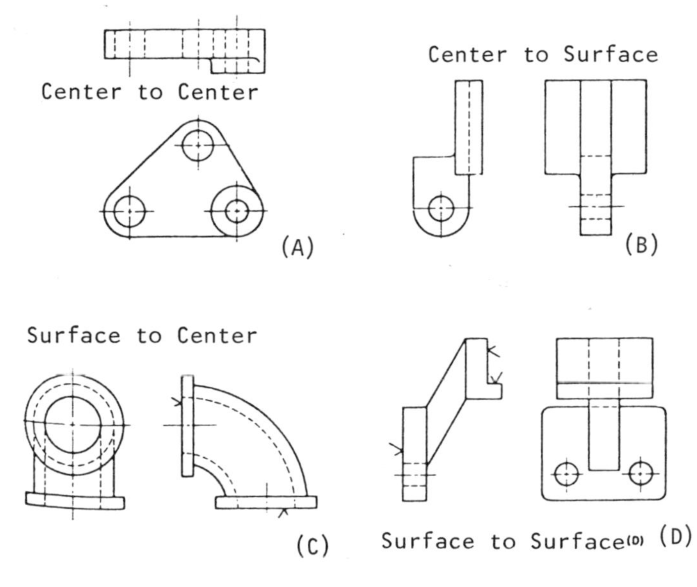

LOCATION DIMENSIONS

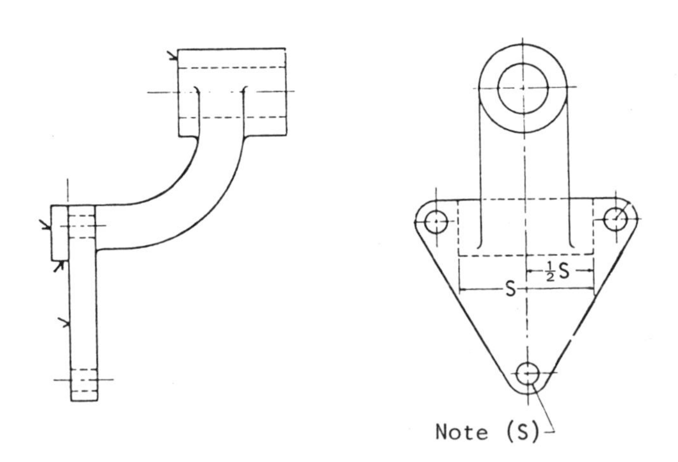

Solution: A. Location Dimensions. As previously stated, in dimensioning an object the first step is to break down the object into elementary shapes and to dimension the shapes. Location dimensions are then placed on the drawing to tie these elementary shapes together. Location dimensions are of several types.

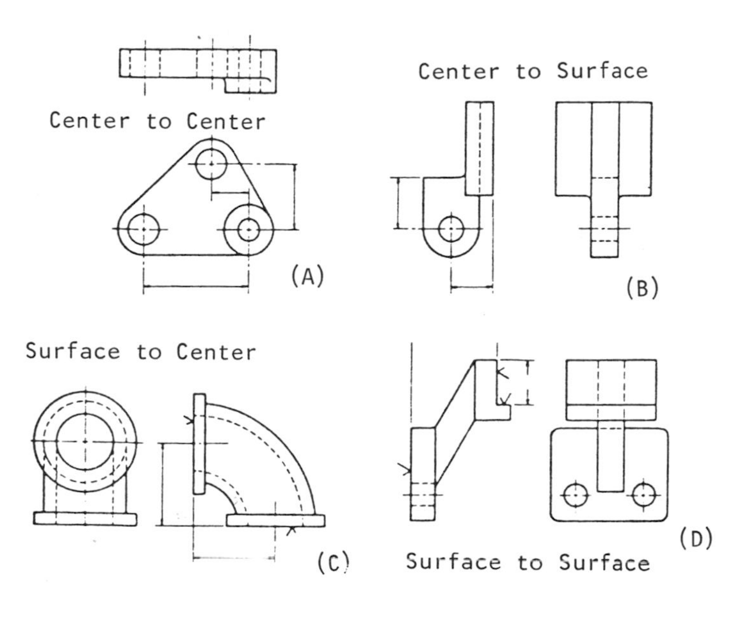

LOCATION DIMENSIONS Fig. 1

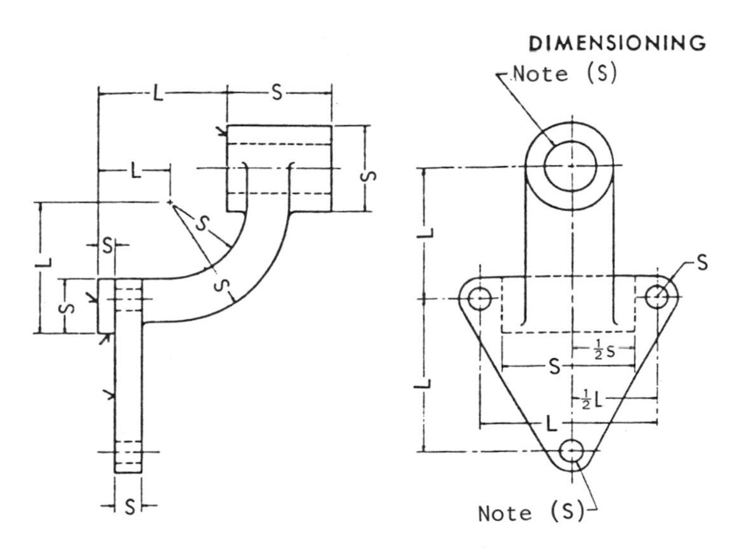

EXAMPLE OF COMPLETELY DIMENSIONED DRAWING. Fig. 2

1. Locating center lines of holes with respect to other center lines (Fig. 1A).

2. Locating surfaces from center lines of holes (Fig. 1B).

3. Locating center lines from finished surfaces (Fig. 1C).

4. Locating finished surfaces with respect to other finished surfaces (Fig. 1D).

Fig. 2 shows a completely dimensioned drawing. Instead of numbers, L has been used to indicate location dimensions and S to indicate size dimensions. It will be noted that many of the rules previously set forth are illustrated in this drawing.

B. Over-All Dimensions. Over-all dimensions should be shown for all drawings. An over-all dimension, which is neither a size dimension nor a location dimension, must show the summation of all intermediate dimensions. The only exception to this rule is that the over-all dimension is never given to pieces having circular ends (Fig. 2).

• PROBLEM 17-9

Draw two views of and dimension the following:

A) A right circular cone with a height of 1 3/4 in. and a base diameter of 1 1/2 in.

B) A frustum of a right circular cone (cone with the top cut off) with a base diameter of 1 1/2 in., a top diameter of 5/8 in. and a height of 1 in.

C) A Cone originally 1 3/4 in. high and 1 1/2 in. in diameter at the base which was cut to give a maximum height of 1 1/4 in. and a minimum height of 3/8 in.

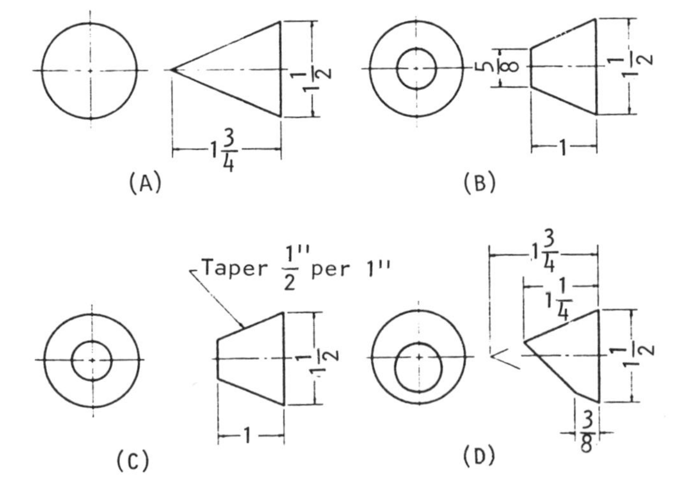

DIMENSIONING OF CONES. Fig. 1

Solution: Cones. Fig. 1 shows the methods used to dimension different types of cones. The general rule

for dimensioning a cone is to show all the dimensions on the view which shows the altitude.

Two different methods of dimensioning the same cone are shown in Figs. 1B and 1C. The choice between the two methods would be dependent solely upon its method of construction in the shop.

Another method of dimensioning a cone which is frequently used is to give the diameter of the base and the angle of the vertex. In this instance it is not necessary to show a diagram of the cone.

• PROBLEM 17-10

The following figure is of an object with a flange of outside diameter 3 3/4 inches and hub of outside diameter 1 1/2 inches. The full length of the object is 2 1/4 inches and the flange length is 7/8 inches. Dimension the figure.

HUB FLANGE

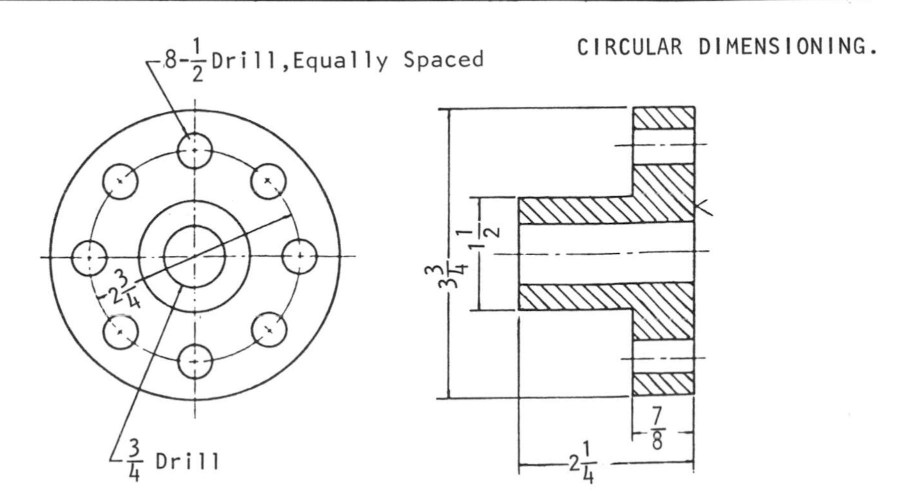

Solution: A good rule to remember in dimensioning cylinders is the following: the only dimension that should appear on a plan view, showing cylinders as circles, is the diameter of the circular center line of bolt circles in the flange of the object, as illustrated in the figure (circular dimensioning). This figure also shows how a piece should be dimensioned when it is built up entirely of cylinders.

• PROBLEM 17-11

Draw two views of and dimension:

A) A 2-inch long cylinder with a 1 1/2-inch diameter.

B) A 2-inch long cylinder with an outside diameter of 1 1/2 inches and an inside drill hole of 3/4 in. diameter.

C) A truncated cylinder of maximum length 2 in., minimum length 3/4 in. and diameter 1 1/2 in.

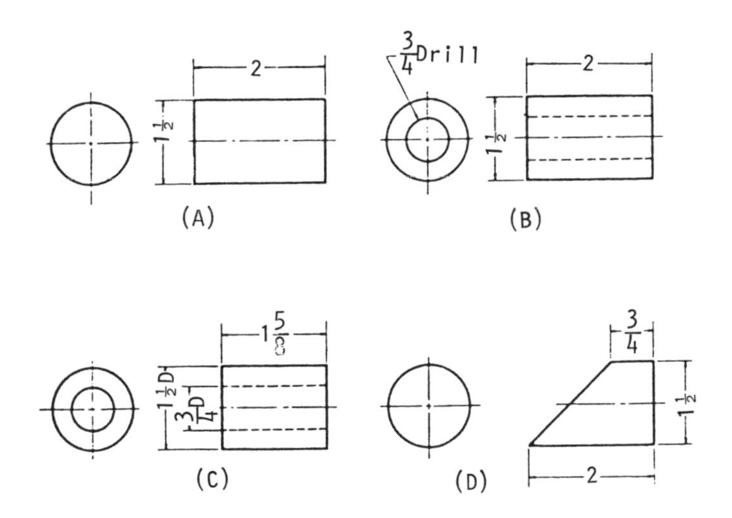

DIMENSIONING OF CYLINDER.

Solution: The general rule for dimensioning cylinders is to locate the dimensions so that the length and diameter appear on the same view. Where a hole or a negative cylinder exists, it is best to dimension by a note when the method of producing the hole is known (Fig. B).

The approved methods of dimensioning cylinders are shown in the figure. The method used in Fig. B is preferred to the one shown in Fig. C. The method illustrated in Fig. C is generally used when the numeral is followed by a D (diameter) and the circular or plan view is omitted.

• PROBLEM 17-12

Draw two views of and dimension:

A) A right pyramid of base ABCD and apex P · AB = DC = 1 in., AD = BC = 1 3/8. The height is 1 1/2 in. and the distance between the projection of P on the base and DC is 11/16 in.

B) A frustum of a pyramid (pyramid with top cut off) with a bottom base ABCD, top plane A'B'C'D', and a height of 7/8 in. AB = DC = 1 in. AD = BC = 1 3/8 in. A'B' = D'C' = 1/2 in. A'D' = B'C' = 5/8 in.

C) An oblique pyramid with base ABCD, apex P, and height of 1 1/2 in. AB = CD = 1 in. AD = BC = 1 3/8 in.

The distance between P' and the projection of P on the plane of ABCD and AD is 1/2 inch and P' to DC is 1 7/8 inches.

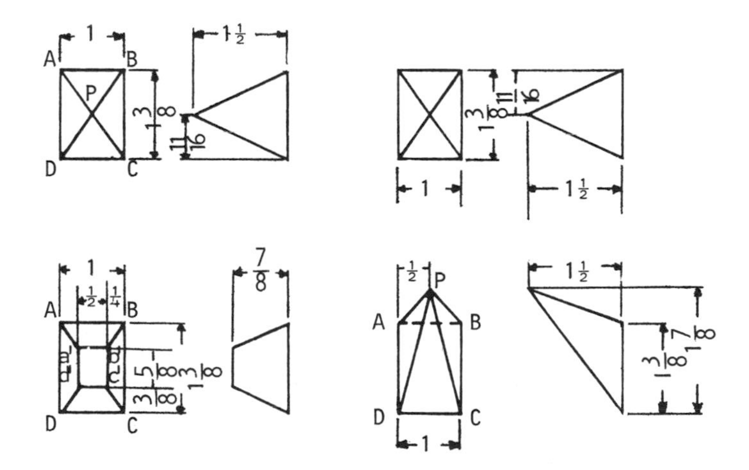

DIMENSIONING OF PYRAMIDS.

Solution: Pyramids are dimensioned as shown in Figure. The general rule to follow in this case is to show the two dimensions of the base in the view where they appear true size, and to show the altitude in the other view.

• PROBLEM 17-13

Draw the three views of a prism and dimension the drawing to these specifications: 2 1/8 inches in length, 1 1/4 in height and 1 inch in width.

Solution: The Figure shows the approved methods of dimensioning a prism. The general rule to remember is to show the length and height in the front view or front elevation, and the depth in one of the other views.

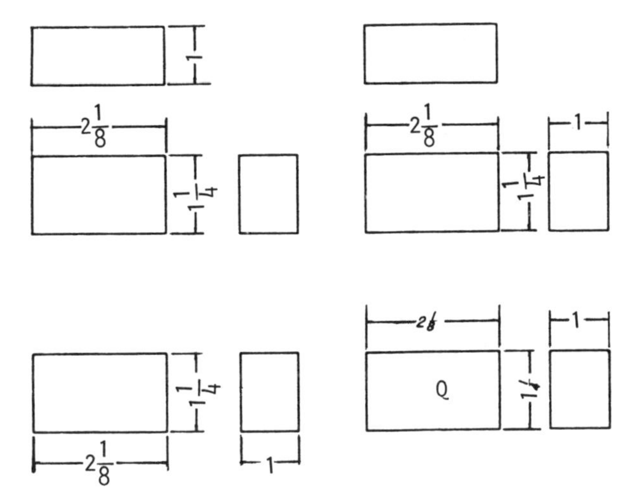

DIMENSIONING OF PRISM.

APPENDIX

Decimal Equivalents of Fractions of an Inch and of a Foot

Fractions of inch or foot	Decimal equivalents	Inch equivalents of foot fractions	Fractions of inch or foot	Decimal equivalents	Inch equivalents of foot fractions	Fractions of inch or foot	Decimal equivalents	Inch equivalents of foot fractions	Fractions of inch or foot	Decimal equivalents	Inch equivalents of foot fractions
	.0052	1/16		.2552	3 1/16		.5052	6 1/16		.7552	9 1/16
	.0104	1/8		.2604	3 1/8		.5104	6 1/8		.7604	9 1/8
1/64	.015625	3/16	17/64	.265625	3 3/16	33/64	.515625	6 3/16	49/64	.765625	9 3/16
	.0208	1/4		.2708	3 1/4		.5208	6 1/4		.7708	9 1/4
	.0260	5/16		.2760	3 5/16		.5260	6 5/16		.7760	9 5/16
1/32	.03125	3/8	9/32	.28125	3 3/8	17/32	.53125	6 3/8	25/32	.78125	9 3/8
	.0365	7/16		.2865	3 7/16		.5365	6 7/16		.7865	9 7/16
	.0417	1/2		.2917	3 1/2		.5417	6 1/2		.7917	9 1/2
3/64	.046875	9/16	19/64	.296875	3 9/16	35/64	.546875	6 9/16	51/64	.796875	9 9/16
	.0521	5/8		.3021	3 5/8		.5521	6 5/8		.8021	9 5/8
	.0573	11/16		.3073	3 11/16		.5573	6 11/16		.8073	9 11/16
1/16	.0625	3/4	5/16	.3125	3 3/4	9/16	.5625	6 3/4	13/16	.8125	9 3/4
	.0677	13/16		.3177	3 13/16		.5677	6 13/16		.8177	9 13/16
	.0729	7/8		.3229	3 7/8		.5729	6 7/8		.8229	9 7/8
5/64	.078125	15/16	21/64	.328125	3 15/16	37/64	.578125	6 15/16	53/64	.828125	9 15/16
	.0833	1		.3333	4		.5833	7		.8333	10
	.0885	1 1/16		.3385	4 1/16		.5885	7 1/16		.8385	10 1/16
3/32	.09375	1 1/8	11/32	.34375	4 1/8	19/32	.59375	7 1/8	27/32	.84375	10 1/8
	.0990	1 3/16		.3490	4 3/16		.5990	7 3/16		.8490	10 3/16
	.1042	1 1/4		.3542	4 1/4		.6042	7 1/4		.8542	10 1/4
7/64	.109375	1 5/16	23/64	.359375	4 5/16	39/64	.609375	7 5/16	55/64	.859375	10 5/16
	.1146	1 3/8		.3646	4 3/8		.6146	7 3/8		.8646	10 3/8
	.1198	1 7/16		.3698	4 7/16		.6198	7 7/16		.8698	10 7/16
1/8	.1250	1 1/2	3/8	.3750	4 1/2	5/8	.6250	7 1/2	7/8	.8750	10 1/2
	.1302	1 9/16		.3802	4 9/16		.6302	7 9/16		.8802	10 9/16
	.1354	1 5/8		.3854	4 5/8		.6354	7 5/8		.8854	10 5/8
9/64	.140625	1 11/16	25/64	.390625	4 11/16	41/64	.640625	7 11/16	57/64	.890625	10 11/16
	.1458	1 3/4		.3958	4 3/4		.6458	7 3/4		.8958	10 3/4
	.1510	1 13/16		.4010	4 13/16		.6510	7 13/16		.9010	10 13/16
5/32	.15625	1 7/8	13/32	.40625	4 7/8	21/32	.65625	7 7/8	29/32	.90625	10 7/8
	.1615	1 15/16		.4115	4 15/16		.6615	7 15/16		.9115	10 15/16
	.1667	2		.4167	5		.6667	8		.9167	11
11/64	.171875	2 1/16	27/64	.421875	5 1/16	43/64	.671875	8 1/16	59/64	.921875	11 1/16
	.1771	2 1/8		.4271	5 1/8		.6771	8 1/8		.9271	11 1/8
	.1823	2 3/16		.4323	5 3/16		.6823	8 3/16		.9323	11 3/16
3/16	.1875	2 1/4	7/16	.4375	5 1/4	11/16	.6875	8 1/4	15/16	.9375	11 1/4
	.1927	2 5/16		.4427	5 5/16		.6927	8 5/16		.9427	11 5/16
	.1979	2 3/8		.4479	5 3/8		.6979	8 3/8		.9479	11 3/8
13/64	.203125	2 7/16	29/64	.453125	5 7/16	45/64	.703125	8 7/16	61/64	.953125	11 7/16
	.2083	2 1/2		.4583	5 1/2		.7083	8 1/2		.9583	11 1/2
	.2135	2 9/16		.4635	5 9/16		.7135	8 9/16		.9635	11 9/16
7/32	.21875	2 5/8	15/32	.46875	5 5/8	23/32	.71875	8 5/8	31/32	.96875	11 5/8
	.2240	2 11/16		.4740	5 11/16		.7240	8 11/16		.9740	11 11/16
	.2292	2 3/4		.4792	5 3/4		.7292	8 3/4		.9792	11 3/4
15/64	.234375	2 13/16	31/64	.484375	5 13/16	47/64	.734375	8 13/16	63/64	.984375	11 13/16
	.2396	2 7/8		.4896	5 7/8		.7396	8 7/8		.9896	11 7/8
	.2448	2 15/16		.4948	5 15/16		.7448	8 15/16		.9948	11 15/16
1/4	.2500	3	1/2	.5000	6	3/4	.7500	9	1	1.0000	12

American Standard Cap Screws

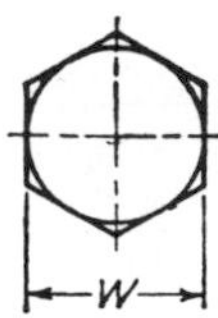

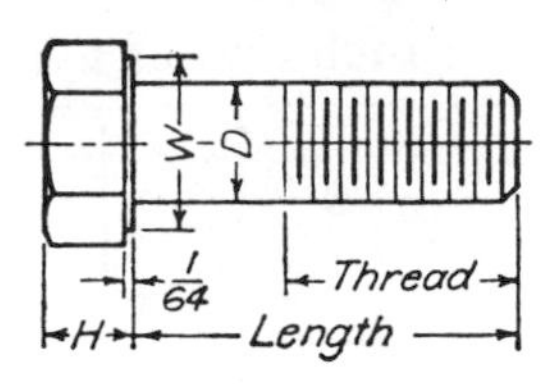

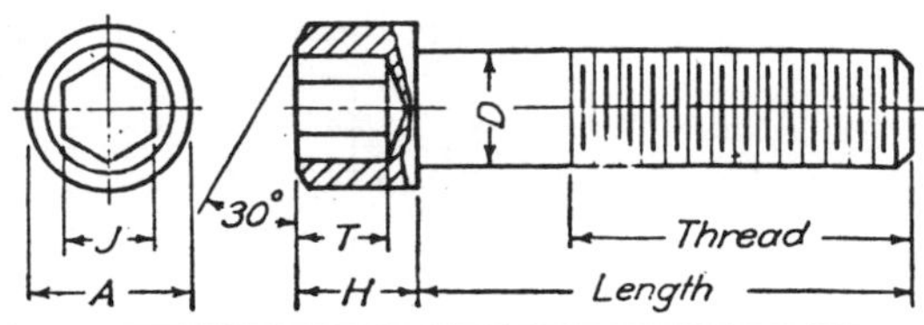

Diameter D	Threads per inch		Hexagon[1]		Socket hexagon[2]			
	Coarse	Fine[3]	Width across flats W	Height of head H	A	H	J	T
1/4	20	28	7/16	3/16	3/8	1/4	3/16	Socket depth is not specified by the standards. Exact depth may be obtained from manufacturers' catalogues. For drawing purposes $T = \frac{3}{4}H$
5/16	18	24	1/2	15/64	7/16	5/16	7/32	
3/8	16	24	9/16	9/32	9/16	3/8	5/16	
7/16	14	20	5/8	21/64	5/8	7/16	5/16	
1/2	13	20	3/4	3/8	3/4	1/2	3/8	
9/16	12	18	13/16	27/64	13/16	9/16	3/8	
5/8	11	18	7/8	15/32	7/8	5/8	1/2	
3/4	10	16	1	9/16	1	3/4	9/16	
7/8	9	14	1 1/8	21/32	1 1/8	7/8	9/16	
1	8	14	1 5/16	3/4	1 5/16	1	5/8	
1 1/8	7	12	1 1/2	27/32	1 1/2	1 1/8	3/4	
1 1/4	7	12	1 11/16	15/16	1 3/4	1 1/4	3/4	

Body-length increments:
- For screw lengths 1/4 to 1 = 1/8"
- For screw lengths 1" to 4" = 1/4"
- For screw lengths 4" to 6" = 1/2"

Thread length:
- Coarse thread: $2D + \frac{1}{2}''$
- Fine thread: $1\frac{1}{2}D + \frac{1}{2}''$

All dimensions in inches.

[1] ASA B18.2 1940.

[2] ASA B18.3 1936.

[3] Not included in American Standards but in common use.

Dimensions[1] of Slotted-head Cap Screws[2,3]

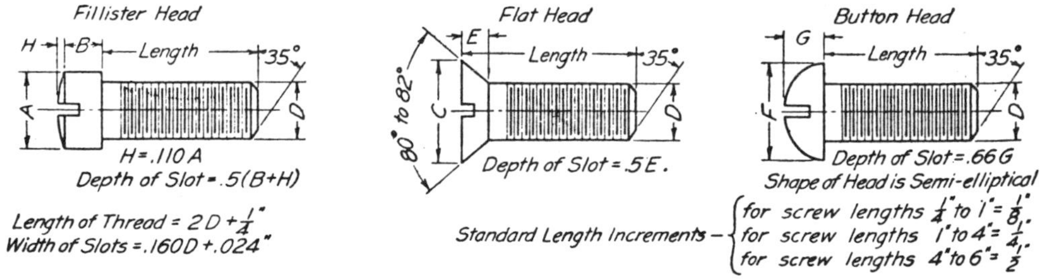

Fillister Head

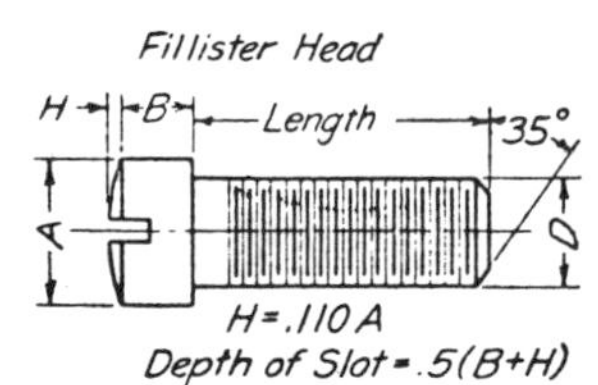

H = .110 A

Depth of Slot = .5(B+H)

Flat Head

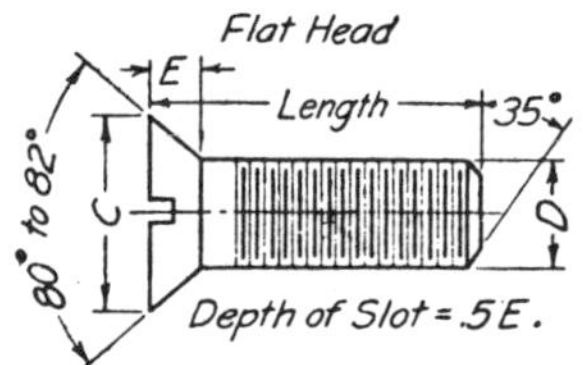

Depth of Slot = .5E.

Button Head

G — Length — 35° — F — D

Depth of Slot = .66 G

Shape of Head is Semi-elliptical

Length of Thread = 2D + 1/4"

Width of Slots = .160D + .024"

Standard Length Increments —
- for screw lengths 1/4" to 1" = 1/8"
- for screw lengths 1" to 4" = 1/4"
- for screw lengths 4" to 6" = 1/2"

Diameter D of screw	A	B	C	E	F	G
1/4	3/8	11/64	1/2	0.146	7/16	3/16
5/16	7/16	13/64	5/8	0.183	9/16	15/64
3/8	9/16	1/4	3/4	0.220	5/8	1/4
7/16	5/8	19/64	13/16	0.220	3/4	5/16
1/2	3/4	21/64	7/8	0.220	13/16	21/64
9/16	13/16	3/8	1	0.256	15/16	25/64
5/8	7/8	27/64	1 1/8	0.293	1	7/16
3/4	1	1/2	1 3/8	0.366	1 1/4	17/32
7/8	1 1/8	19/32				
1	1 15/16	21/32				

All dimensions in inches.

[1] Nominal.

[2] ASA B18c 1930.

[3] Compiled from American Standards.

Dimensions of Machine Screws and Machine-screw and Stove-bolt Nuts[1]

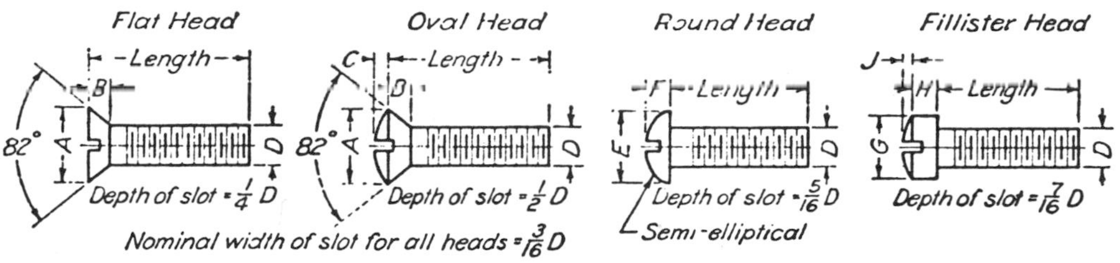

Nominal size	Diameter D	Threads per in. (coarse)	Threads per in. (fine)	A	B	C	E	F	G	H	J
2	0.086	56	64	0.164	0.046	0.041	0.154	0.065	0.132	0.050	0.023
3	0.099	48	56	0.190	0.054	0.048	0.178	0.073	0.153	0.058	0.027
4	0.112	40	48	0.216	0.061	0.054	0.202	0.081	0.175	0.066	0.030
5	0.125	40	..	0.242	0.069	0.061	0.227	0.089	0.198	0.075	0.033
6	0.138	32	40	0.268	0.076	0.067	0.250	0.097	0.217	0.083	0.037
8	0.164	32	36	0.320	0.092	0.080	0.298	0.113	0.260	0.099	0.043
10	0.190	24	32	0.372	0.107	0.094	0.346	0.130	0.303	0.115	0.049
12	0.216	24	28	0.424	0.122	0.107	0.395	0.146	0.344	0.132	0.056
1/4	0.250	20	28	0.492	0.142	0.124	0.458	0.168	0.402	0.153	0.064
5/16	0.3125	18	24	0.618	0.179	0.156	0.574	0.207	0.505	0.193	0.080
3/8	0.375	16	24	0.742	0.215	0.186	0.689	0.247	0.606	0.232	0.096

Nominal size		2	3	4	5	6	8	10	12	1/4	5/16	3/8
Machine-screw[2] and stove-bolt[3] nuts	W	3/16	3/16	1/4	5/16	5/16	11/32	3/8	7/16	7/16	9/16	5/8
	T	1/16	1/16	3/32	7/64	7/64	1/8	1/8	5/32	3/16	7/32	1/4

Dimensions in inches.
[1] Compiled from formulas of American Standards.
[2] Machine-screw nuts are hexagonal.
[3] Stove-bolt nuts are square.

American Phillips-head Machine Screws

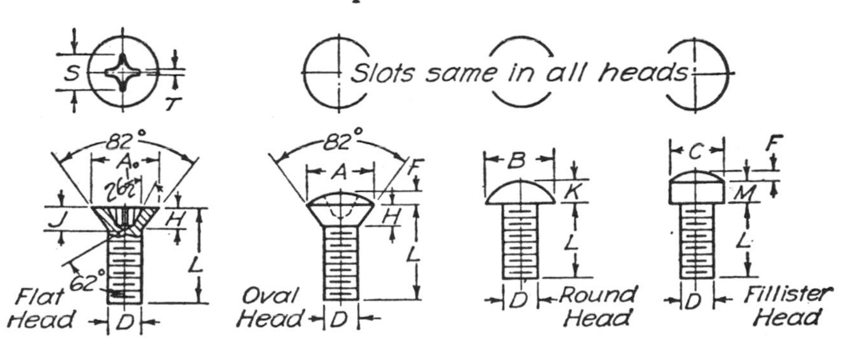

Nominal size[1]	Diameter D	Head major diam A	Head major diam B	Head major diam C	Head height (max) H	Head height (max) K	Head height (max) M	Oval height (max) F	Recess spread (max) S	Recess depth (max) J	Wing thick (max) T
2	0.086	0.172	0.162	0.140	0.051	0.070	0.055	0.029	0.111	0.089	0.020
3	0.099	0.199	0.187	0.161	0.059	0.078	0.063	0.033	0.119	0.097	0.020
4	0.112	0.225	0.211	0.183	0.067	0.086	0.072	0.037	0.127	0.105	0.020
5	0.125	0.232	0.236	0.204	0.075	0.095	0.081	0.041	0.151	0.104	0.027
6	0.138	0.279	0.260	0.226	0.083	0.103	0.089	0.045	0.159	0.112	0.027
8	0.164	0.332	0.309	0.270	0.100	0.119	0.106	0.053	0.175	0.128	0.027
10	0.190	0.385	0.359	0.313	0.116	0.136	0.123	0.061	0.192	0.145	0.027
12	0.216	0.438	0.408	0.357	0.132	0.152	0.141	0.069	0.246	0.165	0.032
1/4	0.250	0.507	0.472	0.414	0.153	0.174	0.163	0.079	0.265	0.187	0.032
5/16	0.3125	0.636	0.591	0.519	0.192	0.214	0.205	0.098	0.305	0.227	0.032
3/8	0.375	0.762	0.708	0.622	0.230	0.254	0.246	0.117	0.384	0.281	0.045

Dimensions in inches.
[1] See machine screw table above for threads per inch.

American Standard Hexagonal Socket Setscrews[1]

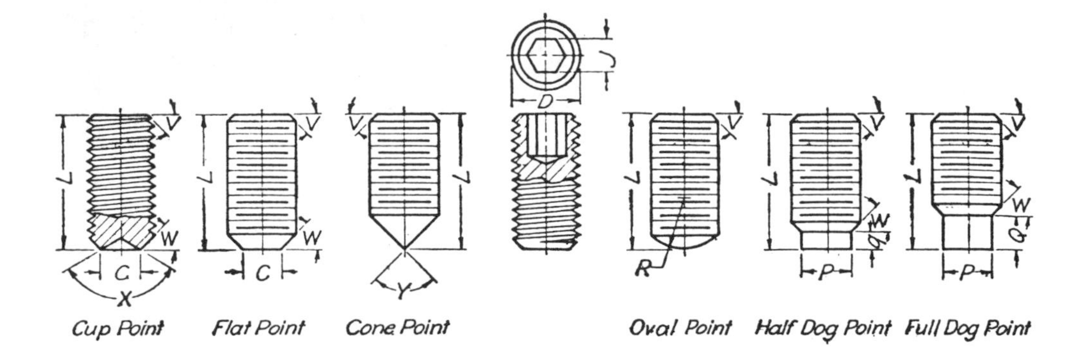

Diameter D	Cup and flat-point diameter C	Oval-point radius R	Cone-point angle Y for these lengths and		Full and half dog points			Socket width J
			Under	Over	Diameter P	Length		
			118° ± 2°	90° ± 2°		Full Q	Half q	
5	1/16	3/32	1/8	3/16	0.083	0.06	0.03	1/16
6	0.069	7/64	1/8	3/16	0.092	0.07	0.03	1/16
8	5/64	1/8	3/16	1/4	0.109	0.08	0.04	5/64
10	3/32	9/64	3/16	1/4	0.127	0.09	0.04	3/32
12	7/64	5/32	3/16	1/4	0.144	0.11	0.06	3/32
1/4	1/8	3/16	1/4	5/16	5/32	1/8	1/16	1/8
5/16	11/64	15/64	5/16	3/8	13/64	5/32	5/64	5/32
3/8	13/64	9/32	3/8	7/16	1/4	3/16	3/32	3/16
7/16	15/64	21/64	7/16	1/2	19/64	7/32	7/64	7/32
1/2	9/32	3/8	1/2	9/16	11/32	1/4	1/8	1/4
9/16	5/16	27/64	9/16	5/8	25/64	9/32	9/64	1/4
5/8	23/64	15/32	5/8	3/4	15/32	5/16	5/32	5/16
3/4	7/16	9/16	3/4	7/8	9/16	3/8	3/16	3/8
7/8	33/64	21/32	7/8	1	21/32	7/16	7/32	1/2
1	19/32	3/4	1	1 1/8	3/4	1/2	1/4	9/16
1 1/8	43/64	27/32	1 1/8	1 1/4	27/32	9/16	9/32	9/16
1 1/4	3/4	15/16	1 1/4	1 1/2	15/16	5/8	5/16	5/8
1 3/8	53/64	1 1/32	1 3/8	1 5/8	1 1/32	11/16	11/32	5/8
1 1/2	29/32	1 1/8	1 1/2	1 3/4	1 1/8	3/4	3/8	3/4
1 3/4	1 1/16	1 5/16	1 3/4	2	1 5/16	7/8	7/16	1
2	1 7/32	1 1/2	2	2 1/4	1 1/2	1	1/2	1

All dimensions in inches.
Chamfer and point angle.
W = 45°, +5°, −0°; draw 45°.
X = 118° ± 5°; draw 120°.
V and Z = 35° + 5°, −0°; draw 45°.
Standard length increments: 1/4″ to 5/8″ by (1/16″); 5/8″ to 1″ by (1/8″); 1″ to 4″ by (1/4″); 4″ to 6″ by (1/2″). Fractions in parentheses show length increments; for example, 5/8″ to 1″ by (1/8″) includes the lengths 5/8″, 3/4″, 7/8″, and 1″.
[1] Compiled from ASA B18.3 1936.

American Standard Square-headed Setscrews

THREADS ARE AMERICAN STANDARD[1]

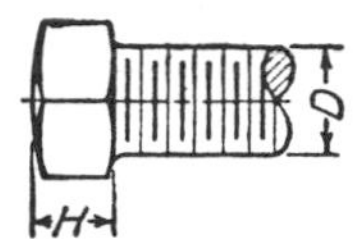

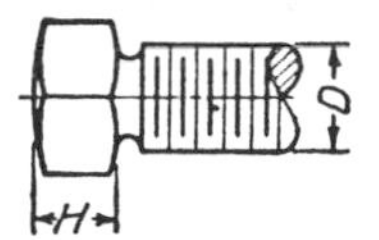

Diameter	1/4	5/16	3/8	7/16	1/2	9/16	5/8	3/4	7/8	1	1 1/8	1 1/4	1 3/8	1 1/2
Width across flats W	1/4	5/16	3/8	7/16	1/2	9/16	5/8	3/4	7/8	1	1 1/8	1 1/4	1 3/8	1 1/2
Height of head H	3/16	15/64	9/32	21/64	3/8	27/64	15/32	9/16	21/32	3/4	27/32	15/16	1 1/32	1 1/8

[1] ASA B18.2 1940.

Dimensions of Wood Screws[1]

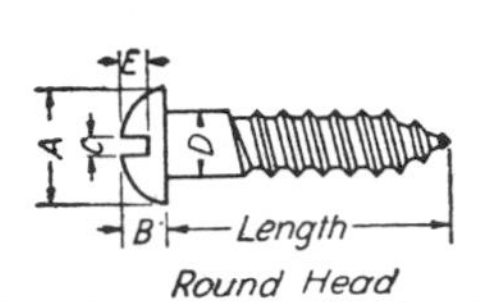

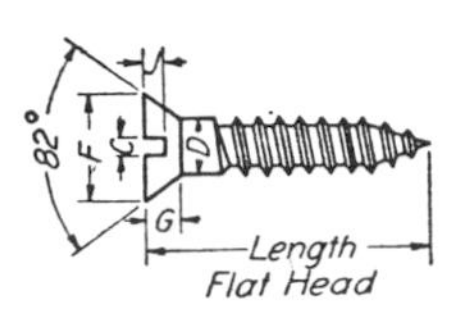

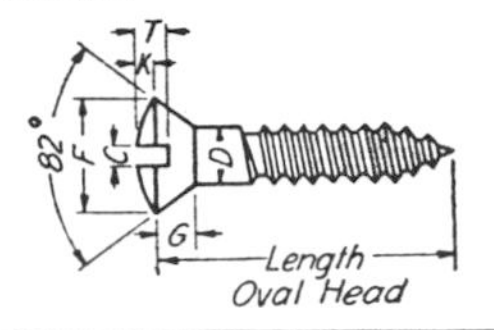

Screw No.	Diameter D	A	B	C	E	F	G	J	K	T
0	0.060	0.106	0.047	0.025	0.034	0.112	0.030	0.012	0.018	0.027
1	0.073	0.130	0.056	0.027	0.038	0.138	0.038	0.015	0.022	0.034
2	0.086	0.154	0.064	0.030	0.042	0.164	0.045	0.019	0.025	0.041
3	0.099	0.178	0.072	0.032	0.046	0.190	0.053	0.022	0.029	0.047
4	0.112	0.202	0.080	0.034	0.050	0.216	0.061	0.025	0.033	0.054
5	0.125	0.228	0.089	0.037	0.054	0.242	0.068	0.028	0.037	0.061
6	0.138	0.250	0.097	0.039	0.058	0.268	0.076	0.031	0.040	0.067
7	0.151	0.274	0.105	0.041	0.062	0.294	0.083	0.034	0.044	0.073
8	0.164	0.298	0.113	0.043	0.066	0.320	0.092	0.037	0.048	0.080
9	0.177	0.322	0.121	0.045	0.070	0.346	0.100	0.040	0.051	0.086
10	0.190	0.346	0.130	0.048	0.075	0.371	0.107	0.043	0.055	0.093
11	0.203	0.370	0.138	0.050	0.078	0.398	0.114	0.046	0.059	0.100
12	0.216	0.395	0.146	0.052	0.083	0.424	0.123	0.049	0.063	0.116
14	0.242	0.443	0.162	0.057	0.091	0.476	0.137	0.056	0.069	0.120
16	0.268	0.491	0.178	0.061	0.099	0.528	0.152	0.062	0.077	0.133
18	0.294	0.539	0.195	0.066	0.107	0.580	0.167	0.068	0.085	0.146

Dimensions in inches. [1] Compiled from ASA B18c 1930.

Parker-Kalon Self-tapping Cap Screws[1]

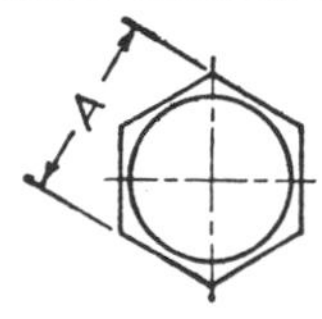

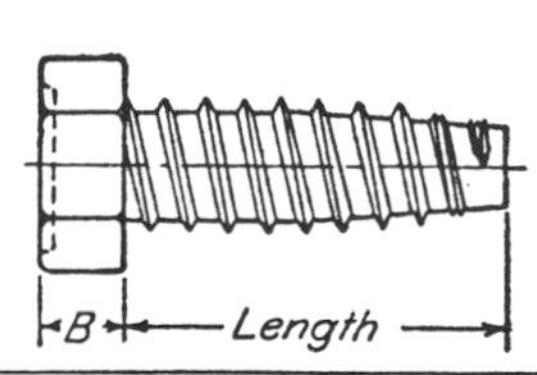

Diameter	A	B	Length L[2]	Drill size	
				Aluminum die castings, etc.	Slate, ebony, asbestos, etc.
6	1/4	5/64	3/16, 3/8, 1/4, 5/16	No. 30	No. 31
8	1/4	7/64	1/4" to 1" by (1/8")	No. 24	No. 26
10	9/32	1/4	3/8" to 1" by (1/8")	No. 16	No. 19
14	3/8	11/64	3/8" to 1 1/2" by (1/8")	15/64	No. 1
5/16	1/2	13/64	1/2" to 1" by (1/8") 1 1/4, 1 1/2	L	9/32
3/8	9/16	9/32	5/8" to 1" by (1/8") 1" to 2" by (1/4")	S	21/64
7/16	5/8	5/16	3/4" to 1" by (1/8") 1" to 2" by (1/4")	Z	X
1/2	3/4	3/8	3/4" to 1" by (1/8") 1" to 2 1/2" by (1/4")	15/32	29/64

[1] Compiled from Parker-Kalon catalogue. [2] Fractions in parentheses show length increments; for example, 1/4" to 1" by (1/8") included the lengths 1/4", 3/8", 1/2", 5/8", 3/4", 7/8", and 1".

Recommended[1] SAE Standard Lock Washers

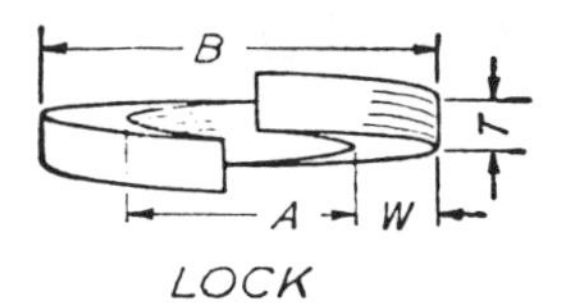

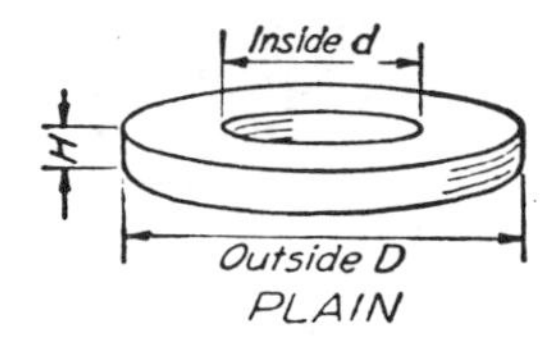

Screw or bolt size, nominal	SAE standard sizes			Lock washers for use with					
	SAE light W × T	SAE standard[2] W × T	SAE heavy W × T	All regular boltheads and nuts, series A W × T	Cap screws, series B W × T	Round-head mach. screws, series C W × T	Fillister-head mach. screws, series D W × T	Mach.-screw and stovebolt nuts, series E W × T	Socket-head cap screws W × T
2*	0.022 × 0.022	1/32 × 0.022	1/32 × 1/32	...	...	...	1/32 × 0.022		
2†	1/32 × 0.022	1/32 × 1/32	3/64 × 1/32	...	...	1/32 × 1/32			
4*	1/32 × 0.022	1/32 × 1/32	3/64 × 1/32	...	...	...	1/32 × 1/32	...	0.022 × 0.022
4†	3/64 × 1/32	1/16 × 1/32	5/64 × 1/32	...	...	3/64 × 1/32			
6*	1/32 × 1/32	3/64 × 1/32	3/64 × 3/64	...	...	...	3/64 × 1/32	...	1/32 × 1/32
6†	1/16 × 1/32	5/64 × 1/32	5/64 × 3/64	...	...	1/16 × 3/64	...	3/32 × 3/64	
8*	3/64 × 1/32	3/64 × 3/64	1/16 × 3/64	...	...	...	1/16 × 3/64	...	3/64 × 3/64
8†	5/64 × 1/32	5/64 × 3/64	3/32 × 3/64	...	...	5/64 × 3/64	...	7/64 × 1/16	
10*	3/64 × 3/64	1/16 × 3/64	1/16 × 1/16	...	...	...	1/16 × 3/64	...	3/64 × 3/64
10†	5/64 × 3/64	3/32 × 3/64	3/32 × 1/16	...	...	3/32 × 1/16	...	7/64 × 1/16	
1/4	3/32 × 3/64	3/32 × 1/16	3/32 × 5/64	9/64 × 5/64	1/8 × 1/16	1/8 × 1/16	3/64 × 1/16	9/64 × 5/64	3/64 × 5/64
5/16	1/8 × 3/64	1/8 × 1/16	1/8 × 3/32	5/32 × 3/32	9/64 × 5/64	9/64 × 5/64	7/64 × 1/16	5/32 × 3/32	3/64 × 5/64
3/8	1/8 × 1/16	1/8 × 3/32	1/8 × 1/8	11/64 × 7/64	5/32 × 3/32	5/32 × 3/32	1/8 × 3/32	11/64 × 7/64	5/64 × 1/8
7/16	5/32 × 1/16	5/32 × 1/8	5/32 × 5/32	13/64 × 1/8	11/64 × 7/64	...	...	...	5/64 × 1/8
1/2	11/64 × 1/16	11/64 × 1/8	11/64 × 11/64	7/32 × 5/32	3/16 × 1/8	...	...	...	7/64 × 11/64
9/16	3/16 × 3/32	3/16 × 1/8	3/16 × 3/16	1/4 × 3/16	13/64 × 1/8				
5/8	13/64 × 3/32	13/64 × 5/32	13/64 × 13/64	17/64 × 3/16	7/32 × 5/32	...	...	...	7/64 × 11/64
3/4	1/4 × 1/8	1/4 × 3/16	1/4 × 1/4	5/16 × 7/32	1/4 × 3/16	...	...	...	7/64 × 3/16
7/8	17/64 × 5/32	17/64 × 3/16	17/64 × 17/64	11/32 × 1/4	5/16 × 7/32	...	...	...	7/64 × 3/16
1	5/16 × 3/16	5/16 × 1/4	5/16 × 5/16	13/32 × 9/32	5/16 × 1/4	...	...	...	1/8 × 13/64
1 1/8	3/8 × 3/16	3/8 × 1/4	3/8 × 3/8	7/16 × 5/16	11/32 × 1/4	...	...	...	5/32 × 5/16
1 1/4	7/16 × 3/16	7/16 × 1/4	7/16 × 5/16	1/2 × 3/8	3/8 × 5/16	...	...	...	7/32 × 5/16
1 3/8	7/16 × 1/4	7/16 × 5/16	7/16 × 3/8	1/2 × 3/8	3/8 × 5/16	...	...	...	
1 1/2	1/2 × 1/4	1/2 × 5/16	1/2 × 3/8	1/2 × 3/8	3/8 × 5/16	...	...	...	7/32 × 5/16

Dimensions in inches. * For fillister-head machine screws. † For round-head machine screws. [1] By Spring Washer Industry. [2] Also called "Regular."

SAE Standard Plain Washers

Nominal size	1/4	5/16	3/8	7/16	1/2	9/16	5/8	11/16	3/4	7/8	1	1 1/8	1 1/4	1 3/8	1 1/2
Inside diameter d	9/32	11/32	13/32	15/32	17/32	19/32	21/32	23/32	13/16	15/16	1 1/16	1 3/16	1 5/16	1 7/16	1 9/16
Outside diameter D	5/8	11/16	13/16	15/16	1 1/16	1 3/16	1 5/16	1 3/8	1 1/2	1 3/4	2	2 1/4	2 1/2	2 3/4	3
Thickness H	1/16	1/16	1/16	1/16	3/32	3/32	3/32	3/32	1/8	1/8	1/8	1/8	5/32	5/32	5/32

Dimensions in inches.

Wire and Sheet-metal Gages

DIMENSIONS IN DECIMAL PARTS OF AN INCH

Number of gage	American or Brown and Sharpe[1]	Washburn & Moen or American Steel & Wire Co.[2]	Birmingham or Stubs iron wire[3]	Music wire[4]	Imperial wire gage[5]	U.S. Std. for plate[6]
0000000	......	0.4900	.....	.....	0.5000	0.5000
000000	0.5800	0.4615	.....	0.004	0.4640	0.4688
00000	0.5165	0.4305	0.500	0.005	0.4320	0.4375
0000	0.4600	0.3938	0.454	0.006	0.4000	0.4063
000	0.4096	0.3625	0.425	0.007	0.3720	0.3750
00	0.3648	0.3310	0.380	0.008	0.3480	0.3438
0	0.3249	0.3065	0.340	0.009	0.3240	0.3125
1	0.2893	0.2830	0.300	0.010	0.3000	0.2813
2	0.2576	0.2625	0.284	0.011	0.2760	0.2656
3	0.2294	0.2437	0.259	0.012	0.2520	0.2500
4	0.2043	0.2253	0.238	0.013	0.2320	0.2344
5	0.1819	0.2070	0.220	0.014	0.2120	0.2188
6	0.1620	0.1920	0.203	0.016	0.1920	0.2031
7	0.1443	0.1770	0.180	0.018	0.1760	0.1875
8	0.1285	0.1620	0.165	0.020	0.1600	0.1719
9	0.1144	0.1483	0.148	0.022	0.1440	0.1563
10	0.1019	0.1350	0.134	0.024	0.1280	0.1406
11	0.0907	0.1205	0.120	0.026	0.1160	0.1250
12	0.0808	0.1055	0.109	0.029	0.1040	0.1094
13	0.0720	0.0915	0.095	0.031	0.0920	0.0938
14	0.0641	0.0800	0.083	0.033	0.0800	0.0781
15	0.0571	0.0720	0.072	0.035	0.0720	0.0703
16	0.0508	0.0625	0.065	0.037	0.0640	0.0625
17	0.0453	0.0540	0.058	0.039	0.0560	0.0563
18	0.0403	0.0475	0.049	0.041	0.0480	0.0500
19	0.0359	0.0410	0.042	0.043	0.0400	0.0438
20	0.0320	0.0348	0.035	0.045	0.0360	0.0375
21	0.0285	0.0317	0.032	0.047	0.0320	0.0344
22	0.0253	0.0286	0.028	0.049	0.0280	0.0313
23	0.0226	0.0258	0.025	0.051	0.0240	0.0281
24	0.0201	0.0230	0.022	0.055	0.0220	0.0250
25	0.0179	0.0204	0.020	0.059	0.0200	0.0219
26	0.0159	0.0181	0.018	0.063	0.0180	0.0188
27	0.0142	0.0173	0.016	0.067	0.0164	0.0172
28	0.0126	0.0162	0.014	0.071	0.0148	0.0156
29	0.0113	0.0150	0.013	0.075	0.0136	0.0141
30	0.0100	0.0140	0.012	0.080	0.0124	0.0125
31	0.0089	0.0132	0.010	0.085	0.0116	0.0109
32	0.0080	0.0128	0.009	0.090	0.0108	0.0102
33	0.0071	0.0118	0.008	0.095	0.0100	0.0094
34	0.0063	0.0104	0.007	0.100	0.0092	0.0086
35	0.0056	0.0095	0.005	0.106	0.0084	0.0078
36	0.0050	0.0090	0.004	0.112	0.0076	0.0070
37	0.0045	0.0085	.....	0.118	0.0068	0.0066
38	0.0040	0.0080	.....	0.124	0.0060	0.0063
39	0.0035	0.0075	.....	0.130	0.0052	
40	0.0031	0.0070	.....	0.138	0.0048	

[1] Recognized standard in the United States for wire and sheet metal of copper and other metals except steel and iron.

[2] Recognized standard for steel and iron wire. Called the "U.S. steel wire gage."

[3] Formerly much used, now nearly obsolete.

[4] American Steel & Wire Company's music or piano wire gage. Recommended by U.S. Bureau of Standards.

[5] Official British Standard.

[6] Legalized U.S. Standard for iron and steel plate, although plate is now always specified by its thickness in decimals of an inch.

Preferred thicknesses for uncoated thin flats metals (under 0.250 in.), ASA B32 1941, gives recommended sizes for sheets.

Widths and Heights of Standard Square- and Flat-stock Keys with Corresponding Shaft Diameters

APPROVED BY AMERICAN STANDARDS ASSOCIATION[1]

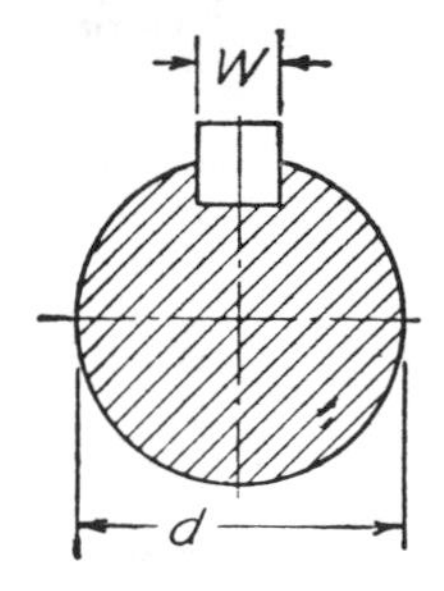

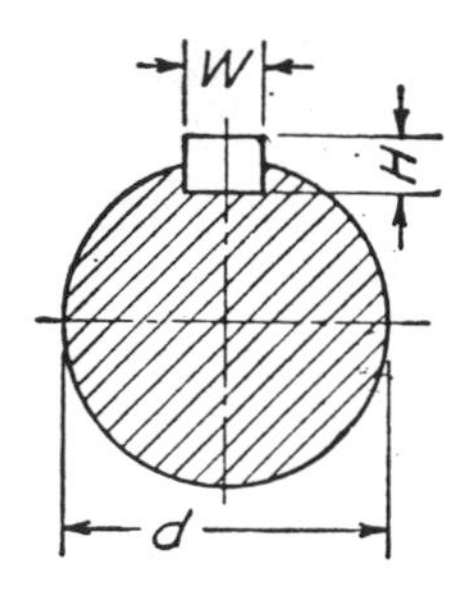

Shaft diameter d (inclusive)	Square-stock keys W	Flat-stock keys, $W \times H$	Shaft diameter d (inclusive)	Square-stock keys W	Flat-stock keys, $W \times H$
1/2 – 9/16	1/8	1/8 × 3/32	2 7/8–3 1/4	3/4	3/4 × 1/2
5/8 – 7/8	3/16	3/16 × 1/8	3 3/8–3 3/4	7/8	7/8 × 5/8
15/16–1 1/4	1/4	1/4 × 3/16	3 7/8–4 1/2	1	1 × 3/4
1 5/16–1 3/8	5/16	5/16 × 1/4			
1 7/16–1 3/4	3/8	3/8 × 1/4	4 3/4–5 1/2	1 1/4	1 1/4 × 7/8
1 13/16–2 1/4	1/2	1/2 × 3/8	5 3/4–6	1 1/2	1 1/2 × 1
2 5/16–2 3/4	5/8	5/8 × 7/16			

Dimensions in inches.

[1] ASA B17.1 1934.

Dimensions of Standard Gib-head Keys, Square and Flat

APPROVED BY AMERICAN STANDARDS ASSOCIATION[1]

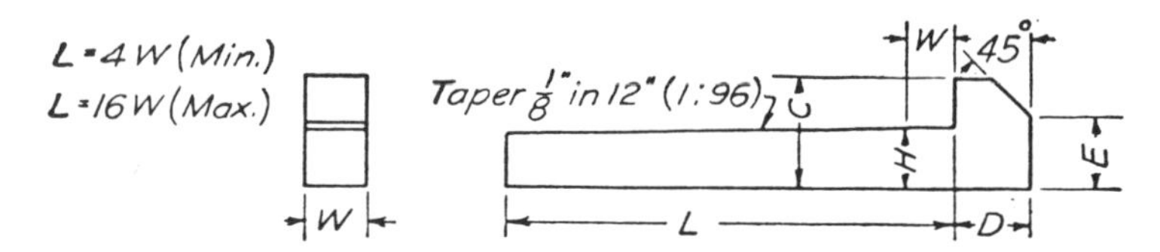

Diameters of shafts	Square type					Flat type				
	Key		Gib head			Key		Gib head		
	W	H	C	D	E	W	H	C	D	E
1/2 – 9/16	1/8	1/8	1/4	7/32	5/32	1/8	3/32	3/16	1/8	1/8
5/8 – 7/8	3/16	3/16	5/16	9/32	7/32	3/16	1/8	1/4	3/16	5/32
15/16–1 1/4	1/4	1/4	7/16	11/32	11/32	1/4	3/16	5/16	1/4	3/16
1 5/16–1 3/8	5/16	5/16	9/16	13/32	13/32	5/16	1/4	3/8	5/16	1/4
1 7/16–1 3/4	3/8	3/8	11/16	15/32	15/32	3/8	1/4	7/16	3/8	5/16
1 13/16–2 1/4	1/2	1/2	7/8	19/32	5/8	1/2	3/8	5/8	1/2	7/16
2 5/16–2 3/4	5/8	5/8	1 1/16	23/32	3/4	5/8	7/16	3/4	5/8	1/2
2 7/8 –3 1/4	3/4	3/4	1 1/4	7/8	7/8	3/4	1/2	7/8	3/4	5/8
3 3/8 –3 3/4	7/8	7/8	1 1/2	1	1	7/8	5/8	1 1/16	7/8	3/4
3 7/8 –4 1/2	1	1	1 3/4	1 3/16	1 3/16	1	3/4	1 1/4	1	13/16
4 3/4 –5 1/2	1 1/4	1 1/4	2	1 7/16	1 7/16	1 1/4	7/8	1 1/2	1 1/4	1
5 3/4 –6	1 1/2	1 1/2	2 1/2	1 3/4	1 3/4	1 1/2	1	1 3/4	1 1/2	1 1/4

Dimensions in inches.

[1] ASA B17.1 1934.

Table of Limits for Cylindrical Fits[1]

Size of hole or external member, inclusive	Clearance fits							
	Class 1 Loose fit				Class 2 Free fit			
	Hole or external member		Shaft or internal member		Hole or external member		Shaft or internal member	
	+		−	−	+		−	−
0–3/16	0.001	0.000	0.001	0.002	0.0007	0.0000	0.0004	0.0011
3/16–5/16	0.002	0.000	0.001	0.003	0.0008	0.0000	0.0006	0.0014
5/16–7/16	0.002	0.000	0.001	0.003	0.0009	0.0000	0.0007	0.0016
7/16–9/16	0.002	0.000	0.002	0.004	0.0010	0.0000	0.0009	0.0019
9/16–11/16	0.002	0.000	0.002	0.004	0.0011	0.0000	0.0010	0.0021
11/16–13/16	0.002	0.000	0.002	0.004	0.0012	0.0000	0.0012	0.0024
13/16–15/16	0.002	0.000	0.002	0.004	0.0012	0.0000	0.0013	0.0025
15/16–1 1/16	0.003	0.000	0.003	0.006	0.0013	0.0000	0.0014	0.0027
1 1/16–1 3/16	0.003	0.000	0.003	0.006	0.0014	0.0000	0.0015	0.0029
1 3/16–1 3/8	0.003	0.000	0.003	0.006	0.0014	0.0000	0.0016	0.0030
1 3/8 –1 5/8	0.003	0.000	0.003	0.006	0.0015	0.0000	0.0018	0.0033
1 5/8 –1 7/8	0.003	0.000	0.004	0.007	0.0016	0.0000	0.0020	0.0036
1 7/8 –2 1/8	0.003	0.000	0.004	0.007	0.0016	0.0000	0.0022	0.0038
2 1/8 –2 3/8	0.003	0.000	0.004	0.007	0.0017	0.0000	0.0024	0.0041
2 3/8 –2 3/4	0.003	0.000	0.005	0.008	0.0018	0.0000	0.0026	0.0044
2 3/4 –3 1/4	0.004	0.000	0.005	0.009	0.0019	0.0000	0.0029	0.0048
3 1/4 –3 3/4	0.004	0.000	0.006	0.010	0.0020	0.0000	0.0032	0.0052
3 3/4 –4 1/4	0.004	0.000	0.006	0.010	0.0021	0.0000	0.0035	0.0056
4 1/4 –4 3/4	0.004	0.000	0.007	0.011	0.0021	0.0000	0.0038	0.0059
4 3/4 –5 1/2	0.004	0.000	0.007	0.011	0.0022	0.0000	0.0041	0.0063
5 1/2 –6 1/2	0.005	0.000	0.008	0.013	0.0024	0.0000	0.0046	0.0070
6 1/2 –7 1/2	0.005	0.000	0.009	0.014	0.0025	0.0000	0.0051	0.0076
7 1/2 –8 1/2	0.005	0.000	0.010	0.015	0.0026	0.0000	0.0056	0.0082

Size of hole or external member, inclusive	Class 3 Medium fit				Class 4 Snug fit			
	Hole or external member		Shaft or internal member		Hole or external member		Shaft or internal member	
	+		−	−	+			−
0–3/16	0.0004	0.0000	0.0002	0.0006	0.0003	0.0000	0.0000	0.0002
3/16–5/16	0.0005	0.0000	0.0004	0.0009	0.0004	0.0000	0.0000	0.0003
5/16–7/16	0.0006	0.0000	0.0005	0.0011	0.0004	0.0000	0.0000	0.0003
7/16–9/16	0.0006	0.0000	0.0006	0.0012	0.0005	0.0000	0.0000	0.0003
9/16–11/16	0.0007	0.0000	0.0007	0.0014	0.0005	0.0000	0.0000	0.0003
11/16–13/16	0.0007	0.0000	0.0007	0.0014	0.0005	0.0000	0.0000	0.0004
13/16–15/16	0.0008	0.0000	0.0008	0.0016	0.0006	0.0000	0.0000	0.0004
15/16–1 1/16	0.0008	0.0000	0.0009	0.0017	0.0006	0.0000	0.0000	0.0004
1 1/16–1 3/16	0.0008	0.0000	0.0010	0.0018	0.0006	0.0000	0.0000	0.0004
1 3/16–1 3/8	0.0009	0.0000	0.0010	0.0019	0.0006	0.0000	0.0000	0.0004
1 3/8–1 5/8	0.0009	0.0000	0.0012	0.0021	0.0007	0.0000	0.0000	0.0005
1 5/8–1 7/8	0.0010	0.0000	0.0013	0.0023	0.0007	0.0000	0.0000	0.0005
1 7/8–2 1/8	0.0010	0.0000	0.0014	0.0024	0.0008	0.0000	0.0000	0.0005
2 1/8–2 3/8	0.0010	0.0000	0.0015	0.0025	0.0008	0.0000	0.0000	0.0005
2 3/8–2 3/4	0.0011	0.0000	0.0017	0.0028	0.0008	0.0000	0.0000	0.0005
2 3/4–3 1/4	0.0012	0.0000	0.0019	0.0031	0.0009	0.0000	0.0000	0.0006
3 1/4–3 3/4	0.0012	0.0000	0.0021	0.0033	0.0009	0.0000	0.0000	0.0006
3 3/4–4 1/4	0.0013	0.0000	0.0023	0.0036	0.0010	0.0000	0.0000	0.0006
4 1/4–4 3/4	0.0013	0.0000	0.0025	0.0038	0.0010	0.0000	0.0000	0.0007
4 3/4–5 1/2	0.0014	0.0000	0.0026	0.0040	0.0010	0.0000	0.0000	0.0007
5 1/2–6 1/2	0.0015	0.0000	0.0030	0.0045	0.0011	0.0000	0.0000	0.0007
6 1/2–7 1/2	0.0015	0.0000	0.0033	0.0048	0.0011	0.0000	0.0000	0.0008
7 1/2–8 1/2	0.0016	0.0000	0.0036	0.0052	0.0012	0.0000	0.0000	0.0008

All dimensions in inches.
[1] Compiled from American Standard ASA B4a 1925.

Table of Limits for Cylindrical Fits[1]—(*Continued*)

Size of hole or external member, inclusive	Interference fits							
	Class 5 Wringing fit				Class 6 Tight fit			
	Hole or external member		Shaft or internal member		Hole or external member		Shaft or internal member	
	+		+		+		+	+
0–3/16	0.0003	0.0000	0.0002	0.0000	0.0003	0.0000	0.0003	0.0000
3/16–5/16	0.0004	0.0000	0.0003	0.0000	0.0004	0.0000	0.0005	0.0001
5/16–7/16	0.0004	0.0000	0.0003	0.0000	0.0004	0.0000	0.0005	0.0001
7/16–9/16	0.0005	0.0000	0.0003	0.0000	0.0005	0.0000	0.0006	0.0001
9/16–11/16	0.0005	0.0000	0.0003	0.0000	0.0005	0.0000	0.0007	0.0002
11/16–13/16	0.0005	0.0000	0.0004	0.0000	0.0005	0.0000	0.0007	0.0002
13/16–15/16	0.0006	0.0000	0.0004	0.0000	0.0006	0.0000	0.0008	0.0002
15/16–1 1/16	0.0006	0.0000	0.0004	0.0000	0.0006	0.0000	0.0009	0.0003
1 1/16–1 3/16	0.0006	0.0000	0.0004	0.0000	0.0006	0.0000	0.0009	0.0003
1 3/16–1 3/8	0.0006	0.0000	0.0004	0.0000	0.0006	0.0000	0.0009	0.0003
1 3/8–1 5/8	0.0007	0.0000	0.0005	0.0000	0.0007	0.0000	0.0011	0.0004
1 5/8–1 7/8	0.0007	0.0000	0.0005	0.0000	0.0007	0.0000	0.0011	0.0004
1 7/8–2 1/8	0.0008	0.0000	0.0005	0.0000	0.0008	0.0000	0.0013	0.0005
2 1/8–2 3/8	0.0008	0.0000	0.0005	0.0000	0.0008	0.0000	0.0014	0.0006
2 3/8–2 3/4	0.0008	0.0000	0.0005	0.0000	0.0008	0.0000	0.0014	0.0006
2 3/4–3 1/4	0.0009	0.0000	0.0006	0.0000	0.0009	0.0000	0.0017	0.0008
3 1/4–3 3/4	0.0009	0.0000	0.0006	0.0000	0.0009	0.0000	0.0018	0.0009
3 3/4–4 1/4	0.0010	0.0000	0.0006	0.0000	0.0010	0.0000	0.0020	0.0010
4 1/4–4 3/4	0.0010	0.0000	0.0007	0.0000	0.0010	0.0000	0.0021	0.0011
4 3/4–5 1/2	0.0010	0.0000	0.0007	0.0000	0.0010	0.0000	0.0023	0.0013
5 1/2–6 1/2	0.0011	0.0000	0.0007	0.0000	0.0011	0.0000	0.0026	0.0015
6 1/2–7 1/2	0.0011	0.0000	0.0008	0.0000	0.0011	0.0000	0.0029	0.0018
7 1/2–8 1/2	0.0012	0.0000	0.0008	0.0000	0.0012	0.0000	0.0032	0.0020

Size of hole or external member, inclusive	Class 7 Medium force fit				Class 8 Heavy force and shrink fit			
	Hole or external member		Shaft or internal member		Hole or external member		Shaft or internal member	
	+		+	+	+		+	+
0–3/16	0.0003	0.0000	0.0004	0.0001	0.0003	0.0000	0.0004	0.0001
3/16–5/16	0.0004	0.0000	0.0005	0.0001	0.0004	0.0000	0.0007	0.0003
5/16–7/16	0.0004	0.0000	0.0006	0.0002	0.0004	0.0000	0.0008	0.0004
7/16–9/16	0.0005	0.0000	0.0008	0.0003	0.0005	0.0000	0.0010	0.0005
9/16–11/16	0.0005	0.0000	0.0008	0.0003	0.0005	0.0000	0.0011	0.0006
11/16–13/16	0.0005	0.0000	0.0009	0.0004	0.0005	0.0000	0.0013	0.0008
13/16–15/16	0.0006	0.0000	0.0010	0.0004	0.0006	0.0000	0.0015	0.0009
15/16–1 1/16	0.0006	0.0000	0.0010	0.0005	0.0006	0.0000	0.0016	0.0010
1 1/16–1 3/16	0.0006	0.0000	0.0012	0.0006	0.0006	0.0000	0.0017	0.0011
1 3/16–1 3/8	0.0006	0.0000	0.0012	0.0006	0.0006	0.0000	0.0019	0.0013
1 3/8–1 5/8	0.0007	0.0000	0.0015	0.0008	0.0007	0.0000	0.0022	0.0015
1 5/8–1 7/8	0.0007	0.0000	0.0016	0.0009	0.0007	0.0000	0.0025	0.0018
1 7/8–2 1/8	0.0008	0.0000	0.0018	0.0010	0.0008	0.0000	0.0028	0.0020
2 1/8–2 3/8	0.0008	0.0000	0.0019	0.0011	0.0008	0.0000	0.0031	0.0023
2 3/8–2 3/4	0.0008	0.0000	0.0021	0.0013	0.0008	0.0000	0.0033	0.0025
2 3/4–3 1/4	0.0009	0.0000	0.0024	0.0015	0.0009	0.0000	0.0039	0.0030
3 1/4–3 3/4	0.0009	0.0000	0.0027	0.0018	0.0009	0.0000	0.0044	0.0035
3 3/4–4 1/4	0.0010	0.0000	0.0030	0.0020	0.0010	0.0000	0.0050	0.0040
4 1/4–4 3/4	0.0010	0.0000	0.0033	0.0023	0.0010	0.0000	0.0055	0.0045
4 3/4–5 1/2	0.0010	0.0000	0.0035	0.0025	0.0010	0.0000	0.0060	0.0050
5 1/2–6 1/2	0.0011	0.0000	0.0041	0.0030	0.0011	0.0000	0.0071	0.0060
6 1/2–7 1/2	0.0011	0.0000	0.0046	0.0035	0.0011	0.0000	0.0081	0.0070
7 1/2–8 1/2	0.0012	0.0000	0.0052	0.0040	0.0012	0.0000	0.0092	0.0080

All dimensions in inches.

[1] Compiled from American Standard ASA B4a 1925.

American 150-lb Malleable-iron Screwed-fitting Standard[1]

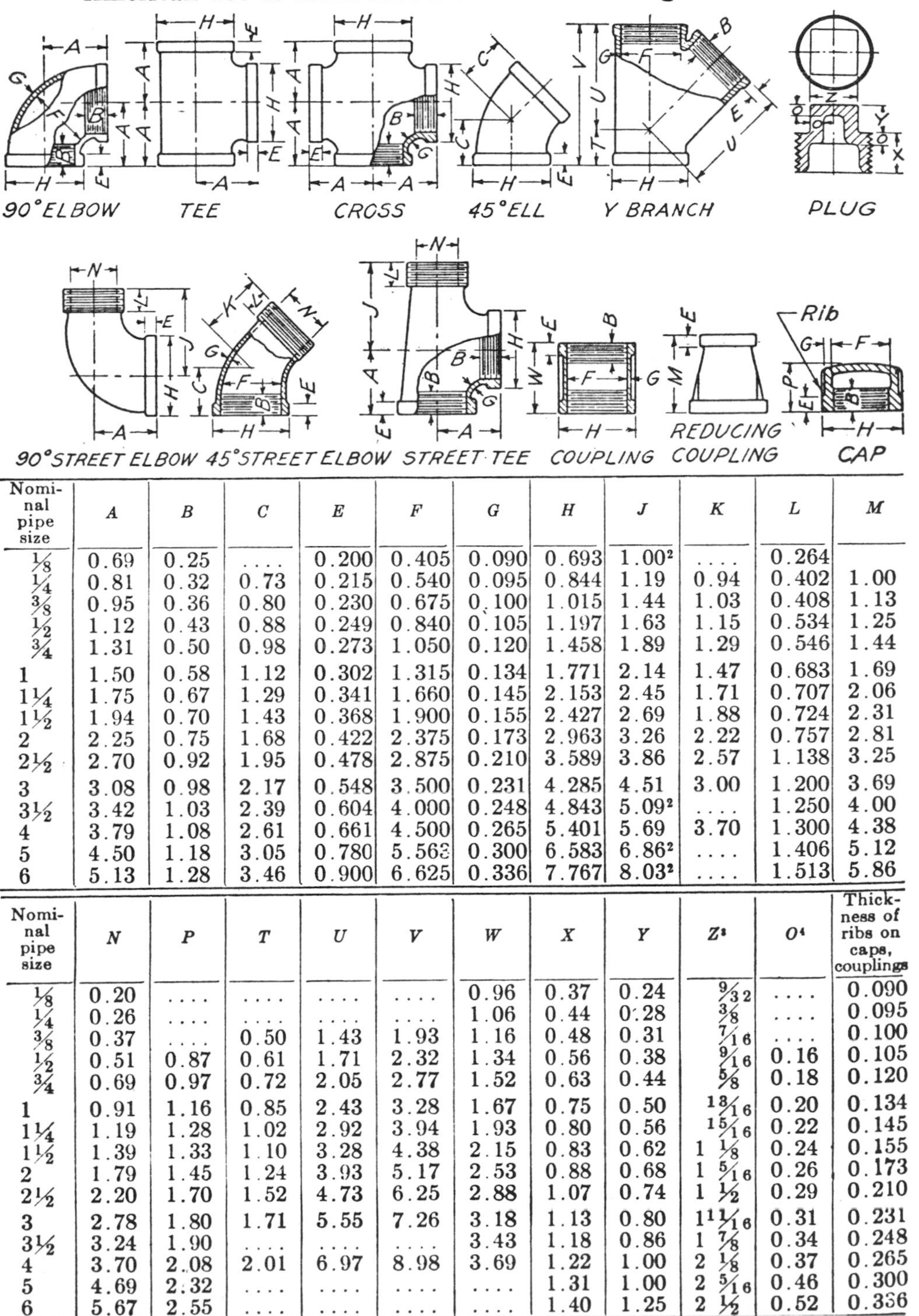

Nominal pipe size	A	B	C	E	F	G	H	J	K	L	M
1/8	0.69	0.25	....	0.200	0.405	0.090	0.693	1.00[2]	....	0.264	
1/4	0.81	0.32	0.73	0.215	0.540	0.095	0.844	1.19	0.94	0.402	1.00
3/8	0.95	0.36	0.80	0.230	0.675	0.100	1.015	1.44	1.03	0.408	1.13
1/2	1.12	0.43	0.88	0.249	0.840	0.105	1.197	1.63	1.15	0.534	1.25
3/4	1.31	0.50	0.98	0.273	1.050	0.120	1.458	1.89	1.29	0.546	1.44
1	1.50	0.58	1.12	0.302	1.315	0.134	1.771	2.14	1.47	0.683	1.69
1 1/4	1.75	0.67	1.29	0.341	1.660	0.145	2.153	2.45	1.71	0.707	2.06
1 1/2	1.94	0.70	1.43	0.368	1.900	0.155	2.427	2.69	1.88	0.724	2.31
2	2.25	0.75	1.68	0.422	2.375	0.173	2.963	3.26	2.22	0.757	2.81
2 1/2	2.70	0.92	1.95	0.478	2.875	0.210	3.589	3.86	2.57	1.138	3.25
3	3.08	0.98	2.17	0.548	3.500	0.231	4.285	4.51	3.00	1.200	3.69
3 1/2	3.42	1.03	2.39	0.604	4.000	0.248	4.843	5.09[2]	....	1.250	4.00
4	3.79	1.08	2.61	0.661	4.500	0.265	5.401	5.69	3.70	1.300	4.38
5	4.50	1.18	3.05	0.780	5.563	0.300	6.583	6.86[2]	....	1.406	5.12
6	5.13	1.28	3.46	0.900	6.625	0.336	7.767	8.03[2]	....	1.513	5.86

Nominal pipe size	N	P	T	U	V	W	X	Y	Z[3]	O[4]	Thickness of ribs on caps, couplings
1/8	0.20	....	....	....	....	0.96	0.37	0.24	9/32	....	0.090
1/4	0.26	....	....	....	....	1.06	0.44	0.28	3/8	....	0.095
3/8	0.37	....	0.50	1.43	1.93	1.16	0.48	0.31	7/16	....	0.100
1/2	0.51	0.87	0.61	1.71	2.32	1.34	0.56	0.38	9/16	0.16	0.105
3/4	0.69	0.97	0.72	2.05	2.77	1.52	0.63	0.44	5/8	0.18	0.120
1	0.91	1.16	0.85	2.43	3.28	1.67	0.75	0.50	13/16	0.20	0.134
1 1/4	1.19	1.28	1.02	2.92	3.94	1.93	0.80	0.56	15/16	0.22	0.145
1 1/2	1.39	1.33	1.10	3.28	4.38	2.15	0.83	0.62	1 1/8	0.24	0.155
2	1.79	1.45	1.24	3.93	5.17	2.53	0.88	0.68	1 5/16	0.26	0.173
2 1/2	2.20	1.70	1.52	4.73	6.25	2.88	1.07	0.74	1 1/2	0.29	0.210
3	2.78	1.80	1.71	5.55	7.26	3.18	1.13	0.80	1 11/16	0.31	0.231
3 1/2	3.24	1.90	....	....	....	3.43	1.18	0.86	1 7/8	0.34	0.248
4	3.70	2.08	2.01	6.97	8.98	3.69	1.22	1.00	2 1/8	0.37	0.265
5	4.69	2.32	....	....	....	....	1.31	1.00	2 5/16	0.46	0.300
6	5.67	2.55	....	....	....	....	1.40	1.25	2 1/2	0.52	0.336

Dimensions in inches. Left-hand couplings have four or more ribs. Right-hand couplings have two ribs.
[1] ASA B16c 1939. Street tee not made in 1/8″ size. [2] Street ell only. [3] These dimensions are the nominal size of wrench (ASA B18.2 1941). Square-head plugs are designed to fit these wrenches. [4] Solid plugs are provided in sizes 1/8 to 3 1/2 in. incl.; cored plugs 1/2 to 3 1/2 in., inclusive. Cored plugs have minimum metal thickness at all points equal to dimension O except at the end of the thread.

PIPE FITTING AND VALVE SYMBOLS

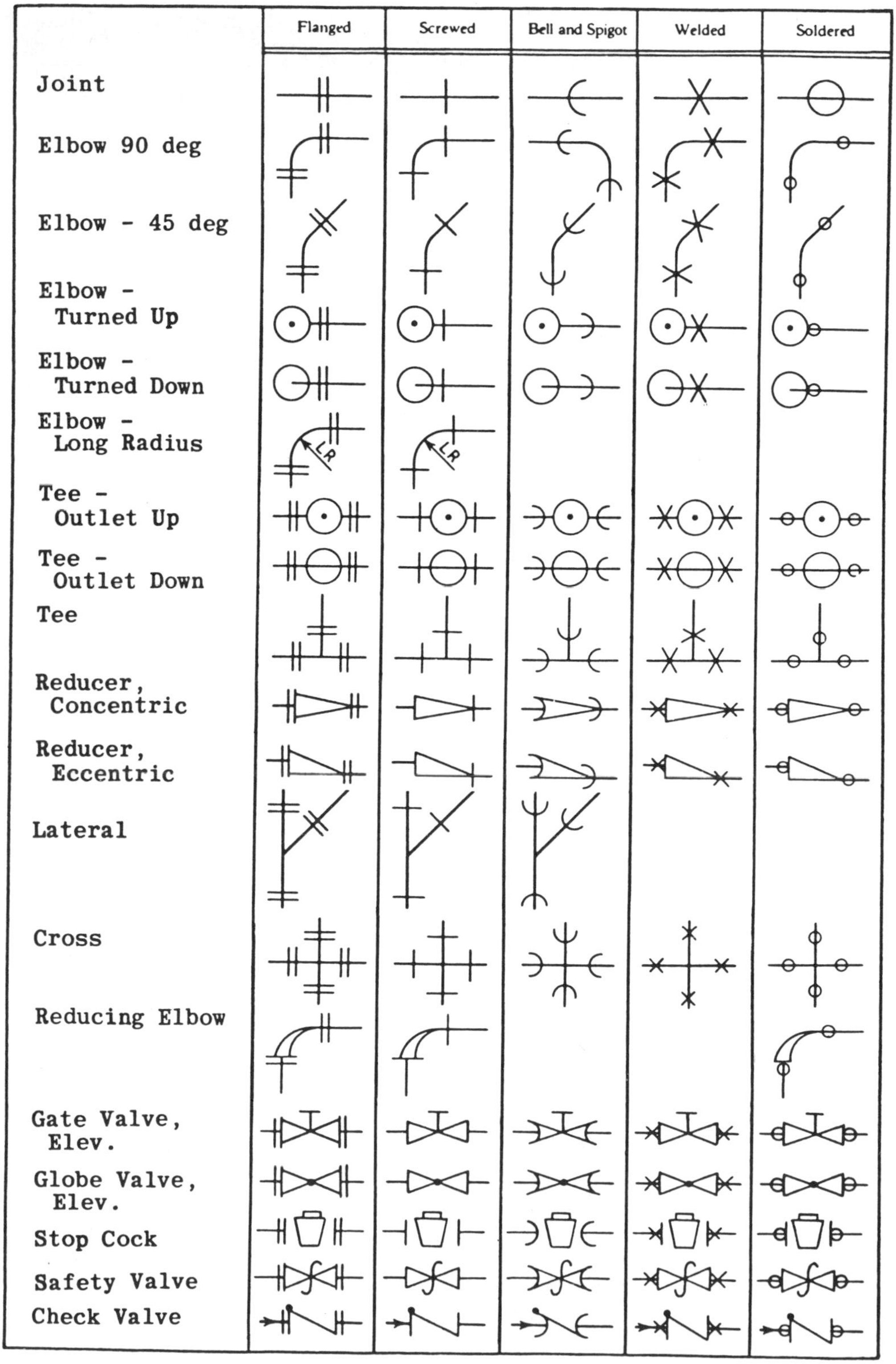

	Flanged	Screwed	Bell and Spigot	Welded	Soldered
Joint					
Elbow 90 deg					
Elbow - 45 deg					
Elbow - Turned Up					
Elbow - Turned Down					
Elbow - Long Radius	LR	LR			
Tee - Outlet Up					
Tee - Outlet Down					
Tee					
Reducer, Concentric					
Reducer, Eccentric					
Lateral					
Cross					
Reducing Elbow					
Gate Valve, Elev.					
Globe Valve, Elev.					
Stop Cock					
Safety Valve					
Check Valve					

Symbols for materials (section)

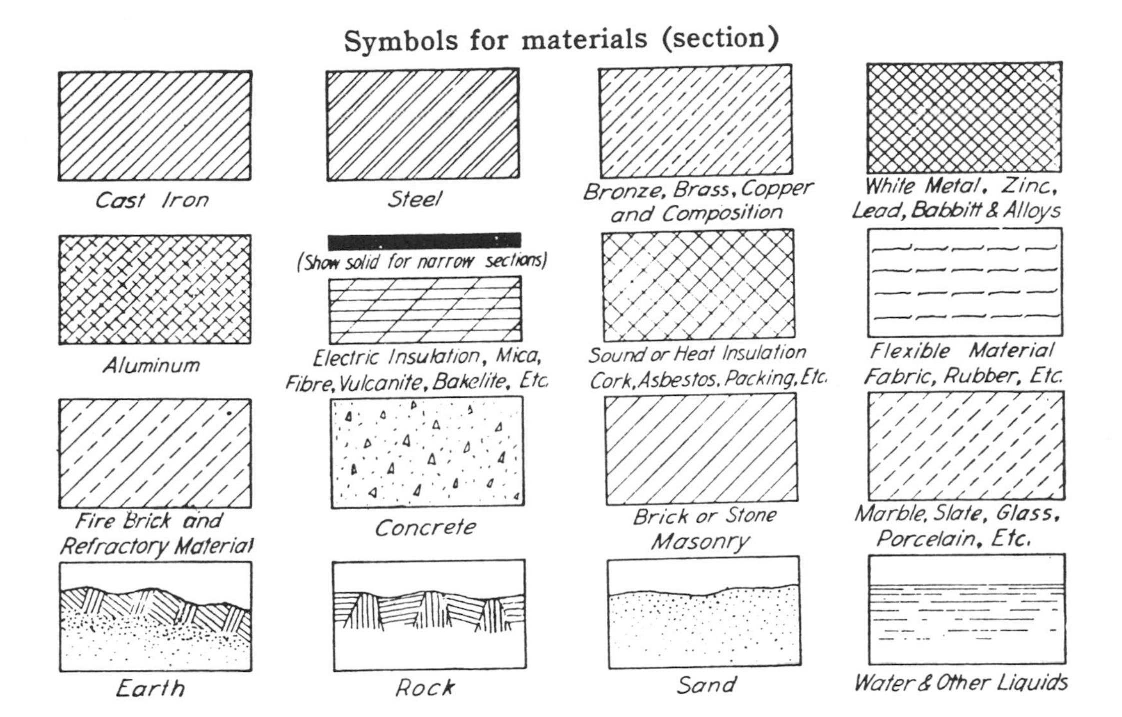

Weights of Materials

METALS

	lb/cu in.
Aluminum alloy, cast	0.099
Aluminum, cast	0.094
Aluminum, wrought	0.097
Babbitt metal	0.267
Brass, cast or rolled	0.303–0.313
Brass, drawn	0.323
Bronze, aluminum cast	0.277
Bronze, phosphor	0.315–0.321
Chromium	0.256
Copper, cast	0.311
Copper, rolled, drawn or wire	0.322
Dowmetal *A*	0.065
Duralumin	0.101

	lb/cu in.
Gold	0.697
Iron, cast	0.260
Iron, wrought	0.283
Lead	0.411
Magnesium	0.063
Mercury	0.491
Monel metal	0.323
Silver	0.379
Steel, cast or rolled	0.274–0.281
Steel, tool	0.272
Tin	0.263
Zinc	0.258

WOOD

	lb/cu in.
Ash	0.024
Balsa	0.0058
Cedar	0.017
Cork	0.009
Hickory	0.0295
Maple	0.025
Oak (white)	0.028
Pine (white)	0.015
Pine (yellow)	0.025
Poplar	0.018
Walnut (black)	0.023

MISCELLANEOUS MATERIALS

	lb/cu ft
Asbestos	175
Bakelite	79.5
Brick, common	112
Brick, fire	144
Celluloid	86.4
Earth, packed	100
Fiber	89.9
Glass	163
Gravel	109
Limestone	163
Plexiglass	74.3
Sandstone	144
Water	62.4

ELECTRICAL SYMBOLS

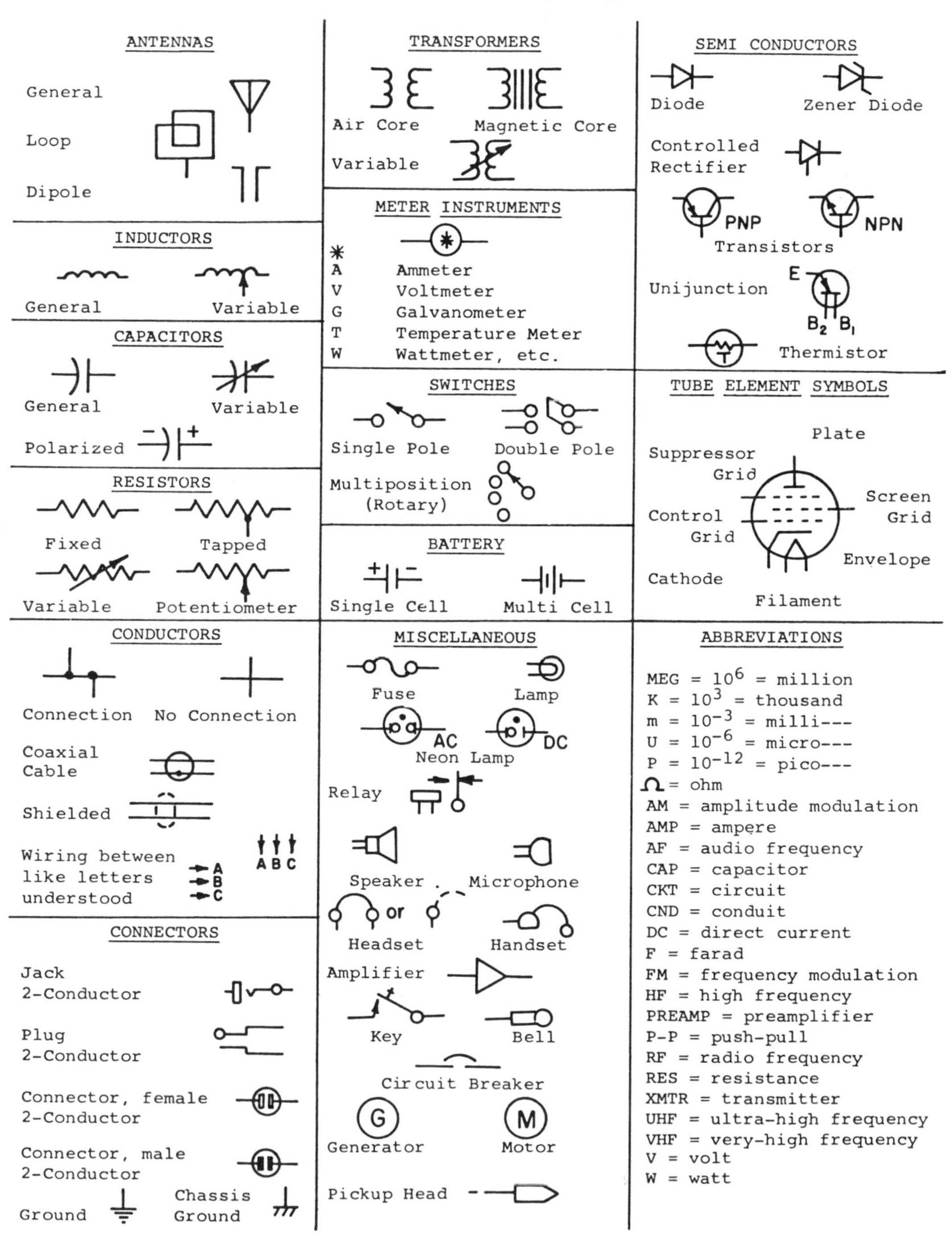

ABBREVIATIONS

MEG = 10^6 = million
K = 10^3 = thousand
m = 10^{-3} = milli---
U = 10^{-6} = micro---
P = 10^{-12} = pico---
Ω = ohm
AM = amplitude modulation
AMP = ampere
AF = audio frequency
CAP = capacitor
CKT = circuit
CND = conduit
DC = direct current
F = farad
FM = frequency modulation
HF = high frequency
PREAMP = preamplifier
P-P = push-pull
RF = radio frequency
RES = resistance
XMTR = transmitter
UHF = ultra-high frequency
VHF = very-high frequency
V = volt
W = watt

AREA AND VOLUME FORMULAS

Plane Figures

Nomenclature

a, b, c, d	—	Lengths of Sides
A	—	Area
d, d_1, d_2	—	Diameters
e, f	—	Lengths of Diagonals
h	—	Vertical Height or Altitude
l, l_1, l_2	—	Length of Arc
L	—	Lateral Length or Slant Height
n	—	Number of Sides
θ	—	Number of Degrees of Arc
p	—	Perimeter
r, r_1, r_2, R	—	Radii

1. RIGHT TRIANGLE

$p = a + b + c$

$c^2 = a^2 + b^2$

$b = \sqrt{c^2 - a^2}$

$A = \frac{ab}{2}$

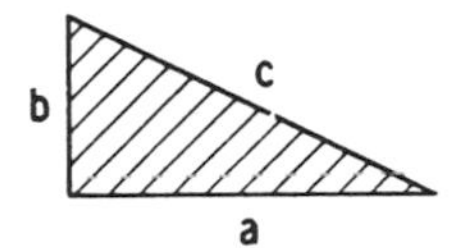

2. EQUILATERAL TRIANGLE

$p = 3a$

$h = \frac{a}{2}\sqrt{3} = .866\,a$

$A = a^2 \frac{\sqrt{3}}{4} = .433\,a^2$

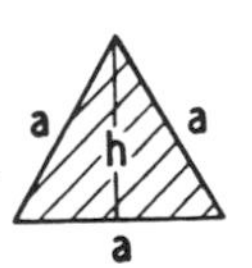

3. GENERAL TRIANGLE

Let $s = \frac{a + b + c}{2}$

$p = a + b + c$

$h = \frac{2}{a}\sqrt{s(s-a)(s-b)(s-c)}$

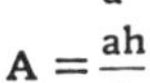

$A = \frac{ah}{2}$

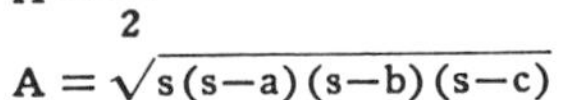

$A = \sqrt{s(s-a)(s-b)(s-c)}$

4. SQUARE

$a = b$

$p = 4a$

$A = a^2 = .5e^2$

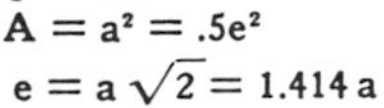

$e = a\sqrt{2} = 1.414\,a$

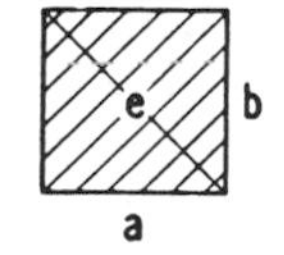

5. RECTANGLE

$p = 2(a + b)$

$e = \sqrt{a^2 + b^2}$

$b = \sqrt{e^2 - a^2}$

$A = ab$

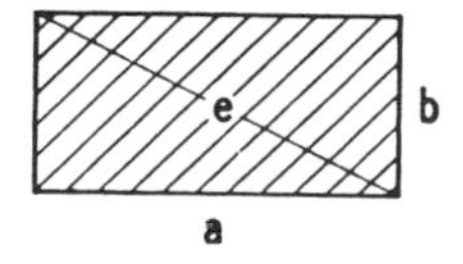

6. GENERAL PARALLELOGRAM OR RHOMBOID; AND RHOMBUS

Rhomboid—opposite sides parallel

$p = 2(a + b)$

$e^2 + f^2 = 2(a^2 + b^2)$

$A = ah$

Rhombus—opposite sides parallel and all sides equal

$a = b$

$p = 4a = 4b$

$e^2 + f^2 = 4a^2$

$A = ah = \frac{ef}{2}$

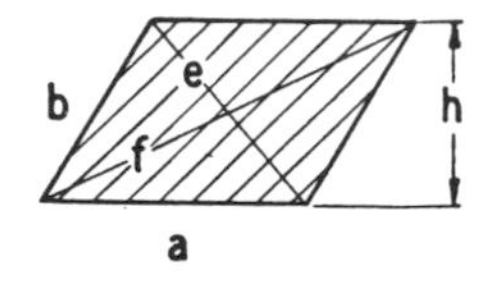

7. TRAPEZOID

$p = a + b + c + d$

$A = \frac{(a + b)}{2} h$

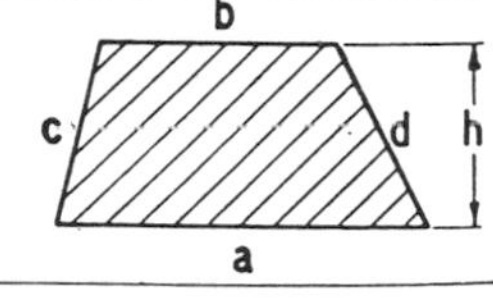

8. TRAPEZIUM

$p = a + b + c + d$

A = Sum of Areas of two major triangles

$A = \frac{(h_1 + h_2)g + fh_1 + jh_2}{2}$

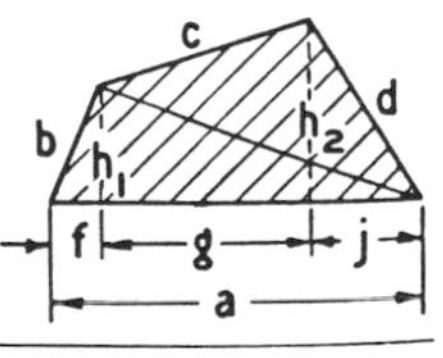

9. REGULAR POLYGON

Let n = number of sides

$p = na$

$a = 2\sqrt{R^2 - r^2}$

$A = \frac{nar}{2} = \frac{na}{2}\sqrt{R^2 - \frac{a^2}{4}}$

= n × Area of each triangle

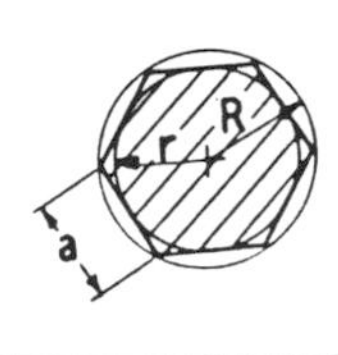

10. CIRCLE

$p = 2\pi r = \pi d = 3.1416d$

$A = \pi r^2 = \frac{\pi d^2}{4} = .7854d^2$

$= \frac{p^2}{4\pi} = .07958p^2$

11. HOLLOW CIRCLE or ANNULUS

$A = \frac{\pi}{4}(d_2^2 - d_1^2) = .7854\,(d_2^2 - d_1^2)$

$= \pi\,(r_2^2 - r_1^2)$

$= \pi \frac{d_1 + d_2}{2}(r_2 - r_1)$

$= \pi\,(r_1 + r_2)(r_2 - r_1)$

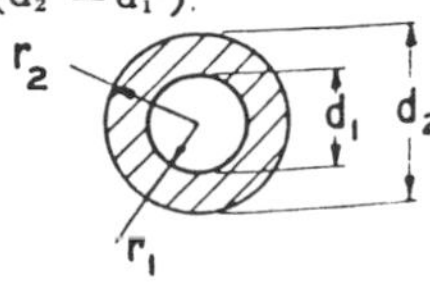

12. SECTOR of CIRCLE

$$l = \frac{\pi r\theta}{180} = \frac{r\theta}{57.3} = .01745r\theta$$
$$= \frac{2A}{r}$$
$$A = \frac{\pi\theta r^2}{360} = .008727\theta r^2$$
$$= \frac{lr}{2}$$

13. SEGMENT of CIRCLE

for $\theta < 90°$

$$A = \frac{r^2}{2}\left(\frac{\pi\theta}{180} - \sin\theta\right)$$

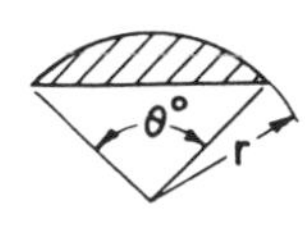

for $\theta > 90°$

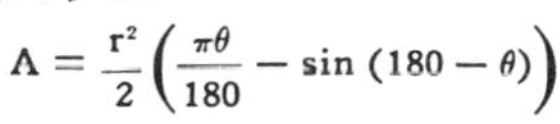

$$A = \frac{r^2}{2}\left(\frac{\pi\theta}{180} - \sin(180 - \theta)\right)$$

for chord rise, etc., see "Properties of Circle"

14. SECTOR of HOLLOW CIRCLE

$$A = \frac{\pi\theta(r_2^2 - r_1^2)}{360}$$
$$A = \frac{r_1 - r_2}{2}(l_1 + l_2)$$

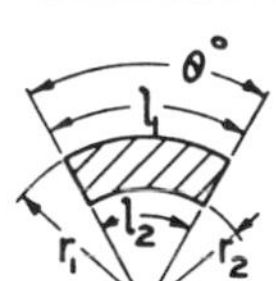

15. FILLET

$A = .215r^2$

or approximately

$$A = \frac{r^2}{5}$$

16. ELLIPSE

$p = \pi(a + b)$ approximately
$= \pi[1.5(a + b) - \sqrt{ab}]$ more nearly

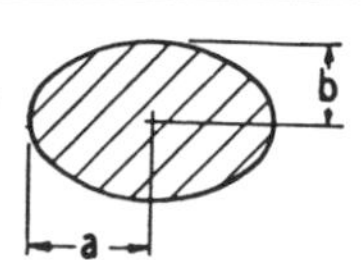

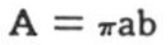

$A = \pi ab$

17. PARABOLA

$$A = \frac{2}{3}ab$$

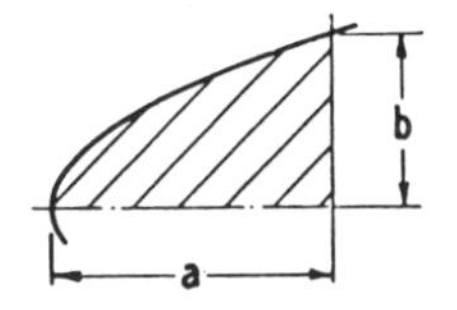

Nomenclature

a, b, c, d	— Lengths of Sides
C	— Length of Chord
A	— Total Area
A_B	— Area of Base
A_L	— Area of Lateral or Convex Surfaces
A_R	— Area of Right Section
A_T	— Area of Top Section
h, h_1, h_2	— Vertical Height or Altitude
h_G	— Vertical Distance between Centers of Gravity of Areas
L, L_1, L_2	— Lateral Length or Slant Height
L_G	— Slant Height between Centers of Gravity of Areas
p	— Perimeter
p_B	— Perimeter of Base
p_R	— Perimeter of Right Section
r, r_1	— Radii
V	— Volume

18. CUBE

$A = 6a^2$
$V = a^3$

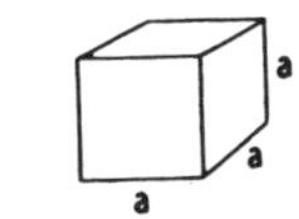

19. PARALLELOPIPED

$A = 2(ab + bc + ac)$
$V = abc$

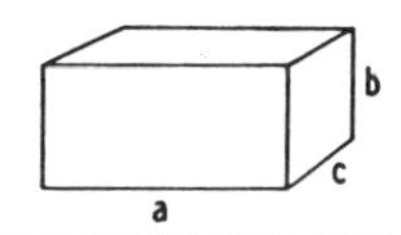

20. GENERAL PRISM AND RIGHT REGULAR PRISM

$A_L = p_R L = p_B h$
$A = A_L + 2A_B$
$V = A_R \times L = A_B h$

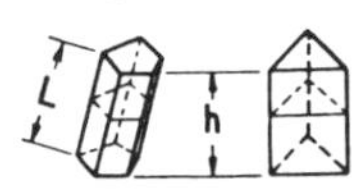

21. FRUSTUM of PRISM

$V = A_B h_G$
$V = A_R L_G$

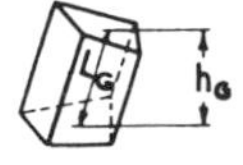

22. RIGHT REGULAR PYRAMID or CONE

$$A_L = \frac{1}{2}p_B L$$
$$V = \frac{1}{3}A_B h$$

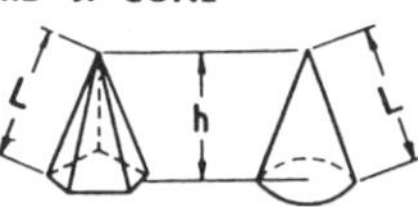

(Cont.)

23. GENERAL PYRAMID or CONE

$$V = \frac{1}{3} A_B h$$

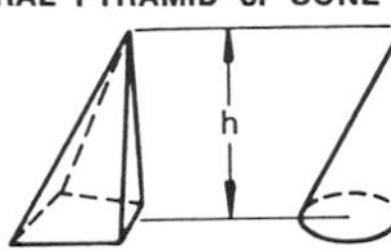

24. FRUSTUM of RIGHT REGULAR PYRAMID or CONE

$$A_L = \frac{1}{2} L(p_B + p_T)$$

$$A = A_L + A_B + A_T$$

$$V = \frac{1}{3} h(A_B + A_T + \sqrt{A_B A_T})$$

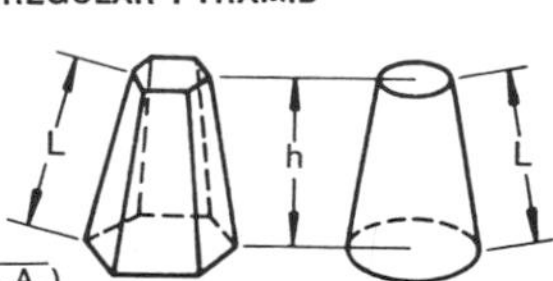

25. FRUSTUM of GENERAL PYRAMID or CONE (PARALLEL ENDS)

$$V = \frac{1}{3} h(A_B + A_T + \sqrt{A_B A_T})$$

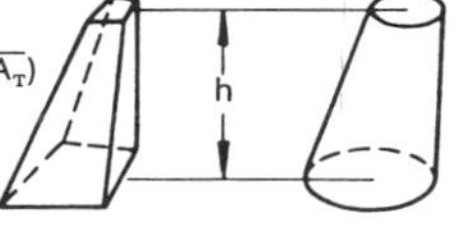

26. RIGHT CIRCULAR CYLINDER

$$A_L = 2\pi rh$$

$$A = 2\pi r(r + h)$$

$$V = \pi r^2 h$$

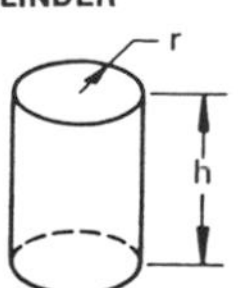

27. GENERAL CYLINDER (ANY CROSS SECTION)

$$A_L = p_B h = p_R L$$

$$A = A_L + 2A_B$$

$$V = A_B h = A_R L$$

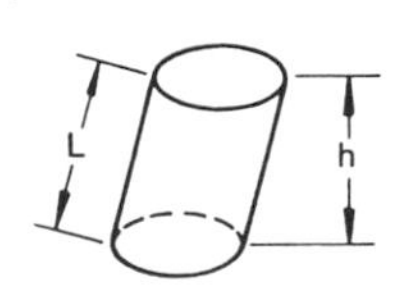

28. FRUSTUM of GENERAL CYLINDER

$$V = \frac{1}{2} A_R(L_1 + L_2)$$

$$V = A_B h_G$$

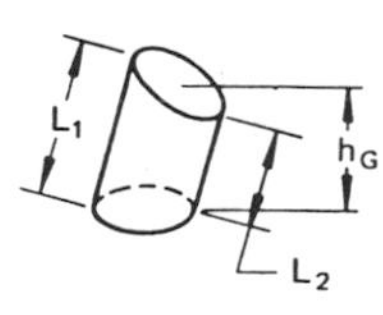

29. FRUSTUM of RIGHT CIRCULAR CYLINDER

$$A_L = \pi r(h_1 + h_2)$$

$$A_T = \pi r\sqrt{r^2 + \left(\frac{h_1 - h_2}{2}\right)^2}$$

$$A_B = \pi r^2$$

$$A = A_L + A_T + A_B$$

$$V = \frac{\pi r^2}{2}(h_1 + h_2)$$

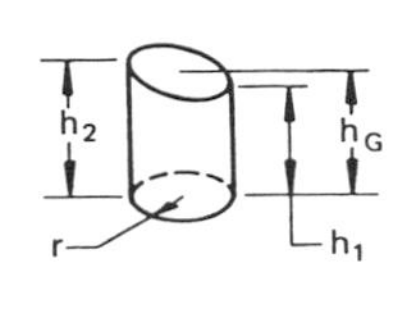

30. SPHERE

$$A = 4\pi r^2 = 12.566r^2$$

$$V = \frac{4}{3}\pi r^3 = 4.189r^3$$

31. SPHERICAL SECTOR

$$A = \frac{\pi r}{2}(4h + C)$$

$$V = \frac{2}{3}\pi r^2 h = 2.0944r^2 h$$

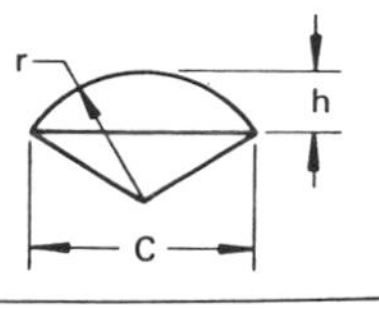

32. SPHERICAL SEGMENT

$$A_T = 2\pi rh = \frac{\pi}{4}(4h^2 + C^2)$$

$$V = \frac{\pi}{3} h^2(3r - h) = \frac{\pi}{24} h(3C^2 + 4h^2)$$

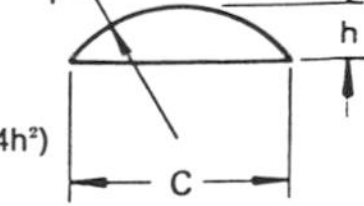

33. SPHERICAL ZONE

$$A_L = 2\pi rh$$

$$A = \frac{\pi}{4}(8rh + a^2 + b^2)$$

$$V = \frac{\pi h}{24}(3C^2 + 3b^2 + 4h^2)$$

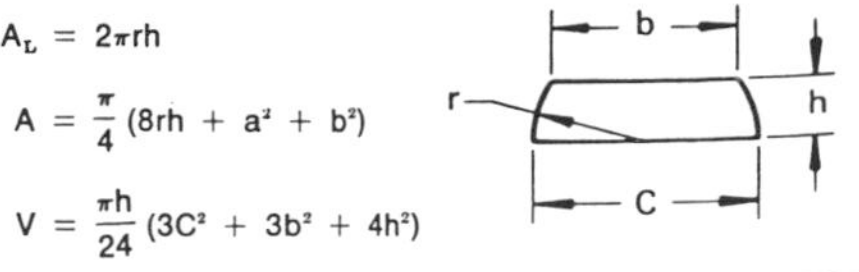

34. TORUS

$$A = 4\pi^2 rr_1$$

$$V = 2\pi^2 r^2 r_1$$

INDEX

Numbers on this page refer to <u>PROBLEM NUMBERS</u>, not page numbers

Numbers on this page refer to PROBLEM NUMBERS, not page numbers

Numbers on this page refer to **PROBLEM NUMBERS**, not page numbers

Numbers on this page refer to **PROBLEM NUMBERS,** not page numbers

Numbers on this page refer to PROBLEM NUMBERS, not page numbers

Numbers on this page refer to PROBLEM NUMBERS, not page numbers

Numbers on this page refer to PROBLEM NUMBERS, not page numbers